Lecture Notes in Artificial Intelligence　　10939

Subseries of Lecture Notes in Computer Science

More information about this series at http://www.springer.com/series/1244

Dinh Phung · Vincent S. Tseng
Geoffrey I. Webb · Bao Ho
Mohadeseh Ganji · Lida Rashidi (Eds.)

Advances in Knowledge Discovery and Data Mining

22nd Pacific-Asia Conference, PAKDD 2018
Melbourne, VIC, Australia, June 3–6, 2018
Proceedings, Part III

 Springer

Editors
Dinh Phung
Deakin University
Geelong, VIC
Australia

Vincent S. Tseng
National Chiao Tung University
Hsinchu City
Taiwan

Geoffrey I. Webb (ID)
Monash University
Clayton, VIC
Australia

Bao Ho
Japan Advanced Institute
   of Science and Technology
Nomi, Ishikawa
Japan

Mohadeseh Ganji
University of Melbourne
Melbourne, VIC
Australia

Lida Rashidi
University of Melbourne
Melbourne, VIC
Australia

ISSN 0302-9743   ISSN 1611-3349   (electronic)
Lecture Notes in Artificial Intelligence
ISBN 978-3-319-93039-8   ISBN 978-3-319-93040-4   (eBook)
https://doi.org/10.1007/978-3-319-93040-4

Library of Congress Control Number: 2018944425

LNCS Sublibrary: SL7 – Artificial Intelligence

Printed on acid-free paper

This Springer imprint is published by the registered company Springer International Publishing AG
part of Springer Nature
The registered company address is: Gewerbestrasse 11, 6330 Cham, Switzerland

PC Chairs' Preface

With its 22nd edition in 2018, the Pacific-Asia Conference on Knowledge Discovery and Data Mining is the second oldest conference and a leading venue in the area of knowledge discovery and data mining (KDD). It provides a prestigious international forum for researchers and industry practitioners to share their new ideas, original and latest research results, and practical development experiences from all KDD-related areas, including data mining, data warehousing, machine learning, artificial intelligence, deep learning, databases, statistics, knowledge engineering, visualization, and decision-making systems.

This year, we received 592 valid submissions, which is the highest number of submissions in the past 10 years. The diversity and reputation of PAKDD were also evident from the various regions from which submissions came, with over 25 different countries, noticeably from North America and Europe. Our goal was to continue to ensure a rigorous reviewing process with each paper assigned to one Senior Program Committee (SPC) member and at least three Technical Program Committee (TPC) members, resulting in an ideal minimum number of reviews of four for each paper. Owing to the unusually large number of submissions this year, we had to increase almost doubling the number of committee members, resulting in 72 SPC members and 330 TPC members. Each valid submission was reviewed by three PC members and meta-reviewed by one SPC member who also led the discussion. This required a total of approximately 2,000 reviews. The program co-chairs then considered recommendations from the SPCs, the submission, and the reviews to make the final decision. Borderline papers were discussed intensively before final decisions were made. In some cases, additional reviews were also requested.

In the end, 164 out of 592 papers were accepted, resulting in an acceptance rate of 27.9%. Among them, 58 papers were selected for long presentation and 107 papers were selected for regular presentation. This year, we introduced a new track in Deep Learning for Knowledge Discovery and Data Mining. This track was particularly popular (70 submissions); however, in the end, the number of papers accepted as the primary category for this track was moderate (six accepted papers), standing at 8.8%. The conference program contained 32 sessions in total. Long presentations were allocated 25 minutes and regular presentations 15 mins. These two types of papers, however, are not distinguished in the proceedings.

We would like to sincerely thank all SPC members, TPC members, and external reviewers for their time, effort, dedication, and services to PAKDD 2018.

April 2018

Dinh Phung
Vincent S. Tseng

General Chairs' Preface

Welcome to the proceedings of the 22nd Pacific-Asia Conference on Knowledge Discovery and Data Mining (PAKDD). This conference has a reputable tradition in bringing researchers, academia, developers, practitioners, and industry together with a focus on the Pacific-Asian regions. This year, PAKDD was held in the wonderful city of Melbourne, Australia, during June 3–6, 2018.

The single most important element of PAKDD is the technical contributions and submissions in the area of KDD. We were very pleased with the number of submissions received this year, which was well close to 600, showing a significant boost in the number of submissions and the popularity of this conference. We sincerely thank the many authors from around the world who submitted their work to the PAKDD 2018 technical program as well as its data competition and satellite workshops. In addition, PAKDD 2018 featured three high-profile keynote speakers: Professor Kate Smith-Miles, Australian Laureate Fellow from Melbourne University; Dr. Rajeev Rastogi, Director of Machine Learning at Amazon; and Professor Bing Liu from the University of Illinois at Chicago. The conference featured three tutorials and five satellite workshops in addition to a data competition sponsored by the Fourth Paradigm Inc. and ChaLean.

We would like to express our gratitude to the contribution of the SPC, TPC, and external reviewers, led by the program co-chairs, Dinh Phung and Vincent Tseng. We would like to thank the workshop co-chairs, Benjamin Fung and Can Wang; the tutorial co-chairs, Wray Buntine and Jeffrey Xu Yu; the competition co-chairs, Wei-Wei Tu and Hugo Jair Escalante; the local arrangements co-chairs, Gang Li and Wei-Luo; the publication co-chairs, Mohadeseh Ganji and Lida Rashidi; the Web and content co-chairs, Trung Le, Uyen Pham, and Khanh Nguyen; the publicity co-chairs, De-Chuan Zhan, Kozo Ohara, Kyuseok Shim, and Jeremiah Deng; and the award co-chairs, James Bailey, Bart Goethals, and Jinyan Li.

We are grateful to our sponsors: Deakin University as the host institution and gold sponsor; Monash University as the gold sponsor, University of Melbourne, Trusting Social, and the Asian Office of Aerospace Research and Development/Air Force Office of Scientific Research as silver sponsors, Springer as the publication sponsor, and the Fourth Paradigm, CodaLab and ChaLearn as the data competition sponsors.

April 2017

Tu-Bao Ho
Geoffrey I. Webb

Organization

Organizing Committee

General Co-chairs

Geoffrey I. Webb	Monash University, Australia
Bao Ho	Japan Advanced Institute of Science and Technology, Japan

Program Committee Co-chairs

Dinh Phung	Deakin University, Australia
Vincent Tseng	National Chiao Tung University, Taiwan

Tutorial Co-chairs

Wray Buntine	Monash University, Australia
Jeffrey Xu Yu	Chinese University of Hong Kong, Hong Kong, SAR China

Workshop Co-chairs

Benjamin Fung	McGill University, Canada
Can Wang	Griffith University, Australia

Data Competition Co-chairs

Wei-Wei Tu	Fourth Paradigm Inc., China
Hugo Jair Escalante	INAOE Mexico, ChaLearn, USA

Publicity Co-chairs

De-Chuan Zhan	Nanjing University, China
Kozo Ohara	Aoyama Gakuin University, Japan
Kyuseok Shim	Seoul National University, South Korea
Jeremiah Deng	University of Otago, New Zealand

Publication Co-chairs

Mohadeseh Ganji	University of Melbourne, Australia
Lida Rashidi	University of Melbourne, Australia

Local Arrangements Co-chairs

Gang Li	Deakin University, Australia
Wei Luo	Deakin University, Australia

Web and Content Co-chairs

Trung Le	Deakin University, Australia
Uyen Pham	Vietnam National University, Vietnam

Award Co-chairs

James Bailey	University of Melbourne, Australia
Bart Goethals	University of Antwerp, Belgium
Jinyan Li	University of Technology Sydney, Australia

Steering Committee

Co-chairs

Ee-Peng Lim	Singapore Management University, Singapore
Takashi Washio	Institute of Scientific and Industrial Research, Osaka University, Japan

Treasurer

Longbing Cao	Advanced Analytics Institute, University of Technology, Sydney, Australia

Members

Ee-Peng Lim	Singapore Management University, Singapore (member since 2006, co-chair 2015–2017)
P. Krishna Reddy	International Institute of Information Technology, Hyderabad (IIIT-H), India (member since 2010)
Joshua Z. Huang	Shenzhen Institutes of Advanced Technology, Chinese Academy of Sciences, China (member since 2011)
Longbing Cao	Advanced Analytics Institute, University of Technology, Sydney (member since 2013)
Jian Pei	Simon Fraser University, Canada (member since 2013)
Myra Spiliopoulou	Otto von Guericke University Magdeburg, Germany (member since 2013)
Vincent S. Tseng	National Chiao Tung University, Taiwan (member since 2014)
Tru Hoang Cao	Ho Chi Minh City University of Technology, Vietnam (member since 2015)
Gill Dobbie	University of Auckland, New Zealand (member since 2016)
Kyuseok Shim	Seoul National University, South Korea

Life Members

Hiroshi Motoda	AFOSR/AOARD and Osaka University, Japan (member since 1997, co-chair 2001–2003, chair 2004–2006, life member since 2006)
Rao Kotagiri	University of Melbourne, Australia (member since 1997, co-chair 2006–2008, chair 2009–2011, life member since 2007, treasury Co-sign since 2006)
Huan Liu	Arizona State University, USA (member since 1998, treasurer 1998–2000, life member since 2012)
Ning Zhong	Maebashi Institute of Technology, Japan (member since 1999, life member since 2008)
Masaru Kitsuregawa	Tokyo University, Japan (member since 2000, life member since 2008)
David Cheung	University of Hong Kong, SAR China (member since 2001, treasurer 2005–2006, chair 2006–2008, life member since 2009)
Graham Williams	Australian National University, Australia (member since 2001, treasurer since 2006, co-chair 2009–2011, chair 2012–2014, life member since 2009)
Ming-Syan Chen	National Taiwan University, Taiwan, ROC (member since 2002, life member since 2010)
Kyu-Young Whang	Korea Advanced Institute of Science and Technology, South Korea (member since 2003, life member since 2011)
Chengqi Zhang	University of Technology Sydney, Australia (member since 2004, life member since 2012)
Tu Bao Ho	Japan Advanced Institute of Science and Technology, Japan (member since 2005, co-chair 2012–2014, chair 2015–2017, life member since 2013)
Zhi-Hua Zhou	Nanjing University, China (member since 2007, life member since 2015)
Jaideep Srivastava	University of Minnesota, USA (member since 2006, life member since 2015)
Takashi Washio	Institute of Scientific and Industrial Research, Osaka University (member since 2008, life member since 2016)
Thanaruk Theeramunkong	Thammasat University, Thailand (member since 2009)

Past Members

Hongjun Lu	Hong Kong University of Science and Technology, Hong Kong, SAR China (member 1997–2005)
Arbee L. P. Chen	National Chengchi University, Taiwan, ROC (member 2002–2009)
Takao Terano	Tokyo Insitute of Technology, Japan (member 2000–2009)

Senior Program Committee

Albert Bifet	Universite Paris-Saclay, France
Andrzej Skowron	University of Warsaw, Poland
Benjamin C. M. Fung	McGill University, Canada
Byung Suk Lee	University of Vermont, USA
Chandan Reddy	Virginia Tech, USA
Chuan Shi	Beijing University of Posts and Telecommunications, China
Dat Tran	University of Canberra, Australia
Dinh Phung	Deakin University, Australia
Eibe Frank	University of Waikato, New Zealand
Feida Zhu	Singapore Management University, Singapore
Gang Li	Deakin University, Australia
Geoff Holmes	University of Waikato, New Zealand
George Karypis	University of Minnesota, USA
Guozhu Dong	Wright State University, USA
Hanghang Tong	City University of New York, USA
Hu Xia	Texas A&M University, USA
Hui Xiong	Rutgers University, USA
Jae-Gil Lee	KAIST, South Korea
James Bailey	University of Melbourne, Australia
Jeffrey Xu Yu	Chinese University of Hong Kong, Hong Kong, SAR China
Jia Wu	Macquarie University, Australia
Jian Pei	Simon Fraser University, Canada
Jianyong Wang	Tsinghua University, China
Jiliang Tang	Michigan State University, USA
Jiuyong Li	University of South Australia, Australia
Joshua Huang	Shenzhen Institutes of Advanced Technology, Chinese Academy of Sciences, China
Kai Ming Ting	Federation University, Australia
Kamalakar Karlapalem	International Institute of Information Technology, Hyderabad, India
Krishna Reddy P.	International Institute of Information Technology, Hyderabad, India
Kyuseok Shim	Seoul National University, South Korea
Latifur Khan	University of Texas at Dallas, USA
Longbing Cao	University of Technology Sydney, Australia
Masashi Sugiyama	University of Tokyo, Japan
Michael Berthold	University of Konstanz, Germany
Ming Li	Nanjing University, China
Min-Ling Zhang	Southeast University, China
Nikos Mamoulis	University of Ioannina, Greece
Niloy Ganguly	IIT, Kharagpur, India
Nitin Agarwal	University of Arkansas at Little Rock, USA

Program Committee

Arnaud Giacometti	François Rabelais University, France
Arnaud Soulet	François Rabelais University, France
Arthur Zimek	University of Southern Denmark, Denmark
Athanasios Nikolakopoulos	University of Minnesota, USA
Bay Vo	Ho Chi Minh City University of Technology, Vietnam
Bettina Berendt	Katholieke Universiteit Leuven, Belgium
Bin Liu	IBM T. J. Watson Research Center, USA
Bing Xue	Victoria University of Wellington, New Zealand
Bo Jin	Dalian University of Technology, China
Bolin Ding	Microsoft Research, USA
Brendon Woodford	University of Otago, New Zealand
Bruno Cremilleux	Université de Caen Normandie, France
Bum-Soo Kim	Korea University, South Korea
Canh Hao Nguyen	Kyoto University, Japan
Carson Leung	University of Manitoba, Canada
Chao Lan	University of Wyoming, USA
Chao Qian	University of Science and Technology of China, China
Chedy Raissi	Inria, France
Chen Chen	Nankai University, China
Chengzhang Zhu	University of Technology Sydney, Australia
Chenping Hou	National University of Defence Technology, China
Chia Hui Chang	National Central University, Taiwan
Choochart Haruechaiyasak	National Electronics and Computer Technology Centre, NECTEC, Thailand
Chuan Shi	Beijing University of Posts and Telecommunications, China
Chulyun Kim	Sookmyung Women's University, South Korea
Chun-Hao Chen	Tamkang University, Taiwan
Dao-Qing Dai	Sun Yat-Sen University, China
Dat Tran	University of Canberra, Australia
David Anastasiu	San José State University, USA
David Taniar	Monash University, Australia
David Tse Jung Huang	University of Auckland, New Zealand
De-Chuan Zhan	Nanjing University, China
Defu Lian	University of Electronic Science and Technology of China, China
Dejing Dou	University of Oregon, USA
De-Nian Yang	Academia Sinica, Taiwan
Dhaval Patel	IBM T. J. Watson Research Center, USA
Dinh Quoc Tran	University of North Carolina at Chapel Hill, USA
Divyesh Jadav	IBM Research, USA
Dragan Gamberger	Rudjer Boskovic Institute, Croatia
Du Zhang	California State University, USA
Duc Dung Nguyen	Institute of Information Technology, Vietnam
Elham Naghizade	University of Melbourne, Australia
Enhong Chen	University of Science and Technology of China, China

Jing Zhang	Nanjing University of Science and Technology, China
Jingrui He	IBM Research, USA
Jingwei Xu	Nanjing University, China
Jingyuan Yang	Rutgers University, USA
Joao Vinagre	LIAAD – INESC Tec, Porto, Portugal
Johannes Bloemer	University of Paderborn, Germany
Jörg Wicker	University of Auckland, New Zealand
Joyce Jiyoung Whang	Sungkyunkwan University, South Korea
Jun Gao	Peking University, China
Jun Luo	Lenovo, Hong Kong, SAR China
Junbin Gao	University of Sydney, Australia
Jundong Li	Arizona State University, USA
Jungeun Kim	KAIST, South Korea
Jun-Ki Min	Korea University of Technology and Education, South Korea
Junping Zhang	Fudan University, China
K. Selçuk Candan	Arizona State University, USA
Keith Chan	Hong Kong Polytechnic University, Hong Kong, SAR China
Kevin Bouchard	Université du Quebec a Chicoutimi, Canada
Khoat Than	Hanoi University of Science and Technology, Vietnam
Ki Yong Lee	Sookmyung Women's University, South Korea
Ki-Hoon Lee	Kwangwoon University, South Korea
Kitsana Waiyamai	Kasetsart University, Thailand
Kok-Keong Ong	La Trobe University, Australia
Kouzou Ohara	Aoyama Gakuin University, Japan
Krisztian Buza	University of Bonn, Germany
Kui Yu	University of South Australia, Australia
Kun-Ta Chuang	National Cheng Kung University, Taiwan
Kyoung-Sook Kim	Artificial Intelligence Research Centre, South Korea
Latifur Khan	University of Texas, USA
Le Wu	Hefei University of Technology, China
Lei Gu	Nanjing University of Post and Telecommunications, China
Leong Hou U	University of Macau, SAR China
Liang Hu	Jilin University, China
Liang Hu	University of Technology Sydney, Australia
Liang Wu	Arizona State University, USA
Lida Rashidi	University of Melbourne, Australia
Lijie Wen	Tsinghua University, China
Lin Liu	University of South Australia, Australia
Lin Wu	University of Queensland, Australia
Ling Chen	University of Technology Sydney, Australia
Lizhen Wang	Yunnan University, China
Long Yuan	University of New South Wales, Australia
Lu Zhang	University of Arkansas, USA

Luiza Antonie   University of Guelph, Canada
Maciej Grzenda   Warsaw University of Technology, Poland
Mahito Sugiyama   National Institute of Informatics, Japan
Mahsa Salehi   Monash University, Australia
Makoto Kato   Kyoto University, Japan
Marco Maggini   University of Siena, Italy
Marzena Kryszkiewicz   Warsaw University of Technology, Poland
Md Zahidul Islam   Charles Sturt University, Australia
Meng Chang Chen   Academia Sinica, Taiwan
Meng Jiang   University of Illinois, USA
Miao Xu   RIKEN, Japan
Michael E. Houle   National Institute of Informatics, Japan
Michael Hahsler   Southern Methodist University, USA
Ming Li   Nanjing University, China
Ming Tang   Chinese Academy of Sciences, China
Ming Yin   Microsoft Research and Purdue University, USA
Mingbo Zhao   Donghua University, China
Min-Ling Zhang   Southeast University, China
Miyuki Nakano   Advanced Institute of Industrial Technology, Japan
Mohadeseh Ganji   University of Melbourne, Australia
Mohit Sharma   Walmart Labs, USA
Mostafa Haghir   Telecom Paristech, France
  Chehreghani
Motoki Shiga   GIFU University, Japan
Muhammad Aamir Cheema   Monash University, Australia
Murat Kantarcioglu   University of Texas at Dallas, USA
Nam Huynh   Japan Advanced Institute of Science and Technology,
                           Japan
Nayyar Zaidi   Monash University, Australia
Ngoc-Thanh Nguyen   Wroclaw University of Technology, Poland
Nguyen Le Minh   Japan Advanced Institute of Science and Technology,
                           Japan
Noseong Park   University of North Carolina at Charlotte, USA
P Sastry   IISc, India
P. Krishna Reddy   International Institute of Information Technology
                           Hyderabad, India
Pabitra Mitra   Indian Institute of Technology Kharagpur, India
Panagiotis Papapetrou   Stockholm University, Sweden
Patricia Riddle   University of Auckland, New Zealand
Peixiang Zhao   Florida State University, USA
Pengpeng Zhao   Soochow University, China
Philippe Fournier-Viger   Harbin Institute of Technology, China
Philippe Lenca   IMT Atlantique, France
Qi Liu   University of Science and Technology of China, China
Qiang Tang   Luxembourg institute of Science and Technology,
                           Luxembourg

Qing Wang	Australian National University, Australia
Qingshan Liu	Nanjing University of Information Science and Technology, China
Ranga Vatsavai	North Carolina State University, USA
Raymond Chi-Wing Wong	Hong Kong University of Science and Technology, Hong Kong, SAR China
Reza Zafarani	Syracuse University, USA
Rong-Hua Li	Shenzhen University, China
Rui Camacho	University of Porto, Portugal
Rui Chen	Samsung Research America, USA
Sael Lee	SUNY, South Korea
Sangkeun Lee	Korea University, South Korea
Sanjay Jain	National University of Singapore, Singapore
Santu Rana	Deakin University, Australia
Sarah Erfani	University of Melbourne, Australia
Satoshi Hara	Osaka University, Japan
Satoshi Oyama	Hokkaido University, Japan
Shanika Karunasekera	University of Melbourne, Australia
Sheng Li	Adobe Research, USA
Shirui Pan	University of Technology Sydney, Australia
Shiyu Yang	University of New South Wales, Australia
Shoji Hirano	Shimane University, Japan
Shoujin Wang	University of Technology Sydney, Australia
Shu Wu	NLPR, China
Shu-Ching Chen	Florida International University, USA
Shuhan Yuan	University of Arkansas, USA
Shuigeng Zhou	Fudan University, China
Sibo Wang	University of Queensland, Australia
Silvia Chiusano	Polytechnic University of Turin, Italy
Simon James	Deakin University, Australia
Songcan Chen	Nanjing University of Aeronautics and Astronautics, China
Songlei Jian	University of Technology Sydney, Australia
Steven Ding	McGill University, Canada
Suhang Wang	Arizona State University, USA
Sunhwan Lee	IBM Research, USA
Sunil Gupta	Deakin University, Australia
Tadashi Nomoto	National Institute of Japanese Literature, Japan
Takehiro Yamamoto	Kyoto University, Japan
Takehisa Yairi	University of Tokyo, Japan
Tanmoy Chakraborty	University of Maryland, College Park, USA
Teng Zhang	Nanjing University, China
Tetsuya Yoshida	Nara Women's University, Japan
Thanh Nguyen	Deakin University, Australia
Thin Nguyen	Deakin University, Australia
Tho Quan	John Von Neumann Institute, Vietnam

Tong Xu	University of Science and Technology of China, China
Toshihiro Kamishima	National Institute of Advanced Industrial Science and Technology, Japan
Trong Dinh Thac Do	University of Technology, Sydney, Australia
Tru Cao	Ho Chi Minh City University of Technology, Vietnam
Tuan-Anh Hoang	Leibniz University of Hanover, Germany
Tzung-Pei Hong	National University of Kaohsiung, Taiwan
Vien Ngo	Queen's University Belfast, UK
Viet Huynh	Deakin University, Australia
Vincenzo Piuri	University of Milan, Italy
Vineeth Mohan	Arizona State University, USA
Vladimir Estivill-Castro	Griffith University, Australia
Wai Lam	Chinese University of Hong Kong, Hong Kong, SAR China
Wang-Chien Lee	Pennsylvania State University, USA
Wei Ding	University of Massachusetts Boston, USA
Wei Kang	University of South Australia, Australia
Wei Liu	UTS, Australia, Australia
Wei Luo	Deakin University, Australia
Wei Shen	Nankai University, China
Wei Wang	University of New South Wales, Australia
Wei Zhang	ECNU, China
Weiqing Wang	University of Queensland, Australia
Wenjie Zhang	University of New South Wales, Australia
Wilfred Ng	HKUST, China
Woong-Kee Loh	Gacheon University, South Korea
Xian Wu	Microsoft Research Asia, China
Xiangfu Meng	Liaoning Technical University, China
Xiangjun Dong	Qilu University of Technology, China
Xiangliang Zhang	King Abdullah University of Science and Technology, Saudi Arabia
Xiangmin Zhou	RMIT University, Australia
Xiangnan He	National University of Singapore, Singapore
Xiangnan Kong	Worcestor Polytechnic Institute, USA
Xiaodong Yue	Shanghai University, China, China
Xiaofeng Meng	Renmin University of China, China
Xiaohui (Daniel) Tao	University of Southern Queensland, Australia
Xiaoying Gao	Victoria University of Wellington, New Zealand
Xin Huang	Hong Kong Baptist University, Hong Kong, SAR China
Xin Wang	University of Calgary, Canada
Xingquan Zhu	Florida Atlantic University, USA
Xintao Wu	University of Arkansas, USA
Xiuzhen Zhang	RMIT University, Australia
Xuan Vinh Nguyen	University of Melbourne, Australia

Xuan-Hieu Phan	University of Engineering and Technology – VNUHN, Vietnam
Xuan-Hong Dang	UC Santa Barbara, USA
Xue Li	University of Queensland, Australia
Xuelong Li	Chinese Academy of Science, China
Xuhui Fan	University of Technology Sydney, Australia
Yaliang Li	University at Buffalo, USA
Yanchang Zhao	CSIRO, Australia
Yang Gao	Nanjing University, China
Yang Song	University of Sydney, Australia
Yang Wang	University of New South Wales, Australia
Yang Yu	Nanjing University, China
Yang-Sae Moon	Kangwon National University, South Korea
Yanjie Fu	Missouri University of Science and Technology, USA
Yao Zhou	Arizona State University, USA
Yasuhiko Morimoto	Hiroshima University, Japan
Yasuo Tabei	RIKEN Centre for Advanced Intelligent Project, Japan
Yating Zhang	RIKEN AIP Centre/NAIST, Japan
Yidong Li	Beijing Jiaotong University, China
Yi-Dong Shen	Chinese Academy of Sciences, China
Yifeng Zeng	Teesside University, UK
Yim-ming Cheung	Hong Kong Baptist University, Hong Kong, SAR China
Ying Zhang	University of New South Wales, Australia
Yi-Ping Phoebe Chen	La Trobe University, Australia
Yi-Shin Chen	National Tsing Hua University, Taiwan
Yong Guan	Iowa State University, USA
Yong Zheng	Illinois Institute of Technology, USA
Yongkai Wu	University of Arkansas, USA
Yuan Yao	Nanjing University, China
Yuanyuan Zhu	Wuhan University, China
Yücel Saygın	Sabancı University, Turkey
Yue-Shi Lee	Ming Chuan University, Taiwan
Yu-Feng Li	Nanjing University, China
Yun Sing Koh	University of Auckland, New Zealand
Yuni Xia	Indiana University – Purdue University Indianapolis (IUPUI), USA
Yuqing Sun	Shandong University, China
Zhangyang Wang	Texas A&M University, USA
Zhaohong Deng	Jiangnan University, China
Zheng Liu	Nanjing University of Posts and Telecommunications, China
Zhenhui (Jessie Li)	Pennsylvania State University, USA
Zhiyuan Chen	University of Maryland Baltimore County, USA
Zhongfei Zhang	Binghamton University, USA
Zhou Zhao	Zhejiang University, China

Sponsors

Contents – Part III

Community Detection and Network Science

Deep Learning Theory and Applications in KDD

Clustering and Unsupervised Learning

Privacy-Preserving and Security

Recommendation and Data Factorization

Social Network, Ubiquitous Data and Graph Mining

Feature Learning and Data Mining Process

Discovering High Utility Itemsets Based on the Artificial Bee Colony Algorithm

Wei Song$^{(\boxtimes)}$ and Chaomin Huang

College of Computer Science and Technology,
North China University of Technology, Beijing 100144, China
songwei@ncut.edu.cn

Abstract. Mining high utility itemsets (HUI) is an interesting research problem in data mining. Recently, evolutionary computation has attracted researchers' attention, and based on the genetic algorithm and particle swarm optimization, new algorithms for mining HUIs have been proposed. In this paper, we propose a new algorithm called HUI mining based on the artificial bee colony algorithm (HUIM-ABC). In HUIM-ABC, a bitmap is used to transform the original database that represents a nectar source and three types of bee. In addition to an efficient bitwise operation and direct utility computation, a bitmap can also be used for search space pruning. Furthermore, the size of discovered itemsets is used to generate new nectar sources, which has a higher chance of producing HUIs than generating new nectar sources at random. Extensive tests show that the proposed algorithm outperforms existing state-of-the-art algorithms.

Keywords: Data mining · High utility itemset · Artificial bee colony
Bitmap · Direct nectar source generation

1 Introduction

Unlike frequent itemset [1], high utility itemset (HUI) emphasizes quantity and profit, and has received increasing attention. Although several algorithms [6, 11, 12] have been proposed, enumerating all HUIs exactly cannot avoid the exponential problem of a very large search space when the number of items or size of the database is large.

Evolutionary computation (EC) methods, such as genetic algorithm (GA) [3] and particle swarm optimization (PSO) [9], are applied for mining HUIs recently. Using an EC technique, discovering most HUIs is a promising solution to the problem of the large search space of all HUIs. Two HUI mining algorithms, HUPE$_{UMU}$-GARM and HUPE$_{WUMU}$-GARM, based on the GA are proposed in [8]. The difference between them is that the minimum utility threshold is not required for the second algorithm. Premature convergence is the main problem of these two algorithms; that is, the two algorithms fall easily into local optima. Lin et al. proposed two algorithms, HUIM-BPSO$_{sig}$ [5] and HUIM-BPSO [4], for mining HUIs based on PSO. According to [4], HUIM-BPSO outperforms HUIM-BPSO$_{sig}$ using an OR/NOR-tree structure.

Unlike the GA and PSO, the artificial bee colony (ABC) algorithm [7] is an algorithm inspired by the foraging behavior of bees. Two distinguishing characteristics

© Springer International Publishing AG, part of Springer Nature 2018
D. Phung et al. (Eds.): PAKDD 2018, LNAI 10939, pp. 3–14, 2018.
https://doi.org/10.1007/978-3-319-93040-4_1

of the ABC algorithm are self-organization and decentralized control. To the best of our knowledge, there has not been any work applying ABC algorithm for mining HUIs.

Using the ABC algorithm, we propose a novel HUI mining algorithm called HUI mining based on the ABC algorithm (HUIM-ABC). First, we model the problem of mining HUIs from the perspective of the ABC algorithm. Second, a bitmap is used for both information representation and search space pruning, which can accelerate the HUI discovery process. Third, instead of randomly generating new candidates, the size information of the discovered HUIs is used for producing new nectar sources. Thus, more HUIs can be discovered within fewer iteration cycles. Extensive experimental results show that the HUIM-ABC algorithm outperforms three existing algorithms based on EC in terms of efficiency, number of results, and convergence speed.

2   Preliminaries

2.1   Problem of HUI Mining

Let $I = \{i_1, i_2,..., i_m\}$ be a finite set of items. Then, set $X \subseteq I$ is called an *itemset*. Let $D = \{T_1, T_2, ..., T_n\}$ be a transaction database. Each transaction $T_i \in D$, with unique identifier *tid*, is a subset of I.

The *internal utility* $q(i_p, T_d)$ represents the quantity of item i_p in transaction T_d. The *external utility* $p(i_p)$ is the unit profit value of item i_p. The *utility* of item i_p in transaction T_d is defined as $u(i_p, T_d) = p(i_p) \times q(i_p, T_d)$. The utility of itemset X in transaction T_d is defined as $u(X, T_d) = \sum_{i_p \in X \wedge X \subseteq T_d} u(i_p, T_d)$. The utility of itemset X in D is defined as $u(X) = \sum_{X \subseteq T_d \wedge T_d \in D} u(X, T_d)$. The *transaction utility* (TU) of transaction T_d is defined as $TU(T_d) = u(T_d, T_d)$.

The *minimum utility threshold* δ, specified by the user, is defined as a percentage of the total TU values of the database, whereas the *minimum utility value* is defined as $min\_util = \delta \times \sum_{T_d \in D} TU(T_d)$. An itemset X is called an HUI if $u(X) \geq min\_util$. Given a transaction database D, the task of HUI mining is to determine all itemsets that have utilities no less than min_util.

The *transaction-weighted utilization* (TWU) of itemset X [6] is the sum of the transaction utilities of all the transactions containing X, which is defined as $TWU(X) = \sum_{X \subseteq T_d \wedge T_d \in D} TU(T_d)$. X is a high transaction-weighted utilization itemset (HTWUI) if $TWU(X) \geq min\_util$. An HTWUI with k items is called a k-HTWUI.

2.2   ABC Algorithm

The main components of the ABC algorithm are nectar sources and artificial bees, and nectar sources (solutions) are refined by the artificial bees iteratively. The value of a nectar source is usually represented by a single number, and there are three types of bee in the ABC algorithm: employed bees, onlooker bees, and scout bees.

Let S_i ($i = 1, 2, ..., SN$) be the ith solution with a D-dimensional vector, where SN is the number of nectar sources that are generated randomly initially.

The number of employed bees equals the number of nectar sources. Every employed bee produces a new solution from the old solution using

$$V_{m,j} = S_{m,j} + \varphi\left(S_{m,j} - S_{n,j}\right) \tag{1}$$

where j is a random dimension index in $\{1, 2, ..., D\}$, n is a randomly selected nectar source differentiating from m, and φ is a random number in the range $[-1, 1]$. If the fitness value of the newly generated solution is better than that of the old solution, the old solution is forgotten and the new solution is memorized. Otherwise, the old solution is kept.

When all employed bees have finished their searching process, they share the fitness (nectar) information of their solution (nectar sources) with the onlookers. The number of onlooker bees is also SN. Each onlooker bee selects one of the memorized nectar sources depending on the fitness value obtained from the employed bees. The probability that a food source will be selected can be obtained from

$$P_i = \frac{fitness_i}{\sum_{j=1}^{SN} fitness_j} \tag{2}$$

where $fitness_i$ is the fitness value of the ith nectar source.

After the nectar source is selected, Eq. 1 is used again by an onlooker bee to generate a new solution. If the fitness value of the new solution is better than that of the old solution, the bee memorizes the new solution and forgets the old solution.

If a nectar source cannot be improved further within a predetermined number of cycles, that nectar source is assumed to be abandoned. Then the nectar source abandoned by the bees is replaced with a new nectar source by one scout bee using

$$S_{m,j} = x_j^{min} + \gamma(x_j^{max} - x_j^{min}) \tag{3}$$

where γ is a random number in $[0, 1]$, and x_j^{min} and x_j^{max} are the lower and upper bounds of dimension j, respectively.

The employed, onlooker, and scout bees' phases repeat until the termination condition is met.

3 Mining HUIs Using the ABC

3.1 Bitmap Item Information Representation

A bitmap is used to transform the original database. The *bitmap* of D is an $n \times m$ Boolean matrix $B(D)$ with entries from the set $\{0, 1\}$. The entry in $B(D)$ that corresponds to transaction T_j $(1 \leq j \leq n)$ and item i_k $(1 \leq k \leq m)$ is denoted by (j, k), which is in the jth row and kth column of $B(D)$. The value of (j, k) is defined by

$$B_{j,k} = \begin{cases} 1, & \text{if } i_k \in T_j, \\ 0, & \text{otherwise.} \end{cases} \tag{4}$$

that is, entry (j, k) of $B(D)$ is one if and only if item i_k is included in transaction T_j; otherwise, it is set to zero.

In $B(D)$, the *bitmap cover* of item i_k, denoted by $Bit(i_k)$, is the kth column vector. This naturally extends to itemsets: the bitmap cover of itemset X is defined as Bit (X) = bitwise-AND$_{i \in X}(Bit(i))$.

In addition to transforming the original database and representing itemset information, a bitmap can also be used for pruning in the HUIM-ABC algorithm.

Definition 1. Let V be a bit vector that corresponds to a nectar source, employed bee, onlooker bee, or scout bee; and X the itemset that V represents. If $Bit(X)$ is only composed of zeros, V is called an *unpromising bit vector* (UPBV); otherwise, V is called a *promising bit vector* (PBV).

Thus, if a newly generated bit vector is an UPBV, the fitness value computation can be avoided. This technique is called the *PBV check* (PBVC) pruning strategy in the HUIM-ABC algorithm.

3.2   Modeling HUI Discovery Using the ABC

After transforming the database into a bitmap, it is natural to encode each nectar source in a bit vector. The length of this vector is equal to the number of 1-HTWUIs. Based on the bit vector, a new nectar source generation in the HUIM-ABC algorithm can be achieved by only changing the value of 1 bit in the old nectar source, either from zero to one, or from one to zero.

To discover HUIs from the transaction database, the utility of the itemset is used as the fitness function directly:

$$fitness(S_i) = u(X) \tag{5}$$

where X is the union of items in the nectar source S_i if its value is set to one. Thus, X is an HUI if $fitness(S_i) \geq min\_util$.

3.3   Direct Nectar Source Generation for Scout Bees

For the standard ABC algorithm, the scout bees search for a new nectar source randomly. Because the search space of HUIs is huge, when using the simple random search strategy, the number of discovered HUIs could be limited within a certain number of cycles, and the computational cost is high. Thus, the question that arises is: can we generate more promising new nectar sources as early as possible so that more HUIs can be discovered within a certain number of cycles and the computational cost can also be lowered? The answer is "yes" by using the sizes of discovered HUIs.

Park et al. [10] indicated that the processes in the initial iterations of the Apriori algorithm dominate the total execution cost; that is, the candidate sets with smaller sizes (number of items) are crucial to improving the performance of the Apriori algorithm. This inspires us to deduce that the resulting itemsets' sizes are not evenly distributed. Thus, the scout bees can search more frequently in new nectar sources that represent certain sizes that could generate more HUIs. This is called the *direct nectar source generation* (DNSG) strategy in the HUIM-ABC algorithm.

Specifically, we set m buckets, denoted by BK_1, BK_2,..., BK_m, and each bucket stores the number of HUIs whose sizes are within a certain range, denoted by $BK_1.ran$, $BK_2.ran$,..., $BK_m.ran$. For example, the first bucket BK_1 stores the number of HUIs with sizes in $BK_1.ran$ (from 1 to 5), and the second bucket BK_2 stores the number of HUIs with sizes in $BK_2.ran$ (from 6 to 10). Let $BK_1.num$, $BK_2.num$,..., $BK_m.num$ be the number of itemsets in BK_1, BK_2,..., BK_m. Once a new HUI X is generated, and the size of X is in $BK_i.ran$, then $BK_i.num$ is incremented by one. It should be noted that the initial number of $BK_1.num$, $BK_2.num$,..., $BK_m.num$ is set to one because some HUIs with certain sizes may not be discovered during the early cycles. If the initial value is zero, then this type of HUI has no chance of being generated as a new nectar source until the cycle for which the corresponding number is not zero.

For each cycle of the ABC algorithm, after the employed bee phase and onlooker bee phase, the probability of generating new nectar sources whose sizes are in $BK_i.ran$ $(1 \leq i \leq m)$ is determined by

$$P_{nec}^i = \frac{BK_i.num}{\sum_{j=1}^{m} BK_j.num} \tag{6}$$

Using the DNSG strategy, the information about the discovered results is used. Thus, the new nectar sources are more promising for HUI discovery than using the simple random approach.

3.4 Algorithm Description

Based on the above discussion, the proposed HUIM-ABC algorithm for mining HUIs is described in Algorithm 1.

Algorithm 1. HUIM-ABC
1   Scan database D once. Delete 1-LTWUIs;
2   Represent the reorganized database using a bitmap;
3   Initialization( );
4   *gen*=1;
5   **while** *gen<=max_cyc* **do**
6   Employed_bees( );
7   Onlooker_bees( );
8   Scout_bees( );
9   *gen*++;
10   **end while**
11   Output all discovered HUIs.

In Algorithm 1, the transaction database is first scanned once to determine the high TWU single items (Step 1). In Step 2, the bitmap representation of the pruned database is constructed. The procedure Initialization (described in Algorithm 2) is called in Step 3. Then, the number of cycles is set to one (Step 4). The main loop (Steps 5–10) repeats the three phases of the employed bees, onlooker bees, and scout bees until the maximum cycle number is reached. The procedures of these three phases are described in Algorithms 3, 4, and 5, respectively. Finally, Step 11 outputs all discovered HUIs.

Algorithm 2. Procedure Initialization( )

  1   Initialize m buckets;
  2   **for** each nectar source S_i **do**
  3   S_i is initialized as a bit vector with length $|1\text{-HTWUI}|$;
  4   If S_i is an UPBV, generate a new S_i by changing 1 bit until it is a PBV;
  5   Get itemset X by unifying of items in S_i if its value is set to 1;
  6   **while** $fitness(S_i){\geq}min\_util$ **do**
  7         $X{\rightarrow}SHUI$;
  8   Update $BK_i.num$ $(1{\leq}i{\leq}m)$ using X;
  9   Initialize a new S_i by changing 1 bit of the original one;
10   If S_i is an UPBV, generate a new S_i by changing 1 bit until it is a PBV;
11   Get itemset X by unifying of items in S_i if its value is set to 1;
12   **end while**
13   Initialize an employed bee EB_i that corresponds to S_i;
14   Initialize an onlooker bee OB_i that corresponds to S_i;
15   **end for**

Algorithm 2 first initializes all buckets (Step 1). Then all nectar sources are initialized individually (Steps 2–15). Each source is first represented by a bit vector whose length is equal to the number of 1-HTWUIs (Step 3). Steps 4 performs the PBVC pruning discussed in Sect. 3.1. Step 5 determines the itemset that corresponds to the enumerating nectar source. The loop from Step 6 to Step 12 repeats until a new nectar source with a fitness value lower than the minimum utility value is generated. Step 7 records the newly discovered itemset. Here $SHUI$ is the set of discovered HUIs. Step 8 updates the number of one bucket. A new nectar source is generated by randomly changing 1 bit of the original vector (Step 9). Steps 10 also performs PBVC pruning. Step 11 determines the itemset that corresponds to the enumerating nectar source. The employed bees and onlooker bees are then initialized in Steps 13 and 14, respectively.

Algorithm 3 is used by the employed bees to generate the HUIs. The main loop from Step 1 to Step 18 processes all nectar sources individually. In Step 2, *count*, which is a parameter to indicate whether the loop (Steps 7–14) is executed, is set to zero. Then, an employed bee EB_i is mapped to the enumerating nectar source S_i in Step 3. A new employed bee is generated in Step 4. Step 5 performs PBVC pruning. Step 6 determines the itemset X that corresponds to EB_i. The loop from Steps 7–14 repeats until a new employed bee with a fitness value lower than the minimum utility value is generated. Step 8 records the newly discovered HUI. Step 9 updates the number of the bucket whose range covers the size of X. Step 10 generates a new employed bee. Step 11 also performs PBVC pruning. Step 12 determines the itemset X that corresponds to EB_i. Then the parameter *count* is incremented by one in Step 13. If the loop (Steps 7–14) is executed, then S_i is updated by the newest EB_i (Steps 15–17); otherwise, $S_i.trial$, which is a parameter that indicates the number of trials that fail to generate a better nectar source, is incremented by one in Step 17.

Algorithm 3. Procedure Employed_bees( )

---

 1   **for** i=1 to SN **do**
 2   count=0;
 3   Map EB_i to S_i;
 4   Generate a new EB_i by changing 1 bit of the original one;
 5   If EB_i is an UPBV, generate a new EB_i by changing 1 bit until it is a PBV;
 6   Get itemset X by unifying of items in EB_i if its value is 1;
 7   **while** $fitness(EB_i) \geq min\_util$ **do**
 8      $X \rightarrow SHUI$;
 9   Update $BK_i.num$ $(1 \leq i \leq m)$ using X;
10   Generate a new EB_i by changing 1 bit of the original one;
11   If EB_i is an UPBV, generate a new EB_i by changing 1 bit until it is a PBV;
12   Get itemset X by unifying of items in EB_i if its value is 1;
13   count++;
14   **end while**
15   **if** count>0 **then**
16   Maps S_i to EB_i;
17   **else** S_i.trial++;
18   **end for**

Algorithm 4. Procedure Onlooker_bees ( )

---

 1   **for** i=1 to SN **do**
 2   count=0;
 3   Map OB_i to S_j using Eq.2;
 4   Generate a new OB_i by changing 1 bit of the original source;
 5   If OB_i is an UPBV, generate a new OB_i by changing 1 bit until it is a PBV;
 6   Get itemset X by unifying of items in OB_i if its value is 1;
 7   **while** $fitness(OB_i) \geq min\_util$ **do**
 8      $X \rightarrow SHUI$;
 9   Update $BK_i.num$ $(1 \leq i \leq m)$ using X;
10   Generate a new OB_i by changing 1 bit of the original one;
11   If OB_i is an UPBV, generate a new OB_i by changing 1 bit until it is a PBV;
12   Get itemset X by unifying the items in OB_i if its value is 1;
13   count++;
14   **end while**
15   **if** count>0 **then**
16   Maps S_j to OB_i;
17   **else** S_j.trial++;
18   **end for**

Algorithm 4 is used by the onlooker bees to generate the HUIs of the HUIM-ABC algorithm. It is similar to Algorithm 3, except for Step 3; that is, instead of a one-to-one mapping between employed bees and nectar sources, an onlooker bee maps to a nectar source using the roulette wheel selection mechanism.

Algorithm 5. Procedure Scout_bees( )
1   **for** i=1 to SN **do**
2   **if** $S_i.trial \geq LT$ **then** // LT is the maximum number of trials
3   Initialize a new nectar source S_i using the DNSG strategy;
4   If S_i is an UPBV, generate a new S_i by changing 1 bit until it is a PBV;
5   Get itemset X by unifying of items in S_i if its value is 1;
6   **while** $fitness(S_i) \geq min\_util$ **do**
7         $X \rightarrow SHUI$;
8   Update $BK_i.num$ ($1 \leq i \leq m$) using X;
9   Generate a new S_i by changing 1 bit of the original one;
10   If S_i is an UPBV, generate a new S_i by changing 1 bit until it is a PBV;
11   Get itemset X by unifying of items in S_i if its value is 1;
12   **end while**
13   **end if**
14 **end for**

Algorithm 5 examines the nectar sources individually. If the number of trial times of one nectar source reaches the maximum number of trials (Step 2), then Step 3 uses the DNSG discussed in Sect. 3.3 to generate a new nectar source. Steps 4–12 are similar to the corresponding steps in Algorithms 3 and 4, with the function of recording the discovered HUIs and generating a new promising nectar source.

4   Performance Evaluation

We evaluate the performance of our HUIM-ABC algorithm and compare it with the HUPE$_{UMU}$-GARM [8], HUIM-BPSO$_{sig}$ [5], and HUIM-BPSO [4] algorithms.

4.1   Experimental Environment and Datasets

The experiments were performed on a supercomputer with 16-Core 2.00 GHz CPU, 48 GB memory, and running on 64-bit Microsoft Windows 7. Our programs were written in Java. Four datasets, downloaded from the SPMF data mining library [2], were used for evaluation, and their characteristics are presented in Table 1.

Table 1. Characteristics of the datasets

Dataset	Average transaction length	Number of items	Number of transactions
Chess	37	76	3,196
Mushroom	23	119	8,124
Accidents_10%	34	469	34,018
Connect	43	130	67,557

Similar to the work in [4], only 10% of the total Accident dataset was used for our experiment. For all experiments, the termination criterion was set to 2,000 iterations, the number of nectar sources was set to 10, and the number of buckets was 10.

4.2   Running Time

Figure 1 shows the execution time comparisons for the four datasets.

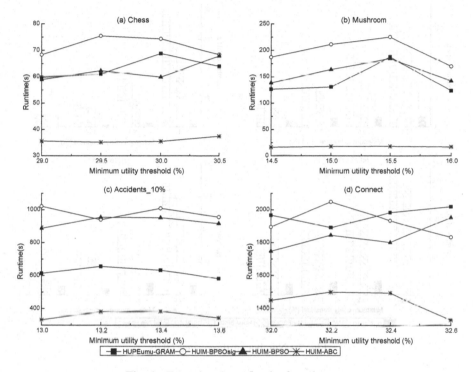

Fig. 1. Execution times for the four datasets

As shown in Fig. 1, the HUIM-ABC algorithm was always faster than the other three algorithms. In particular, the HUIM-ABC algorithm demonstrated relatively steady execution times on the Mushroom dataset, it was 7.19 times faster than HUPE$_{UMU}$-GARM, 8.06 times faster than HUIM-BPSO, and had an order of magnitude faster than HUIM-BPSO$_{sig}$, on average, for this dataset. The reason for the high performance of the HUIM-ABC algorithm can be explained by the use of bitmap representation. In addition to efficient bitwise operations, the PBVC pruning strategy can avoid the unnecessary computation of fitness values as early as possible. Furthermore, the real utility of PBVs can be verified from the recorded transactions rather than by resorting to using the entire database.

4.3   Number of Discovered HUIs

Because the EC-based HUI mining algorithm cannot ensure the discovery of all itemsets within a certain number of cycles, we also compared the number of discovered HUIs. The Two-Phase algorithm [6] was used to discover the actual and complete HUIs from the four datasets. The comparison results are shown in Fig. 2.

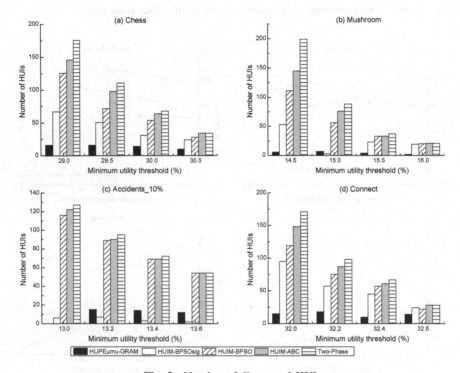

Fig. 2. Number of discovered HUIs

As shown in Fig. 2, the HUIM-ABC algorithm always discovered more HUIs than the other three EC-based algorithms. On average, the HUIM-ABC algorithm discovered 91.34%, 87.10%, 96.65%, and 91.59% of the total number of HUIs on the Chess, Mushroom, Accidents_10%, and Connect datasets, respectively. When the minimum utility threshold was high, for example, 30.5% for Chess, the HUIM-ABC algorithm discovered all the HUIs.

4.4   Convergence

The convergence performance results of the four algorithms are shown in Fig. 3.

For this set of experiments, we can observe that the convergence speed of $HUPE_{UMU}$-GARM was lower than that of the other three EC-based algorithms. This is because the GA-based algorithm suffered from the combination explosion problem in the evolution process composed of selection, crossover, and mutation. For the two PSO-based algorithms, HUIM-BPSO_{sig} and HUIM-BPSO, the latter demonstrated

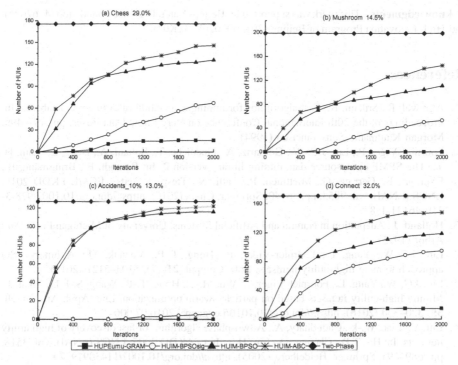

Fig. 3. Convergence performance comparison

better convergence speed because the OR/NOR-tree structure used by HUIM-BPSO avoided invalid combinations of particles. Although HUIM-BPSO demonstrated similar convergence performance to the HUIM-ABC algorithm on Accidents_10%, the HUIM-ABC algorithm always converged faster than HUIM-BPSO. The main reason is that the DNSG strategy of the HUIM-ABC algorithm generated new nectar sources by making use of the discovered HUIs rather than completely at random.

5   Conclusions

In this paper, we proposed an HUI mining algorithm called the HUIM-ABC algorithm based on the ABC algorithm. In the HUIM-ABC algorithm, the problem of HUI discovery was modeled from the perspective of the ABC algorithm. A bitmap was used for information representation. UPBVs could be detected by bitwise operations efficiently. Thus, the useless operation of utility calculation could be avoided. Furthermore, the sizes of discovered HUIs were recorded by different buckets. Based on this, new nectar sources were more likely to be generated within the range of most discovered results' sizes. Thus, more HUIs were mined within limited iteration cycles composed of employed bees, onlooker bees, and scout bees. As a result of the efficient strategies and optimizations introduced, the HUIM-ABC algorithm outperformed existing state-of-the-art HUI mining algorithms based on EC.

Acknowledgments. This work was supported by Beijing Natural Science Foundation (4162022), and High Innovation Program of Beijing (2015000026833ZK04).

References

1. Agrawal, R., Srikant, R.: Fast algorithms for mining association rules in large databases. In: Proceedings of the 20th International Conference on Very Large Data Bases, pp. 487–499. Morgan Kaufmann, San Francisco (1994)
2. Fournier-Viger, P., Lin, J.C.-W., Gomariz, A., Gueniche, T., Soltani, A., Deng, Z., Lam, H. T.: The SPMF open-source data mining library version 2. In: Berendt, B., Bringmann, B., Fromont, É., Garriga, G., Miettinen, P., Tatti, N., Tresp, V. (eds.) ECML PKDD 2016. LNCS (LNAI), vol. 9853, pp. 36–40. Springer, Cham (2016). https://doi.org/10.1007/978-3-319-46131-1_8
3. Holland, J.: Adaptation in Natural and Artificial Systems. University of Michigan Press, Ann Arbor (1975)
4. Lin, J.C.-W., Yang, L., Fournier-Viger, P., Hong, T.-P., Voznak, M.: A binary PSO approach to mine high-utility itemsets. Soft. Comput. **21**(17), 5103–5121 (2017)
5. Lin, J.C.-W., Yang, L., Fournier-Viger, J., Wu, M.-T., Hong, T.-P., Wang, S.-L.L., Zhan, J.: Mining high-utility itemsets based on particle swarm optimization. Eng. Appl. Artif. Intell. **55**, 320–330 (2016). https://doi.org/10.1016/j.engappai.2016.07.006
6. Liu, Y., Liao, W.-k., Choudhary, A.: A two-phase algorithm for fast discovery of high utility itemsets. In: Ho, T.B., Cheung, D., Liu, H. (eds.) PAKDD 2005. LNCS (LNAI), vol. 3518, pp. 689–695. Springer, Heidelberg (2005). https://doi.org/10.1007/11430919_79
7. Karaboga, D.: An idea based on honey bee swarm for numerical optimization. Technical report, Erciyes University, Engineering Faculty, Computer Engineering Department (2005)
8. Kannimuthu, S., Premalatha, K.: Discovery of high utility itemsets using genetic algorithm with ranked mutation. Appl. Artif. Intell. **28**(4), 337–359 (2014). https://doi.org/10.1080/08839514.2014.891839
9. Kennedy, J., Eberhart, R.: Particle swarm optimization. In: Proceedings of the IEEE International Conference on Neural Networks, pp. 1942–1948. IEEE Press, New York (1995). https://doi.org/10.1109/icnn.1995.488968
10. Park, J.S., Chen, M.S., Yu, P.S.: An effective hash-based algorithm for mining association rules. In: Proceedings of the 1995 ACM SIGMOD International Conference on Management of Data, pp. 175–186. ACM, New York (1995). https://doi.org/10.1145/223784.223813
11. Song, W., Zhang, Z., Li, J.: A high utility itemset mining algorithm based on subsume index. Knowl. Inf. Syst. **49**(1), 315–340 (2016). https://doi.org/10.1007/s10115-015-0900-1
12. Tseng, V.S., Shie, B.-E., Wu, C.-W., Yu, P.S.: Efficient algorithms for mining high utility itemsets from transactional databases. IEEE Trans. Knowl. Data Eng. **25**(8), 1772–1786 (2013). https://doi.org/10.1109/TKDE.2012.59

A Scalable and Efficient Subgroup Blocking Scheme for Multidatabase Record Linkage

Thilina Ranbaduge[1]([✉]), Dinusha Vatsalan[1,2], and Peter Christen[1]

[1] Research School of Computer Science, The Australian National University,
Canberra, ACT, Australia
thilina.ranbaduge@anu.edu.au
[2] Data61, CSIRO, Eveleigh, NSW, Australia

Abstract. Record linkage is a commonly used task in data integration to facilitate the identification of matching records that refer to the same entity from different databases. The scalability of multidatabase record linkage (MDRL) is significantly challenged with the increase of both the sizes and the number of databases that are to be linked. Identifying matching records across subgroups of databases is an important aspect in MDRL that has not been addressed so far. We propose a scalable subgroup blocking approach for MDRL that uses an efficient search over a graph structure to identify similar blocks of records that need to be compared across subgroups of multiple databases. We provide an analysis of our technique in terms of complexity and blocking quality. We conduct an empirical study on large real-world datasets that shows our approach is scalable with the size of subgroups and the number of databases, and outperforms an existing state-of-the-art blocking technique for MDRL.

Keywords: Entity resolution · Apriori · Depth-first · P-partite graph · Cliques

1 Introduction

Many organisations, including government agencies, businesses, and research centres, collect vast quantities of data on a daily basis [3]. To improve the efficiency and effectiveness of decision making, organisations increasingly require data from different databases to be integrated. Multidatabase record linkage (MDRL) is the process of identifying records that match (i.e. correspond to the same entities) across multiple databases [4]. The process of linking records across different databases is also known as 'data linkage' or 'entity resolution' [3].

This work was funded by the Australian Research Council under Discovery Projects DP130101801 and DP160101934. The authors would also like to thank Vassilios Verykios for his valuable feedback.

© Springer International Publishing AG, part of Springer Nature 2018
D. Phung et al. (Eds.): PAKDD 2018, LNAI 10939, pp. 15–27, 2018.
https://doi.org/10.1007/978-3-319-93040-4_2

A real-world example of MDRL would be a health surveillance system that continuously links data from hospitals and pharmacies. The data collected from these sources can facilitate the investigation of geographical and temporal effects of diseases, or adverse drug reactions in certain patient groups [2]. Such analyses require the linkage across subgroups of hospital and pharmacy databases collected at different locations since the linkage across all databases would not be sufficient to identify subsets of matching records such as cancer patients who visited a certain number of hospitals in a country, but not all hospitals.

In a MDRL context, potentially each record from one database needs to be compared with all records in all other databases to determine if a set of records corresponds to the same entity or not [15]. This becomes computationally expensive as the number of record pair comparisons grows exponentially with the number of databases to be linked [12]. To overcome this issue, blocking is generally applied in the linkage process [10]. Blocking reduces the record comparison space by grouping similar records, that likely correspond to true matches, based on the values of a set of attributes into the same block, while inserting records that likely correspond to non-matches into different blocks.

To identify subsets of matching records in multiple databases, it must be possible to link records from subgroups of databases. Though various blocking techniques have been developed for MDRL [12,15], these techniques cannot identify similar blocks across subgroups of databases because of two reasons. (1) Existing blocking techniques for MDRL are only capable of generating candidate blocks across all the databases, or for subgroups of a specific size [12], and (2) the application of a blocking technique multiple times for linking subgroups of different sizes is computationally infeasible due to the large number of potential subgroup combinations that need to be considered. This makes subgroup linkage for MDRL currently not scalable with an increasing number of databases.

We propose a subgroup blocking approach for MDRL which can efficiently identify blocks of records within a user specific range of subgroup sizes. Assuming d databases to be linked, we introduce two parameters, g_α and g_β, with $2 \leq g_\alpha \leq g_\beta \leq d$, to specify the minimum and maximum number of databases, respectively, that are to be included in a subgroup. Our approach accepts as input the sets of blocks generated from the databases that are to be linked, and it generates a set of candidate block tuples ($CBTs$) for each subgroup size from g_α to g_β. The records across the blocks in each CBT can then be compared in more detail [3].

To generate $CBTs$ that need to be compared across subgroups of different sizes, we first arrange the sets of blocks from different databases to be linked into a graph structure $\mathbf{G}$. We then introduce two constraints based algorithms for the generation of $CBTs$ by traversing over $\mathbf{G}$. Additionally, our approach allows for subgroups where some of the databases are fixed such that these databases must appear in every subgroup combination that is generated by our approach.

Contributions: We propose (1) a scalable subgroup blocking approach for MDRL that can be used under different real-world blocking scenarios, and (2) two constraint based graph traversal algorithms to generate candidate block tuples for subgroups of different sizes. (3) We analyse our subgroup blocking approach in terms of complexity and blocking quality, and (4) empirically evaluate the approach using large real-world databases with millions of records to validate its efficiency and effectiveness under different subgroup blocking scenarios. The results show that our subgroup blocking approach outperforms a state-of-the-art MDRL blocking approach [12] in terms of efficiency with no loss in effectiveness.

2   Related Work

Papadakis et al. [10] recently provided a survey of blocking techniques that have been proposed for record linkage. Most of these techniques are limited to linking two databases, and only few techniques have been developed for MDRL. Sadinle and Fienberg [15] proposed a probabilistic technique for linking multiple databases by extending the seminal work of Fellegi and Sunter [5]. This technique uses standard blocking [3] to group records into blocks, however it can only be used to match records across all the databases that are being linked.

Kong et al. [9] recently proposed an unsupervised technique to link records from multiple heterogeneous databases. This approach uses locality sensitive hashing (LSH) [7] to block each database to improve efficiency when generating candidate record tuples. The authors then adapted [5] to calculate the likelihood of a candidate record tuple being a match or a non-match based on several attributes. This approach, however, does not scale with the number of databases due to the large number of probability calculations required for each record tuple to be compared, and it cannot perform subgroup matching across databases.

Ranbaduge et al. [12] proposed a distributed blocking technique for privacy-preserving MDRL. This approach allows each owner to block its database independently by conducting a local clustering over their database to generate blocks. These blocks are then hashed using LSH [7] to identify the blocks that need to be compared across databases. While this approach can identify the blocks that need to be compared for a single subgroup size, it has to be run repeatedly for subgroups of different sizes, making the approach neither efficient nor effective in terms of identifying subgroup block tuples of different sizes.

In contrast to these existing multidatabase blocking approaches, our approach generates in one single run all candidate block tuples for subgroups of different sizes. We also allow the user to specify the minimum and maximum size of the subgroups, g_α and g_β, that are to be generated. The approach efficiently generates the most similar candidate block tuples by applying a constraints based pruning technique over a graph structure that is created based on the generated blocks. Fu et al. [6] proposed a graph based approach to match households across time in historical census data, while a MDRL meta-blocking approach recently proposed by Ranbaduge et al. [11] also uses a graph structure to remove redundant record pair comparisons. However, both these approaches are not capable of performing subgroup blocking across databases.

3 Subgroup Blocking Process

Let us assume d (≥ 2) de-duplicated databases are to be linked. We use the notation $\mathbf{D}_A, \mathbf{D}_B, \mathbf{D}_C$, and so on to represent different databases, while A_i, B_j, C_k and so on to represent the blocks generated for each corresponding database. The aim of our approach is to generate candidate block tuples ($CBTs$) for subgroup combinations across these d databases. A CBT is a tuple of blocks which consists of a maximum of one block per database, and blocks from at least two databases. As illustrated in Fig. 1, the approach consists of three steps:

1. *Potential Candidate Grouping:* As we discuss in Sect. 3.1, the block description pairs ($BDPs$) of each database are grouped into a set of candidate groups (**CG**) based on the similarities between their block representatives.
2. *Candidate Graph Generation:* A candidate graph **G** is constructed based on the generated **CG**. This requires an iteration over each group in the **CG** to create vertices and edges in **G**. Then weights (w) are calculated for each edge in **G**, as we discuss in detail in Sect. 3.2.
3. *Subgroup Candidate Generation:* $CBTs$ are generated for each subgroup combination using **G**. As we describe in Sect. 3.3, a *weight threshold* (w_t) is used to remove low weighted edges ($w < w_t$) in **G** to ensure the block pairs that have a low similarity are excluded from the CBT generation process.

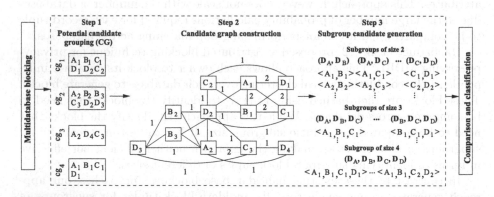

Fig. 1. Overview of our subgroup blocking approach with its three main steps. First, in step 1 the set of candidate groups (**CG**) is generated. A graph is constructed in step 2, where each block description pair (BDP) in a candidate group $cg \in \mathbf{CG}$ becomes a vertex. Two vertices are connected by an edge if they occur in the same cg. Each edge is assigned with a weight which in this example is the number of $cg \in \mathbf{CG}$ that contain a given pair of $BDPs$. In step 3, candidate block tuples ($CBTs$) are identified for subgroups that need to be compared. In this example, (g_α, g_β) is set to $(2, 4)$.

The two user defined parameters, g_α and g_β, with $2 \leq g_\alpha \leq g_\beta \leq d$, specify the minimum and maximum number of databases that are to be included into subgroup combinations, respectively. As an optional parameter, the user can also define the set of fixed databases (F) that must be included in every subgroup that is generated. Based on (g_α, g_β), our approach is capable of generating $CBTs$ for the following three scenarios with d databases in a MDRL context:

1. $g_\alpha = g_\beta = d$: This setting gives the linkage between all databases only, i.e. only sets of blocks that are to be compared across all databases are generated.
2. $2 \leq g_\alpha \leq g_\beta < d$: This setting gives all possible linkages for subgroups with at least g_α to at most g_β databases, i.e. $CBTs$ are generated for every subgroup combination between sizes g_α and g_β across all databases.
3. $F = \{\mathbf{D}_x, \mathbf{D}_y, \cdots, \mathbf{D}_z\}$, $|F| \leq g_\alpha \leq g_\beta < d$: This setting generates $CBTs$ for subgroups with size at least g_α to a maximum size of g_β out of d databases, where databases $\mathbf{D}_x, \mathbf{D}_y, \cdots, \mathbf{D}_z$ must appear in every subgroup.

As shown in Fig. 1, for example, if $F = \{\mathbf{D}_A\}$ and $(g_\alpha, g_\beta) = (2, 3)$ then the subgroup combinations that will be considered in our approach are, for subgroups of size 2: $(\mathbf{D}_A, \mathbf{D}_B)$, $(\mathbf{D}_A, \mathbf{D}_C)$, and $(\mathbf{D}_A, \mathbf{D}_D)$; and for subgroups of size 3: $(\mathbf{D}_A, \mathbf{D}_B, \mathbf{D}_C)$, $(\mathbf{D}_A, \mathbf{D}_B, \mathbf{D}_D)$, and $(\mathbf{D}_A, \mathbf{D}_C, \mathbf{D}_D)$.

To perform the linkage across subgroups of databases, as a prerequisite, first each database needs to be blocked. Any blocking technique [3,10] can be used to generate the set of blocks for each database, as long as the same technique is used on all d databases. We assume each database is blocked independently, as this provides flexibility and efficiency over the block generation process [12].

After blocking is completed, a *block description pair* (BDP) is generated for each block from each database. Each BDP (b, b_{rep}) consists of a block identifier (b) and a block representative (b_{rep}). A b_{rep} can be generated in different forms, such as a Min-Hash signature [7], a Bloom filter [12], or a phonetic encoding [3], as long as the same technique is used on all databases to generate the b_{rep}s for all blocks. The set of generated $BDPs$ of each database is then added to an overall set of $BDPs$, $\mathbf{B}$, which is used as input in our MDRL subgroup blocking approach. Hence, we assume this local multidatabase blocking phase to be a *black box*. We next describe the three steps of our approach in more detail.

3.1 Potential Candidate Grouping

In step 1 of our approach we identify the potential candidates among the sets of blocks of each database by grouping the $BDPs$ in $\mathbf{B}$ into a set of candidate groups $(\mathbf{CG})$. The grouping technique is based on the similarities calculated between the corresponding b_{rep}s that have been generated. For example, a Jaccard based LSH [7] technique can be used with b_{rep}s based on Min-Hash signatures, where blocks that hash to the same bucket become candidates and each bucket is considered as a candidate group (cg) that is added to the overall set $\mathbf{CG}$.

Each $cg \in \mathbf{CG}$ helps to identify the candidate blocks that need to be considered for comparison. If a pair of blocks appears in multiple cgs it is more likely that these blocks are more similar. As in Fig. 1 (step 1), for example, the pair (A_1, C_1) is more likely to be similar compared to (A_1, C_2), because (A_1, C_1)

Algorithm 1. Apriori Candidate Generation	**Algorithm 2.** Depth-first Candidate Generation

Input:
- **G**:   Undirected candidate graph
- F: Set of fixed databases
- d: Number of databases
- w_t: Weight threshold
- g_α, g_β: Minimum and maximum subgroup size

Output:
- **SBT**: Inverted index of subgroup block tuples

1: **SBT** $\leftarrow \{\}$, $K \leftarrow \{\}$
2: $k \leftarrow 2$
3: **if** $k < g_\alpha$ **then:**
4:     $K.\text{add}(\text{getCombinations}(k, g_\alpha, d, F))$
5: $K.\text{add}(\text{getCombinations}(g_\alpha, g_\beta, d, F))$
6: **foreach** $S \in K[2]$ **do:**
7:     $E \leftarrow \text{identifyEdges}(\mathbf{G}, S, w_t))$
8:     $C_2.\text{add}(E)$
9: **SBT**$[2] \leftarrow C_2$
10: $k \leftarrow 3$
11: **while** $k \leq g_\beta$ **and** $C_{k-1} \neq \emptyset$ **do:**
12:     $C_k \leftarrow \text{getCliques}(\mathbf{G}, K[k], w_t, C_{k-1})$
13:     **if** $g_\alpha \leq k$ **then:**
14:         **SBT**$[k] \leftarrow C_k$
15:     $k \leftarrow k + 1$
16: **return SBT**

Input:
- **G**:   Undirected candidate graph
- F: Set of fixed databases
- d: Number of databases
- w_t: Weight threshold
- g_α, g_β: Minimum and maximum subgroup size
- **B**:   Set of block description pairs

Output:
- **SBT**: Inverted index of subgroup block tuples

1: **SBT** $\leftarrow \{\}$, $K \leftarrow \{\}$
2: $K.\text{add}(\text{getCombinations}(g_\alpha, g_\beta, d, F))$
3: **foreach** $k \in K.keys()$ **do:**
4:     **foreach** $S \in K[k]$ **do:**
5:         **SBT**$[k].\text{add}(\text{genCandidates}(\mathbf{G}, S, w_t, \mathbf{B}))$
6: **return SBT**

Function genCandidates($\mathbf{G}, S, w_t, \mathbf{B}$):
7: **if** $|S| = 2$ **then:**
8:     **return** identifyEdges($\mathbf{G}, S, w_t$)
9: **else:**
10:     $\mathbf{C} \leftarrow []$
11:     $\mathbf{D}, L_{BDP} \leftarrow \text{getDBWithMinBlocks}(\mathbf{B}, S)$
12:     **foreach** $(b_i, b_{rep}) \in L_{BDP}$ **do:**
13:         $L_v \leftarrow \text{getNeighbours}(\mathbf{G}, \{S - \mathbf{D}\}, w_t, b_i)$
14:         $\mathbf{C} \leftarrow \text{genCandidates}(\mathbf{G}, \{S - \mathbf{D}\}, w_t, L_v)$
15:         $\mathbf{C}.\text{add}(\text{updateCandidates}(C, b_i))$
16:     **return C**

occurs in two cgs while (A_1, C_2) only occurs in one. Hence, this grouping reduces the overall number of block comparisons since only the blocks in a cg will be compared next in the linkage process. Reducing the number of block comparisons therefore reduces comparisons between records that are unlikely to be similar.

3.2   Candidate Graph Construction

In step 2 we construct the candidate graph $\mathbf{G} = (V, E)$ from **CG**, where **G** is an undirected d-partite graph [1]. The construction of **G** requires a pass over the **CG** where each BDP that appears in a $cg \in \mathbf{CG}$ becomes a vertex $v \in V$ in **G**.

An edge $e_{i,j} \in E$ is created between two vertices, v_i and v_j, if their corresponding BDPs (BDP_i and BDP_j) appear in the same cg, with the constraint that edges are created only between BDPs from different databases. As shown in Fig. 1 (step 2), a weight $w_{i,j}$ is calculated for each $e_{i,j}$. The weight $w_{i,j}$ of edge (BDP_i, BDP_j) can be computed in different ways based on the generated b_{rep}s, such as the similarity between the corresponding block representatives b_{rep}^i and b_{rep}^j, or the normalised cardinality of (BDP_i, BDP_j) which is $w_{i,j} = |\{cg : \forall_{cg \in \mathbf{CG}} (BDP_i, BDP_j) \in cg\}| / |\mathbf{CG}|$, where $| \cdot |$ represents the cardinality of a given set. These weights are used in the next step to generate the CBTs.

3.3   Subgroup Candidate Generation

In step 3 of our approach, CBTs are generated for each subgroup combination required. For a given subgroup of size g_α a CBT contains a maximum of one block identifier per database, and identifiers from at least g_α databases. Each CBT is a *clique* $c \in \mathbf{G}$, where each $c \subseteq V$, such that all pairs of vertices in c must be connected by an edge, i.e., $\forall v_i, v_j \in c : (v_i, v_j) \in E$. For generating CBTs we propose two candidate generation algorithms, as detailed below:

- *Apriori based Candidate Generation*: CBTs for subgroup sizes from g_α to g_β are generated using an Apriori based breadth-first search over $\mathbf{G}$ [1,8].
- *Depth-first based Candidate Generation*: CBTs for subgroup sizes from g_α to g_β are generated using a depth-first traversal through graph $\mathbf{G}$ [1,14].

Apriori based Candidate Generation (ACG): The proposed ACG approach is outlined in Algorithm 1. In lines 3 to 5, the function *getCombinations()* generates all the required subgroup combinations from g_α to g_β to be considered in the candidate generation process which are then added to an inverted index K using the subgroup sizes as keys. For example, with $(g_\alpha, g_\beta) = (2,3)$ of a linkage between databases $\mathbf{D}_A, \mathbf{D}_B$, and $\mathbf{D}_C$, $K[2]$ contains the list of subgroup combinations $(\mathbf{D}_A, \mathbf{D}_B)$, $(\mathbf{D}_A, \mathbf{D}_C)$, and $(\mathbf{D}_B, \mathbf{D}_C)$, while $K[3]$ contains the subgroup combination $(\mathbf{D}_A, \mathbf{D}_B, \mathbf{D}_C)$. If $g_\alpha > 2$ (line 3) ACG needs to generate the set of subgroups of sizes 2 to g_α because an Apriori based iterative approach [1,8] is used to identify cliques $(CBT$s) of size k from the cliques of size $k-1$ that were identified in the previous iteration (starting from pairs, i.e. $k = 2$).

To control the number of CBTs generated for each subgroup, we use a constraint, named as *weight threshold* (w_t), on the weight $w_{i,j}$ of each $e_{i,j} \in E$, which specifies the minimum weight that each $e_{i,j}$ must have in order to be considered in the CBT generation. w_t helps to control the density of $\mathbf{G}$ by efficiently pruning block pairs that have a low similarity. In practice, different w_ts can be specified for different subgroup combinations depending on user requirements.

The function *identifyEdges()* generates the trivial cliques of size $k = 2$ for each subgroup S which are the set of edges $E \in \mathbf{G}$ that satisfy w_t (line 7). In line 12, the function *getCliques()* traverses through $\mathbf{G}$ to identify all cliques of size k that satisfy w_t. These cliques are then added to the set C_k. Following the Apriori principle [1,8], in lines 11 to 15, ACG continues until k reaches g_β or no cliques were generated in the previous iteration (i.e. $C_{k-1} = \emptyset$). These generated CBTs are added to an inverted index **SBT** using subgroup combinations as keys.

Depth-first based Candidate Generation (DCG): The proposed DCG approach uses an iterative deepening depth-first search algorithm [14], as detailed in Algorithm 2. DCG generates CBTs from size g_α to g_β by incrementally expanding the size of subgroups. DCG uses multi-branch recursion that allows $\mathbf{G}$ to be searched progressively for similar blocks from the corresponding databases of a subgroup combination until the required CBT size is reached.

Similar to ACG, DCG starts with generating subgroup combinations for all databases by using the function *getCombinations()*. These combinations are then added to K (line 2 of Algorithm 2). For each subgroup S in K the recursive function *genCandidates()* is called to generate the set of CBTs (in lines 3 to 5). Similar to Algorithm 1, the function *identifyEdges()* is used to get the set of edges from $\mathbf{G}$ for each subgroup of size $|S|$ that satisfy the threshold w_t (line 8).

For subgroup sizes greater than 2, the function $genCandidates()$ first selects the database $\mathbf{D}$ with the minimum number of blocks using a function $getDBWithMinBlocks()$. This function returns $\mathbf{D}$ and its list of BDPs L_{BDP} in $\mathbf{B}$. This selection minimises the number of recursive branches in the CBT generation (line 11). Next, for each (b_i, b_{rep}) in L_{BDP} the function $getNeighbours()$ retrieves the neighbouring vertices for the remaining set of databases that connect with b_i (line 13). Those pairs of vertices that have an edge weight greater than or equal to w_t are added to the list of neighbouring vertices L_v.

For each b_i in L_{BDP} the function $genCandidates()$ is called recursively with the list L_v, w_t, and the set of remaining databases as inputs (line 14). Each of these recursive calls returns a list of block tuples C, where each tuple in C is updated with the current processed b_i using the function $updateCandidates()$ (line 15). This allows DCG to progressively generate CBTs until they reach the required size k. These CBTs are finally added to $\mathbf{SBT}$ (line 5).

4   Analysis of Subgroup Blocking

We analyse our approach in terms of complexity and blocking quality. We neither consider the block generation nor the comparison and classification techniques since they are outside the scope of our approach. Let us assume d databases are to be linked and the blocks of all these databases are added into the set $\mathbf{B}$.

Complexity: Though the generation of the set of candidate groups ($\mathbf{CG}$) in step 1 depends on the grouping technique used, it would require a complexity of $O(|\mathbf{B}|)$. We assume each candidate group $cg \in \mathbf{CG}$ contains d BDPs from different databases. In step 2, $\mathbf{CG}$ is used to construct the graph $\mathbf{G}$. This requires to iterate over each $cg \in \mathbf{CG}$ to add a vertex v to $\mathbf{G}$, and to create edges between vertices if they share the same cg. A cg with d BDPs generates $d(d-1)/2$ edges in $\mathbf{G}$. Hence, the construction of $\mathbf{G}$ has a complexity of $O(|\mathbf{CG}| \cdot d^2)$.

In ACG the CBT generation would require a complexity of $O(n^{g_\alpha})$ [8] if each of the g_α databases generates $n = |\mathbf{B}|/d$ BDPs. In line 10 of Algorithm 1, the generation of cliques of size k depends on the number of $(k-1)$ size cliques, C_{k-1}, generated previously. Hence, the complexity of ACG is $O(|E| + \sum_{g_\beta=3}^{g_\beta} \binom{g_\alpha}{g_\beta}|C_{g_\alpha-1}|^2)$, where $|E|$ is the number of edges (pairs of blocks) in $\mathbf{G}$. This becomes computationally infeasible when n, and g_β are increasing.

Based on the g_α and g_β settings, DCG requires to generate CBTs for $n_c = \sum_{g_\alpha=2}^{g_\beta} \binom{d}{g_\alpha}$ subgroup combinations. For each combination, DCG uses a multi-branch recursion to generate CBTs. In the function $genCandidates()$ of Algorithm 2, at each recursion a database $\mathbf{D}$ out of g_α databases with the minimum number of vertices is selected. Without loss of generality, let us assume the number of BDPs selected for $\mathbf{D}$ is m. m defines the maximum recursion branch factor of $\mathbf{D}$. The total number of vertex traversals in $\mathbf{G}$ for a given subgroup combination of size g_α can be calculated as $g_\alpha \cdot m + (g_\alpha - 1) \cdot m^2 + \cdots + 2 \cdot m^{g_\alpha-1} + m^{g_\alpha}$ which is $\sum_{i=1}^{g_\alpha}(g_\alpha + 1 - i) \cdot m^i$. Hence, for n_c subgroup combinations DCG has a complexity of $O(\sum_{g_\alpha=2}^{g_\beta} \sum_{i=1}^{g_\alpha} \binom{d}{g_\alpha}(g_\alpha+1-i) \cdot m^i)$. However, DCG would require a complexity of $O(n_c \cdot (|\mathbf{B}|/d)^{g_\alpha})$ if $\mathbf{G}$ is a complete graph with $\forall e \in E : e.w \geq w_t$.

Blocking Quality: In step 1 the effectiveness of the candidate grouping depends on the b_{rep}s and grouping technique used on those b_{rep}s. In step 2 the density of the graph **G** depends on the number of blocks in each $cg \in$ **CG** from different databases to be linked. A large number of blocks in a cg will increase the effectiveness of CBT generation as more edges are created in **G**.

In step 3 of our approach, the weight threshold w_t provides a trade-off between the quality and efficiency of the CBT generation process. The threshold w_t is used to prune edges (block pairs) with weights lower than w_t from the candidate generation process. A lower w_t will generate more cliques (CBTs) as more edges are considered in the CBT generation process for a given subgroup. This will increase the number of true matches as more block tuples are generated as cliques to be compared in the comparison and classification step, which will improve the effectiveness of the overall linkage. However, a lower w_t will potentially increase the overall runtime and space requirements of our approach because more edges are considered for a given subgroup combination.

Table 1. Datasets use in our experimental evaluation. 'Dataset Size (min-max)' is the minimum and maximum number of records in the databases of a dataset, and 'Avg. overlap' is the average percentage of records matched across the databases in a dataset.

Datasets	Number of databases (d)	Dataset size (min-max)	Avg. overlap	Provenance
NC-CLN	16	5,614,747–7,453,886	90%	Real
NC-DRT	16	72,903–1,308,796	20%	Real
NC-SYN	10	5,000–1,000,000	50%	Synthetic
UKCD	6	17,033–31,059	5,000 records	Real

5   Experiments and Discussion

For evaluation purposes we use two real-world datasets as outlined in Table 1. NC contains registration records of around 8 million voters from the US state of North Carolina (available from: http://dl.ncsbe.gov/). We use *given name*, *surname*, *city*, and *zipcode* as the blocking key attributes, as these are commonly used for record linkage [3,13].

We use 16 voter databases, collected at different points in time with two months interval between each pair of databases. The records in these databases can be grouped into three categories: (1) *exact matching*: those records (about the same person) that are exactly matching with each other, (2) *unique*: those records that are only appearing in one database, and (3) *updated*: those records where at least one attribute value has changed across two consecutive databases.

We use three different variations of the NC datasets, named as NC-CLN, NC-DRT, and NC-SYN. For NC-CLN we extracted *unique* and *exact matching* records from each database. Due to the skewness of *exact matching* records NC-CLN is only used to evaluate the scalability of our approach. For NC-DRT we extracted *unique* and *updated* records from each database. Since each update in an attribute value is considered as a modification in a record, NC-DRT is used to evaluate the blocking quality of our approach. To evaluate our approach with different levels of data quality, we use NC-SYN that contains 10 synthetic databases, as used in and provided by [12], which was created by extracting records from the original NC dataset. Some of these databases included corrupted records, where the corruption levels were set to 20% and 40% [12].

UKCD is another real dataset used in and provided by [6], consisting of census records collected from the years 1851 to 1901 in 10 year intervals for the town of Rawtenstall and surrounds in the United Kingdom. It contains approximately 150,000 records of 32,000 households with partial gold standard data (records manually linked by domain experts) for testing. Both NC and UKCD have been used for the evaluation of various other RL approaches [6,11,12] and we are not aware of any other available large real-world datasets that contain records from more than two databases that could be used to evaluate MDRL.

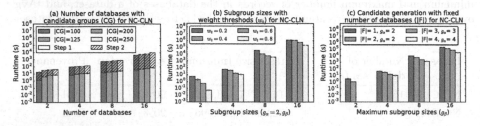

Fig. 2. The average runtime required with different (a) number of databases; (b) weight thresholds (w_t); and (c) number of fixed databases ($|F|$) for the NC-CLN datasets.

For comparison we use the state-of-the-art MDRL blocking technique proposed in [12] (named HDC for *Hashing based Distributed Clustering*) as this is the only existing technique we are aware of that can be used for subgroup blocking. For the prerequisites of our blocking approach, we use the same steps and parameter settings as used in HDC. Each database is blocked using a hierarchical clustering approach to generate an average of 100 to 1,000 blocks per database. Next, Min-Hash signatures are generated for each block as b_{rep}s. As in HDC, we hash these b_{rep}s into a set of buckets using locality sensitive hashing (LSH) [7] in step 1. Each bucket is added to the set **CG** as a candidate group. To measure the similarity between the b_{rep}s in step 2 we use the Jaccard coefficient [3].

We evaluated the complexity using runtime and memory, while the blocking quality was measured using pairs completeness (PC) and reduction ratio (RR) [3]. PC was calculated as the ratio of the number of matched records against the total number of true matched records across all databases. RR measures the reduction in the number of compared record pairs against the total number of record pairs. All experiments were conducted on a server with 64-bit Intel Xeon (2.4 GHz) CPUs, 128 GBytes of memory, and Ubuntu 14.04. We implemented all approaches using Python (version 2.7), and to allow repeatability the programs and test datasets are available from the authors.

Discussion: As shown in Fig. 2(a), the average runtime required for steps 1 and 2 of our approach increases linearly with the number of databases d. We noted that the average runtime also increases linearly with the number of candidate groups ($|\mathbf{CG}|$), which suggests that more edges are being generated in the graph $\mathbf{G}$. The average runtime decreases with an increase in the weight threshold (w_t) because the edges with lower similarity between their corresponding b_{rep}s are not considered in the CBT generation (see Fig. 2(b)). However, the runtime increases linearly with the size of subgroups as more combinations are considered in the CBT generation while the runtime decreases when more databases are included in the set of fixed databases F as shown in Fig. 2(c).

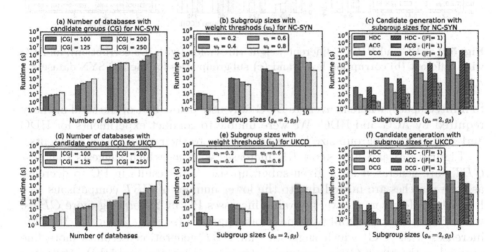

Fig. 3. Runtime results, where plots (a) to (c) and (d) to (f) show the results for the NC-SYN and UKCD datasets, respectively. Plots (a) and (d) show the total runtime with different number of databases, and (b) and (e) show the average runtime with different weight thresholds for different subgroup sizes. Plots (c) and (f) show the average runtime required for DCG compared with HDC and ACG.

Figures 3(a) and (d) show the total runtime increases exponentially as the number of subgroup combinations grows exponentially with d, while the runtime grows linearly with d for a given subgroup size. Similar to Fig. 2 the average runtime decreases with an increase in w_t as shown in Fig. 3(b) and (e). Figures 3(c) and (f) show DCG requires less runtime compared to HDC and ACG, which suggests that DCG is more efficient for CBT generation. However, ACG is still competitive with DCG if the number of blocks from each database remains small. We also measured the memory required for the CBT generation (not shown due

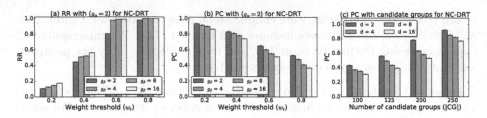

Fig. 4. (a) RR and (b) PC for different weight thresholds for different subgroup sizes, and (c) PC with different number of candidate groups (|**CG**|) for the NC-DRT datasets.

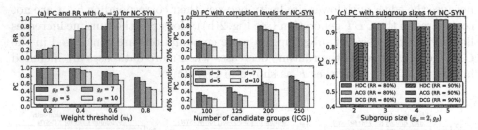

Fig. 5. (a) PC and RR for different weight thresholds (with 0% corruption), and PC with different (b) corruption levels and (c) subgroup sizes for the NC-SYN datasets.

to limited space) where in average DCG only uses below 10% of the total memory required by ACG and HDC. We were unable to conduct experiments for HDC and ACG with subgroup size larger than 5 due to their memory requirements.

Figures 4(a) and 5(a) show RR increases with w_t which suggests that less *CBT*s are generated for a given subgroup size. This results in PC to decrease as true matches are missed due to the lower number of *CBT* comparisons (see Fig. 4(b)). However, a lower w_t value increases PC by generating more *CBT*s, which increases the overall runtime of our approach (see Figs. 2 and 3). Also, PC increases with |**CG**| which suggests that *CBT* generation becomes more fine grained as the graph **G** becomes more dense (see Figs. 4(c) and 5(b)). After step 3 of our approach, we applied the same ranking algorithm as used in HDC to compare ACG and DCG with HDC [12]. HDC ranks the *CBT*s for comparison according to an approximation of RR. As shown in Fig. 5(c), we observed that ACG and DCG achieve the same PC as HDC which suggests that our approach can perform subgroup blocking more efficiently with no loss in effectiveness.

6   Conclusions and Future Work

We proposed a subgroup blocking approach for multidatabase record linkage (MDRL) based on a graph structure that is used for generating candidate block tuples using cliques. The evaluation on real datasets showed that our approach is scalable with the size of subgroups and it outperforms an existing MDRL blocking approach in terms of subgroup blocking. In future we aim to adapt pattern growth methodologies [1] and parallelisation into our blocking approach.

References

1. Aggarwal, C., Wang, H.: Managing and Mining Graph Data. Springer, New York (2010). https://doi.org/10.1007/978-1-4419-6045-0
2. Boyd, J., Ferrante, A., O'Keefe, C., et al.: Data linkage infrastructure for cross-jurisdictional health-related research in Australia. BMC Health Serv. Res. **12**, 480 (2012)
3. Christen, P.: Data Matching. Springer, Heidelberg (2012). https://doi.org/10.1007/978-3-642-31164-2
4. Elmagarmid, A., Ipeirotis, P., Verykios, V.: Duplicate record detection: a survey. IEEE TKDE **19**, 1–16 (2007)
5. Fellegi, I., Sunter, A.: A theory for record linkage. JASA **64**, 1183–1210 (1969)
6. Fu, Z., Christen, P., Zhou, J.: A graph matching method for historical census household linkage. In: Tseng, V.S., Ho, T.B., Zhou, Z.-H., Chen, A.L.P., Kao, H.-Y. (eds.) PAKDD 2014. LNCS (LNAI), vol. 8443, pp. 485–496. Springer, Cham (2014). https://doi.org/10.1007/978-3-319-06608-0_40
7. Indyk, P., Motwani, R.: Approximate nearest neighbors: towards removing the curse of dimensionality. In: Theory of Computing (1998)
8. Inokuchi, A., Washio, T., Motoda, H.: An apriori-based algorithm for mining frequent substructures from graph data. In: Zighed, D.A., Komorowski, J., Żytkow, J. (eds.) PKDD 2000. LNCS (LNAI), vol. 1910, pp. 13–23. Springer, Heidelberg (2000). https://doi.org/10.1007/3-540-45372-5_2
9. Kong, C., Gao, M., Xu, C., Qian, W., Zhou, A.: Entity matching across multiple heterogeneous data sources. In: ACM DASFAA (2016)
10. Papadakis, G., Svirsky, J., et al.: Comparative analysis of approximate blocking techniques for entity resolution. VLDB Endow. **9**, 684–695 (2016)
11. Ranbaduge, T., Vatsalan, D., Christen, P.: Scalable block scheduling for efficient multi-database record linkage. In: IEEE ICDM (2016)
12. Ranbaduge, T., Vatsalan, D., Christen, P., Verykios, V.: Hashing-based distributed multi-party blocking for privacy-preserving record linkage. In: Bailey, J., Khan, L., Washio, T., Dobbie, G., Huang, J.Z., Wang, R. (eds.) PAKDD 2016. LNCS (LNAI), vol. 9652, pp. 415–427. Springer, Cham (2016). https://doi.org/10.1007/978-3-319-31750-2_33
13. Randall, S., Ferrante, A., Boyd, J., Semmens, J.: The effect of data cleaning on record linkage quality. BMC Med. Inform. Decis. Mak. **13**, 64 (2013)
14. Russell, S., Norvig, P.: Artificial Intelligence: A Modern Approach (2009)
15. Sadinle, M., Fienberg, S.: A generalized Fellegi-Sunter framework for multiple record linkage with application to homicide record systems. JASA **108**, 385–397 (2013)

Efficient Feature Selection Framework for Digital Marketing Applications

Wei Zhang$^{(\boxtimes)}$, Shiladitya Bose, Said Kobeissi, Scott Tomko, and Chris Challis

Adobe Systems, Mclean, USA
wzhang@adobe.com

Abstract. Digital marketing strategies can help businesses achieve better Return on Investment (ROI). Big data and predictive modelling are key to identifying these specific customers. Yet the very rich and mostly irrelevant attributes(features) will adversely affect the predictive modelling performance, both computationally and qualitatively. So selecting relevant features is a crucial task for marketing applications. The feature selection process is very time consuming due to the large amount of data and high dimensionality of features. In this paper, we propose to reduce the computation time through regularizing the feature search process using expert knowledge. We also combine the regularized search with a generative filtering step, so we can address potential problems with the regularized search and further speed up the process. In addition, a progressive sampling and coarse to fine selection framework is built to further lower the space and time requirements.

1 Introduction

In order to effectively use their marketing budgets, more and more companies are resorting to digital marketing to segment and target their customers. By exploiting big data and utilizing predictive modelling techniques, such as logistic regression and random forest [2], companies can build all kinds of predictive models to help them identify customers of interest, e.g., these who are likely to purchased a product or those likely to churn. Usually each predictive model is trained using a large amount of user data so that the predictive model is generalizable to unseen data. In addition, each user is characterized by hundreds of attributes, including both categorical and numeric attributes, e.g., "geo-location" ("United States" or "Brazil") and "referring domain" ("www.google.com" or "www.yahoo.com"), "Revenue" and "Number of times activating a device", etc. These "attributes" serve as "features" for building predictive models and we use them interchangeably in this paper. The set of attributes is very comprehensive and covers all aspects of user visits, so that there are always discriminative features for building different predictive models. On the other hand, most of the attributes are redundant for each individual prediction task, i.e., the prediction output is usually determined by only a small number of influential attributes. Therefore, it is not a good idea to use all available features to build predictive models. The

D. Phung et al. (Eds.): PAKDD 2018, LNAI 10939, pp. 28–39, 2018.
https://doi.org/10.1007/978-3-319-93040-4_3

large number of irrelevant features will lead to unnecessary heavy computational cost, as well as causing overfitting, thus leading to inferior results [3].

In order to avoid using irrelevant features, one can ask users to select good (relevant) features for the predictive module to work with. However, it is extremely difficult for ordinary users to understand the high dimensional data and know a *priori* which features are relevant for each prediction task. Usually only a domain expert can choose a sensible set of features. The requirement of user input is a significant hindrance for general users. Moreover, even inputs from domain experts have no guarantee to be good enough. Therefore, we would like to do automatic feature selection to alleviate the problem. This is a challenging task because we are dealing with a huge amount of data with hundreds of millions of records, in addition to the large number of features. In this paper, we propose a novel feature selection framework to address the problem. First, we use semantic ranking to encode expert knowledge and guide the feature selection process, so it is more efficient. Second, we introduce a progressive sampling and coarse to fine selection framework to further reduce the space and time complexity.

The rest of the paper is organized as follows. Section 2 discusses related work. Section 3 describes the overall feature selection framework. Section 4 presents the semantic ranking guided feature selection algorithm. Section 5 presents the progressive sampling framework. We discuss experiments in Sect. 6 and conclude the paper in Sect. 7.

2 Related Work

Feature selection is a fundamental problem for data mining and machine learning tasks. Past works can be found in several review papers [5,9,13,19]. And it is still being actively investigated [22,23]. In general, feature selection methods can be divided into three categories: wrappers, filters and embedded [5].

Wrapper methods search for the best subset of features by scoring features using modelling algorithms (e.g., logistic regression). The brute force method of evaluating all possible feature combinations is computationally prohibitive. Greedy search strategies, including sequential forward search and backward elimination, are much more efficient and exhibit robustness against overfitting [5], although they can lead to local optima. The forward search procedure grows the selected feature set sequentially, adding one feature at each time. The backward elimination procedure starts with all features and progressively eliminates inferior features. Note even the greedy search will have heavy computational cost when the number features is large. The Minimum Redundancy Maximum Relevance(MRMR) [16] is a well known greedy algorithm based on Mutual information(MI). Along this line of research, Nguyen *et al.* [15] proposed an approximate global Mutual information based approach. More recently, Berrendero *et al.* [1] proposed to define relevance and redundancy in the MRMR algorithm using a distance correlation association measure.

Filter methods select features based on general measures, such as the correlation and mutual information between each feature and the dependent variable [17,21]. They are computationally more efficient than wrapper methods, because wrapper methods usually run the learning module many times before finding a good subset of features. Yet filter methods do not cater to any specific prediction model, so they usually lead to lower prediction performance. Hybrid approaches [6,7,11] exploit multiple feature selection algorithms to combine their advantages. For example, Hsu et al. [6] propose to use the filter approach to select a candidate set then refine them by the wrapper approach.

In this paper, we perform supervised feature selection as our data is labeled (e.g., order placed, customer churned). Unsupervised feature selection is also an important research field. Authors in [4,12] addressed the feature selection problem based on selecting features that minimize the reconstruction error of the data matrix. Tang et al. [20] used spectral clustering for multi-view data. Shao et al. [18] proposed an online unsupervised multi-view feature selection algorithm to deal with streaming data.

3   Overall Framework

We propose to incorporate expert knowledge into the greedy forward search process, as shown in Algorithm 1. The expert knowledge is represented by a semantic categorization scheme. A generative filtering step is applied to each category of features before doing greedy search, which bears some resemblance to the hybrid approach [6,7]. The generative filtering step is used mainly to eliminate extremely irrelevant features, which otherwise might be selected because of the categorization based search (explained in Sect. 4.2). Therefore we can use a very conservative criteria for filtering, without concerning that predictive features might be filtered out. The emphasize is on the semantic classification guided search algorithm, which lowers both the computational cost and the likelihood of selecting inferior features. On average, the combination of the knowledge guided search and the generative filtering yields almost 10× speedup over the standard forward search algorithm. In order to further reduce the computation and especially the space requirement, we introduce a progressive sampling and coarse to fine selection framework, which will be described in Algorithm 2. It uses Algorithm 1 as its core feature selection module in both the coarse selecting and the refining stages.

4   Feature Exploration Using Semantic Ranking and Generative Filtering

The greedy forward search algorithm goes as follows. Starting from an empty set $S = \{\}$, it first evaluates every single feature, then selects the one with the best predictive power and adds it to S as the 1st selected feature. Then from the remaining candidate features, it selects the 2nd feature which achieves the best predictive power together with the 1st feature. The process repeats until some stopping criteria are met, such as reaching the desired number of features. Let N be the number of original features, n be the number of selected features, assuming the cost of building and evaluating a model using k features is $C(k)$, the complexity of the forward search algorithm is:

$$T_{FS} = N \times C(1) + (N-1) \times C(2) + \ldots + (N - n + 1) \times C(n)$$

$N \times C(1)$ is the computational cost of selecting the 1st features, because each one of the N features has to be evaluated to determine which is the best. Then there are $N - 1$ remaining features to be selected from, each time the model is built and evaluated using 2 features (the 1st and one of the $N - 1$ features), thereby the computational cost is $(N - 1) \times C(2)$ for selecting the 2nd feature, so on and so forth. Usually $N \gg n$, so $N - n + 1 \approx N$, thus the complexity is approximately:

$$T_{FS} \approx N \times (C(1) + C(2) + \ldots + C(n)) \tag{1}$$

So the time complexity of the forward search algorithm is approximately linear to N.[1]

Besides the large number of features, the number of instances (visits in our dataset) in the marketing data is huge, easily on the order of hundreds of millions. Fortunately, it is not necessary to use the entire dataset to build a predictive model, instead we take a sample of data to do it. However, to ensure the sample has a reasonable coverage of the original data distribution and that the predictive model can be generalizable, we found empirically the sample size needs to be on the order of half million. Given this large sample size, the computational cost is rather high. For example, using a single computing node, an implementation based on the standard forward search algorithm took ~20 min to select from 25 features and build a logistic regression model. It will take hours to process hundreds of features, which is certainly not desirable to our customer. Thus, we introduce a semantic feature classification scheme to reduce the computational cost.

4.1 The Semantic Ranking Guided Feature Selection Algorithm

Although the complexity of the forward search algorithm is linear to N in general, digital marketing applications exhibit special properties which we can utilize to enhance the greedy search. As a result, we can reduce the time complexity without sacrificing the predictive performance. We first classify features into semantically meaningful categories and then do selection progressively. In addition, we use generative feature quality measures to eliminate features which are highly irrelevant. Algorithm 1 is a high level description of our feature selection algorithm.

---

[1] $C(n)$ depends on the modelling algorithm. It is linear w.r.t. n for many commonly used algorithms such as logistic regression and random forest [8]. In this case, the complexity term $C(1) + C(2) + \ldots + C(n) \propto n^2$. Without loss of generality, we use Eq. 1 to represent the complexity. The derivation in Sect. 4.1 holds either way.

Algorithm 1. The feature selection algorithm

---

1. Let $S = \{\}$, $F =$ the set of all features.
2. Classify F into subsets, $F = \{F_1, F_2, \ldots, F_k\}$, using the designed semantic based classifier, k is the number of semantic categories.
3. for $i = 1, 2, \ldots, k$
   (a) Estimate the feature relevance for all features in F_i based on correlation, then remove highly irrelevant features, and form a new subset F_i^*.
   (b) Do forward search within F_i^*:
       i. Select the feature f^* from F_i^*, so that the model Φ^* build using $\{S, f^*\}$ gives the best performance among all models built using $\{S, f\}, \forall f \in F_i^*$.
       ii. Check if Φ^* is better than the previous model (built using S), using model quality measures such as BIC:
           A. If not, stop selection within this category, $i = i + 1$, go to 3.
           B. If yes, $S = \{S, f^*\}$, remove f^* from F_i^*, go to 3b.
4. Output S as the set of selected features.

Semantic Based Feature Ranking. Based on our experiences of dealing with a large number of customer datasets, we have observed that there are direct relationships between customer behaviors (for example, conversion, revenue, etc.) and the segments they belong to. Some classes of attributes can clearly segment customers. For example, it is reasonable to expect that a customer's traits (e.g. "age", "gender" and "geo-location") will affect him/her purchasing a certain product. Attributes like "Browsers", "Operating systems" and "screen type" (mobile/desktop) also segment the customer population fairly well, although usually not as discriminatively as customer traits. Therefore, we propose to classify attributes based on their likelihood of segmenting the customer population, which in turn has a direct impact on the predictive power of these attributes with regard to prediction outputs such as conversion or revenue.

Designing the Semantic Classifier. We have formalized this observation and defined the following semantic classes for grouping attributes, ordered based on their likelihood of segmenting a customer population.

1. Visitor traits (attributes describing visitor traits like Age, Gender etc.)
2. Visitor sources (Browser, Operating System, mobile etc.)
3. Social channels (attributes describing interaction via social channels like Facebook, Twitter)
4. Visitor actions (attributes related to visitor actions, like click, view, purchase etc.)
5. Temporal (time related attributes.)
6. Others (attributes which cannot be classified such as customer defined attributes.)

Classification Method. We use a dictionary based approach to solve the classification problem, where we have identified a list of keywords for each semantic class. For example, "visitor traits" contains keywords such as "age", "gender", "city" and "zip". So if an attribute name contains the keyword "gender", it is classified to the class "visitor traits". The dictionary is designed by studying digital marketing glossaries in the literature. An attribute can belong to multiple classes at the same time. For instance, if we consider the attribute "Mobile Purchase" it will match successfully to both keywords "mobile" and "purchase". Thus it will be classified to the class "visitor actions" (Purchase) and "visitor source" (Mobile) simultaneously. Ultimately "Mobile Purchase" will be processed in the "visitor source" category because it has a higher rank. Note as the keywords in our dictionary may occur in various forms in the attribute names, the dictionary only keeps the stem [10] of each keyword. For example, "clicked" and "clicks" both appears in attribute names. They are converted to their stem "click" in the dictionary, so that we can successfully match attribute names to their corresponding keywords.

Computation Reduction Through Categorization. We now explain how the computational cost can be reduced by categorizing features. Suppose that we can divide the original feature set into two subsets of equal size $N/2$, and select $n/2$ features respectively from each subset. In the end, we will still have n features selected. Yet the time complexity becomes:

$$T \approx N/2 \times (C(1) + C(2) + \ldots + C(n/2))$$
$$+N/2 \times (C(n/2+1) + \ldots + C(n)) \tag{2}$$

The first part of the equation is the complexity of selecting $n/2$ features from the 1st subset, which has the size $N/2$. The 2nd part of the equation represents the complexity of adding additional $n/2$ features from the 2nd subset to the selected set. Equation 2 can be simplified as:

$$T \approx N/2 \times (C(1) + C(2) + \ldots + C(n)) = T_{FS}/2 \tag{3}$$

So we can reduce the computation by half in this case. Dividing features into more categories can bring more reduction in computation. The reduction will ultimately be determined by how features are categorized. For example, if 99% of features are put into one category, which is like no categorization at all, the reduction would be trivial. Our feature classification scheme has 6 classes and the numbers of features among categories are roughly balanced, so we avoid this situation in practice. On average, our classification scheme brings a $\sim 3\times$ speedup.

The semantic classification based search can also be thought of as using the expert knowledge to regularize the feature search process. Remember that we are taking a sample of data to process. When using a sample instead of the entire data, there is always some risk of performance degrading (in this case, meaning selecting suboptimal features). Based on the expert knowledge, we put features which are likely to be more influential in higher categories, so that they are more likely to be selected, thereby reducing the likelihood of selecting inferior features by the greedy search. The combination of human knowledge and the computer based search is one *key* to reaching our desired performance.

4.2   Combining with Generative Filtering for Better Performance

The user has the freedom of setting a wide variety of prediction targets, such as the probability of placing an order, or conversion. We cannot guarantee that the features in the top semantic category are good for all tasks. For some particular tasks, it is possible that none of the features in the 1st category is predictive. However, if we only do category based forward search, we will always select some features from the 1st category, even though they may not be good features for the particular task.

In order to address this problem, we apply a generative filtering process before performing forward search within each category. We use a correlation filter [17] to check the relevance of each feature, and screen out highly irrelevant features. As a result, when the features in the top category are irrelevant for a particular task, they will be removed and not affect the performance. This filtering step not only helps us to address the potential problem of selecting inferior features, it also lowers the computational cost since the number of candidate features is reduced after filtering. Although we set the threshold very conservatively, usually over 60% of the features can be eliminated as they are irrelevant.

5   Progressive Sampling and Feature Selection Framework

In order to build a predictive model, we have to query a cluster of servers to obtain the training data. Let M represents the number of samples that we use to build models. Usually M has to be large enough so that the model can be generalizable to unseen data. We typically set $M = 2.5 \times 10^5$. It may vary depending on tasks, but we use it as an example. N is the number of features to select from, which is also the number of attributes that we need to query from servers. In the case of $N = 1000$ attributes, the size of the returned table is $M \times N = 1000 \times 2.5 \times 10^5 = 2.5 \times 10^8$. Each table entry is represented by 8 byte, so that is 2 Gb of data to be queried and transferred. Clearly, querying too many attributes will put a big burden on servers. In addition, transferring the data back to the local workstation will also take more time and user experiences will deteriorate.

5.1   Coarse to Fine Implementation

In order to mitigate the problem, we propose a progressive sampling and coarse to fine selection framework as described in Algorithm 2. The computational cost for building and evaluating a model is typically linear (or superlinear) to the number of samples M. For example, the time complexity of logistic regression is $O(MN)$ (depending on the implementation, can also be $O(MN^2)$) [14], while

---

Algorithm 2. The coarse to fine framework

---

1. Querying servers with a reduced number of M^* samples, and use these samples to do feature selection and select a subset of features.
    (a) $M^* = M \times K/N$, where K is a constant, usually $M^* \ll M$.
    (b) Feed those samples to the feature selection module as described in Algorithm 1.
    (c) A subset of attributes will be selected. Denote the number of selected attributes as N^*, usually $N^* \ll N$.
2. Regular sampling for final feature selection and model building.
    (a) Query the server for M samples, but only for the selected N^* attributes.
    (b) Feed these samples again to the feature selection module and select predictive features from these N^* features.
    (c) Use the final set of features to build the output predictive model.

---

the time complexity of random forest is $O(NM \log(M))$ [8]. So the reduction in the number of samples will significantly reduce the computation.

In Algorithm 2, we empirically set $K = 50$. The choice of K is based on two factors: the number of the original features and the resulting M^*. M^* can not be too small as we need to build a meaningful model using these samples. For instance, if $N = 250$, in the 1st round of query, we will request $250000 \times 50/250 = 50000$ samples, which is a decent number of samples and good for our purpose. In the 1st stage of feature selection, the threshold for the correlation based filtering is set even more conservatively to reduce the risk of removing good features (because of partial data). Typically more than 85% of the features are passed to the guided search in this stage. The 1st stage will screen out around 80% of the features on average. The remaining features all show effectiveness in the 1st round of feature screening. This is like an expert preselecting a subset of candidate features after going through a portion of the data.

With the reduced number of samples in the 1st stage, the data is not as representative as that in the 2nd stage. If we use them for model building, the model will be suboptimal. Nevertheless, this is just the coarse selection; we only use it for eliminating the most unrelated attributes, not for building the final model. The final model, which is built in the 2nd stage, is reliable because it utilizes a large number of data and relevant features.

5.2   Time and Space Reduction Through Progressive Sampling

By coarse sampling with fewer data samples, we got a partial view of the data. Although not as comprehensive as the regular sample, the partial view is good enough for eliminating most non-predictive attributes. This is due to the fact that most attributes are virtually irrelevant for each individual prediction task. The reason is that our attribution set is overly comprehensive to cover all aspects of customer visits. Consequently, it is enough to identify most irrelevant attributes with a partial view of data (which still contains a significant number of samples, e.g., $M^* = 50000$ when $N = 250$ and $K = 50$). Therefore, in the 1st stage, the coarse sampling and feature screening process can help to eliminate most features. Then in the 2nd stage, we perform the full fledged sampling and feature selection from these selected (good) attributes, so we can build a truthful model.

Table 1 shows the speed improvement when building a logistic regression model, when setting $K = 50$. For simplicity, we omit the Big O notation, because they have the same constant factor. For example, for the proposed framework, when $N = 300$, the time complexity in the 1st stage is:

$$M^*N = M \times K/N \times N = MK = 50M$$

On average around $300 \times 20\% = 60$ features will get through the 1st screening stage. Then the time complexity in the 2nd stage will be $\sim 60M$, so the overall time complexity is roughly $50M + 60M = 110M$. While for the baseline case, which does feature selection directly without progressive sampling, the time complexity is $MN = 300M$. When building other models such as the decision tree and random forest, the reduction can be even higher since their time complexities are superlinear with regard to the number of samples.

Table 1. Time complexity of the proposed coarse to fine framework vs. the baseline (no progressive sampling), where N is the number of attributes and M is the number of data points (visit records). Note the Big O notation is omitted, since they have the same constant factors.

N	Baseline	Proposed	Computation reduction
100	100M	$\sim$70M	30%
200	200M	$\sim$90M	55%
300	300M	$\sim$110M	60%
500	500M	$\sim$150M	70%

Improvement on the space complexity is more significant. Because the 2nd stage is separate from the 1st stage and uses less space, the maximum space usage only happens in the 1st stage, which is $O(KM)$. While the space complexity for the baseline is $O(NM)$. N is usually several hundred as opposed to the typical setting of $K = 50$.

6   Experiments and Discussion

We have carried out extensive experiments with the proposed feature selection framework using Adobe.com web traffic data. The test dataset contains 42 million visit records. The number of attributes are typically 200–500, including: (1). 200 predefined attributes, such as "Page Views", "Revenues" and "Geo-country"; (2) hundreds of ad hoc attributes, which are customer defined and different across users. They can have arbitrary names and meaning, for example a name can be "A Specific Search Event". In our experiments, 80% of data are used for training predictive models and the remaining 20% are used for testing. We use the F-measure, which is defined as $\frac{2 \times (precision \times recall)}{(precision + recall)}$, as the performance metric. The higher the F-measure, the better the performance, ranging from 0 to 1. We compared the proposed algorithm with the baseline forward search algorithm, using the same initial feature set. The resulting F-measures were usually quite similar. On average, the difference is less than 0.03. On balanced data samples, the F-measures are typically in the range of [0.6, 0.9] depending on different prediction tasks. So the difference is less than 5% of the F-measure. The selected features usually have significant overlaps although they are not always the same. As we mentioned before, the features are very comprehensive and redundant. So it is not surprising that different combinations of features lead to models with comparable performances. In addition, we are processing a sample of data rather than the entire data. This also contributes to some variations in the selected set of features, because data samples will be different each time. Nevertheless, what we really care about is the predictive performance. Our experiments show that the resulting model is as good as the model produced by the standard algorithm. The proposed Algorithm 1, which combines the knowledge guided greedy search and the generative filtering, is on average almost 10 times as fast as the baseline greedy forward search algorithm. The speed and space improvements due to the progressive sampling framework vary by the initial number of features, as explained in Sect. 5.2. In a typical scenario of $N = 300$, we get approximately a 2× increase in speed and an 80% reduction in the space requirement.

7   Conclusion

In this paper, we introduce an efficient feature selection framework for doing predictive analytics in the digital marketing domain. We show that the computational cost of the greedy feature search can be further reduced by feature classification. First, we use the expert knowledge to categorize features and regularize the feature search process. Thus instead of the blind search, the features which are more likely to be influential are evaluated before other features. This helps us to retain good features and speed up the process. Second, we use generative filtering to alleviate the possible problem of category based search, and reduce the computation as well. In addition, a progressive sampling and coarse to fine selection framework is running on top of the core feature selection algorithm, which effectively reduces the time and space complexities. The proposed feature selection framework enables us to do feature selection from a large set of attributes on a big amount of data. This greatly enhances our user experiences for using the powerful predictive capability.

References

1. Berrendero, J.R., Cuevas, A., Torrecilla, J.L.: The mRMR variable selection method: a comparative study for functional data. J. Stat. Comput. Simul. **86**(5), 891–907 (2016)
2. Breiman, L.: Random forests. Mach. Learn. **45**, 5–32 (2001)
3. Deng, K.: Omega: On-line memory-based general purpose system classifier. Ph.D. dissertation, Carnegie Mellon University (1998)
4. Farahat, A.K., Ghodsi, A., Kamel, M.S.: An efficient greedy method for unsupervised feature selection. In: ICDM, pp. 161–170 (2011)
5. Guyon, I., Elisseeff, A.: An introduction to variable and feature selection. J. Mach. Learn. **3**, 1157–1182 (2003)
6. Hsu, H.H., Hsieh, C.W., Lu, M.D.: Hybrid feature selection by combining filters and wrappers. Expert Syst. Appl. **38**(7), 8144–8150 (2011)
7. Huda, S., Yearwood, J., Stranieri, A.: Hybrid wrapper-filter approaches for input feature selection using maximum relevance-minimum redundancy and artificial neural network input gain measurement approximation. In: ACSC, pp. 43–52 (2011)
8. Iyer, K.: Computational complexity of data mining algorithms used in fraud detection. Ph.D. dissertation, Pennsylvania State University (2005)
9. Kotsiantis, S.: Feature selection for machine learning classification problems: a recent overview. Artif. Intell. Rev. **42**, 1–20 (2011)
10. Kroeger, P.R.: Analyzing Grammar: An Introduction. Cambridge University Press, Cambridge (2005)
11. Lee, C.P., Leu, Y.: A novel hybrid feature selection method for microarray data analysis. Appl. Soft Comput. **11**(1), 208–213 (2011)
12. Mahdokht, M., Yan, Y., Cui, Y., Dy, J.: Convex principal feature selection. In: Proceedings of the 2010 SIAM International Conference on Data Mining, pp. 619–628 (2010)
13. Manikandan, P., Venkateswaran, C.J.: Feature selection algorithms: literature review. Smart Comput. Rev. **4**(3) (2014)
14. Minka, T.P.: A comparison of numerical optimizers for logistic regression. Unpublished draft (2003)
15. Nguyen, X.V., Chan, J., Romano, S., Bailey, J.: Effective global approaches for mutual information based feature selection. In: KDD, pp. 512–521 (2014)
16. Peng, H., Long, F., Ding, C.: Feature selection based on mutual information criteria of max-dependency, max-relevance, and min-redundancy. IEEE Trans. Pattern Anal. Mach. Intell. **27**(8), 1226–1238 (2005)
17. Senliol, B., Gulgezen, G., Yu, L., Cataltepe, Z.: Fast correlation based filter (FCBF) with a different search strategy. In: 23rd International Symposium on Computer and Information Sciences, pp. 1–4 (2008)
18. Shao, W., He, L., Lu, C., Wei, X., Yu, P.: Online unsupervised multi-view feature selection. In: ICDM, pp. 1203–1208 (2016)
19. Tang, J., Alelyani, S., Liu, H.: Feature selection for classification: a review (2014)
20. Tang, J., Hu, X., Gao, H., Liu, H.: Unsupervised feature selection for multi-view data in social media. In: SDM, pp. 270–278 (2013)

21. Torkkola, K.: Feature extraction by non-parametric mutual information maximization. J. Mach. Learn. Res. **3**, 1415–1438 (2003)
22. Venkateswara, H., Lade, P., Lin, B., Ye, J., Panchanathan, S.: Efficient approximate solutions to mutual information based global feature selection. In: ICDM, pp. 1009–1014 (2015)
23. Vinzamuri, B., Padthe, K.K., Reddy, C.K.: Feature grouping using weighted l1 norm for high-dimensional data. In: ICDM, pp. 1233–1238. IEEE (2016)

Dynamic Feature Selection Algorithm Based on Minimum Vertex Cover of Hypergraph

Xiaojun Xie and Xiaolin Qin[✉]

College of Computer Science and Technology,
Nanjing University of Aeronautics and Astronautics, Nanjing, China
{xiexj,qinxcs}@nuaa.edu.cn

Abstract. Feature selection is an important pre-processing step in many fields, such as data mining, machine learning and pattern recognition. This paper focuses on dynamically updating a subset of features with new samples arriving and provides a hypergraph model to deal with dynamic feature selection problem. Firstly, we discuss the relationship between feature selection of information system and minimum vertex cover of hypergraph, and feature selection is converted to a minimum vertex cover problem based on this relationship. Then, an algorithm for generating induced hypergraph from information system is presented, the induced hypergraph can be divided into two part: the original induced hypergraph and the added hypergraph with new samples arriving. Finally, a novel dynamic feature selection algorithm based on minimum vertex cover of hypergraph is proposed, and this algorithm only needs a small amount of computation. Experiments show that the proposed method is feasible and highly effective.

Keywords: Feature selection · Hypergraph · Minimum vertex cover
Dynamic reduct

1 Introduction

In rough set theory, feature selection is also called attribute reduction [1,2], the target of feature selection is to find a minimal feature subset from a problem domain while retaining a suitably high accuracy in representing the original features. Many static feature selection approaches have been developed in rough set [1–4]. In recent years, dynamic feature selection has attracted many scholars, which focus on dynamic data set.

Dynamic data set environment can be categorized along the following three situations: variation of features, variation of feature values and variation of samples. With variation of features, Wang et al. [5] developed a dimension incremental algorithm for dynamic data sets based on information entropy. Shu and Shen [6] presented an incremental attribute reduction algorithm based on incomplete decision system, which is based on the analysis of the update mechanism

© Springer International Publishing AG, part of Springer Nature 2018
D. Phung et al. (Eds.): PAKDD 2018, LNAI 10939, pp. 40–51, 2018.
https://doi.org/10.1007/978-3-319-93040-4_4

of the positive region in the case of the feature changed. With variation of feature values, Shu and Shen [7] presented the update mechanism of the positive region when the feature values of multiple samples vary simultaneously, and proposed an incremental attribute reduction based on positive region. Xie and Qin [8] proposed three update strategies of inconsistency degree for dynamic incomplete decision systems and established the framework of the incremental attribute reduction algorithm. With variation of samples, Fan et al. [9] proposed a dynamic attribute reduction algorithm, when a new sample is added. Liang et al. [10] proposed a group incremental attribute reduction algorithm, when a group of sample are added. Shu and Qian [11] presented a positive region update mechanism with respect to the adding and deleting of samples in incomplete decision systems and a dynamic attribute reduction algorithm is proposed. Yang et al. [12] proposed two incremental algorithms for attribute reduction with fuzzy rough sets are presented for one incoming sample and multiple incoming samples.

The graph-based method has been used to solve feature selection problem for static data [13], it is more effective to handle the large-scale data. The previous algorithm based on graph theory is not designed to handle dynamic data sets with sample arriving where one sample or multiple samples arrive successively. At the arrival of new samples, this algorithm needs to generate new induced hypergraph and re-compute a minimum vertex cover from the whole new dataset, which includes the accumulated samples and new incoming samples. To the best of our knowledge, graph-based feature selection for dynamic data with sample arriving has not yet been discussed so far, which is the motivation for this paper. We apply graph-based method in feature selection and propose a dynamic feature selection algorithm for dynamic data with sample arriving.

The rest of the paper is organized as follows. In Sect. 2, several basic concepts about rough set theory and graph theory are introduced. In Sect. 3, we provide an algorithm for generating the induced hypergraph and a method for updating the induced hypergraph with sample arriving, then a dynamic feature selection algorithm based on minimum vertex cover of hypergraph is proposed. In Sect. 4, Experimental comparisons are performed to show the effectiveness of our proposed algorithm. Finally, the conclusions are presented in Sect. 5.

2   Preliminaries

This section introduces some basic concepts and definitions about rough set and graph theory.

In rough set theory, data sets represented by a table called the information system, which is composed of a 4-tuple $S = (U, A, V, f)$, where $U = \{x_1, x_2, \cdots, x_{|U|}\}$ is the universe, a finite nonempty set of objects, the attribute set is $A = C \cup D$, C is a set of condition attributes and D is a set of decision attributes, V is the domain of attribute $a \in A$, $f: U \times (C \cup D) \to V$ is an information function.

Table 1. An exemplary information system

U	a_1	a_2	a_3	a_4	D
x_1	1	0	1	0	1
x_2	1	0	1	0	2
x_3	1	1	1	0	3
x_4	0	1	0	0	2
x_5	0	1	1	1	2

Let $S = (U, A, V, f)$ be an information system, given a subset of attributes $B \subseteq A$, $IND(B) = \{(x,y) \in U \times U | \forall a \in B \wedge f(x,a) = f(x,b)\}$ is the indiscernibility relation generated by B. $U/B = \{[x]_B | x \in U\}$ is the classification induced by B, where $[x]_B = \{y | \forall a \in B \wedge f(x,a) = f(y,a)\}$ is called an equivalence class of x with respect to B. The positive region of d with respect to B is expressed as $POS_B(D) = \bigcup_{X \in U/D} B_-(X)$, where $B_-(X)$ denotes the B-lower approximation of X, i.e., $B_-(X) = \{x \in U | [x]_B \subseteq X\}$ and $U/B = \{B_1, B_2, \cdots, B_{|U/B|}\}$ is the classification induced by decision attributes D.

Given an information system $S = (U, A, V, f)$, B is a condition attribute set, the discernibility matrix is defined as follows.

$$m_{ij} = \begin{cases} \{a \in C : f(x_i, a) \neq f(x_j, a) \wedge f(x_i, D) \neq f(x_j, D)\}, x_i, x_j \in U_{POS} \\ \{a \in C : f(x_i, a) \neq f(x_j, a)\}, x_i \in U_{POS}, x_i \in U' \\ \emptyset, else \end{cases}$$

where $U_{POS} = POS_C(D)$ and $U' = delrep(U - U_{POS})$, which denotes a set of deleting the repetitive objects in $U - U_{POS}$.

A discernibility function of an information system $S = (U, C \cup D, V, f)$ is a Boolean function of $|C|$ Boolean variables $a_1^*, a_2^*, \cdots, a_{|C|}^*$ corresponding to the attribute set $a_1, a_2, \cdots, a_{|C|}$, i.e., $f\left(a_1^*, a_2^*, \cdots, a_{|C|}^*\right) = \wedge \{\vee M | M \in \mathbf{M}, M \neq \emptyset\}$, where $\vee M$ is disjunction of all attributes in $\mathbf{M}$. An attribute set $B \subseteq C$ is a reduct of S $iff \wedge_{a_i \in B} a_i^*$ is a prime implicant of the discernibility function f_S. All the candidate reduct of S is denoted by $RED(S)$.

A hypergraph is a pair $H = (V, E)$, where $V = \{v_1, v_2, \cdots, v_{|V|}\}$ is the set of vertices and $E = \{e_1, e_2, \cdots, e_{|E|}\}$ is the set of hyperedges. A vertex cover of H is a subset $K \subseteq V$ such that $\forall e_i \in E$ has at least one endpoint in K. A minimum vertex cover is a vertex cover with the least number of vertices.

Given a hypergraph $H = (V, E)$, the function f_H for H is a Boolean function of $|V|$ Boolean variables $v_1^*, v_2^*, \cdots, v_{|V|}^*$ corresponding to the vertices $v_1, v_2, \cdots, v_{|V|}$, i.e., $f\left(v_1^*, v_2^*, \cdots, v_{|V|}^*\right) = \wedge \{\vee N(e_i) | e_i \in E\}$, where $N(e_i)$ is a set of vertices connected by the hyperedge e_i. A subset $K \subseteq V$ is a minimum vertex cover of H $iff \wedge_{v_i \in K} v_i^*$ is a prime implicant of the Boolean function f_H. $COV(H)$ is a set of all minimal vertex covers of H.

Based on above, attribute reduction of an information system and minimum vertex cover of a hypergraph both can be obtained via Boolean formula. Therefore, finding a reduct of an information system can be translated into a minimum vertex cover problem of a hypergraph.

3 Dynamic Feature Selection Algorithm Based on Minimum Vertex Cover of Hypergraph

3.1 The Induced Hypergraph

Definition 1. *Given an information system $S = (U, C \cup D, V, f)$, the discernibility matrix is $\mathbf{M} = m_{ij}$ and $M^* = \{M \in \mathbf{M} | M - \emptyset\}$. The induced hypergraph from S is $H = (V, E)$, where $V = C$ and $E = M^*$.*

In Definition 1, each attribute in an information system is a vertex in the induced hypergraph and each non-empty element of the discernibility matrix is a hyperedge of the induced hypergraph. Obviously, non-empty elements M_i^* and M_j^* corresponding to the hyperedges e_i and e_j, if $M_i^* = M_j^*$ that hyperedges e_i and e_j are the same hyperedge. According to $E = M^*$, we can get $f_S = f_H$, if all the candidate reduct of S is $RED(S)$ and all minimal vertex covers of H is $COV(H)$, then $RED(S) = COV(H)$. In other words, if attributes B is a reduct of S, then vertices K corresponding to attributes B is a minimum vertex cover of the induced hypergraph of S, and if the vertices K is a minimum vertex cover of the induced hypergraph of H, that the attributes B corresponding to vertices K is a reduct of S. This is the connection between the reduct of information system and the minimum vertex cover of the induced hypergraph.

Algorithm 1. The detailed process of generating the induced hypergraph from an information system

---

Input: An information system $S = (U, C \cup D, V, f)$
Output: The induced hypergraph $H = (V, E)$
1 Compute the discernibility matrix $\mathbf{M} = m_{ij}$;
2 Initialize hypergraph $H = (V, E)$, where $V = C$ and $E = \emptyset$;
3 **for** *each $M \in \mathbf{M}$ and $M \neq \emptyset$* **do**
4   Add the hyperedge e (corresponding to M) into E, $E = E \cup \{e\}$;
5   **if** *the hyperedge e is a redundant hyperedge* **then**
6 | Remove the hyperedge e, $E = E - \{e\}$;
7   **end**
8 **end**
9 Return the induced hypergraph $H = (V, E)$;

Proposition 1. *Given an information system $S = (U, C \cup D, V, f)$, the induced hypergraph from S is $H = (V, E)$, the discernibility matrix is $\mathbf{M} = m_{ij}$, $M^* = \{M \in \mathbf{M} | M \neq \emptyset\}$ and two non-empty elements $M_i^*, M_j^* \in M^*$ corresponding to the hyperedges $e_i, e_j \in E$. If the vertices $K \subseteq V$ cover the hyperedge e_i and $M_i^* \subseteq M_j^*$, then the vertices K must cover the hyperedge e_j.*

Proof. If $M_i^* = M_j^*$, then hyperedges e_i and e_j are the same hyperedge. The vertices $K \subseteq V$ cover the hyperedge e_i, obviously, vertices $K \subseteq V$ must cover the hyperedge e_j. If $M_i^* \neq M_j^*$, i.e., $M_i^* \subset M_j^*$ and the vertices $K \subseteq V$ cover the hyperedge e_i. Therefore, $\exists v_k \in M_i^*$, that $v_k \in K$ and $M_i^* \subset M_j^*$, i.e., $v_k \in M_i^* \subset M_j^*$, hence, the vertices K must cover the hyperedge e_j. This completes the proof. $\qquad\Box$

According to Proposition 1, in the process of generating the induced hypergraph from an information system $S = (U, C \cup D, V, f)$, if the hyperedges $e_i, e_j \in E$ corresponding to the non-empty elements $M_i^*, M_j^* \in M^*$ and $M_i^* \subseteq M_j^*$, then the hyperedge e_j is a redundant hyperedge. The redundant hyperedge e_j can be deleted or merge into the hyperedge e_i. The detailed process of generating the induced hypergraph from an information system $S = (U, C \cup D, V, f)$ is presented in Algorithm 1.

In Algorithm 1, the induced hypergraph can be obtained. Based on this induced hypergraph, the reduct of the information system is the minimum vertex cover of this induced hypergraph.

Example 1. Table 1 is an exemplary information system $S = (U, C \cup D, V, f)$, where $U = \{x_1, x_2, x_3, x_4, x_5\}$ and $C = \{a_1, a_2, a_3, a_4\}$. Through calculating, $POS_C(D) = \{x_3, x_4, x_5\}$ is obtained, then the the discernibility matrix is
$$M = \begin{bmatrix} \emptyset & \{a_1, a_3\} & \{a_1, a_4\} & \{a_2\} \\ \{a_1, a_3\} & \emptyset & \emptyset & \{a_1, a_2, a_3\} \\ \{a_1, a_4\} & \emptyset & \emptyset & \{a_1, a_2, a_4\} \end{bmatrix}.$$
According to Algorithm 1, we can generate the induced hypergraph $H = (V, E)$, where $E = \{\{a_2\}, \{a_1, a_3\}, \{a_1, a_4\}\}$. The induced hypergraph is shown in Fig. 1(a), the minimum vertex cover of this induced hypergraph is $K = \{a_1, a_2\}$. Therefore, the reduct of the information system in Table 1 is $R = \{a_1, a_2\}$.

3.2   Updating Minimum Vertex Cover of Hypergraph

An information system changes over time, and the database has not been set up from the beginning. Some new samples arrive into an information system, the reduct of this information system may change. Obviously, we can generate the new induced hypergraph and find the new minimum vertex cover of it. This method is easily taken into consideration, but it is not a recommendable method. How to dynamically update the minimum vertex cover is the task confronting us. Based on the above analysis, the first problem we need to solve is how to dynamically update the induced hypergraph.

Definition 2. *Given an information system* $S_n = (U, C \cup D, V, f)$, *the induced hypergraph from* S *is* $H_n = (V, E)$, x' *is a new sample arrive into* S_n, *the new information system is* $S_{n+1} = (U \cup \{x'\}, C \cup D, V, f)$ *and the new induced hypergraph is expressed as* $H_{n+1} = (V, E \cup E_{ad})$, *where* E_{ad} *is new hyperedges generated by new sample* x', $E_{ad} = \bigcup_{y_i \in U} \{a \in C | f(y_i, a) \neq f(x', a) \wedge f(y_i, d) \neq f(x', d)\}$.

In Definition 2, adding new hyperedges to the original hypergraph H_n may produce some redundant hyperedges in H_n. These redundant hyperedges are expressed as E_{de}, hence, the original induced hypergraph is updated to $H_n = (V, E - E_{de})$. In hyperedges E_{ad}, if $e \in E_{ad}$ is a redundant hyperedge, we remove it. Removing all redundant hyperedges in E_{ad}, then the new induced hypergraph is updated to $H_{n+1} = (V, (E - E_{de}) \cup E'_{ad})$, where $E'_{ad} \subseteq E_{ad}$ is the non-redundant hyperedges in E_{ad}.

Proposition 2. *Given an information system $S_n = (U, C \cup D, V, f)$, the induced hypergraph from S_n is $H_n = (V, E)$, x' is a new sample arrive into S_n. If $\exists y \in U - POS_C(D)$ and $(x', y) \in IND(C)$, and H_{n+1} is the new induced hypergraph from, then $H_{n+1} = H_n$.*

Proof. According to Definition 2, M is the non-empty element in original discernibility matrix $\mathbf{M}$, E_{ad} is new hyperedges generated by new sample x' and $\exists y \in U - POS_C \wedge (x', y) \in IND(C)$, each $e \in E_{ad}$, the non-empty element M_e corresponding to e must satisfy $\exists M' \in \mathbf{M} \wedge M' = M_e$, that is to say e is a redundant hyperedge in H_n, i.e., $E'_{ad} = \emptyset$ and $E_{de} = \emptyset$. Because the new induced hypergraph is $H_{n+1} = (V, (E - E_{de}) \cup E'_{ad})$, $H_{n+1} = (V, E)$ is obtained. Obviously, $H_{n+1} = H_n$, this completes the proof. $\qquad\square$

Proposition 3. *Given an information system $S_n = (U, C \cup D, V, f)$, x' is a new sample arrive into S_n, the new induced hypergraph $H_{n+1} = (V, (E - E_{de}) \cup E'_{ad})$, the original induced hypergraph $H_n = (V, E - E_{de})$, K_1 is a minimum vertex cover of H_n. H_{n+1} is divided into two parts: the original part $H_n = (V, E - E_{de})$ and the added part $\Delta H_{n+1} = (V, E''_{ad})$, where $E''_{ad} = E'_{ad} - \{e \in E'_{ad} | e \in N(K_1)\}$, $e \in N(K_1)$ indicates that vertices K_1 cover hyperedge e. If K_2 is a minimum vertex cover of ΔH_{n+1}, then $K = K_1 \cup K_2$ is a minimum vertex cover of H_{n+1}.*

Proof. Suppose that vertices K cannot cover at least one hyperedge in H_{n+1}, i.e., vertices K cannot cover hyperedge $e \in E_{n+1}$. If this hyperedge $e \in E - E_{de}$, then vertices K_1 cannot cover e in H_n, i.e., K_1 is not a minimum vertex cover of H_n, this assumption is invalid. If this hyperedge $e \in E''_{ad}$, then vertices K_2 cannot cover e in ΔH_{n+1}, i.e., K_2 is not a minimum vertex cover of ΔH_{n+1}, this assumption is also invalid in this case. Therefore, K is a minimum vertex cover of H_{n+1}. This completes the proof. $\qquad\square$

Based on Propositions 2 and 3, if $E'_{ad} \neq \emptyset$, then $K = K_1$. Updating minimum vertex cover of H_{n+1} can be divided into two steps. Firstly, we update the original induced hypergraph H_n and update its minimum vertex cover. Then, we obtain the added hypergraph ΔH_{n+1} and compute its minimum vertex cover.

3.3   Dynamic Feature Selection Algorithm

As mentioned earlier, we can dynamically update the reduct, when some new samples are added into an information system, a dynamic feature selection algorithm is presented in the following.

In Algorithm 2, the time complexity of Step 3 is $O\left(|U|\right)$ in the worst case, Steps 5–11 update the original induced hypergraph and its minimum vertex cover, the time complexity of updating the original induced hypergraph is $O\left(|U|\right)$ in the worst case and the time complexity of updating the minimum vertex cover is $O\left(|R_1|\right)$, Steps 12–14 initialize the added hypergraph and compute its minimum vertex cover, the time complexity of initializing the added hypergraph is $O\left(|U|\right)$ and the time complexity of computing the minimum vertex cover of the added hypergraph is $O\left(|C|\,|E''_{ad}|\right)$. So in the worst case, the time complexity of Algorithm 2 is $O\left(|U'|\cdot|C|\cdot|U|\right)$.

Algorithm 2. Dynamic feature selection algorithm based on minimum vertex cover of hypergraph (DFSMVC)

Input: An information system $S = (U, C \cup D, V, f)$, the induced
   hypergraph H_n, the reduct of original information system R_1, the
   new samples U';
Output: A new reduct R.
1 Initialize the original induced hypergraph H_n and $R = R_1$;
2 **for** *each* $x' \in U'$ **do**
3 Compute the new hyperedges E'_{ad};
4 **if** $E_{ad} \neq \emptyset$ **then**
5   Compute E_{de} and E''_{ad};
6   Update the original induced hypergraph H_n;
7   **for** *each* $v \in R$ **do**
8   **if** $N\left(v\right) = \emptyset$ **then**
9     $R = R - \{v\}$;
10   **end**
11   **end**
12   Initialize the added hypergraph $\Delta H_{n+1} = (V, E''_{ad})$;
13   Compute the minimum vertex cover of ΔH_{n+1} is R_2;
14   $R = R \cup R_2$;
15 **end**
16 **end**
17 return the new reduct R;

Example 2. Table 2 is an information system $S = (U_0, C \cup D, V, f)$, with some new samples U' are added, the reduct of original information system is $R_1 = \{a_1, a_2\}$, the detailed process of dynamic feature selection based on the minimum vertex cover of hypergraph is as follows.

According to Example 1, we can see that the original induced hypergraph is $H_1 = (V, E_1)$, where $E_1 = \{\{a_2\}, \{a_1, a_3\}, \{a_1, a_4\}\}$, the minimum vertex cover of H_1 is $K_1 = \{a_1, a_2\}$. Figure 1 displays the hypergraphs in the different states.

☐ Adding the new sample x_6 to the information system, the original induced hypergraph becomes $H_2 = (V, E_2)$. Because of $(x_1, x_6) \in IND\left(C\right)$, so $H_2 = H_1$, i.e., the minimum vertex cover of H_2 is $K_2 = K_1 = \{a_1, a_2\}$.

Table 2. Original information system of Table 1 and the new added samples U'

	U	a_1	a_2	a_3	a_4	D
U_0	x_1	1	0	1	0	1
	x_2	1	0	1	0	2
	x_3	1	1	1	0	3
	x_4	0	1	0	0	2
	x_5	0	1	1	1	2
U'	x_6	1	0	1	0	3
	x_7	1	1	0	1	3
	x_8	0	1	0	0	3
	x_9	0	1	0	1	3

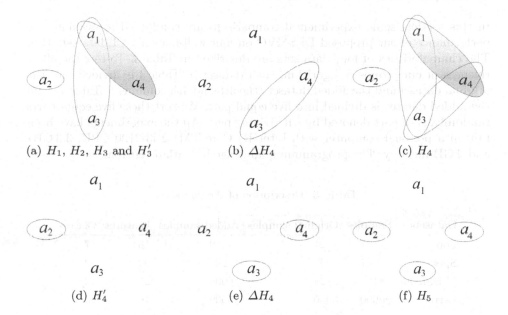

(a) H_1, H_2, H_3 and H'_3 (b) ΔH_4 (c) H_4

(d) H'_4 (e) ΔH_4 (f) H_5

Fig. 1. The hypergraphs in different states

☐ Adding the new sample x_7 to the information system, the original induced hypergraph becomes $H_3 = (V, E_3)$, After calculation, we can get $E_{ad} = \{\{a_2, a_3, a_4\}, \{a_1, a_3\}, \{a_1, a_4\}\}$, then $E'_{ad} = \emptyset$ is obtained. Hence, $H_3 = H_2$ and $K_3 = K_2 = \{a_1, a_2\}$ is a minimum vertex cover of H_3.

☐ Adding the new sample x_8 to the information system, the original induced hypergraph becomes $H_4 = (V, E_4)$. After calculation, we can get $E_{ad} = \{\{a_2, a_3\}, \{a_3, a_4\}\}$, then $E'_{ad} = E''_{ad} = \{\{a_3, a_4\}\}$ and $E_{de} = \emptyset$ are obtained. Therefore, the original induced hypergraph is unchanged $H'_3 = H_3$ and the added hypergraph $\Delta H_4 = (V, \{\{a_3, a_4\}\})$, the minimal vertex of ΔH_4

is $\Delta K_4 = a_3$ or $\Delta K_4 = a_4$. According to this result, the minimum vertex cover of H_4 is $K_4 = K_3 \cup \Delta K_4 = \{a_1, a_2, a_3\}$ or $\{a_1, a_2, a_4\}$.

☐ Adding the new sample x_9 to the information system, the original induced hypergraph becomes $H_5 = (V, E_5)$. After calculation, we can get $E_{ad} = \{\{a_2, a_3, a_4\}, \{a_3\}, \{a_4\}\}$, then $E_{de} = \{\{a_1, a_3\}, \{a_1, a_4\}, \{a_3, a_4\}\}$ and $E'_{ad} = E''_{ad} = \{\{a_3\}, \{a_4\}\}$ are obtained. Therefore, the original induced hypergraph is $H'_4 = (V, E_4 - E_{de})$, the minimum vertex cover of H'_4 is $K_4 = K_4 - \{a_1, a_3, a_4\} = \{a_2\}$, the minimum vertex cover of added hypergraph $\Delta H_5 = (V, \{\{a_3\}, \{a_4\}\})$ is $\Delta K_5 = \{a_3, a_4\}$. Therefore, the minimum vertex cover of H_5 is $K_5 = K_4 \cup \Delta K_5 = \{a_2, a_3, a_4\}$, $R = \{a_2, a_3, a_4\}$ is a new reduct of the new information system.

4   Experimental Analysis

In this section, some experimental comparisons are conducted to evaluate the performance of our proposed DFSMVC on four well-known UCI data sets [14]. The characteristics of four data sets are described in Table 3. To test the effectiveness of our proposed algorithm, each dataset in Table 3 is divided into the original dataset and the added dataset (the 3th and 4th columns of Table 3), and the added dataset is divided into five equal part. We sort these five equal parts randomly, each part denoted by ith arriving part. All the experiments have been ran on a personal computer with Inter(R) Core(TM) 2 i3-2120 CPU, 3.3 GHz and 4 GB memory. The programming language is Matlab R2016a.

Table 3. Description of the datasets.

Datasets	Samples	Original samples	Added samples	Features	Classes
Zoo	101	21	80	16	7
Spect	267	67	200	22	2
Mushroom	8124	2124	6000	22	2
Letter	20000	5000	15000	16	26

We evaluate the feasibility of DFSMVC from the following two aspects. One is to compare the runtime of DFSMVC with that of the non-incremental algorithms, i.e., POSR in [4] and INCONR in [8]. The average computational time (s) and the number of reduct are recorded and the experiment results shown in Table 4. The other is to compare the classification accuracy of attributes selected by POSR and INCONR. The classification accuracy is conducted on the selected attribute reducts found by the three algorithms with classifier linear SVM, J48 and 3NN. All of the classification accuracies are obtained with 10-fold cross validation. In 10-fold cross validation, a given data set is randomly divided into 10 nearly equally sized subsets, of these 10 subsets, 9 subsets are used as training set, a single subset is retained as testing set to assess the classification accuracy.

Experimental results on four data sets are listed in Fig. 2, where Raw denotes the accuracies of classifiers on data sets with original feature set.

Table 4. Comparison of reduct size and computation time in six data set

Data set		DFSMVC		POSR		INCONR	
		Size	Time	Size	Time	Size	Time
Zoo	1th	4	0.031	5	0.218	4	0.046
	2th	4	0.028	5	0.234	4	0.052
	3th	4	0.044	4	0.187	4	0.031
	4th	4	0.037	5	0.296	5	0.045
	5th	5	0.043	5	0.281	5	0.059
Spect	1th	11	0.087	10	0.468	11	0.093
	2th	15	0.069	15	0.453	15	0.145
	3th	15	0.089	15	0.625	15	0.192
	4th	17	0.128	17	0.796	17	0.218
	5th	17	0.119	17	1.012	17	0.256
Mushroom	1th	1	2.591	1	6.187	1	5.156
	2th	3	2.971	4	16.171	4	7.514
	3th	3	3.424	4	26.484	4	9.846
	4th	4	3.698	4	38.859	4	11.252
	5th	4	4.192	5	52.391	4	14.531
Letter	1th	9	7.203	9	67.797	9	12.871
	2th	9	9.641	9	113.578	9	16.976
	3th	10	13.297	10	197.469	10	30.981
	4th	10	15.082	10	304.078	10	58.098
	5th	11	18.688	11	473.828	11	79.872

As shown in Table 4, the reduction results of these three algorithms are almost identical, and the computational time of our proposed algorithm DFSMVC can find a reduct in a much shorter time. The main reason is that our proposed algorithm DFSMVC can avoid some recalculation. Figure 2 displays the more detailed changes of the classification accuracy of features selected by DFSMVC, POSR and INCONR. The four curves of classification accuracy are non-monotonic with samples arriving and the trend of the curves DFSMVC, POSR and INCONR are basically the same. The classification accuracy of our proposed algorithm is higher than the other two algorithms in most cases.

Based on above analysis, we can conclude that our proposed algorithm is feasible and highly effective.

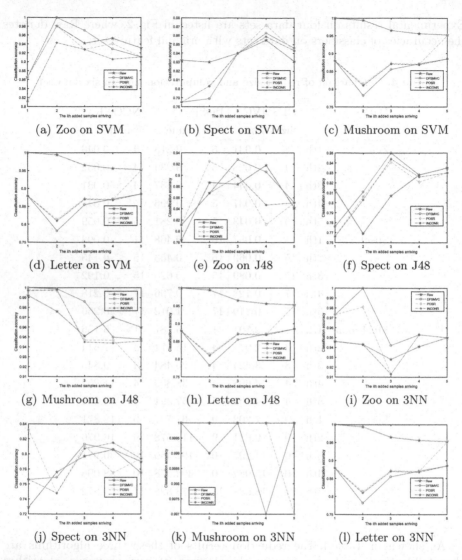

Fig. 2. Classification accuracy changes with respect to samples continuously arriving on classifier SVM, J48 and 3NN

5   Conclusions

Toward dynamic information system with samples arriving, we apply graph-based method for dynamic feature selection problem. Firstly, we introduce an induced hypergraph from the information system and transform the feature selection problem into a minimum vertex cover of this induced hypergraph. Then, we divide the new hypergraph into original party and added party with new sample arriving. Finally, the update mechanism of minimum vertex cover of new hypergraph is established and the dynamic feature selection algorithm based on minimum vertex cover of hypergraph is proposed, we only need a small amount of computation for computing minimum vertex cover of the added hypergraph and updating minimum vertex cover of the original induced hypergraph in our proposed algorithm. The experimental results show that, the proposed algorithm can obtain a reduct with a comparable classification accuracy in a much shorter time. This paper focus on streaming samples, graph-based dynamic feature selection method for streaming features is our future work.

Acknowledgments. The research was supported by The National Natural Science Foundation of China (grant nos. 61373015, 41301047, 61300052).

References

1. Swiniarski, R.W., Skowron, A.: Rough set methods in feature selection and recognition. Pattern Recogn. Lett. **24**(6), 833–849 (2003)
2. Degang, C., Changzhong, W., Qinghua, H.: A new approach to attribute reduction of consistent and inconsistent covering decision systems with covering rough sets. Inf. Sci. **177**(17), 3500–3518 (2007)
3. Qian, Y., Liang, J., Pedrycz, W., Dang, C.: Positive approximation: an accelerator for attribute reduction in rough set theory. Artif. Intell. **174**(9), 597–618 (2010)
4. Xu, Z.Y., Liu, Z.P., Yang, B.R.: A quick attribute reduction algorithm with complexity of $\max(O(|C||U|), O(|C|^2|U/C|))$. Chin. J. Comput. **29**(3), 391–399 (2006)
5. Wang, F., Liang, J., Qian, Y.: Attribute reduction: a dimension incremental strategy. Knowl.-Based Syst. **39**, 95–108 (2013)
6. Shu, W., Shen, H.: Updating attribute reduction in incomplete decision systems with the variation of attribute set. Int. J. Approx. Reason. **55**(3), 867–884 (2014)
7. Shu, W., Shen, H.: Incremental feature selection based on rough set in dynamic incomplete data. Pattern Recogn. **47**(12), 3890–3906 (2014)
8. Xie, X., Qin, X.: A novel incremental attribute reduction approach for dynamic incomplete decision systems. Int. J. Approx. Reason. **93**, 443–462 (2018)
9. Fan, Y.N., Tseng, T.L.B., Chern, C.C., Huang, C.C.: Rule induction based on an incremental rough set. Expert Syst. Appl. **36**(9), 11439–11450 (2009)
10. Liang, J., Wang, F., Dang, C., Qian, Y.: A group incremental approach to feature selection applying rough set technique. IEEE Trans. Knowl. Data Eng. **26**(2), 294–308 (2014)
11. Shu, W., Qian, W.: An incremental approach to attribute reduction from dynamic incomplete decision systems in rough set theory. Data Knowl. Eng. **100**, 116–132 (2015)
12. Yang, Y., Chen, D., Wang, H., Tsang, E.C., Zhang, D.: Fuzzy rough set based incremental attribute reduction from dynamic data with sample arriving. Fuzzy Sets Syst. **312**, 66–86 (2017). Theme: Fuzzy Rough Sets
13. Chen, J., Lin, Y., Lin, G., Li, J., Zhang, Y.: Attribute reduction of covering decision systems by hypergraph model. Knowl.-Based Syst. **118**, 93–104 (2017)
14. Lichman, M.: UCI Machine Learning Repository (2013)

Feature Selection for Multiclass Binary Data

Kushani Perera[1](✉), Jeffrey Chan[2], and Shanika Karunasekera[1]

[1] University of Melbourne, Melbourne, VIC 3010, Australia
bperera@student.unimelb.edu.au, karus@unimelb.edu.au
[2] RMIT University, Melbourne, VIC 3000, Australia
jeffrey.chan@rmit.edu.au

Abstract. Feature selection in binary datasets is an important task in many real world machine learning applications such as document classification, genomic data analysis, and image recognition. Despite many algorithms available, selecting features that distinguish all classes from one another in a multiclass binary dataset remains a challenge. Furthermore, many existing feature selection methods incur unnecessary computation costs for binary data, as they are not specifically designed for binary data. We show that exploiting the symmetry and feature value imbalance of binary datasets, more efficient feature selection measures that can better distinguish the classes in multiclass binary datasets can be developed. Using these measures, we propose a greedy feature selection algorithm, *CovSkew*, for multiclass binary data. We show that *CovSkew* achieves high accuracy gain over baseline methods, upto ∼40%, especially when the selected feature subset is small. We also show that *CovSkew* has low computational costs compared with most of the baselines.

1 Introduction

Binary datasets are commonly used for machine learning tasks in many domains including, text and document classification [1], image recognition [2] and gene analysis [3]. They include either data collected in binary format or non-binary data binarized for various purposes such as reducing data transmission costs and reducing processing costs in image matching [4]. Most of these datasets are large, with thousands of features [1,5] and require the removal of irrelevant features before using them for machine learning tasks. Feature selection is preferred to feature extraction or projection (e.g.: Principal Component analysis, deep learning) for removing irrelevant features because the latter can be hard to interpret [6]. Therefore, computationally efficient and accurate feature selection algorithms for binary datasets have wide applicability across many domains.

Despite many feature selection methods available in the literature [7–10], selecting a high quality feature subset with low computation costs still remains a challenge. However, the special properties of binary data, such as symmetry, makes it easier to achieve this goal for binary data. For example, measures equivalent to the commonly used measures such as mutual information can be

© Springer International Publishing AG, part of Springer Nature 2018
D. Phung et al. (Eds.): PAKDD 2018, LNAI 10939, pp. 52–63, 2018.
https://doi.org/10.1007/978-3-319-93040-4_5

	d_1 d_2 d_3 d_4	d_5 d_6 d_7 d_8	d_9 d_{10} d_{11} d_{12}
Cat	1 1 1 0	0 0 0 0	0 0 0 0
Compiler	0 0 0 0	1 1 0 0	0 0 0 0
Classifier	0 0 0 0	0 0 1 1	0 0 0 0
Dog	0 1 1 1	0 0 0 0	0 0 0 0
Class	Z Z Z Z	C C C C	P P P P

(a) Dataset 1

	d_1 d_2 d_3 d_4	d_5 d_6 d_7 d_8	d_9 d_{10} d_{11} d_{12}
Cat	1 1 1 0	0 0 0 0	0 0 0 0
Dog	0 1 1 1	0 0 0 0	0 0 0 0
Virus	1 1 0 0	1 0 0 0	0 0 0 0
Milk	1 1 1 0	0 0 0 0	0 0 0 0
Fish	0 0 1 1	0 0 0 0	0 0 0 0
Class	Z Z Z Z	C C C C	P P P P

(b) Dataset 2

Fig. 1. Example text document datasets. Row: a term, Column: a document, Class: document type, 1/0: Presence/absence of the term

computed with less computation costs in the case of binary data, considering only the probability distribution of a one feature value [11]. However, *utilisation of binary data properties for feature selection is a less investigated problem and feature selection methods which are specifically designed for binary data are rare.*

Previous works are either limited to two classes or text data for binary data [5,11] or are general feature selection methods [8,9]. They improve global prediction accuracy across all the classes and do not consider the accuracy for individual classes. Therefore, they do not provide good class separation across all classes in multiclass data, resulting in low overall accuracy. However, many binary datasets in domains such as document classification and character recognition are multiclass and require predicting each class equally well.

Example 1: Consider selecting three features from the binary dataset in Fig. 1a. In this dataset, each document (d_i) is categorized into three types (classes), Zoology (Z), Computer Science (C) and Physics (P). The rows represent the feature vector, the words appearing in the documents. Each feature value represents the presence (1) or absence (0) of a word, in the given document.

The best three features are {Cat, Compiler, Classifier}, because "Cat" distinguishes 75% of instances in zoology class, "Compiler" and "Classifier" together distinguish all the instances in the computer science class. Only one Zoology document, d_4, remains undistinguishable from the physics documents. This feature subset includes more features ("Compiler" and "Classifier") for distinguishing class C, than for class Z ("Cat"), which shows that different classes require different numbers of features to distinguish its instances from the rest.

However, *existing feature selection methods do not consider these different feature requirements of different classes.* As a result, all feature selection methods discussed above, select {Cat, Dog, Compiler} as the best three features (see Sect. 4 for details). In that case, 50% class C instances (d_7 and d_8) remain undistinguishable from class P instances. Selecting features in a one vs. all approach using the same measures, can improve the class separability quality to some extent [12]. However, selecting the classes in a round-robin method does not guarantee it [6,12]. For example, in the previous example, this also results in the same feature subset, {Cat, Dog, Compiler} as other methods. Therefore,

local feature selection measures, that focus on features that distinguish individual
classes are required for multiclass binary datasets.

Our Contribution

To select a better quality feature subset in multiclass binary datasets, we propose
two feature selection measures and a feature selection algorithm, which address
the shortcomings of the existing measures. Also, our proposed measures are
specifically designed for binary data, and therefore have lower computational
costs. The intuition is that *in a binary dataset, feature selection is possible by
considering only the distribution of one feature value.* We propose:

- Measures, *skewness* and *total coverage*, to accurately measure a feature's predictive power in a feature value imbalanced binary dataset.
- A feature selection algorithm, *CovSkew*, which uses the proposed measures,
  to gain high prediction accuracy for all the classes in a binary dataset, with
  a minimal number of features.

2 Related Work

We select filter feature selection methods over wrapper and embedded methods [13] because filter methods are classifier independent and incur low computational costs. Many filter methods, such as minimum Redundancy Maximum
Relevancy (mRMR) [8,10], are designed for non-binary data and incur unnecessary computational costs for binary data [11]. Class-dependent density based
feature elimination (CDFE) is an efficient feature selection measure catered for
binary datasets, yet limited to two class classification problems. Distinguishing
Feature Selector (DFS) [5], Gini index (GINI) [7], maximum Information Gain
(IG) [9] and χ^2 [5] are efficient algorithms ($nlog(n)$ computational complexity,
n = number of features) which are commonly used for binary data. However,
they ignore the feature redundancy and do not ensure class separability in multiclass data, resulting in poor prediction accuracy. Algorithms which consider
the class separability are designed for non-binary data and select the same number of features for each class, irrespective of the different feature requirements
for different classes [6,12]. *In summary, selecting a high quality feature subset
with low computational costs in multiclass binary data remains an open research
challenge.*

3 Preliminary Concepts

In this section and in Table 1, we introduce some new terms, used in the paper.

**Definition 1. *Sparse Value:* For a feature f in a binary dataset, consider n_0
to be the number of 0 s and n_1 to be the number of 1 s in the feature. If $n_0 > n_1$
then 1 is the sparse value and 0 is the common value and vice versa.**

Table 1. Frequently used definitions

D	Binary dataset	f	A single feature $f \in F$
C	Set of all classes in D	t	A single instance $t \in T$
F	Set of all features in D	T_c	Set of instances with class label c, $T_c \subseteq T$
T	Set of all instances in D	$k_{t,f}$	Feature value of instance t for feature f
c	A single class $c \in C$	S	Selected feature subset, $S \subseteq F$

We represent the sparse value with 1 and the common value with 0.

Definition 2. *Given an instance $t_i \in T$ and $S = \{f_1, \cdots, f_n\}$ is the selected feature subset, **the feature value assignment for t_i with respect to S** is defined as $v_{i,S} = [k_{t_i,f_1}, \cdots k_{t_i,f_n}]$.*

As described in Definition 3, we count the number of instance pairs in the dataset with different class labels and out of them, count the number of pairs with the same feature value assignment. We repeat the count for the instance pairs with same class labels.

Definition 3. *Let $t_i, t_j \in T$ ($i \neq j$), c_x the class label of t_x and $v_{x,S}$ is the feature value assignment of t_x with respect to S. Let $T_1 = $ set of all $\{t_i, t_j\}$ pairs where $c_i \neq c_j$, $T_2 = $ set of all $\{t_i, t_j\}$ pairs where $c_i = c_j$, $T_3 = $ set of all $\{t_i, t_j\}$ pairs where $v_{i,S} = v_{j,S}$.*

1. **The proportion of different class instances which have the same feature value assignment** *is defined as $pdiff(S) = \frac{|T_1 \cap T_3|}{|T_1|}$.*
2. **The proportion of same class instances which have the same feature value assignment** *is defined as $psame(S) = \frac{|T_2 \cap T_3|}{|T_2|}$.*

4   Problem Formulation

Our objective is similar to the one proposed in [14], hence we use it as the basis. Our objective is to (1) maximise the proportion of same class instances which have the same feature value assignment and (2) minimise the proportion of different class instances which have the same feature value assignment. That is to select a feature subset $S \subseteq F$ to

$$\max \Big(psame(S) - pdiff(S) \Big) \text{ s.t. } |S| = n \tag{1}$$

where $n \in \mathbb{Z}^+$, and $n \leq |F|$. In binary datasets the instances with same feature value assignments are the ones with a zero Hamming distance [4]. Figure 2a shows an example for an ideal six feature dataset according to the above objective.

Many existing feature selection methods fail to satisfy the above objective for multiclass binary datasets. We demonstrate this using DFS algorithm [5],

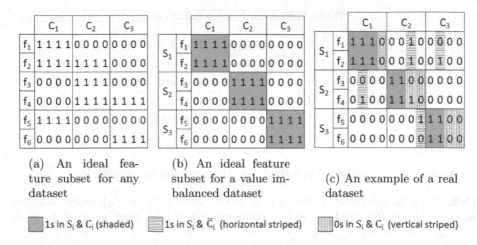

(a) An ideal feature subset for any dataset

(b) An ideal feature subset for a value imbalanced dataset

(c) An example of a real dataset

Fig. 2. Selecting features in a multiclass binary dataset. a column: an instance, a row (f_i): a feature, C_i: a class, $\bar{C}_i$: all classes excluding C_i, S_i: feature subset selected to distinguish C_i from the other classes

which has shown comparatively good accuracy for binary data, as a representative to solve Example 1. For each feature f, DFS assigns a score, $DFS(f) = \sum_{i=1}^{m} \frac{P(c_i|t)}{P(t|c_i) + P(t|\bar{c}_i) + 1}$, and selects the top k features with highest score. t is any of the two feature values of f, c_i is a class and m is the number of classes. DFS scores for features "Cat", "Compiler", "Classifier" and "Dog" are 0.8, 0.67, 0.67 and 0.8, respectively. Therefore, the selected feature subset, $S' = \{$Cat, Dog, Compiler$\}$. For S', $pdiff(S') = 8/48$, $psame(S') = 9/18$ and $psame(S')$ - $pdiff(S') = 0.33$. Similar to DFS, other methods, such as mRMR [7–9], also select the same feature subset. However, the optimal feature subset according to the above objective is $S = \{$Cat, Compiler, Classifier$\}$, because it results in $pdiff(S) = 4/48$ and $psame(S) = 11/18$ and $psame(S)$ - $pdiff(S) = 0.53$.

5 Our Approach

As selecting an optimal feature subset according to the objective in Eq. (1) is NP-hard [14], heuristic methods are required to find a nearly optimal feature subset. However, selecting a feature at a time to directly achieve the objective results in low prediction accuracy due to many local optimums. A more effective method for a feature value imbalanced binary dataset is to select a feature subset S_i for each class C_i, to distinguish c_i from the rest. As there are only two values in a binary dataset, this is possible by selecting an S_i which has a high sparse value density in C_i, compared to the other classes. Figure 2b shows the ideal feature subset, $S = \bigcup_{i=1}^{3} S_i$, we get in this one vs. all approach. To measure the sparse value density, we propose two new measures: *skewness* and *total coverage*, whose maximisation also achieves the above objective. Using these measures, we propose *CovSkew*, a heuristic feature selection algorithm.

5.1   Measuring the Sparse Value Distribution

In this section, we discuss our proposed measures, skewness and total coverage in detail. The proof of the theorem is omitted due to space limitations.

Definition 4. *Skewness of feature f in class c is given by:*

$$skewness(f, c) = P(C = c | f = 1) = \frac{No.\ of\ sparse\ values\ in\ f\ and\ c}{No.\ of\ all\ sparse\ values\ in\ f} \quad (2)$$

When a feature's sparse values are scattered across many classes the *skewness* is decreased. For example, in Fig. 1b, $skewness(Cat, Z) = 1$ and $skewness(Virus, Z) = 2/3$. For a feature subset, S_c, with multiple features, the total skewness of all features ($\sum_{f \in S_c} skewness(f, c)$) is considered.

Definition 5. *Given a feature subset S, and 0 and 1 are the common and sparse values of a feature, respectively, the total coverage of S within class c is:*

$$tcov(S, c) = \prod_{t \in T_c} \left(\sum_{f_j \in S} k_{t, f_j} + 1 \right) \quad (3)$$

Total coverage measures *the proportion of the class instances for which a feature subset has sparse values*. Total coverage of a *feature subset within a class increases with* (1) *the total number of sparse values its features have within the class.* (2) *the distribution of sparse values of features across the class instances.* For example, in Fig. 1b, {Cat, Dog} has a higher number of sparse values than {Cat, Fish} in Z (6 vs. 5) and $tcov(\{Cat, Dog\}, Z) = 36$ and $tcov(\{Cat, Fish\}, Z) = 24$. {Cat, Dog} has the same number of sparse values (6) as {Cat, Milk}, yet has a wider distribution of sparse values across the class instances. $tcov(\{Cat, Dog\}, Z) = 36$ and $tcov(\{Cat, Milk\}, Z) = 27$.

Theorem 1. *Given that S_i is a feature subset in a feature value imbalanced binary dataset, C_i is the ith class and $S = \bigcup_{i=1}^{|C|} S_i$ where $S_x \cap S_y = \emptyset$ for $x \neq y$, maximising $\sum_{f \in S_i} skewness(f, C_i)$ and $tcov(S_i, C_i)$ for each C_i is equivalent to maximising $psame(S) - pdiff(S)$.[1]*

As shown in Fig. 2c, selecting an S_i, with a high total skewness for C_i is equivalent to decreasing the corresponding horizontal striped areas in other classes. Maximising the total coverage is equivalent to minimising the corresponding vertical striped areas in C_i. Maximum total skewness of S_i, is achieved when all its features' sparse values are concentrated within only C_i. Maximum total coverage within C_i is achieved when all the features in S_i have sparse values for all the class instances. This results in no striped areas in Fig. 2c. *Selecting an S_i for each C_i with maximum total skewness and maximum total coverage results in the ideal feature subset $S = \bigcup_{i=1}^{3} S_i$ in Fig. 2b, in which $psame(S) = 1$ and $pdiff(S) = 0$.* Theorem 1 describes this phenomenon.

---

[1] https://sites.google.com/view/kushani/publications.

5.2   New Feature Selection Objective

As selecting a feature subset S to maximise total skewness and coverage within each class is analogous to maximising $psame(S)$ - $pdiff(S)$, we reformulate the feature selection objective as follows. *For a given class c, our objective is to select a feature subset $S_c \subseteq F$ to*

$$\max_{S_c} \left(tcov(S_c, c) \cdot \sum_{f \in S_c} skewness(f, c) \right) \text{ s.t. } |S_c| = n_c.$$

where $n_c \in \mathbb{Z}^+$, and $n_c \leq |F|$. We refer to this product of the two measures as the *predictive power* of S_c for class c $(predpow(S_c, c) = tcov(S_c, c) \cdot \sum_{f \in S_c} skewness(f, c))$, as it measures S_c's ability to distinguish c from the rest. Considering all the classes, our objective is to maximise the *total predictive power* of all classes. To ensure that each class is distinguished equally well, we also maximise the *minimal predictive power*. Therefore, given that N is the number of features to be selected and S_c is the best feature subset for c, the new objective is to select $S \subseteq F$ to

$$\max_{S} \left(\sum_{c \in C} predpow(S_c, c) + predpow(S_x, x) \right) \text{ s.t. } |S| = N$$

where $S = \bigcup_{c=1}^{|C|} S_c$, $S_i \cap S_j = \emptyset, i \neq j$ and $x = \text{argmin}_{c \in C} \, predpow(S_c, c)$.

5.3   A Greedy Feature Selection Approach

As selecting a feature subset to maximise the total coverage is still NP-hard, we propose a heuristic algorithm, *CovSkew*, to achieve the new objective. *CovSkew* iteratively selects a class c such that

$$c = \underset{x \in C}{\text{argmin}} \left(\frac{1}{|T_x| \cdot |S_x|} predpow(S_x, x) \right) \tag{4}$$

where S_x is the already selected feature subset for class x and also selects a feature f for c such that

$$f = \underset{y \notin S_c}{\text{argmax}} \left(skewness(y, c)^\alpha \cdot tcov(S_c \cup y, c) \right) \tag{5}$$

where S_c is the already selected feature subset for class c and α a user defined parameter (Line 3–4 in Algorithm 1). f is moved from the unselected feature subset, U, to S_c and the selected feature subset for all classes, S (Line 5 in Algorithm 1). The steps are repeated until the required feature count is obtained. In Eq. (4), to prevent the class imbalance problem, we normalise the class predictive power over the number of class instances. To prevent the bias towards the number of features selected for each class, we normalise it over the number of features in the subset. In Eq. (5), maximising the skewness of an individual feature maximises the total skewness of the selected feature subset because the skewness of a feature is independent from other features. We refer to $\frac{1}{|T_x| \cdot |S_x|} predpow(S_x, x)$ as the *class score* of c and $skewness(y, c)^\alpha \cdot tcov(S_c \cup y, c)$ the *feature score* of f within c.

Algorithm 1. *CovSkew* algorithm

---

input : Dataset (D), Requested Feature Count $(reqFeaCount)$
output: Selected feature subset (S)

1 $feaCount \leftarrow 0$; $U \leftarrow$ Set of features in D; $C \leftarrow$ Set of classes in D;
2 **while** $feaCount < reqFeaCount$ **do**
3 $\quad$ $c \leftarrow \text{argmin}_{x \in C} \frac{1}{|T_x| \cdot |S_x|} predpow(S_x, x)$ (Equation (4));
4 $\quad$ $f \leftarrow \text{argmax}_{y \notin S_c} skewness(y, c)^\alpha \cdot tcov(S_c \cup y, c)$ (Equation (5));
5 $\quad$ $S \leftarrow S + f$; $S_c \leftarrow S_c + f$; $U \leftarrow U$ - f;
6 $\quad$ $feaCount$++;
7 **end**
8 **return** S;

Example 1 Revisited: Consider solving Example 1 with *CovSkew*. Assume α = 1. As all class scores are initially zero, it randomly selects class Z and feature "Cat", which gives the highest feature score for Z (8). Class score of Z is now 2. Next it selects "Compiler", which has the highest feature score for class C (4). Class score of C is now 1. No features have sparse values for class P, therefore no features are selected for P. Out of Z and C, C has the minimum class score and "Classifier" gives the highest maximum feature score for C (16). Therefore, the selected feature subset by *CovSkew* is {Cat, Compiler, Classifier}, which is the optimal feature subset according to the objective in Eq. (1).

Computational Complexity Analysis: Assume n and f, are the number of instances and features, n_c the number of instances in a single class, and s the number of selected features. As a feature's skewness is independent from other features, *skewness computation is performed only once, with a time complexity linear to f ($O(nf)$). The total coverage and class predictive power computation in a single iteration has only $O(fn_c)$ complexity, which is significantly low. This is because these measures consider only the instances of a single class. As the algorithm runs s iterations, the total time complexity is $O(sf(n_c) + nf)$.

6 Evaluation

Datasets: We evaluate *CovSkew*'s accuracy and performance using publicly available, real datasets. A summary of the datasets, which is presented in Table 2, shows that the datasets have a wide variation in terms of the number of features, instances and classes. For continuous data, zero threshold binarization is applied.

Experimental Setup: We design three sets of experiments using above datasets. Default $\alpha = 2$ and the classifier used is Support Vector Machine (SVM), as they give the best average accuracy for all datasets. We have also evaluated with different classifiers and settings (Logistic Regression and Naive Bayes), for which all the classifiers give similar accuracy results as SVM, but due to space

Table 2. Dataset description. n: # features, c: # classes, m: # instances

Dataset	Type	n	c	m	Description
StudentLife (SL) [15]	Continuous	7,483	39	117,000	Wi-Fi fingerprint data
UJIIndoorLoc (UJ) [16]	Continuous	520	3	21,048	Wi-Fi fingerprint data
Citeseer (CS) [1]	Binary	3,703	6	3,312	Text document data
Cora (CO) [1]	Binary	1,433	7	2,708	Text document data
WebKB (WK) [1]	Binary	1,703	5	877	Text document data
Terrorist Attack (TA) [1]	Binary	106	5	1,293	Terrorist attack data
Genomic (GN) [17]	Continuous	12,532	11	175	Genomic data

limits we leave their discussion to further work. The experiments are performed on a Core i7, 2.60 GHz computer with 16 GB RAM.

- **Experiment 1:** Measures the classification accuracy obtained for the datasets with selected features. The aim is to evaluate the prediction accuracy of the feature selection algorithm.
- **Experiment 2:** Feature selection algorithm is executed 100 times to evaluate its performance.
- **Experiment 3:** Performs feature selection for different α values (1–5). The aim is to test the effect of the α parameter value on classification accuracy.

Baselines: As baseline feature selection methods, we use DFS [5], GINI [7], CDFE [11], which are limited for binary data, IG [9] and mRMR [8], which have achieved good accuracy and performance for both binary and non-binary data. Among them, DFS has shown to have good prediction accuracy over many existing binary feature selection methods and CDFE and GINI consider the distribution of one feature value, similar to our method. CDFE method is extended for multiclass problems using the one vs. all approach. "Unselect" means no feature selection is performed and the complete dataset is used.

Evaluation Criteria: The classifier's prediction accuracy on the dataset with selected features is considered as the prediction accuracy of the feature selection algorithm. Using the 10-fold cross validation method, we compute the F1-score for each class and report the *average F1-score* for all the classes (*AVGF*), Macro-F1 in other terms [5]. The performance is evaluated in terms of *average runtime* for 100 algorithm executions. We also report the 95% confidence intervals.

6.1 Experimental Results

Figures 3, 4 and 5 show the results for Experiment 1, 2 and 3. *For all the selected feature numbers in all the datasets, CovSkew shows higher or same accuracy compared to all the baselines*, with only ~5 exceptions in TA and UJ datasets.

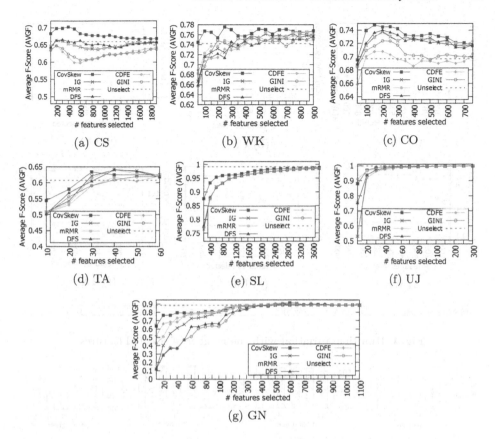

Fig. 3. SVM classification accuracy variation with the number of selected features

CovSkew's average and maximum accuracy gains over baselines are ∼5% and ∼40%, respectively. *Compared to baselines, CovSkew shows a significantly higher accuracy for small numbers of selected features.* CovSkew also shows a higher or same accuracy as the "Unselect" case for most datasets. The second best accuracies are for DFS and mRMR, however, *CovSkew's* running time is significantly lower than theirs (∼50 times lower for GN dataset). *CovSkew* has a higher running time than IG and GINI, yet has a high accuracy gain over them (∼8% average accuracy gain than GINI). Figure 5 shows that different datasets have different accuracy variation patterns with α value and both higher and lower accuracies are possible than with default α. In Fig. 5, AUC is the Area Under the Curve in AVGF vs. number of features graph (up to 50% of all features).

6.2   Evaluation Insights

CovSkew shows good classification accuracy for all the datasets compared to all the baselines (∼5% average and ∼40% maximum accuracy gains). The accuracy gain is higher when the selected feature subset is small, which shows that

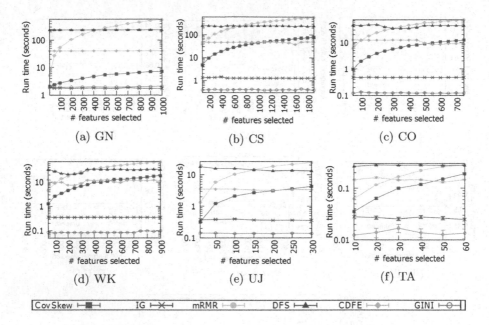

Fig. 4. Run time variation with the number of selected features

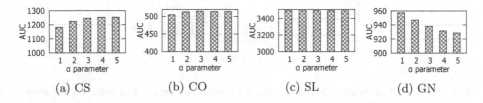

Fig. 5. Average accuracy (AUC for 50% of features) for different α values

CovSkew gains good accuracy with a minimal number of features. By optimising the α parameter for the dataset, higher accuracy can be obtained than the reported results. *CovSkew* has lower running time compared to most of the baselines (~50 times lower than mRMR and DFS for GN dataset). Although the running time of *CovSkew* is higher than of IG and GINI, *CovSkew* has a higher accuracy than these methods. The run time difference is smaller for small numbers of selected features, which is acceptable given that *CovSkew* gives good accuracy even for small numbers of selected features.

7   Conclusion

We propose two novel feature selection measures to select a better quality feature subset in multiclass binary datasets to improve the classification accuracy. The measures are based on the sparse value distribution of the dataset, therefore computationally efficient. Using these measures, we propose *CovSkew*, a feature selection algorithm. We show that compared to existing algorithms, *CovSkew* achieves high prediction accuracy, especially when the selected feature subset is small. The experimental results also show that *CovSkew* has a lower running time compared to most of the baselines. Future directions of this work include automatically computing the optimal α for a dataset, using dataset properties.

Acknowledgements. This work is supported by the Australian Government under the Australian Postgraduate Award.

References

1. Sen, P., Namata, G.M., Bilgic, M., et al.: Collective classification in network data. AI Mag. **29**(3), 93–106 (2008)
2. Juan, A., Vidal, E.: Bernoulli mixture models for binary images. In: Proceedings of 17th IEEE ICPR, vol. 3, pp. 367–370 (2004)
3. Shmulevich, I., Zhang, W.: Binary analysis and optimization-based normalization of gene expression data. Bioinformatics **18**(4), 555–565 (2002)
4. Calonder, M., Lepetit, V., Ozuysal, M., et al.: Brief: computing a local binary descriptor very fast. TPAMI **34**(7), 1281–1298 (2012)
5. Uysal, A.K., Gunal, S.: A novel probabilistic feature selection method for text classification. Knowl.-Based Syst. **36**, 226–235 (2012)
6. Pereira, R.B., Plastino, A., Zadrozny, B., et al.: Categorizing feature selection methods for multi-label classification. AI Rev. **49**, 1–22 (2016)
7. Park, H., Kwon, S., Kwon, H.C.: Complete Gini-Index Text (GIT) feature-selection algorithm for text classification. In: SEDM 2010, pp. 366–371. IEEE (2010)
8. Peng, H., Long, F., Ding, C.: Feature selection based on mutual information criteria of max-dependency, max-relevance, and min-redundancy. IEEE TPAMI **27**(8), 1226–1238 (2005)
9. Yang, Y., Pedersen, J.O.: A comparative study on feature selection in text categorization. ICML **97**, 412–420 (1997)
10. Herman, G., Zhang, B., Wang, Y., et al.: Mutual information-based method for selecting informative feature sets. Pattern Recogn. **46**(12), 3315–3327 (2013)
11. Javed, K., Babri, H.A., Saeed, M.: Feature selection based on class-dependent densities for high-dimensional binary data. IEEE TKDE **24**(3), 465–477 (2012)
12. Forman, G.: A pitfall and solution in multi-class feature selection for text classification. In: Proceedings of the 21st ICML, p. 38. ACM (2004)
13. Xiang, S., Shen, X., Ye, J.: Efficient nonconvex sparse group feature selection via continuous and discrete optimization. Artif. Intell. **224**, 28–50 (2015)
14. Davies, S., Russell, S.: NP-completeness of searches for smallest possible feature sets. In: Proceedings of the AAAI Fall Symposium on Relevance, pp. 37–39 (1994)
15. Wang, R., Chen, F., Chen, Z., et al.: StudentLife: assessing mental health, academic performance and behavioral trends of college students using smartphones. In: Proceedings of the ACM Ubicomp, pp. 3–14 (2014)
16. Torres-Sospedra, J., Montoliu, R., Martínez-Usó, A., et al.: UJIIndoorLoc: a new multi-building and multi-floor database for WLAN fingerprint-based indoor localization problems. In: IPIN 2014, pp. 261–270. IEEE (2014)
17. Su, A.I., Welsh, J.B., Sapinoso, L.M., et al.: Molecular classification of human carcinomas by use of gene expression signatures. Cancer Res. **61**(20), 7388–7393 (2001)

Scalable Model-Based Cascaded Imputation of Missing Data

Jacob Montiel[1]($\boxtimes$), Jesse Read[2], Albert Bifet[1], and Talel Abdessalem[1,3]

[1] LTCI, Télécom ParisTech, Université Paris-Saclay, 75013 Paris, France
{jacob.montiel,albert.bifet}@telecom-paristech.fr,
talel.abdessalem@enst.fr
[2] LIX, École Polytechnique, 91120 Palaiseau, France
jesse.read@polytechnique.edu
[3] UMI CNRS IPAL & National University of Singapore, Singapore, Singapore

Abstract. Missing data is a common trait of real-world data that can negatively impact interpretability. In this paper, we present CASCADE IMPUTATION (CIM), an effective and scalable technique for automatic imputation of missing data. CIM is not restrictive on the characteristics of the data set, providing support for: Missing At Random and Missing Completely At Random data, numerical and nominal attributes, and large data sets including highly dimensional data sets. We compare CIM against well-established imputation techniques over a variety of data sets under multiple test configurations to measure the impact of imputation on the classification problem. Test results show that CIM outperforms other imputation methods over multiple test conditions. Additionally, we identify optimal performance and failure conditions for popular imputation techniques.

Keywords: Classification · Missing data · Imputation

1 Introduction

Missing data is a common phenomenon in real-world applications. It can be introduced during data collection by human manipulation or by sensor failures, or by hardware/software failures during data storage/transmission. The average amount of missing data is estimated in a range between 5% and 20% [24,26]. Missing data has a negative effect on performance of supervised learning methods, according to [1], ratios between 5–15% require the usage of sophisticated methods while above 15% of missing values can compromise data interpretation.

Missing data *mechanisms* describe the underlying nature of this phenomenon and are classified into three major categories [18]: I. *Missing Completely At Random* (MCAR), the events behind missing values are independent of observable variables and the missing values themselves. II. *Missing At Random* (MAR), missingness can be explained by observable variables. III. *Missing Not At Random* (MNAR), when data is not MCAR nor MAR and the reason for missingness

© Springer International Publishing AG, part of Springer Nature 2018
D. Phung et al. (Eds.): PAKDD 2018, LNAI 10939, pp. 64–76, 2018.
https://doi.org/10.1007/978-3-319-93040-4_6

of data is related to the value of the missing data itself. Consider X the matrix of input attributes and y the corresponding labels, let $D = (X, y)$. For the observed values (not missing) D_{obs}, let $D_{obs} = (X_{obs}, y)$. A *missingness* matrix M with the same shape as X indicates if a value is missing on X by setting its ijth entry to 1. It follows that:

$$\text{MCAR}: \quad P(M|D) = P(M) \tag{1}$$

$$\text{MAR}: \quad P(M|D) = P(M|D_{obs}) \tag{2}$$

In the presence of missing data, three approaches are usually considered [9]: (i) Discard instances with missing values, (ii) let the learning algorithm deal with missing values, (iii) impute (fill) missing data. Imputation is usually the recommended approach. However, manual imputation of MCAR and MAR data is a time consuming process and requires deep understanding of the data and the phenomena that it describes. On the other hand, manual imputation is recommended for MNAR data, given that data specialists are more likely to identify the reasons behind this type of missing data and the proper way to handle it. Additionally, current trends in data generation and collection have shifted the data archetype in the Machine Learning community to larger and more complex data. Under this scenario, imputing data manually is impractical, therefore *scalable* automatic imputation solutions are required for real-world applications.

One of these applications is *Classification*, a type of Supervised Learning, where a model h is generated from labeled data (X_{train}, y_{train}). This model is applied to unlabeled data $X_{predict}$ to predict the corresponding class $y_{predict} = h(X_{predict})$ where $X_{train} \neq X_{predict}$. Two classes are considered in *binary* classification, $y \in \{0, 1\}$, while $K > 2$ classes are used in *multi-class* classification, $y \in \{0, 1, \ldots, K\}$. For both *binary* and *multi-class* classification only one class is assigned per instance.

The contributions of this paper are the following:

- A new scalable and effective model-based imputation method that casts the imputation process as a set of classification/regression tasks.
- Different to well established imputation techniques, the proposed method is non-restrictive on the type of missing data to process, supporting:
  • MAR and MCAR missing data mechanisms.
  • Numerical and Nominal data.
  • Small to large data sets, including high dimensional data.
- Since the proposed method does not require additional tools to the ones for Classification, implementing a pipeline *imputation+classification* is straightforward.
- We provide a comprehensive evaluation of different imputation methods, identifying optimal operation conditions as well as failure conditions.

The rest of the paper is organized as follows. Section 2 provides an overview of related work. Section 3 describes our imputation method. Test methodology is described in Sect. 4 and results are discussed in Sect. 5. Section 6 presents our conclusions and future directions.

2   Related Work

Various approaches have been proposed for missing data imputation. Basic methods rely on simple statistical values such as *mean* [18,20] and *covariance* [18]. Other methods *reconstruct* incomplete data incrementally: *kNN* based methods [2,13,27] impute data based on the neighborhood of a missing value and show good results when the data is small. However, these methods suffer the computational burden of the kNN algorithm making scalability an issue. *Self-Organizing Map Imputation* (SOMI) [8] is a Neural-Network model that first ignores missing data when finding patterns, and then imputes missing data based on the weights of activation nodes of the previously estimated patterns. *Expectation Maximization Imputation* (EMI) [5] performs imputation based on the Expectation-Maximization algorithm. EMI imputes values based on mean and covariance during the expectation step, then updates them in the maximization step, this process continues until convergence. Kernel methods [21,22,29] build imputation models based on kernel functions. These non-parametric methods usually consist of two stages: kernel function selection and bandwidth adjustment. *Fuzzy C-Means Imputation* (FCMI) [16] and *k-Means clustering* (KMC) [28], use clusters generated from non-missing data. While some imputations methods use the entire data set (EMI, Mean Imputation, Most Frequent Value), others use only sections of it (kNNI, FCMI, KMI).

Table 1. Related imputation methods categorization

		(a)	(b)	(c)	(d)	(e)	(f)
Data Type	Numerical	✓	✓	✓	✗	✓	✓
	Nominal	✗	✓	✓	✓	✓	✗
Mechanism	MAR	✗	✗	✗	✗	✗	✗
	MCAR	✓	✓	✓	✓	✓	✓
Performance evaluation	Classification	✓	✗	✗	✓	✗	✗
	Regression	✓	✗	✗	✗	✗	✗
	Imputation error[a]	✗	✓	✓	✗	✓	✓
Data set size[b]		S	S-L	S-L	S-M	S	S
Missing values ratio[c]		S-L	S	S	M-L	M-L	M-L

[a] Respect to complete data set.
[b] (*Instances* × *Attributes*) → Small [<200K], Medium [200K–600K], Large [>600K]
[c] Small [<10%], Medium: [10%–25%], Large [>25%]

In the following, we summarize relevant imputation methods and categorize them in Table 1: (**a**) *Locally Linear Reconstruction* (LLR) [13] determines the number of neighbors k and the weights given to the neighbors in kNN learning. Imputation is limited to numerical values. Imputed data sets are used to train classification/regression models. LLR assumes k to be 'sufficiently large'

which represents a compromise for large data sets. (**b**) A combination of EMI
with Decision Trees (DMI) and Decision Forest (SiMI) are proposed in [23]. The
goal is to identify segments of data where instances have higher similarity and
attribute correlation. A tree-based algorithm identifies partitions in the data,
given that leaves are sets of mutually exclusive instances. Imputation of numerical and nominal data is performed via EMI and Majority Class respectively. (**c**)
FEMI [24] uses the General Fuzzy C-Means (GFCM) [15] clustering algorithm to
find most similar instances for imputation via EMI. FEMI supports imputation
of numerical and nominal data. The required number of k clusters is manually
set. (**d**) A model-based approach is provided in [26]. Imputation of nominal data
is performed via a classification approach, based only on observed values. (**e**) A
non-parametric iterative imputation method for numerical and nominal values is
proposed in [29]. This method uses a mixed-kernel-based estimator and applies
a grid search strategy for selecting the optimal bandwidth. Imputed values are
used to impute missing values in subsequent iterations. (**f**) An iterative model-
based imputation method is presented in [25]. Imputation of numerical data is
carried by iteratively applying regression functions, first using observed data and
then including imputed data, until the difference between predictive values falls
bellow a user defined threshold.

3   Proposed Method

In this paper we present C̲ASCADE I̲MPUTATION (CIM), a model-based incremental imputation method. CIM casts the imputation process as a set of classi-
fication/regression tasks where unobserved values are imputed on a supervised
learning fashion. This approach is supported by the underlying presence of high
correlation/interaction between attributes [3, 10, 17]. In the following, we describe
the main steps performed by CIM, see Fig. 1.

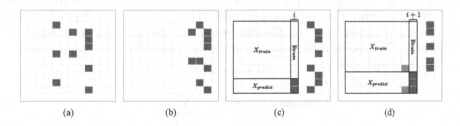

Fig. 1. CIM Steps. (1a) Original data with missing values marked in red. (1b) Updated
positions after sorting attributes by count of missing values. (1c–1d) Imputation iterations, repeated until data set is complete. Imputed values in green are used in following
iterations of the algorithm. (Color figure online)

Given an incomplete data set $D = (X, y)$, we want to find the corresponding
imputed data set $D' = (X', y)$. Figure 1a shows the original positions of missing

data (in red) in X. First, CIM splits X column-wise, keeping complete data to the left and incomplete data to the right. Columns are sorted in incremental order depending on the amount of missing values. Figure 1b shows the updated column order after sorting.

CIM iterates trough the columns with missing values, incrementally performing imputation via predictive models, hence cascade imputation. For each column with missing values i, CIM trains a classification model if the attribute is nominal or a regression model if it is numerical. On each iteration i, CIM sorts rows in X, placing all instances where the value of i is known (observed) at the top and instances with unknown (missing) values at the bottom. Figure 1c shows the updated row positions after sorting.

Columns $0 \to i$ correspond to the setup of a classification/regression problem. Known inputs X_{train} and their responses y_{train} are used to train a model h_i. Imputed values result from applying the corresponding model $imputed = h_i(X_{predict})$. This process is repeated until all attributes with missing values have been processed. Imputed values (in green) are used on following iterations of the algorithm, Fig. 1d.

Additionally, CIM calculates for each instance j a *missingness* weight w_j, ranging from 0 (all values missing) to 1 (no missing values). In practice, $0 < w_j \leq 1$. w_j represents the level of noise that may be introduced by the imputation process, and is used in the final classification task where it assigns higher importance to complete instances over imputed ones.

Although CIM can be paired with different classifiers/regressors, we propose two configurations based on popular algorithms in the research community [4]. CIM-LR uses Logistic Regression and Linear Regression, while CIM-RF uses Random Forest. Logistic and Linear Regression are both types of generalized linear models which solve

$$y = \phi(X\beta) + \epsilon \tag{3}$$

where y is continuous in the regression case; while for classification, y is the probability of a categorical outcome, a two states variable for the most basic case. Random Forest, a type of Ensemble Trees, creates n independent and fully grown Decision Trees T_i. For regression, the average value of the trees is calculated. For classification, a majority vote is applied for the class predictions $C_i(x) = k$ of each tree.

$$\text{Regression,} \quad f(x) = \frac{1}{n} \sum_{i=1}^{n} T_i(x) \tag{4}$$

$$\text{Classification,} \quad C(x) = \arg\max_{k} \sum_{i=1}^{n} (C_i(x) = k) \tag{5}$$

Resources required by CIM are upper bounded by the classification task on the complete attributes set X. On each independent iteration, a subset $X_{train} \in X$ is used. The number of iterations m required is limited by the number of attributes d where $m \leq d$. If all attributes have at least one missing value, then

CIM starts the cascade immediately after the attribute with less missing values. For each iteration, if an instance in X_{train} has missing values, it is removed; if a missing values is in $X_{predict}$, it is replaced with zero. Once the cascade ends, CIM imputes the first attribute. The worst case scenario is to have missing values in all the attributes of a high dimensional data set. We discuss this scenario in the test section.

4 Methodology

We are interested in the impact of imputation on classification. In order to thoroughly evaluate our proposed method, we use 10 data sets (Table 2) from a variety of domains and with the following characteristics: **Classification type**: Our tests focus on binary and multi-class classification. Although we use multi-label data sets, we test binary classification by learning and evaluating only one label (first label by default). **Data type**: Numerical, nominal or a combination of both. **Size**: Defined by the number of instances × the number of attributes. Data sets range from small ($< 200K$ values) to large ($> 600K$ values). Tests fall into the following categories:

Table 2. Data sets.

Name	Domain	Instances	Attributes Num.	Attributes Nom.	Source	Classification
Adult	Demography	48,842	6	8	UCI Rep.	Binary
Census-IKDD[a]	Demography	299,285	7	33	UCI Rep.	Binary
Music	Music	593	72	0	MEKA Rep.	Binary[b]
Enron	Text	1,702	0	1,001	MEKA Rep.	Binary[b]
Genbase	Biology	661	0	1,186	MULAN Rep.	Binary
Llog	Text	1,460	0	1004	MEKA Rep.	Binary[b]
Medical	Text	978	0	1,449	MEKA Rep.	Binary[b]
Scene	Image	2,407	294	0	MEKA Rep.	Binary[b]
Yeast	Biology	2,417	103	0	MEKA Rep.	Binary[b]
Covtype	Biology	581,012	10	44	UCI Rep.	Multi-class

[a] Census-Income (KDD)
[b] Multi-label data set. We use only the first class-label to test binary classification.

1. **Imputation-Classification tests.** We compare performance of classifiers trained on imputed data with multiple ratios of missing data vs the (baseline) performance of a classifier trained using complete data. This test is performed as follows:
   (a) **Generate missing data.** First, we remove all original missing values from each data set. Then, for each complete set we generate incomplete versions with 4 missing values ratios (5%, 10%, 25% and 50%) and 2

mechanisms (MCAR and MAR). For MCAR, we draw at random a number from a uniform distribution $f_U(v_{i,j}) = x_{i,j}$ for each value $v_{i,j}$ in the data set, if $x_{i,j} \le t$ then we mark the value as missing. The threshold t is defined by the ratio of missing values. For MAR, we draw at random a pair of attributes (A, B) and a threshold $t_A \in A$. A value B_j is marked as missing if $A_j \le t_A$. This process is repeated until the ratio of missing values is reached. We generate 10 different versions of each configuration for a total of $10 \times 4 \times 2$ incomplete sets for each complete data set.

(b) **Impute missing data.** We compare CIM against 4 well established imputation techniques. *Constant Imputation* (CONSTANT) where a constant value is used to fill missing values. We use 0 for numerical values and define a 'missing' class for nominal values. *Simple Imputation* (SIMPLE), fills missing values using the *mean value* for numerical attributes and the *most-frequent value* for nominal attributes. *Expectation-Maximization Imputation* (EMI) [5,12,14] is an iterative method with two steps. Expectation (E), where values are imputed based on observed values. And Maximization (M), where imputed values are evaluated and updated if necessary according to the data distribution. The EM algorithm converges to imputed values consistent with the observed distribution. *k-Nearest Neighbor Imputation* (KNNI) [2] uses the neighborhood of a missing value to estimate the corresponding imputation value. Defining the optimal k value is challenging and has important implications on performance at the cost of computational burden [6,19]. In our tests we use $k = 3$, as a compromise given the range of data set sizes. Although parameter tuning can improve performance of predictive models, it would increase the complexity of CIM. In our tests, we set the classifiers/regressors in CIM-LR and CIM-RF to default values, using 100 trees for Random Forest.

(c) **Use imputed data for classification.** We train classification models using the imputed data and compare the performance of these models against the baseline performance of models generated using complete data. In order to control the complexity of our tests we use Logistic Regression and Random Forest as final classifiers. We use 10-fold cross validation for a total of $90 \times 2 \times 10$ tests for each data set.

2. **Scalability test.** To test the scalability of CIM, we focus on three large data sets, Adult, Covtype and Llog. We compare the two versions of CIM against EMI and KNNI. SIMPLE and CONSTANT are not included given their low complexity. For each complete data set D, we create subsets $D_i \in D$ corresponding to 5%, 10%, 25%, 50%, 75% and 100% of the complete data set sizes. Then, for each complete subset D_i we generate 10 incomplete versions with 5%, 10%, 25% and 50% missing values ratios using MCAR and MAR. In total, we measure imputation time for $6 \times 10 \times 5 \times 2$ incomplete sets. Reported times are the average of imputing each incomplete sets for each combination of set size, missing values ratio and missingness mechanism.

The data sets in Table 2 present different degrees of class imbalance. In this context, accuracy can be misleading [11], therefore we use two metrics that

account for the actual amount of correctly classified instances, namely: Area
Under the Receiver Operating Characteristic Curve (AUROC) [7] to evaluate
performance of binary classifiers and F1-Score for multi-class classifiers.

Given the multiple configurations in our test setup, we define an *Overall Performance Ranking* to simplify the interpretation of results. Since we are interested in robust methods, we focus on the general behavior of imputation methods
over multiple configurations rather than on isolated cases. We use the Root Mean
Square Error (RMSE) to measure the performance difference between a classifier
trained on complete data (z_{base}), and the same classifier trained on imputed data
(z_i), where i corresponds to n missing values ratios. The RMSE is calculated as:

$$RMSE = \sqrt{\frac{1}{n} \sum_{i=1}^{n} (z_{base} - z_i)^2} \tag{6}$$

Then, we rank each imputation method based on its RMSE within a test set.
One test set contains results grouped by [data-set, supervised learning algorithm,
missing data mechanism] for n missing values ratios. We award points based on
the following criteria:

3: *Top*: If the method is the best performer and there are no *ties*.
2: *Tie*: If multiple methods fall in a 5% band from the *top* performance.
1: *Runner-up*: Methods bellow 5% from the *top*.
0: If the imputation method *fails* for all the files in the test set.
0: If the method is the last among non-*fail* methods and the difference with
the best *runner-up* is larger than 1x the difference between *top* and best
runner-up.

5   Discussion of Experimental Results

Tests results for Imputation-Classification tests are shown in Fig. 2 for Logistic
Regression and Random Forest. Due to space limitations, we only show results
for 3 data sets with one metric, AUC ROC for binary data sets and F1-Score
for multi-class data sets. Test results indicate that the operating range (conditions under which imputation is successful) of EMI and KNNI is rather small,
and they fail under multiple test configurations. Table 3 shows the number of
successful imputations over 40 incomplete data sets for each mechanism. KNNI
fails when the number of values in the data set is large, regardless of the missing
data mechanism or missing values ratio. On the contrary, when the number of
values is small, KNNI performance is in the top-tier. On the other hand, EMI
is sensitive to high dimensionality, and success ratio drops as the missing values
ratio increases, especially on MCAR data. This indicates that EMI is sensitive
to data size, missingness mechanism and ratio of missing values. In contrast,
CIM-LR, CIM-RF, CONSTANT and SIMPLE methods successfully impute data
for all test configurations. However, CONSTANT's performance is mostly inconsistent with worst performance on small data sets and on MCAR. SIMPLE performs

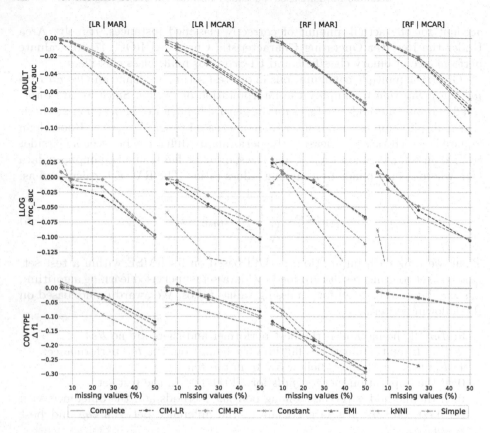

Fig. 2. Performance of classification models trained on imputed data vs baseline (complete).

remarkably well through our tests. It is interesting that such simplistic solution performs in overall better than EMI and κNNI across multiple test configurations. Finally, the two versions of CIM show the best overall performance, being CIM-RF the top performer.

The ranking of overall performance across multiple test configurations is available in Table 4. Top performer is CIM-RF, followed by CIM-LR and SIMPLE. Notice that optimal performance is achieved by CIM without using parameter tuning for the internal classification/regression tasks. Although CONSTANT imputes data successfully for all tests, its overall performance is low. κNNI is next as it shows good performance with small to medium size data sets but fails on large data sets. The worst performer in our tests is EMI given its narrow operating range. However, it is important to remark its good performance when the missing values ratio and the data set size are small.

The impact of the supervised learning algorithm on performance is also important to consider. In Table 4 we see that Logistic Regression and CIM-RF are a good combination for both MAR and MCAR. Results show that Random

Table 3. Number of successful imputations (out of 40) for each test set.

	MAR										MCAR									
	Adult	Census-IKDD	Music	Enron	Genbase	Llog	Medical	Scene	Yeast	Covtype	Adult	Census-IKDD	Music	Enron	Genbase	Llog	Medical	Scene	Yeast	Covtype
CIM-LR	40	40	40	40	40	40	40	40	40	40	40	40	40	40	40	40	40	40	40	40
CIM-RF	40	40	40	40	40	40	40	40	40	40	40	40	40	40	40	40	40	40	40	40
Constant	40	40	40	40	40	40	40	40	40	40	40	40	40	40	40	40	40	40	40	40
EMI	40	40	11	0	0	0	0	15	36	0	31	31	10	0	0	0	0	0	10	0
kNNI	0	0	40	0	0	0	0	0	40	0	0	0	40	0	0	0	0	0	40	0
Simple	40	40	40	40	40	40	40	40	40	40	40	40	40	40	40	40	40	40	40	40

Table 4. Overall performance ranking over multiple test configurations. Larger numbers are better. Top performer is CIM-RF, followed by CIM-LR and SIMPLE.

Classifier	Mechanism	CIM-LR	CIM-RF	CONSTANT	EMI	kNNI	SIMPLE
Logistic regression	MAR	13	**17**	14	1	2	12
	MCAR	15	**21**	5	0	4	10
Random forest	MAR	16	11	13	1	2	**18**
	MCAR	13	**18**	7	0	3	16

Forest does a better job at exploiting imputed data from CIM-LR and SIM-PLE. In particular, we see a significant boost in performance when SIMPLE is paired with Random Forest. This explains the close gap between CIM-LR and SIMPLE in the overall performance ranking. Covtype with more than 31M values provides insight on the suitability of sophisticated imputation methods for big data sets. The top performer in this case is CIM-LR, followed closely by SIM-PLE and CIM-RF. This suggests that SIMPLE represents a good compromise for *extremely* large data sets, given the computational burden of sophisticated imputation methods.

Scalability test results are shown in Fig. 3 for Adult, Llog and Covtype. We observe that CIM takes longer to impute MCAR data. This is expected given that missing values are equally distributed across the data attributes, requiring more iterations of the cascade. CIM is faster on MAR data with low missing values ratio, but as the ratio increases ($\geq 25\%$) there are more attributes to process and imputation time gets closer to the time for MCAR. Notice that for kNNI and EMI, imputation time increases along with the missing values ratio, while the contrary happens for CIM-LR and CIM-RF. This is explained by the fact that there is less training data as the number of missing values increases, which results in less training time for the classification/regression models within CIM. In Adult, kNNI is the slowest imputation method. On the other hand, Llog presents the worst case scenario for CIM given its high dimensionality, especially with MCAR data. Nonetheless, notice that the gap between CIM and kNNI decreases as data size and missing values ratio increase. For Covtype, kNNI is again the slowest to impute data and starts to fail at 25% of the original data set size, while CIM-LR and CIM-RF successfully impute data for all subset sizes. EMI fails to impute data for all subsets of Llog and Covtype.

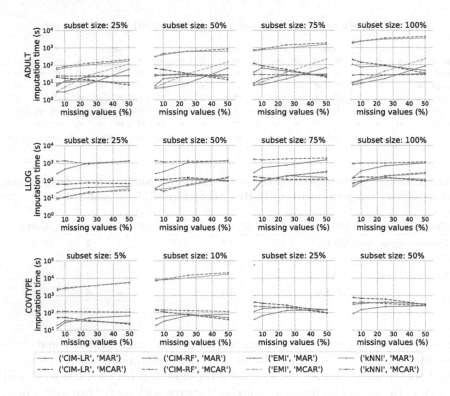

Fig. 3. Scalability test results for CIM, EMI and κNNI. CIM imputation time decreases with larger missing values ratios, while EMI and κNNI are weak against large data set sizes and ratio of missing values.

Test results show that CIM performs well on a variety of conditions, expanding beyond the operating range of EMI and κNNI. An additional consideration is the straightforward implementation of an *imputation+classification* pipeline using CIM, given that it does not require extra tools to the ones used for Classification.

6   Conclusions

We presented CIM, an effective and scalable imputation method. CIM imputes both numerical and nominal data and mitigates the impact of MAR and MCAR data on binary and multi-class classification. Test results show that CIM performs well over a wide range of missing values ratios and does not require parameter tuning to achieve optimal performance. CIM is scalable to large data sets, including highly dimensional data sets, a limitation of well established methods such as EMI and κNNI. Additionally, CIM always results in an imputed data set, something that is not guaranteed by EMI. Implementing an *imputation+classification* pipeline is straightforward given that CIM does not require additional tools to the ones related to Classification.

In future work we will improve the selection of training instances in the cascade; by accounting for the correlation between attributes we aim to reduce the computational burden from unrelated attributes while maintaining data interpretability. A natural extension of the method we propose is a multi-imputation solution based on multi-label classification and regression techniques, in an ensemble to improve robustness.

References

1. Acuña, E., Rodriguez, C.: The treatment of missing values and its effect on classifier accuracy. Classif. Clust. Data Min. Appl. **1995**, 639–647 (2004)
2. Batista, G.E.A.P.A., Monard, M.C.: A study of k-nearest neighbour as an imputation method. Frontiers in Artificial Intelligence and Applications **87**, 251–260 (2002)
3. Brown, G., Pocock, A., Zhao, M.J., Lujan, M.: Conditional likelihood maximisation: a unifying framework for mutual information feature selection. J. Mach. Learn. Res. **13**, 27–66 (2012)
4. Caruana, R., Niculescu-Mizil, A.: An empirical comparison of supervised learning algorithms. In: Proceedings of the 23rd international conference on Machine learning vol. C, no. 1, pp. 161–168 (2006)
5. Dempster, A., Laird, N., Rubin, D.B.: Maximum likelihood from incomplete data via the EM algorithm. J. R. Stat. Soc. Ser. B Methodol. **39**(1), 1–38 (1977)
6. Duda, R.O., Hart, P.E., Stork, D.G.: Pattern Classification. John Wiley & Sons, Hoboken (2012)
7. Fawcett, T.: An introduction to ROC analysis. Pattern Recogn. Lett. **27**(8), 861–874 (2006)
8. Fessant, F., Midenet, S.: Self-organising map for data imputation and correction in surveys. Neural Comput. Appl. **10**, 300–310 (2002)
9. Friedman, J., Hastie, T., Tibshirani, R.: The Elements of Statistical Learning: Data Mining, Inference, and Prediction. Springer, New York (2009)
10. Guyon, I., Elisseeff, A.: An Introduction to variable and feature selection. J. Mach. Learn. Res. (JMLR) **3**(3), 1157–1182 (2003)
11. He, H., Garcia, E.A.: Learning from imbalanced data. IEEE Trans. Knowl. Data Eng. **21**(9), 1263–1284 (2009)
12. Honaker, J., King, G., Blackwell, M.: Amelia ii: A program for missing data. J. Stat. Softw. **45**(1), 1–47 (2011)
13. Kang, P.: Locally linear reconstruction based missing value imputation for supervised learning. Neurocomputing **118**, 65–78 (2013)
14. King, G., Honaker, J., Joseph, A., Scheve, K.: Analyzing incomplete political science data. Am. Polit. Sci. Rev. **85**(1269), 49–69 (2001)
15. Lee, M., Pedrycz, W.: The fuzzy c-means algorithm with fuzzy p-mode prototypes for clustering objects having mixed features. Fuzzy Sets Syst. **160**(24), 3590–3600 (2009)
16. Li, D., Deogun, J., Spaulding, W., Shuart, B.: Towards missing data imputation: a study of fuzzy k-means clustering method. In: Tsumoto, S., Słowiński, R., Komorowski, J., Grzymała-Busse, J.W. (eds.) RSCTC 2004. LNCS (LNAI), vol. 3066, pp. 573–579. Springer, Heidelberg (2004). https://doi.org/10.1007/978-3-540-25929-9_70

17. Li, J., Cheng, K., Wang, S., Morstatter, F., Trevino, R.P., Tang, J., Liu, H.: Feature selection: a data perspective. J. Mach. Learn. Res. **50**, 1–73 (2016)
18. Little, R.J., Rubin, D.B.: Statistical Analysis with Missing Data. John Wiley & Sons, Hoboken (2002)
19. Maier, M., Hein, M., Von Luxburg, U.: Optimal construction of k-nearest-neighbor graphs for identifying noisy clusters. Theoret. Comput. Sci. **410**, 1749–1764 (2009)
20. Mundfrom, D.J., Whitcomb, A.: Imputing missing values: the effect on the accuracy of classification (1998)
21. Qin, Y., Zhang, S., Zhu, X., Zhang, J., Zhang, C.: POP algorithm: Kernel-based imputation to treat missing values in knowledge discovery from databases. Expert Systems with Applications **36**(2, Part 2), 2794–2804 (2009)
22. Racine, J., Li, Q.: Nonparametric estimation of regression functions with both categorical and continuous data. J. Econom. **119**(1), 99–130 (2004)
23. Rahman, M.G., Islam, M.Z.: Missing value imputation using decision trees and decision forests by splitting and merging records: two novel techniques. Knowl. Based Syst. **53**, 51–65 (2013)
24. Rahman, M.G., Islam, M.Z.: Missing value imputation using a fuzzy clustering-based EM approach. Knowl. Inf. Syst. **46**, 389–422 (2015)
25. Richman, M.B., Trafalis, T.B., Adrianto, I.: Missing data imputation through machine learning algorithms. In: Haupt, S.E., Pasini, A., Marzban, C. (eds.) Artificial Intelligence Methods in the Environmental Sciences, pp. 153–169. Springer, Dordrecht (2009). https://doi.org/10.1007/978-1-4020-9119-3_7
26. Su, X., Greiner, R., Khoshgoftaar, T.M., Napolitano, A.: Using classifier-based nominal imputation to improve machine learning. In: Huang, J.Z., Cao, L., Srivastava, J. (eds.) PAKDD 2011. LNCS (LNAI), vol. 6634, pp. 124–135. Springer, Heidelberg (2011). https://doi.org/10.1007/978-3-642-20841-6_11
27. Wang, L., Fu, D.M.: Estimation of missing values using a weighted k-nearest neighbors algorithm. In: Proceedings - 2009 International Conference on Environmental Science and Information Application Technology ESIAT 2009 vol. 3, no. 2, pp. 660–663 (2009)
28. Zhang, C., Qin, Y., Zhu, X., Zhang, J., Zhang, S.: Clustering-based missing value imputation for data preprocessing. In: 2006 IEEE International Conference on Industrial Informatics, pp. 1081–1086. IEEE (2006)
29. Zhu, X., Zhang, S., Jin, Z., Zhang, Z., Xu, Z.: Missing value estimation for mixed-attribute data sets. IEEE Trans. Knowl. Data Eng. **23**(1), 110–121 (2011)

On Reducing Dimensionality of Labeled Data Efficiently

Guoxi Zhang[1,3]([✉]), Tomoharu Iwata[2], and Hisashi Kashima[1,3]

[1] Graduate School of Informatics, Kyoto University, Kyoto, Japan
guoxi@ml.ist.i.kyoto-u.ac.jp, kashima@i.kyoto-u.ac.jp
[2] NTT Communication Science Laboratories, Kyoto, Japan
iwata.tomoharu@lab.ntt.co.jp
[3] Riken Center for Advanced Intelligence Project, Tokyo, Japan

Abstract. We address the problem of reducing dimensionality for labeled data. Our objective is to achieve better class separation in latent space. Existing nonlinear algorithms rely on pairwise distances between data samples, which are generally infeasible to compute or store in the large data limit. In this paper, we propose a parametric nonlinear algorithm that employs a spherical mixture model in the latent space. The proposed algorithm attains grand efficiency in reducing data dimensionality, because it only requires distances between data points and cluster centers. In our experiments, the proposed algorithm achieves up to 44 times better efficiency while maintaining similar efficacy. In practice, it can be used to speedup k-NN classification or visualize data points with their class structure.

1 Introduction

Dimensionality reduction is an important task in machine learning. Although the canonical objective for reducing data dimensionality is the perseverance of data similarity, we argue that manifesting class structure is also desirable. Intuitively, the idea is to project high-dimensional data points to a low-dimensional latent space, such that points from the same class locate at nearby regions, and they are distant from points of other classes. We refer to this idea as collapsing classes.

Collapsing classes expedites k-nearest neighbor (k-NN) classification. k-NN is a competitive algorithm in applications such as document classification [8], kinship verification [16] and animal behavioral classification [1]. Yet its time complexity for classifying a point is linear in the size of training data and data dimensionality, which prohibits application on large-scale datasets. Because the neighborhood of classes are explicitly considered, by collapsing classes we can accelerate k-NN while retaining classification performance for k-NN.

Besides, it also reduces memory consumption of k-NN. k-NN requires storing the entire training data, which is problematic for large-scale datasets or real-time applications. The idea of collapsing classes is closely related to several data compression algorithm for k-NN. Stochastic Neighbor Compression (SNC) [12] and Learning Discriminative Projections and Prototypes (LDPP) [21] project training data to a few class-specific reference vectors, which are considered as proxies for data. The idea of collapsing classes is incorporated to ensure the discriminability of reference vectors.

Yet another use case is data visualization. Visualizations facilitate inspection of data. Algorithms such as t-SNE [17] and Elastic Embedding [2] take as input pairwise distances of data. When proper similarity metric is unavailable, it is challenging to generate satisfactory visualizations. For labeled data we may instead visualize their class information with discriminative dimensionality reduction techniques. Collapsing classes is advantageous for this purpose because it improves readability by emphasizing separation of classes.

However, existing algorithms for collapsing classes are not satisfactory when for large-scale datasets. Algorithms such as Neighborhood Component Analysis (NCA) [6] and Maximally Collapsing Metric Learning algorithm (MCML) [5] compute pairwise distances of data, which means they are of quadratic space and time complexity. Meanwhile, the Parametric Embedding (PE) algorithm [10] computes only distances between data points and reference vectors of classes and has better efficiency. Nevertheless, PE takes as input posterior probability distribution of labels given data and outputs embedding vectors of data directly. It is not a tempting choice when such distribution is unavailable, or when embeddings of new data points are constantly queried.

In this paper, we propose an efficient algorithm called Nonlinear Parametric Embedding (NPE) for collapsing classes. NPE takes as input pairs of features and labels, and outputs a nonlinear projection function parametrized by neural networks. NPE computes only distance between data points and reference vectors of classes, thus it is more efficient than nonlinear extensions of NCA (NNCA) and that of MCML (NMCML). Meanwhile, NPE is more generic than PE. PE takes as input posterior distribution of classes given features, and it outputs embedding vectors of data points. In many cases, estimating such distribution is itself a challenging task. Moreover, PE determines representations of test points by invoking optimization routines, which is in general more time-consuming than applying a neural network.

We evaluate NPE on five datasets in terms of efficiency and efficacy. Our results show that compared to existing nonlinear methods, NPE can be up to 44 times faster. The contribution of this paper are three-folds: 1. we address the efficiency issue when collapsing class with nonlinear transformation, which is rarely considered in literature; 2. we propose a parametric model and a learning algorithm which solve the task efficiently; 3. we extensively evaluate the proposed method.

The rest of this paper is organized as the following. Section 2 briefly reviews the related literature. Section 3 describes the proposed algorithm. Section 4 describes our evaluation of the proposed algorithm. In the final section we conclude this paper.

2   Related Work

2.1   Metric Learning

Metric learning is concerned with learning a metric for high-dimensional data. Often, by learning a projection matrix and restricting its rank, one can reduce the dimensionality of data at the same time [5,6]. Nonlinear metric learning incorporates non-linearity to capture higher-order correlations among feature dimensions, which can be achieved either by kernel density estimation [7] or nonlinear distance measure [11]. In this paper, we consider incorporating nonlinearity with neural networks, which is highly flexible and generic in practice.

2.2   Nonlinear Algorithms for Collapsing Classes

Tow linear algorithms for collapsing classes have been extended to nonlinear setting. In [20] the authors propose a nonlinear extension of NCA (NNCA) using feedforward neural networks. In [18] the authors propose nonlinear MCML (NMCML) and compare performance of NNCA and NMCML on image recognition tasks. However, as explained below, the high time complexity of NNCA and NMCML hampers applying them to large datasets.

NMCML assumes a non-parametric distribution over data points in latent space. Denote by N the number of data points. A Student-t distribution with degree of freedom α is centered at each point in latent space. Each point $f(x_j)$ is associated with a Categorical distribution, modeling the identity of its nearest neighbor. This distribution is parameterized by q_{ij}:

$$q_{ij} = \begin{cases} \frac{(1+\alpha^{-1}\|f(x_i)-f(x_j)\|^2)^{-\frac{1+\alpha}{2}}}{\sum_{k \neq i}(1+\alpha^{-1}\|f(x_i)-f(x_k)\|^2)^{-\frac{1+\alpha}{2}}} & ,j \neq i, \\ 0 & ,j = i. \end{cases} \quad (1)$$

When classes in data are ideally collapsed, a point and its nearest neighbor should belong to the same classes. A sufficient specification of q_{ij} for this ideal case, p_{ij} is:

$$p_{ij} \propto \begin{cases} 1, & c_j = c_i, j \neq i, \\ 0, & \text{otherwise.} \end{cases} \quad (2)$$

NMCML minimizes the sum of KL divergence between Eq. 2 and Eq. 1 so as to collapse classes in data. Dropping the term for entropy of p_{ij}, its objective function is:

$$C(f) = -\sum_i \sum_{j \in C_k, j \neq i} p_{ij} \log(q_{ij}). \quad (3)$$

Meanwhile, NNCA also utilizes Eq. 1, but it tries to maximize accuracy of 1-NN classification.

In fact, Eq. 1 manifests the flaws of NNCA and NMCML. The optimization problems involved in NNCA and NMCML have to solved iteratively, and q_{ij} is re-computed in every iteration. Computing q_{ij} requires computing the pairwise distances between projected points, which is of quadratic time complexity. This high complexity hinders applying NNCA and NMCML on large datasets. Moreover, it can be infeasible to load the entire distance matrix of large-scale datasets into memory. In such cases objective functions of NNCA and NMCML can only be optimized in a stochastic batch optimization setting. However, in that case q_{ij} is not properly normalized, causing large variance in optimization. Therefore, there is still need for efficient nonlinear algorithm for collapsing classes.

2.3 Parametric Embedding

Parametric Embedding (PE) is an algorithm for visualizing posterior probability distribution [10]. It takes as input conditional probability of class given features $p(c_k|x_i)$, and outputs coordinates z_i for each data point and a reference vector ϕ_k for each class k. It assumes a spherical Gaussian mixture model in latent space, and learns z_i by minimizing the KL divergence between $p(c_k|x_i)$ and $p(c_k|z_i)$. This parametric formulation gives rise to high efficiency, as only the distance between data points and class centers are iteratively computed. However, a proper posterior distribution has to be estimated prior to appling PE. Moreover, determining low-dimensional representations of new data can be time consuming, since optimization procedures must be invoked.

3   Nonlinear Parametric Embedding

In this section we describe the proposed algorithm. Denote by D the dimensionality of data and by K the number of classes. We assumes each data point is associated with one class, so the data samples can be represented as $\{(x_i, c_i)\}$, where $i \in \{1, 2, \ldots, N\}$, $x_i \in \mathbf{R}^D$, and $c_i \in \{1, 2, \ldots, K\}$. We seek for a nonlinear transformation $f(\cdot)$ to project data points onto a latent space with dimensionality D', where $D' < D$. $f(\cdot)$ is parameterized with feedforward neural network.

The proposed method, Nonlinear Parametric Embedding (NPE), learns a reference vector for each class in latent space. Denote the reference vector for k^{th} class as ϕ_k, where $k \in \{1, 2, \ldots, K\}$ and $\phi_k \in \mathbf{R}^{D'}$. The Euclidean distance between $f(x_i)$ and ϕ_k depicts how likely x_i belongs to class k in latent space. Intuitively, to collapse class, NPE pulls $f(x_i)$ to ϕ_{c_i} and pushes away reference vectors of other classes.

The class structure in data are modeled with a mixture of K Student's t-distribution with uniform class probability. The kth mixture component is centered at ϕ_k. The conditional probability of x_i belonging to class k in latent space is:

$$w(c_i = k|x_i) = \frac{(1 + \alpha^{-1} \|f(x_i) - \phi_k\|^2)^{-\frac{1+\alpha}{2}}}{\sum_{\ell=1}^{K}(1 + \alpha^{-1} \|f(x_i) - \phi_\ell\|^2)^{-\frac{1+\alpha}{2}}}, \tag{4}$$

where α is the degrees of freedom of the Student's t-distribution. We choose the Student's t-distribution because it enhances class separation [17,18]. We set α to $D' - 1$ throughout this paper, as suggested by [18].

To collapse classes, NPE reduces the distance between the projected point $f(x_i)$ and the reference vector of the corresponding class ϕ_{c_i} and increases the distances between $f(x_i)$ and ϕ_j where $j \neq c_i$. It minimizes the KL divergence between the one-hot encoding of label c_i, $v(c_i|x_i)$ and $w(c_i|x_i)$. Specifically, $v(c_i|x_i)$ is a K-dimensional vector with c_i^{th} element equals to one and all other elements equal to zero. This corresponds to the case where data points are perfectly classified in latent space. Dropping the term for entropy of $v(c_i|x_i)$, the objective function of NPE is:

$$C(f, \{\phi_k\}) = -\sum_{i}^{N}\sum_{k=1}^{K} v(c_i = k|x_i)\log(w(c_i = k|x_i)). \tag{5}$$

One may notice the connection between Eq. 5 and the soft-max output layer used in neural networks for classification tasks. Indeed, NPE can be considered as a discriminative probabilistic classifier parameterized by neural network. It relaxes PE's dependency over posterior distribution by estimating it jointly with reducing data dimensionality. Note that training PE on $\{(x_i, c_i)\}$ yields a trivial case, in which projections of data points belonging to some class locate very close to the reference vector of that class. The correlation between classes are totally discarded.

The derivatives of C with respective to $f(x_i)$ and derivatives with respective to ϕ_k are the following:

$$
\begin{aligned}
\frac{\partial C}{\partial f(x_i)} &= \sum_{k=1}^{K} \beta_{k,i} \frac{1+\alpha}{\alpha} \frac{(f(x_i) - \phi_k)}{(1 + \alpha^{-1}\|f(x_i) - \phi_k\|^2)}, \\
\frac{\partial C}{\partial \phi_k} &= \sum_{i-1}^{N} \beta_{k,i} \frac{1+\alpha}{\alpha} \frac{(\phi_k - f(x_i))}{(1 + \alpha^{-1}\|f(x_i) - \phi_k\|^2)},
\end{aligned}
\tag{6}
$$

where $\beta_{k,i} = v(c_i = k|x_i) - w(c_i = k|x_i)$. From Eq. 6 we can make two observations. On the one hand, $\beta_{k,i} \geq 0$ only if $k = c_i$, and $\beta_{k,i} < 0$ otherwise. That is to say, minimizing the objective function poses attraction between $f(x_i)$ and ϕ_k, and repulsion between $f(x_i)$ and ϕ_j where $j \neq k$. On the other hand, since $f(x_i) - \phi_k$ are computed for $k \in 1, 2, \ldots, K$ and for $i \in 1, 2, \ldots, N$, the computational complexity in one iteration of optimization is $O(NK)$. In many real-word datasets, $K << N$, and NPE will be much more efficient than existing algorithms.

4   Evaluation

4.1   Experiment Settings

Datasets. We use five datasets to evaluate NPE: MNIST [14], 20 newsgroups [13], isolet [4], sensorless and satimage. Statistics of datasets are shown in

Table 1. MNIST contains images of handwritten digits. In experiments we converting pixel values to $[0, 1]$. 20 newsgroups dataset (20NEWS) contains newsgroups posts on 20 topics. We remove stop-words and 100 most frequent words. We then select 10,000 most frequent words as vocabulary and vectorize documents using tf-idf algorithm. Isolet is a dataset for spoken letter recognition. Satimage dataset consists of multi-spectral satellite images. Sensorless dataset contains data extracted from electric current drive signals. We use the scaled version of satimage dataset and sensorless dataset provided by LIBSVM data repository [3]. For isolet we use the version provided by UCI Machine Learning Repository [15].

Table 1. Dataset statistics

Dataset	# of Train	# of Test	D	K
MNIST	60,000	10,000	784	10
SENSORLESS	43,881	14,628	48	11
20NEWS	11,314	7,532	10,000	20
ISOLET	6,238	1,559	617	26
SATIMAGE	4,435	2,000	36	6

Evaluation Method. The extent to which classes are collapsed is can be revealed with k-NN test accuracy. A k-NN classifier is constructed on representations of training data, and it is then used to classify representations of test data. Its performance reflects how well data points from different classes are separated. Because the appropriate latent dimensionality and number of neighbors used in k-NN may vary for different datasets, we perform experiments with k in k-NN as $1, 2, 5, 10, 15, 20, 50, 100$ on all five datasets and perform experiments with latent dimensionality $2, 5, 10, 20, 30, 40, 50, 100$ on three relatively high-dimensional datasets MNIST, 20NEWS, and ISOLET.

To evaluate efficiency of algorithms, we compare the time that algorithms need to converge. We first smooth the loss values with the moving average index of window size 10. From the smoothed loss values we calculate the total loss drop in training. We report as execution time the elapsed time when loss drop reaches 99% of the corresponding total loss drop. Efficiency is evaluated with dimensionality of latent space equals to 30.

Alternative Methods. In experiments we compare performance of the proposed NPE with NNCA and NMCML. To demonstrate how collapsing class facilitate classification, we also include k-NN classification results on embeddings generated by PCA, latent discriminant analysis (LDA) and using k-NN algorithm directly, which we denote as "Direct" in results. On account of nonconvex optimization we repeat evaluation of NPE, NMCML and NNCA for ten times and report the mean values.

Implementation Details. We use five layers of feedforward networks to parameterize nonlinear mappings. The first four layers have the same number of latent units. For MNIST, ISOLET, 20NEWS, we set it to 512; for SENSORLESS and SATIMAGE we set it to 128. We set D' to 30 for evaluation of dimensional reduction and two for visualization purpose. The final layer is an affine transformation with the number of latent units equals to D'. For the first three layers we use the Rectified Linear Unit [19] as activation function, and use the hyperbolic tangent function for the fourth layer.

We add batch normalization [9] to the output of every layer in order to use large batch size as 512 in training. The reason to select such a large batch size is that NNCA and NMCML are vulnerable to noise introduced in mini-batch optimization, despite merit of stochastic optimization in training neural networks. All networks are trained using stochastic gradient descent with momentum, using 10^{-2} as learning rate. Optimizations are executed for 500 epoch to ensure convergence. To palliate over-fitting problem, we use weight decay with rate 10^{-6} and add drop-out after the third layer with rate 0.5.

4.2 Results

Efficiency. Figure 1 shows execution time of NPE, NMCML and NNCA. In all cases NPE is much faster than NNCA and NMCML by more than a magnitude. Specifically, on large datasets such as MNIST and SENSORLESS, NPE only takes two and five minutes, whereas NMCML and NNCA take more than one hour. These results demonstrate NPE's superiority on large-scale datasets.

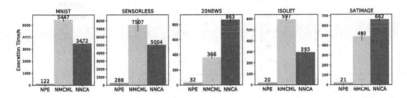

Fig. 1. Execution time of NPE, NMCML and NNCA on five datasets. The minimum and maximum speedup of the proposed method (NPE) with respective to NMCML are 11.4x on 20NEWS and 44.6x on MNIST, respectively. The minimum and maximum speedup of NPE with respective to NNCA is 14.8x on ISOLET and 31.5x on SATIMAGE, respectively.

Dimensionality Reduction. In Table 2 we show the test accuracy of 10-NN classification on five datasets. On all datasets, the three nonlinear methods have better performance than PCA, LDA and using kNN directly. These results support our claim that using nonlinear transformation to reduce dimensionality benefits kNN classification. The proposed method outperforms NMCML and NNCA on 20NEWS, ISOLET and SATIMAGE datasets. Particularly, on 20NEWS superiority of NPE is significant. These results demonstrate efficacy of NPE.

Table 2. 10-NN test accuracy on five datasets. Except for direct method, the dimensionality of latent space is set to 30. The proposed method, NMCML and NNCA have similar performance on four of five datasets, and they are better than PCA, LDA, and using *k*-NN directly.

Dataset	NPE	NMCML	NNCA	Direct	PCA	LDA
MNIST	0.981	**0.985**	0.981	0.967	0.974	0.919
SENSORLESS	0.999	0.998	**0.999**	0.989	0.989	0.949
20NEWS	**0.661**	0.574	0.573	0.414	0.565	0.539
ISOLET	**0.959**	0.955	0.952	0.913	0.896	0.949
SATIMAGE	**0.908**	0.905	0.905	0.895	0.895	0.87

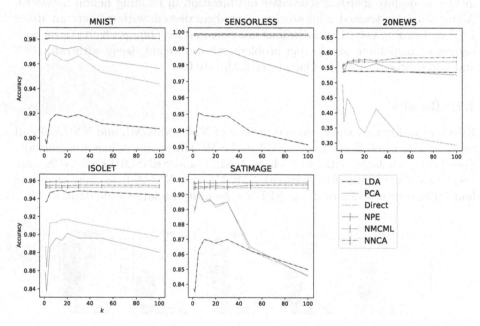

Fig. 2. *k*-NN test accuracy for different *k*. The proposed method (NPE), NMCML and NNCA are note sensitive to choice of *k*, indicating well class separation in the latent space.

In Fig. 2 we show test accuracy of *k*-NN with different values of *k* when dimensionality of latent space is set to 30. For NPE, NMCML and NNCA we draw the standard errors. It is shown in the figure that performance of all the three nonlinear methods is not sensitive to *k*, demonstrating well class separation in latent space.

In Fig. 4 we show test accuracy of 10-NN classification on five datasets, with different latent dimensionality. Nonlinear method becomes beneficial when dimensionality of latent space is greater than five. On MNIST and ISOLET, the three nonlinear methods being compared have similar performance. On 20NEWS

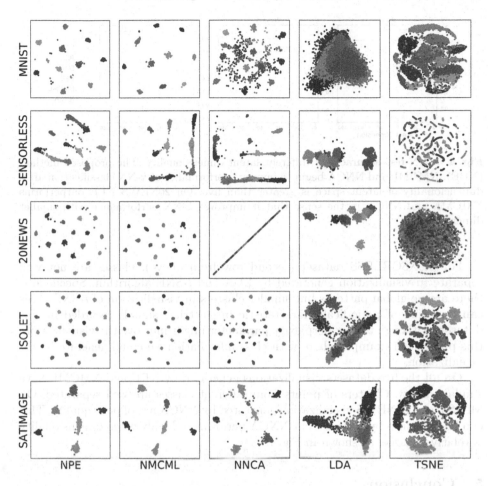

Fig. 3. Visualization. The proposed method (NPE) and NMCML generate similar visualizations. The separation of classes makes visualization easy to read. NNCA fails to generate good visualization on MNIST and 20NEWS. In visualizations generated by LDA classes are not well separated, so we might not determine class of test points easily. t-SNE fails to generate global patterns in visualization on SENSORLESS.

dataset, the proposed method achieves high test accuracy at much lower dimensionality. Because computational complexity of k-NN grows linearly in dimensionality of data, being able to generate good embeddings at low dimensionality makes NPE more practical on large datasets.

Visualization. In Fig. 3 we present visualization of training data, generated by NNCA, NPE, NMCML, LDA and t-SNE. Due to memory limit, for MNIST and SENSORLESS datasets we sample $10,000$ data points to use tSNE algorithm. The perplexity parameter of tSNE is set to 50.

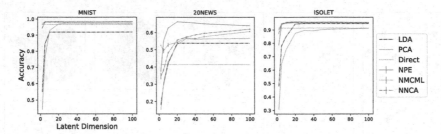

Fig. 4. k-NN test accuracy for different latent dimensionality. The proposed method (NPE), NMCML and NNCA begin to improve performance of k-NN classification after dimensionality of latent space is greater than five. On 20NEWS NPE outperforms NMCML and NNCA in the sense that it improves k-NN performance with smaller dimensionality.

For SENSORLESS dataset (second row from above), classes are not well separate in visualization generated by LDA and t-SNE algorithm. Specifically, there are no global patterns but smaller clusters in visualization of t-SNE. This can be result of periodical characteristic in electrical signals recorded in this dataset. Although tuning perplexity parameter might improve the visualization, this provide an example where with class information we could generate better visualization.

On all the five datasets visualizations generated by NPE and NMCML have similar quality. Clusters of points from different classes are well separated. On MNIST and 20NEWS visualization generated by NNCA are of poor quality. This can be explained by the fact that NNCA performs well only when dimensionality is relatively large, as shown in Fig. 4.

5   Conclusion

In this paper we address the idea of collapsing classes for reducing dimensionality of labeled data, and propose an efficient algorithm that learns nonlinear transformation from high-dimensional space to low-dimensional space. The proposed NPE algorithm utilizes parametric formulation in latent space and thus is more efficient than algorithms with nonparametric formulation. We demonstrate the efficacy and efficiency of the proposed Nonlinear Parametric Embedding algorithm with experiments on five multi-class classification dataset. Specifically, the proposed method can be up to 44.6 times faster. In addition, it can achieve best class separation with relatively low dimensionality. Future directions include extension to other form of data, such as multi-label datasets or continuous labels.

References

1. Bidder, O.R., Campbell, H.A., Gómez-Laich, A., Urgé, P., Walker, J., Cai, Y., Gao, L., Quintana, F., Wilson, R.P.: Love thy neighbour: automatic animal behavioural classification of acceleration data using the k-nearest neighbour algorithm. PLoS One **9**(2), e88609 (2014)
2. Carreira-Perpinán, M.A.: The elastic embedding algorithm for dimensionality reduction. In: ICML 2010, pp. 167–174 (2010)
3. Chang, C.C., Lin, C.J.: LIBSVM: a library for support vector machines. ACM Trans. Intell. Syst. Technol. (TIST) **2**(3), 27 (2011)
4. Fanty, M., Cole, R.: Spoken letter recognition. In: Advances in Neural Information Processing Systems, pp. 220–226 (1991)
5. Globerson, A., Roweis, S.T.: Metric learning by collapsing classes. In: Advances in neural information processing systems, pp. 451–458 (2006)
6. Goldberger, J., Hinton, G.E., Roweis, S.T., Salakhutdinov, R.R.: Neighbourhood components analysis. In: Saul, L.K., Weiss, Y., Bottou, L. (eds.) Advances in Neural Information Processing Systems, vol. 17, pp. 513–520. MIT Press, Cambridge (2005). http://papers.nips.cc/paper/2566-neighbourhood-components-analysis.pdf
7. He, Y., Mao, Y., Chen, W., Chen, Y.: Nonlinear metric learning with kernel density estimation. IEEE Trans. Knowl. Data Eng. **27**(6), 1602–1614 (2015)
8. Huang, G., Guo, C., Kusner, M.J., Sun, Y., Sha, F., Weinberger, K.Q.: Supervised word mover's distance. In: Lee, D.D., Sugiyama, M., Luxburg, U.V., Guyon, I., Garnett, R. (eds.) Advances in Neural Information Processing Systems, vol. 29, pp. 4862–4870. Curran Associates, Inc., New York (2016). http://papers.nips.cc/paper/6139-supervised-word-movers-distance.pdf
9. Ioffe, S., Szegedy, C.: Batch normalization: accelerating deep network training by reducing internal covariate shift. In: International Conference on Machine Learning, pp. 448–456 (2015)
10. Iwata, T., Saito, K., Ueda, N., Stromsten, S., Griffiths, T.L., Tenenbaum, J.B.: Parametric embedding for class visualization. In: Saul, L.K., Weiss, Y., Bottou, L. (eds.) Advances in Neural Information Processing Systems, vol. 17, pp. 617–624. MIT Press, Cambridge (2005). http://papers.nips.cc/paper/2556-parametric-embedding-for-class-visualization.pdf
11. Kedem, D., Tyree, S., Sha, F., Lanckriet, G.R., Weinberger, K.Q.: Non-linear metric learning. In: Pereira, F., Burges, C.J.C., Bottou, L., Weinberger, K.Q. (eds.) Advances in Neural Information Processing Systems 25, pp. 2573–2581. Curran Associates, Inc., New York (2012). http://papers.nips.cc/paper/4840-non-linear-metric-learning.pdf
12. Kusner, M., Tyree, S., Weinberger, K., Agrawal, K.: Stochastic neighbor compression. In: International Conference on Machine Learning, pp. 622–630 (2014)
13. Lang, K.: Newsweeder: learning to filter netnews. In: Proceedings of the 12th International Conference on Machine Learning, vol. 10, pp. 331–339 (1995)
14. LeCun, Y., Bottou, L., Bengio, Y., Haffner, P.: Gradient-based learning applied to document recognition. Proc. IEEE **86**(11), 2278–2324 (1998)
15. Lichman, M.: UCI machine learning repository (2013). http://archive.ics.uci.edu/ml
16. Lu, J., Zhou, X., Tan, Y.P., Shang, Y., Zhou, J.: Neighborhood repulsed metric learning for kinship verification. IEEE Trans. Pattern Anal. Mach. Intell. **36**(2), 331–345 (2014)

17. van der Maaten, L., Hinton, G.: Visualizing data using t-SNE. J. Mach. Learn. Res. **9**(Nov), 2579–2605 (2008)
18. Min, M.R., Maaten, L., Yuan, Z., Bonner, A.J., Zhang, Z.: Deep supervised t-distributed embedding. In: Proceedings of the 27th International Conference on Machine Learning (ICML 2010), pp. 791–798 (2010)
19. Nair, V., Hinton, G.E.: Rectified linear units improve restricted Boltzmann machines. In: Proceedings of the 27th International Conference on Machine Learning (ICML 2010), pp. 807–814 (2010)
20. Salakhutdinov, R., Hinton, G.E.: Learning a nonlinear embedding by preserving class neighbourhood structure. In: International Conference on Artificial Intelligence and Statistics, pp. 412–419 (2007)
21. Villegas, M., Paredes, R.: Dimensionality reduction by minimizing nearest-neighbor classification error. Pattern Recogn. Lett. **32**(4), 633–639 (2011)

Using Metric Space Indexing
for Complete and Efficient Record
Linkage

Özgür Akgün[1]([⊠]), Alan Dearle[1], Graham Kirby[1], and Peter Christen[2]

[1] School of Computer Science, University of St Andrews,
St Andrews, Scotland
{ozgur.akgun,alan.dearle,graham.kirby}@st-andrews.ac.uk
[2] Research School of Computer Science, The Australian National University,
Canberra, Australia
peter.christen@anu.edu.au

Abstract. Record linkage is the process of identifying records that refer
to the same real-world entities in situations where entity identifiers are
unavailable. Records are linked on the basis of similarity between common attributes, with every pair being classified as a link or non-link
depending on their similarity. Linkage is usually performed in a three-
step process: first, groups of similar candidate records are identified using
indexing, then pairs within the same group are compared in more detail,
and finally classified. Even state-of-the-art indexing techniques, such as
locality sensitive hashing, have potential drawbacks. They may fail to
group together some true matching records with high similarity, or they
may group records with low similarity, leading to high computational
overhead. We propose using *metric space indexing* (MSI) to perform
complete linkage, resulting in a parameter-free process combining indexing, comparison and classification into a single step delivering complete
and efficient record linkage. An evaluation on real-world data from several domains shows that linkage using MSI can yield better quality than
current indexing techniques, with similar execution cost, without the
need for domain knowledge or trial and error to configure the process.

Keywords: Entity resolution · Data matching · Similarity search
Blocking

1 Introduction

Record linkage, also known as entity resolution, data matching and duplicate
detection [4], is the process of identifying and matching records that refer to the
same real-world entities within or across datasets. The entities to be linked are
often people (such as patients in hospital or customers in business datasets), but
record linkage can also be applied to link consumer products or bibliographic
records [4]. Record linkage is commonly challenged by the lack of unique entity

© Springer International Publishing AG, part of Springer Nature 2018
D. Phung et al. (Eds.): PAKDD 2018, LNAI 10939, pp. 89–101, 2018.
https://doi.org/10.1007/978-3-319-93040-4_8

identifiers (keys) in the datasets to be linked, which prevents the use of a database join. Instead, the linkage of records requires the comparison of the common attributes (or fields) that are available within the datasets, for example the names, addresses and dates of birth of individuals.

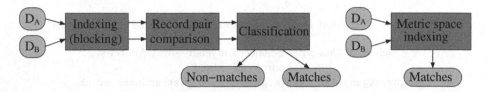

Fig. 1. Overview of the steps of the traditional record linkage process (left side) and our proposed metric space indexing based approach (right side), as described in Sect. 1, where records from two datasets, $\mathbf{D}_A$ and $\mathbf{D}_B$, are being linked.

To overcome data quality issues such as typographical errors and variations (common in name and address values [4]), approximate string comparison functions (e.g. edit distance, the Jaro-Winkler comparator, or Jaccard similarity [4]) are used to compare record pairs, leading to a vector of similarities (one similarity per attribute compared) for each pair. These are used to classify the record pairs into *links* (where it is assumed both records correspond to the same real-world entity) and *non-links* (where they are assumed to correspond to different entities). Various classification methods have been employed in record linkage [4,10], ranging from simple threshold-based to sophisticated clustering, supervised classification, and active learning approaches [30].

Besides a lack of unique entity identifiers, and data quality issues, linkage is also challenged by dataset scale [10]. To avoid full pair-wise comparison of all possible record pairs (quadratic in the dataset sizes), blocking techniques, commonly known as *indexing* [5], are used. These split the datasets into smaller blocks in an efficient way, grouping together records that are likely to correspond to the same entity. Only records within blocks are then compared in detail.

While indexing allows efficient linkage of large datasets [10], scalability is at the cost of reduced linkage quality, because potentially matching record pairs are ignored, leading to lower recall [4]. Indexing techniques, discussed in more detail later, range from simple phonetic based blocking [4] and sorting of the datasets [11] to locality sensitive hashing based techniques [18,29], and unsupervised [17,26] and supervised [1,22] learning of optimal blocking schemes.

Traditional linkage systems that perform indexing prior to comparison and classification (on the left in Fig. 1) add a further complexity. Indexing, comparison and classification are often conducted using algorithms and parameters selected using domain expertise, followed by manual assessment of the linkage outcomes [4]. If the resulting link quality is too low for a certain application, the process is repeated with different parameter settings or algorithms, giving a time-consuming iterative process [13]. The choice of an appropriate indexing technique as well as suitable parameter settings (including which attributes to use in indexing) will significantly affect the final linkage outcome.

We focus on approaches using a similarity threshold to classify links. These are fundamentally limited by the extent to which true matching records are similar, and true non-matches are dissimilar—this is dataset-dependent. Within this domain, we define a technique to be *complete* if it guarantees to find all record pairs within the specified threshold. Many indexing techniques are incomplete, since they reduce computational cost at the expense of potentially overlooking some true matches. By definition, incomplete techniques yield lower recall than complete ones. Conversely, and counter-intuitively, complete techniques can yield lower precision with some datasets. This is discussed further in Sect. 3.

Metric space indexing (MSI) is a complete technique with lower computational cost than a brute force approach. It allows indexing, comparison and classification to be combined into a single step (on the right in Fig. 1), making the process simpler, more efficient and more effective than incomplete approaches.

The motivation for this work is the Digitising Scotland project [9], which aims to transcribe and link all civil registration events recorded in Scotland between 1856 and 1973. This dataset will include around 14 million birth records, 11 million death records and 4 million marriage records.

Contribution: Our primary contribution is the novel application of MSI to achieve complete and efficient record linkage, without the need for complex parameter tuning. We evaluate our approach on several real-world datasets and demonstrate its advantages over existing indexing techniques for record linkage.

2 Related Work

We review relevant work in the areas of indexing for record linkage (for recent surveys see [5,25]), and metric space indexing [31]. Techniques to link records have been investigated for over five decades [12,24], with scalability being an ongoing challenge as datasets grow in size and complexity. Traditional blocking [5] uses a set of attributes (a *blocking key*) to insert records with the same value(s) in their blocking key into the same block. Only records within the same block are compared to each other. To overcome variations and misspellings, the values can be phonetically encoded using functions such as Soundex, NYSIIS, or Double-Metaphone [4]. These convert a string into a code according to its pronunciation, assigning the same code to similar sounding names (such as 'Gail' and 'Gayle'). Multiple blocking keys may also be used to deal with missing attribute values.

A different approach uses sorted neighbourhoods [23], where the datasets are sorted according to a *sorting key* (usually a concatenation of several attribute values). A sliding window is moved over the datasets and only records within the window are compared. Techniques that adaptively shrink or expand the window size based on the characteristics of the sorting key values have been shown to improve both linkage efficiency and quality [11].

These techniques are heuristics, requiring domain knowledge, such as the choice of appropriate blocking or sorting keys. Poor choices of blocking attributes result in records being inserted into inappropriate blocks, and thus true matches being missed, giving *incomplete* linkage. Conversely, many pairs compared in a block may have low similarity, being non-matches, giving *inefficient* linkage.

Locality sensitive hashing (LSH), proposed for efficient nearest-neighbour search in high-dimensional spaces [16], has been used for record linkage indexing. Attribute values are hashed multiple times, and blocks are created from those records that share some hash values. *HARRA* [18] is a linkage approach based on MinHash [3] and LSH which blocks, compares, and then merges linked records iteratively. [29] evaluates two LSH variations, concluding that to get good results, they must be tuned to the datasets. This requires good ground truth data which may be unavailable in real-world applications or expensive to obtain.

Metric space indexing (MSI) techniques [31] support similarity search. They require a distance measure between records, with certain properties including the *triangle inequality* [31]. Similarity search operations include *range-search(q, d)*, identifying all records within a distance d of a query record q; *nearest-neighbour(q)*, returning the record with smallest distance to q; and *nearest-n(q, n)*, returning the n closest records to q. Here we choose one MSI structure, the M-tree [6], and investigate its efficacy for record linkage. The M-tree is dynamically balanced. Every node contains a reference to a record being indexed, a pointer to its parent, the distance to its parent, and the node's radius. The radius of a node is the distance from it to its furthest child. For a parent node with radius r, all its children may be visualised as being contained within a ball of radius r from it.

A linkage method using R-trees [15] was described in [20], demonstrating that high linkage quality can be achieved using Jaccard similarity. [6] shows that M-trees are almost always more efficient than R-trees, hence their use here.

3 Approach

We address the following general linkage problem: for two datasets D_A and D_B, we wish to find, for each record in D_A, all the records in D_B that match it with regard to a certain distance threshold d (i.e. have a distance of d or less). We compare several linkage algorithms: traditional blocking, an incomplete similarity search method, LSH-MinHash, and a complete method, M-tree. We also use a complete brute force technique as a baseline, though this can only feasibly be applied to our smallest dataset. All experiments have a number of parameters to configure the search space and algorithm behaviour, including the distance function and the threshold, d, specifying the maximum distance for two records to be classified as a link (i.e. referring to the same entity). We focus on a single distance function in these experiments, to constrain the experimental space. In Sect. 5 we return to the selection of alternative distance functions.

Brute force: Every record in $\mathbf{D}_A$ is compared with every record in $\mathbf{D}_B$. Each pair is classified as a link if the distance between the records is less than or equal to the threshold d. This always finds all links, with complexity $O(|\mathbf{D}_A| \cdot |\mathbf{D}_B|)$.

Traditional Blocking: The parameters are the set of blocking keys and (optionally) the phonetic encodings applied to each attribute. These are selected as described in [5], exploiting knowledge of the domain and of the data, and chosen with the intention of giving the best possible results. Each record in $\mathbf{D}_A$ is placed into the appropriate block based on its blocking key value. The algorithm then iterates over the records in $\mathbf{D}_B$, and for each one compares it with each of the records from $\mathbf{D}_A$ in the block with the same blocking key value.

LSH-MinHash: The parameters for LSH-Minhash are [3] *shingle size* (l_{ss}), *band size* (l_{bs}) and *number of bands* (l_{nb}). First, the attributes of each record in $\mathbf{D}_A$ are concatenated, and the result *shingled* into a set of n-grams with $n = l_{ss}$. Next, a set of deterministically generated hash functions is applied to each n-gram in the set and the smallest result (the MinHash) of each hash application is added to a signature for the record. The number of hashes used, and thus the size of the signature, is set to $l_{nb} \times l_{bs}$. Finally, the signature is split into l_{nb} bands and the values from each band are hashed again to create a number of keys. The original record is added to a map associated with each of the keys. To perform linkage, the algorithm iterates over the records in $\mathbf{D}_B$. Each record is hashed as described above, to obtain a set of keys. Each key is looked up in the data structure, and the associated records from $\mathbf{D}_A$ added to the result set. Finally, the record from $\mathbf{D}_B$ is compared in turn with each record in the result set, with the pair being classified as a link or non-link based on their distance.

In some circumstances, incomplete approaches such as traditional blocking and LSH-MinHash can yield higher precision than complete techniques. This can occur when a significant number of non-matches nonetheless have high similarity. In this situation, the fact that an incomplete technique omits consideration of some potential links can serve to improve precision, since a classification decision based on a certain similarity threshold is incorrect for high-similarity non-matches. By definition, recall can never be higher for incomplete techniques.

M-tree: The linkage algorithm has no additional parameters. As with *LSH-MinHash*, each record in $\mathbf{D}_A$ is inserted into an M-tree. To perform linkage, the algorithm iterates over each record $\mathbf{b} \in \mathbf{D}_B$. A *range-search(b, d)* operation is performed on the M-tree, passing the distance threshold d as the second parameter. All the returned records are directly classified as links.

4   Experiments and Results

We now describe the datasets and method used in our evaluation[1]. We used three datasets from two domains in our experiments, as summarised in Table 1. The first is *Cora* [21], which contains 1,295 records that refer to 112 machine learning publications. Cora is commonly used as a benchmark dataset in the literature for assessing linkage algorithms. Ground truth is provided via a unique *paper_id* identifier of the form "blum1993". In this experiment linkage is performed over the same set of records (i.e. a de-duplication [4]).

Table 1. Characteristics of datasets used in the experiments.

Dataset name(s)	Records in dataset $\mathbf{D}_A$	Records in dataset $\mathbf{D}_B$	Number of true matching pairs	Entities linked
Cora	1,295	1,295	17,184	Publication–Publication
Isle of Skye	17,612	12,284	2,900	Birth–Death
Kilmarnock	38,430	23,714	8,300	Birth–Death

The other two datasets are historical Scottish records of vital events (birth, marriages and deaths), one registered on the *Isle of Skye*, a rural district, and the other records from *Kilmarnock*, an industrial town. These datasets were created, curated and linked by historical demographers [27,28]. Both include the names and genders of individuals and their parents. Ground truth was generated by the demographers based on their extensive domain knowledge.

In all of our experiments we use a single distance metric: the sum of the attribute-level Levenshtein [19] edit distances.

4.1   Cora Results

We perform linkage on the Cora dataset using all approaches presented in this paper: brute force, traditional blocking, LSH and M-tree, using several selected configurations for blocking and LSH. The distance threshold is varied between 0 and 250[2]. For traditional blocking, the following attributes are used individually as blocking keys: *author*, *title*, *venue*, *location*, *publisher* and *year*. We also use a combined blocking key comprising all attributes.

Figure 2 shows the precision, recall, and F-measure [4] for various thresholds[3]. As expected, low thresholds give high precision and low recall, and the reverse for high thresholds. Brute force and M-tree give identical results, as expected.

---

[1] Experimental data, additional figures and source code can be downloaded from: http://github.com/digitisingscotland/pakdd2018-metric-linkage.

[2] Relatively high Levenshtein edit distances are included since Cora contains a number of low-similarity true matches.

[3] Noting that recent research identifies some problematic aspects with using the F-measure to compare record linkage procedures at different similarity thresholds [14].

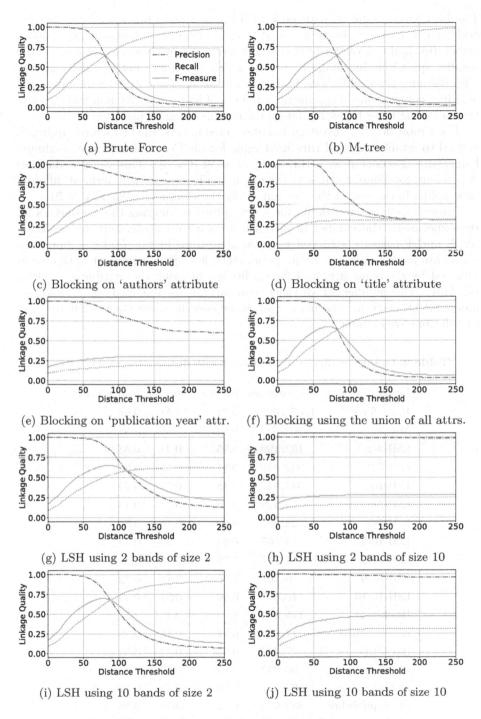

(a) Brute Force

(b) M-tree

(c) Blocking on 'authors' attribute

(d) Blocking on 'title' attribute

(e) Blocking on 'publication year' attr.

(f) Blocking using the union of all attrs.

(g) LSH using 2 bands of size 2

(h) LSH using 2 bands of size 10

(i) LSH using 10 bands of size 2

(j) LSH using 10 bands of size 10

Fig. 2. Linkage results on the Cora dataset.

The best linkage quality, with an F-measure of around 0.7, is achieved by several linkers, including brute force, M-tree, blocking on *authors*, blocking on all attributes, and two of the LSH configurations. All of these give similar overall results, apart from blocking on *authors*, which gives much better quality at very high distance thresholds. This is due to the incomplete nature of the approach, avoiding comparisons of significant numbers of high-similarity non-matches and thus avoiding these becoming false positives and keeping precision high.

For a more detailed investigation of selected linkers, the brute force approach is used to establish a good threshold value for the Cora dataset. The maximum F-measure is observed at a threshold value of $d = 70$. This value is dataset-dependent; for different datasets the maximum F-measure will occur at different thresholds. In the rest of this section we fix the threshold value at $d = 70$.

Table 2 shows greater detail for selected linkers, showing the parameters for the experiment, the number of distance comparisons made, and the precision, recall and F-measure achieved by each algorithm. In the *Linker* column the algorithm name is followed by its parameters: for LSH the number of the bands followed by the band size, and for traditional blocking the attributes used for blocking. The number of distance comparisons is reported as a machine-independent proxy for execution cost, since code profiling shows that distance calculations are dominant.

Table 2. Linkage quality on Cora dataset with distance threshold $d = 70$.

Linker	Comparisons	Precision	Recall	F-measure
Brute force	$1,677,025$	0.84	0.57	0.68
M-tree	$902,693$	0.84	0.57	0.68
LSH-2-2	$192,199$	0.95	0.47	0.63
LSH-5-2	$342,849$	0.91	0.55	0.69
LSH-10-2	$513,947$	0.88	0.57	0.69
LSH-2-5	$14,329$	0.99	0.28	0.43
LSH-5-5	$22,057$	0.99	0.36	0.53
LSH-10-5	$26,167$	0.98	0.40	0.57
LSH-2-10	$4,711$	1.00	0.15	0.27
LSH-5-10	$6,501$	1.00	0.19	0.32
LSH-10-10	$10,627$	0.99	0.27	0.43
Block-year	$115,893$	0.99	0.35	0.51
Block-authors	$11,039$	0.94	0.16	0.28
Block-title	$27,407$	0.95	0.42	0.58
Block-venue	$36,647$	0.85	0.29	0.44
Block-location	$1,009,957$	0.83	0.43	0.57
Block-publisher	$833,079$	0.85	0.44	0.58
Block-combined	$1,214,269$	0.84	0.56	0.67

M-tree yields the same linkage quality as brute force, although using a significantly lower number of comparisons. This is as expected, since both techniques are complete. Several of the incomplete linkers give similar quality, for example *LSH-2-2*, *LSH-5-2*, *LSH-10-2* and *Block-combined*. These, and a number of other incomplete linkers, give better precision than the complete techniques. This is due to high-similarity non-matches, as discussed in Sect. 3. Although several of the incomplete linkers give as good quality as M-tree, and in some cases at lower cost, this is offset by the need to select appropriate configuration parameters. Some other linkers give very poor results.

4.2 Demographic Dataset Results

Birth records were linked to death records, separately for the Skye and Kilmarnock datasets, using M-tree and a range of LSH configurations. It was not computationally feasible to run the brute force linker. Of the incomplete linkers, LSH was selected as it gave slightly better results for Cora. The shingle size was set to $l_{ss} = 2$ for all the LSH experiments reported, as this was found to give good results and LSH was not especially sensitive to this parameter. Results for other shingle sizes are omitted from this paper for brevity. A lower range of distance thresholds was explored, based on domain knowledge of the datasets.

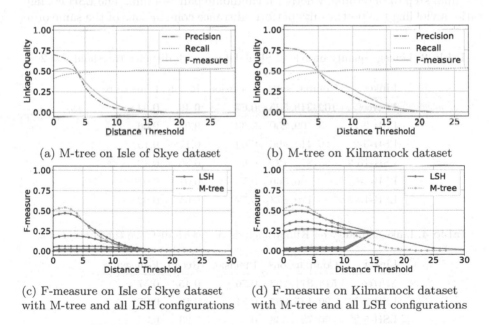

(a) M-tree on Isle of Skye dataset   (b) M-tree on Kilmarnock dataset

(c) F-measure on Isle of Skye dataset   (d) F-measure on Kilmarnock dataset
with M-tree and all LSH configurations   with M-tree and all LSH configurations

Fig. 3. Linkage results on the demographic datasets.

Figure 3 plots (a) and (b) show the M-tree precision, recall, and F-measure for various thresholds. In both datasets, the best F-measure values are obtained with

a low distance threshold of $d = 2$. Plots (c) and (d) compare the F-measure curves
for M-tree with those obtained from a range of LSH configurations. The best
F-measure value for M-tree is higher than that of any of the LSH configurations,
for both datasets. This demonstrates both the competitiveness of M-tree with
respect to linkage quality, and its important characteristic of being parameter-
free—the linkage quality is obtained without the need to tune for the dataset.

Tables 3 and 4 show greater detail for selected linkers. In both cases the F-
measure achieved is better for M-tree than any of the LSH linkers. The better
linkage quality achieved by M-tree is largely due to recall for M-tree being much
higher than for any of the LSH configurations. In most cases, LSH out-performs
M-tree in terms of precision. More significantly, LSH linkage quality is heavily
dependent on the configuration parameters. For plausible settings for the number
of bands and band size, F-measure varies from 0.01 (extremely poor) to 0.47
(relatively good) for Skye and from 0.03 to 0.49 for Kilmarnock. In both cases
LSH-10-2 performs best, but since this is data-dependent there is no guarantee
that these parameters would work well with another dataset.

The number of distance comparisons varies dramatically among the various
linkers. M-tree always performs the most comparisons, since they are intrinsic
to the *range-search* algorithm. The core part of the LSH linker performs Jaccard
similarity comparisons and hashing; distance comparisons are only performed in
the final step to determine whether a candidate pair is a link. The LSH configurations yielding the best results perform distance comparisons of the same order

Table 3. Linkage quality on Isle of Skye dataset with distance threshold $d = 2$.

Linker	Comparisons	Precision	Recall	F-measure
M-tree	$102, 318, 525$	0.65	0.46	0.54
LSH-2-2	$3, 109, 250$	0.63	0.03	0.06
LSH-5-2	$10, 412, 496$	0.64	0.11	0.19
LSH-10-2	$53, 874, 127$	0.68	0.36	0.47
LSH-5-5	$36, 566$	0.76	0.01	0.01
LSH-10-5	$129, 873$	0.72	0.01	0.02

Table 4. Linkage quality on Kilmarnock dataset with distance threshold $d = 2$.

Linker	Comparisons	Precision	Recall	F-measure
M-tree	$514, 871, 153$	0.76	0.45	0.57
LSH-2-2	$99, 145, 887$	0.81	0.16	0.27
LSH-5-2	$130, 721, 338$	0.79	0.23	0.36
LSH-10-2	$177, 168, 848$	0.79	0.36	0.49
LSH-5-5	$239, 368$	0.84	0.01	0.02
LSH-10-5	$855, 431$	0.87	0.02	0.03

of magnitude as M-tree. This indicates that the good LSH linkers return many candidates beyond the distance threshold. Thus, in order to get good results, LSH tends towards a brute force search over the candidate results. Despite this, LSH is faster due to the efficiency of the hashing process.

5 Conclusions and Future Work

In this paper we have demonstrated the efficacy of MSI in achieving complete and efficient record linkage, without the need for complex parameter tuning. In conclusion, this claim deserves some careful unpacking. It is always possible to achieve high quality linkage using a brute force approach. However the quadratic complexity of this approach prevents its practical application for datasets of even moderate size. We have shown that MSI techniques such as M-tree can deliver high precision, high recall results that are the same as those delivered by brute force. Furthermore this is achieved with fewer distance comparisons, and consistently without the need for complex parameter tuning.

We contrast this to traditional blocking and LSH-based approaches. Their major drawback is that whilst they can produce extremely good results, they can also produce extremely poor results. It was our observations of low recall given by these approaches that originally led us to experiment with M-trees.

We note that the good results obtained by both traditional blocking and LSH are partly due to the fact that (in the limit) they tend towards brute force as the number of records in the blocks increase. A second, unexpected, result is that illustrated in Table 2, namely that incomplete approaches such as LSH can in some cases yield higher precision than that achieved by a complete method such as M-tree. This is due to the incomplete linker masking the inability of a classifier based solely on record similarity to correctly classify high-similarity non-matches or low-similarity matches.

We have focused on a single distance function: the sum of attribute-level Levenshtein distances. This gives a straightforward intuition of record-level distances, but it is more common to normalise metrics to the range 0–1.

Many distance functions, such as Jaro-Winkler, are not metric, and therefore cannot be used with MSI techniques. Care must be taken to preserve metric properties when combining metrics over individual fields, as is highlighted in [2], since this may yield a function that is not metric. It is then possible for MSI techniques to yield results that are subtly incorrect.

Different metrics can give different distributions of inter-record distances, which can affect both the linkage results and the number of comparisons made, and hence the efficiency of the algorithm. Datasets with low variation in inter-record distances are said to have high *intrinsic dimensionality*, and tend to require high numbers of comparisons. The Scottish vital event datasets [28], combined with Levenshtein-based metrics, appear to have high intrinsic dimensionality. We are now, therefore, investigating the application of various different metrics to this domain, including Jensen-Shannon, Cosine and Structured Entropic Distance, as described in [7]. We are also investigating the applicability of a novel technique for dimensionality reduction [8].

Acknowledgements. This work was supported by ESRC grants ES/K00574X/2 "Digitising Scotland" and ES/L007487/1 "Administrative Data Research Centre—Scotland".

We thank Alice Reid of the University of Cambridge and her colleagues, especially Ros Davies and Eilidh Garrett, for the work undertaken on the Kilmarnock and Isle of Skye databases.

References

1. Bilenko, M., Kamath, B., Mooney, R.J.: Adaptive blocking: learning to scale up record linkage. In: IEEE ICDM, Hong Kong, pp. 87–96 (2006)
2. Bo, L., Yujian, L.: A normalized Levenshtein distance metric. IEEE Trans. Pattern Anal. Mach. Intell. **29**, 1091–1095 (2007). https://doi.org/10.1109/TPAMI.2007. 1078
3. Broder, A.: On the resemblance and containment of documents. In: IEEE Compression and Complexity of Sequences, Salerno, Italy, pp. 21–29 (1997)
4. Christen, P.: Data Matching - Concepts and Techniques for Record Linkage, Entity Resolution, and Duplicate Detection. Springer, Heidelberg (2012). https://doi.org/ 10.1007/978-3-642-31164-2
5. Christen, P.: A survey of indexing techniques for scalable record linkage and deduplication. IEEE TKDE **24**(9), 1537–1555 (2012)
6. Ciaccia, P., Patella, M., Rabitti, F., Zezula, P.: Indexing metric spaces with M-tree. In: Italian Symposium on Advanced Database Systems 1997, pp. 67–86 (1997)
7. Connor, R.: A tale of four metrics. In: Amsaleg, L., Houle, M.E., Schubert, E. (eds.) SISAP 2016. LNCS, vol. 9939, pp. 210–217. Springer, Cham (2016). https://doi. org/10.1007/978-3-319-46759-7_16
8. Connor, R., Vadicamo, L., Rabitti, F.: High-dimensional simplexes for supermetric search. In: Beecks, C., Borutta, F., Kröger, P., Seidl, T. (eds.) SISAP 2017. LNCS, pp. 96–109. Springer, Cham (2017). https://doi.org/10.1007/978-3-319-68474-1_7
9. Dibben, C., Williamson, L., Huang, Z.: Digitising Scotland (2012). http://gtr.rcuk. ac.uk/projects?ref=ES/K00574X/2
10. Dong, X.L., Srivastava, D.: Big data integration. Synth. Lect. Data Manag. **7**(1), 1–198 (2015)
11. Draisbach, U., Naumann, F., Szott, S., Wonneberg, O.: Adaptive windows for duplicate detection. In: IEEE ICDE, Washington, DC, pp. 1073–1083 (2012)
12. Fellegi, I.P., Sunter, A.B.: A theory for record linkage. J. Am. Stat. Assoc. **64**(328), 1183–1210 (1969)
13. Fisher, J., Wang, Q.: Unsupervised measuring of entity resolution consistency. In: IEEE ICDM DINA Workshop, pp. 218–221 (2015)
14. Hand, D., Christen, P.: A note on using the F-measure for evaluating record linkage algorithms. Stat. Comput. **28**(3), 539–547 (2018)
15. Hjaltason, G.R., Samet, H.: Incremental distance join algorithms for spatial databases. SIGMOD Rec. **27**(2), 237–248 (1998)
16. Indyk, P., Motwani, R.: Approximate nearest neighbors: towards removing the curse of dimensionality. In: ACM TOC, Dallas, pp. 604–613 (1998)
17. Kejriwal, M., Miranker, D.P.: An unsupervised algorithm for learning blocking schemes. In: IEEE ICDM, Dallas, pp. 340–349 (2013)
18. Kim, H., Lee, D.: HARRA: fast iterative hashed record linkage for large-scale data collections. In: EDBT, Lausanne, pp. 525–536 (2010)

19. Levenshtein, V.: Binary codes capable of correcting deletions, insertions and reversals. Cybern. Control Theory **10**, 707–710 (1966)
20. Li, C., Jin, L., Mehrotra, S.: Supporting efficient record linkage for large data sets using mapping techniques. World Wide Web **9**(4), 557–584 (2006)
21. McCallum, A.: Cora dataset: cora.csv (2017). https://doi.org/10.3886/E4728V1
22. Michelson, M., Knoblock, C.A.: Learning blocking schemes for record linkage. In: AAAI, Boston (2006)
23. Monge, A.E., Elkan, C.P.: The field-matching problem: algorithm and applications. In: ACM SIGKDD, Portland, pp. 267–270 (1996)
24. Newcombe, H., Kennedy, J., Axford, S., James, A.: Automatic linkage of vital records. Science **130**(3381), 954–959 (1959)
25. Papadakis, G., Svirsky, J., Gal, A., Palpanas, T.: Comparative analysis of approximate blocking techniques for entity resolution. PVLDB **9**(9), 684–695 (2016)
26. Ramadan, B., Christen, P.: Unsupervised blocking key selection for real-time entity resolution. In: Cao, T., Lim, E.-P., Zhou, Z.-H., Ho, T.-B., Cheung, D., Motoda, H. (eds.) PAKDD 2015. LNCS (LNAI), vol. 9078, pp. 574–585. Springer, Cham (2015). https://doi.org/10.1007/978-3-319-18032-8_45
27. Reid, A., Garrett, E., Davies, R., Blaikie, A.: Scottish census enumerators' books: Skye, Kilmarnock, Rothiemay and Torthorwald, 1861–1901. Economic and Social Data Service (2006)
28. Reid, A., Davies, R., Garrett, E.: Nineteenth-century Scottish demography from linked censuses and civil registers: a 'sets of related individuals' approach. History Comput. **14**(1–2), 61–86 (2002)
29. Steorts, R.C., Ventura, S.L., Sadinle, M., Fienberg, S.E.: A comparison of blocking methods for record linkage. In: Domingo-Ferrer, J. (ed.) PSD 2014. LNCS, vol. 8744, pp. 253–268. Springer, Cham (2014). https://doi.org/10.1007/978-3-319-11257-2_20
30. Wang, Q., Vatsalan, D., Christen, P.: Efficient interactive training selection for large-scale entity resolution. In: Cao, T., Lim, E.-P., Zhou, Z.-H., Ho, T.-B., Cheung, D., Motoda, H. (eds.) PAKDD 2015. LNCS (LNAI), vol. 9078, pp. 562–573. Springer, Cham (2015). https://doi.org/10.1007/978-3-319-18032-8_44
31. Zezula, P., Amato, G., Dohnal, V., Batko, M.: Similarity Search: The Metric Space Approach. Springer, Boston (2010). https://doi.org/10.1007/0-387-29151-2

Dimensionality Reduction via Community Detection in Small Sample Datasets

Kartikeya Bhardwaj[✉] and Radu Marculescu

Carnegie Mellon University, Pittsburgh, PA, USA
kbhardwa@andrew.cmu.edu, radum@cmu.edu

Abstract. Real world networks constructed from raw data are often characterized by complex community structures. Existing dimensionality reduction techniques, however, do not take such characteristics into account. This is especially important for problems with low number of samples where the *curse of dimensionality* is particularly significant. Therefore, in this paper, we propose *FeatureNet*, a novel community-based dimensionality reduction framework targeting small sample problems. To this end, we propose a new method to directly construct a network from high-dimensional raw data while explicitly revealing its hidden community structure; these communities are then used to learn low-dimensional features using a representation learning framework. We show the effectiveness of our approach on eight datasets covering application areas as diverse as handwritten digits, biology, physical sciences, NLP, and computational sustainability. Extensive experiments on the above datasets (with sizes mostly between 100 and 1500 samples) demonstrate that FeatureNet significantly outperforms (*i.e.*, up to 40% improvement in classification accuracy) ten well-known dimensionality reduction methods like PCA, Kernel PCA, Isomap, SNE, t-SNE, *etc.*

1 Introduction

Many Artificial Intelligence (AI) and Machine Learning (ML) problems use very high-dimensional datasets. This high dimensionality leads to the *curse of dimensionality* where the performance of ML models decreases with increasing data dimensions [9]. The curse of dimensionality is particularly profound if in addition to the large number of features, the problem also suffers from the availability of a low number of samples [5,9,10,16]. Specifically, as theoretically established in [5,9,10,16], in order to obtain high classification performance in high-dimensional spaces, the number of samples must also be very large (*e.g.*, $\sim 10^5$ samples). Therefore, for problems with a low number of samples (say, 100–1500 samples), extracting a set of useful features from the high-dimensional data is a very challenging task [23]. Consequently, in this paper, we address the well-known dimensionality reduction problem specifically in a *high dimensions and low sample-size* setting. This important class of problems has many engineering and scientific applications such as on-device mobile applications, remote sensing, fMRI processing, sustainability, finance, and biological systems.

© Springer International Publishing AG, part of Springer Nature 2018
D. Phung et al. (Eds.): PAKDD 2018, LNAI 10939, pp. 102–114, 2018.
https://doi.org/10.1007/978-3-319-93040-4_9

Dimensionality reduction can also be seen as an *automatic feature extraction* problem which yields low-dimensional features from the initial high-dimensional raw data. Indeed, a major goal in AI and representation learning is to enable machines to learn such useful, low-dimensional features *automatically* from the raw data rather than using manually engineered features. Towards this automatic feature learning, deep learning has emerged for vision, speech, and natural language processing (NLP) applications [3]. However, deep learning often needs enormous training datasets ($\sim 10^5$–10^6 samples) and results in significant overfitting for problems with small sample-size and high dimensions [4]. Therefore, for small sample-size problems, new dimensionality reduction techniques are needed to automatically extract useful features from the initial high-dimensional data.

Several outstanding dimensionality reduction techniques exist ranging from linear methods such as Principal Component Analysis (PCA), Probabilistic PCA (PPCA), to graph-based non-linear techniques like Maximum Variance Unfolding (MVU) [21], Isomap [19], *etc.* (see [12] for a review). Other techniques include Kernel PCA, deep learning autoencoders [8], stochastic proximity embedding (SPE) [1], stochastic neighbor embedding (SNE) [7] and t-distributed SNE (t-SNE) [11]. The graph-based techniques build a *neighborhood graph* to learn a lower-dimensional embedding [12]. For example, Isomap builds a neighborhood graph based on a *fixed* parameter which controls the neighborhood size or the number of nearest-neighbors of each node in the graph.

In real world, however, networks constructed from raw data often possess complex characteristics such as *communities* (*i.e.*, groups of tightly connected nodes) and *structural equivalence* (*i.e.*, nodes with similar roles in network, *e.g.*, hubs) [17]. Therefore, such network characteristics must be accounted for while computing network neighborhoods for dimensionality reduction as they can lead to more accurate feature learning. Specifically, the neighborhood of a given node must depend on the community it belongs to. By contrast, prior techniques like Isomap assume a rigid (fixed) neighborhood for all nodes in the network. Similarly, stochastic graph-based methods such as SNE/t-SNE use a fixed parameter called perplexity which measures the effective number of neighbors for each node. Hence, the complex community structure hidden within the raw data has *not* been explicitly taken into account in prior dimensionality reduction methods.

Recently, representation learning has been proposed in the context of learning features on networks while accounting for community structure, *e.g.*, node2vec [6], DeepWalk [15], community preserving embedding [20], LINE [18], *etc.* We refer to this problem space as *"Representation Learning on Networks"* throughout the paper. However, the networks considered in this prior art do *not* come from high-dimensional raw data, but rather from social networks (*e.g.*, blogs, Youtube, Flickr), authorship networks or Wikipedia webpage networks. Hence, the prior representation learning on networks research does *not* directly address the problem of dimensionality reduction. In contrast, we argue that by capturing communities and structural equivalence, ideas from "representation learning on networks" problem space can have significant implications for dimensionality reduction. Therefore, in this paper, we address the following two **key questions**:

1. Can representation learning on networks have more general implications in dimensionality reduction if we leverage the hidden communities in raw data?
2. If so, how can we best construct a network from high-dimensional data to optimally capture its latent communities for dimensionality reduction?

To answer these questions, we propose *FeatureNet*, a new community-based dimensionality reduction framework. We further contribute a new method to construct a network directly from the raw data while explicitly revealing its hidden communities; this enables us to employ network representation learning ideas to learn *low-dimensional* community- and structural equivalence-based features from this network, thereby reducing the dimensions of the dataset.

We evaluate our proposed approach on five very diverse application areas ranging from handwritten digit recognition, biology, physical sciences, NLP, to computational sustainability. As mentioned earlier, our datasets are relatively small with sizes mostly between 100 and 1500 samples. This is because, *automatic feature engineering* for relatively small datasets still remains an important problem as deep learning models often lead to overfitting for such datasets.

To summarize, we make the following **key contributions**:

1. We propose FeatureNet, a novel community-based dimensionality reduction framework. We also propose a new method to construct a network directly from high-dimensional raw data, thereby revealing its hidden communities explicitly. To the best of our knowledge, we are the first to employ community-based representation learning ideas for dimensionality reduction.
2. We evaluate FeatureNet on eight datasets spanning five diverse real-world applications like handwritten digit recognition, biology, physical science, NLP, and computational sustainability. Our new sustainability datasets can be used by research community to further benchmark dimensionality reduction.
3. We further compare FeatureNet against ten most notable dimensionality reduction techniques such as PCA, deep learning autoenoders, t-SNE, Isomap, *etc.* Overall, the proposed FeatureNet significantly outperforms (in terms of accuracy) all of these techniques on the above diverse datasets by 3%–40%.
4. Finally, we introduce a new challenging computational sustainability problem as a case study: *Given high-dimensional Carbon Emissions data, how can we learn optimal low-dimensional features to best classify the GDP growth of nations?* Again, FeatureNet achieves the state-of-the-art performance.

Next, we review the related work reported in the literature.

2   Related Work

As mentioned earlier, prior techniques such as node2vec, DeepWalk, community-preserving embedding, LINE, *etc.* explicitly require a network as an input [6,15, 18,20]. In contrast, we start with high-dimensional (raw) data that does not exist in predefined network forms and learn the network structure directly from the raw data to explicitly reveal its latent communities. For example, consider

Arcene, a high-dimensional cancer benchmark dataset, where each sample comes from a patient and features specify the abundance of certain proteins[1]. This dataset does not have a predefined network structure like in social networks. Hence, methods like node2vec, LINE, *etc.* are *not* a natural choice. This is where our key contribution lies – in representing *any kind of dataset* as a network which reveals hidden communities in raw data; we then use this network for dimensionality reduction using ideas from community-based feature learning.

To summarize, prior work in representation learning on networks focuses only on network-based classification tasks (*e.g.*, classify interests of a blogger based on communities/homophily in a blog social network). Our work, however, truly generalizes this "network representation learning" space to *any* classification problem which has high-dimensional data and does *not* restrict it only to network classification tasks (see Supplementary Sect. 1 for detailed related work[2]).

3   Proposed Approach

Given a classification problem $\{X, y\}$, let $X \in \mathbb{R}^{n \times p}$ denote the original dataset with n samples and p features, while $y \in \mathbb{R}^{n \times 1}$ denotes the labels. Also, let $x^{(i)} \in \mathbb{R}^{p \times 1}$ be the i-th sample in X. Then, dimensionality reduction is a function $f : \mathbb{R}^{n \times p} \rightarrow \mathbb{R}^{n \times d}$, where d is the number of features in the reduced space ($d << p$). Since prior neighborhood graph-based methods [11,19] do not take communities into account, this results in loss of important network information.

Let $\mathcal{X}$ be the low-dimensional mapping of X, and $\mathrm{x}^{(i)} \in \mathbb{R}^{d \times 1}$ be the i-th sample of $\mathcal{X}$ (*i.e.*, the reduced representation of initial $x^{(i)}$). Then, the problem is to find $\mathcal{X}$ which accounts for the latent community structure and structural equivalences hidden within the raw data. Hence, unlike established techniques such as Isomap, the network neighborhood for each sample in our approach is *not* fixed, but rather takes communities and structural equivalences into account. To find such a mapping, therefore, we maximize the probability of observing a certain neighborhood $\mathcal{N}$ of sample $x^{(i)}$, conditional on its low-dimensional representation, as well as on its latent community structure and structural equivalences:

$$\max_{\mathcal{X} = \{\mathrm{x}^{(i)} | i = 1:n\}} \sum_{x^{(i)} \in X} \log Pr(\mathcal{N}(x^{(i)}) | \mathrm{x}^{(i)}, \mathcal{C}(x^{(i)}), \mathcal{S}(x^{(i)})) \qquad (1)$$

where, $\mathcal{C}(x^{(i)})$ and $\mathcal{S}(x^{(i)})$ are latent variables containing information about communities and structural equivalence of sample $x^{(i)}$ hidden within the raw data.

To solve problem (1), we propose FeatureNet, a two stage solution for dimensionality reduction which: (*i*) transforms the raw data into a network space to *explicitly* reveal data's inherent communities, and (*ii*) performs representation learning on this network (see Fig. 1). Next, we present our proposed network construction technique to reveal hidden communities naturally.

---

[1] https://archive.ics.uci.edu/ml/datasets/Arcene.
[2] Supplementary material available at: https://goo.gl/LvkmjB.

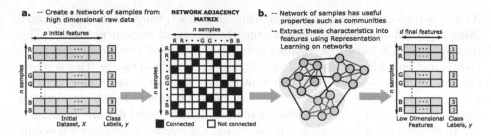

Fig. 1. Complete flow of FeatureNet: (a) First, construct a network of samples using the proposed correlations-based method to explicitly reveal hidden communities in raw data (Sect. 3: Step 1). (b) Next, use representation learning on this network to find the community-based low-dimensional features (Sect. 3: Step 2).

Step 1: Proposed K-τ Method for Network Construction

We create a network directly from raw data as follows: (i) construct a correlation-based network (τ-step), and (ii) improve the density of network communities (K-step). For the τ-step, we transform the initial high-dimensional data into a correlation-based network of samples (*i.e.*, each sample now becomes a node). This first step is a mapping $l : \mathbb{R}^{n \times p} \to \mathbb{R}^{n \times n}$, which yields $\mathcal{G} = l(X)$. Here, $\mathcal{G} \in \mathbb{R}^{n \times n}$ is the adjacency matrix of the network of samples with elements:

$$\mathcal{G}^{ij} = \begin{cases} c(x^{(i)}, x^{(j)}) & \text{if } c(x^{(i)}, x^{(j)}) \geq \tau \\ 0 & \text{if } c(x^{(i)}, x^{(j)}) < \tau \text{ or } i = j \end{cases} \tag{2}$$

where, $c(\cdot, \cdot)$ is the Pearson's correlation function, and τ is a threshold used on $c(\cdot, \cdot)$ to remove weakly correlated links from the network.

Threshold (τ-step): Setting a higher τ removes the noise from the network by encouraging connections only among samples of the same class (*intra-class links*) and not among samples of different classes (*inter-class links*). Ideally, a network should mostly have intra-class links and not many inter-class links. To elaborate, we consider MNIST handwritten digit dataset where each sample has 784 features. Since our focus is on relatively small datasets, we randomly select 1000 samples (100 images for each digit 0–9) from the MNIST database.

Next, we create a Pearson's correlation-based network of samples from this 1000×784 dataset using Eq. 2 and a threshold $\tau = 0.7$. Figure 2(a) illustrates the adjacency matrix of this network. Clearly, the ten diagonal clusters in Fig. 2(a) represent the intra-class links, thus revealing the hidden communities of each digit. Moreover, using a high threshold of 0.7, most of the *noisy links* in the network (*i.e.*, the inter-class links) are removed. We note, however, that a too high threshold can also result in some samples getting completely disconnected from the network and very sparse communities (see digits 2 and 5 in Fig. 2(a) zoomed inset). To overcome this problem, we introduce a network density parameter, K.

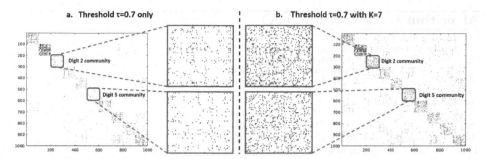

Fig. 2. Adjacency matrix for MNIST dataset network. (a) Threshold $\tau = 0.7$ removes noise from the network and reveals a clear community structure (*i.e.*, the diagonal clusters). (b) Introducing a density parameter K fixes the problem of sparse communities without adding significant noise and yields reliable low-dimensional representations.

Network Density (K-step): To connect the disconnected nodes and increase the density of network communities, we next connect each sample to its corresponding K highest correlated samples; *i.e.*, after the thresholding step, if a sample $x^{(i)}$ has less than K links, we connect it to samples $x^{(j)}$'s until it has K links. Here, samples $x^{(j)}$'s are selected based on the K-highest correlations. This step is a variant of the K-nearest neighbor approach to handle correlations rather than euclidean distances (*i.e.*, instead of K neighbors with minimum distances, we use K neighbors with maximum correlations). As shown in Fig. 2(b), introducing a network density parameter of $K = 7$, (i) connects all disconnected nodes, and (ii) increases the density of diagonal clusters significantly without too much additional noise (see the zoomed-in inset of Fig. 2(b)). Hence, the threshold and density steps yield a K-τ method-based network of samples, $\mathcal{G}$.

To summarize, our proposed approach creates a network from raw data using two parameters: Threshold τ and density K, which provide a tradeoff between the noise and the density of communities. Best K and τ can be selected via cross-validation and a simple grid search which would generate an optimal neighborhood for each sample. Hence, in our approach, the neighborhood of each sample is not rigid, rather is determined *automatically* by its community structure. Explicitly revealing these hidden communities in raw data, therefore, enables the use of community-based representation learning in dimensionality reduction.

Step 2: Community-Based Representation Learning
The network of samples, $\mathcal{G}$, often possesses characteristics such as communities and structural equivalence. Once in the network space, problem (1) reduces to:

$$\max_{\mathcal{X}=\{x^{(i)}|i=1:n\}} \sum_{v \in \mathcal{V}(\mathcal{G})} \log Pr(\mathcal{N}_{\mathcal{G}}^{R}(v)|x^{(i)}) \tag{3}$$

where, $\mathcal{V}(\mathcal{G})$ denotes the set of nodes in network $\mathcal{G}$, and sample $x^{(i)}$ is now represented by a node v in the network of samples. Finding the neighborhood of $x^{(i)}$, $\mathcal{N}(x^{(i)})$, now becomes the problem of finding the network neighborhood

---

Algorithm 1. FeatureNet(X, τ, K)

1: **Input:** Raw Data $X \in \mathbb{R}^{n \times p}$, τ, K
2: **Output:** Low-dimensional representation $\mathcal{X} \in \mathbb{R}^{n \times d}$
      —— **K-τ method to reveal communities**——
3:   $A \leftarrow$ corr(X') /* Pairwise correlations b/w samples */
4:   $\mathcal{G} \leftarrow A$, $\mathcal{G}(\mathcal{G} < \tau) \leftarrow 0$ /* Remove links below threshold */
5:   NODES $\leftarrow \mathcal{V}(\mathcal{G})$
6:   **for all** i in NODES **do**
7:     $a^{(i)} \leftarrow A(i, :)$, $g^{(i)} \leftarrow \mathcal{G}(i, :)$ /* i^{th} row of A (G) */
8:     **while** $\|g^{(i)}\|_0 < K$ **do**   /* # links for node $i < K$ */
9:       $\{m, j\} \leftarrow$ max$(a^{(i)})$, $\mathcal{G}^{ij} \leftarrow m$, $g^{(i)} \leftarrow \mathcal{G}(i, :)$
10:      $a^{ij} \leftarrow \phi$ /* Remove max element from $a^{(i)}$ */
11:     **end while**
12:   **end for**

13: $\mathcal{X} \leftarrow h(\mathcal{G})$ /* Learn low-dimensional representation by solving problem (3). Use node2vec search strategy to explicitly account for communities revealed by $\mathcal{G}$ */

$\mathcal{N}_{\mathcal{G}}^{R}(v)$ of node v in $\mathcal{G}$. This network neighborhood can be found using a strategy R, which can account for the latent community structure $\mathcal{C}(x^{(i)})$ and structural equivalence $\mathcal{S}(x^{(i)})$. However, note that, this is precisely the skip-gram objective which the recent research on network representation learning aims to optimize [6,13,15]. Therefore, once the high-dimensional raw data is transformed into a network which *explicitly* reveals the community structure, the final low-dimensional representation can be learned using techniques such as node2vec [6]. Specifically, node2vec acts as a mapping $h : \mathbb{R}^{n \times n} \rightarrow \mathbb{R}^{n \times d}$, which yields $\mathcal{X} = h(\mathcal{G})$. The $n \times d$ matrix $\mathcal{X}$ contains the final low-dimensional features based on hidden communities in raw data. Algorithm 1 shows these two stages of FeatureNet. For more information on the classic word2vec skip-gram objective [13] and node2vec search strategy [6], please refer to Supplementary Sect. 2.

4 Experimental Setup and Results

4.1 Experimental Setup

We implement the K-τ method in MATLAB, while node2vec-neighborhood search, optimization, and the subsequent classification are all carried out in Python. We use *one-vs-rest* logistic regression with L2 regularization and a broad range of inverse regularization strength parameter, $C \in \{10^{-2}, 10^{-1}, \ldots, 10^4\}$ for multi-class classification. Of note, node2vec parameters (return parameter p, and in-out parameter q) which control a trade-off between communities and structural equivalence, are optimized via a grid search on $p, q \in \{0.25, 0.75, 0.9, 1.5, 2, 4\}$. Finally, the two parameters of FeatureNet (K, τ) are also optimized using a grid search. τ is varied in steps of 0.05 from 0.6 to 0.95, while K varies from 1 to 9. The best parameter values are selected using 10-fold cross-validation (CV).

To show the effectiveness of FeatureNet on many applications, we conduct experiments on eight datasets coming from five very different application areas as summarized in Table 1. Our focus in this paper is on dimensionality reduction for relatively small datasets which explains why the sample sizes are mostly between 100 and 1500 in Table 1. Reuters subset data is used to analyze the scalability of our approach. Table 1 contains five benchmarks from UCI ML repository[3].

Table 1. Characteristics of the datasets

Area	Dataset	X ($n \times p$)	#classes	Area	Dataset	X ($n \times p$)	#classes
Cancer (Bio.)	Arcene	100 × 10000	2	Comp. Sust.	CE-GDP 1980	1015 × 1095	17
Phys. Sci.	Musk1	476 × 166	2	Comp. Sust.	CE-GDP 1990	1420 × 1095	16
Hand Digits	MNIST	1000 × 784	10	Comp. Sust.	CE-GDP 2000	1448 × 1095	11
NLP	CNAE-9	1079 × 856	9	NLP	Reuters (subset)	5946 × 18933	65

Table 2. 10-fold CV F_1-Macro and F_1-Micro (Accuracy) scores for UCI benchmarks ($d = 16$): Best six prior methods shown.

Dimension reduction	Arcene		Musk1		MNIST		Dimension reduction	CNAE-9	
	F_1-Macro	F_1-Micro	F_1-Macro	F_1-Micro	F_1-Macro	F_1-Micro		F_1-Macro	F_1-Micro
PCA	0.7596	0.76	0.7502	0.7563	0.8069	0.808	PCA	0.855	0.852
PPCA	0.7681	0.7699	0.7502	0.7563	0.805	0.806	PPCA	0.855	0.8535
KPCA (Poly.)	0.7493	0.75	0.7034	0.7037	0.6655	0.679	SNE	0.8728	0.8721
SPE	0.6124	0.62	0.7354	0.7436	0.7872	0.789	SPE	0.8527	0.8535
t-SNE	0.7312	0.7399	0.709	0.7142	0.8837	0.884	t-SNE	0.8068	0.81
Isomap	0.6386	0.64	0.726	0.7331	0.8338	0.8349	Isomap	0.8317	0.8331
FeatureNet	**0.8164**	**0.82**	**0.829**	**0.834**	**0.9128**	**0.913**	FeatureNet	**0.9220**	**0.923**

Table 1 also shows three datasets from the computational sustainability domain in which quantitatively inferring economic growth from anthropogenic carbon emissions remains an active area of research [14]. Here, we make a twofold contribution: First, we propose the following new computational sustainability problem: *"Given multiple years of daily carbon emissions (CE) data across the world, can we correctly classify the Gross Domestic Product (GDP) growth of different regions?"* Second, we contribute three new datasets to further benchmark dimensionality reduction. The datasets are compiled using a carbon dioxide database [2] and the World Bank [22] data (see Supplementary Sect. 3).

Finally, we compare our approach against ten[4] well-established dimensionality reduction techniques: (1) PCA, (2) PPCA, (3) Polynomial Kernal PCA (KPCA - Poly.), (4) KPCA – gaussian kernel, (5) Linear Discriminant Analysis (LDA), (6) SPE, (7) Deep Autoencoders, (8) SNE, (9) t-SNE, and (10) Isomap. We used a dimensionality reduction toolbox [12] for these techniques.

---

[3] http://archive.ics.uci.edu/ml/index.php.
[4] For the ease of presentation, we will report only the top six performers.

4.2   Results

UCI Machine Learning Repository Benchmarks. In our experiments, we reduce the dimensions of each dataset from initial p features to $d = 16$ features. We then conduct logistic regression on the reduced features and report its 10-fold CV F_1-Macro and F_1-Micro scores. Note that, F_1-Micro scores have the same interpretation as classification accuracy for multiclass classification problems. Table 2 presents these results for FeatureNet and the best six traditional techniques for all UCI datasets. As shown, our proposed FeatureNet *significantly* outperforms all six (and, implicitly, all ten!) prior techniques.

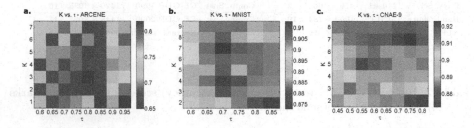

Fig. 3. F_1-Micro for varying FeatureNet parameters (K, τ): (a) Arcene, (b) MNIST, and (c) CNAE-9. Red (blue) indicates higher (lower) accuracy. For all datasets, FeatureNet outperforms prior methods for many combinations of K and τ. (Color figure online)

For Arcene, FeatureNet achieves a F_1-Micro of 0.82 improving over the best performing PPCA method by 6.5%. Arcene is a challenging dataset because 3000 out of its 10000 features are 'probes' with no predictive power. This shows that our proposed FeatureNet is able to handle such noisy datasets. Next, for Musk1, we achieve an improvement of 10.27% in F_1-Micro scores over the best traditional methods – PCA and PPCA. Similarly, for MNIST, we observe an improvement of 3.28% in F_1-Micro over the best performing t-SNE technique. Recall that, we are only using 1000 samples for MNIST and not all 60,000 images for training. In fact, all datasets used in the present work are "relatively small" with number of samples mostly between 100 and 1500. This is why, deep learning-based autoencoders do not perform very well and, as expected, overfit the data.

Finally, for the CNAE-9 dataset (NLP), we improve the F_1-Micro by 5.83% over the best performing SNE method. CNAE-9 is a business description text data for certain companies classified according to economic sectors. Each document is processed using standard NLP techniques (*e.g.*, stop-word removal, stemming, *etc.*) and is converted to a term frequency vector. This results in a very sparse dataset wherein 99.22% of the raw data is all zeros. In summary, our results demonstrate that FeatureNet can handle dimensionality reduction problems on very diverse applications and can also handle noisy and sparse datasets. Similar improvements are observed for F_1-Macro scores.

Table 3. 10-fold CV F_1-Micro (Accuracy) for CE-GDP problems ($d = 16$): Six best prior methods are shown.

Years	PCA	PPCA	KPCA (Poly.)	KPCA (Gauss.)	t-SNE	Isomap	FeatureNet
1980	0.5123	0.6137	0.4108	0.4059	0.398	0.5241	**0.86**
1990	0.6098	0.6661	0.483	0.4845	0.5366	0.6443	**0.8161**
2000	0.625	0.6609	0.4544	0.4171	0.4682	0.6153	**0.8411**

Empirical Evaluation of FeatureNet in the K-τ Parameter Space.
Figure 3 shows the impact of varying density K (y-axis) and threshold τ (x-axis) for various UCI datasets (see Supplementary Fig. S2(a) for Musk1 dataset). As shown, FeatureNet outperforms the traditional methods for several combinations of K and τ (see orange/red portions in Fig. 3). For instance, for MNIST, CNAE-9 (Fig. 3(b, c)) and Musk1 (Fig. S2(a)), almost any combination of parameters gives a high classification accuracy. Indeed, for Arcene (Fig. 3(a)), we observe that only a few parameter combinations give high performance (*e.g.*, for $\tau = 0.95$ and all K values). A possible reason for FeatureNet's behavior for Arcene could be due to the additional noise in this dataset. We leave the theoretical analysis of stability of FeatureNet as a future work (*e.g.*, analyzing impact of noise, *etc.*).

Why We Achieve Performance Gains? As mentioned before, the parameters τ and K control the tradeoff between noise in the network and density of communities. Consider the case $\tau = 0.85$ and varying K's for MNIST (*i.e.*, the rightmost column of Fig. 3(b)). For a high threshold of 0.85, the diagonal communities are even more sparse than those shown in Fig. 2 where the threshold used was only 0.7 (see also supplementary Fig. S1). Now, if we increase the density K, the F_1-Micro increases from 0.873 for $K = 2$, to 0.904 for $K = 5$, to 0.902 for $K = 9$ (probably too much noise for $K = 9$). This clearly demonstrates the tradeoff between the noise and density, and how it can affect the model performance. Therefore, our K-τ method for network construction successfully captures the best tradeoff and thus yields a high classification accuracy. Hence, our results show that choosing a good network construction approach for revealing hidden communities in data is very important for obtaining higher performance.

Computational Sustainability – A Case Study and New AI Datasets.
Table 3 shows F_1-Micro for the competitive methods across the three years for the CE-GDP datasets. As evident, FeatureNet significantly outperforms the best PPCA method by 40.13%, 22.51%, and 27.26% for 1980, 1990, and 2000, respectively. We also observed results like Fig. 3 for the CE-GDP datasets (see supplementary Fig. S2(b)). Moreover, Fig. S3 shows results for varying the number of target dimensions from $d = 16$ to 32. Again, FeatureNet outperforms other techniques for all d. Therefore, the CE-GDP datasets can also be used by the ML community to benchmark dimensionality reduction.

Finally, the K-τ network shown in Fig. 4(a) for CE-GDP 2000 dataset demonstrates that FeatureNet models the hidden communities with significantly different sizes very accurately, thus explaining the excellent performance of FeatureNet (see Fig. S4 for the 1980 network). Hence, fixed neighborhood size or perplexity methods (*e.g.*, Isomap, t-SNE) cannot capture such massive heterogeneity in raw data's community structure. To show this, we vary the *fixed* neighborhood size for Isomap in Fig. 4(b). As shown, FeatureNet is far superior (with F_1-Micro nearly 0.9 for $d = 32$) to Isomap for all neighborhood sizes.

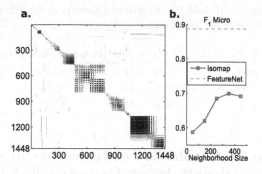

Fig. 4. (a) K-τ network for CE-GDP 2000 shows communities with very different sizes that are accurately modeled by FeatureNet. (b) Varying fixed neighborhood size in Isomap and other methods cannot capture such variable size communities ($d = 32$).

Note on Scalability. In order to analyze the scalability of FeatureNet, we consider a subset of Reuters-21578 dataset in which documents with multiple category labels were removed. This yielded 8293 documents from 65 classes with 18933 distinct terms. Of the total 8293 documents, we focus on the given training dataset[5] of 5946 documents and report 10-fold CV classification F_1-Micro after reducing its dimensions from 18933 to 16. We compare FeatureNet with some of the top performers from the above experiments – SPE, PCA, and t-SNE as these were amongst the only few techniques that were able to finish execution in a reasonable time (*e.g.*, about 2–4 h) using reasonable computational resources (an 8-core Intel i7 desktop).

For relatively small datasets like MNIST, the number of links is not very big (*e.g.*, 6669 links for 1000 nodes). However, for larger datasets like Reuters, the number of links can increase rapidly (719, 080 links for ($\tau = 0.7$, $K = 30$) case and 2.1 million links for ($\tau = 0.5$, $K = 50$) case; see Table S1). Figure S5 shows the diagonal communities for the Reuters $\tau = 0.7$ and $K = 30$ case ($\approx$700,000 links), whereas Fig. S6 shows the same for $\tau = 0.5$ and $K = 50$ (>2.1M links; to create this network, MATLAB takes only 10 s and up to 7 GB of memory.). Clearly, the diagonal communities of the former are significantly more sparse than those of the latter. Consequently, our proposed FeatureNet successfully reduced the dimensions and finished executing for the former but not for the latter. In terms of the classification accuracy, F_1-Micro for SPE, PCA, and t-SNE were 0.725, 0.82 and 0.823 respectively, whereas FeatureNet again significantly outperformed these techniques with a F_1-Micro score of 0.867 (5.34% improvement). These results demonstrate that currently FeatureNet can indeed

---

[5] See details at http://www.cad.zju.edu.cn/home/dengcai/Data/TextData.html.

scale up to large datasets provided their networks contain several hundreds of thousands of links. However, optimizing FeatureNet to handle datasets which result in more than several million links, is left for future research.

5 Conclusion and Future Work

We have proposed FeatureNet, a new community-based dimensionality reduction framework for small sample problems. To this end, we have proposed a new technique to construct a network from any general raw data while revealing its hidden communities. Community-based low-dimensional features are then learned using a representation learning framework. We have demonstrated the effectiveness of FeatureNet across five very different application domains ranging from handwritten digit recognition, biology, physical science, NLP, to computational sustainability. We have further shown that FeatureNet significantly outperforms many well-known dimensionality reduction techniques such as PCA, PPCA, deep autoencoders, t-SNE and Isomap. This ultimately shows how representation learning ideas can have huge implications for dimensionality reduction.

As a future work, we plan to develop even stronger algorithms and parallelization techniques to scale FeatureNet to hundred-thousand samples/features. Finally, we plan to provide an in-depth theoretical analysis for FeatureNet.

References

1. Agrafiotis, D.K.: Stochastic proximity embedding. J. Comput. Chem. **24**(10), 1215–1221 (2003)
2. Andres, R.: Monthly Fossil-Fuel CO2 Emissions: Mass of Emissions Gridded by One Degree Latitude by One Degree Longitude. CDIAC, U.S.A. (2013)
3. Bengio, Y., et al.: Representation learning: a review and new perspectives. IEEE Trans. PAMI **35**(8), 1798–1828 (2013)
4. Friedman, J., Hastie, T., Tibshirani, R.: The Elements of Statistical Learning. Springer Series in Statistics, vol. 1. Springer, New York (2001). https://doi.org/10.1007/978-0-387-21606-5
5. Fukunaga, K., Hayes, R.R.: Effects of sample size in classifier design. IEEE Trans. Pattern Anal. Mach. Intell. **11**(8), 873–885 (1989)
6. Grover, A., Leskovec, J.: node2vec: scalable feature learning for networks. In: KDD, pp. 855–864. ACM (2016)
7. Hinton, G., Roweis, S.: Stochastic neighbor embedding. In: NIPS, vol. 15 (2002)
8. Hinton, G.E., Salakhutdinov, R.R.: Reducing the dimensionality of data with neural networks. Science **313**(5786), 504–507 (2006)
9. Hughes, G.: On the mean accuracy of statistical pattern recognizers. IEEE Trans. Inf. Theory **14**(1), 55–63 (1968)
10. Jain, A.K., Duin, R.P.W., Mao, J.: Statistical pattern recognition: a review. IEEE Trans. Pattern Anal. Mach. Intell. **22**(1), 4–37 (2000)
11. Van der Maaten, L., Hinton, G.: Visualizing data using t-SNE. J. Mach. Learn. Res. **9**(Nov), 2579–2605 (2008)
12. Maaten, V.D.: Dimensionality reduction: a comparative review. J. Mach. Learn. Res. **10**, 66–71 (2009)

13. Mikolov, T., Chen, K., Corrado, G., Dean, J.: Efficient estimation of word representations in vector space. arXiv preprint arXiv:1301.3781 (2013)
14. Moss, R.H., et al.: The next generation of scenarios for climate change research and assessment. Nature **463**(7282), 747 (2010)
15. Perozzi, B., Al-Rfou, R., Skiena, S.: DeepWalk: online learning of social representations. In: KDD, pp. 701–710. ACM (2014)
16. Raudys, S.J., et al.: Small sample size effects in statistical pattern recognition. IEEE Trans. PAMI **13**(3), 252–264 (1991)
17. Steinhaeuser, K., et al.: Multivariate and multiscale dependence in the global climate system revealed through complex networks. Clim. Dyn. **39**, 889–895 (2012)
18. Tang, J., et al.: LINE: large-scale information network embedding. In: Proceedings of the 24th International Conference on World Wide Web, pp. 1067–1077 (2015)
19. Tenenbaum, J.B., et al.: A global geometric framework for nonlinear dimensionality reduction. Science **290**(5500), 2319–2323 (2000)
20. Wang, X., Cui, P., Wang, J., Pei, J., Zhu, W., Yang, S.: Community preserving network embedding. In: AAAI, pp. 203–209 (2017)
21. Weinberger, K.Q., Saul, L.K.: An introduction to nonlinear dimensionality reduction by maximum variance unfolding. In: AAAI 2006, pp. 1683–1686 (2006)
22. World Bank: GDP Growth Data (%) (2017). http://data.worldbank.org/
23. Yamada, M., et al.: High-dimensional feature selection by feature-wise kernelized Lasso. Neural Comput. **26**(1), 185–207 (2014)

An Interaction-Enhanced Feature Selection Algorithm

Xiaochuan Tang[(✉)], Yuanshun Dai, Yanping Xiang, and Liang Luo

School of Computer Science and Engineering,
University of Electronic Science and Technology of China, Chengdu 611731, China
xiaochuantang@std.uestc.edu.cn, ydai@uestc.edu.cn

Abstract. Feature selection is a crucial pre-processing step in machine learning and data mining. A popular approach is based on information theoretic measures. Most of the existing methods used low-dimensional mutual information terms that are ineffective in detecting high-order feature interactions. To fill this gap, we employ higher-order interactions for feature selection. We first relax the assumptions of MI-based methods to allow for higher-order interactions. A direct calculation of the interaction terms is computationally expensive. We use four-dimensional joint mutual information, a computationally efficient measure, to estimate the interaction terms. We also use the 'maximum of the minimum' nonlinear approach to avoid the overestimation of feature significance. Finally, we arrive at an effective feature selection method that makes use of higher-order interactions. To evaluate the performance of the proposed method, we compare it with seven representative feature selection methods, including RelaxMRMR, JMIM, IWFS, CIFE, MIFS, MIM, and reliefF. Experimental results on eighteen benchmark data sets demonstrate that higher-order interactions are effective in improving MI-based feature selection.

Keywords: Feature selection · Mutual information
Feature interaction

1 Introduction

In machine learning and data mining, the data under processing have become increasingly larger, both in terms of the number of features and the number of instances. Such data significantly increase the computation time and memory requirements for data analytics. Moreover, irrelevant and redundant features widely exist in the data, which not only increase computational burden but also have negative impact on learning models. Towards this issue, a straightforward solution is feature selection. Feature selection methods can be broadly classified as wrapper [1], filter [2,3] and embedded [4]. Feature selection does not alter the physical meaning of the original features. The readability and interpretability of feature selection is better than feature extraction [5].

In this paper, we focus on studying mutual information (MI) based feature selection. The goal of MI-based feature selection is to select a feature subset that

© Springer International Publishing AG, part of Springer Nature 2018
D. Phung et al. (Eds.): PAKDD 2018, LNAI 10939, pp. 115–125, 2018.
https://doi.org/10.1007/978-3-319-93040-4_10

maximizes the Multidimensional Joint Mutual Information (MJMI) between the selected features and the class label. However, accurate estimation for MJMI entails estimating a high-dimensional joint probability distribution, which is a long-standing challenge in statistics. Many MI-based feature selection methods address this issue by making a set of assumptions to decompose MJMI into low-dimensional MI terms. Brown et al. [6] proposed a unifying framework for MI-based feature selection methods. The framework consisted of two two-dimensional and one three-dimensional MI terms, i.e., relevancy, redundancy and conditional redundancy. Many well-known methods are special cases of this framework, such as MIM [7], mRMR [3], MIFS [8], and CIFE [9]. One limitation of these methods is that they neglected some important higher-dimensional MI terms.

Recently, studies showed that higher-dimensional MI terms can improve the performance of feature selection [10]. Our study is inspired by the following works. First, a comparative study [6] showed that the Joint Mutual Information (JMI) method achieved a good trade-off between accuracy and stability. The reason was that JMI balanced the relevancy/redundancy terms and included the conditional redundancy. The drawback of JMI was that it allowed the overestimation of the significance of some features. Bennasar et al. [11] used the 'maximum of the minimum' method to overcome this problem. Their method was referred to as Joint Mutual Information Maximization (JMIM). Our method not only use the 'maximum of the minimum' method, but also employ a higher-dimensional JMI term. Second, Vinh et al. [10] studied higher-order dependencies and proposed a new feature selection method (RelaxMRMR). It introduced a three-dimensional MI term to estimate redundancy. The author identified a relaxed assumption of conditional independence as a theoretical underpinning for RelaxMRMR. We further relaxed the assumptions to allow higher-dimensional MI terms. Third, Zeng et al. [12] showed that three-way interaction can improve the performance of feature selection. Shishkin et al. [13] proposed a modification of the Conditional Mutual Information Maximization (CMIM) method [14] to identify interactions. It used greedy algorithm to reduce computational complexity and binary representatives to reduce sample complexity. Therefore, higher-order interactions have great potential for improving the performance of feature selection. Inspired by these works, we propose a new feature selection method that takes into account higher-order interactions.

Our contributions are summarized as follows: (1) We identify a relaxed assumption, which allows us to decompose the MI-based feature selection into a sum of interactions. (2) We propose a new feature selection method that takes into account higher-order feature interactions. We conduct comprehensive experiments to show the effectiveness of the proposed method.

2   Proposed Method for Feature Selection

In this section, a new MI-based feature selection method is proposed. We first introduce some concepts on information measures for feature interaction. Then, a preliminary method that directly takes into account higher-order interactions is proposed. Finally, we propose an improved feature selection method.

2.1   An Information Measure for Feature Interaction

This section introduces an information measure for evaluating feature interaction. To begin with, some necessary notations are introduced.

The *mutual information* between a feature X and a class label Y quantifies the information shared by X and Y. It is defined as,

$$I(X;Y) = \sum_{x_i \in X} \sum_{y_j \in Y} p(x_i, y_j) log \frac{p(x_i, y_j)}{p(x_i)p(y_j)} = H(X) + H(Y) - H(X, Y) \quad (1)$$

where $H(\cdot)$ denotes *entropy*.

Definition 1 (Interaction information). *Interaction information quantifies the information shared by multi-variables [15]. We also adopt interaction information to quantify feature interactions. It is defined as,*

$$I(S) \triangleq - \sum_{T \subseteq S} (-1)^{|S|-|T|} H(T) \quad (2)$$

where $S = \{X_{i_1}, \cdots, X_{i_s}\}$ is a subset of features, and $T = \{X_{j_1}, \cdots, X_{j_t}\}$ is a subset of S. $I(S) = I(X_{i_1}; \cdots; X_{i_s})$ is the interaction information of all the features in S, where the semicolon ';' refers to an interaction information. $H(T) = H(X_{j_1}, \cdots, X_{j_t})$ is the joint entropy of all the features in T, where the comma ',' refers to a joint variable.

The following equation shows the connection between three-dimensional joint mutual information and interaction information.

$$I(X_i, X_j; Y) = I(X_i; Y) + I(X_j; Y) + I(X_i; X_j; Y) \quad (3)$$

2.2   An Interaction Based Feature Selection Method

In this section, we first identify a relaxed assumption, and then propose a interaction based feature selection method.

For a feature set $X = \{X'_1, \cdots, X'_N\}$ of N features, the feature selection process selects the most informative feature subset $S = \{X_1, \cdots, X_k\} \subset X$, where $k < N$. Ideally, the MI-based methods should maximize the joint mutual information between the selected features S and the class label Y:

$$S_{opt} = arg \max_{S \subset X} I(S; Y) \quad (4)$$

However, exhaustive search of the optimal feature subset S is impractical. A popular sub-optimal search strategy is the sequential forward selection (SFS) [13]. The MI-based methods usually use low-dimensional MI terms to estimate $I(S; Y)$. Balagani and Phoha [16] proposed a set of assumptions to decompose high-dimensional dependency $I(S;Y)$ into a sum of very low-dimensional MI terms (E.g., relevancy $I(X_i;Y)$ and redundancy $I(X_i;X_j)$). The assumptions are: (1) The selected features are independent; (2) The selected features are conditionally independent given the candidate feature; (3) Each feature independently influences the class label. Vinh et al. [10] relaxed assumption (2) as follows: the selected features are conditionally independent given the candidate feature and any feature $X_j \in S$.

However, studies have shown that higher-order interactions are also informative. For example, in Natural Language Processing, the N-gram language model [17] is widely used to capture term dependencies. The n-gram of size one (unigram), two (bigram) and three (trigram) are popular. Trigram consists of three features, which can be viewed as a special case of four-way interaction. Therefore, we make a relaxed assumption to enable higher-order interactions.

Assumption 1. *Given a candidate feature $X_k \in X \setminus S_k$, a selected feature $X_i \in S_k$, and a class label Y, then each feature in set S_i independently interact with the given variables:*

$$I(S_i; X_i; X_k; Y) = \sum_{j=1}^{i-1} I(X_j; X_i; X_k; Y) \tag{5}$$

where $S_i = \{X_1, \cdots, X_{i-1}\}$ are the selected features before X_i.

By this assumption, we make a balance between accuracy and efficiency. As n increases, the accuracy of an n-way interaction based feature selection method increases. However, the estimation of n-way interaction information become less reliable. Since it requires much more data and computation.

Based on Assumption 1, the MI-based feature selection problem of Eq. 4 is written as,

$$\begin{aligned}
&arg \max_{X_k \in X \setminus S_k} I(S_k \cup X_k; Y) \\
&= arg \max_{X_k \in X \setminus S_k} I(X_k; Y) + \sum_{i=1}^{m} I(X_i; X_k; Y) + \sum_{i=2}^{m} \sum_{j=1}^{i-1} I(X_i; X_j; X_k; Y)
\end{aligned} \tag{6}$$

Proof. Let $X_k \in X \setminus S_k$ be a candidate feature, and $S_k = \{X_1, \cdots, X_{k-1}\}$ be a feature subset that contains all the selected features before the kth round.

Using Eq. 3, the MI-based feature selection problem can be transformed into:

$$\begin{aligned}
I(S_k \cup X_k; Y) &= I(S_k, X_k; Y) = I(S_k; X_k; Y) + I(S_k; Y) + I(X_k; Y) \\
&= I(S_{k-1}, X_{k-1}; X_k; Y) + I(S_k; Y) + I(X_k; Y) \\
&= I(S_{k-1}; X_{k-1}; X_k; Y) + I(S_{k-1}; X_k; Y) + I(X_{k-1}; X_k; Y) + I(X_k; Y) + \Omega \\
&= I(S_{k-1}; X_{k-1}; X_k; Y) + I(S_{k-2}; X_{k-2}; X_k; Y) + I(S_{k-2}; X_k; Y) + I(X_{k-2}; X_k; Y) \\
&\quad + I(X_{k-1}; X_k; Y) + I(X_k; Y) + \Omega \\
&= \cdots = \sum_{i=2}^{k} I(S_i; X_i; X_k; Y) + \sum_{i=1}^{k} I(X_i; X_k; Y) + I(X_k; Y) + \Omega
\end{aligned}$$

$$\tag{7}$$

Under Assumption 1, Eq. 7 is turned to Eq. 8:

$$\sum_{i=2}^{k} \sum_{j=1}^{i-1} I(X_i; X_j; X_k; Y) + \sum_{i=1}^{k} I(X_i; X_k; Y) + I(X_k; Y) + \Omega \qquad (8)$$

where Ω is a constant w.r.t X_k. Therefore, we finish the proof.

However, the goal function in Eq. 6 has $O(|S|^2)$ interaction terms. So the computational complexity is high. We employ four-dimensional joint mutual information to address the drawbacks of Eq. 6. The proposed method is referred to as Four-way Interaction Maximization (FIM). The goal function is written as:

$$J_{FIM}(X_k) = arg \min_{X_i, X_j \in S} I(X_i, X_j, X_k; Y) \qquad (9)$$

where $X_k \in X \setminus S$ is a candidate feature, and S contains the selected features before X_k.

Theorem 1. *The four-way joint mutual information $I(X_1, X_2, X_3; Y)$ equals to a sum of interaction information of all the feature subsets of its component features:*

$$I(X_1, X_2, X_3; Y) = \sum_{i=1}^{3} I(X_i; Y) + \sum_{i=1}^{2} \sum_{j=i+1}^{3} I(X_i; X_j; Y) + I(X_1; X_2; X_3; Y)$$

$$\qquad (10)$$

where X_1, X_2, X_3 and Y are random variables.

Proof. Using Eq. 1, the left-hand side of Eq. 10 can be transformed to

$$I(X_1, X_2, X_3; Y) = H(X_1, X_2, X_3) + H(Y) - H(X_1, X_2, X_3, Y) \qquad (11)$$

Using Eqs. 1 and 2, the right-hand side of Eq. 10 can be transformed to,

$$\sum_{i=1}^{3} I(X_i; Y) + \sum_{i=1}^{2} \sum_{j=i+1}^{3} I(X_i; X_j; Y) + I(X_1; X_2; X_3; Y)$$

$$= \sum_{i=1}^{3} \{H(X_i) + H(Y) - H(X_i, Y)\} - \sum_{i=1}^{2} \sum_{j=i+1}^{3} \{H(X_i) - H(X_j) - H(Y) + H(X_i, X_j)$$

$$+ H(X_i, Y) + H(X_j, Y) - H(X_i, X_j, Y)\} + \{H(X_1) + H(X_2) + H(X_3) + H(Y)$$

$$- H(X_1, X_2) - H(X_1, X_3) - H(X_1, Y) - H(X_2, X_3) - H(X_2, Y) - H(X_3, Y)$$

$$+ H(X_1, X_2, X_3) + H(X_1, X_2, Y) + H(X_1, X_3, Y) + H(X_2, X_3, Y) - H(X_1, X_2, X_3, Y)\}$$

$$= H(Y) + H(X_1, X_2, X_3) - H(X_1, X_2, X_3, Y)$$

Therefore, Eq. 10 holds.

Theorem 1 shows that the four-dimensional joint mutual information equals to the sum of all the interactions of its component variables. The goal function in Eq. 9 has many advantages. First, a direct computation of $I(X_i, X_j, X_k; Y)$ avoids the calculation of a large number of interaction terms in Eq. 6. An effective method for estimating $I(X_i, X_j, X_k; Y)$ is the histogram method based on the frequency of occurrences in the data [18]. Second, the goal function in Eq. 9 employs the 'maximum of the minimum' method [11] to avoid the over estimation of some features, such as redundant features.

Algorithm 1. The FIM feature selection algorithm

Input: The original features $\{X_1, \cdots, X_n\}$; the number of features to be selected k.
Output: The selected feature subset S.
1: Initialize $S = \Phi$; $T = \{X_1, X_2, \cdots, X_n\}$.
2: **for** $i = 1$ to k **do**
3:    **for** $j = 1$ to $n - i$ **do**
4:        Calculate $J_{FIM}(T_j)$ using Eq. 9.
5:    **end for**
6:    $z = \max_{X_t \in T}(J_{FIM}(X_t))$.
7:    $S = S \cup z$.
8:    $T = T \setminus z$.
9: **end for**

For filter methods, conventional search strategies are forward and backward search. The proposed method FIM is a filter method with forward search, which is summarized in Algorithm 1.

2.3   Complexity Analysis

We analyze the computational complexity of the proposed method FIM. Recall that the input data set $D \in R^{M \times N}$ has M instances and N features, and the number of features to be selected is k. The complexity of mRMR, JMI and other similar methods are $O(k^2 MN)$ [10].

FIM and RelaxMRMR generally have higher complexity since more MI terms are taken into account. Compared with mRMR, RelaxMRMR needs an additional loop over the selected feature subset to compute $I(X_i; X_k | X_j)$. The time complexity for RelaxMRMR is $O(k^3 MN)$. Similarly, the time complexity for FIM is $O(k^3 MN)$. The reason is that, compared with JMI, FIM needs an additional loop over the selected features to compute $I(X_i, X_j, X_k; Y)$. In future works, we plan to use parallel computing and quantum computing technologies to accelerate FIM.

3   Experiments

We conduct experiments to evaluate the performance of the proposed FIM method. The experiments are performed on eighteen data sets, including thirteen

UCI data sets and five microarray data sets [3,10]. Table 1 shows that the data sets cover a wide range of conditions, i.e., different number of features, instances and classes.

Table 1. Dataset summary.

Dataset	#Features	#Instances	#Class
Wine	13	178	3
Segment	19	2310	7
Cardio	21	2126	3
Waveform	21	5000	3
Parkinsons	22	195	2
Steel	27	1941	7
Breast	30	569	2
Ionosphere	33	351	2
Landsat	36	6435	6
Spambase	57	4601	2
Musk	166	476	2
Semeion	256	1593	10
Arrhythmia	257	430	2
Lung	325	73	7
Colon	2000	62	2
Lymphoma	4026	96	9
Leukemia	7129	73	2
NCI60	9996	60	10

For classification, we use SVM with linear kernel, k-Nearest Neighbor (kNN) with 3 neighbors, and Decision Tree (DT) with the CART algorithm. All the classifiers are available in the Matlab Statistics and Machine Learning Toolbox. The classification accuracy is used to compare the performance of the feature selection methods.

The experimental settings are as follows. For each data set, the 10-fold cross-validation (or leave-one-out cross validation if the number of instances < 100) is used to divide the data into train/test sets. For each partition of the data, do the following feature selection and classification process: (1) each feature selection method selects a subset of $k = 50$ features based on the train data. (2) The train/test data are updated according to the selected features (from 1 to k features). Then, each classifier is trained on the train data, and classification accuracy is obtained by applying the trained classifier to the test data. Finally, we compute the *mean ± std* classification accuracy across a range of feature size (from 1 to k). We also obtain a plot of the classification accuracy vs. the number of selected features.

The proposed FIM method is compared with seven popular feature selection methods. Specifically, FIM is compared with some well-known MI-based feature selection methods, including RelaxMRMR [10], JMIM [11], IWFS [12], CIFE, MIFS, and MIM [6]. To make a more comprehensive comparison, FIM is also compared with a *non-MI-based* method, i.e., reliefF [2]. All the features selection methods are implemented in Matlab/C++. Specifically, the MI terms of the MI-based methods are implemented using a C++ implementation called Mutual Information Toolbox[1]. Each continuous feature is discretized into five equal-size bins [10]. The *relieff* function is a Matlab built-in function for reliefF.

3.1 Overall Performance

Table 2 shows the average classification accuracy of all the feature selection methods. The last row of each table shows the win-tie-loss values, which are used to indicate that FIM performs better (+), equally well (=), or worse (−) than the competitor. It is obtained by the one-sided paired t-test at the 5% significance level.

In general, for all the three classifiers, FIM achieves better performance than other methods in over half of the cases. While in about ten percent of the cases, one of the competitors wins. In the remaining cases, the performance of FIM and the other approaches are similar. Therefore, FIM performs better or equally well than all the other methods in about ninety percent of the cases. The methods that take into account feature interactions (e.g., reliefF, JMIM, and RelaxM-RMR) generally have better performance than others.

When kNN and Decision Tree are used as classifier, for each of the competing feature selection method, the number of wins is bigger than the number of loss. FIM tend to outperform other methods. When the classifier is SVM, FIM outperforms all the competing method except RleaxMRMR. One possible reason is that SVM does not make full use of the four-way interactions.

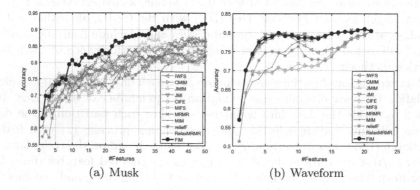

(a) Musk   (b) Waveform

Fig. 1. Performance comparison with respect to #features.

---

[1] http://home.penglab.com/proj/mRMR/.

Table 2. Classification accuracy ($mean\% \pm std$) comparison of different feature selection methods.

Dataset	FIM	MIM	MIFS	CIFE	IWFS	JMIM	RelaxMRMR	reliefF
kNN								
Wine	75.4±6.5	72.4±2.2(=)	79.8±6.8(-)	74.5±5.8(=)	72.1±2.9(+)	73.8±7.4(+)	72.1±5.3(+)	70.8±1.5(+)
Parkinsons	83.4±1.1	84.6±2.1(-)	82.1±1.1(+)	82.1±1.3(+)	83.2±1.1(=)	82.9±0.8(+)	83.3±0.8(=)	83.0±2.3(=)
Ionosphere	86.7±2.2	85.4±1.4(+)	86.7±3.0(=)	86.4±2.7(=)	85.5±3.0(+)	86.6±1.8(=)	86.8±1.9(=)	85.2±1.9(+)
Breast	92.1±0.8	91.9±1.1(=)	91.8±0.8(+)	92.0±0.7(=)	91.9±0.8(+)	92.3±0.8(-)	91.8±0.9(+)	89.6±6.0(+)
Segment	92.9±8.4	89.9±8.4(+)	89.8±7.8(+)	88.3±8.6(+)	92.1±8.2(+)	92.8±8.3(+)	92.6±8.3(=)	87.9±14.7(+)
Cardio	89.0±3.8	88.9±3.8(=)	84.4±5.3(+)	84.1±5.4(+)	88.9±4.2(=)	89.2±3.8(=)	88.2±4.0(+)	88.7±3.2(=)
Steel	43.6±11.6	44.5±7.2(=)	28.6±10.8(+)	29.0±10.1(+)	35.4±17.6(+)	47.2±10.4(-)	44.6±11.8(=)	32.2±8.7(+)
Musk	81.6±6.3	75.2±4.7(+)	78.1±7.1(+)	78.4±7.5(+)	79.3±6.2(+)	79.2±5.9(+)	80.0±7.6(+)	77.2±6.6(+)
Waveform	76.7±7.9	73.7±10.0(+)	71.2±8.4(+)	70.9±8.5(+)	75.8±7.4(+)	76.8±8.0(=)	76.6±7.8(=)	72.8±8.6(+)
Arrhythmia	59.2±2.5	58.5±3.0(+)	58.9±1.7(=)	58.2±1.6(+)	55.7±1.4(+)	59.6±2.7(=)	59.2±2.2(=)	59.2±3.6(=)
Landsat	88.1±8.4	87.3±8.5(+)	87.3±8.4(+)	87.3±8.6(+)	87.8±8.3(+)	87.9±8.3(+)	88.0±8.5(=)	85.4±11.6(+)
Spambase	78.9±9.1	77.9±9.2(+)	78.9±11.4(=)	76.9±12.7(+)	76.9±9.0(+)	78.2±9.7(+)	82.5±10.9(-)	79.9±6.2(=)
Semeion	68.0±18.1	54.6±18.2(+)	70.3±17.7(-)	67.4±16.8(=)	58.8±13.9(+)	64.1±17.1(+)	63.7±16.2(+)	48.7±18.6(+)
Lung	77.1±10.9	67.6±11.0(+)	74.0±13.5(+)	75.1±13.0(+)	65.2±9.9(+)	74.1±11.0(+)	76.0±11.2(+)	69.6±11.4(+)
Colon	84.5±2.3	78.6±3.0(+)	78.4±2.9(+)	77.5±6.1(+)	79.6±3.3(+)	81.1±2.0(+)	83.9±1.9(=)	84.3±2.5(=)
Leukemia	94.8±4.1	95.3±4.0(-)	88.2±3.6(+)	89.5±3.2(+)	94.3±4.3(=)	95.6±3.9(-)	94.7±4.0(=)	95.2±4.1(=)
NCI60	47.4±9.8	46.4±10.0(+)	28.6±4.5(+)	34.8±3.0(+)	24.5±5.1(+)	47.8±6.5(=)	46.8±5.9(=)	39.8±11.3(+)
Lymphoma	85.6±8.9	73.0±10.8(+)	75.3±6.7(+)	81.5±9.9(+)	66.5±5.7(+)	83.8±9.9(+)	87.1±9.9(-)	77.6±12.5(+)
Win/Tie/Loss	-	12/4/2	13/3/2	13/5/0	15/3/0	10/5/3	6/10/2	12/6/0
SVM								
Wine	93.6±4.7	92.7±4.5(=)	93.3±4.5(=)	93.9±4.7(=)	93.3±4.5(=)	93.7±4.7(=)	93.7±4.7(=)	91.4±8.1(+)
Parkinsons	89.3±2.0	87.8±1.9(+)	89.3±2.2(=)	89.1±2.3(=)	89.0±2.5(=)	87.6±1.8(+)	89.7±2.1(=)	89.1±2.1(=)
Ionosphere	84.4±2.7	85.0±2.0(=)	85.1±2.6(-)	85.1±2.5(-)	84.4±2.7(=)	84.6±2.1(-)	84.6±2.2(=)	82.6±2.4(+)
Breast	95.2±1.0	94.5±1.2(+)	93.6±1.1(+)	94.2±0.9(+)	93.8±0.8(+)	94.8±1.0(+)	95.0±1.1(=)	93.7±5.3(=)
Segment	94.5±7.9	91.6±8.5(+)	94.6±8.0(=)	92.3±8.6(+)	94.8±8.0(-)	94.6±8.0(=)	94.5±7.9(=)	93.8±11.4(=)
Cardio	88.8±7.1	88.9±6.8(=)	70.5±13.5(+)	71.7±13.1(+)	89.0±7.5(=)	89.4±6.0(=)	77.8±8.4(+)	76.4±10.1(+)
Steel	69.1±6.7	64.8±6.6(+)	69.3±6.5(=)	68.5±7.4(=)	69.1±6.6(=)	67.9±6.8(+)	69.9±6.9(-)	66.6±6.7(+)
Musk	84.5±7.1	76.0±3.6(+)	76.1±5.4(+)	76.4±5.4(+)	80.2±6.0(+)	80.3±5.7(+)	78.7±5.8(+)	76.1±7.5(+)
Waveform	77.6±5.3	75.1±6.7(+)	72.3±5.1(+)	72.2±5.1(+)	76.1±5.0(+)	77.3±5.3(+)	77.5±5.4(=)	73.5±6.2(+)
Arrhythmia	59.5±3.0	59.2±3.0(=)	56.5±2.8(+)	63.2±1.5(-)	56.0±3.2(+)	59.4±2.5(=)	60.8±2.9(-)	60.7±3.5(-)
Landsat	87.8±5.5	87.1±5.5(+)	86.9±5.4(+)	86.9±5.5(+)	87.5±5.4(+)	87.5±5.4(+)	87.6±5.5(+)	84.8±9.5(+)
Spambase	83.6±4.4	83.4±4.7(=)	85.4±3.9(-)	86.0±4.0(-)	86.1±4.6(-)	83.6±4.3(=)	88.8±3.9(-)	83.5±4.7(=)
Semeion	73.3±15.3	60.4±15.6(+)	77.6±15.7(-)	75.1±15.0(-)	75.1±15.0(-)	65.9±12.9(+)	69.8±14.8(+)	55.1±17.1(+)
Lung	60.3±6.8	58.7±7.0(+)	60.2±7.8(=)	59.0±7.1(+)	59.8±8.4(=)	60.1±8.1(=)	62.4±8.3(-)	59.9±6.0(=)
Colon	78.5±2.2	74.6±3.1(+)	77.8±4.5(=)	75.9±3.4(+)	75.4±4.1(+)	75.4±3.9(+)	80.8±2.3(-)	75.7±3.4(+)
Leukemia	94.9±2.0	96.1±1.2(-)	86.2±4.4(+)	87.6±3.0(+)	88.1±4.9(+)	94.9±2.0(=)	92.5±1.9(+)	93.7±1.7(+)
NCI60	30.4±3.9	33.2±4.6(-)	19.9±6.8(+)	31.4±4.8(=)	20.4±7.7(+)	38.8±3.8(-)	32.2±4.6(-)	30.3±3.2(=)
Lymphoma	71.6±3.1	71.2±5.9(=)	58.9±6.1(+)	68.4±4.2(+)	54.7±4.9(+)	71.5±3.3(-)	74.2±4.0(-)	68.8±7.0(+)
Win/Tie/Loss	-	10/6/2	9/6/3	10/4/4	10/6/2	8/9/1	5/6/7	11/6/1
Decision Tree								
Wine	89.8±5.0	90.4±5.3(=)	90.0±5.1(=)	89.5±4.8(=)	89.7±5.1(=)	89.5±4.9(=)	89.1±4.8(+)	88.1±6.3(+)
Parkinsons	85.5±1.8	84.9±1.7(=)	86.1±1.9(=)	86.4±1.9(-)	85.3±1.6(=)	85.6±1.9(=)	86.2±1.6(-)	84.6±2.1(+)
Ionosphere	87.0±1.7	85.6±2.0(+)	87.6±2.1(=)	87.4±1.9(=)	87.9±2.7(-)	86.9±1.7(=)	87.1±1.6(=)	87.6±3.0(=)
Breast	93.2±1.2	92.2±1.2(+)	92.7±1.0(+)	92.7±1.0(+)	92.7±1.2(+)	92.3±1.0(+)	92.9±1.0(=)	91.2±5.9(+)
Segment	93.9±7.4	90.8±7.9(+)	93.9±7.5(=)	91.9±7.8(+)	93.8±7.4(=)	93.9±7.4(=)	93.9±7.4(=)	93.3±10.6(=)
Cardio	91.1±2.9	91.2±2.9(=)	88.4±3.5(+)	88.5±3.6(+)	91.6±3.5(-)	91.2±2.9(=)	90.6±3.4(+)	91.3±3.1(=)
Steel	71.3±3.7	68.1±4.4(+)	71.2±4.3(=)	69.5±4.9(+)	71.2±3.8(=)	71.5±3.7(=)	71.6±4.1(=)	70.9±3.7(=)
Musk	78.6±3.8	75.9±3.8(+)	74.1±2.6(+)	75.7±3.4(+)	76.4±3.1(+)	77.2±3.7(+)	76.8±3.8(+)	75.1±4.3(+)
Waveform	72.7±5.7	69.3±7.7(+)	68.0±6.2(+)	67.6±6.3(+)	72.0±5.4(+)	72.2±5.7(+)	72.5±5.7(+)	69.4±6.8(+)
Arrhythmia	57.1±1.9	56.9±3.3(=)	56.3±1.1(+)	61.5±1.8(-)	51.3±1.1(+)	56.5±2.6(+)	55.5±1.5(+)	58.2±1.3(-)
Landsat	84.9±4.7	84.2±4.7(+)	83.5±4.6(+)	83.5±4.6(+)	84.7±4.6(+)	84.7±4.6(+)	84.7±4.7(=)	81.9±9.1(+)
Spambase	90.9±2.7	90.7±2.6(+)	88.8±2.6(+)	88.0±3.0(+)	91.0±2.7(-)	90.9±2.6(=)	91.2±2.8(-)	91.0±3.2(=)
Semeion	66.2±11.0	57.5±13.6(+)	67.7±10.8(-)	67.4±10.8(-)	60.9±9.3(+)	64.1±10.9(+)	65.3±10.6(+)	50.5±13.4(+)
Lung	51.4±4.9	50.7±8.0(=)	49.1±6.3(+)	54.7±6.7(-)	49.4±5.3(+)	50.7±4.9(=)	53.5±5.8(=)	51.2±4.5(=)
Colon	72.5±4.5	72.4±3.4(=)	72.7±3.0(=)	77.9±3.0(-)	83.8±1.4(-)	70.9±4.2(+)	75.5±3.1(=)	75.3±4.7(=)
Leukemia	93.2±1.2	94.4±0.0(-)	91.7±0.4(+)	91.9±0.6(+)	93.1±0.2(=)	93.0±1.0(=)	93.1±0.4(+)	93.9±1.7(-)
NCI60	30.9±2.9	38.0±5.2(-)	28.6±4.3(+)	38.1±2.4(-)	27.2±2.9(+)	37.5±4.4(-)	39.2±3.2(-)	43.5±7.9(-)
Lymphoma	62.8±3.3	61.8±5.1(=)	64.8±1.6(-)	64.8±3.2(-)	57.6±1.1(+)	62.1±5.1(=)	62.1±4.1(+)	67.2±7.1(-)
Win/Tie/Loss	-	9/7/2	10/6/2	9/2/7	9/5/4	7/10/1	7/6/5	7/6/5

For the data sets with high dimensionality and small sample size (such as Lymphoma, NCI60, Leukemia, Colon, and Lung), the estimation error of the four-way joint mutual information could be high. Interestingly, kNN still achieves considerably higher accuracy than SVM and Decision Tree. One possible reason is that kNN makes better use of the selected interactions.

3.2 Performance with Respect to the Number of Features

We compare the performance of feature selection methods with increasing number of features. Fig. 1 shows the classification accuracy of SVM. The data sets are Musk and Waveform, which vary in the number of features and instances. We can see that FIM (round marker) is competitive to other methods, especially for the Musk data set. A possible explanation is that higher-order interactions can improve the performance of feature selection.

The performance is also affected by the data size. Large data size is necessary for a better estimation of MI terms. Therefore, FIM tends to perform better on the data sets with large instance/feature ratio.

4   Discussion and Conclusion

We presented a new feature selection method FIM that took into account two-through four-way interaction between features and the class label. Theoretic derivation showed that two- through four-way interactions can be merged into four-dimensional joint mutual information, which is computationally more efficient. Thus, instead of directly calculating the interaction information terms, FIM calculated joint mutual information to reduce computational complexity. In addition, the problem of overestimating the feature significance widely exist in almost all methods that used the cumulative sum of MI terms. FIM employed the 'maximum of the minimum' method to address this problem. Experiments on eighteen benchmark data sets demonstrated that FIM was effective in identifying informative features.

Acknowledgments. The authors would like to thank the anonymous reviewers for their careful reading of this paper and for their constructive comments and suggestions. This work is supported by the National Natural Science Foundation of China under Grant no. 61602094.

References

1. Kohavi, R., John, G.H.: Wrappers for feature subset selection. Artif. Intell. **97**(1–2), 273–324 (1997)
2. Robnik-Siknja, M., Kononeko, I.: Theoretical and empirical analysis of ReliefF and RReliefF. Mach. Learn. **53**(1–2), 23–69 (2003)
3. Peng, H., Long, F., Ding, C.: Feature selection based on mutual information: criteria of max-dependency, max-relevance, and min-redundancy. PAMI **27**(8), 1226–1238 (2005)

4. Nie, F., Huang, H., Cai, X., Ding, C.H.: Efficient and robust feature selection via joint l2,1-norms minimization. In: NIPS, pp. 1813–1821 (2010)
5. Hinton, G.E.: Reducing the dimensionality of data with neural networks. Science **313**(5786), 504–507 (2006)
6. Brown, G., Pocock, A., Zhao, M.J., Luján, M.: Conditional likelihood maximisation: a unifying framework for information theoretic feature selection. JMLR **13**(1), 27–66 (2012)
7. Lewis, D.D.: Feature selection and feature extraction for text categorization. In: Proceedings of Speech and Natural Language Workshop, pp. 212–217 (1992)
8. Battiti, R.: Using mutual information for selecting features in supervised neural net learning. IEEE Trans. Neural Netw. **5**(4), 537–550 (1994)
9. Lin, D., Tang, X.: Conditional infomax learning: an integrated framework for feature extraction and fusion. In: Leonardis, A., Bischof, H., Pinz, A. (eds.) ECCV 2006. LNCS, vol. 3951, pp. 68–82. Springer, Heidelberg (2006). https://doi.org/10.1007/11744023_6
10. Vinh, N.X., Zhou, S., Chan, J., Bailey, J.: Can high-order dependencies improve mutual information based feature selection? Pattern Recogn. **53**, 46–58 (2016)
11. Bennasar, M., Hicks, Y., Setchi, R.: Feature selection using joint mutual information maximisation. Expert Syst. Appl. **42**(22), 8520–8532 (2015)
12. Zeng, Z., Zhang, H., Zhang, R., Yin, C.: A novel feature selection method considering feature interaction. Pattern Recogn. **48**(8), 2656–2666 (2015)
13. Shishkin, A., Bezzubtseva, A., Drutsa, A., Shishkov, I., Gladkikh, E., Gusev, G., Serdyukov, P.: Efficient high-order interaction-aware feature selection based on conditional mutual information. In: NIPS, pp. 4637–4645 (2016)
14. Fleuret, F.: Fast binary feature selection with conditional mutual information. JMLR **5**(8), 1531–1555 (2004)
15. Jakulin, A.: Machine learning based on attribute interactions. Ph.D. thesis, pp. 1–252 (2005)
16. Balagani, K.S., Phoha, V.V.: On the feature selection criterion based on an approximation of multidimensional mutual information. PAMI **32**(7), 1342–1343 (2010)
17. Brown, P.F., Desouza, P.V., Mercer, R.L., Pietra, V.J.D., Lai, J.C.: Class-based n-gram models of natural language. Comput. Linguist. **18**(4), 467–479 (1992)
18. Scott, D.W.: Multivariate Density Estimation. Wiley Series in Probability and Statistics, 2nd edn. Wiley, Hoboken (2015)

An Extended Random-Sets Model
for Fusion-Based Text Feature Selection

Abdullah Semran Alharbi[1,2]([✉]), Yuefeng Li[1], and Yue Xu[1]

[1] School of EECS, Queensland University of Technology, Brisbane, QLD, Australia
{y2.li,yue.xu}@qut.edu.au
[2] Department of CS, Umm Al-Qura University, Mecca, Saudi Arabia
asaharbi@uqu.edu.sa

Abstract. Selecting features that represent a specific corpus is important for the success of many machine learning and text mining applications. In information retrieval (IR), fusion-based techniques have shown remarkable performance compared to traditional models. However, in text feature selection (FS), popular models do not consider the fusion of the taxonomic features of the corpus. This research proposed an innovative and effective extended random-sets model for fusion-based FS. The model fused scores of different hierarchal features to accurately weight the representative words based on their appearance across the documents in the corpus and in several latent topics. The model was evaluated for information filtering (IF) using TREC topics and the standard RCV1 dataset. The results showed that the proposed model significantly outperformed eleven state-of-the-art baseline models in six evaluation metrics.

Keywords: Feature selection · Data fusion · Topic modelling
Term weighting · Extended Random Set

1 Introduction

Over the last three decades, fusion-based techniques have been effective in IR [9, 30]. These techniques have shown that by combining different representations of documents and queries instead of using a single IR model, search systems outputs and ranking and scoring algorithms can maintain better results with substantial improvements [9]. However, applying similar techniques to FS is still limited. FS plays a crucial role in improving the accuracy and scaling down the complexity of many classifiers [19]. Improvements are achieved by selecting some relevant features and discarding irrelevant ones [17].

Analogous to most data fusion models that reward highly ranked documents in retrieved lists [4], FS models reward highly representative features. Thus, the *scoring function* is the most important component in a text FS model. It gives weights to features and specify how informative those features are to a corpus

© Springer International Publishing AG, part of Springer Nature 2018
D. Phung et al. (Eds.): PAKDD 2018, LNAI 10939, pp. 126–138, 2018.
https://doi.org/10.1007/978-3-319-93040-4_11

that describes a specific topic of interest [12]. Most low-level, term-based[1] fusion FS models apply the early fusion concept [31] by combining different terms' scores (frequencies) heuristically [9] based on the flat Bag-of-Words (BoW) representation. Adopting a representation where no relationships between features were assumed made these models susceptible to noise and caused them to suffer from polysemy and synonymy [17].

Alternatively, phrase-based, pattern-based and topic-based FS models adopted the late fusion strategy [31]. They combine the term weights (scores) based on the relationships between the high-level features (phrases, patterns, topics or a mixture of them) and the documents in a specific corpus. However, phrase-based models suffer from a low frequency of specific phrases [18], and pattern-based techniques do not assume that a corpus can discuss multiple subjects or themes [12]. Topic-based techniques, such as the probabilistic latent semantic analysis (pLSA) [13] and latent Dirichlet allocation (LDA) [7], can alleviate the polysemy problem [13] and are built on the assumption that documents can exhibit more than one topic [7].

LDA is the most popular unsupervised topic modelling technique with multiple applications. However, LDA calculates term weights on a document-by-document basis using the hierarchical local topics-document probability distributions and global (corpus level) term-topics assignments [12]. It does not automatically consider the sub-hierarchal features of a document, such as its paragraphs-topics distributions, or the features higher up in the hierarchy that represent the whole corpus. Thus, term weights assigned by the LDA term scoring function do not accurately reflect the importance of these terms in their local documents or the corpus. Recent studies [3,5,6,12] have confirmed that the LDA scoring function negatively influenced the LDA's FS performance.

Relevant terms can be identified in a specific corpus by fusing various instances (evidence) of these terms in different representations [31]. At the corpus level, the global statistics of terms, such as document frequency (df), are important pieces of evidence that represent terms more discriminatively [15]. Nevertheless, in IR, representing words using global term-weighting schemes could not give better retrieval results [20] because term global statistics could not reveal the term's local, document-level importance [22], and neither can the LDA. This research asked whether there was a method to significantly fuse the LDA's hierarchal features with corpus statistical features (particularly df) to overcome their limitations in representing the local and global importance of terms.

This research aimed to develop an effective, fusion-based text FS model called (SIF2)[2]. SIF2 adopts a complex hierarichal representation for the corpus (documents, their paragraphs, latent topics from the paragraphs and all terms in the corpus). SIF2 then models the complicated and imprecise relationships between these hierarichal features using multiple Extended Random-Sets (ERS) to effectively weight terms. The model provides an elegant way to combine the

---

[1] Words, keywords and terms are used interchangeably in this paper.

[2] SIF stands for **S**election of **I**nformative **F**eatures, and the '2' refers to the utilisation of both local and global statistics.

advantages of both topic modelling and terms' global statistics. The experimental results showed that the framework was highly effective and significantly outperformed state-of-the-art FS methods in IF regardless of the fusion technique or the type of text features they utilised.

This research made two major contributions: (a) a new fusion-based model for FS that combines the advantages of unsupervised learning (topic modelling) and corpus statistics and (b) a new method for adapting the LDA for feature weighting by modelling it with multiple ERSs.

2   Related Work

FS techniques that adopt the early fusion strategy are efficient and were developed based on sophisticated mathematical and statistical weighting theories [18]. Popular examples are TF*IDF, Information Gain, the Gini-index, the Chi-Square (X^2) [15], BM25 [26], the ranking SVM [14], Rocchio, LASSO, Mutual Information and many others. These methods use low-level terms, which makes them noise sensitive and causes them to suffer from the problems of polysemy and synonymy [17]. Also, because they use the BoW representation, they ignore the order of words in documents, and consequently, they lose the semantic relationships between these words.

Many late fusion models have been developed to overcome the limitations of term-based methods. Phrase-based approaches use n-grams, but they do not show encouraging results in FS [12] because phrases are statistically inferior to words and can be noisy [18]. Patterns are more frequent than phrases and are used to mine specific features [1,2]. However, patterns are sensitive to noise and redundant. A topic can be discovered by topic modelling techniques like pLSA [13] and LDA [7]. However, using LDA and pLSA for relevant FS does not show encouraging results [6,12]. For better performance, researchers in [12, 17,18,29] integrated different types of high-level features, but these mix-based models could inherit the limitations of each feature and could be expensive.

3   Problem Formulation and Background

Let assume that a researcher maintains a corpus of long documents D^+ that describe a particular topic of interest that might also have multiple sub-topics. For further investigation, the researcher wants to enrich the corpus by collecting documents from the Web. To achieve this goal, the researcher needs a model that can select and accurately weight terms to effectively describe the corpus. The weighted terms are used to gather relevant documents.

We assumed that the corpus $D^+ = \{d_1, d_2, d_3, \ldots, d_x\}$ has X documents that are related to a particular topic of interest, which is different from a latent topic. A document d consists of a set of paragraphs S and a paragraph g consists of a bag of words. The set of all paragraphs in the corpus is $G = \cup_{d \in D^+}\{g | g \in d\}$ and $S \subseteq G$. The set of all unique words in G is $\Omega = \{w_1, w_2, w_3, \ldots, w_K\}$, where $K = |\Omega|$. SIF2 uses the LDA to discover a set of latent topics T from G where V

is the number of topics. The LDA is an effective model to discover hidden topics from a corpus, but it does not show sufficient performance in FS.

3.1 Latent Dirichlet Allocation and Limitations

LDA describes a topic $t_j \in T$ as a probability distribution over all words in Ω using $p(w_i|t_j)$, in which $\sum_i^{|\Omega|} p(w_i|t_j) = 1$ where $1 \leq j \leq V$ and $w_i \in \Omega$. Also, LDA describes a document d by a probabilistic mixture of topics using $p(t_j|d)$. All hidden variables, $p(w_i|t_j)$ and $p(t_j|d)$, are inferred statistically by the Gibbs sampler [28]. As a result, and based on T, the local score (probability) of word w_i in a document d can be estimated by $p(w_i|d) = \sum_{j=1}^{V} p(w_i|t_j)p(t_j|d)$.

For every topic, estimating $p(w_i|d)$ requires the fusion of two hierarchal features, the word-topic assignment $p(w_i|t_j)$ and the topic-document distribution $p(t_j|d)$. However, we argue that using only these features made the LDA ineffective for selecting informative terms in a specific corpus (see LDA's results in Table 1). Adapting the LDA to estimate words informativeness has two challenges: (a) how to fuse other hierarchal features for a better weight estimation for words, and (b) how to decide the optimal number for hidden topics.

3.2 Extending Random-Sets

A random set (RS) is an arbitrary entity that contains a subset of values selected from existing space [24]. Random sets can measure uncertainty when inaccurate data are used for some decision analysis applications [25]. To effectively fuse the local score of word w in document d, we proposed the set-valued function $\Gamma : T \rightarrow 2^{\Omega}$ from T onto Ω. Let us call T the evidence space, and P is a probability function specified on T. Then, we call the (P, Γ) pair a RS [24]. Γ can also be extended as $\xi :: T \rightarrow 2^{\Omega \times [0,1]}$ and is called an extended set-valued mapping [16] given that $\sum_{(w,p) \in \xi(t)} p = 1$ for each $t \in T$.

4 The Proposed SIF2 Model

Distinct hierarchal entities and their relationships to each other can affect the term scoring in a specific corpus. As can be seen in Fig. 1, SIF2 uses four entities, which are the corpus documents D^+, their paragraphs G, the LDA latent topics T and the corpus keywords Ω. SIF2 also models the complex relationships between these entities using the ERS theory to fuse and, thus, generalise the LDA weight of a local term to a global one that can be combined with a more representative global statistic.

SIF2 proposed two ERSs Γ_1 and Γ_2 and their inverse to model the one-to-many relationships between the used entities. In every ERS, including its inverse, a function is used to describe a specific relationship and to assign a score that represents the rank of the targeted entity. Then, a new term scoring function is developed by integrating the proposed ERSs.

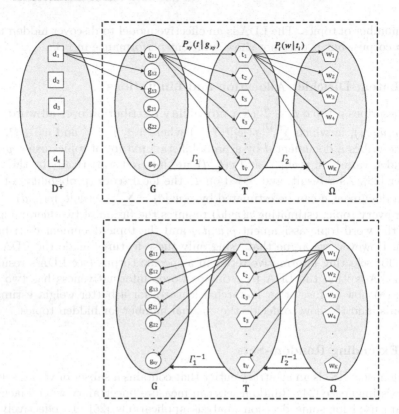

Fig. 1. The proposed SIF2 model and the mapping of Γ and Γ^{-1}.

4.1   Fusing Hierarchical Features

Given a topic $t \in T$, let us define a probability function P that satisfies $\sum_{t \in T} P(t) = 1$. Thus, we can call the pair (ξ, P) an extended Random-Set. Also, for every $t_i \in T$, we define $P_i(w|t_i)$ as a conditional probability function on the set of keywords Ω such that the mapping $\Gamma(t_i) = \{w | w \in \Omega, P_i(w|t_i) > 0\}$. Further, Γ^{-1} is the inverse function of Γ and we define it as $\Gamma^{-1} : \Omega \to 2^T$; $\Gamma^{-1}(w) = \{t \in T | w \in \Gamma(t)\}$. Thus, to estimate the fused local score of word w, the extended set-valued mapping is used to define the probability function $pr_d(w)$ on Ω such that $pr_d :: \Omega \to \mathbb{R}$ as follows:

$$pr_d(w) = \sum_{t_i \in \Gamma^{-1}(w)} (P(t_i) \times P_i(w|t_i)) \tag{1}$$

where $pr_d(w)$ is the fused local score of word w at the document d level.

The ERS Γ_1 defines the conditional probability function $P_{xy}(t|g_{xy})$ on the set of paragraphs G to describe the one-to-many relationship between a paragraph and a topic as $\Gamma_1 : G \to 2^{T \times [0,1]}$; $\Gamma_1(g_{xy}) = \{(t_1, P_{xy}(t_1|g_{xy})), \ldots\}$. Similarly, as a topic can have many terms, Γ_2 defines $P_i(w|t_i)$ on T as another conditional probability function that estimates the probability of a word based on its appearance in each topic $t_i \in T$ as $\Gamma_2: T \rightarrow 2^{\Omega \times [0,1]}$; $\Gamma_2(t_i) = \{(w_1, P_i(w_1|t_i)), \ldots\}$.

Inversely, as a topic also can appear in one or more paragraphs that belong to a certain document, Γ_1^{-1} is proposed to describe such relationship using the $P_t(t_i)$ function in which a subset of paragraphs S will only be mapped to its document d as $\Gamma_1^{-1}(t) = \{g_{xy}|t \in \Gamma_1(g_{xy}), g_{xy} \in S\}$. Similarly, as a word w in a specific document can occur in multiple topics, Γ_2^{-1} is also proposed to govern this relationship using the probability function $pr_d(w)$ as $\Gamma_2^{-1}(w) = \{t|w \in \Gamma_2(t)\}$.

Fusing Topic Scores. To calculate the fused weight of word w in document d_x, Γ_1^{-1} and Γ_2^{-1} are used to estimate two probabilistic scores. The first score is the topic marginal probability $P_t(t_i)$ for every topic appears in paragraph $g_y \in d_x$ by fusing the topic-paragraph distribution $P_{xy}(t_i|g_{xy})$ to estimate its topic-document marginal probability in which we assume $P_G(g_{xy}) = \frac{1}{N}$ where $N = |S|$ as follows:

$$P_t(t_i) = \sum_{g_{xy} \in \Gamma_1^{-1}(t_i)} (P_G(g_{xy}) \times P_{xy}(t_i|g_{xy}))$$

$$= \frac{1}{N} \sum_{g_{xy} \in \Gamma_1^{-1}(t_i)} P_{xy}(t_i|g_{xy}) \qquad (2)$$

where $P_{xy}(t_i|g_{xy})$ is estimated by LDA, and g_{xy} denotes to paragraph y of document x.

Fusing Local Word Scores. To calculate the second probability, for each topic $t_i \in T$, the conditional probability $P_i(w|t_i)$ needs to be estimated for the word w in topic t_i (which was estimated by LDA in this study). Thus, the fused, word weight at a document d level can be estimated by substituting $P_t(t_i)$ in Eq. 1 by its formula in Eq. 2 as follows:

$$pr_d(w) = \sum_{t_i \in \Gamma_2^{-1}(w)} (P_t(t_i) \times P_i(w|t_i))$$

$$= \sum_{t_i \in \Gamma_2^{-1}(w)} \left[\left(\frac{1}{N} \sum_{g_{xy} \in \Gamma_1^{-1}(t_i)} P_{xy}(t_i|g_{xy}) \right) \times P_i(w|t_i) \right]$$

$$= \frac{1}{N} \sum_{t_i \in \Gamma_2^{-1}(w)} \left[P_i(w|t_i) \times \left(\sum_{g_{xy} \in \Gamma_1^{-1}(t_i)} P_{xy}(t_i|g_{xy}) \right) \right] \qquad (3)$$

Fusing Global Word Scores. As the LDA does not explicitly provide a global score for a word at a corpus level, we estimated the global score for a word w

as the sum of its $pr_d(w)$ in every document $d_i \in D^+$. Finally, the fused, global word score $sr(w)$ at the corpus level is estimated as follows:

$$sr(w) = df(w) \cdot \sum_{w \in d_i, d_i \in D^+} pr_{d_i}(w) \tag{4}$$

where $df(w)$ is the document frequency of word w, and $pr_{d_i}(w)$ is the fused score of w in document d_i.

5   Evaluation

The evaluation experiments aimed to show (a) how the integration of late fusion hierarchical LDA features and early fusion global statistics (e.g., df) can accurately weight terms and effectively represent a specific corpus, and (b) how sensitive this integration can be to some hyper-parameters, mainly the number of latent topics and top-selected features (top-n). This research hypothesised that a late fusion of topical taxonomic features could provide an effective term weight, especially after it was fused with an appropriate global statistic. We used an IF system-based methodology for evaluating this hypothesis.

5.1   Dataset and Evaluation Measures

The Reuters Corpus Volume 1 (RCV1) dataset and its TREC[3] relevance judgments were used to evaluate SIF2 and the baseline models. RCV1 has 806,791 documents that were grouped into 100 collections that discuss 100 different topics of interest. However, we used the first 50 collections only because they were assessed at NIST [27] by subject-matter experts. Fifty collections are sufficient for a reliable experiment [8]. Every collection has a set of training and testing documents and each set has some relevant D^+ and irrelevant D^- documents. Each document is a news story that has some paragraphs (a title and several text XML elements). The SIF2 model was trained and tested on these paragraphs after their terms were stemmed and the stop-words were removed.

Six metrics were used to measure the performance of the SIF2 model and the baselines. The measures are the average precision of the top-20 ranked documents (P@20), break-even point (b/p), 11-point interpolated average precision (IAP), mean average precision (MAP), interpolated precision averages at 11 standard recall levels (11-point) and F-score (F_1). More information about these measures can be found in [21]. The Student's t-test was used to analyse the significance of the difference between the results of the SIF2 and the baselines.

5.2   Baseline Models and Settings

For an extensive evaluation, we compared the performance of the SIF2 model to major FS models. We selected eleven baseline models that were grouped based on the fusion strategies they adopted.

---

[3] http://trec.nist.gov/.

Early Fusion Models. SVM [14] and Okapi BM25 [26] were selected to represent the early fusion category. They are state-of-the-art, supervised, term-based models. Since IF can be considered another type of binary classification problem, the rank-based SVM was used in this research similarly to how it was used in [2]. BM25 is a popular document ranking algorithm in IR, and we set its experimental parameters to $b = 0.75$ and $k_1 = 1.2$ as recommended in [21].

Late Fusion Models. Nine state-of-the-art FS models were selected for this category. They use different high-level features. The standard n-Grams model, where $n = 3$, was selected because it was the best value reported in [10] for RCV1. PDS [32] and SCSP [1] represented the pattern-based FS models. PDS and SCSP utilise frequent patterns and closed sequential patterns, respectively. LDA [7] and its predecessor, pLSA [13], were used to represent the topic-based group. pLSA and LdaDoc were trained on D^+ while LdaPara was trained on D^+ paragraphs. Four mixed-based models were selected for the experiment. PBTM-FP [10], PBTM-FCP [10] and SPBTM [11] discovered frequent patterns, closed frequent patterns and significant matched patterns, respectively. The Topical N-Grams(TNG) [29] used n-Gram-based topics to represent the corpus.

The MALLET toolkit [23] was used to implement all the topic-based models including our SIF2, but for the pLSA, the Lemur toolkit[4] was used instead. As recommended in [28], we set the hyper-parameters β and α for the LDA-based models to (0.01) and $(50/V)$ respectively, and the Gibbs sampler iterations to 1000. The default value (1000) was accepted for the pLSA.

5.3 Experimental Design

To demonstrate the validity of our evaluation hypothesis, we used our SIF2 model as a FS step for an IF system based on TREC filtering track [27]. We did extensive experiments on the RCV1 50 assessed collections and their TREC relevance judgements. For each collection, we trained the SIF2 on the paragraphs of D^+ in the training set. We used LDA to generate 10 topics, as in [12], but unlike [12], SIF2 was insensitive to this hyper-parameter (see Fig. 3). Then, SIF2 was used to score and rank terms using Eq. 4, and a top-$n(n = 15)$ feature was selected as a query Q to the IF testing system. The IF system used a new document from the testing set and decided its relevance to the corpus. Determining the value of n was experimental, and the SIF2's performance was stable even with higher n features (see Fig. 3). A similar process was also separately applied to each baseline. If the six metrics of the IF system's results were significantly better than the baseline's, then we could say that our SIF2 outperformed a baseline model.

5.4 Results

The experimental results of the SIF2 and the baseline models (averaged over the 50 collections) are presented in Table 1 and Fig. 2. Table 1 results are grouped

---

[4] https://www.lemurproject.org/.

Table 1. SIF2 results compared with the baselines.

	P@20	b/p	MAP	$F_{\beta=1}$	IAP
SIF2	**0.594**	**0.494**	**0.522**	**0.485**	**0.541**
SVM	0.457	0.408	0.409	0.421	0.435
BM25	0.445	0.407	0.407	0.414	0.428
Improvement%	**+30.1**	**+20.9**	**+27.5**	**+15.2**	**+24.4**
SPBTM	0.527	0.448	0.456	0.445	0.478
PDS	0.496	0.430	0.444	0.439	0.464
LdaPara	0.492	0.414	0.442	0.437	0.468
PBTM-FP	0.470	0.402	0.427	0.423	0.449
PBTM-FCP	0.489	0.420	0.423	0.422	0.447
SCSP	0.480	0.407	0.420	0.423	0.442
LdaDoc	0.457	0.391	0.400	0.413	0.434
pLSA	0.423	0.386	0.379	0.392	0.404
TNG	0.447	0.360	0.372	0.386	0.394
n-Gram	0.401	0.342	0.361	0.386	0.384
Improvement%	**+12.7**	**+10.1**	**+14.4**	**+8.9**	**+13.2**

Table 2. p-values of SIF2 compared with the best baselines.

	P@20	b/p	MAP	$F_{\beta=1}$	IAP
SVM	1.50E−05	2.03E−04	1.31E−06	6.43E−06	2.32E−06
SPBTM	1.10E−02	2.98E−02	1.26E−03	4.81E−03	1.74E−03

based on the fusion strategy utilised by the baseline model, and the improvement% is the difference between the SIF2's performance and the best result of the baseline model. Any improvement greater than 5.0% was considered significant.

Table 1 demonstrates that our SIF2 model outperformed the baselines in all metrics. Compared to the best baseline in the early fusion category (SVM), SIF2 was significantly better on average by a minimum increase of 15.2% and a maximum of 30.1%. Compared to SPBTM, which was the best baseline in the late fusion group, SIF2 was significantly better on average by a minimum increase of 8.9% and a maximum of 14.4%. The 11-points measure in Fig. 2 illustrated the superiority of SIF2 and confirmed the significant increases presented in Table 1. The t-test *p-values* in Table 2 also confirmed the statistical significance (*p-value* < 0.05) of the improvements in the SIF2 model.

5.5   Discussion

As seen in Table 1 and Fig. 2, adopting the late fusion strategy for taxonomic topical features enabled a significant improvement over all other types of FS methodologies. The results showed that using topic modelling can be effective for FS and more efficient than mix-based techniques. Also, modelling the complex relationships between all entities in the taxonomy using the ERS made it more effective to adapt LDA feature weights and accurately estimate local and global term weights. This enabled their combination with representative global statistics.

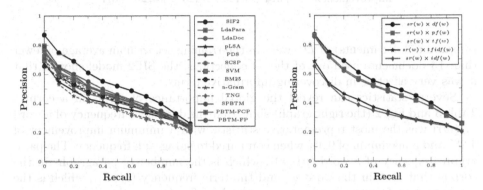

Fig. 2. 11-point result of SIF2 compared with baselines (left), and SIF2 combined with other global statistics (right).

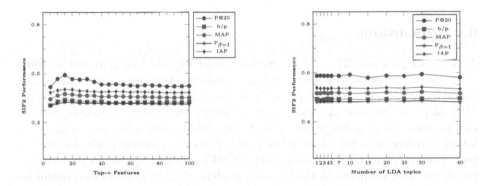

Fig. 3. SIF2 sensitivity to the selected top-n features (left) and to the number of latent topics (right).

As seen in Fig. 3, adopting a combination of topic modelling and global statistics made the SIF2 insensitive to the number of latent topics, which is one of the major advantages of using such an approach. Selecting the number of best features that represent a specific collection was, and still is, challenging. However, SIF2 maintained a stable performance after selecting the top-15 terms from the vocabulary, which is another important advantage. Overall, even though most of the documents in the RCV1 were in the testing set, and only a small number

Table 3. Equation 4 using other global statistics.

	P@20	b/p	MAP	$F_{\beta=1}$	IAP
$sr(w) \times df(w)$	**0.594**	**0.494**	**0.522**	**0.485**	**0.541**
$sr(w) \times pf(w)$	0.543	0.460	0.485	0.465	0.510
$sr(w) \times tf(w)$	0.532	0.454	0.480	0.462	0.505
$sr(w) \times tfidf(w)$	0.390	0.344	0.349	0.373	0.376
$sr(w) \times idf(w)$	0.357	0.328	0.332	0.361	0.359
Improvement%	**+9.4**	**+7.4**	**+7.6**	**+4.2**	**+6.1**

(minimum 3 documents) of D^+ were in the training set with an average of fewer than 13 documents used out of the 50 collections, the SIF2 model showed that it was very effective in discovering informative terms.

Several statistics can reveal the global importance of terms in a corpus. Table 3 and Fig. 2 (the right graph) show that the document frequency of a word ($df(w)$) was the most representative statistic with a minimum improvement of 4.2% and a maximum of 9.4% when compared to paragraph frequency. The paragraph frequency of a word ($pf(w)$), which is the number of paragraphs in the corpus that contain the word w, and the term frequency ($tf(w)$), which is the total number of times a word w appears in the collection, are the next representative global statistics, as they underperformed the df. The term frequency-inverse document frequency $tfidf(w)$ and the inverse document frequency ($idf(w)$) [15] scored the worst results, and should not be combined with LDA for FS.

6   Conclusions

This paper presents SIF2, an innovative model for selecting informative terms from specific corpora. The model extends random-sets to fuse hierarchical LDA-based features and to accurately weight terms on a document-by-document basis. SIF2 also combines the aggregated terms' weights with their document frequencies to estimate a global score. This fused global score effectively reflects the informativeness of a term to the key topic of interest discussed in a specific corpus. The experimental results showed that SIF2 attained significant performance improvements compared to the baseline models. SIF2 demonstrates a promising methodology for combining the advantages of unsupervised topic modelling and corpus statistics.

References

1. Albathan, M., Li, Y., Xu, Y.: Using extended random set to find specific patterns. In: WI 2014, vol. 2, pp. 30–37. IEEE (2014)
2. Algarni, A., Li, Y.: Mining specific features for acquiring user information needs. In: Pei, J., Tseng, V.S., Cao, L., Motoda, H., Xu, G. (eds.) PAKDD 2013 Part I. LNCS (LNAI), vol. 7818, pp. 532–543. Springer, Heidelberg (2013). https://doi.org/10.1007/978-3-642-37453-1_44

3. Alharbi, A.S., Li, Y., Xu, Y.: Integrating LDA with clustering technique for relevance feature selection. In: Peng, W., Alahakoon, D., Li, X. (eds.) AI 2017. LNCS (LNAI), vol. 10400, pp. 274–286. Springer, Cham (2017). https://doi.org/10.1007/978-3-319-63004-5_22

4. Anava, Y., Shtok, A., Kurland, O., Rabinovich, E.: A probabilistic fusion framework. In: CIKM 2016, pp. 1463–1472. ACM (2016)

5. Bashar, M.A., Li, Y.: Random set to interpret topic models in terms of ontology concepts. In: Peng, W., Alahakoon, D., Li, X. (eds.) AI 2017. LNCS (LNAI), vol. 10400, pp. 237–249. Springer, Cham (2017). https://doi.org/10.1007/978-3-319-63004-5_19

6. Bashar, M.A., Li, Y., Gao, Y.: A framework for automatic personalised ontology learning. In: WI 2016, pp. 105–112. IEEE (2016)

7. Blei, D.M., Ng, A.Y., Jordan, M.I.: Latent Dirichlet allocation. J. Mach. Learn. Res. **3**, 993–1022 (2003)

8. Buckley, C., Voorhees, E.M.: Evaluating evaluation measure stability. In: SIGIR 2000, pp. 33–40. ACM (2000)

9. Croft, W.B.: Combining approaches to information retrieval. In: Croft, W.B. (ed.) Advances in Information Retrieval. INRE, vol. 7, pp. 1–36. Springer, Boston (2002). https://doi.org/10.1007/0-306-47019-5_1

10. Gao, Y., Xu, Y., Li, Y.: Pattern-based topic models for information filtering. In: ICDM 2013, pp. 921–928. IEEE (2013)

11. Gao, Y., Xu, Y., Li, Y.: Topical pattern based document modelling and relevance ranking. In: Benatallah, B., Bestavros, A., Manolopoulos, Y., Vakali, A., Zhang, Y. (eds.) WISE 2014 Part I. LNCS, vol. 8786, pp. 186–201. Springer, Cham (2014). https://doi.org/10.1007/978-3-319-11749-2_15

12. Gao, Y., Xu, Y., Li, Y.: Pattern-based topics for document modelling in information filtering. IEEE TKDE **27**(6), 1629–1642 (2015)

13. Hofmann, T.: Unsupervised learning by probabilistic latent semantic analysis. Mach. Learn. **42**(1–2), 177–196 (2001)

14. Joachims, T.: Optimizing search engines using clickthrough data. In: KDD 2002, pp. 133–142. ACM (2002)

15. Lan, M., Tan, C.L., Su, J., Lu, Y.: Supervised and traditional term weighting methods for automatic text categorization. IEEE TPAMI **31**(4), 721–735 (2009)

16. Li, Y.: Extended random sets for knowledge discovery in information systems. In: Wang, G., Liu, Q., Yao, Y., Skowron, A. (eds.) RSFDGrC 2003. LNCS (LNAI), vol. 2639, pp. 524–532. Springer, Heidelberg (2003). https://doi.org/10.1007/3-540-39205-X_87

17. Li, Y., Algarni, A., Albathan, M., Shen, Y., Bijaksana, M.A.: Relevance feature discovery for text mining. IEEE TKDE **27**(6), 1656–1669 (2015)

18. Li, Y., Algarni, A., Zhong, N.: Mining positive and negative patterns for relevance feature discovery. In: KDD 2010, pp. 753–762. ACM (2010)

19. Li, Y., Li, T., Liu, H.: Recent advances in feature selection and its applications. Knowl. Inf. Syst. **53**, 1–27 (2017)

20. Macdonald, C., Ounis, I.: Global statistics in proximity weighting models. In: Web N-gram Workshop. p. 30. Citeseer (2010)

21. Manning, C.D., Raghavan, P., Schütze, H.: Introduction to Information Retrieval. Cambridge University Press, Cambridge (2008)

22. Maxwell, K.T., Croft, W.B.: Compact query term selection using topically related text. In: SIGIR 2013, pp. 583–592. ACM (2013)

23. McCallum, A.K.: Mallet: a machine learning for language toolkit (2002)

24. Molchanov, I.: Theory of Random Sets. Springer, Heidelberg (2006). https://doi.org/10.1007/1-84628-150-4
25. Nguyen, H.T.: Random sets. Scholarpedia **3**(7), 3383 (2008)
26. Robertson, S., Zaragoza, H.: The Probabilistic Relevance Framework: BM25 and Beyond. Now Publishers Inc., Breda (2009)
27. Robertson, S.E., Soboroff, I.: The TREC 2002 filtering track report. In: TREC, vol. 2002, p. 5 (2002)
28. Steyvers, M., Griffiths, T.: Probabilistic topic models. Handb. Latent Semant. Anal. **427**(7), 424–440 (2007)
29. Wang, X., McCallum, A., Wei, X.: Topical n-grams: phrase and topic discovery, with an application to information retrieval. In: ICDM 2007, pp. 697–702. IEEE (2007)
30. Wu, S.: Data Fusion in Information Retrieval. Springer, Heidelberg (2012)
31. Zhang, S., Balog, K.: Design patterns for fusion-based object retrieval. In: Jose, J.M., Hauff, C., Altıngovde, I.S., Song, D., Albakour, D., Watt, S., Tait, J. (eds.) ECIR 2017. LNCS, vol. 10193, pp. 684–690. Springer, Cham (2017). https://doi.org/10.1007/978-3-319-56608-5_66
32. Zhong, N., Li, Y., Wu, S.T.: Effective pattern discovery for text mining. IEEE TKDE **24**(1), 30–44 (2012)

Attribute Reduction Algorithm Based on Improved Information Gain Rate and Ant Colony Optimization

Jipeng Wei[1], Qianjin Wei[2(✉)], and Yimin Wen[2]

[1] School of Computer Science and Information Safety,
Guilin University of Electronic Technology, Guilin 541004, China
wei61457@163.com, wei_qj@guet.edu.cn
[2] Guangxi Key Laboratory of Trusted Software,
Guilin University of Electronic Technology, Guilin 541004, China
ymwen@guet.edu.cn

Abstract. Solving minimal attribute reduction (MAR) in rough set theory is a NP-hard and nonlinear constrained combinatorial optimization problem. Ant colony optimization (ACO), a new intelligent computing method, takes strategies of heuristic search, which is characterized by a distributed and positive feedback and it has the advantage of excellent global optimization ability for handling combinatorial optimization problems. Having considered that the existing information entropy and information gain methods fail to help to select the optimal minimal attribute every time, this paper proposed a novel attribute reduction algorithm based on ACO. Firstly, the algorithm adopts an improved information gain rate as heuristic information. Secondly, each ant solves a problem of minimum attributes reduction and then conduct redundancy test to each selected attribute. What's more, redundant detection of all non-core attributes in the optimal solution will be performed in each generation. The result of the experiment on several datasets from UCI show that the proposed algorithms are more capable of finding the minimum attribute reduction and can faster converge and at the same time they can almost keep the classification accuracy, compared with the traditional attribute reduction based on ACO algorithm.

Keywords: Rough set · Ant colony optimization · Information gain rate
Attribute reduction

1 Introduction

With the rapid development of Internet and cloud computing technology, the scale of data grows exponentially. There are many important information hidden behind massive data. Facing the massive data, how to extract the key and effective information is the current research hotspot. Rough set theory (RST) was proposed by Polish scientist Pawlak in 1982 [1], which provides a mathematical methodology to deal with imprecision and uncertainty in information system. The method of using rough set

D. Phung et al. (Eds.): PAKDD 2018, LNAI 10939, pp. 139–150, 2018.
https://doi.org/10.1007/978-3-319-93040-4_12

theory to reduce the dimension and feature selection of data has been widely used in many fields such as data mining, machine learning. Attribute reduction is one of the most important research contents, and also the key step of knowledge acquisition in rough set theory. However, relying solely on rough set theory can't solve the problem well. Therefore, how to integrate attribute reduction algorithms with various intelligent methods according to the actual situation is one of the future development directions [2].

Intelligent computing is also called "soft computing", which is inspired by the laws of nature (Biology), and invent inventions according to their principles and structures. People have designed the tabu search, genetic algorithm, particle swarm optimization algorithm, which have been widely used [3–5]. In the 1990s, the Italian scholar Dorigo, Maniezzo, Colorni, etc., by simulating the behavior of the natural ant search path, proposed a new simulation evolutionary algorithm ant colony algorithm [6, 7]. Ant colony optimization (ACO) is a general heuristic algorithm developed in recent years [8]. This method is used to solve the TSP problem, distribution problem and has achieved good results. It has a good global optimization ability in combinatorial optimization problems, heuristic search and parallel distributed computing. Besides, this method is easy to combine with other problems.

The method of attribute reduction based on the traditional rough set theory is mainly based on the positive region method proposed by Pawlak [1]. Miao proposed attribute reduction algorithm based on information entropy [9]. The algorithm based on discernibility matrix proposed by Skowron [10] and other forms of methods [11–13]. In [14], it pointed out that traditional attribute reduction algorithm can't guarantee the MAR of the decision table. Searching for minimal attribute reduction has been proved to be a NP-hard problem [15]. Concerning the ACO algorithm has a good optimization ability in dealing with combinatorial optimization problems. Jensen and Shen [16] first introduced the ant colony algorithm into the field of attribute reduction in rough sets, and proposed an attribute reduction algorithm based on ACO. Ke et al. [17] proposes the ACOAR algorithm based on the attribute positive region as heuristic information, and each ant randomly selects the starting node. Literature [18, 19] uses information entropy and information gain rate respectively as the heuristic information, and proposes an attribute reduction algorithm based on ACO. But it is easy for both to add redundant attributes to the reduced set as selected attributes. In [20], the ant colony optimization algorithm is applied to deal with the attribute reduction of continuous attributes.

In this research, a new ant colony optimization algorithm for attribute reduction is proposed. An improved information gain rate designed by document [21] is used as heuristic information and add redundancy judgment to the process of attribute selection in each ant. At the same time, the non-core attributes redundancy detection is performed for each generation optimal result. The experiment is validated by selecting data sets in UCI and the reduction results verified by C4.5 and Naive Bayes. Which can verify that the algorithm has good attribute optimization ability and fast convergence ability.

This paper is organized as follows. Section 2 introduces the basic concepts and definitions of RST. In Sect. 3, the principle of ant colony optimization algorithm is introduced and three improved algorithms are proposed to overcome the shortcomings mentioned above. Experimental results and comparative analysis are presented in Sect. 4. Section 5 gives the conclusion of this paper.

2 Preliminary

2.1 Rough Set Theory

This section recalls some basic and essential definitions from RST that are used for attribute reduction. Detailed description and formal definitions of the theory can be found in [1].

Definition 1. In RST, a decision system S is defined as $S = (U, C \cup D, V, f)$, where U called universe, is a nonempty set of finite objects. C and D are the set of conditional attributes and the set of decision attributes. V is the union of feature domains such that $V = \cup_{a \in C} V_a$ for V_a denotes the value domain of feature a, any $a \in A$ determines a function $f_a : U \to V_a$, where V_a is the set of values of a. For an attributes set $B \subseteq C$, there is an associated indiscernibility relation $IND(B)$:

$$IND(B) = \{(x, y) \in U \times U \mid \forall a \in B, f(x, a) = f(y, a)\} \tag{1}$$

$U/IND(B)$ called an equivalent partition called U. The equivalence class includes multiple equivalence classes, each of which is called a Knowledge Granule.

Definition 2. (Upper and Lower Approximation of Set) Given an information table $S = (U, C \cup D, V, f)$, for a subset $X \subseteq U$ and equivalence relation $IND(B)$ on domain U. The subsets X-*lower* and X-*upper* approximation of R defined as:

$$\underline{R}(X) = \{x \mid (\forall x \in U) \wedge ([x]_R \subseteq X)\} = \cup \{Y \mid (\forall Y \in U / R) \wedge (Y \subseteq X)\} \tag{2}$$

$$\overline{R}(X) = \{x \mid (\forall x \in U) \wedge ([x]_R \cap X \neq \emptyset)\} = \cup \{Y \mid (\forall Y \in U / R) \wedge (Y \cap X \neq \emptyset)\} \tag{3}$$

2.2 Information Representation in Decision Table

In attribute reduction problem, different attributes contain different information. The task of attribute reduction is to find those conditions attributes that contain as much information as possible about decision attributes. For this purpose, Shannon's information theory [22] provides us with a possible way to measure the information of data set with entropy and mutual information.

Definition 1. Let $S = (U, C \cup D, V, f)$ be a decision table. For any $B \subseteq C$, let $IND(B) = \{X_1, X_2, \cdots, X_n\}$ denote the partition induced by equivalence relation IND (B). Similarly $IND(D) = \{Y_1, Y_2, \cdots, Y_m\}$. The information entropy $H(B)$ of feature set B is defined as:

$$H(B) = -\sum_{i=1}^{n} p(X_i) \log_2 p(X_i) \tag{4}$$

The conditional entropy of D conditioned to B is defined as:

$$H(D|B) = -\sum_{i=1}^{n} p(X_i) \sum_{j=1}^{m} p(Y_j | X_i) \log_2 p(Y_j | X_i) \tag{5}$$

Where $p(X_i) = |X_i|/|U|, p(Y_j|X_i) = p|X_i \cap Y_j|/|X_i|$, $1 \leq i \leq n$, $1 \leq j \leq m$, $|X_i|$ is the cardinality of X_i.

Definition 2. Let $S = (U, C \cup D, V, f)$ be a decision table, with $B \subseteq C$. $IND(B)$ and $IND(D)$ represents equivalence relations of conditions and decisions attribute. The mutual information between B and D is defined as:

$$I(B; D) = H(D) - H(D|B) \tag{6}$$

Definition 3. [8] Let $S = (U, C \cup D, V, f)$ be a decision table. For every $a \in C$, If $I(C - \{a\}; D) < I(C; D)$, then a is a core attribute of S. The integration of all core attributes is called attribute core. Attribute core can be used as the starting point of reduction computation.

Definition 4. Let $S = (U, C \cup D, V, f)$ be a decision table. For any $B \in C$ *of attribute, and any attribute* $a \in C - B$, the significance of attribute a with respect to B and D is defined as:

$$\text{sgn}(a, B, D) = I(B \cup \{a\}; D) - I(B; D) \tag{7}$$

The significance of feature attributes can be used as heuristic information in greedy algorithms to compute the minimal attribute reduction.

3   Ant Colony Optimization for Attribute Reduction

3.1   Ant Colony Optimization Principle

In nature, ants can leave pheromones on the path they walk through. Ants in the process of movement to perceive the existence and strength of pheromones are more likely to move toward higher intensity pheromones. It is through the exchange of information that the ants can realize the goal of searching for food.

Based on this positive feedback mechanism to find the shortest path. Inspired by the behavior of real ants, Dorigo and Caro proposed ACO to solve combinatorial optimization problems [3]. Jensen and Shen (2003) propose a method for feature selection based on rough sets and ACO (JSACO) [6]. The algorithm idea is as follows: given a colony of k artificial ants randomly to select the initial node (attribute). Then, the next attribute is selected based on pheromones and heuristic information. During every iteration k, each ant searches for a set of attribute reduction and updates pheromone according to the optimal attribute path. The algorithm stops iterating when a termination condition is met.

3.2 Local Solution

In ACO, we need first to find a local solution, and then obtain the global optimal solution according to the positive feedback mechanism of the ant colony. Ants randomly selects the initial attribute, and then select the next probability of the largest attribute according to the probability formula. The formula is defined as follows:

$$p_{ij}^k(t) = \frac{\tau_{ij}^\alpha \eta_{ij}^\beta(t)}{\sum\limits_{l \in allowed_k} \tau_{il}^\alpha \eta_{il}^\beta(t)}, \; j \in allowed_k \tag{8}$$

Where k and t denote the number of ants and iterations, respectively. $\alpha > 0$ and $\beta > 0$ are two parameters which determine the relative importance of the pheromone trail and heuristic information. If α is larger than β, the ant select the attribute path when the main consideration pheromone trails. If β is much larger than α, ants will select those edges with higher heuristic information in a greedy manner. α and β should be chosen in the range 0–1 and be determined by experimentation. $allowed_k$ denotes the set of conditional attributes that have not yet been selected. τ_{ij} and β_{ij} are the pheromone value and heuristic information of choosing attribute j when at attribute i.

A construction process is terminated by one of the following two conditions:

(1) $I(C; D) = I(R_k; D)$, where R_k is local attribute solution constructed by the k-th ants.

(2) The cardinality of the current solution is larger than that of temporary minimal attribute reduction.

The first condition to stop the search means that the current reduction set of attributes has achieved the reduction effect. If the local solution attribute set cardinality is less than the current global optimal solution, the local solution is taken as a new global optimal solution. The second condition implies that the cardinality of attribute reduction set for the k-th ant is greater than the cardinality of the current global optimal solution.

3.3   Pheromone Updating

After each ant has find a set of reduction, which represents the completion of an iteration. The pheromone on each edge should be updated according to the following rule:

$$\tau_{ij}(t+1) = \rho\tau_{ij}(t) + \Delta\tau_{ij}(t) \tag{9}$$

Parameter $\tau_{ij}(t)$ is the amount of pheromone on a given edge (i, j) at iteration t, and $\tau_{ij}(t+1)$ is the amount of pheromone on a given edge (i, j) at next iteration. $\rho(0 \leq \rho \leq 1)$ represents constant used to simulate the evaporation of pheromone, and $\Delta\tau_{ij}(t)$ is the amount of pheromone deposited, typically given by:

$$\Delta\tau_{ij}(t) = \begin{cases} \sum \frac{q}{|R(t)|} & \text{if edge (i,j) has been traversed} \\ 0 & \text{otherwise} \end{cases} \tag{10}$$

Where parameter q is a given constant, $|R(t)|$ is the cardinality of the minimal attribute reduction at iteration t.

3.4   The Proposed Algorithm

RSFSACO algorithm [19] uses information gain as heuristic (7), which tends to choose attributes that contain more values. From the perspective of information theory, it is the property that tends to choose a more chaotic value. With the information gain rate as the heuristic information, there is a preference for the attributes with less number of values. Therefore, the first contribution of this paper is to use the improved information gain rate [21] as heuristic information (11). This not only considers the increment of mutual information after adding the selected attributes, but also takes the information entropy of the selected attribute itself into account.

The heuristic information formula is as follows:

$$\eta_{r,p} = \text{sgn}(r, \{Core \cup p\}, D)/H(D|r) \tag{11}$$

Suppose *Core* denote the core of C with respect to D in decision table. All ants start from *Core* set, and the given ant is currently at node p. The next selected attribute $r \in C - \{Core + p\}$.

Algorithm IGRARACO (Improved-information Gain Rate Attribute Reduction based on ACO)

Input: a decision table $S = (U, C \cup D, V, f)$ and parameters

Output: a minimal feature reduction R_{min} and its cardinality L_{min}

1: Initial $R_{min} = C, L_{min} = |C|, iteration = 0$, η_{ij} and τ_{ij}, $1 \leq i, j \leq |C|$;

2: Compute $I(C; D)$ and $Core$;

3: **For** every $c \in C$ **do**

4:      **If** $I(C - \{c\}; D) < I(C; D)$ **then** $Core = \{\cup c\}$;

5:      **end If**

6: **end For**

7: **For** $iteration \leftarrow 1$ to $maxcycle$ **do**

8:      **For** every $k \in Ant$ **do**

9:          $R_k = Core, L_k = |Core|$, Select a feature $a_k \in \{C - Core\}$ random

10:          $R_k = \{R_k \cup a_k\}, L_k = L_k + 1$;

11:      **Do:**

12:          Calculate $I(R_k; D)$;

13:          Select next attribute $b_k \in \{C - R_k\}$ by formula (8);

14:          Calculate $I(R_k \cup \{b_k\}; D)$;

15:          **If** $I(R_k; D) \neq I(R_k \cup \{b_k\}; D)$ **then** $R_k = \{R_k \cup b_k\}, L_k = L_k + 1$;

16:          **end If**

17:      **Until** $I(R_k; D) = I(C; D)$ or $L_k > L_{min}$;

18:      **end For**

19:      **For** every $x_k \in R_k - Core$ **do**

20:          **If** $I(R_k - \{x_k\}; D) = I(C; D)$ **then** $R_k = R_k - \{x_k\}$;

21:          **else** select next attribute;

22:          **end If**

23:      **end For**

24:      **If** $I(R_k; D) = I(C; D)$ and $L_k < L_{min}$ **then** $R_{min} = R_k, L_{min} = L_k$;

25:      **end If**

26:      Update $\tau_{ij}(t+1) = \rho \tau_{ij}(t) + \Delta \tau_{ij}(t)$;

27: **end For**

28: **Output** R_{min} and L_{min};

Whether traditional attribute reduction algorithm or an attribute reduction are based on ACO algorithm, they usually only focus on the choice of heuristic information and how to optimize and improve it. However, the redundant detection of selected attributes is ignored, which can easily lead to the addition of redundant attributes to the reduction set. Therefore, the second contribution of this paper is to make redundant judgement for each ant's attribute selection process, which can effectively avoid adding redundant attributes to the reduction result. Although the optimal solution of each generation has been obtained by the above method, there may still be redundant attributes, that is, the local optimal solution is not the global optimal solution. The third contribution is to redundantly detect the optimal solution of each generation again, thus eliminating the possibility of redundancy due to the random selection of the initial attributes and bringing the result closer to global optimal.

It should be pointed out that the purpose of the article is to find the MAR, that is, the fewer the number of attributes in the reduction set is, the better it can be. The comparison between the local optimal solution and the global optimal solution can be expressed as the comparison of reduction length.

4   Experimental Analysis

4.1   Comparison with Other Methods

In order to verify the effectiveness of the proposed algorithm, IEACO [18] and RSFSACO [19] are compared experimentally, and the effects of dimension reduction are compared as well. The performance of the algorithm is evaluated as follows: (1) Comparison of minimum reduction capability; (2) The change of fitness value in iterative updating process is used to evaluate the convergence speed of the algorithm. The algorithm is tested on a personal computer running Windows10 with 2.60 GHZ processor and 8 GB memory. In our experiments, we set the parameters $\alpha = 1$, $\beta = 0.01$, $= 0.9$, $q = 0.1$ and the initial pheromone η was set to 0.5 with a small random perturbation added, the number of ants was 1.5 times the number of attributes. Each dataset is tested for twenty times and the halting condition is reaching the maximum cycle or getting the same attribute reduction under five consecutive iterations. We also compare with other metaheuristic algorithm GenRSAR [3]. The experiments are carried out on four UCI datasets, Zoo, Audiology, Soybean and Vote. The experimental results are shown in Tables 1 and 2.

Table 1. Comparison of performances between RSFSACO, IEACO and IGRARACO

Data set	Inst.	Feat	Min-Redu	RSFSACO		IEACO		IGRARACO	
				Redu	Time	Redu	Time	Redu	Time
Audiology	200	70	13	$14^{15}15^5$	78	20	25.8	13	76
Breastcancer	699	10	4	4	5.42	$14^{12}5^8$	1.11	4	6.7
Chess-king	3196	37	29	$29^8 30^{12}$	1563	36	2226	29	3753
Monk1	124	7	3	3	<0.1	$3^{13}4^7$	<0.1	3	<0.1
Monk3	122	7	4	4	<0.1	4	<0.1	4	<0.1
Mushroom	8124	23	4	5	2431	$4^9 5^{11}$	973	$4^{15}5^5$	2675
Vote	435	17	9	9	5.91	$12^6 13^{12}$	2.78	9	7.4
Wine	178	14	4	4	1.57	4	0.19	4	2.1
Zoo	101	17	5	$5^2 6^{18}$	0.53	$6^8 7^{11}8^1$	0.2	5	0.5
Glass	214	10	5	$5^{14}6^6$	1.73	$5^4 6^{12}7^4$	0.25	5	1.9
Soybean	307	36	9	$15^{19}20^1$	59	$11^1 12^{19}$	5.8	$9^7 10^{13}$	41

Table 2. Comparison of performances between GenRSAR and IGRARACO

Data set	Inst.	Feat	Min-Redu	GenRSAR		IGRARACO	
				Redu	Time	Redu	Time
Zoo	101	17	5	$5^5 6^{14} 7^1$	0.8	5	0.5
Soybean	307	36	9	$11^1 12^3, >13^{16}$	6.3	$9^7 10^{11} 11^2$	41
Audiology	200	70	13	$15^1 16^1 17^{18}$	29.4	13	76
Vote	435	17	9	$9^5 10^{15} 11^5$	10.3	9	7.4

From the results, we can see that IGRARACO outperforms other algorithms with respect to ability of finding optimal reduction. For example, the results in IGRAR-ACO, IEACO and GenACO all contain redundant attributes for dataset Audiology, Soybean. For other datasets, although the comparison algorithm can get the MAR, but some results are not optimal, with redundant attributes. The reason is that the redundancy attribute is added to the reduction set in the process of attribute selection and also lacks redundant detection of the reduction results. At the same time, we find that the running time of IGRARACO is greater than other algorithms since in the process of computing, we have redundant detection of each attribute and result.

4.2 Analysis of Convergence Rate

In order to compare the optimization ability and convergence speed of the algorithm and IEACO and RSFSACO algorithm in the reduction process. We compare the fitness values of three algorithms in the experiment, and calculate the fitness value by the weight of attributes and the weighted length of attributes [16]. The fitness calculation is as follows:

$$fitenss = \lambda \frac{I(R_k; D)}{I(C; D)} + (1 - \lambda) \frac{|C| - |R_k|}{|C|} \tag{12}$$

Experiments are carried out by using multiple data sets in UCI. In order to show the effectiveness of the algorithm more clearly, the variable design is the optimal attribute reduction after each ant search. Number of searches = iteration times * ant number. The datasets are Audiology, Breastcancer, Glass, Vote, Zoo and Mushroom. The fitness value are shown in Fig. 1.

The convergence rates of IGRARACO and the GenRSAR algorithm are analyzed in Fig. 2. The fitness value of each generation is used to reflect the change trend of fitness value, and the number of iterations is set to 500 times. The datasets are Soybean and Zoo.

From the convergence rate analysis Figs. 1 and 2 can be seen that the algorithm designed in this research has faster convergence rate than other comparison algorithms. IGRARACO has the ability to quickly converge in locating the optimal solution, which is obtained in fewer iterations through the detection of selected attributes and reduction results In most cases, a satisfying solution can be achieved within five iterations.

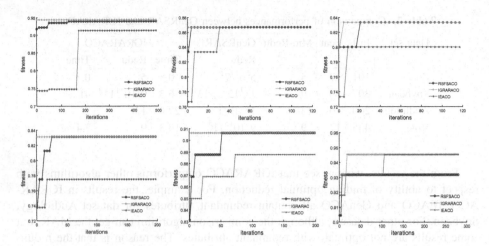

Fig. 1. The change of fitness value over successive iterations for IGRARACO, RSFSACO and IEACO.

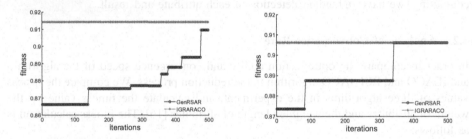

Fig. 2. The change of fitness value for IGRARACO and GenRSAR.

The performance of the proposed algorithm is evaluated by training C4.5 classifier and Naive Bayes classifier on the original data sets and the reduced data sets. We have used the implementation WEKA software [23]. Classification accuracies uses 10-fold-cross validation approach for validation, the results of experiments are shown in Tables 3 and 4.

Table 3. Classification accuracy before and after reduction obtained by C4.5 classifier

Data set	Before reduction	IGRARACO reduction	IEACO reduction	RSFSACO reduction
Audiology	77.8761	73.8938	73.8938	74.3363
Breastcancer	94.8498	95.4220	95.4220	95.4220
Chess-king	99.4368	99.2157	99.4368	99.2157
Mushroom	100	100	100	100
Vote	96.3218	96.3218	96.3218	96.3218
Wine	92.1348	91.5730	91.5730	91.5730
Zoo	92.0792	94.0594	96.0396	94.0594
Glass	66.8224	69.6262	69.6262	58.8785

Table 4. Classification accuracy before and after reduction obtained by Naive Bayes classifier

Data set	Before reduction	IGRARACO reduction	IEACO reduction	RSFSACO reduction
Audiology	73.4513	73.8938	73.8938	69.0265
Breastcancer	95.7082	94.2775	94.2775	94.2775
Chess-king	87.8911	89.8592	87.8911	89.8592
Mushroom	95.8272	98.6091	98.6091	98.5229
Vote	90.1149	92.3333	91.2644	92.3333
Wine	96.6292	94.3820	94.3820	94.3820
Zoo	95.0495	90.0990	89.1089	90.0990
Glass	48.5981	47.1963	47.1963	53.7383

From the experimental results obtained, we conclude that the three algorithms have little differences in classification quality after most data reduction and the reduced attribute results can ensure the classification of the original data is valid. Most of the reducts found by IGRARACO exhibit higher classification accuracy.

5 Conclusion

This paper discusses the shortcomings of conventional ACO attribute reduction. The results of attribute reduction in these algorithms are often fail to find minimal attribute reduction. Meanwhile, they need to iterate many times to get an appropriate result. These techniques usually fail to find optimal reductions, as no perfect heuristic can guarantee optimality [19].

We proposed a novel attribute reduction based on rough set and ACO. ACO has the ability to quickly converge. It has a strong search capability in combinatorial optimization problem to find minimal reduction. Our algorithm has the following characteristics: (a) Its heuristic information use an improved information gain rate as attribute selected significance; (b) Attribute redundancy judgment is carried out in each ant attribute selection process; (c) Redundant attributes are deleted for each generation of optimal attribute reduction results. Experimental results on real datasets demonstrate the effectiveness of our method to attribute reduction.

At the same time, through the experimental results we found that when the data set is larger, the algorithm need much longer time and the experiment lacks large data sets for verification. Therefore, the next step is to combine with parallel technology to deal with attribute reduction in large-scale datasets.

Acknowledgements. This work has been supported by National Natural Science Foundation of China (61363029, 61572146, U1711263), Science Foundation of Guangxi Key Laboratory of Trusted Software (kx201515), and the Foundation of Guangxi Educational Committee (KY2015YB105).

References

1. Pawlak, Z.: Rough sets. Int. J. Comput. Inf. Sci. **11**(5), 341–356 (1982)
2. Ding, H., Ding, S.F., Li-Hua, H.U.: Research progress of attribute reduction based on rough sets. Comput. Eng. Sci. **32**(6), 92–94 (2010). (in Chinese)
3. Jensen, R., Shen, Q.: Semantics-preserving dimensionality reduction: rough and fuzzy-rough-based approaches. IEEE Trans. Knowl. Data Eng. **16**(12), 1457–1471 (2004)
4. Hedar, A.R., Wang, J., Fukushima, M.: Tabu search for attribute reduction in rough set theory. Soft. Comput. **12**(9), 909–918 (2008)
5. Zhai, J.H., Liu, B., Zhang, S.: A feature selection approach based on rough set relative classification information entropy and particle swarm optimization. CAAI Trans. Intell. Syst. **12**(3), 397–404 (2017). in Chinese
6. Dorigo, M., Maniezzo, V., Colorni, A.: Ant system: optimization by a colony of cooperating agents. IEEE Trans. Syst. Man. Cybern. B **26**(1), 29–41 (1996)
7. Liao, T., Stützle, T., Oca, M.A.M.D., et al.: A unified ant colony optimization algorithm for continuous optimization. Eur. J. Oper. Res. **234**(3), 597–609 (2014)
8. Duan, H., Wang, D., Yu, X.: Review on research progress in ant colony algorithm. Chin. J. Nat. **28**(2), 102–105 (2006). in Chinese
9. Miao, D., Wang, J.: Information representation of the concepts and operations in rough set theory. J. Softw. **10**(2), 113–116 (1999). (in Chinese)
10. Skowron, A., Rauszer, C.: The discernibility matrices and functions in information systems. In: Slowinski, R. (ed.) Intelligent Decision Support. Handbook of Applications and Advances of the Rough Set Theory, pp. 331–362. Kluwer, Dordrecht (1992)
11. Yao, Y.: Three-way decisions with probabilistic rough sets. Inf. Sci. **180**(3), 341–353 (2010)
12. Lu, J., Li, D., Zhai, Y., et al.: A model for type-2 fuzzy rough sets. Inf. Sci. **328**(C), 359–377 (2016). in Chinese
13. Qian, Y., Li, S., Liang, J., et al.: Pessimistic rough set based decisions: a multigranulation fusion strategy. Inf. Sci. **264**(6), 196–210 (2014)
14. Wang, J., Miao, D.: Analysis on attribute reduction strategies of rough set. J. Comput. Sci. Technol. **13**(2), 189–192 (1998). in Chinese
15. Wong, S.K.M., Ziarko, W.: On optimal decision rules in decision tables. Bull. Pol. Acad. Sci. Math. **33**(11), 693–696 (1985)
16. Jensen, R., Shen, Q.: Finding rough set reducts with ant colony optimization. In: Proceedings of 2003 UK Workshop on Computational Intelligence, pp. 15–22 (2003)
17. Ke, L., Feng, Z., Ren, Z.: An efficient ant colony optimization approach to attribute reduction in rough set theory. Pattern Recogn. Lett. **29**(9), 1351–1357 (2008)
18. Chen, Y., Chen, Y.: Attribute Reduction Algorithm Based on Information Entropy and Ant Colony Optimization. J. Chin. Comput. Syst. **36**(3), 586–590 (2015). in Chinese
19. Chen, Y., Miao, D., Wang, R.: A rough set approach to feature selection based on ant colony optimization. Elsevier Science Inc. (2010)
20. Chebrolu, S., Sanjeevi, S.G.: Attribute reduction on continuous data in rough set theory using ant colony optimization metaheuristic. In: Proceedings of International Symposium on Women in Computing and Informatics. ACM, New York, pp. 17–24 (2015)
21. Yan, Y., Yang, H.: Knowledge reduction algorithm based on mutual information. J. Tsinghua Univ. **42**(2), 1903–1906 (2007). in Chinese
22. Shannon, C.E., Weaver, W., Hajek, B., et al.: The mathematical theory of communication. Phys. Today **3**(9), 31–32 (1950)
23. Hall, M., Frank, E., Holmes, G., et al.: The WEKA data mining software: an update. ACM SIGKDD Explor. Newsl. **11**(1), 10–18 (2009)

Efficient Approximate Algorithms for the Closest Pair Problem in High Dimensional Spaces

Xingyu Cai[1], Sanguthevar Rajasekaran[1(✉)], and Fan Zhang[2]

[1] Department of CSE, University of Connecticut, Storrs, CT, USA
{xingyu.cai,sanguthevar.rajasekaran}@uconn.edu
[2] Zhejiang University, Hangzhou, China
fanzhang@zju.edu.cn

Abstract. The Closest Pair Problem (CPP) is one of the fundamental problems that has a wide range of applications in data mining, such as unsupervised data clustering, user pattern similarity search, etc. A number of exact and approximate algorithms have been proposed to solve it in the low dimensional space. In this paper, we address the problem when the metric space is of a high dimension. For example, the drug-target or movie-user interaction data could contain as many as hundreds of features. To solve this problem under the ℓ_2 norm, we present two novel approximate algorithms. Our algorithms are based on the novel idea of projecting the points into the real line. We prove high probability bounds on the run time and accuracy for both of the proposed algorithms. Both algorithms are evaluated via comprehensive experiments and compared with existing best-known approaches. The experiments reveal that our proposed approaches outperform the existing methods.

Keywords: Closest pair · High dimension · Approximate algorithms

1 Introduction

Similarity search has been widely used in data mining. Example applications include finding the similarity between user patterns from online merchant transactions, analysis of social media connections, unsupervised data clustering, knowledge discovery from semantic data, etc. Two of the fundamental problems in data mining are finding the Nearest Neighbor (NN) and finding the Closest Pair (CP). These two problems are closely related. For instance, CP could be seen as an extension of NN, which requires more computation and thus is more challenging. In general, multi-feature data could be modeled as points in a high dimensional metric space. Among all the different similarity measurement metric, ℓ_p norm is commonly used. In this paper, ℓ_2 norm, or Euclidean distance, is employed as it is one of the most widely applicable measurements.

The Closest Pair Problem (CPP) we are addressing is that of identifying the closest pair of points from a given set of N points $\in \Re^m$ when m is not small.

© Springer International Publishing AG, part of Springer Nature 2018
D. Phung et al. (Eds.): PAKDD 2018, LNAI 10939, pp. 151–163, 2018.
https://doi.org/10.1007/978-3-319-93040-4_13

This classical problem has been studied extensively [19]. A straightforward algorithm for solving this problem takes $O(N^2 m)$ time, where m is the dimension of the input space. Research works have been carried out in different domains for different purposes to solve this problem in an efficient way. In 1979, Fortune and Hopcroft presented a deterministic algorithm with a run time of $O(N \log \log N)$ assuming that the floor operation takes $O(1)$ time [8]. In [1], another divide-and-conquer deterministic algorithm was introduced. Later improvements include [9,11,17,18]. In his seminal paper, Rabin proposed a randomized algorithm with an expected run time of $O(N)$ [7] (where the expectation is in the space of all possible outcomes of coin flips made in the algorithm). Rabin's algorithm also used the floor function as a basic operation. In 1995, the sieve method was proposed to eliminate points in a randomized way such that the actual comparison of the remaining candidates could be dramatically reduced [14]. A sample-based randomized approach was proposed in [6] in 1997 to solve several issues existing in [7]. Yao has proven a lower bound of $\Omega(N \log N)$ on the algebraic decision tree model (for any dimension) [21]. All these algorithms assume a constant dimensional space (i.e., $m = O(1)$), and the run times are exponentially dependent on the dimension, making them not applicable for a dimension of several hundreds.

In recent years, database applications have driven the research on CPP. By exploring the connection between CPP and matrix multiplication, Indyk [10] has presented an $O(N^{(w+3)/2)}$ time algorithm for CPP in ℓ_1 and ℓ_∞ norms, where $O(N^w)$ is the time needed to multiply two $N \times N$ matrices. This algorithm is not applicable for ℓ_2 norm. Corral et al. [3] have provided a method that uses tree data structures. Besides exact algorithms, Lopez and Liao [15] have provided an approximate algorithm to address this problem by making copies of the original data and employing random shifting on each copy. More recently, Locally Sensitivity Hashing (LSH) has gained attention in solving the NN problem. As a consequence, approximate algorithms based on LSH for CPP are proposed in the literature. Datar [5] has proposed a sub-quadratic time algorithm using LSH that solves the c-approximate problem (output neighbors that are no further than c times the distance between the nearest neighbors). Later Tao [20] improved Datar's algorithm and extended it to out-of-core CPP, where the I/O costs are optimized. The comparison in [20] shows that their algorithm outperforms the methods in [3,15]. These algorithms mainly focus on the NN problem, or address the problem for efficiency in I/O cost, making them fit for out-of-core computation with many applications such as in database query processing. However, they are not very suitable for in-memory computation. Also, the construction of special data structures (such as LSB tree in [20]) will bring significant overhead for in-core tasks. In addition, the approximate methods in this domain are addressing the c-approximate problem that introduces a relaxation factor c.

Mueen et al., have presented an elegant exact algorithm called MK for the CPP [16]. Though this algorithm was originally proposed to solve a special case of the CPP, known as the time series motif mining problem, it can be used to solve the CPP very well. Although MK is an $O(N^2 m)$ time algorithm, it improves the performance of the brute-force algorithm in practice using the triangular inequality and a technique called early-abandoning. MK is a deterministic algorithm that always finds the closest pair. To the best of our knowledge, MK is still one of the best performing algorithms for high dimensional CPP in practice, even though it is no longer the state-of-the-art choice for time series motif mining problem. In this paper we use MK as the baseline to evaluate our proposed algorithms. To provide a fair comparison, we use the original MK code that is publicly available in http://alumni.cs.ucr.edu/mueen/MK/. The code for our proposed approaches can be found at https://github.com/TideDancer/ACPP.git.

In this paper we present two approximate algorithms for the CPP. One of them revisits the divide-and-conquer approach but modifies it to high dimensional settings. The other uses a novel idea in random projection: The original Johnson-Lindenstrauss Lemma shows the existence of a random projection of $O(\log N)$ dimension that preserves all pairwise distances with a high probability. For the CPP we only have to preserve the distance between the closest pair. We use random projection of points into 1D. We show that if we perform this projection $O(\log N)$ times, then the distance between the closest pair will be preserved at least once with a high probability. Note that although the proposed algorithms are sequential, all these algorithms along with MK, could be easily parallelized due to the independence of their subroutines.

The rest of this paper is organized as follows: In Sect. 2, we present two approximate algorithms for the high dimensional CPP. Running time and accuracy bounds are proved for both of the approaches (refer to Appendix for details). Comprehensive experiments are carried out to evaluate the performance of both algorithms in Sect. 3, and some conclusions are provided at the end.

2 Proposed Approximate Algorithms

Two approximate approaches for the high dimensional CPP are provided: ACP-P and ACP-D. Both algorithms always keep an upper bound δ_u on the distance δ^* between the closest pair of points. The common initial step for both algorithms is to obtain an upper bound on δ^* by picking a random sample of $\sqrt{N}$ points and identifying the distance between the closest pair of points in the sample. Even a brute-force algorithm will only take $O(N)$ time for doing this.

2.1 ACP-D

The divide-and-conquer algorithm of [1] performs well on low dimensional (e.g., 2D and 3D) data with a run time of $O(N \log N)$. Its performance degrades significantly on high dimensional data since the run time has an exponential dependence on the dimension. The divide-and-conquer algorithm proceeds by partitioning the input into two using the median along one of the dimensions. The closest pairs are recursively found for each of the two parts. Followed by this, we have to find the closest among the cross-part pairs. When the input is

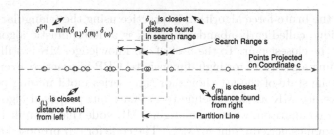

Fig. 1. Illustration of search within a range s around the partition line along a particular coordinate

from 2D (i.e., $m = 2$), the number of cross-pairs that have to be considered is proved to be $O(N)$. When the input is from an m-dimensional space, the number of candidate cross-pairs to be considered goes up to $O(N \times 3^m)$. This number can be $\Omega(N^2)$ or worse. Thus the performance could be as bad as that of the brute force algorithm.

Algorithm 1. ACP-D

Input: N points $p_i \in \Re^m$ $(1 \leq i \leq N)$, brute-force subset size T, search range constant α. Initialize left $= 1$, right $= N$, depth $= 1$

Output: function   ACP-D(depth, left, right) finds the closest pair (l, r) with the smallest Euclidean distance $D(l, r)$

1: len = right - left + 1
2: **if** depth = 1 **then**
3:    Randomly select one coordinate $c^{(\mathcal{H})}$, $c^{(\mathcal{H})} \in [1, d]$;
4:    Sort the points based on values $p_i[c^{(\mathcal{H})}]$ along the coordinate $c^{(\mathcal{H})}$;
5: **end if**
6: **if** len $\leq T$ **then**
7:    Use brute-force to find best-so-far $d(i, j)$ where left $\leq i \leq j \leq$ right;
8:    **if** $d(i, j) < D(l, r)$ **then**
9:       $l = i$; $r = j$; $D(l, r) = d(i, j)$;
10:   **end if**
11:   **return** $D(l, r)$;
12: **else**
13:   mid = left + len/2;
14:   ACP-D(depth+1, left, mid);
15:   ACP-D(depth+1, mid+1, right);
16:   Obtain two sets of indices: $S = \{p_i\}, i < \text{mid}, p_{mid} - p_i \leq \alpha D / \sqrt{N}$ and $S' = \{p_j\}, j > \text{mid}, p_j - p_{mid} \leq \alpha D / \sqrt{N}$;
17:   **for** $i \in S$ **do**
18:      **for** $j \in S'$ **do**
19:         **if** dist$(p_i, p_j) < D(l, r)$ **then**
20:            $l = i$; $r = j$; $D(l, r) = $ dist(p_i, p_j);
21:         **end if**
22:      **end for**
23:   **end for**
24: **end if**

In this section we propose an enhanced divide-and-conquer algorithm. The idea is to choose the candidate cross-part pairs appropriately. Here again we partition the input into two and recursively find the closest pair in each part. To find the closest cross-part pair we do the following: Let $\mathcal{H}$ be the hyperplane that partitions the input into two (based on the median along one of the coordinates). We have to consider all pairs of the form (a, b) where a is one side of the hyperplane $\mathcal{H}$ and b is on the other side of $\mathcal{H}$. Instead of checking all such pairs we only consider pairs where a and b are on different sides of $\mathcal{H}$ but within a distance of s. We refer to s as the search range (see Fig. 1). This procedure is repeated by partitioning along different coordinates to increase the chances of finding the closest pair.

To begin with, ACP-D randomly chooses a coordinate to do partition. It then recursively finds the closest pair's distance from the left and the right partitions (denote the distances as $\delta_{(L)}, \delta_{(R)}$). Next we look at all the points that reside within a search range s around the partition hyperplane $\mathcal{H}$ along this coordinate, and find the closest pair (distance as $\delta_{(s)}$) among these candidates. Followed by this we update the pair with the minimum distance denoted as $\delta^{(\mathcal{H})} = \min(\delta_{(L)}, \delta_{(R)}, \delta_{(s)})$. The detailed pseudocode of ACP-D is given in Algorithm 1. An illustration of searching within a range s is shown in Fig. 1. We establish the following theorem and provide the proof in Appendix. This theorem offers a probabilistic bound on success rate and run time of ACP-D. To boost the success rate, we repeat ACP-D and output the closest pair seen as the closest pair of points.

Theorem 1. *Let $p[c]$ represent vector p's c-th element, or equivalently p's c-coordinate value. Assume that the coordinate values in each dimension are uniformly distributed and let the spread length of points be $r = \max_i p_i[c^{(\mathcal{H})}] - \min_j p_j[c^{(\mathcal{H})}]$ on the partition coordinate $c^{(\mathcal{H})}$. Use a search range of $s = \sqrt{\alpha}\frac{\delta^{(\mathcal{H})}}{\sqrt{m}}$. As long as $r^2 = \Omega(N)$ where m is the dimension, ACP-D algorithm's expected run time will be $T(N) = O(N \log N)$, with a high probability.*

Corollary 1. *We have the following probability bound on the run time:*

$$Prob\{T(N) \geq (\beta + 1)\alpha(\delta^{(\mathcal{H})})^2 N \log N\} \leq e^{-\beta}, \text{ for any } \beta > 0.$$

2.2 ACP-P

Random projection lemma [12] states that pairwise distances are closely preserved in a random $O(\log N)$-dimensional space with a high probability. In this paper we prove that, if we repeat projecting the input points from $\Re^m$ to $\Re^d$ randomly $(d < m)$ a total of k times, as long as kd satisfies a certain condition, the closest pair's distance will be closely preserved in at least one of the projections, with a high probability. In addition, $d = 1$ would significantly reduce the computation cost. We exploit this property in the ACP-P algorithm.

After the projection, all the pairs in the projected space that are within a distance of $\delta^{(P)} = (1 + \epsilon)\delta_u$ (in $\Re^d$) needs to be identified, where ϵ is a small constant. For the case of $d > 1$, identifying these close pairs in $\Re^d$ still remains a difficult task. One can use hyper-sphere centered at each point with a radius of $\delta^{(P)}$, and check if there are other points in the hyper-sphere. However, this might be even harder than directly computing all pairwise distances in $\Re^d$, which takes

$O(N^2 d)$ running time. On the other hand, if in 1-D space ($d = 1$), the hypersphere becomes left and right intervals, making the job of identifying close points within an interval of $\delta^{(P)}$ extremely easy. To be specific, one can use any sorting algorithm to first sort all the projected points because all the points are identified by a scalar value in 1-D space. Then a scanning from left to right is performed and all the adjacent points within a certain range are detected. In total it only requires an $O(N \log N)$ running time. In fact, sorting could be replaced by a griding approach to identify pairs within an interval (see Appendix 4.4). After identifying these pairs, the Euclidean distance between each pair is computed in $\Re^m$. The pair with the least distance is kept and δ_u is updated.

Algorithm 2. ACP-P

Input: N points in $\Re^m$: $p_1, p_2, \ldots, p_N$.
Output: The closest pair of input points.

1: $j = 1$
2: **repeat**
3:    Randomly generate a projection vector $\Phi \in \Re^{1 \times n}$
4:    **for** $i = 1$ to N **do**
5:        $p_i' = \Phi p_i^T$
6:    **end for**
7:    Sort $p_1', p_2', \ldots, p_N'$;
8:    **for** $i = 1$ to N **do**
9:        Identify the interval that p_i' belongs to;
10:   **end for**
11:   **for** every interval **do**
12:       Generate all possible pairs from the points that have fallen into this interval. These are candidate pairs;
13:   **end for**
14:   For each candidate pair compute the distance in $\Re^m$ and pick the pair with the least distance. Let this distance be δ_j;
15:       $j = j + 1$;
16:   **until** $j = k$
17:   Find $\delta_o = \min\{\delta_1, \delta_2, \ldots, \delta_k\}$;
18:   **return** δ_o

The above projection-identification process is repeated k times. We show (in Appendix) that if $kd = \Theta(\log N)$, then the closest pair would come within a distance of $(1 + \epsilon)\delta_u$ in the projected space at least once with a high probability. Note that in the original Johnson-Lindenstrauss Lemma, $d = O(\log N)$. **The reason that we can push the limit to $d = 1$ is because we only have to preserve the distance between the closest pair, and not all pairwise distances.** We provide the following theorems and the corresponding proofs (in Appendix).

Theorem 2. *Let the closest pair have a distance of δ^*. If we repeat the random projection $\Re^m \to \Re^d$ for a total of k times, then the probability that $(\delta^{(P)})^2 < (1 + \epsilon)(\delta^*)^2$ at least once is high (i.e., $\geq 1 - N^{-\alpha}$ where α is some constant), as long as $dk \geq \frac{4\alpha}{\epsilon^2 - \epsilon^3} \log N$, for any $\epsilon \in [0, 1]$.*

Corollary 2. *Let $d = 1$. In each iteration of ACP-P, let the projected points be quantized with intervals of length $2(1 + \epsilon)\delta_u$. The probability that the closest* *pair (in $\Re^m$) will fall into the same interval is $\geq \frac{1}{2}[1 - e^{-(\epsilon^2 - \epsilon^3)/4}]$. This in turn means that the number of iterations taken by ACP-P to identify the closest pair of points with a high probability is $k = O\left(\frac{\alpha \log N}{\epsilon^2 - \epsilon^3}\right)$.*

In practice, we have found that the sorting based implementation in 1-D does not introduce an observable overhead. A detailed pseudocode of ACP-P that employs sorting is given in Algorithm 2. It is worth pointing out that the key difference between ACP-P and the LSH method used in [20] is after projection. The linear search based on the sorted list of points is much more efficient to identify each points' close neighbors, rather than a grid scheme using hashset technique. In our experiments we have realized that neither C++/boost hashset nor google's hashset could achieve desirable in-core performance, making the method in [20] not suitable for in-memory computations. Besides, the probability bound analyses in Appendix are also different.

3   Experiments

We have conducted experiments to evaluate the performance of ACP-P and ACP-D against MK. We have employed an Intel Xeon E5 CPU @ 3.2 GHz machine. The experiments have been performed on synthetic datasets, with different numbers of points and dimensions. Coordinate values have been generated uniformly randomly from the range: $[0, 1000]$. The following values have been used: $N = 10, 20, 30, 40, 50 \times 10^3$ and $m = 128, 256, 512, 1024$ and 2048.

To further boost the success rate, in each run we repeat the approximate algorithms Q times and output the best among them. Q is designed as $Q_{\text{ACP-D}} = h\frac{N}{10 \times 10^3}$ and $Q_{\text{ACP-P}} = h(\frac{N}{10 \times 10^3})^2$ for ACP-D and ACP-P, respectively. N is the input size and h is the hyper parameter. For instance if $N = 30k, h = 2$, then $Q = 6$ for ACP-D and $Q = 18$ for ACP-P. We perform 10 runs and provide the average running time, average rank and the hit rate (i.e., the fraction of the number of times the closest pair is found in 10 runs). Clearly, the larger the h, the better is the accuracy and the worse is the run time.

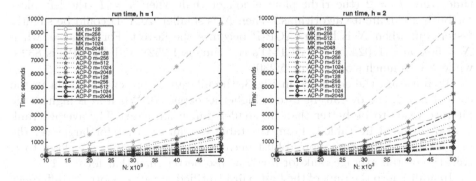

Fig. 2. Run time comparison

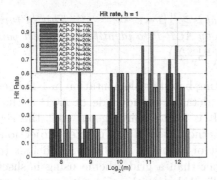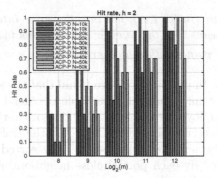

Fig. 3. Hit rate comparison

Table 1. Average rank (the smaller the better)

	$m = 128$		$m = 256$		$m = 512$		$m = 1024$		$m = 2048$	
$h = 1$	ACP-D	ACP-P	ACP-D	ACP-P	ACP-D	ACP-P	ACP-D	ACP-P	ACP-D	ACP-P
$N = 10k$	2.7	7.1	4	6.2	4.4	8.5	4	6.6	3.9	7.9
$N = 20k$	1.7	4.7	2	3.7	3.5	4.9	3.6	5	3.2	5.5
$N = 30k$	2.1	1.9	1.7	2.7	1.9	2.3	1.5	3.2	1.8	2.9
$N = 40k$	1.5	1.9	1.2	3.3	2.4	2	1.3	2.9	2.1	2.2
$N = 50k$	1.6	1.5	1.4	2.6	1.2	1.4	1.7	2	2	3.5
$h = 2$	ACP-D	ACP-P	ACP-D	ACP-P	ACP-D	ACP-P	ACP-D	ACP-P	ACP-D	ACP-P
$N = 10k$	1.6	4.8	2.5	3.7	1.7	5.5	2.2	5.8	3.5	5.7
$N = 20k$	1.5	1.9	1.2	3.3	1.8	2.4	1.5	4.6	2.2	3.1
$N = 30k$	1	1.2	1	2.1	1.2	1.5	1.5	2	1.3	1.6
$N = 40k$	1.1	1.2	1	1.4	1.1	1.1	1.3	1.7	1.3	1.3
$N = 50k$	1	1	1.1	1.1	1.1	1.2	1	1.8	1	1.4

Figure 2 shows the run time comparison. Clearly, for all settings, ACP-D and ACP-P are significantly faster than the MK algorithm. As expected, the run time when $h = 2$ (the right plot) is longer than when $h = 1$ (the left plot) for both approximate algorithms. When N is smaller, ACP-D could be slightly faster, but when N is larger, ACP-P becomes the fastest. For instance, when $N = 50k, m = 1,024, h = 1$, ACP-D's run time is 1,252 s and ACP-P's is 832 s, while MK is much slower using 5,230 s.

To illustrate the accuracy, in Fig. 3, the hit rate is presented. Again in the case of $h = 2$, the overall hit rate is higher as expected. When m is higher, the hit rate tends to be better than in smaller dimension cases. The average rank is also provided in Table 1. From the table we can see that for larger N, the proposed algorithms are more robust because the average ranks are closer to 1. And the overall average rank for $h = 2$ is also better than that for $h = 1$.

In addition to the rank of the best pair identified, we also report the difference between the output pair's distance and the true closest pair's distance. We define the distance ratio as $\rho = d/d^*$, where d^* is the distance between the closest pair of points and d is the distance between the output pair of points. In Table 2, we show the mean and variance of ρ when $h = 2$, and demonstrate that for our synthetic dataset, the distance ratio is very close to 1 with a small variance. This also proves the robustness of our proposed approximate algorithms.

Table 2. Distance ratio ρ (mean and standard deviation) when $h = 2$

$h = 2$	$m = 128$		$m = 256$		$m = 512$		$m = 1024$		$m = 2048$	
Mean	ACP-D	ACP-P	ACP-D	ACP-P	ACP-D	ACP-P	ACP-D	ACP-P	ACP-D	ACP-P
N = 10k	1.010	1.034	1.004	1.008	1.002	1.012	1.001	1.004	1.006	1.006
N = 20k	1.008	1.015	1.000	1.002	1.002	1.004	1.001	1.003	1.005	1.007
N = 30k	1.000	1.004	1.000	1.002	1.000	1.001	1.000	1.001	1.001	1.002
N = 40k	1.001	1.003	1.000	1.012	1.000	1.000	1.001	1.001	1.001	1.001
N = 50k	1.000	1.000	1.002	1.002	1.000	1.000	1.000	1.005	1.000	1.001
Std	ACP-D	ACP-P	ACP-D	ACP-P	ACP-D	ACP-P	ACP-D	ACP-P	ACP-D	ACP-P
N = 10k	0.011	0.029	0.004	0.003	0.002	0.006	0.001	0.002	0.002	0.004
N = 20k	0.015	0.015	0.000	0.003	0.003	0.003	0.001	0.003	0.005	0.005
N = 30k	0.000	0.011	0.000	0.002	0.001	0.002	0.000	0.001	0.002	0.003
N = 40k	0.004	0.006	0.000	0.014	0.001	0.001	0.001	0.002	0.001	0.001
N = 50k	0.000	0.000	0.007	0.007	0.001	0.001	0.000	0.005	0.000	0.001

4   Conclusions

In this paper we have offered two approximate algorithms for solving the CPP. Both of them are based on the idea of converting high dimensional search into line search. We provide theoretical bounds on the run time and prove the accuracy of ACP-D. For ACP-P, we exploit random projections but push the limit to 1-D space because we only identify the closest pair rather than preserving all pairwise distances. A theoretical analysis is also provided. In the experiments, we perform comprehensive simulations to evaluate both the run time and the accuracy for the proposed approximate algorithms. The results reveal that our approach runs much faster than the state-of-the-art method while still keeping a very good accuracy. Our algorithms could be easily parallelized for further speedups.

Acknowledgments. This work has been partly supported by NSF Grants 1447711 & 1743418 to SR; and the National Natural Science Foundation of China Grants 61472357 & 61571063 to FZ.

Appendix

4.1   Proof of Theorem 1

Proof. In the recursive algorithm ACP-D, the initial data size is N and each recursion splits the input into two halves. Recursion ends when the size is below the threshold T. The brute-force or MK algorithm can be used when the final input size is $\leq T$. Thus the total number K of recursions is $K = \log(N/T)$.

In each iteration, the closest pair (p_a, p_b) may not be captured only if this pair is neither in the left half nor the right half, and also not captured by the search range in the middle. So the probability of failure in this iteration equals the probability that p_a, p_b is split into left and right halves (denote this probability as $P_{(sep)}$), multiplied by probability that (p_a, p_b) is missed in the search range s (denote this probability as $P_{(miss)}$) conditioned on $P_{(sep)}$. Denote iteration k with a superscript such as $P_{(sep)}^k$.

Assume a uniform distribution of points' coordinate values. In the randomly chosen coordinate $c^{(\mathcal{H})}$, denote p_M as the index of the median point along this coordinate. Then, for the first iteration,

$$P_{(sep)}^1 = \mathrm{Prob}\{p_a[c^{(\mathcal{H})}], p_b[c^{(\mathcal{H})}] \text{ are split by } p_M[c^{(\mathcal{H})}]\} = 1/2.$$

In the second iteration, probability that p_a, p_b are split again in both halves is

$$P_{(sep)}^2 = 2 \times (\mathrm{Prob}\{p_a[c^{(\mathcal{H})}], p_b[c^{(\mathcal{H})}] \text{in same half}\} \times \mathrm{Prob}\{\text{They are separated}\})$$

$$= 2 \times \left(\frac{1}{4} \times \frac{1}{2}\right) = \frac{1}{4}.$$

Thus we can easily see that $P_{(sep)}^k = \frac{1}{2^k}$.

Next let us compute $P_{(miss)}$. Let the distance between the closest pair (p_a, p_b) be δ^* and let $\delta^{(D)} = \min(\delta_{(L)}, \delta_{(R)})$. Clearly, $\delta^{(D)} \geq \delta^*$. Since there are m dimensions, the expected contribution of δ^* to coordinate $c^{(\mathcal{H})}$ is $E[l_{c^{(\mathcal{H})}}(p_a, p_b)] = \delta^*/\sqrt{m}$, which means there are at least $(m - \frac{m}{\alpha})$ coordinates for which $l_c(p_a, p_b) \leq \sqrt{\alpha}\frac{\delta^{(D)}}{\sqrt{m}}$, $\alpha > 1$.

If we set the search range as $s = \sqrt{\alpha}\frac{\delta^{(D)}}{\sqrt{m}}$, meaning we check pairs (p_i, p_j) such that $p_i[c] \in [p_M[c] - s, p_M[c]]$ and $p_j[c] \in [p_M[c], p_M[c] + s]$, then choosing a random coordinate would give $\mathrm{Prob}\{\text{Capture the closest pair}| P_{(sep)}^1\} \geq 1 - 1/\alpha$. As a result, for any iteration k, the conditional probability of failure is

$$P_{(miss)}^k\{\text{Fail} \mid P_{(sep)}^k\} = (1 - \mathrm{Prob}[\text{Capture}| P_{(sep)}^k]) \leq 1/\alpha.$$

So the probability of failure through all the recursions is

$$\mathrm{Prob}[\text{Fail}] = \sum_{k=1}^{K} P_{(sep)}^k P_{(miss)}^k \leq \frac{1}{\alpha} \sum_{k=1}^{\log N/T} \frac{1}{2^k} \leq \frac{1}{\alpha}.$$

Thus we arrive at the following Lemma:

Lemma 1. *Assuming a uniform distribution of points, if we set the search range as $s = \sqrt{\alpha}\frac{\delta^{(\mathcal{D})}}{\sqrt{m}}, \alpha > 1$, where $\delta^{(\mathcal{D})}$ is minimum of the closest distances returned by the left and the right halves, ACP-D will find the closest pair with a probability of $\geq 1 - 1/\alpha$. Q.E.D.*

Assuming a uniform distribution and letting the spread length of points be $r = \max_i p_i[c^{(\mathcal{H})}] - \min_j p_j[c^{(\mathcal{H})}]$ on the chosen coordinate, within search range s, the expected number of points residing in will be sN/r. Thus the expected number of pairs that we need to compute distances for would be $(sN/r)^2$. Therefore, the expected running time will be

$$T(N) = 2T(\frac{N}{2}) + \tilde{O}(\frac{s^2N^2}{r^2}m) = 2T(\frac{N}{2}) + \tilde{O}(\frac{\alpha(\delta^{(\mathcal{D})})^2N^2}{r^2})$$

assuming that each distance computation takes $O(m)$ time. As long as $r^2 = \Omega(N)$ ACP-D algorithm's expected run time will be:
   $T(N) = 2T(N/2) + \tilde{O}(\alpha(\delta^{(\mathcal{D})})^2N) = \tilde{O}(N \log N)$ as α and $\delta^{(\mathcal{D})}$ are some constants.   □

4.2   Proof of Corollary 1

Proof. Applying the probabilistic recurrence relationship [13] and the revised version [2], under the same assumptions of Theorem 1, for a positive β, we have

$$\text{Prob}\{T(N) \geq (\beta + 1)\alpha(\delta^{(\mathcal{D})})^2N \log N\} \leq e^{-\beta}.$$

□

4.3   Proof of Theorem 2

Proof. In the proof for the JL lemma [4], the following lemma is also given:

Lemma 2. *For any fixed vector $v \in \Re^m$, projection matrix $\Phi : \Re^m \to \Re^d$ with i.i.d. Gaussian entries, i.e., $\Phi_{ij} = \frac{1}{\sqrt{d}}N(0,1)$, the following statements are true:*
$E[||\Phi v||^2] = ||v||^2$ and $Prob[||\Phi v||^2 \geq (1 + \epsilon)||v||^2] \leq e^{-(\epsilon^2 - \epsilon^3)d/4}$, *as well as*
$Prob[||\Phi v||^2 \leq (1 - \epsilon)||v||^2] \leq e^{-(\epsilon^2 - \epsilon^3)d/4}$.

For simplicity in both analysis and implementation, we use Gaussian projections. Let the closest pair (in $\Re^m$) be (p_a, p_b) with a distance of δ^*. Using the second equation of the above theorem, we obtain:

$$\text{Prob}\{||\Phi(p_b - p_a)||^2 \geq (1 + \epsilon)||(p_b - p_a)||^2\} \leq e^{-(\epsilon^2 - \epsilon^3)d/4}.$$

Let the distance between p_a and p_b in $\Re^d$ be $\delta^{(\mathcal{P})}$. The above equation becomes:

$$\text{Prob}\{(\delta^{(\mathcal{P})})^2 \geq (1 + \epsilon)\delta^{*2}\} \leq e^{-(\epsilon^2 - \epsilon^3)d/4}.$$

The probability that the above event happens at least once in the k iterations is:

$$\text{Prob}\{(\delta^{(\mathcal{P})})^2 < (1+\epsilon)\delta^{*2}]\} \geq 1 - e^{-(\epsilon^2-\epsilon^3)dk/4}.$$

We want this to be a high probability. By high probability we mean a probability of $\geq (1 - N^{-\alpha})$, α being a probability parameter (normally assumed to be a constant ≥ 1). We want $1 - e^{-(\epsilon^2-\epsilon^3)dk/4} \geq 1 - N^{-\alpha}$. This happens when:

$$e^{-(\epsilon^2-\epsilon^3)dk/4} \leq N^{-\alpha} \Rightarrow (\epsilon^2 - \epsilon^3)dk/4 \geq \alpha \log N \Rightarrow dk \geq \frac{4\alpha}{\epsilon^2 - \epsilon^3} \log N.$$

$\square$

4.4   Proof of Corollary 2

Proof. Projecting the input points into real numbers, i.e., $d = 1$, has a great computational advantage. To identify pairs within a specific distance in $O(N)$ time, we use quantization technique. Let the minimum and maximum projected values be m_1 and m_2, respectively. We partition the range $[m_1, m_2]$ into intervals of length $L = 2(1 + \epsilon)\delta_u$ each. Each such interval has an integer index (starting from 1). We also extend the range $[m_1, m_2]$ by a random number r in the range $[0, (1 + \epsilon)\delta_u]$. Specifically, we use the range: $[m_1 - r, m_2]$. Then for each point p_i we identify the interval that it falls into as: $\text{ID}(p_i) = \lceil (p_i/L) \rceil$. This can be done in a total of $O(N)$ time for all the points. For each interval, we generate all possible pairs from out of the points that belong to this interval. From out of all of these candidate pairs we pick the one with the least distance (in $\Re^m$). Clearly, the probability that two points that are within a distance of $\leq (1+\epsilon)\delta_u$ (in $\Re^m$) will fall into the same interval is $\geq 1/2$. Thus the number of iterations k can be computed in the same manner as above. $\square$

References

1. Bentley, J.L., Shamos, M.I.: Divide-and-conquer in multidimensional space. In: Proceedings of the Eighth Annual ACM Symposium on Theory of Computing, pp. 220–230. ACM (1976)
2. Chaudhuri, S., Dubhashi, D.: Probabilistic recurrence relations revisited. Theoret. Comput. Sci. **181**(1), 45–56 (1997)
3. Corral, A., Manolopoulos, Y., Theodoridis, Y., Vassilakopoulos, M.: Closest pair queries in spatial databases. In: ACM SIGMOD Record, vol. 29, pp. 189–200. ACM (2000)
4. Dasgupta, S., Gupta, A.: An elementary proof of a theorem of Johnson and Lindenstrauss. Random Struct. Algorithms **22**(1), 60–65 (2003)
5. Datar, M., Immorlica, N., Indyk, P., Mirrokni, V.S.: Locality-sensitive hashing scheme based on p-stable distributions. In: Proceedings of the Twentieth Annual Symposium on Computational Geometry, pp. 253–262. ACM (2004)
6. Dietzfelbinger, M., Hagerup, T., Katajainen, J., Penttonen, M.: A reliable randomized algorithm for the closest-pair problem. J. Algorithms **25**(1), 19–51 (1997)
7. Rabin, M.: Probabilistic algorithms (1976)

8. Fortune, S., Hopcroft, J.: A note on Rabin's nearest-neighbor algorithm. Inf. Process. Lett. **8**(1), 20–23 (1979)
9. Ge, Q., Wang, H.-T., Zhu, H.: An improved algorithm for finding the closest pair of points. J. Comput. Sci. Technol. **21**(1), 27–31 (2006)
10. Indyk, P., Lewenstein, M., Lipsky, O., Porat, E.: Closest pair problems in very high dimensions. In: Díaz, J., Karhumäki, J., Lepistö, A., Sannella, D. (eds.) ICALP 2004. LNCS, vol. 3142, pp. 782–792. Springer, Heidelberg (2004). https://doi.org/10.1007/978-3-540-27836-8_66
11. Jiang, M., Gillespie, J.: Engineering the divide-and-conquer closest pair algorithm. J. Comput. Sci. Technol. **22**(4), 532–540 (2007)
12. Johnson, W.B., Lindenstrauss, J.: Extensions of Lipschitz mappings into a Hilbert space. Contemp. Math. **26**(189–206), 1 (1984)
13. Karp, R.M.: Probabilistic recurrence relations. J. ACM (JACM) **41**(6), 1136–1150 (1994)
14. Khuller, S., Matias, Y.: A simple randomized Sieve algorithm for the closest-pair problem. Inf. Comput. **118**(1), 34–37 (1995)
15. Lopez, M.A., Liao, S.: Finding k-closest-pairs efficiently for high dimensional data (2000)
16. Mueen, A., Keogh, E., Zhu, Q., Cash, S., Westover, B.: Exact discovery of time series motifs. In: Proceedings of the 2009 SIAM International Conference on Data Mining, pp. 473–484. SIAM (2009)
17. Pereira, J.C., Lobo, F.G.: An optimized divide-and-conquer algorithm for the closest-pair problem in the planar case. J. Comput. Sci. Technol. **27**(4), 891–896 (2012)
18. Preparata, F.P., Shamos, M.: Computational Geometry: An Introduction. Springer, Heidelberg (2012). https://doi.org/10.1007/978-1-4612-1098-6
19. Shamos, M.I., Hoey, D.: Closest-point problems. In: 16th Annual Symposium on Foundations of Computer Science, pp. 151–162. IEEE (1975)
20. Tao, Y., Yi, K., Sheng, C., Kalnis, P.: Efficient and accurate nearest neighbor and closest pair search in high-dimensional space. ACM Trans. Database Syst. (TODS) **35**(3), 20 (2010)
21. Yao, A.C.-C.: Lower bounds for algebraic computation trees of functions with finite domains. SIAM J. Comput. **20**(4), 655–668 (1991)

Efficient Compression Technique
for Sparse Sets

Rameshwar Pratap[1(✉)], Ishan Sohony[2], and Raghav Kulkarni[3]

[1] Wipro Technologies, Bangalore, India
rameshwar.pratap@gmail.com
[2] PICT, Pune, India
ishangalbatorix@gmail.com
[3] Chennai Mathematical Institute, Chennai, India
kulraghav@gmail.com

Abstract. Recent growth in internet has generated large amount of data over web. Representations of most of such data are high-dimensional and sparse. Many fundamental subroutines of various data analytics tasks such as clustering, ranking, nearest neighbour scales poorly with the data dimension. In spite of significant growth in the computational power performing such computations on high dimensional data sets are infeasible, and at times impossible. Thus, it is desirable to investigate on compression algorithms that can significantly reduce dimension while preserving similarity between data objects. In this work, we consider the data points as sets, and use Jaccard similarity as the similarity measure. Pratap and Kulkarni [10] suggested a compression technique for high dimensional, sparse, binary data for preserving the Inner product and Hamming distance. In this work, we show that their algorithm also works well for Jaccard similarity. We present a theoretical analysis of compression bound and complement it with rigorous experimentation on synthetic and real-world datasets. We also compare our results with the state-of-the-art "min-wise independent permutation [6]", and show that our compression algorithm achieves almost equal accuracy while significantly reducing the compression time and the randomness.

1 Introduction

We are at the dawn of a new age. An age in which the availability of raw computational power and massive data sets gives machines the ability to learn, leading to the first practical applications of Artificial Intelligence. The human race has generated more amount of data in the last 2 years than in the last couple of decades, and it seems like just the beginning. As we can see, practically everything we use on a daily basis generates enormous amounts of data and in order to build smarter, more personalised products, it is required to analyse these datasets and draw logical conclusions from it. Therefore, performing computations on big data is inevitable, and efficient algorithms that are able to deal with large amounts of data, are the need of the day. We would like to emphasize

© Springer International Publishing AG, part of Springer Nature 2018
D. Phung et al. (Eds.): PAKDD 2018, LNAI 10939, pp. 164–176, 2018.
https://doi.org/10.1007/978-3-319-93040-4_14

that many of these datasets are high dimensional and sparse – the number of possible attributes in the dataset are large, however, only a small number of them are present in most of the data points. Sparsity is also quite common in web documents, text, audio, video data. Therefore, it is desirable to investigate the compression techniques that can compress the dimension of the data while preserving the similarity between data objects.

In this work, we focus on sparse, binary data, which can also be considered as sets, and the underlying similarity measure as Jaccard similarity. Given two sets A and B the Jaccard similarity between them is denoted as $JS(A, B)$ and is defined as $JS(A, B) = |A \cap B|/|A \cup B|$. Jaccard similarity is popularly used to determine whether two documents are similar. Broder [3] showed that this problem can be reduced to set intersection problem via *shingling*[1]. For example: two documents A and B first get converted into two shingles S_A and S_B, then similarity between these two documents is defined as $JS(A, B) = |S_A \cap S_B|/|S_A \cup S_B|$. Experiments validate that high Jaccard similarity implies that two documents are similar. Broder *et al.* [5,6] suggested a technique to compress a collection of sets while preserving the Jaccard similarity between every pair of sets. For a set U of binary vectors $\{\mathbf{u_i}\}_{i=1}^n \subseteq \{0,1\}^d$, their technique includes taking a random permutation of $\{1, 2, \ldots, d\}$ and assigning a value to each set which maps to minimum under that permutation. Throughout this paper, we represent sets as binary vectors.

Theorem 1 (Minhash [5,6]). *Let π be a permutations over $\{1, \ldots, d\}$, then for a set $\mathbf{u} \subseteq \{1, \ldots d\}$ $h_\pi(\mathbf{u}) = \arg\min_i \pi(i)$ for $i \in \mathbf{u}$. Then,*

$$\Pr[h_\pi(\mathbf{u}) = h_\pi(\mathbf{v})] = \frac{|\mathbf{u} \cap \mathbf{v}|}{|\mathbf{u} \cup \mathbf{v}|}.$$

1.1 Revisiting Compression Scheme of [10]

Pratap and Kulkarni [10] suggested a compression scheme for binary data that compress the data while preserving both hamming distance and inner product. A major advantage of their scheme is that the compression-length depends only on the sparsity of the data and is independent of the dimension of data. We briefly revisit their compression scheme. Consider a set of n binary vectors in d-dimensional space, then, given a binary vector $\mathbf{u} \in \{0,1\}^d$, their scheme compresses it into a N-dimensional binary vector (say) $\mathbf{u}' \in \{0,1\}^N$ as follows, where N to be specified later. It randomly assigns each bit position (say) $\{i\}_{i=1}^d$ of the original data to an integer $\{j\}_{j=1}^N$. Further, to compute the j-th bit of the compressed vector $\mathbf{u}'$ we check which bits positions have been mapped to j, and compute the parity of bits located at those positions, and assign it to the j-th bit position. Figure 1 illustrates this with an example. In continuation of their analogy we call it as BCS.

---

[1] A document is a string of characters. A k-shingle for a document is defined as a contiguous substring of length k found within the document. For example: if our document is $abcd$, then shingles of size 2 are $\{ab, bc, cd\}$.

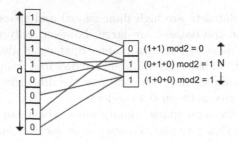

Fig. 1. Binary compression scheme (BCS) of [10]

1.2 Our Result

Using the above mentioned compression scheme, we are able to prove the following compression guarantee for the Jaccard similarity.

Theorem 2. *Consider a set* U *of binary vectors* $\{u_i\}_{i=1}^n \subseteq \{0,1\}^d$ *with maximum number of* 1 *in any vector is at most* ψ, *and* $\epsilon > 0$. *If we set* $N = O(\psi^2 \log^2 n)$, *and compress them into a set* U' *of binary vectors* $\{u_i'\}_{i=1}^n \subseteq \{0,1\}^N$ *using* BCS. *Then for all* $u_i, u_j \in U$, *the following is true with probability at least* $1 - 2/n$,

$$(1 - \epsilon)\mathrm{JS}(u_i, u_j) \leq \mathrm{JS}(u_i', u_j') \leq (1 + \epsilon)\mathrm{JS}(u_i, u_j).$$

Remark 1. A major benefit (as also mentioned in [10]) of BCS is that it also works well in the streaming setting. The only prerequisites are an upper bound on the sparsity ψ, and the number of data points.

Parameters for Evaluating a Compression Scheme

The quality of a compression algorithm can be evaluated on the following parameters: (1) *Randomness* is the number of random bits required for compression, (2) *Compression time* is the time required for compression, (3) *Compression length* is the dimension of data after compression, (4) The amount of *space* required to store the compressed matrix. Ideally, the compression length and the compression time should be as small as possible while maintaining a desired accuracy.

1.3 Comparison Between BCS and Minhash and Its Variants

We evaluate the quality of our compression scheme with minhash on the parameters stated earlier.

Randomness: One of the major advantages of BCS is the reduction in the number of random bits required for compression. We quantify it below.

Lemma 1. *Let a set of n d dimensional binary vectors, which get compressed into a set of n vectors in N dimension via minhash and BCS, respectively. Then, the amount of random bits required for BCS and minhash are $O(d \log N)$ and $O(Nd \log d)$, respectively.*

Compression Time: BCS is significantly faster than minhash algorithm in terms of compression time. This is because, generation of random bits requires a considerable amount of time. Thus, reduction in compression time is proportional to the reduction in the amount of randomness required for compression. Also, for compression length N, minhash scans the vector N times - once for each permutation, while BCS just requires a single scan.

Space Required for Compressed Data: Minhash compression generates an integer matrix as opposed to the binary matrix generated by BCS. Therefore, the space required to store the compressed data of BCS is significantly less as compared to minhash.

Search Time: Binary form of our compressed data leads to a significantly faster search as efficient bitwise operations can be used.

In Sect. 3, we numerically quantify the advantages of BCS on the later three parameters *via* experimentations on synthetic and real-world datasets.

Li *et al.* [7] presented "b-bit minhash" an improvement over Broder's minhash by reducing the compression size. They store only a vector of last b-bits of the corresponding binary representation of hash values. However, this approach reduces the accuracy. If we compare BCS with b-bit minhash, then we have same the advantage as of minhash in savings of randomness and compression time. We also get benefited in the search time, as our sketch is more succinct than [7].

Mitzenmacher [9] suggest a compression technique – Oddsketch – that gives a compression for Jaccard similarity. Oddsketch is similar to BCS in the sense that mapping (random bucketing and parity) is the same in both of these approaches, and after compression both gives binary sketch. However, a major difference between them is that Oddsketch is build on the top of minhash signatures of the input sets, while BCS is directly applied on the input data. For BCS sparsity assumption over data is required, while there is no such assumption is needed for Oddsketch. Our analysis technique is simple and gives a bound which is quadratic in the sparsity, and is based on simple *birthday-paradox* type analysis, while Oddsketch analysis is tight, and is based on concentration inequalities.

1.4 Applications of Our Result

In cases of high dimensional, sparse data, BCS can be used to improve numerous applications where currently minhash is used. We discuss a few of them below.

Faster Ranking/De-duplication of Documents: Given a corpus of documents and a set of query documents, ranking documents from the corpus based on similarity with the query documents is an important problem in information-retrieval. This also helps in identifying duplicates, as documents that are ranked high with respect to the query documents, share high similarity. Broder [4] suggested an efficient de-duplication technique for documents – by converting documents to *shingles*; defining the similarity of two documents based on their Jaccard similarity; and then using minhash sketch to efficiently detect near-duplicates. As most the datasets are sparse, BCS can be more effective than minhash on the parameters stated earlier.

Scalable Clustering of Documents: Clustering is one of the fundamental information-retrieval problems. Broder *et al.* [2] suggested an approach to cluster data objects that are similar. The approach is to partition the data into shingles; defining the similarity of two documents based on their Jaccard similarity; and then via minhash generate a sketch of each data object. These sketches preserve the similarity of data objects. Thus, grouping these sketches gives a clustering on the original documents. However, when documents are high dimensional such as webpages, minhash sketching approach might not be efficient. Again exploiting the sparsity, BCS can be more effective.

Beyond above applications, minhash compression has been widely used in applications like spam detection [3], all pair similarity [1]. As in most of these cases, data objects are sparse, BCS can provide almost accurate and more efficient solutions to these problems. We experimentally validate the performance of BCS for ranking experiments on UCI [8] "BoW" dataset, and achieved significant improvements over minhash. We discuss this in Subsection 3.2. Similarly, other mentioned applications can also be validated.

Organization of the Paper: Below, we first present some necessary notations that are used in the paper. In Sect. 2, we first revisit the results of [10], then building on it we give a proof on the compression bound for Jaccard similarity. In Sect. 3, we complement our theoretical results *via* extensive experimentation on synthetic as well as real-world datasets. Finally, in Sect. 4 we conclude our discussion and state some open questions.

Notations	
N	Dimension of the compressed data
ψ	Upper bound on the number of 1's in binary data
$\mathbf{u}[i]$	i-th bit position of vector $\mathbf{u}$
$\mathrm{JS}(\mathbf{u}, \mathbf{v})$	Jaccard similarity between binary vectors $\mathbf{u}$ and $\mathbf{v}$
$d_\mathrm{H}(\mathbf{u}, \mathbf{v})$	Hamming distance between binary vectors $\mathbf{u}$ and $\mathbf{v}$
$\langle \mathbf{u}, \mathbf{v} \rangle$	Inner product between binary vectors $\mathbf{u}$ and $\mathbf{v}$

2   Analysis

We first revisit the results of [10] which discuss compression bounds for hamming distance and inner product, and then building on it, we give a compression bound for Jaccard similarity. We start with discussing the intuition and a proof sketch of their result. Consider two binary vectors $\mathbf{u}, \mathbf{v} \in \{0,1\}^d$, we call a bit position *"active"* if at least one of the vector between $\mathbf{u}$ and $\mathbf{v}$ has value 1 in that position. Further, given the sparsity bound ψ, there can be at most 2ψ active positions between $\mathbf{u}$ and $\mathbf{v}$. Then let via BCS, they compressed into binary vectors $\mathbf{u}', \mathbf{v}' \in \{0,1\}^N$. In the compressed version, we call a bit position *"pure"* if the number of active positions mapped to it is at most one, and *"corrupted"* otherwise. The contribution of pure bit positions in $\mathbf{u}', \mathbf{v}'$ towards hamming distance (or inner product similarity), is exactly equal to the contribution of the bit positions in $\mathbf{u}, \mathbf{v}$ which get mapped to the pure bit positions. Further, the deviation of hamming distance (or inner product similarity) between $\mathbf{u}'$ and $\mathbf{v}'$ from that of $\mathbf{u}$ and $\mathbf{v}$, corresponds to the number of corrupted bit positions shared between $\mathbf{u}'$ and $\mathbf{v}'$. Figure 2 illustrate this with an example, and the lemma below analyse it.

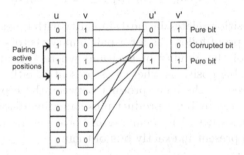

Fig. 2. Illustration of pure/corrupted bits in BCS.

Lemma 1 (Lemma 14 **of** [10]**).** *Consider two binary vectors* $\mathbf{u}, \mathbf{v} \in \{0,1\}^d$, *which get compressed into vectors* $\mathbf{u}', \mathbf{v}' \in \{0,1\}^N$ *using the BCS, and suppose* ψ *is the maximum number of 1 in any vector. Then for an integer* $r \geq 1$, *and* $\epsilon \in (0,1)$, *probability that* $\mathbf{u}'$ *and* $\mathbf{v}'$ *share more than* ϵr *corrupted positions is at most* $\left(2\psi/\sqrt{N}\right)^{\epsilon r}$.

The lemma below generalise the above result for a set of n binary vectors, and suggest a compression bound so that any pair of compressed vectors share only a very small number of corrupted bits, with high probability.

Lemma 2 (Lemma 15 **of** [10]**).** *Consider a set* U *of* n *binary vectors* $\{\mathbf{u_i}\}_{i=1}^n \subseteq \{0,1\}^d$, *which get compressed into a set* U' *of binary vectors* $\{\mathbf{u_i'}\}_{i=1}^n \subseteq \{0,1\}^N$ *using the BCS. Then for any positive integer* r, *and* $\epsilon \in (0,1)$,

- *if* $\epsilon r > 3 \log n$, *and we set* $\mathrm{N} = O(\psi^2)$, *then probability that for all* $\mathbf{u_i'}, \mathbf{u_j'} \in \mathrm{U}'$ *share more than* ϵr *corrupted positions is at most* $1/n$.

– If $\epsilon r < 3\log n$, and we set $N = O(\psi^2 \log^2 n)$, then probability that for all $\mathbf{u}_i', \mathbf{u}_j' \in U'$ share more than ϵr corrupted positions is at most $1/n$.

After compressing binary data *via* BCS, the hamming distance between any pair of binary vectors can not increase. This is due to the fact that compression doesn't generate any new 1 bit, which could increase the hamming distance from the uncompressed version. In the following, we recall the main result of [10], which holds due the above fact and Lemma 2.

Theorem 3 (Theorem 1, 2 of [10]). *Consider a set* U *of binary vectors* $\{\mathbf{u}_i\}_{i=1}^n \subseteq \{0,1\}^d$, *a positive integer* r, *and* $\epsilon \in (0,1)$. *If we set* $N = O(\psi^2 \log^2 n)$, *and compress them into a set* U' *of binary vectors* $\{\mathbf{u}_i'\}_{i=1}^n \subseteq \{0,1\}^N$ *using BCS. Then for all* $\mathbf{u}_i, \mathbf{u}_j \in U$,

– *if* $d_H(\mathbf{u}_i, \mathbf{u}_j) \leq r$, *then* $\Pr[d_H(\mathbf{u}_i', \mathbf{u}_j') \leq r] = 1$,
– *if* $d_H(\mathbf{u}_i, \mathbf{u}_j) \geq (1+\epsilon)r$, *then* $\Pr[d_H(\mathbf{u}_i', \mathbf{u}_j') \leq r] < \frac{1}{n}$.

For inner product, the following is true with probability at least $1 - 1/n$,

$$(1 - \epsilon)\langle \mathbf{u}_i, \mathbf{u}_j \rangle \leq \langle \mathbf{u}_i', \mathbf{u}_j' \rangle \leq (1 + \epsilon)\langle \mathbf{u}_i, \mathbf{u}_j \rangle.$$

The following proposition relates Jaccard similarity with inner product and hamming distance. The proof follows as for a pair binary vectors their Jaccard similarity is the ratio of the number of positions where 1 is appearing together, with the number of bit positions where 1 is present in either of them. Clearly, numerator is captured by the inner product between those pair of vectors, and denominator is captured by inner product plus hamming distance between them – number of positions where 1 is occurring in both vectors, plus the number of positions where 1 is present in exactly one of them.

Proposition 4. *For any pair of vectors* $\mathbf{u}, \mathbf{v} \subseteq \{0,1\}^d$, *we have* $JS(\mathbf{u}, \mathbf{v}) = \langle \mathbf{u}, \mathbf{v} \rangle / (\langle \mathbf{u}, \mathbf{v} \rangle + d_H(\mathbf{u}, \mathbf{v}))$

A proof of Theorem 2 follows by combining the results of Proposition 4, and Theorem 3, and applying probability union bound analysis.

3   Experimental Evaluation

We performed our experiments on a machine having the following configuration: CPU: Intel(R) Core(TM) i5 CPU @ 3.2 GHz x 4; Memory: 8 GB 1867 MHz DDR3; OS: macOS Sierra 10.12.5; OS type: 64-bit. We performed our experiments on synthetic and real-world datasets, we discuss them one-by-one as follows:

3.1   Results on Synthetic Data

We performed two experiments on synthetic dataset and showed that it preserve all-pair-similarity, that is, given a set of n binary vectors in d-dimensional space with the sparsity bound ψ, we showed that after compression Jaccard similarity between every pair of vector is preserved. We performed experiments on dataset consisted of 1000 vectors in 100000 dimension. Throughout synthetic data experiments, we calculate the accuracy *via* Jaccard ratio, that is, if the set $\mathcal{O}$ denotes the ground truth result, and the set $\mathcal{O}'$ denotes our result, then the accuracy of our result is calculated by the Jaccard ratio between the sets $\mathcal{O}$ and $\mathcal{O}'$ – that is $\mathrm{JS}(\mathcal{O}, \mathcal{O}') = |\mathcal{O} \cap \mathcal{O}'|/|\mathcal{O} \cup \mathcal{O}'|$. To reduce the effect of randomness, we repeat the experiment 10 times and took the average.

Dataset Description: We generated 1000 binary vectors in dimension 100000 such that the sparsity of each vector is at most ψ. If we randomly choose binary vectors respecting the sparsity bound, then most likely every pair of vector will have similarity 0. Thus, we had to deliberately generate some vectors having high similarity. We generated 200 pairs whose similarity is high. To generate such a pair, we choose a random number (say s) between 1 and ψ, then we randomly select those many position (in dimension) from 1 to 100000, set 1 in both of them, and set remaining to 0. Further, for each of the vector in the pair, we choose a random number (say s') from the range 1 to $(\psi - s)$, and again randomly sample those many positions from the remaining positions and set them to 1. This gives a pair of vectors having similarity at least $\frac{s}{s+2s'}$ and respecting the sparsity bound. We repeat this step 200 times and obtain 400 vectors. For each of the remaining 600 vectors, we randomly choose an integer from the range 1 to ψ, choose those many positions in the dimension, set them to 1, and set the remaining positions to 0. Thus, we obtained 1000 vectors of dimension 100000, which we used as an input matrix.

Data Representation: We can imagine synthetic dataset as a binary matrix of dimension 100000 × 1000. However, for ease and efficiency of implementation, we use a compact representation which consist of a list of lists. The number of lists is equal to the number of vectors in the binary matrix, and within each list we just store the indices (co-ordinate) where 1s are present. We use this list as an input for both BCS and minhash.

Evaluation Metric: We performed two experiments on synthetic dataset – (1) fixed sparsity while varying compression length, and (2) fixed compression length while varying sparsity. We present these experimental results in Fig. 3. In both of these experiments, we compare and contrast the performance BCS with minhash on *accuracy, compression time,* and *search time* parameters. All-pair-similarity experiment result requires a quadratic search – generation of all possible candidate pairs and then pruning those whose similarity score is high, and the corresponding search time is the time required to compute all such pairs.

In order to calculate the accuracy on a given support threshold value, we first run a simple brute-force search algorithm on the entire (uncompressed) dataset, and obtain the ground truth result. Then, we calculate the Jaccard ratio between our algorithm's result/ minhash's result, with the corresponding exact result, and compute the accuracy.

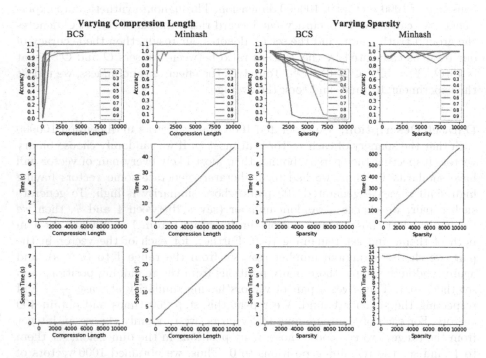

Fig. 3. Experiments on synthetic data: (1) fixed sparsity $\psi = 200$ and varying compression length, and (2) varying sparsity and fixed compression lenght 5000.

Insight: In Fig. 3, we plot the result of BCS and minhash for all-pair-similarity. For this experiment, we fix the sparsity $\psi = 200$ and generate the datasets as stated above. We compress the datasets using BCS and minhash for a range of compression lengths from 50 to 10000. It can be observed that BCS performs remarkably well on the parameters of compression time and search time. Our compression time remains almost constant at 0.2 s in contrast to the compression time of minhash, which grows linearly to almost 50 s. On an average, BCS is 90 times faster than minhash. Also accuracy for BCS and minhash is almost equal above compression length 300, but in the window of 50–300 minhash performs slightly better than BCS. Further, the search-time on BCS is also significantly less than minhash for all compression lengths. On an average search-time is 75 times less than the corresponding minhash search-time.

In Fig. 3, we plot the result of BCS and minhash for all-pair-similarity. For this experiment, we generate datasets for different values of sparsity ranging from 50 to 10000. We compress these datasets using BCS and minhash to a fixed value of compression length 5000. In all-pair-similarity, when sparsity value is below 2200, average accuracy of BCS is above 0.85. It starts decreasing after that value, at sparsity value is 7500, the accuracy of BCS stays above 0.7, on most of the threshold values. The compression time of BCS is always below 2 s while compression time of minhash grows linearly with sparsity – on an average compression time of BCS is around 550 times faster than the corresponding minhash compression time. Further, we again significantly reduce search time – on an average our search-time is 91 times less than minhash.

3.2   Results on Real-World Data

Dataset Description: We compare the performance of BCS with minhash on the task of retrieving top-ranked elements based on Jaccard similarity. We performed this experiment on publicly available high dimensional sparse dataset of UCI machine learning repository [8] (described in Table 1). These datasets are binary "BoW" representation of the corresponding text corpus. We consider each of these datasets as a binary matrix, where each document corresponds to a binary vector, that is if a particular word is present in the document, then the corresponding entry is 1 in that position, and it is 0 otherwise. For ENRON and NYTimes we take a uniform sample of 10000 documents from their corpus.

Table 1. Real-world dataset description

Data Set	No. of points	Dimension	Sparsity
NYTimes news articles	10000	102660	871
Enron Emails	10000	28102	2021
NIPS full papers	1500	12419	914
KOS blog entries	3430	6906	457

Evaluation Metric: We split the dataset in two parts 90% and 10% – the bigger partition is use to compress the data, and is referred as the *training partition*, while the second one is use to evaluate the quality of compression and is referred as *querying partition*. We call each vector of the querying partition as query vector. For each query vector, we compute the vectors in the training partition whose Jaccard similarity is higher than a certain threshold (ranging from 0.1 to 0.9). We first do this on the uncompressed data inorder to find the underlying ground truth result – for every query vector compute all vectors that are similar to them. Then we compress the entire data, on various values of compression lengths, using our compression scheme/minhash. For each query vector, we calculate the accuracy of BCS/minhash by taking Jaccard ratio between the

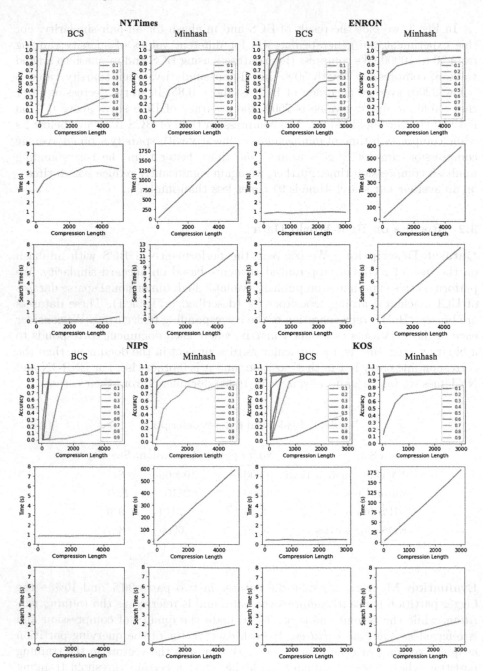

Fig. 4. Experiments on real-world datasets [8].

set outputted by BCS/minhash, on various values of compression length, with set outputted a simple linear search algorithm on entire data. This gives us the accuracy of compression of that particular query vector. We repeat this for every vector in the querying partition, take the average, and we plot the average accuracy for each value in support threshold and compression length. We also note down the corresponding compression time on each of the compression length for both BCS and minhash. Search time is time required to do a linear search on the compressed data, we compute the search time for each of the query vector and take the average in the case of both BCS and minhash.

Insights: We plot experiments of real world dataset [8] in Fig. 4, and found that performance of BCS is similar to its performance on synthetic datasets. NYTimes is the sparsest among all other dataset, so the performance of BCS is relatively better as compare to other datasets. For NYTIMES dataset, on an average BCS is 135 times faster than minhash, and search time for BCS is 25 times less than search time for minhash. For BCS accuracy starts dropping below 0.9 when data is compressed below compression length 300. For minhash, accuracy starts dropping below compression compression length 150. Similar pattern is observed for ENRON dataset as well, where BCS is 268 times faster than minhash, and a search on the compressed data obtained from BCS is 104 times faster than search on data obtained from minhash. KOS and NIPS are dense, low dimensional datasets. However here also, for NIPS, our compression time is 271 times faster and search-time is 90 times faster as compared to minhash. For KOS, our compression time is 162 times faster and search time is 63 times faster than minhash.

To summarise, BCS is significantly faster than minhash in terms of both - compression time and search time while giving almost equal accuracy. Also, the amount of randomness required for BCS is also significantly less as compared to minhash. However, as sparsity is increased, accuracy of BCS starts decreasing slightly as compared to minhash.

4 Concluding Remarks and Open Questions

We showed that BCS is able to compress sparse, high-dimensional binary data while preserving the Jaccard similarity. It is considerably faster than the "state-of-the-art" minhash permutation, and also maintains almost equal accuracy while significantly reducing the amount of randomness required. Moreover, the compressed representation obtained from BCS is in binary form, as opposed to integer in case of minhash, due to which the space required to store the compressed data is reduced, and consequently leads to a faster search on the compressed representation. Another major advantage of BCS is that its compression bound is independent of the dimensions of the data, and only grows polynomially with the sparsity and poly-logarithmically with the number of data points. We present a theoretical proof of the same and complement it with rigorous and extensive experimentations. Our work leaves the possibility of several open questions – improving the compression bound of our result, and extending it to other similarity measures.

References

1. Bayardo, R.J., Ma, Y., Srikant, R.: Scaling up all pairs similarity search. In: Proceedings of the 16th International Conference on World Wide Web, WWW 2007, Banff, Alberta, Canada, 8–12 May 2007, pp. 131–140 (2007)
2. Broder, A., Glassman, S., Nelson, C., Manasse, M., Zweig, G.: Method for clustering closely resembling data objects. US Patent 6,119,124, 12 September 2000
3. Broder, A.Z.: On the resemblance and containment of documents. In: Proceedings of the Compression and Complexity of SEQUENCES 1997 (Cat. No.97TB100171), pp. 21–29, June 1997
4. Broder, A.Z.: Identifying and filtering near-duplicate documents. In: Giancarlo, R., Sankoff, D. (eds.) CPM 2000. LNCS, vol. 1848, pp. 1–10. Springer, Heidelberg (2000). https://doi.org/10.1007/3-540-45123-4_1
5. Broder, A.Z.: Min-wise independent permutations: theory and practice. In: Montanari, U., Rolim, J.D.P., Welzl, E. (eds.) ICALP 2000. LNCS, vol. 1853, pp. 808–808. Springer, Heidelberg (2000). https://doi.org/10.1007/3-540-45022-X_67
6. Broder, A.Z., Charikar, M., Frieze, A.M., Mitzenmacher, M.: Min-wise independent permutations (extended abstract). In: Proceedings of the Thirtieth Annual ACM Symposium on the Theory of Computing, Dallas, Texas, USA, 23–26 May 1998, pp. 327–336 (1998)
7. Li, P., Mahoney, M.W., She, Y.: Approximating higher-order distances using random projections. In: UAI 2010, Proceedings of the Twenty-Sixth Conference on Uncertainty in Artificial Intelligence, Catalina Island, CA, USA, 8–11 July 2010, pp. 312–321 (2010)
8. Lichman, M.: UCI machine learning repository (2013)
9. Mitzenmacher, M., Pagh, R., Pham, N.: Efficient estimation for high similarities using odd sketches. In: 23rd International World Wide Web Conference, WWW 2014, Seoul, Republic of Korea, 7–11 April 2014, pp. 109–118 (2014)
10. Pratap, R., Kulkarni, R.: Similarity preserving compressions of high dimensional sparse data. CoRR, abs/1612.06057 (2016)

It Pays to Be Certain: Unsupervised Record Linkage via Ambiguity Minimization

Anna Jurek[✉] and Deepak P.

Queen's University Belfast, Belfast, UK
a.jurek@qub.ac.uk, deepaksp@acm.org

Abstract. Record linkage (RL) is a process of identifying records that refer to the same real-world entity. Many existing approaches to RL apply supervised machine learning (ML) techniques to generate a classification model that classifies a pair of records as either linked or non-linked. In such techniques, the labeled data helps guide the choice and relative importance to similarity measures to be employed in RL. Unsupervised RL is therefore a more challenging problem since the quality of similarity measures needs to be estimated in the absence of linkage labels. In this paper we propose a novel optimization approach to unsupervised RL. We define a scoring technique which aggregates similarities between two records along all attributes and all available similarity measures using a weighted sum formulation. The core idea behind our method is embodied in an objective function representing the overall ambiguity of the scoring across a dataset. Our goal is to iteratively optimize the objective function to progressively refine estimates of the scoring weights in the direction of lesser overall ambiguity. We have evaluated our approach on multiple real world datasets which are commonly used in the RL community. Our experimental results show that our proposed approach outperforms state-of-the-art techniques, while being orders of magnitude faster.

1 Introduction

RL, also referred to as data matching or entity resolution, is the task of finding records that correspond to the same entity from one or more data sources. Given two data sources, each pair of records can be classified into one of two classes: linked and non-linked. Table 1 shows a simple example of RL. The table contains records from two bibliographic data sources, viz., DBLP and ACM digital library. The aim is to identify those pairs of records referring to the same publications, which in this case are (ACM1, DB1) and (ACM2, DB2). Any other pairs of records should be identified as non-linked. If records have error-free and unique identifiers, such as social security numbers, RL is a straightforward process that can be easily performed by the standard database join operation. In many practical scenarios, however, such a unique identifier does not exist and the linkage process needs to be performed by approximate matching of the corresponding fields of two records. It is also notable that the same data can be

© Springer International Publishing AG, part of Springer Nature 2018
D. Phung et al. (Eds.): PAKDD 2018, LNAI 10939, pp. 177–190, 2018.
https://doi.org/10.1007/978-3-319-93040-4_15

Table 1. An example of RL.

ID	Title	Authors	Venue
ACM1	A compact B-tree	Peter Bumbulis, Ivan T. Bowman	International conference on management of data
ACM2	A theory of redo recovery	David Lomet, Mark Tuttle	International conference on management of data
DB1	A compact B-tree	Peter Bumbulis, Ivan Bowman	SIGMOD conference
DB2	A theory of redo recovery	Mark R. Tuttle, David B. Lomet	SIGMOD conference
DB3	The nimble integration engine	Denise Draper Alon Y. Halevy Daniel S. Weld	SIGMOD conference

represented in different ways in different data sources due to factors such as different conventions, typographical errors, missing and out of date values. This makes similarity matching and aggregation of similarity scores to perform record linkage, a challenging task.

Efficiency and Effectiveness in RL: The problem of RL can be seen as comprising two main fields of research which are: (1) developing time-efficient algorithms for RL [21] and (2) efforts on developing techniques for effective link discovery [6,7,23]. The former focuses on improving the turnaround time for record linkage through heuristically avoiding comparison between records that hold a low apriori chance of getting linked. The space of candidate record pairs for record linkage is evidently quadratic in the size of the datasets. This quadratic space is often pruned out through indexing and filtering, collectively referred to as blocking methodologies. The latter field of research, that towards effective RL, focuses on the orthogonal problem of accurately determining which among compared candidate pairs are to be labelled as linked/non-linked. In this work, we will focus on the second research problem, which is the development of models for accurately determining the linkage status of record pairs.

Effective Record Linkage: Techniques for effective RL may be seen as comprising two major streams, based on whether labelled data is exploited for the task. The large majority of techniques for effective RL rely on the usage of a training dataset comprising record pairs that are labelled as linked/non-linked to learn a classifier, thus treating it as a supervised learning problem. In these techniques, each pair of records is represented as a similarity vector representing a set of numeric similarities, each calculated with a similarity measure on a pair of field values of the two records. The task of RL is then considered as a binary classification problem over similarity vectors [6]. The second category address RL in the absence of training data. Traditionally, this task has been addressed by replacing training data with assistance from a domain expert, who would handcraft bespoke domain-specific rules that aid determining the linkage likelihood of a candidate record pair [7,23]. Drawing up rules for record linkage requires deep topical expertise in the domain, an impractical or costly proposition in many scenarios. This makes unsupervised machine learning for RL, the task that targets to tackle RL without the aid of training data or a domain expert, a promising avenue of research in RL. It is notable that unsupervised machine learning for RL is much more challenging than the supervised or expert-assisted variants; this explains the relative dearth of research.

Our Contribution: In this paper, we address the problem of unsupervised RL and propose a novel method for unsupervised scoring of record pairs modeling the task as an optimization problem. *Our core idea is that a good record linkage method would be able to make conclusive decisions on the linkage of most pairs of records, if not all; we look for methods that can achieve conclusive decisions, while staying strictly within the space of RL models that make linkage decisions on a weighted sum aggregate of similarity scores.* Accordingly, we outline a model for the ambiguity of record linkage, and progressively refine the weightings associated with similarity measures in the direction of reducing overall ambiguity. Through an empirical analysis over multiple real-world datasets, we illustrate that our method is able to outperform existing methods.

2   Related Work

We briefly survey recent RL methods under two separate heads.

Semi-supervised Record Linkage. In semi-supervised learning, a small set of labeled instances and a large set of unlabeled instances are used in the training process. A popular approach to semi-supervised RL is that using active learning (AL) [1]. AL identifies highly informative instances for manual labeling that are later used for training classification models. In [18] the instances that are not assigned to the same class by majority of the classifiers are selected for manual labeling. A different approach, where a set of similarity vectors are ranked and those in the middle (ambiguous) region are selected for manual labeling, is proposed in [4]. In the work presented in [24], all the record pairs are clustered by their similarity vectors and randomly selected similarity vectors from each cluster are selected for manual labeling. Depending on the output of the manual labeling, similarity vectors in each cluster are automatically labeled as linked or non-linked, or the cluster is further divided into sub-clusters. The system reported in [12] takes as input a small set of training examples, referred to as seeds, to initially train the classification model, which is then used over unseen data.

Unsupervised Record Linkage. In [6], k-means clustering is used to predict the status of a small set of similarity vectors (seeds). Following this, the seeds are used as training set for a supervised learning algorithm. Automatic seed selection, referred to as nearest based, was applied with the self-training process in [2]. In [13] an entity matching algorithm is proposed, which allows to identify best k results for a user-specific scoring function. Unsupervised approaches to RL based on maximizing the value of pseudo f-measure were investigated in [15,16]. Pseudo f-measure, an unsupervised variant of the f-measure, is formulated using the assumption that while different records often represent the same entity in different repositories, distinct record within one dataset is expected to denote distinct entity. It can be calculated using sets of unlabeled records. The idea is to find the decision rule for record matching which maximizes the value of the pseudo f-measure applying genetic programming [16] or hierarchical search [15]. In more recent work the authors proposed to address the problem of unsupervised record linkage using graphical models [20] and multi view ensemble self-learning [10].

Discussion. While semi-supervised learning significantly reduces the number of manually labeled examples required for generating a classification model, it still requires a certain amount of human input in the training process. Methods that require labeled data for RL are not applicable in many real-world situations; in particular for privacy preserving RL, where the data is private and confidential [22]. While unsupervised methods such as [2] do not require any labeled data, there is much gap to close between them and supervised methods in terms of accuracy.

3   Problem Definition

Consider a dataset of relational records $\mathcal{R} = \{r_1, r_2, \ldots, r_n\}$ where each record comprises values it takes for attributes from a schema $\mathcal{A} = [a_1, a_2, \ldots, a_m]$. Accordingly, we can represent a record r_i as $[r_{i1}, r_{i2}, \ldots, r_{im}]$ where r_{ij} is the value that the ith record takes, for the jth attribute in the schema. For each attribute $a_j \in \mathcal{A}$, we use $\mathcal{S}_j$ to denote the set of similarity measures that are available for the attribute. There could be many similarity measures available for each attribute type [5]. Examples of common similarity measures include Jaccard, and inverses of edit-distance, or L_1 and L_2 distances. Thus, $\mathcal{S}_j : dom(a_j) \times dom(a_j) \rightarrow \mathbb{R}$, where $dom(a_j)$ denotes the domain of the attribute a_j. Here, we address the task of *unsupervised record linkage scoring*, that of leveraging $\mathcal{R}$ and $\mathcal{S}_j$s to learn a scoring method for pairs from $\mathcal{R}$, the score quantifying the likelihood that both records relate to the same entity. Notationally:

$$[\mathcal{R}, \{\mathcal{S}_1, \ldots, \mathcal{S}_m\}] \xrightarrow[\text{Learning}]{\text{Unsupervised}} RLS : \mathcal{R} \times \mathcal{R} \rightarrow \mathbb{R} \qquad (1)$$

Thus, $RLS(r_i, r_j)$ would be a numeric score directly related to the likelihood that r_i and r_j relate to the same entity. A few points are in order; first, unlike the bulk of literature in supervised record linkage [3] - i.e., pairs of records that are known to be linked or not-linked - we make use of no such labeled information, and thus address the unsupervised problem. Second, in the interest of retaining generality, we do not necessitate that the scoring by RLS needs to be in $[0, 1]$, or have a probabilistic or possibilistic semantics since that would require corresponding semantics on the similarity measures as well. In other words, for an accurate estimate of RLS, we only expect that pairs that are scored higher are more likely to be linked to the same entity than those are scored lower, i.e., that the relative ordering of pairs on RLS scores is meaningful. Third, *record linkage scoring* is a direct building block for the record linkage problem of classifying pairs of records as either linked or not linked. Applying an appropriate threshold to the RLS scores would yield an intuitive solution to the record linkage problem; one with the pairs that score above the threshold marked as linked, and others as not-linked.

3.1 Evaluating Record Linkage Scoring

As is the case with any unsupervised machine learning task, we would like to evaluate the quality of RLS against gold-standard labeled data. This is done by checking the relative ordering between record pairs which are known to be 'linked' and pairs that are known to be 'not linked'. With a threshold on RLS scores yielding a record linkage method, the precision, recall and f-measure on the linked and non-linked classes can be measured on varying values of the threshold. However, most record linkage datasets are very unbalanced [3] with a much larger fraction of unlinked records (recall that this was also the case even in the small example outlined in Table 1). This lopsided distribution makes it easier for RLS methods to achieve high precision and recall on the unlinked class. Thus, rank-aware measures that can incentivize RLS methods that put linked record pairs at the top of the ordering would help better evaluate the quality of RLS methods. Accordingly, we outline two simple evaluation measures below. Let $\mathcal{L}$ and $\mathcal{U}$ be the set of labeled data comprising linked record pairs and unlinked pairs respectively. Our evaluation measure is then:

$$ARL(RLS, \mathcal{L}, \mathcal{U}) = average\{Rank_{\mathcal{L}, \mathcal{U}}(RLS, l) | l \in \mathcal{L}\}$$

$$MRL(RLS, \mathcal{L}, \mathcal{U}) = median\{Rank_{\mathcal{L}, \mathcal{U}}(RLS, l) | l \in \mathcal{L}\}$$

where $Rank_{\mathcal{L}, \mathcal{U}}(RLS, l)$ denotes the rank of the record pair $l \in \mathcal{L}$ in the decreasing RLS-score ordering of record pairs in $(\mathcal{L} \cup \mathcal{U})$. Since our gold standard labellings may not be comprehensive, $(\mathcal{L} \cup \mathcal{U}) \subseteq \mathcal{R} \times \mathcal{R}$. Thus, $ARL(\ldots)$ and $MRL(\ldots)$ measures the mean and median ranks of the record pairs in $\mathcal{L}$ in the RLS score ordering. Since we would like to see the record pairs in $\mathcal{L}$ at the top of the ordering, numerically lower values of ARL and MRL are desirable. These metrics differ in their character in that ARL is affected by all changes in orderings of record-pairs, whereas MRL is less sensitive to outliers at the ends of the ordering.

4 Our Method

4.1 The Scoring Formulation

Consider a record pair $p_{xy} = (r_x, r_y)$ where $\{r_x, r_y\} \subseteq \mathcal{R}$. The similarity between the two records in p_{xy} can be measured along each of the m attributes in $\mathcal{A}$.

Further, for each attribute $a_j \in \mathcal{A}$, similarity between the records in p_{xy} can be measured using each of the similarity measures in $\mathcal{S}_j$. Thus, there are $\sum_{j=1}^{m} |\mathcal{S}_j|$ signals of similarity that are available for each record pair, across the similarity measures. Given a set of weights to associate with each of these similarity signals, collectively denoted as $\mathcal{W}$, our method aggregates these similarities using a linear aggregation (weighted sum) into a single value yielding the RLS scores:

$$RLS_{\mathcal{W}}(r_x, r_y) = \sum_{a_j \in \mathcal{A}} \sum_{S \in \mathcal{S}_j} w_{jS}^2 \times S(r_x.a_j, r_y.a_j) \qquad (2)$$

with $S(r_x.a_j, r_y.a_j)$ denoting the similarity measure between the values for attribute a_j on r_x and r_y, as measured using the similarity measure $S \in \mathcal{S}_j$. The square of the weights, $w_{jS} \in \mathcal{W}$, is used in the formulation for optimization convenience to automatically disallow negative weights. Thus, the crux of our method is in learning the set of weights, $\mathcal{W}$, so that the $RLS_{\mathcal{W}}$ scores estimate the linkage likelihood effectively.

4.2   Developing the Objective Function

Our interest is in ensuring that the set of weights, $\mathcal{W}$, leads to an $RLS_{\mathcal{W}}$ scoring that is similar to an "ideal" RLS scoring. Within our unsupervised setting, the notion of ideal-ness needs to be outlined without the luxury of knowing information such as the balance of the $\mathcal{L}$-$\mathcal{U}$ split in $\mathcal{R}$. Thus, we choose to go with a simple goal motivated by *unambiguity* - we would like to learn an estimate of $\mathcal{W}$ such that the resultant weighted-sum based $RLS_{\mathcal{W}}$ scoring can decidedly determine whether each pair of records from $\mathcal{R}$ is to be linked or not. In other words, for every pair of records, we would like the $RLS_{\mathcal{W}}$ scoring to be either close to the lower extreme (unlinked) or close to the higher extreme (linked), avoiding the (ambiguous) bay between the extremes as much as possible. Since it is evidently impractical to enforce this strictly for all record pairs in $\mathcal{R}$ for such a $\mathcal{W}$ may not even exist, we use an approach of iteratively optimizing an objective function to progressively refine $\mathcal{W}$ in the direction of lesser overall ambiguity.

We now outline an objective function. Consider a record-pair $p_{xy} = (r_x, r_y)$ and an estimate of weights $\mathcal{W}$, the ambiguity for p_{xy} may be modeled as follows:

$$AMB_{\mathcal{W}}(r_x, r_y) = min\{RLS_{\mathcal{W}}(r_x, r_y) - \rho, \tau - RLS_{\mathcal{W}}(r_x, r_y)\} \qquad (3)$$

where the $RLS_{\mathcal{W}}$ scores for all record-pairs in $\mathcal{R}$ reside in $[\rho, \tau]$, ρ being the lower extreme and τ being the upper extreme for the $RLS_{\mathcal{W}}$ scores. Informally, $AMB_{\mathcal{W}}(r_x, r_y)$ measures the extent to which the RLS score computed using $\mathcal{W}$ deviates from either ends. The min aggregation ensures that scores in the extremes, i.e., both $RLS_{\mathcal{W}}(r_x, r_y) = \rho$ and $RLS_{\mathcal{W}}(r_x, r_y) = \tau$, would bring the ambiguity score down to zero. This is a desirable condition since these extreme scorings indicate that $RLS_{\mathcal{W}}$ is in no way uncertain about the linkage status for the record pairs in p_{xy}. Analogously, as the $RLS_{\mathcal{W}}$ score moves into the gulf between the extremes, the $AMB_{\mathcal{W}}$ correspondingly goes up.

In most practical cases, similarity functions return non-negative values since negative values for similarity do not make much practical sense. This non-negativity assumption makes 0.0 an intuitive lower bound (ρ) for $RLS_\mathcal{W}$ scores. The upper extrem for the $RLS_\mathcal{W}$ scores (τ) is set as 1. Incorporating it, we refine the ambiguity notion as:

$$AMB_\mathcal{W}(r_x, r_y) = min\{RLS_\mathcal{W}(r_x, r_y), \tau - RLS_\mathcal{W}(r_x, r_y)\} \tag{4}$$

We would like to aggregate this across record-pairs in $\mathcal{R}$ to arrive at a notion of overall ambiguity of $\mathcal{W}$ as follows:

$$AMB_\mathcal{W}(\mathcal{R}) = \sum_{\{r_x, r_y \in \mathcal{R}, x \neq y\}} min\{RLS_\mathcal{W}(r_x, r_y), \tau - RLS_\mathcal{W}(r_x, r_y)\} \tag{5}$$

Connecting back to our original goal of reducing ambiguity, our task is simply to learn a set of weights $\mathcal{W}$ that minimizes the overall ambiguity across the dataset.

$$\mathcal{W}^* = \arg\min_\mathcal{W} \; AMB_\mathcal{W}(\mathcal{R}) \tag{6}$$

Our intent in the optimization approach is to ensure that the initial estimates of $\mathcal{W}$ are re-balanced over iterations to orient them towards those similarity signals that can play a role in reducing overall ambiguity. Specifically, all the weights increasing (or decreasing) together do not benefit us much since that would mostly change the range of $RLS_\mathcal{W}$ rather than altering the ordering among pairs. In order to be robust and to avoid such cases, we use an add-to-one constraint in our optimization approach.

$$\sum_{a_j \in \mathcal{A}} \sum_{S \in S_j} w_{jS} = 1 \tag{7}$$

This enforces that all the weights in $\mathcal{W}$ sum up to 1.0. It may be noted that the optimization problem in Eq. 6, in combination with the constraint in Eq. 7, while simple to state, involves searching over all possible settings of $\mathcal{W}$ such that its components add up to 1.0. This is evidently a massive search space, making brute force search impossible. In the next section, we will outline a gradient co-ordinate descent formulation.

4.3   Optimization Formulation

The objective to be minimized (from Eq. 6) can be written by expanding $RLS_\mathcal{W}$:

$$\sum_{\{r_x, r_y \in \mathcal{R}, x \neq y\}} min\{\sum_{a_j \in \mathcal{A}} \sum_{S \in S_j} w_{jS}^2 \times S_{xyj}, \tau - \sum_{a_j \in \mathcal{A}} \sum_{S \in S_j} w_{jS}^2 \times S_{xyj}\} \tag{8}$$

where S_{xyj} is a shorthand for $S(r_x.a_j, r_y.a_j)$. The min aggregation function, being not differentiable, does not easily yield to optimization. We observe that exponentiation can be used to approximate the min aggregation (similar to another approximation [17]).

$$min\{a, b\} \approx \frac{1}{\phi} log\Big(exp(\phi a) + exp(\phi b) \Big) \tag{9}$$

This approximation holds for high negative values of ϕ for any numbers a and b. In this paper we set $\phi = -50$. The inner multiplication of a and b separately with ϕ enables spacing out the two terms due to the large numeric value of ϕ; observe that $|a - b| << |\phi a - \phi b|$. For cases where $a > b$ holds, ϕa would be much smaller than ϕb, given that ϕ is a large negative value. Consequently, $exp(\phi a)$ would be much lesser than $exp(\phi b)$, making their sum much closer to the latter than the former. Thus, the log of their sum would be closer to ϕb, making the entire term in the RHS of Eq. 9 a good approximation of b. It is notable that this approximation works reasonably well when $a = b$ too, since the RHS would reduce to $a + log(2)/\phi$; the second term is a very small term, due to having a numerically large ϕ in the denominator, thus yielding a good approximation for min. We simply apply this approximation to re-write our objective:

$$\frac{1}{\phi} \sum_{\{r_x, r_y \in \mathcal{R}, x \neq y\}} log\Bigg(exp\Big[\phi \sum_{a_j \in \mathcal{A}} \sum_{S \in \mathcal{S}_j} w_{jS}^2 S_{xyj} \Big] + exp\Big[\phi \Big(\tau - \sum_{a_j \in \mathcal{A}} \sum_{S \in \mathcal{S}_j} w_{jS}^2 S_{xyj} \Big) \Big] \Bigg) \tag{10}$$

Towards optimizing this more convenient and differentiable objective function, we adopt a gradient descent approach, optimizing for one variable within $\mathcal{W}$, at a time. Consider the variable $w_{j'S'}$; the partial derivative, $\frac{\partial AMB_{\mathcal{W}}(\mathcal{R})}{\partial w_{j'S'}}$ is then

$$\sum_{\{r_x, r_y \in \mathcal{R}, x \neq y\}}$$

$$\frac{2w_{j'S'} S'_{xyj'} \Big(exp\Big[\phi \sum_{a_j \in \mathcal{A}} \sum_{S \in \mathcal{S}_j} w_{jS}^2 S_{xyj} \Big] - exp\Big[\phi \Big(\tau - \sum_{a_j \in \mathcal{A}} \sum_{S \in \mathcal{S}_j} w_{jS}^2 S_{xyj} \Big) \Big] \Big)}{\Big(exp\Big[\phi \sum_{a_j \in \mathcal{A}} \sum_{S \in \mathcal{S}_j} w_{jS}^2 S_{xyj} \Big] + exp\Big[\phi \Big(\tau - \sum_{a_j \in \mathcal{A}} \sum_{S \in \mathcal{S}_j} w_{jS}^2 S_{xyj} \Big) \Big] \Big)} \tag{11}$$

The update for $w_{j'S'}$ follows gradient descent[1], using a learning rate, μ.

$$w_{j'S'} = w_{j'S'} - \mu \times \frac{\partial AMB_{\mathcal{W}}(\mathcal{R})}{\partial w_{j'S'}} \tag{12}$$

4.4   Overall Approach

Our iterative approach targets arriving at a good estimate of $\mathcal{W}$ by updating each w_{jS} in turn using the values of $\mathcal{W}$ from the previous iteration, followed by re-normalizing them to sum up to 1.0 within each iteration. This approach

---

[1] https://en.wikipedia.org/wiki/Gradient_descent.

Algorithm 1: Our Unsupervised Record Linkage Scoring Method

    input   : Set of Records $\mathcal{R}$ and similarity functions $\{\ldots, \mathcal{S}_j, \ldots\}$
    output : Set of weights $\mathcal{W} = \{\ldots, w_{jS}, \ldots\}$ to define the scoring $RLS_{\mathcal{W}}$

1 Initialize all w_{jS}s uniformly satisfying $\sum_{a_j \in \mathcal{A}} \sum_{S \in \mathcal{S}_j} w_{jS} = 1$;
2 While (iterations limit not reached and not converged)
3   Initialize a new set of weights $\mathcal{W}'$
4    $Sum_{\mathcal{W}'} \leftarrow 0$;
5    $\forall\, w_{jS} \in \mathcal{W}$
6       $w'_{jS} \leftarrow w_{jS} - \mu \times \frac{\partial AMB_{\mathcal{W}}(\mathcal{R})}{\partial w_{jS}}$;
7       $Sum_{\mathcal{W}'} = Sum_{\mathcal{W}'} + w'_{jS}$;
8    $\forall\, w'_{jS} \in \mathcal{W}$
9       $w'_{jS} \leftarrow \frac{w'_{jS}}{Sum_{\mathcal{W}'}}$;
10   $\mathcal{W} \leftarrow \mathcal{W}'$;
11 Output $\mathcal{W}$;

---

is outlined in Algorithm 1. Line 6 denotes the gradient descent update, whereas Line 9 performs the normalization.

Complexity: The update in Eq. 12 has terms for each record pair to be evaluated hidden within the slope term, i.e., $\frac{\partial AMB_{\mathcal{W}}(\mathcal{R})}{\partial w_{jS}}$, and needs to be run for each similarity measure for each attribute (notice the iteration over w_{jS}s). From a computational perspective, consider the construction of the slope term (Ref. Eq. 12); it may be observed that the inner term in the numerator (difference between exponentiated terms) and the inner term in the denominator (sum of exponentiated terms) are both independent of j' and S', and thus, can be computed once per record pair. This makes the full complexity $\mathcal{O}(\sum_{j=1}^{m} |\mathcal{S}_j| \times \mathcal{P})$ per iteration where $\mathcal{P}$ is the number of record pairs. It may be noted that the $\sum_{j=1}^{m} |\mathcal{S}_j|$ term is small, there being only a handful of attributes and a handful of similarity measures, making the complexity largely dependent on the size of $\mathcal{P}$. We will show later that our method stabilizes to reasonable accuracy in 100s of iterations in our experimental section. Coming to $\mathcal{P}$, though the number of possible record pairs in $\mathcal{R}$ is quadratic in $|\mathcal{R}|$, typical record linkage scenarios use efficient blocking strategies to rule out a large fraction of record pairs from being considered for linkage determination making $\mathcal{P} << |\mathcal{R}|^2$. Usage of better blocking strategies would benefit our method since they reduce $\mathcal{P}$, leading to our method running faster.

5 Experiments and Results

5.1 Experimental Setup

In the experiments, we use gold standard linkage labellings to evaluate the methods. It may be emphasized here that the gold standard labellings were used only for the evaluation purposes. All experiments were run on a workstation with Intel(R) Core(TM) i7-4790 CPU @ 3,60 GHz processor, 16 GB (RAM) and 64-bit Windows 7. As with a typical RL approach, we perform blocking as the first step to reduce the number of record pairs for each of the RL methods. Any blocking method could be employed, blocking being orthogonal to the task we evaluate; we used the recently proposed unsupervised blocking scheme learner [11] in our evaluation.

Baselines. As baselines for our method, we used two unsupervised RL methods [15,16] based on the pseudo f-measure. To compare the two methods against our approach we used the similarity scoring configurations output by each of the method to rank all record pairs. Following this, we use ARL and MRL to evaluate each of the rankings. All baseline parameters were set to the values recommended in respective papers.

Data. The experiments were conducted with three real world datasets commonly used for evaluating RL methods: Restaurant, Cora and ACM-DBLP. The Restaurant dataset contains 864 restaurant records (with 112 pairs of matching records), each with five fields, including name, address, city, phone and type. The Cora dataset is a collection of 1,295 (with 14,184 pairs of matching records) citations to computer science papers. Each citation is represented by 4 fields (author, title, venue, year). The ACM-DBLP is a bibliographic datasets of Computer Science bibliography records represented by four attributes (author, title, venue, year). The total number of entity pairs is 6,001,104.

Parametrization of the algorithms. One parameter needs to be set to run our method, which is the number of iterations. The reported results were obtained for number of iterations equals to 200 for each of the 3 datasets. For each of the evaluated methods we applied five commonly used similarity measures for RL, namely Jaro [9], Winkler [25], Jaccard [8], Q-Gram [19], and Levenshtein edit distance [14].

5.2   Comparative Evaluation of Record Pairs Ordering

Effectiveness. Table 2 lists the results of the comparative evaluation of our method against the genetic algorithm and linear classifiers baselines, among the latest methods for unsupervised ML-based record linkage. For each of the evaluation measures, ARL and MRL, lower values are better, as seen in Sect. 3.1. Recall measures the fraction of correctly linked record pairs among the top $|\mathcal{L}|$ pairs; thus, this is the recall of the method if the top-$\mathcal{L}$ pairs according to the scoring were output as the result set. We additionally have also included the values of the measures that can be achieved by a perfect scoring, one that puts all linked record pairs at the top, followed by the unliked pairs; this implicitly has a recall of 1.0. This enables understanding the gap between our method and the best possible scoring. It can be observed that our method obtained better result than two baseline methods for Cora and DBLP-ACM datasets. For the Restaurant dataset, our method was outperformed by the genetic algorithm

Table 2. Evaluation over rank-aware metrics and recall (best numbers highlighted).

	Genetic algorithm			Linear classifier			Our method			Perfect scoring	
	AR$\mathcal{L}$	MR$\mathcal{L}$	Rec.	AR$\mathcal{L}$	MR$\mathcal{L}$	Rec.	AR$\mathcal{L}$	MR$\mathcal{L}$	Rec.	AR$\mathcal{L}$	MR$\mathcal{L}$
Restaurant	**59.6**	**56**	**0.92**	79.42	67	0.72	65.2	61	0.89	55.5	55.5
Cora	8514.5	7006.5	0.79	9396.2	8858.5	0.75	**8077.3**	**6836.5**	**0.82**	6756	6756
DBLP-ACM	1475.2	1444.5	0.65	1769.6	1009.5	0.84	**1100**	**926.5**	**0.85**	851	851

Table 3. The execution time: *hours: minutes: seconds* (best numbers highlighted).

	Genetic algorithm	Linear classifier	Our method
Restaurant	00:19:42	30:26:06	**00:15:07**
Cora	17:34:01	22:55:34	**00:07:27**
DBLP-ACM	30:26:47	72:00:00+	**02:14:58**

based approach, which is able to navigate the search space using the randomized approach effectively for the small dataset. However, as expected, our method, due to it's rather highly focused search along the space of solutions, is able to achieve better effectiveness over large datasets (and their correspondingly larger solution spaces). This character is well pronounced in the large DBLP-ACM dataset where our method is able to make massive improvements.

Efficiency. For each of the evaluated methods we measure their execution time, baselines implemented according to our best understanding from the respective papers. The run times of each of the methods are reported in Table 3. We can observe that our method was able to perform the linkage process significantly faster.

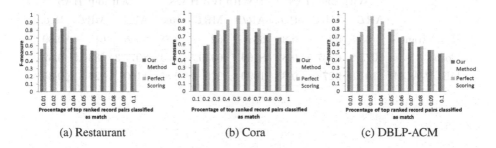

| (a) Restaurant | (b) Cora | (c) DBLP-ACM |

Fig. 1. F-measure obtained with different values of cut off point.

5.3   Further Analysis of Our Method

F-Measure at Different Cut-offs. For usage of our record linkage scoring method in a practical scenario, one would need to apply a cut-off point in the ranked list, so that record pairs above that cut-off could be regarded as linked, and those below may be considered as unlinked. The recall (for the linked class) reported in the previous section is precisely equal to the recall when the cut-off point is chosen after $\mathcal{L}$ records; at this cut-off, the precision and recall are equivalent due to the usage of the same denominator. However, information about $\mathcal{L}$ is part of the gold-standard data and is not available within a realistic record linkage setting. Thus, we evaluate the recall, precision and F-measure for the linked class over varying cut-offs, to illustrate the effectiveness trends of our methods at varying cut-offs. We plot the F-measure achieved by our method against that of the perfect ordering for each of the datasets in Fig. 1, with the choice of the percentage of all record pairs in the dataset used for the cut-off point indicated in the X-axis. The range of cut-off points studied differ for the Cora dataset due to the higher ratio of matching records. Our method is seen to trail the perfect scoring quite closely in some parts of the space, with the gap widening, though not significantly, at other parts. This further illustrates the effectiveness of our method.

Resilience to Noisy Similarity Measures: We now study the resilience of our method to *highly ambigious* similarity measures (abbreviated to HASs), those that hold no utility in separating the linked and unlinked record pairs. We conduct this study through the usage of similarity measures that simply sample from a normal distribution centered at 0.5, the value midway between the linkage favoring extreme of 0.0 and the other extreme of 1.0; we use a standard deviation of 0.2. Table 4 lists the results of our method when one or two HASs are added to the dataset. The results show that the effectiveness deteriorations of our methods in the presence of HASs are decidedly miniscule.

Table 4. Evaluation of resilience to ambigious similarities.

	With one HAS			With two HASs			Without HAS		
	AR$\mathcal{L}$	MR$\mathcal{L}$	Rec.	AR$\mathcal{L}$	MR$\mathcal{L}$	Rec.	AR$\mathcal{L}$	MR$\mathcal{L}$	Rec.
Restaurant	65.5	62	0.88	65.5	62	0.88	65.2	61	0.89
Cora	8073.8	6837.5	0.82	8169.7	6939.5	0.81	8077.3	6836.5	0.82
DBLP-ACM	1101.8	927	0.85	1099	926	0.85	1100	926.5	0.85

6   Conclusions

In this paper we addressed the problem of unsupervised RL and proposed a novel approach to RL, which models the task as an optimization problem. Our optimization formulation searches for RL methods that use a weighted sum scoring to determine linkages between records, favoring those that are less ambiguous overall, in the linkage decisions they make. Our experimental results indicate that our method is highly effective in making accurate linkage decisions, while

also being orders of magnitude faster than existing approaches, especially on
large datasets. We also illustrated that our method is fairly accurate as an RL
method at different cut-off points, and that our optimization approach is exceed-
ingly robust to noisy similarity measures.

References

1. Arasu, A., Gotz, M., Kaushik, R.: On active learning of record matching packages.
   In: SIGMOD, pp. 783–794. ACM (2010)
2. Christen, P.: Automatic record linkage using seeded nearest neighbour and support
   vector machine classification. In: SIGKDD, pp. 151–159. ACM (2008)
3. Christen, P.: Data Matching: Concepts and Techniques for Record Linkage, Entity
   Resolution, and Duplicate Detection. Springer, Heidelberg (2012). https://doi.org/
   10.1007/978-3-642-31164-2
4. Cohen, W.W., Richman, J.: Learning to match and cluster large high-dimensional
   data sets for data integration. In: SIGKDD, pp. 475–480. ACM (2002)
5. Deepak, P., Deshpande, P.M.: Operators for Similarity Search: Semantics Techniques and Usage Scenarios. Springer, Heidelberg (2015). https://doi.org/10.1007/
   978-3-319-21257-9
6. Elfeky, M.G., Verykios, V.S., Elmagarmid, A.K.: TAILOR: a record linkage toolbox. In: 2002 Proceedings of 18th International Conference on Data Engineering,
   pp. 17–28. IEEE (2002)
7. Isele, R., Bizer, C.: Learning expressive linkage rules using genetic programming.
   Proc. VLDB Endow. 5(11), 1638–1649 (2012)
8. Jaccard, P.: Distribution de la flore alpine dans le bassin des dranses et dans
   quelques régions voisines. Bull. Soc. Vaud. Sci. Nat. 37, 241–272 (1901)
9. Jaro, M.A.: Advances in record-linkage methodology as applied to matching the
   1985 census of Tampa, Florida. J. Am. Stat. Assoc. 84(406), 414–420 (1989)
10. Jurek, A., Hong, J., Chi, Y., Liu, W.: A novel ensemble learning approach to
    unsupervised record linkage. Inf. Syst. 71, 40–54 (2017)
11. Kejriwal, M., Miranker, D.P.: An unsupervised algorithm for learning blocking
    schemes. In: IEEE 13th International Conference on Data Mining (ICDM), pp.
    340–349 (2013)
12. Kejriwal, M., Miranker, D.P.: Semi-supervised instance matching using boosted
    classifiers. In: Gandon, F., Sabou, M., Sack, H., d'Amato, C., Cudré-Mauroux,
    P., Zimmermann, A. (eds.) ESWC 2015. LNCS, vol. 9088, pp. 388–402. Springer,
    Cham (2015). https://doi.org/10.1007/978-3-319-18818-8_24
13. Lee, S., Lee, J., Hwang, S.W.: Efficient entity matching using materialized lists.
    Inf. Sci. 261, 170–184 (2014)
14. Levenshtein, V.I.: Binary codes capable of correcting deletions, insertions, and
    reversals. In: Soviet Physics Doklady, vol. 10, pp. 707–710 (1966)
15. Ngomo, A.C.N., Lyko, K.: Unsupervised learning of link specifications: determin-
    istic vs. non-deterministic. In: Proceedings of the 8th International Conference on
    Ontology Matching, vol. 1111, pp. 25–36. CEUR-WS.org (2013)
16. Nikolov, A., d'Aquin, M., Motta, E.: Unsupervised learning of link discovery con-
    figuration. In: Simperl, E., Cimiano, P., Polleres, A., Corcho, O., Presutti, V. (eds.)
    ESWC 2012. LNCS, vol. 7295, pp. 119–133. Springer, Heidelberg (2012). https://
    doi.org/10.1007/978-3-642-30284-8_15

17. Deepak, P.: MixKMeans: clustering question-answer archives. In: 2016 Conference on Empirical Methods in Natural Language Processing. In: EMNLP, pp. 1576–1585 (2016)
18. Sarawagi, S., Bhamidipaty, A.: Interactive deduplication using active learning. In: SIGKDD, pp. 269–278. ACM (2002)
19. Shannon, C.E.: A mathematical theory of communication. ACM SIGMOBILE Mob. Comput. Commun. Rev. **5**(1), 3–55 (2001)
20. Steorts, R.C., Hall, R., Fienberg, S.E.: A Bayesian approach to graphical record linkage and deduplication. J. Am. Stat. Assoc. **111**(516), 1660–1672 (2016)
21. Steorts, R.C., Ventura, S.L., Sadinle, M., Fienberg, S.E.: A comparison of blocking methods for record linkage. In: Domingo-Ferrer, J. (ed.) PSD 2014. LNCS, vol. 8744, pp. 253–268. Springer, Cham (2014). https://doi.org/10.1007/978-3-319-11257-2_20
22. Vatsalan, D., Christen, P., Verykios, V.S.: A taxonomy of privacy-preserving record linkage techniques. Inf. Syst. **38**(6), 946–969 (2013)
23. Wang, J., Li, G., Yu, J.X., Feng, J.: Entity matching: how similar is similar. Proc. VLDB Endow. **4**(10), 622–633 (2011)
24. Wang, Q., Vatsalan, D., Christen, P.: Efficient interactive training selection for large-scale entity resolution. In: Cao, T., Lim, E.-P., Zhou, Z.-H., Ho, T.-B., Cheung, D., Motoda, H. (eds.) PAKDD 2015. LNCS (LNAI), vol. 9078, pp. 562–573. Springer, Cham (2015). https://doi.org/10.1007/978-3-319-18032-8_44
25. Winkler, W.E.: String comparator metrics and enhanced decision rules in the fellegi-sunter model of record linkage, pp. 354–359 (1990)

Community Detection and Network Science

Consensus Community Detection in Multilayer Networks Using Parameter-Free Graph Pruning

Domenico Mandaglio, Alessia Amelio, and Andrea Tagarelli[✉]

DIMES - University of Calabria, 87036 Rende, CS, Italy
{d.mandaglio,a.amelio,tagarelli}@dimes.unical.it

Abstract. The clustering ensemble paradigm has emerged as an effective tool for community detection in multilayer networks, which allows for producing consensus solutions that are designed to be more robust to the algorithmic selection and configuration bias. However, one limitation is related to the dependency on a co-association threshold that controls the degree of consensus in the community structure solution. The goal of this work is to overcome this limitation with a new framework of ensemble-based multilayer community detection, which features parameter-free identification of consensus communities based on generative models of graph pruning that are able to filter out noisy co-associations. We also present an enhanced version of the modularity-driven ensemble-based multilayer community detection method, in which community memberships of nodes are reconsidered to optimize the multilayer modularity of the consensus solution. Experimental evidence on real-world networks confirms the beneficial effect of using model-based filtering methods and also shows the superiority of the proposed method on state-of-the-art multilayer community detection.

1 Introduction

Multilayer networks are pervasive in many fields related to network analysis and mining [2,8]. Particularly, *community detection in multilayer networks* (ML-CD) has attracted lot of attention in the past few years, as witnessed by a relatively large corpus of studies (see, e.g., [7] for a survey).

An effective approach to ML-CD corresponds to *aggregation methods*, whose goal is to infer a community structure by combining information from community structures separately obtained on each of the layers [16–18]. A special class of such methods resembles theory on *clustering ensemble* [6,15]: given a set of clusterings as different groupings of the input data, a *consensus* criterion function is optimized to induce a single, meaningful solution that is representative of the input clusterings. A key advantage of using a consensus clustering approach is that the inconvenience of guessing the "best" algorithm selection and parametrization is avoided, and hence consensus results will be more robust and show higher quality when compared to single-algorithm clustering.

© Springer International Publishing AG, part of Springer Nature 2018
D. Phung et al. (Eds.): PAKDD 2018, LNAI 10939, pp. 193–205, 2018.
https://doi.org/10.1007/978-3-319-93040-4_16

Despite the well-recognized benefits of using the consensus/ensemble clustering paradigm, its exploitation to ML-CD is, surprisingly, relatively new in the literature [9,16,18]; actually, to the best of our knowledge, only the most recent of these works goes beyond the use of a clustering ensemble approach as a black-box tool for ML-CD, by proposing the first well-principled formulation of the *ensemble-based community detection* (EMCD) problem. Indeed, in [16], aggregation is not limited at node membership level, but it also accounts for intra-community and inter-community connectivity; moreover, the consensus function is optimized via *multilayer modularity* analysis, instead of being simply based on the sharing of a certain minimum percentage of clusters in the ensemble.

The EMCD method proposed in [16] relies on a *co-association-based consensus clustering scheme*, i.e., the consensus clusters are derived from a co-association matrix built to store the fraction of clusterings in which any two nodes are assigned to the same cluster. Low values in this matrix would reflect unlikely consensus memberships, i.e., noise, and hence should be removed; to this purpose, the matrix is subjected to a filtering step based on a user-specified parameter of minimum co-association, θ. Unfortunately, setting an appropriate θ for a given input network is a challenging task, since too low values will lead to few, large communities, while too high values will lead to many, small communities. Moreover, this approach generally fails to consider properties related to node distributions and linkage in the network.

In this work, we aim to overcome the above issue, by proposing a new EMCD framework featuring a *parameter-free identification of consensus clusters* from which the *consensus community structure* will be induced. Our idea is to exploit a recently developed class of *graph-pruning methods based on generative models*, which are designed to filter out "noisy" edges from weighted graphs. A key advantage of these pruning models is that they do not require any user-specified parameter, since they enable edge-removal decisions by computing a statistical p-value for each edge based on a null model defined on the node degree and strength distributions. We originally introduce these models to multilayer community detection and propose an adaptation to multilayer networks.

Another limitation of EMCD is that the community membership of nodes remains the same through the process of detecting the modularity-driven consensus community structure. In this work, we also address this point, by defining a three-stage process in the EMCD scheme, which iteratively seeks to improve the multilayer modularity of the consensus community structure based on intra-community connectivity refinement, community partitioning, and relocation of nodes from a community to a neighboring one.

Two main findings are drawn from experimental results obtained on real-world multiplex networks: (i) some of the model-filters are effective in simplifying an input multilayer network to support improved community detection, and (ii) our proposed framework outperforms state-of-the-art multilayer community detection methods according to modularity and silhouette quality criteria.

In the rest of the paper, we provide background on generative-model-based filters and on the existing EMCD method (Sect. 2). Next, we present our proposed framework (Sect. 3). Experimental evaluation and results are discussed in Sects. 4 and 5. Section 6 concludes the paper.

2 Background

2.1 Generative Models for Graph Pruning

Pruning is a graph simplification task aimed at detecting and removing irrelevant or spurious edges in order to unveil some hidden property/structure of the network, such as its organization into communities. A simple technique adopted in weighted graphs consists in removing all edges having weight below a predetermined, global threshold. Besides the difficulty of choosing a proper threshold for the input data, this approach tends to remove all ties that are weak at network level, thus discarding local properties at node level.

A relatively recent corpus of study addresses the task of filtering out "noisy" edges from complex networks based on *generative null models*. The general idea is to define a null model based on node distribution properties, use it to compute a p-value for every edge (i.e., to determine the statistical significance of properties assigned to edges from a given distribution), and finally filter out all edges having p-value above a chosen significance level, i.e., keep all edges that are least likely to have occurred due to random chance.

Methods following the above general approach have been mainly conceived to deal with weighted networks, so that the node degree and/or the node strength (i.e., the sum of the weights of all incident edges) are used to generate a model that defines a random set of graphs resembling the observed network. One of the earliest methods is the *disparity* filter [14], which evaluates the strength and degree of each node locally. This filter however introduces some bias in that the strength of neighbors of a node are discarded. By contrast, a global null model is defined with the *GloSS* filter [13], as it preserves the whole distribution of edge weights. The null model is, in fact, a graph with the same topological structure of the original network and with edge weights randomly drawn from the empirical weight distribution. Unlike disparity and GloSS, the null model proposed by Dianati [1] is maximum-entropy based and hence unbiased. Upon it, two filters are defined: the *marginal likelihood filter* (*MLF*), which is a linear-cost method that assigns a significance score to each edge based on the marginal distribution of edge weights, and the *global likelihood filter*, which accounts for the correlations among edges. While performing similarly, the latter filter is more costly than MLF; moreover, both consider the strength of nodes, but not their degrees. Recently, Gemmetto et al. [5] proposed a maximum-entropy filter, *ECM*, for keeping only *irreducible edges*, i.e., the filtered network will retain only the edges that cannot be inferred from local information. ECM employs a null model based on the canonical maximum-entropy ensemble of weighted networks having the same degree and strength distribution as the real network [11]. Due to space limits, we report details of the MLF, GloSS and ECM filters in the **Online Appendix** available at http://people.dimes.unical.it/andreatagarelli/emcd/.

2.2   Ensemble-Based Multilayer Community Detection

Let $G_{\mathcal{L}} = (V_{\mathcal{L}}, E_{\mathcal{L}}, \mathcal{V}, \mathcal{L})$ be a *multilayer network* graph, with set of layers $\mathcal{L} = \{L_1, \ldots, L_\ell\}$ and set of entities $\mathcal{V}$. Each layer corresponds to a given type of entity relation, or edge-label. For each pair of entity in $\mathcal{V}$ and layer in $\mathcal{L}$, let $V_{\mathcal{L}} \subseteq \mathcal{V} \times \mathcal{L}$ be the set of entity-layer pairs representing that an entity is located in a layer. The set $E_{\mathcal{L}} \subseteq V_{\mathcal{L}} \times V_{\mathcal{L}}$ contains the undirected links between such entity-layer pairs. For every layer $L_i \in \mathcal{L}$, V_i and E_i denote the set of nodes and edges, respectively. Also, the inter-layer edges connect nodes representing the same entity across different layers (monoplex assumption).

Given a multilayer network $G_{\mathcal{L}}$, an *ensemble of community structures* for $G_{\mathcal{L}}$ is a set $\mathcal{E} = \{\mathcal{C}_1, \ldots, \mathcal{C}_\ell\}$, such that each $\mathcal{C}_h$ (with $h = 1..\ell$) is a community structure of the layer graph G_h. This ensemble could be obtained by applying any non-overlapping community detection algorithm to each layer graph.

Given an ensemble of community structures for a multilayer network, the problem of ensemble-based multilayer community detection (EMCD) is to compute a *consensus community structure*, as a set of communities that are representative of how nodes were grouped and topologically-linked together over the layer community structures in the ensemble. In order to determine the community membership of nodes in the consensus structure, a *co-association*-based scheme is defined over the layers, to detect a clustering solution (i.e., the consensus) that conforms most to the input clusterings. Given $G_{\mathcal{L}}$, and $\mathcal{E}$ for $G_{\mathcal{L}}$, the *co-association matrix* $\mathbf{M}$ is a matrix with size $|\mathcal{V}| \times |\mathcal{V}|$, whose (i, j)-th entry is defined as $|m_{ij}|/\ell$, where m_{ij} is the set of communities shared by $v_i, v_j \in \mathcal{V}$, under the constraint that the two nodes are linked to each other [16].

EMCD is modeled in [16] as an optimization problem in which the *consensus community structure* solution is optimal in terms of *multilayer modularity*, and is to be discovered within a hypothetical space of consensus community structures that is delimited by a "topological-lower-bound" solution and by a "topological-upper-bound" solution, for a given *co-association threshold* θ. Intuitively, the topological-lower-bound solution may be poorly descriptive in terms of multilayer edges that characterize the internal connectivity of the communities, whereas the topological-upper-bound solution may contain superfluous multilayer edges connecting different communities. The modularity-optimization-driven consensus community structure produced by the method in [16], dubbed M-EMCD, hence produces a solution that is ensured to have higher modularity than both the topologically-bounded solutions.

3   EMCD and Parameter-Free Graph Pruning

As previously discussed, the EMCD framework has one model parameter, i.e., the co-association threshold θ, which allows the user to control the degree of consensus required to every pair of nodes in order to appear in the same consensus

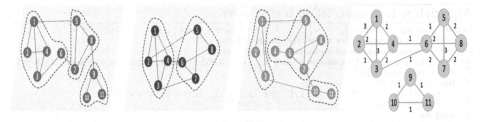

Fig. 1. Community structures (denoted by dotted curves) on a 3-layer network, and corresponding co-association graph.

community. Given a selected value for θ and any two nodes v_i, v_j, we say that their community linkage, expressed by $M(v_i, v_j)$, is considered as meaningful to put the nodes in the same consensus community iff $M(v_i, v_j) \geq \theta$.

However, choosing a fixed value of θ equally valid for all pairs of nodes raises a number of issues. First, there is an intrinsic difficulty of guessing the "best" threshold — since too low values will lead to few, large communities, while too high values will lead to many, small communities. Second, the approach ignores any property of the input network, and consequently a single-shot choice of θ may fail to capture the natural structure of communities. Of course, to overcome the two issues in practical cases, one could always try different choices of the parameter and finally select the best-performing one (e.g., in terms of modularity, as done in [16]), but it is clear that the approach does not scale for large networks.

It would instead be desirable to evaluate the significance of the co-associations by taking into account the topology of the multilayer network, so that a relatively low value of co-association might be retained as meaningful provided that it refers to node relations that make sense only for certain layers, while on the contrary, a relatively high value of co-association could be discarded if it corresponds to the linkage of nodes that have high degree and co-occur in the same community in many layers — in which case, the co-association could be considered as superfluous in terms of community structure.

In order to fulfill the above requirement, we define a *parameter-free approach to EMCD* that exploits the previously discussed pruning models. Since such models are only designed to work with (monoplex) weighted graphs, our key idea is to first infer a *weighted graph representation of the co-association matrix* associated to a multilayer network and its ensemble of community structures, and then apply a pruning model on it to retain only meaningful co-associations.

Definition 1 (Co-association graph).   *Given a multilayer graph $G_{\mathcal{L}}$, an ensemble $\mathcal{E}$ of community structures defined over it, and associated co-association matrix $\boldsymbol{M}$, we define the co-association graph $G_M = \langle V_M, E_M, w \rangle$ as an undirected weighted graph such that $V_M = \mathcal{V}, E_M = \{(v_i, v_j) \mid m_{ij} \neq \emptyset, w_{ij} = |m_{ij}|\}$.*

Below is an example of how the pruning of the co-association graph based on a user-specified threshold could lead to poorly meaningful consensus communities.

Algorithm 1. Co-association matrix filtering

Input: Multilayer graph $G_{\mathcal{L}} = (V_{\mathcal{L}}, E_{\mathcal{L}}, \mathcal{V}, \mathcal{L})$, ensemble of community structures $\mathcal{E} = \{\mathcal{C}_1, \ldots, \mathcal{C}_\ell\}$
(with $\ell = |\mathcal{L}|$), generative model for graph pruning WGP.
Output: Filtered co-association matrix $\mathbf{M}$ for $G_{\mathcal{L}}$ and $\mathcal{E}$.
1: Let α be a statistical significance level (i.e., $\alpha = 0.05$) {*Co-association matrix initialization*}
2: $\mathbf{M} \leftarrow matrix(|\mathcal{V}|, |\mathcal{V}|)$
3: **for** $(i, j) \in \mathbf{M}$ **do**
4:     $m_{ij} \leftarrow \{h \mid L_h \in \mathcal{L} \wedge \exists C \in \mathcal{C}_h, \mathcal{C}_h \in \mathcal{E}, \ s.t. \ v_i, v_j \ in \ C \wedge (v_i, v_j) \in E_h\}$
5:     $M(i, j) \leftarrow |m_{ij}|/\ell$
6: **end for**
7: $G_M = \langle V_M, E_M, w \rangle \leftarrow$ build_coassociation_graph$(G_{\mathcal{L}}, \mathbf{M})$ {*Using Def. 1*}
8: $(e, \gamma_{ij})_{e = (v_i, v_j) \in E_M} \leftarrow$ compute_pValues(G_M, WGP) {*Using Def. 2*}
9: **for** $(v_i, v_j) \in E_M$ **do**
10:     **if** $\gamma_{ij} \geq \alpha$ **then** $M(i, j) \leftarrow 0$ {*Null hypothesis cannot be rejected*}
11: **return** $\mathbf{M}$

Example 1. Consider the 3-layer network and associated co-association graph in Fig. 1. Focusing on the community membership of nodes, consider the following settings of a cutting threshold θ. For any $\theta \leq 1/3$, all edges will be kept (as the minimum valid weight is 1) and hence the co-association graph will be partitioned into the two communities corresponding to its two connected components, i.e., $\{1, .., 8\}$ and $\{9, 10, 11\}$; setting $1/3 < \theta \leq 2/3$ will lead to $\{1, .., 4\}$, $\{5, .., 8\}$, and $\{9\}, \{10\}, \{11\}$; finally, for $2/3 < \theta \leq 1$, the communities will be $\{1, 2, 3\}$, $\{5, 7\}$ and all the other nodes as singletons. It should be noted that no setting of θ can enable the identification of the three "natural" consensus communities, i.e., $\{1, .., 4\}$, $\{5, .., 8\}$, and $\{9, 10, 11\}$.

Definition 2 (Co-association hypothesis testing). *Given a co-association graph $G_M = \langle V_M, E_M, w \rangle$, let WGP denote a statistical inference method whose generative null model is parametric w.r.t. node degree and strength distributions in G_M. We define the co-association hypothesis testing as a parametric testing based on WGP, whose null hypothesis for every observed edge is that its weight has been generated by mere chance, given the empirical strength and degree distributions, and the associated p-value is the probability that the null model produces a weight equal to or greater than the observed edge weight. If the p-value is lower than a desired significance level, then the null hypothesis can be rejected, which implies that the co-association of the two observed nodes is considered as statistically meaningful.*

Algorithm 1 shows the general scheme of creation of the co-association matrix, for a given multilayer network and associated ensemble of community structures, and its filtering based on the co-association hypothesis testing.

Enhanced M-EMCD (M-EMCD*). We propose an enhanced version of M-EMCD that has two main advantages w.r.t. the early M-EMCD method in [16]: (1) it incorporates parameter-free pruning of the co-association matrix described in Algorithm 1, and (2) it fixes the inability of the early M-EMCD in reconsidering the community memberships of nodes during the consensus optimization.

Algorithm 2 shows the pseudo-code of our proposed enhanced M-EMCD, dubbed M-EMCD*. Initially, the filtered co-association matrix computed by

Algorithm 2. Enhanced Modularity-driven Ensemble-based Multilayer Community Detection (M-EMCD*)

Input: Multilayer graph $G_{\mathcal{L}} = (V_{\mathcal{L}}, E_{\mathcal{L}}, \mathcal{V}, \mathcal{L})$, ensemble of community structures $\mathcal{E} = \{C_1, \ldots, C_\ell\}$ (with $\ell = |\mathcal{L}|$), generative model for graph pruning WGP.
Output: Consensus community structure C^* for $G_{\mathcal{L}}$.
1: $\mathbf{M} \leftarrow$ co-associationMatrixFiltering($G_{\mathcal{L}}, \mathcal{E}, WGP$)   {*Algorithm 1*}
2: $C_{lb} \leftarrow$ CC-EMCD($G_{\mathcal{L}}, \mathbf{M}$) {*Compute topological-lower-bound consensus community structure*}
3: $C^* \leftarrow C_{lb}$
4: **repeat**
5:     **for** $L_i \in \mathcal{L}$ **do**
6:         $Q \leftarrow Q(C^*)$
        {*Refine intra-community connectivity of C_j*}
7:         **for** $C_j \in C^*$ **do**
8:             $\langle C'_j, Q'_j \rangle \leftarrow$ update_community(C^*, C_j, L_i)
9:         $j^* \leftarrow \arg\max Q'_j$
10:         **if** $Q'_{j*} > Q$ **then** $Q \leftarrow Q'_{j*}$, $C^* \leftarrow C^* \setminus C_j \cup C'_{j*}$
        {*Refine inter-community connectivity between C_{j*} and each of its neighbors*}
11:         **for** $C_h \in N(C_{j*})$ **do**
12:             $\langle C_h^{IC}, Q_h^{IC} \rangle \leftarrow$ update_community_structure(C^*, C_{j*}, C_h, L_i)
13:             $\langle C_h^R, Q_h^R \rangle \leftarrow$ relocate_nodes(C^*, C_{j*}, C_h)
14:             $\langle C_h, Q_h \rangle \leftarrow \arg\max\{Q_h^{IC}, Q_h^R\}$
15:             $h^* \leftarrow \arg\max Q_h$
16:             **if** $Q_{h*} > Q$ **then**
17:                 $Q \leftarrow Q_{h*}$, $C^* \leftarrow C_{h*}$
18:                 **if** $Q_{h*} = Q_{h*}^R$ **then** $\langle C_h, Q_h \rangle \leftarrow$ update_community_structure(C^*, C_{j*}, C_{h*}, L_i)
19:                 **else** $\langle C_h, Q_h \rangle \leftarrow$ relocate_nodes(C^*, C_{j*}, C_{h*})
20:                 **if** $Q_h > Q$ **then** $Q \leftarrow Q_h$, $C^* \leftarrow C_h$
        {*Evaluate partitioning of C_{j*} into smaller communities*}
21:         $\langle C'_s, Q'_s \rangle \leftarrow$ partition_community(C^*, C_{j*})
22:         **if** $Q'_s > Q$ **then** $Q \leftarrow Q'_s$, $C^* \leftarrow C^* \setminus C_{j*} \cup C'_s$
23:     **end for**
24: **until** $Q(C^*)$ cannot be further maximized
25: **return** C^*

---

a selected model-filter WGP is provided as input to CC-EMCD, which computes the initial (i.e., lower-bound) consensus community structure (Line 2) [16]. This is iteratively improved in a three-stage modularity-optimization process: (i) refinement of connectivity internal to a selected community, (ii) refinement of connectivity between the community and its neighbors also involving relocation of nodes, and (iii) partitioning of the community.

The within-community connectivity refinement step (Lines 7–10) consists in seeking in the current solution C^* the community C_{j*} whose internal connectivity modification leads to the best modularity gain. The internal refinement of a community C_j, applied to the layer L_i, is performed by function update_community (Line 8) which tries to add as many edges of type L_i as possible between nodes belonging to C_j, i.e., the set of edges in E_i whose end-nodes are both in C_j and are not present in the current solution C^*. The function then returns the modified C_j and the updated modularity.

Once identified the community C_{j*} at the previous step, the algorithm tries to relocate nodes from C_{j*} to its neighbor communities $N(C_{j*})$ and/or to refine its external connectivity with them (Lines 11–20). The inter-community connectivity refinement is carried out by function update_community_structure (Line 12) which, for any layer L_i and neighbor communities C_j, C_h, evaluates the resulting

Table 1. Main features of real-world multiplex network datasets used in our evaluation.

	#entities ($|\mathcal{V}|$)	#edges	#layers (ℓ)		#entities ($|\mathcal{V}|$)	#edges	#layers (ℓ)
AUCS [7]	61	620	5	FF-TW-YT [2]	6 407	74 836	3
EU-Air [7]	417	3 588	37	London [19]	369	441	3
FAO-Trade [4]	214	318 346	364	VC-Graders [19]	29	518	3

modularity of adding and/or removing edges of type L_i in the current consensus $\mathcal{C}^*$ between C_j, C_h, compatibly with the set of edges of L_i in the original graph. The relocation of one node at a time from C_{j*} to a neighbor community C_h is evaluated by relocate_nodes (Line 13) until there is no further improvement in modularity. The ordering of node examination is determined by a priority queue that gives more importance to nodes having more edges (of any type) towards C_h than edges linking them to nodes in their current community in $\mathcal{C}^*$.

The step of partitioning of C_{j*} into smaller communities is carried out by function partition_community (Line 21). While this can in principle refer to the use of any (multilayer) modularity-optimization-based community detection method, we choose here to focus on the membership of nodes, and hence to devise this step in the simplified scenario of flattened representation of the consensus community C_{j*}, i.e., a weighted monoplex graph with all and only the nodes belonging to C_{j*} and weights expressing the number of layers on which two nodes are linked in $\mathcal{C}^*$. Upon this representation, we apply a graph partitioning method based on modularity optimization (cf. Sect. 4) and finally maintain the resulting partitioning only if it led to an improvement in modularity.

4   Evaluation Methodology

Datasets. We used six networks for our evaluation (Table 1), which are among the most frequently used in relevant studies in multilayer community detection.

Competing Methods. We selected four of the most representative methods for multilayer community detection: *Generalized Louvain* (GL) [12], *Multiplex Infomap* (M-Infomap) [3], *Principal Modularity Maximization* (PMM) [17], and the consensus clustering approach in [9] (hereinafter denoted as ConClus). Note that the latter two are aggregation-based methods; in particular, ConClus is a simple approach for consensus clustering in weighted networks.

Assessment Criteria and Setting. We employed the *multilayer modularity* defined in [16], the *multilayer silhouette* defined in [16], and *NMI* [15].

To generate the ensemble for each evaluation network, following the lead of the study in [16], we used the serial version of the Nerstrand algorithm [10], a very effective and efficient method for discovering non-overlapping communities in (single-layer) weighted graphs via modularity optimization. We also used Nerstrand for the community-partitioning step in our M-EMCD*.

Table 2. Size and modularity (upper table) and silhouette (bottom table) of lower-bound (CC-EMCD) and M-EMCD*consensus (in brackets, when applicable, the increments over M-EMCD), with or without model-filters.

	CC-EMCD modularity				M-EMCD* modularity				M-EMCD* #communities			
	θ-based	MLF	ECM	GloSS	θ-based	MLF	ECM	GloSS	θ-based	MLF	ECM	GloSS
AUCS	0.60	0.68	0.66	0.21	0.86 (+0.03)	0.91	0.91	0.25	14	13	18	52
EU-Air	0.73	0.60	0.60	0.07	0.91	0.91	0.90	0.09	274	39	45	397 (-2)
FAO-Trade	0.74	0.59	0.30	0.20	1.00	1.00	0.99 (+0.29)	0.99 (+0.56)	41 (+1)	1 (-2)	11 (+3)	40 (-17)
FF-TW-YT	0.48	0.44	0.44	0.05	0.73 (+0.12)	0.94	0.94	0.05	119 (+33)	115	133	5134
London	0.89	0.85	0.85	0.41	0.90	0.97	0.97	0.49 (+0.06)	45	46	46	340 (-3)
VC-Graders	0.22	0.33	0.27	-0.01	0.88 (+0.54)	0.44	0.43	0.03 (-0.01)	3 (-8)	16	17	26 (-1)

	CC-EMCD silhouette				M-EMCD* silhouette			
	θ-based	MLF	ECM	GloSS	θ-based	MLF	ECM	GloSS
AUCS	0.07	0.23	0.28	0.14	0.37 (+0.01)	0.38	0.40	0.15
EU-Air	0.01	0.16	0.18	-0.05	0.09	0.27	0.30	0.04 (-0.02)
FAO-Trade	-0.06	0.01	0.02	0.01	0.08	1.00 (+0.91)	0.06 (-0.05)	0.06 (-0.05)
FF-TW-YT	0.00	0.06	0.06	0.03	0.00 (-0.04)	0.15	0.12	0.03
London	0.14	0.06	0.06	0.03	0 18	0.20	0.20	0.12 (+0.04)
VC-Graders	0.24	0.20	0.21	0.05	0.52 (+0.23)	0.24	0.28	0.83 (+0.77)

As concerns the competing methods, we used the default setting for GL and M-Infomap. We varied the number of communities in PMM from 5 to 100 with increments of 5, and finally selected the value corresponding to the highest modularity. Also, we equipped ConClus with Nerstrand (for the generation of the clusterings), set n_p to the number of layers, and varied θ in the full range (with step 0.01) to finally select the value that determined the consensus clusters with the highest average NMI w.r.t. the initial ensemble solutions.

More details about the evaluation networks and the competing methods can be found at http://people.dimes.unical.it/andreatagarelli/emcd/.

5 Results

5.1 Impact of Model-Filters on M-EMCD*

For every network, we analyzed size, modularity and silhouette of the consensus solution obtained before (i.e., at lower-bound CC-EMCD) and at convergence of the optimization performed by M-EMCD*, when using either global threshold θ pruning or one among MLF, ECM, and GloSS; in the former case, the value of modularity refers to the consensus solution corresponding to the best-performing θ value. Results are reported in Table 2 and discussed next. At the end of this section, we also mention aspects related to time performance evaluation.

Size of Consensus Solutions. MLF and ECM tend to produce similar number of communities. By contrast, GloSS is in general much more aggressive than the other models, which causes proliferation of communities in the co-association graph. Also, the final solution by M-EMCD*can differ in size from the initial consensus by CC-EMCD, due to the optimization of modularity.

Modularity Analysis. Looking at the modularity results, besides the expected improvement by M-EMCD*over CC-EMCD in all cases, the following remarks stand out. First, MLF and ECM again behave similarly in most cases, while GloSS reveals to be much weaker; this is clearly also dependent on the tendency by GloSS of heavily pruning the co-association graph, as discussed in the previous analysis on the size of consensus solutions. Second, using MLF or ECM leads to higher modularity w.r.t. the best-performing global threshold, in all networks but VC-Graders. This would support the beneficial effect deriving from the use of a model-filter for the co-association graph matrix; note however that such results should be taken with a grain of salt, since modularity is computed on differently prunings of the same network. Also, FAO-Trade deserves a special mention, since its much higher multigraph density (13.97) and dimensionality (i.e., number of layers) (cf. Table 1) also caused a densely connected co-association graph, with average degree of 74, average path length of 1.67, clustering coefficient of 0.64, and 1 connected component. This makes FAO-Trade a difficult testbed for a community detection task, which explains the outcome reported in Table 2: 11 consensus communities are produced when using ECM, 41 and 40 with θ-based approach and GloSS, respectively, with most of them singletons and disconnected, and even 1 community for MLF.

It is worth noting that most of the performance gains by M-EMCD*over M-EMCD are obtained for θ-based pruning, but not for model-filter pruning. This would suggest the ability of M-EMCD*of achieving high quality consensus even when a refined model-filter would not be used.

Silhouette and NMI Analysis. In terms of silhouette, the use of model-filter pruning is beneficial to both CC-EMCD and M-EMCD*consensus solutions, where the latter achieve significantly higher silhouette in most cases. Among the filters, again MLF and ECM tend to perform closely—with a slight prevalence of ECM—and better than GloSS (except for VC-Graders, where the number of communities is close to the number of nodes in the co-association graph).

We also measured the NMI of M-EMCD and M-EMCD*model-filter consensus solutions vs. the corresponding solutions obtained by θ-based pruning (results not shown). NMI was found very high (above 0.8, up to 1.0) in EU-Air, AUCS, and VC-Graders, around 0.60–0.70 in FF-TW-YT and London, and around 0.40–0.50 in FAO-Trade. Overall, this indicates that the model-filter pruning has similar capabilities as the best θ-based pruning in terms of community membership, though with the advantage of not requiring parameter selection.

Time Performance Analysis. Considering the execution time of model-filter pruning (results not shown), ECM is in general more costly than GloSS, and this in turn more costly than MLF. This gap—at least one order of magnitude—of ECM against the other two filters can be explained since its higher requirements due to its capability of preserving both degree and strength distributions. Details are reported at http://people.dimes.unical.it/andreatagarelli/emcd/.

Table 3. Increments of number of communities, modularity, silhouette and NMI of M-EMCD*solutions, by varying model-filters, w.r.t. corresponding solutions obtained by GL, PMM, M-Infomap, and ConClus.

	Gains by M-EMCD* vs. GL											
	#communities			Modularity			Silhouette			NMI w.r.t. θ-based pruning		
	MLF	ECM	GloSS	MLF	ECM	GloSS	MLF	ECM	GloSS	MLF	ECM	GloSS
AUCS	+6	+8	+48	+0.09	+0.08	-0.39	+0.11	+0.10	-0.01	+0.06	+0.21	+0.47
EU-Air	-23	-27	+364	+0.12	+0.11	-0.23	+0.26	+0.29	+0.08	+0.51	+0.48	+0.3
FAO-Trade	-5	+4	+30	+0.53	+0.60	+0.70	+0.97	+0.07	+0.07	-0.55	-0.28	+0.21
FF-TW-YT	+111	+130	+5131	+0.29	+0.27	-0.29	-0.07	-0.10	-0.05	+0.02	+0.05	+0.4
London	+23	+23	+318	+0.05	+0.05	-0.42	+0.08	+0.08	-0.30	-0.14	-0.13	-0.06
VC-Graders	0	+2	+18	-0.23	-0.26	-0.40	+0.15	+0.21	+0.71	+0.3	+0.31	+0.08

	Gains by M-EMCD* vs. PMM											
	#communities			Modularity			Silhouette			NMI w.r.t. θ-based pruning		
	MLF	ECM	GloSS	MLF	ECM	GloSS	MLF	ECM	GloSS	MLF	ECM	GloSS
AUCS	-1	+4	+38	+0.43	+0.29	0.00	+0.12	+0.13	-0.04	0.24	+0.26	+0.18
EU-Air	-47	-41	+311	+0.66	+0.65	+0.04	+0.30	+0.33	+0.12	+0.61	+0.61	+0.47
FAO-Trade	-39	-29	0	+0.91	+0.90	+0.90	+1.02	+0.06	+0.07	-0.61	-0.4	+0.06
FF-TW-YT	+104	+122	+5123	+0.66	+0.60	-0.03	-0.14	-0.15	-0.12	-0.1	-0.11	-0.13
London	+1	+1	+295	+0.26	+0.28	0.00	+0.03	+0.03	-0.02	+0.06	+0.07	+0.16
VC-Graders	+1	+2	+11	-0.05	-0.01	-0.13	+0.25	+0.27	+0.95	+0.24	+0.2	-0.29

	Gains by M-EMCD* vs. M-Infomap											
	#communities			Modularity			Silhouette			NMI w.r.t. θ-based pruning		
	MLF	ECM	GloSS	MLF	ECM	GloSS	MLF	ECM	GloSS	MLF	ECM	GloSS
AUCS	+4	+4	+45	+0.18	+0.23	-0.12	+0.17	+0.11	+0.11	+0.48	+0.46	+0.38
EU-Air	-255	-251	+167	+0.38	+0.37	-0.20	+0.35	+0.37	+0.18	+0.74	+0.74	+0.56
FAO-Trade	0	+10	+39	+1.00	0.00	+0.99	+2.00	+1.06	+1.06	0	+0.22	+0.66
FF-TW-YT	+113	+130	+5132	+0.20	+0.24	-0.53	-0.15	-0.15	-0.23	+0.4	+0.3	+0.23
London	+37	+38	+338	+0.52	+0.52	+0.05	+0.21	+0.20	+0.12	+0.39	+0.4	+0.84
VC-Graders	+15	+16	+25	-0.49	-0.50	-0.58	+1.24	+1.28	+1.83	+0.66	+0.64	+0.47

	Gains by M-EMCD* vs. ConClus											
	#communities			Modularity			Silhouette			avg NMI of ensemble		
	MLF	ECM	GloSS	MLF	ECM	GloSS	MLF	ECM	GloSS	MLF	ECM	GloSS
AUCS	+5	+9	+42	+0.33	+0.38	-0.26	+0.13	+0.17	-0.11	-0.03	+0.00	+0.03
EU-Air	-25	-18	+323	+0.71	+0.71	-0.07	+0.23	+0.27	+0.06	-0.05	-0.04	+0.20
FAO-Trade	-16	-11	+21	+0.59	+0.77	+0.74	+0.92	-0.02	-0.01	-0.55	-0.27	+0.01
FF-TW-YT	+17	+74	+4885	+0.48	+0.47	-0.33	+0.15	+0.12	+0.02	-0.06	-0.04	+0.18
London	+16	+21	+298	+0.15	+0.14	-0.30	+0.09	+0.10	-0.01	+0.01	+0.02	+0.12
VC-Graders	+10	+10	+20	+0.21	+0.24	-0.20	+0.09	+0.11	+0.68	+0.02	-0.04	-0.14

5.2 Evaluation with Competing Methods

Table 3 summarizes the increments in terms of size, modularity, silhouette (Table 2), and NMI of M-EMCD*solutions w.r.t. the corresponding solutions obtained by each of the competitors, by varying model-filters. For the NMI evaluation, we distinguished two cases: the one, valid for GL, PMM, or M-Infomap, whereby the reference community structure is the solution obtained by the method in case of θ-based pruning, with θ selected according to the best-modularity performance; the other one, valid for ConClus, whereby we computed the average NMI over the layer-specific community structures.

This comparative analysis was focused on the impact of using the various model-filters on the methods' performance. To this end, for every network and model-filter, we first generated an ensemble of layer-specific community structures via Nerstrand, then we built the co-association graph and applied the filter, finally we removed from the original multilayer network the edges pruned by the model-filter, before providing it as input to each of the competing methods.

One general remark is that M-EMCD*equipped with MLF or ECM outperforms all competing methods in terms of both modularity and silhouette, and tends to produce more communities, with very few exceptions. Concerning NMI results for the first three methods, again the increments by M-EMCD*are mostly positive, thus implying that model-filter pruning appears to be more beneficial, w.r.t. a global threshold based pruning approach, for M-EMCD*than GL, followed by PMM and M-Infomap. Also, it is interesting to observe that, with the exception of FAO-Trade for MLF and ECM, M-EMCD*has average NMI of ensemble comparable to or even better than ConClus, whose performance values are optimal in terms of NMI (i.e., the parameter threshold corresponded to the best NMI over each network).

6   Conclusion

We proposed a new framework for consensus community detection in multilayer networks. This is designed to enhance the modularity-optimization process w.r.t. existing EMCD method. Moreover, by exploiting parameter-free generative models for graph pruning, our framework overcomes the dependency on a user-specified threshold for the global denoising of the co-association graph.

References

1. Dianati, N.: Unwinding the hairball graph: pruning algorithms for weighted complex networks. Phys. Rev. E **93**, 012304 (2016)
2. Dickison, M.E., Magnani, M., Rossi, L.: Multilayer Social Networks. Cambridge University Press, UK (2016)
3. Domenico, M.D., Lancichinetti, A., Arenas, A., Rosvall, M.: Identifying modular flows on multilayer networks reveals highly overlapping organization in interconnected systems. Phys. Rev. X **5**, 011027 (2015)
4. Domenico, M.D., Nicosia, V., Arenas, A., Latora, V.: Structural reducibility of multilayer networks. Nature Commun. **6**, 6864 (2015)
5. Gemmetto, V., Cardillo, A., Garlaschelli, D.: Irreducible network backbones: unbiased graph filtering via maximum entropy. arXiv (June 2017)
6. Gullo, F., Tagarelli, A., Greco, S.: Diversity-based weighting schemes for clustering ensembles. In: Proceedings of SIAM Data Mining, pp. 437–448 (2009)
7. Kim, J., Lee, J.: Community detection in multi-layer graphs: a survey. SIGMOD Rec. **44**(3), 37–48 (2015)
8. Kivela, M., Arenas, A., Barthelemy, M., Gleeson, J.P., Moreno, Y., Porter, M.A.: Multilayer networks. J. Complex Netw. **2**(3), 203–271 (2014)
9. Lancichinetti, A., Fortunato, S.: Consensus clustering in complex networks. Sci. Rep. **2**, 336 (2012)
10. LaSalle, D., Karypis, G.: Multi-threaded modularity based graph clustering using the multilevel paradigm. J. Parallel Distrib. Comput. **76**, 66–80 (2015)

11. Mastrandrea, R., Squartini, T., Fagiolo, G., Garlaschelli, D.: Enhanced reconstruction of weighted networks from strengths and degrees. New J. Phys. **16**, 043022 (2014)
12. Mucha, P.J., Richardson, T., Macon, K., Porter, M.A., Onnela, J.P.: Community structure in time-dependent, multiscale, and multiplex networks. Science **328**(5980), 876–878 (2010)
13. Radicchi, F., Ramasco, J.J., Fortunato, S.: Information filtering in complex weighted networks. Phys. Rev. E **83**, 046101 (2011)
14. Serrano, M.A., Boguna, M., Vespignani, A.: Extracting the multiscale backbone of complex weighted networks. PNAS **106**(16), 6483–6488 (2009)
15. Strehl, A., Ghosh, J.: Cluster ensembles – a knowledge reuse framework for combining multiple partitions. J. Mach. Learn. Res. **3**, 583–617 (2003)
16. Tagarelli, A., Amelio, A., Gullo, F.: Ensemble-based community detection in multilayer networks. Data Min. Knowl. Discov. **31**(5), 1506–1543 (2017)
17. Tang, L., Wang, X., Liu, H.: Uncovering groups via heterogeneous interaction analysis. In: Proceedings of IEEE ICDM, pp. 503–512 (2009)
18. Tang, L., Wang, X., Liu, H.: Community detection via heterogeneous interaction analysis. Data Min. Knowl. Discov. **25**, 1–33 (2012)
19. Zhang, H., Wang, C., Lai, J., Yu, P.S.: Modularity in complex multilayer networks with multiple aspects: a static perspective. CoRR abs/1605.06190 (2016)

Community Discovery Based on Social Relations and Temporal-Spatial Topics in LBSNs

Shuai Xu[1], Jiuxin Cao[1,2(✉)], Xuelin Zhu[1], Yi Dong[1], and Bo Liu[1]

[1] School of Computer Science and Engineering, Southeast University, Nanjing, China
{xushuai7,jx.cao,zhuxuelin,dongyi,bliu}@seu.edu.cn
[2] School of Cybersecurity, Southeast University, Nanjing, China

Abstract. Community discovery is a comprehensive problem associating with sociology and computer science. The recent surge of Location-Based Social Networks (LBSNs) brings new challenges to this problem as there is no definite community structure in LBSNs. This paper tackles the multidimensional community discovery in LBSNs based on user check-in characteristics. Communities discovered in this paper satisfy two requirements: frequent user interaction and consistent temporal-spatial pattern. Firstly, based on a new definition of dynamic user interaction, two types of check-ins in LBSNs are distinguished. Secondly, a novel community discovery model called SRTST is conceived to describe the generative process of different types of check-ins. Thirdly, the Gibbs Sampling algorithm is derived for the model parameter estimation. In the end, empirical experiments on real-world LBSN datasets are designed to validate the performance of the proposed model. Experimental results show that SRTST model can discover multidimensional communities and it outperforms the state-of-the-art methods on various evaluation metrics.

Keywords: Community discovery · LBSNs · Temporal-spatial topics
Topic model

1 Introduction

Community discovery has long been a hot issue in online social network research as it contributes valuable knowledge to practical scenarios, such as user behavior prediction and product recommendation. However, with the increasing prevalence of Location-Based Social Networks (LBSNs) in recent years, new challenges have been brought to traditional community discovery methods since definite community definition is missing in such heterogeneous networks. Abundant user check-in records generated in LBSNs provide a new perspective for community discovery, as a consequence, additional knowledge needs to be mined reasonably considering both user-user and user-location relations.

In our real life, a community can be viewed as a multidimensional cluster [1], where people should have close social relations and similar behavior characteristics including temporal and spatial preferences. Therefore, multiple factors

© Springer International Publishing AG, part of Springer Nature 2018
D. Phung et al. (Eds.): PAKDD 2018, LNAI 10939, pp. 206–217, 2018.
https://doi.org/10.1007/978-3-319-93040-4_17

should be considered when addressing community discovery problem in LBSNs. In this paper, the concept of community is tightly related to four elements, i.e. **WHO, WHEN, WHERE** and **WHAT**. Among the four elements, **WHO** refers to the friend(s) that a user chooses to go out with. It is the reflection of user social relationship. A community considering 'WHO' can keep the close social links within itself. **WHEN** refers to the time when a user conducts his check-in. It is the reflection of user temporal pattern. Users in a community should have consistent temporal pattern. **WHERE** refers to the place where a user plans to go. It is the reflection of user spatial pattern. Users who frequently check-in within the same area should be deemed in the same community, and a community considering 'WHERE' can keep consistent spatial pattern within itself. **WHAT** refers to what the place is when a user decides to visit. It is the reflection of user location preference coming from the user's interest. In this sense, users in a community should have consistent location preference.

Unlike the concepts of W4 proposed by studies [15,16] which focus on modeling individual user's behavior, our concept is devoted to characterizing communities in LBSNs. Communities discovered in this paper satisfy two requirements: frequent user interaction and consistent temporal-spatial pattern, thus can describe user clustering in LBSNs more properly and comprehensively.

Inspired by the generative process of different types of check-ins, this paper proposes a novel community discovery model based on **S**ocial **R**elations and **T**emporal-**S**patial **T**opics (SRTST). We show that SRTST model is able to identify multidimensional communities using real-world LBSN datasets. Besides, we demonstrate its effectiveness and superiority from the perspective of topic modeling, community structure and temporal-spatial consistency, respectively.

2   Related Works

Community Discovery in LBSNs. Up to now, studies on community discovery in LBSNs are relatively rare compared to that in traditional online social networks. Some basic assumptions within LBSN community include that users in the same community should live near each other [3] or have the same topic and location preference [11]. In [5], a new measure of partition quality named 'spatial nearness modularity' is proposed to ensure that communities are not only tightly-knit in terms of topology, but also spatially near in geographic distance. In [11], the user-venue check-in network and user/venue attributes are integrated into an edge-centric coclustering framework to detect overlapping communities. Similarly, based on user check-in records, research in [12] proposes an edge cutting algorithm in order to identify community structure in LBSNs.

Topic Model for Community Discovery. Recently, topic modeling has been widely accepted in community discovery as it is able to detect semantic relations. The common practice is to introduce community related variables into traditional topic model such as LDA and construct a brand new model, enabling specific variables in community to be probability distributions. In Twitter network, a mixed membership stochastic block model is proposed so that each user has a probability distribution over communities [8]. In citation network, as edges between authors are generated by citation relations and similar research interests, a topic model called *Author-Topic-Community* is presented to depict the generative process of citation edges [7]. Several other studies try to find communities in content sharing networks considering topology and content information at the same time [9].

So far, research that applies topic models for community discovery in LBSNs is rare. Although existing studies such as [6,14] can discover community structure in LBSNs through topic models, they can not satisfy the community requirements raised in this paper. The desirable community not only requires close user relations, but also reflects consistency in temporal and spatial preference. Hence, new mechanism is urgently needed to solve the multi-requirement community discovery problem.

3   Community Discovery Model

3.1   Preliminaries

Conventionally, friendship relation in social networks lacks discrimination of closeness, and each friend is treated equally. However, the closeness of friends is unfixed in real life. So the static social relation is unsatisfying when user closeness needs to be described. To emphasize the real situation, we propose a dynamic user interaction definition by combining friendship and user check-in records.

Definition 1 *User interaction. There is one user interaction between u_1 and u_2 when (1) u_1 and u_2 are online friends, and (2) u_1 and u_2 visit the same location in the same windowed time.*

In this paper, week mode [4] is adopted to transfer the absolute time into windowed time. In this sense, we formalize time t as the windowed time $r(t)$, and $r(t) \in \{1, 2, 3, \ldots 168\}$ represents each of the 168 h in a week.

We also give a formalized definition of user check-in:

Definition 2 *User check-in. A check-in is denoted by $o = < u, v, f, t >$, which means user u visits location v at time t with friends f. According to whether f is empty, check-ins are further classified into two types. Check-ins of the **first type** correspond to nonempty set f, which means visiting a place along with friends; check-ins of the **second type** correspond to empty set f, which means visiting a place alone.*

To discover communities from massive check-ins, we need a static snapshot of the global LBSN dataset, which is defined as follows:

Definition 3 *Network Snapshot. The check-in set in a windowed time t_w is treated as the network snapshot of t_w, denoting by $sn_{t_w} = \{< u, v, f, t > | r(t) = t_w\}$.*

Based on Definition 3, the whole LBSN dataset can be denoted by $SN = \{sn_1, sn_2, sn_3, \ldots sn_T\}$, where $T = 168$ as there are 168 time windows under week mode.

3.2 Basic Idea of SRTST Model

Traditional topic modeling methods mainly pay attention to latent semantic topics while ignoring the reason that forms user community. Unlike existing methods, we consider not only temporal-spatial topics, but also the role of communities in the real situation. We take all the four aspects into account: frequent user interaction, consistent temporal pattern, consistent spatial pattern and similar location preference.

With regard to the first type of check-ins (See Fig. 1), we describe the generative process as follows. User u firstly chooses the companion friend set f from community c which u belongs to. Then the activity topic z is considered according to user u and friends f (community c). Next, based on geography and transportation factors, the visiting area, i.e. the region r, is determined. Finally, user u chooses the destination location v considering both topic z and region r. This process involves topic, region as well as community, bringing spatial pattern, location preference and social relations into the model.

With regard to the second type of check-ins (See Fig. 1), the generative process is described as follows. User u firstly decides the purpose of visiting, i.e. the topic z. Then the visiting area, i.e. the region r, is determined. In the end, destination location v is chosen combing topic z and region r. This process involves topic and region, bringing spatial pattern and location preference into the model.

To sum up, in the overall generative process of two types of check-ins, we take social relations and temporal-spatial topics into consideration in each time window.

3.3 Model Construction

According to the above analysis, a Bayesian graphical model for community discovery based on **Social Relations** and **Temporal-Spatial Topics** (SRTST) is conceived. Figure 1 illustrates the structure of our model, and the notations of the model parameters are listed in Table 1. In SRTST model, all the check-ins are divided into T parts according to the time windows. Check-ins in each time window are further divided into two types in order to describe the different generative processes separately. By jointly considering the two types of check-ins, we believe SRTST model can better depict the role of user community in LBSNs.

4 Parameter Estimation

To obtain the posterior distribution of the latent variables conditioned on the existing data, we use Gibbs Sampling for parameter estimation. The updating rules for probability distributions $\{\lambda, \theta, \varphi, \theta', \varphi', \pi\}$ are given as follows.

I. λ, i.e. $P(u|c)$. The probability of user u under community c is:

$$\lambda_{c,u} = \frac{n_u^{(c)} + n_{f.u}^{(c)} + \gamma_u}{\sum_{u' \in U}(n_{u'}^{(c)} + n_{f.u'}^{(c)} + \gamma_{u'})} \tag{1}$$

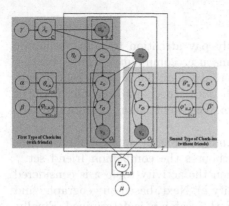

Fig. 1. Graphical Representation of SRTST model. User u and location v are observable variables, which are represented in gray circles. Community c, topic z and region r are latent variables. Subscript o in some symbols represents the current check-in $o = <u, v, f, t>$.

Table 1. Notations of Parameters

Parameters	Interpretation
O_1, O_2	check-in set of the first type and second type respectively
N_t	number of check-ins within windowed time t
$o.f$	set of friends with check-in o
λ	probability distribution $P(u|c)$
π	probability distribution $P(v|z, r)$
θ	probability distribution $P(z|u, c)$
θ'	probability distribution $P(z|u)$
φ	probability distribution $P(r|u, c, z)$
φ'	probability distribution $P(r|u, z)$
γ, α, α'	Dirichlet prior distribution to λ, θ and θ'
β, β', μ	Dirichlet prior distribution to φ, φ' and π
η_t	the initial uniform distribution parameter of the model

where $n_u^{(c)}$ is the number of times that user u belongs to community c; $n_{f.u}^{(c)}$ is the number of times that user u belongs to community c as a companion friend; γ_u is the Dirichlet prior to λ_u.

II. θ, i.e. $P(z|c, u)$. The probability of topic z under community c and user u is:

$$\theta_{c,u,z} = \frac{n_z^{(c,u)} + \alpha_z}{\sum_{z' \in Z}(n_{z'}^{(c,u)} + \alpha_{z'})} \tag{2}$$

where $n_z^{(c,u)}$ is the number of times that topic z belongs to community-user pair $<c, u>$; α_z is the Dirichlet prior to θ_z.

III. φ, i.e. $P(r|c, u, z)$. The probability of region r under community c, user u and topic z is:

$$\varphi_{c,u,z,r} = \frac{n_r^{(c,u.z)} + \beta_r}{\sum_{r' \in R}(n_{r'}^{(c,u,z)} + \beta_{r'})} \tag{3}$$

where $n_r^{(c,u,z)}$ is the number of times that region r belongs to community-user-topic tuple $<c, u, z>$; β_r is the Dirichlet prior to φ_r.

IV. θ', i.e. $P(z|u)$. The probability of topic z under user u is:

$$\theta'_{u,z} = \frac{n_z^{(u)} + \alpha'_z}{\sum_{z' \in Z}(n_{z'}^{(u)} + \alpha'_{z'})} \tag{4}$$

where $n_z^{(u)}$ is the number of times that topic z belongs to user u; α'_z is the Dirichlet prior to θ'_z.

V. φ', i.e. $P(r|u, z)$. The probability of region r under user u and topic z is:

$$\varphi'_{u,z,r} = \frac{n_r^{(u,z)} + \beta'_r}{\sum_{r' \in R}(n_{r'}^{(u,z)} + \beta'_{r'})} \tag{5}$$

where $n_r^{(u,z)}$ is the number of times that region r belongs to user-topic pair $<u, z>$; β'_r is the Dirichlet prior to φ'_r.

VI. π, i.e. $P(v|z, r)$. The probability of location v under topic z and region r is:

$$\pi_{z,r,v} = \frac{n_v^{(z,r)} + \mu_v}{\sum_{v' \in V}(n_{v'}^{(z,r)} + \mu_{v'})} \tag{6}$$

where $n_v^{(z,r)}$ is the number of times that location v belongs to topic-region pair $<z, r>$; μ_v is the Dirichlet prior to π_v.

The detailed Gibbs Sampling algorithm for parameter estimation in SRTST model is shown in Algorithm 1. Theoretically, the time complexity of Algorithm 1 is $O(I \times T \times |SN_t| \times (|C| + |Z| + |R|))$, where I is the predefined number of iterations.

Algorithm 1. Gibbs Sampling Algorithm for SRTST model

Input: $|C|$, $|Z|$, $|R|$, α, β, α', β', γ, μ, η_t

Output: λ, θ, θ', φ, φ', π

1: $I = Number\ of\ Iterations$;
2: /* Initialization for all check-ins in T time windows. */
3: **for** $t = 1$ to T **do**
4:      **for** each $o \in SN_t$ **do**
5:         $c, z, r \sim uniform()$;
6:         assign $\langle c, z, r \rangle$ to o;
7:      **end for**
8: **end for**
9: /* Iteration */
10: **for** $t = 1$ to I **do**
11:      **for** $t = 1$ to T **do**
12:         **for** each $o \in SN_t$ **do**
13:            **if** $o \in O_1$ **then**
14:            $c \sim \frac{n_{-o,u_o}^{(c_o)} + n_{f.u_o}^{(c_o)} + \gamma_{u_o}}{\sum_{u \in U}(n_{-o,u}^{(c_o)} + n_{f.u}^{(c_o)} + \gamma_u)} \frac{n_{-o,z_o}^{(c_o,u_o)} + \alpha_{z_o}}{\sum_{z \in Z}(n_{-o,z}^{(c_o,u_o)} + \alpha_z)} \frac{n_{-o,r_o}^{(c_o,u_o,z_o)} + \beta_{r_o}}{\sum_{r \in R}(n_{-o,r}^{(c_o,u_o,z_o)} + \beta_r)}$;
15:            $z \sim \frac{n_{-o,z_o}^{(c_o,u_o)} + \alpha_{z_o}}{\sum_{z \in Z}(n_{-o,z}^{(c_o,u_o)} + \alpha_z)} \frac{n_{-o,r_o}^{(c_o,u_o,z_o)} + \beta_{r_o}}{\sum_{r \in R}(n_{-o,r}^{(c_o,u_o,z_o)} + \beta_r)} \frac{n_{-o,v_o}^{(z_o,r_o)} + \mu_{v_o}}{\sum_{v \in V}(n_{-o,v}^{(z_o,r_o)} + \mu_v)}$;
16:            $r \sim \frac{n_{-o,r_o}^{(c_o,u_o,z_o)} + \beta_{r_o}}{\sum_{r \in R}(n_{-o,r}^{(c_o,u_o,z_o)} + \beta_r)} \frac{n_{-o,v_o}^{(z_o,r_o)} + \mu_{v_o}}{\sum_{v \in V}(n_{-o,v}^{(z_o,r_o)} + \mu_v)}$;
17:         **end if**
18:         **if** $o \in O_2$ **then**
19:            $c \sim \frac{n_{-o,u_o}^{(c_o)} + n_{f.u_o}^{(c_o)} + \gamma_{u_o}}{\sum_{u \in U}(n_{-o,u}^{(c_o)} + n_{f.u}^{(c_o)} + \gamma_u)}$;

20:     $$z \sim \frac{n_{-o,z_o}^{(u_o)}+\alpha'_{z_o}}{\sum_{z \in Z}(n_{-o,z}^{(u_o)}+\alpha'_z)} \frac{n_{-o,r_o}^{(u_o,z_o)}+\beta'_{r_o}}{\sum_{r \in R}(n_{-o,r}^{(u_o,z_o)}+\beta'_r)} \frac{n_{-o,v_o}^{(z_o,r_o)}+\mu_{v_o}}{\sum_{v \in V}(n_{-o,v}^{(z_o,r_o)}+\mu_v)};$$

21:     $$r \sim \frac{n_{-o,r_o}^{(u_o,z_o)}+\beta'_{r_o}}{\sum_{r \in R}(n_{-o,r}^{(u_o,z_o)}+\beta'_r)} \frac{n_{-o,v_o}^{(z_o,r_o)}+\mu_{v_o}}{\sum_{v \in V}(n_{-o,v}^{(z_o,r_o)}+\mu_v)};$$

22:         **end if**
23:         assign $\langle c, z, r \rangle$ to o;
24:     **end for**
25:     **end for**
26:     /* Update parameters */
27:     update paremeters λ, θ, θ', φ, φ', π according to $Eq.(1) \sim (6)$.
28: **end for**
29: **return** λ, θ, θ', φ, φ', π;

---

5 Experiments

The experiments are conducted in the PC machine with Intel Core i7-6700 processor, 24G RAM and 64bit Ubuntu operating system. The program is mainly coded using Python.

5.1 LBSN Datasets and Evaluation Metrics

Foursquare. The Foursquare dataset was collected by Zhou et al. in [17]. It consists of three parts: social relations (undirected), check-in records and location categories. Each check-in is associated with a location as well as the location category. For the sake of alleviating sparseness, users with less than 10 check-ins and locations which have been visited less than 5 times are removed.

Brightkite. The Brightkite dataset was collected by Cho et al. in [2]. It contains check-in data of users across the United States. In order to facilitate the experiments, we only use a subset of the original dataset covering the New York city area specified by a bounding box $(-74.257159, 40.915568, -73.572489, 40.576913)$. The user network in this dataset is also undirected. Although the location category information is originally missing in the check-in data, the venue category system of Brightkite is almost the same to that in Foursquare[1]. Table 2 summarizes the detailed statistics of the two datasets.

Table 2. Statistics of LBSN datasets.

Dataset	# Users	# Edges	# Check-ins	# Locations	# Avg. check-ins per user
Foursquare	6,291	120,154	788,208	19,905	125.3
Brightkite	1,308	9,248	112,993	17,618	86.4

---

[1] https://developer.foursquare.com/docs/resources/categories.

Evaluation Metrics. We use different metrics from three aspects to evaluate the performance of the community discovery model.

The first metric is **perplexity**, which is often used to evaluate a topic model [10]. Theoretically, a lower perplexity value corresponds to a better topic model for a given dataset. It should be noted that both datasets are split into two parts according to the size ratio 9:1, and the metric perplexity is computed using the latter 10% check-in data.

The second metric is **modularity**, which is widely used to measure the topological quality of community structure [5]. Theoretically, a larger modularity value indicates a better community discovery model.

The third metric is designed to evaluate the **temporal-spatial consistency** within the discovered communities. Specifically, we verify the consistency from two aspects, i.e. temporal consistency and spatial consistency.

5.2 Parameter Configuration

As there are 10 top categories for Foursquare locations, and the number of latent variables is normally set to a multiple of 5 [13], so we set $|Z| = 10$ in SRTST model on both datasets. Besides, as the check-ins in our datasets are made in New York, and there are 5 boroughs in New York city, so we set $|R| = 5$. With regard to the hyperparameters γ, α, α', β, β' and π, for simplicity, we choose a fixed value 0.1 according to [13]. With regard to the number of communities $|C|$ and the number of iterations $|I|$, we plot the relation between perplexity and these two variables using both datasets. We find that perplexity decreases dramatically when $|C| = 10$, $I = 1000$ on Foursquare dataset, and $|C| = 15$, $I = 1200$ on Brightkite dataset. As a result, we set $|C| = 10$, $I = 1000$ for Foursquare dataset, and $|C| = 15$, $I = 1200$ for Brightkite dataset, respectively.

5.3 Comparison Algorithms

State-of-the-art algorithms for comparison in the experiments include TURCM [10], W4 [15], ASTC [13] and UCGT [14]. All these approaches are constructed based on topic models, combining community (except W4), topic, region as well as location. As for W4, we add a latent community variable in the original graphical model so as to serve community discovery in LBSNs. The hyper-parameters of comparison algorithms are set at their suggested values. Besides, we fix $|R| = 5$ and $|Z| = 10$ for all the methods on both datasets. With regard to the number of communities, we set $|C| = 10$ and $|C| = 15$ for Foursquare dataset and Brightkite dataset as mentioned in Sect. 5.2.

5.4 Experimental Results

Perplexity. The comparison results on perplexity with different iterations are shown in Fig. 2. As we can see, SRTST model has lower perplexity than other methods on both datasets when the metric is to converge, which proves the superiority of SRTST model in terms of topic modeling.

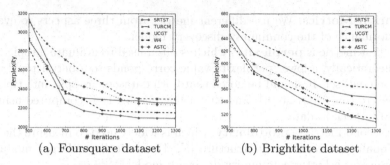

(a) Foursquare dataset   (b) Brightkite dataset

Fig. 2. Perplexity comparison results on two datasets.

Modularity. Due to the sparseness of edges in the datasets, the overall modularity for the entire network is very small. Besides, the modularity values of different methods are pretty close. In order to distinguish the performance of different methods, we report the modularity values of different community size k^2 ranging from 10 to 100. Figure 3 displays the final results. From Fig. 3(a), we observe that the modularity curve of SRTST model fluctuates in the very beginning, and it gradually levels off as k increases. When $k > 20$, the curve of SRTST model is almost always in the highest place compared with other methods. From Fig. 3(b), we observe a similar advantage of SRTST model, even if the numeric values of modularity are fairly small. Since user network in Brightkite dataset is much sparser than that in Foursquare dataset, the discovered community structure may not be desirable in terms of topology. As a result, the modularity values of various models on Brightkite dataset are smaller than that on Foursquare dataset. In a nutshell, the comparison results above indicate the superiority of SRTST model in terms of community structure.

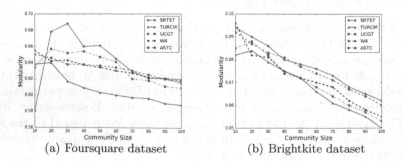

(a) Foursquare dataset   (b) Brightkite dataset

Fig. 3. Modularity comparison results on two datasets.

---

[2] We extract the top-k nodes based on user membership distribution from the corresponding community.

Temporal-Spatial Consistency. We firstly verify the temporal consistency within the communities discovered by SRTST model in an intuitive way. Figure 4 shows the check-in heatmaps of three communities corresponding to nightlife, professionals and entertainments respectively. Limited by paper space, only the results based on Foursquare dataset are reported. For community 1, check-ins are mainly concentrated during 18:00PM to 23:00PM throughout a week, and only a small amount of check-ins take place in the daytime. Besides, there are more check-ins distributed at weekends than that in weekdays, indicating the inherent temporal pattern of nightlife activities. For community 2, check-ins taking place during 10:00AM to 20:00PM are more intensive than other time. In addition, check-ins in weekdays occupy a greater proportion compared with weekends. These characteristics well correspond to professional circumstances. In the end, for community 3, check-ins are densely distributed in the early evening of weekdays and daytime of weekends, indicating this community probably corresponds to entertainments.

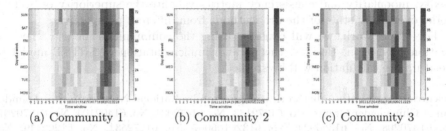

(a) Community 1   (b) Community 2   (c) Community 3

Fig. 4. Check-in heatmaps of three communities on Foursquare dataset. Darker colors represent more check-ins, and vice versa.

Secondly, we evaluate the spatial consistency within the discovered communities. To quantify the spatial consistency, an entropy based measure proposed in [6] is adopted. Specifically, the entropy of venue category distribution within each community is calculated. As a general rule, a lower entropy value indicates a more consistent category distribution, namely, a more consistent location preference within a community. The minimum, maximum as well as the average entropy values of all communities are compared in Table 3. Note that location category is originally missing in Brightkite dataset, so we only report the results on Foursquare dataset. As Table 3 shows, the SRTST model consistently has lower entropy in terms of minimum, maximum as well as average values than other models. This means the discovered communities by SRTST model are more spatially consistent than that by other models, which further reflects the superiority of the proposed model.

In summary, as the above experimental results indicate, the SRTST model proposed in this paper can discover multidimensional communities in LBSNs, and it outperforms state-of-the-art methods with regard to topic modeling, community structure as well as temporal-spatial consistency.

Table 3. Entropy comparison results using Foursquare dataset.

	Min. Entropy	Max. Entropy	Avg. Entropy
SRTST	**2.756**	**2.904**	**2.806**
TURCM	2.930	3.122	2.980
UCGT	2.782	3.064	2.917
W4	2.763	2.934	2.816
ASTC	2.797	3.036	2.962

6   Conclusion

In this paper, we explore the multidimensional community discovery in LBSNs. A probability model called SRTST for community discovery based on social relations and temporal-spatial topics is proposed. Extensive experiments on perplexity, modularity and consistency metrics validate the superiority of SRTST model over many state-of-the-art methods from different perspectives.

For future work, we will consider using the communities for user check-in location prediction. Besides, the parallel implementation of SRTST model to speed up the calculation will be studied.

Acknowledgements. This work is supported by National Natural Science Foundation of China under Grants, No. 61772133, No. 61472081, No. 61402104, No. 61370207, No. 61370208, No. 61300024, No. 61320106007, No. 61272531, No. 61202449, No. 61272054. Collaborative Innovation Center of Wireless Communications Technology, Collaborative Innovation Center of Social Safety Science and Technology, Jiangsu Provincial Key Laboratory of Network and Information Security under Grants No. BM2003201, and Key Laboratory of Computer Network and Information Integration of Ministry of Education of China under Grants No. 93K-9.

References

1. Akbari, M., Chua, T.S.: Leveraging behavioral factorization and prior knowledge for community discovery and profiling. In: Proceedings of the Tenth ACM International Conference on Web Search and Data Mining, pp. 71–79. ACM (2017)
2. Cho, E., Myers, S.A., Leskovec, J.: Friendship and mobility: user movement in location-based social networks. In: Proceedings of the 17th ACM SIGKDD International Conference on Knowledge Discovery and Data Mining, pp. 1082–1090. ACM (2011)
3. Fang, Y., Cheng, R., Li, X., Luo, S., Hu, J.: Effective community search over large spatial graphs. Proc. VLDB Endow. **10**(6), 709–720 (2017)
4. Gao, H., Tang, J., Hu, X., Liu, H.: Modeling temporal effects of human mobile behavior on location-based social networks. In: Proceedings of the 22nd ACM International Conference on Conference on Information & Knowledge Management, pp. 1673–1678. ACM (2013)

5. Hannigan, J., Hernandez, G., Medina, R.M., Roos, P., Shakarian, P.: Mining for spatially-near communities in geo-located social networks. In: Association for the Advancement of Artificial Intelligence-Social Networks and Social Contagion: Web Analytics and Computational Social Science, Arlington, VA, pp. 15–17, November 2013

6. Joseph, K., Tan, C.H., Carley, K.M.: Beyond local, categories and friends: clustering foursquare users with latent topics. In: Proceedings of the 2012 ACM Conference on Ubiquitous Computing, pp. 919–926. ACM (2012)

7. Li, C., Cheung, W.K., Ye, Y., Zhang, X.: The Author-topic-community model: a generative model relating authors' interests and their community structure. In: Zhou, S., Zhang, S., Karypis, G. (eds.) ADMA 2012. LNCS (LNAI), vol. 7713, pp. 753–765. Springer, Heidelberg (2012). https://doi.org/10.1007/978-3-642-35527-1_62

8. Li, W., Ahn, S., Welling, M.: Scalable MCMC for mixed membership stochastic blockmodels. In: Artificial Intelligence and Statistics, pp. 723–731 (2016)

9. Natarajan, N., Sen, P., Chaoji, V.: Community detection in content-sharing social networks. In: Proceedings of the 2013 IEEE/ACM International Conference on Advances in Social Networks Analysis and Mining, pp. 82–89. ACM (2013)

10. Sachan, M., Contractor, D., Faruquie, T.A., Subramaniam, L.V.: Using content and interactions for discovering communities in social networks. In: Proceedings of the 21st International Conference on World Wide Web, pp. 331–340. ACM (2012)

11. Wang, Z., Zhang, D., Zhou, X., Yang, D., Yu, Z., Yu, Z.: Discovering and profiling overlapping communities in location-based social networks. IEEE Trans. Syst. Man Cybern.: Syst. 44(4), 499–509 (2014)

12. Wang, Z., Zhou, X., Zhang, D., Yang, D., Yu, Z.: Cross-domain community detection in heterogeneous social networks. Pers. Ubiquitous Comput. 18(2), 369–383 (2014)

13. Yang, B., Manandhar, S.: Community discovery using social links and author-based sentiment topics. In: 2014 IEEE/ACM International Conference on Advances in Social Networks Analysis and Mining (ASONAM), pp. 580–587. IEEE (2014)

14. Yin, H., Hu, Z., Zhou, X., Wang, H., Zheng, K., Nguyen, Q.V.H., Sadiq, S.: Discovering interpretable geo-social communities for user behavior prediction. In: 2016 IEEE 32nd International Conference on Data Engineering (ICDE), pp. 942–953. IEEE (2016)

15. Yuan, Q., Cong, G., Ma, Z., Sun, A., Thalmann, N.M.: Who, where, when and what: discover spatio-temporal topics for twitter users. In: Proceedings of the 19th ACM SIGKDD International Conference on Knowledge Discovery and Data Mining, pp. 605–613. ACM (2013)

16. Yuan, Q., Cong, G., Zhao, K., Ma, Z., Sun, A.: Who, where, when, and what: a nonparametric Bayesian approach to context-aware recommendation and search for twitter users. ACM Trans. Inf. Syst. (TOIS) 33(1), 2 (2015)

17. Zhou, T., Cao, J., Liu, B., Xu, S., Zhu, Z., Luo, J.: Location-based influence maximization in social networks. In: Proceedings of the 24th ACM International on Conference on Information and Knowledge Management, pp. 1211–1220. ACM (2015)

A Unified Weakly Supervised Framework for Community Detection and Semantic Matching

Wenjun Wang[1,2], Xiao Liu[1,2], Pengfei Jiao[1,2], Xue Chen[1,2], and Di Jin[1,2(✉)]

[1] School of Computer Science and Technology, Tianjin University,
Tianjin 300350, China
{wjwang,xiaoxiao,pjiao,jindi}@tju.edu.cn, chenxuemail@163.com
[2] Tianjin Key Laboratory of Advanced Networking (TANK), Tianjin 300350, China

Abstract. Due to the sparsity of network, some community detection methods only based on topology often lead to relatively low accuracy. Although some methods have been proposed to improve the detection accuracy by using few known semi-supervised information or node content, the research of community detection not only pursues the enhancement of community accuracy, but also pays more attention to the semantic description for communities. In this paper, we proposed a unified nonnegative matrix factorization framework simultaneously for community detection and semantic matching by integrating both semi-supervised information and node content. The framework reveals two-fold community structures as well as their coupling relationship matrix, which helps to identify accurate community structure and at the same time assign specific semantic information to each community. Experiments on some real networks show that the framework is efficient to match each community with specific semantic information, and the performance are superior over the compared methods.

Keywords: Community detection · Nonnegative matrix factorization
Semi-supervised learning · Semantic matching

1 Introduction

The complex network is constituted by a group of entities and their interactive relationships. These direct or indirect interactions can partition the network into several functional communities, making which interact densely in each community and sparsely between them. For example, the protein network is partitioned into different functional units via the interaction among protein molecule. Therefore, the identification of these communities is helpful to understand how the network works and how the functional unit interacts. However, for many networks in real world, due to its community structure is very vague, it is very difficult to identify solely using the observed interactions. How to integrate the structural and semantic information to identify more accurate community structure and

© Springer International Publishing AG, part of Springer Nature 2018
D. Phung et al. (Eds.): PAKDD 2018, LNAI 10939, pp. 218–230, 2018.
https://doi.org/10.1007/978-3-319-93040-4_18

simultaneously assign an appropriate semantic description to each community is a worthy studying heat.

The early community detection methods only use network topology, including hierarchical clustering [1], spectral clustering [2], modularity optimization [3,4] and methods based on generative model [5]. However, for networks with sparse connections and vague community structure, these methods almost fail to accurately identify its community structure.

In order to uncover the vague community hidden in networks, it is necessary to exploit additional available prior information, and some semi-supervised community detection methods [6–10] have been proposed. Specifically, combined with both node labels and pairwise constraints, Eaton and Mansbach proposed a semi-supervised spin-model for community detection, which penalizes the term that violates the guidance and rewards the term that agrees with the guidance [6]. Based on latent space graph regularization, Yang et al. utilized must-link constraints to derive a unified semi-supervised community detection framework [8]. Zhang et al. directly used the pairwise constraints to modify the adjacency matrix of networks, and proposed a semi-supervised community detection framework [9,11]. Considering that the heterogeneity of node degree and community size may lower the utilization of prior constraints, Liu et al. developed a semi-supervised NMF community detection method with node popularity [10]. Indeed, the integration of semi-supervised prior information and network topology plays a vital role in assisting to reveal the vague community structure, but for very sparse networks, the semi-supervised prior cannot be effectively used, and usually has lower utilization. Moreover, it ignores the specific semantic of each community.

In addition, the node contents are often available. For example, a user of a social network often has a person profile with content information such as age, male, education background and profession; a paper in citation network often provides some contents information including author, title, abstract and key words. It is generally assumed that nodes of more similar contents information are more likely to belong to the same community. Therefore, node contents have been widely used to guide the community detection and depict the community semantic [12–14]. The early content-based methods handle the network topologies and content separately, and most of the methods just use node contents to improve the community detection accuracy and compensate the insufficiency of sparse topology. For example, by combining the user similarity, message similarity and user interaction, Pei et al. proposed a nonnegative matrix tri-factorization clustering framework to identify the community structure in a social network [15]. Recently, some researchers often use node contents to describe the semantic explanation for community, so as to further understand why some certain nodes belong to the same community, and what characteristics the community owns. From the perspective of content propagation, Liu et al. combined the topological structure as well as the content information to detect the community structure, and adopted the stable status of random walk to describe the semantic information of communities [16]. By integrating network topology and semantic information of nodes, Wang *et al.* proposed a novel nonnegative matrix factorization (NMF) model [17], and by defining two sets of parameters, the community membership matrix and community attribute matrix respectively, to infer the community structure and its corresponding semantic interpretation.

However, most of these newly proposed methods have three potential problems. Firstly, users tend to form a community due to their interactions. For sparse network, the relatively vague community structure is difficult to accurately identify, and node contents cannot assign appropriate semantic topic for each community when the identified community structure is wrong. Secondly, they generally believe that network topology and node content share the same community membership, but there may be more than one semantic topic for each community. Therefore, although the above methods can identify accurate community structure, they cannot assign correct semantic interpretation to a community. Finally, most of the existing methods utilize network topology and node contents separately, ignoring the relation between topology and content.

In this paper, for sparse networks we propose a unified weakly supervised framework for community detection and semantic matching (WSCDSM). Firstly, we incorporate network topology with must-link prior to derive an accurate topology-driven community (TC) membership, and then utilize node content information to obtain a semantic-driven community (SC) membership. Finally, by introducing a coupling matrix to portray the matching relation between TC and SC community structure, we integrate the above two process into WSCDSM framework to simultaneously detect community structure and match semantic. In our framework, two types of auxiliary information are seamlessly integrated to reveal the vague community structure and help to understand the practical semantic of communities. Consequently, the prior information and node contents are not only more effectively utilized, but also can complement some missing information of each other. We adopt an iterative method to train the TC (SC) community membership and its coupling relationship. Experimental results on several real networks validate that the proposed framework not only improves, as expected, the detection accuracy of vague communities, but also assign an appropriate semantic interpretation to each community.

The contributions of this work are as follows:

(1) Integrating with topological and content information as well as semi-supervised prior, we proposed a unified framework simultaneously for community detection and semantic matching. In this framework, we introduce coupling matrix to depict the relationship between community and semantic topic. Besides, it can also adjust the semantic information of each community.
(2) On the basis of using semi-supervised prior to improve the community accuracy, our proposed framework can integrate content information to compensate the insufficiency of topological information, and further assign more appropriate semantic information to each community.
(3) Our proposed framework is superior over the compared methods in most cases, and the improvement is more obvious on vary sparse network.

2   Proposed WSCDSM Framework

Considering an undirected attributed graph $\mathcal{G} = (\mathbf{V}, \mathbf{E}, \mathbf{S})$ of n nodes $\mathbf{V}$ and e edges $\mathbf{E}$, which often can be represented by a binary-valued adjacency matrix $\mathbf{A} \in \mathbb{R}^{n \times n}$ and an attribute matrix $\mathbf{S} \in \mathbb{R}^{n \times m}$ where m indicates the dimension of attributes each node has. $a_{ij} = 1$ if there is an edge between nodes v_i and v_j and $s_{ij} = 1$ if node v_i has the j-th attribute, and 0 otherwise. Our main task of this paper is to partition the network $\mathcal{G}$ into k communities with well matched semantic interpretation, and the goal is twofold:

(1) Partition the nodes into TC communities based on network topology and must-link prior, and separate the nodes into SC clusters based on nodes content;

(2) Finding the best matching relationship between the two type communities so as to best describe and understand the practical meaning of each community.

2.1   Modeling TC Communities

In this subsection, we utilize must-link constraint to derive an accurate TC community structure. Must-link constraint is a kind of commonly used prior information, which depicts whether two nodes belong to the same community and is helpful to improve the accuracy of community structure. We random select a few of must-link constraints and denote them as $\mathcal{C}_{ml}$. The corresponding must-link constraint matrix $\mathbf{M} \in \mathbf{R}^{n \times n}$ is defined as:

$$(\mathbf{M})_{ij} = \begin{cases} 1, \, if \, i = j, \\ 2, \, if \, (v_i, v_j) \in \mathcal{C}_{ml}, \\ 0, \, others. \end{cases}$$

Assume the TC community membership of all nodes in the network to be $\mathbf{H} \in \mathbf{R}^{n \times k}$, and h_{iz} represents the propensity that node v_i belongs to the z-th TC community. If two nodes belong to the same community, it is often believed that they have similar community membership and close with each other in their geometrical distance. In order to keep this property, we use the following graph regularization to incorporate the must-link constraint for helping reveal the TC community structure:

$$\min \sum_{ij} \|\mathbf{h}_i - \mathbf{h}_j\|^2 M_{ij}$$
$$s.t. \ \mathbf{H} \geq 0. \tag{1}$$

2.2   Modeling SC Communities

Define the semantic driven community membership to be $\mathbf{W} \in \mathbf{R}^{n \times k}$ where w_{ir} denotes the propensity that node v_i belongs to the r-th SC community. For each SC community, it carries some common semantic information which are summarized from the nodes' contents. On one hand, nodes in the same community usually have common contents. For another, if the contents of a node are highly similar to the semantic information of one SC community, the node may belong to this SC community with a high propensity. Therefor, nodes of the similar content may have high propensity constitute one SC community. Assume the common semantic matrix to be $\mathbf{C} \in \mathbf{R}^{m \times k}$, and $\mathbf{c}_r$ is the contents distribution of community r. Then for a node v_i, its propensity belonging to the r-th SC community can be written as:

$$W_{ir} = \mathbf{s}_i \cdot \mathbf{c}_r$$

where $\mathbf{s}_i$ represents the contents of node v_i.

In addition, we realize that each node has multiple contents, but only a small number of contents are relevant to each community and most of contents are background information. For this case, we adopt an l_1 norm to keep the sparse semantic interpretation of each community. Further more, in order to keep the balance of these sparse contents, it needs to impose a constraint $\sum_{r=1}^{k} \|\mathbf{c}(:,r)\|_1^2$ on $\mathbf{C}$. We can derive the SC community detection model as follows:

$$\min \|\mathbf{W} - \mathbf{SC}\|_F^2 + \xi \sum_{r=1}^{k} \|\mathbf{c}(:,r)\|_1^2 \qquad (2)$$
$$s.t. \ \ \mathbf{C} \geq \mathbf{0}.$$

2.3   The Unified Model: Matching TC with SC Communities

According to the above defined TC community membership $\mathbf{H}$ and SC community membership $\mathbf{W}$, we introduce a coupling matrix $\boldsymbol{\Lambda} \in \mathbf{R}^{k \times k}$ to measure how to match semantic information with topological communities, and simultaneously use the relationship of this three matrices to generate the observed network.

In our proposed WSCDSM framework, for any node v_i, it generates a link with node v_j based on the following rule:

(1) According to the SC community structure, node v_i has one kind of common content l with propensity w_{il};
(2) Then the l-th SC community assign its semantic information to the k-th TC community with coupling probability λ_{lk};
(3) As a result, the probability of existing a link between node v_i with common content l and node v_j of the k-th TC community is $w_{il}\lambda_{lk}h_{jk}$.

Summing over all the l and k, we derive the expect number of edge between nodes v_i and v_j is:

$$\widehat{a}_{ij} = \sum_{lk} w_{il}\lambda_{lk}h_{jk}.$$

Using the square error to measure the difference between expected and observed network, it can be further written in matrix formulation:

$$\min \|\mathbf{A} - \mathbf{W}\boldsymbol{\Lambda}\mathbf{H}^T\|_F^2 \qquad (3)$$
$$s.t. \ \ \mathbf{W}, \boldsymbol{\Lambda}, \mathbf{H} \geq \mathbf{0}.$$

By combining the model (3) with the models to derive TC community (1) and SC community (2), we obtain our proposed WSCDSM framework as follows:

$$
\min \|\mathbf{W} - \mathbf{SC}\|_F^2 + \xi \sum_{r=1}^{k} \|\mathbf{c}(:,r)\|_1^2 + \alpha \|\mathbf{A} - \mathbf{W\Lambda H}^T\|_F^2
$$
$$
+ \frac{\mu}{2} \sum_{ij} \|\mathbf{h}_i - \mathbf{h}_j\|^2 M_{ij} + \gamma \|\mathbf{\Lambda}\|_1 \tag{4}
$$
$$
s.t. \ \mathbf{W}, \mathbf{H}, \mathbf{\Lambda}, \mathbf{C} \geq \mathbf{0}
$$

where the parameters α and μ are, respectively, used to adjust the contribution of network topology and must-link prior. The parameter ξ and γ respectively control the sparsity of community common contents and coupling relationship.

3   Optimization

Due to the objective function in (4) is not convex with respect to $\mathbf{W}$, $\mathbf{H}$, $\mathbf{\Lambda}$ and $\mathbf{C}$, it is unreasonable to find its global minimum. Here we use an iteration algorithm to derive the update rule for each matrix by fixing other matrices.

Firstly, the update of $\mathbf{W}$ can be realized by optimizing the following W-subproblem with $\mathbf{H}$, $\mathbf{\Lambda}$ and $\mathbf{C}$ fixed:

$$
\min \|\mathbf{W} - \mathbf{SC}\|_F^2 + \alpha \|\mathbf{A} - \mathbf{W\Lambda H}^T\|_F^2
$$
$$
s.t. \ \mathbf{W} \geq \mathbf{0}. \tag{5}
$$

For the problem (5), we introduce a Lagrange multiplier matrix $\mathbf{\Psi}$ for the constraint $\mathbf{W} \geq \mathbf{0}$, and set the derivative of $\mathcal{L}$ with respect to $\mathbf{W}$ to $\mathbf{0}$, we obtain:

$$
2\mathbf{W} - 2\mathbf{SC} - 2\alpha \mathbf{AH\Lambda}^T + 2\alpha \mathbf{W\Lambda H}^T \mathbf{H\Lambda}^T + \mathbf{\Psi} = \mathbf{0}.
$$

Using the KKT condition $\Psi_{ik} w_{ik} = 0$, we obtain the following update rule for $\mathbf{W}$:

$$
W_{ik} \leftarrow W_{ik} \cdot \frac{(\alpha \mathbf{AH\Lambda}^T + \mathbf{SC})_{ik}}{(\alpha \mathbf{W\Lambda H}^T \mathbf{H\Lambda}^T + \mathbf{W})_{ik}}, \tag{6}
$$

Similarly, the update rules for $\mathbf{H}$ and $\mathbf{\Lambda}$ are as follows:

$$
H_{ik} \leftarrow H_{ik} \cdot \frac{(\alpha \mathbf{A}^T \mathbf{W\Lambda} + \mu \mathbf{MH})_{ik}}{(\alpha \mathbf{H\Lambda}^T \mathbf{W}^T \mathbf{W\Lambda} + \mu \mathbf{QH})_{ik}}, \tag{7}
$$

$$
\mathbf{\Lambda} \leftarrow \mathbf{\Lambda} \cdot \frac{\alpha \mathbf{W}^T \mathbf{AH}}{\alpha \mathbf{W}^T \mathbf{W\Lambda H}^T \mathbf{H} + \gamma \mathbf{E}}, \tag{8}
$$

where $\mathbf{E}$ is a $k \times k$ matrix with all element to be 1, and $\mathbf{Q}$ is a $n \times n$ diagonal matrix ($q_{ii} = \sum_{j} M_{ij}$ and $q_{ij} = 0$ if $i \neq j$).

As for the common content matrix $\mathbf{C}$, it is equivalent to the problem of Wang et al. [17]. The corresponding update rule for $\mathbf{C}$ is:

$$
\mathbf{C} \leftarrow \mathbf{C} \cdot \frac{\mathbf{S}_{new}^T \mathbf{W}_{new}}{\mathbf{S}_{new}^T \mathbf{S}_{new} \mathbf{C}}, \tag{9}
$$

where $\mathbf{S}_{new} = \begin{pmatrix} \mathbf{S} \\ \sqrt{\xi}\mathbf{e}_{1\times m} \end{pmatrix}$, $\mathbf{W}_{new} = \begin{pmatrix} \mathbf{W} \\ \mathbf{0}_{1\times k} \end{pmatrix}$ and $\mathbf{e}_{1\times m}$ is a row vector with all elements equal to 1, $\mathbf{0}_{1\times k}$ is a zero vector.

4   Experimental Results

We evaluate our WSCDSM framework on several real networks with well known communities to validate its accuracy of community detection, and on an online music system *Last.fm* to visualize the semantic information of communities.

4.1   The Performance of Community Detection

The real networks used in the experiments are shown in Table 1.

Table 1. Some real-world networks used.

Dataset	Nodes (n)	Edges (e)	Attributes (m)	Communities (k)
Cora	2708	5429	1433	7
Citeseer	3312	4732	3706	6
Texas	187	328	1703	5
Cornell	195	304	1703	5
Washington	230	446	1703	5
Wisconsin	265	530	1703	5

The Cora and Citeseer networks are both paper citation networks with nodes representing publications and edges denoting that one publication is cited by the other publication. The other four networks are all webpage citation networks where nodes representing webpages gathered from four different universities and edges denoting that one webpage is cited by the other webpage. The node attributes of all six networks are binary-valued word attributes indicating whether each word in the vocabulary is present (indicated by 1) or absent (indicated by 0).

In order to validate the efficiency of prior information and content information for community detection, we compare with the following four types of methods: the first type is only topology-based SNMF method [18]; the second type is only attribute-based SMR method [19] and the third type is two methods based on both network topology and node content, including SCI [17] and NEMBP [20]. In addition, we also compare with one method extracted from our WSCDSM framework, but it ignores the coupling matrix and only combines with must-link constraint. This method is denoted as MLNMF.

In the specific experiments, the number of communities is set to be the same as the ground truth specified. During each experiment, we iterate 2000 times and run 20 times. As for the parameter setting, we set $\alpha = 10$, $\mu = 20$, $\xi = 100$, $\gamma = 5$

for Cora and Citeseer networks and $\gamma = 0.5$ for the other four small networks. For the comparative methods, their parameters are set to be their default values.

In this paper, we only focus on the detection of disjoint community structure, and adopt the normalized mutual information (NMI) and accuracy (AC) to measure the performance of all methods against the ground truth. The results of our WSCDSM framework as well as other 5 comparative methods on Cora and Citeseer networks are shown in Tables 2 and 3, and on the remaining networks are shown in Figs. 1 and 2. From the Tables 2 and 3 and Figs. 1 and 2, we find that due to the sparsity of network and vagueness of community structure, the method only based on topology (SNMF) or content (SMR) almost fail to accurately identify its community structure. However, the detection accuracy can be further improved by integrating both topology and content. In our WSCDSM framework, we believe that the content and topology don't share the same community structure, and on the basis of using few semi-supervised prior to improve the accuracy of community detection, content information can be more effectively utilized to make up for the insufficiency of topology. Therefore, WSCDSM framework outperforms the other five comparative methods on most of networks, especially for Cora and Cornell networks, the improvement is more obvious. Although the randomness of prior information causes that the results of WSCDSM are not always higher than NEMBP on Wisconsin network, it will achieve superior performance when proper prior information is integrated.

Table 2. Comparative results in terms of NMI, and the best results are in bold.

Information used	Method	Cora			Citeseer		
Only topology	SNMF	0.1994			0.0403		
Only content	SMR	0.0078			0.0032		
Topology+Content	SCI	0.1780			0.0922		
	NEMBP	0.4408			0.2427		
Topology+Prior	MLNMF	2%	5%	8%	2%	5%	8%
		0.3159	0.3239	0.3451	0.2664	0.278	0.3081
Topology+Prior+Content	WSCDSM	**0.5254**	**0.7522**	**0.8083**	**0.3532**	**0.4297**	**0.4435**

Table 3. Comparative results in terms of AC, and the best results are in bold.

Information used	Method	Cora			Citeseer		
Only topology	SNMF	0.4173			0.2539		
Only content	SMR	0.3002			0.2111		
Topology+Content	SCI	0.4169			0.3442		
	NEMBP	0.5757			0.4951		
Topology+Prior	MLNMF	2%	5%	8%	2%	5%	8%
		0.4088	0.4106	0.4387	0.4109	0.4233	0.4598
Topology+Prior+Content	WSCDSM	0.5373	**0.7692**	**0.7906**	0.4761	**0.5136**	**0.5444**

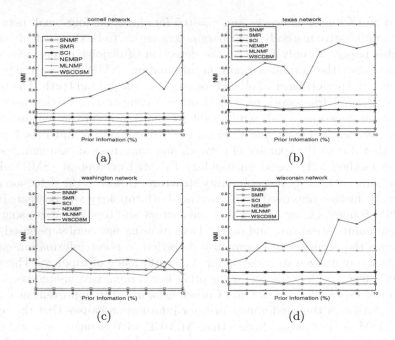

Fig. 1. Comparative results in terms of NMI on (a) Cornell network; (b) Texas network; (c) Washington network; (d) Wisconsin network.

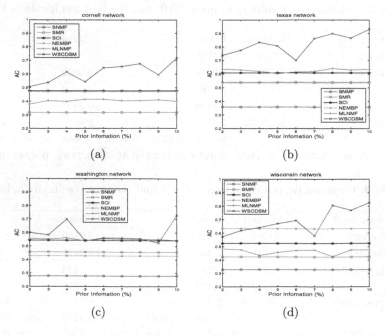

Fig. 2. Comparative results in terms of AC on (a) Cornell network; (b) Texas network; (c) Washington network; (d) Wisconsin network.

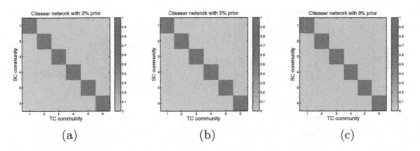

Fig. 3. The coupling relationship between TC and SC community structure on Citeseer network. (a) 2% prior used; (b) 5% prior used; (c) 8% prior used.

Based on the above results, we believe that the higher detection accuracy of TC community structure, the better matching of TC communities and semantic information. Due to the limited space, here we only take Citeseer network for an example to present the better matching between TC and SC community structure, as shown in Fig. 3. We find that each community has different semantic explanations with each other, and the semantic matching is robust to the increase of prior information.

4.2 The Matching Between Semantic and Communities

The *Lsat.fm* system contains 1892 users, and each user has 11,946 dimensional contents, including a list of most-listened-musical to artists and tag assignments, i.e. [user, tag, artist] tuples. Due to the *Lsat.fm* network has no ground truth with respect to the community label of node, we use *Louvain* method [3] as did in Wang et al. [17], but we set the number of communities to be 31, and the corresponding community structure is regarded as the ground truth.

The coupling relationship and semantic information of some communities are presented in Fig. 4. From the Fig. 4(a), we find that our WSCDSM framework can match most TC communities with one specific semantic topic, and only several TC communities have two or three semantic topics. Besides, there are also few communities that they have no semantic topic, which demonstrates the content information of such communities maybe background words. Figure 4(b) depicts a community of only one topic related to Britney Spears, a legend and amazing singer in Louisiana, USA. Her music often has characteristics of "pop", "dance", "rnb" and "electronic". An example community of two topics are shown in Fig. 4(c), this community is composed by a group fans who like "rock" and "heavy metal" two styles of music, and among which the style of "rock" music contains hard rock, classic rock and progressive rock. A community of three topics are illustrated in Fig. 4(d), which is characterized by three types of music including "synthpop", "new wave" and "electronic". For these three types music, Depeche Mode, a representative band, is very popular and active in British. Based on the above analysis, we find that our WSCDSM framework can relatively accurately match the community structure and semantic information.

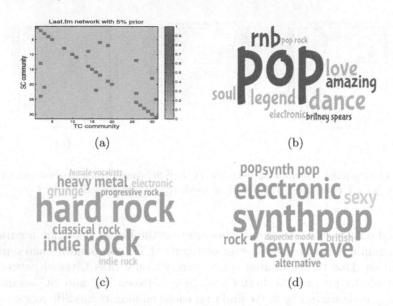

Fig. 4. The matching relationship between SC and TC community, as well as some examples of community interpretation on *Lsat.fm* network. (a) coupling relationship; community with (b) one topic; (c) two topics; (d) three topics.

5   Conclusion

In this paper, we proposed a unified weakly supervised framework simultaneously for community detection and semantic matching. In our framework, the semi-supervised information is firstly utilized to improve the community accuracy. Then by introducing a coupling matrix, the node content information is used to adjust the TC community structure and simultaneously match semantic interpretation for each community. The results on several real networks demonstrated that, for one thing, integrating with few percentage of must-link prior our framework can improve the accuracy of community detection. For another, under the guidance of coupling matrix, the TC community and SC community structure can realize fully interaction with each other, and further derive a well semantic description for communities.

Acknowledgments. This work was supported by the Major Project of National Social Science Fund(14ZDB153), the major research plan of the National Natural Science Foundation (91746205,91746107,91224009,51438009), and the Natural Science Foundation of China (61772361).

References

1. Girvan, M., Newman, M.E.J.: Community structure in social and biological networks. Proc. Natl. Acad. Sci. **99**(12), 7821–7826 (2002)
2. White, S., Smyth, P.: A spectral clustering approach to finding communities in graphs. In: Proceedings of the 2005 SIAM International Conference on Data Mining, Society for Industrial and Applied Mathematics, pp. 274–285 (2005)
3. Blondel, V.D., Guillaume, J.L., Lambiotte, R.: Fast unfolding of communities in large networks. J. Stat. Mech.: Theory Exp. **2008**(10), P10008 (2008)
4. Zhang, S., Zhao, H.: Normalized modularity optimization method for community identification with degree adjustment. Phys. Rev. E **88**(5), 052802 (2013)
5. He, D., Liu, D., Jin, D., Zhang, W.: A stochastic model for the detection of heterogeneous link communities in complex networks. In: Proceedings of the 29th AAAI Conference on Artificial Intelligence, Palo Alto, California, USA, pp. 130–136. AAAI Press (2015)
6. Eaton, E., Mansbach, R.: A spin-glass model for semi-supervised community detection. In: Proceedings of the 26th AAAI Conference on Artificial Intelligence, pp. 900–906 (2012)
7. Ma, X., Gao, L., Yong, X.: Semi-supervised clustering algorithm for community structure detection in complex networks. Physica A: Stat. Mech. Appl. **389**(1), 187–197 (2010)
8. Yang, L., Cao, X., Jin, D.: A unified semi-supervised community detection framework using latent space graph regularization. IEEE Trans. Cybern. **45**(11), 2585–2598 (2015)
9. Zhang, Z.Y.: Community structure detection in complex networks with partial background information. EPL (europhys. lett.) **101**(4), 48005 (2013)
10. Liu, X., Wang, W., He, D.: Semi-supervised community detection based on nonnegative matrix factorization with node popularity. Inf. Sci. **381**, 304–321 (2017)
11. Zhang, Z.Y., Sun, K.D., Wang, S.Q.: Enhanced community structure detection in complex networks with partial background information. Sci. Rep. **3**, 3241 (2013)
12. Ruan, Y., Fuhry, D., Parthasarathy, S.: Efficient community detection in large networks using content and links. In: Proceedings of the 22nd International Conference on World Wide Web, pp. 1089–1098. ACM (2013)
13. Pool, S., Bonchi, F., Leeuwen, M.: Description-driven community detection. ACM Trans. Intell. Syst. Technol. (TIST) **5**(2), 28 (2014)
14. Yang, T., Jin, R., Chi, Y.: Combining link and content for community detection: a discriminative approach. In: Proceedings of the 15th ACM SIGKDD International Conference on Knowledge Discovery and Data Mining, pp. 927–936. ACM (2009)
15. Pei, Y., Chakraborty, N., Sycara, K.: Nonnegative matrix tri-factorization with graph regularization for community detection in social networks. In: Twenty-Fourth International Joint Conference on Artificial Intelligence (2015)
16. Liu, L., Xu, L., Wangy, Z.: Community detection based on structure and content: a content propagation perspective. In: 2015 IEEE International Conference on Data Mining (ICDM), pp. 271–280. IEEE (2015)
17. Wang, X., Jin, D., Cao, X.: Semantic community identification in large attribute networks. In: AAAI, pp. 265–271 (2016)
18. Wang, F., Li, T., Wang, X.: Community discovery using nonnegative matrix factorization. Data Min. Knowl. Discov. **22**(3), 493–521 (2011)

19. Hu, H., Lin, Z., Feng, J.: Smooth representation clustering. In: Proceedings of the IEEE Conference on Computer Vision and Pattern Recognition, pp. 3834–3841 (2014)
20. He, D., Feng, Z., Jin, D.: Joint identification of network communities and semantics via integrative modeling of network topologies and node contents. In: Thirty-First AAAI Conference on Artificial Intelligence (2017)

Tapping Community Memberships and Devising a Novel Homophily Modeling Approach for Trust Prediction

Pulkit Parikh[1(✉)], Manish Gupta[1,2], and Vasudeva Varma[1]

[1] IIIT-Hyderabad, Hyderabad, India
pulkit.parikh@research.iiit.ac.in, {manish.gupta,vv}@iiit.ac.in
[2] Microsoft, Hyderabad, India
gmanish@microsoft.com

Abstract. Prediction of trust relations between users of social networks is critical for finding credible information. Inferring trust is challenging, since user-specified trust relations are highly sparse and power-law distributed. In this paper, we explore utilizing community memberships for trust prediction in a principled manner. We also propose a novel method to model homophily that complements existing work. To the best of our knowledge, this is the first work that mathematically formulates an insight based on community memberships for unsupervised trust prediction. We propose and model the hypothesis that a user is more likely to develop a trust relation within the user's community than outside it. Unlike existing work, our approach for encoding homophily directly links user-user similarities with the pair-wise trust model. We derive mathematical factors that model our hypothesis relating community memberships to trust relations and the homophily effect. Along with low-rank matrix factorization, they are combined into *chTrust*, the proposed multi-faceted optimization framework. Our experiments on the standard Ciao and Epinions datasets show that the proposed framework outperforms multiple unsupervised trust prediction baselines for all test user pairs as well as the low-degree segment, across evaluation settings.

1 Introduction

Trust relations between users are crucial for finding reliable information amidst massive volumes of social media data. While some web portals enable users to explicitly specify trust relations, these relations are usually very sparse [11], resulting in the need for trust inference. The objective of trust prediction is to estimate the likelihood of a trust relation for any ordered user pair in the network, using a relatively tiny number of user-specified trust relations and available auxiliary information such as the users' item ratings in the case of product review datasets. Explicit (given) trust relations on social media follow a power law distribution [11,13]. This combined with the huge ratio of user pairs without trust relations to those with trust relations makes trust inference challenging.

© Springer International Publishing AG, part of Springer Nature 2018
D. Phung et al. (Eds.): PAKDD 2018, LNAI 10939, pp. 231–243, 2018.
https://doi.org/10.1007/978-3-319-93040-4_19

Prior work on trust prediction can be categorized into unsupervised [2,3,7,11, 15] and supervised [6,8,16] approaches. Supervised techniques typically perform classification for an ordered user pair, represented by features extracted from the network structure and sometimes contextual data such as users' item ratings or reviews. Most unsupervised approaches employ trust propagation [2,3,7] or low-rank matrix factorization [11,15]. The success of propagation-based methods, which strongly rely on explicit connections, is jeopardized by the sparsity and power-law distribution of available trust relations [11]. Social theories, namely homophily and status theory, have been modeled for trust prediction in [11,15] respectively using matrix factorization.

In this paper, we investigate how community membership information can be tapped for the task of inferring trust relations. To the best of our knowledge, this is the first work that mathematically models an insight based on community memberships for trust prediction through an optimization framework. We hypothesize that a user's likelihood of forming a trust relation within the user's community is higher than (or equal to) that of trusting a user outside the community. Communities can be extracted either by clustering users on the basis of users' item ratings or from the available sparse trust network using any well-performing community detection technique. We then encode this insight into a matrix factorization based optimization framework through the derivation of a mathematical factor.

Moreover, we formulate a new method for modeling the homophily effect [12], which suggests that similar user pairs are more likely to have trust relations than dissimilar ones. Unlike in [11], which also performs homophily-based trust prediction, user-user similarities and the pair-wise trust model are directly linked in our optimization. Our approach to modeling homophily for learning the matrix factorization based trust model imposes costs directly in accordance with the degrees to which the pair-wise user similarities and the corresponding trust estimates as per the model follow the homophily effect. Moreover, our homophily-based factor comprises terms involving similar user pairs as well as dissimilar ones. We combine it with a complementary term from [11], matrix factorization and the proposed community-based factor to create a versatile, unsupervised trust prediction framework named *chTrust*.

Our main contributions are as follows. (1) We provide the hypothesis that a user's within-community trust propensity is higher than the across-community trust propensity and propose an approach for mathematically modeling it for inferring trust relations. (2) We devise a new formulation for modeling homophily that directly measures the extents to which the pair-wise trust estimates as per the matrix factorization model and the corresponding user-user similarity values conform to the homophily theory. We combine it and the community-based approach into *chTrust*, our multi-faceted unsupervised trust prediction framework. (3) Our results on two real-world datasets find our framework outperforming multiple unsupervised trust prediction baselines across evaluation settings, for the entire set of test user pairs as well as the low-degree segment.

2   Related Work

Binary classification based trust prediction approaches [6,16] use features extracted from network structure and/or users' items ratings or reviews. Besides the obvious drawback of requiring labeled data, classification-based methods face a severe class imbalance issue, because usually the user pairs without specified trust relations hugely outnumber the pairs with trust relations.

Unsupervised methods of trust prediction are usually based on trust propagation [2,3,7] or low-rank matrix factorization [11,15]. The efficacy of propagation-based methods may get adversely affected by the sparsity of available trust relations and the power law distribution (most users having few trustees and trustors) [11]. Tang et al. [11] capitalize on the social theory of homophily to predict trust by using low-rank matrix factorization to represent the matrix of trust relations. The homophily effect conveys that similar users are more likely to form trust relations. In [15], trust inference using matrix factorization is performed by tapping status theory, which suggests that a user is more likely to trust users with high statuses than those with low statuses. Our work falls in this direction of research where we hypothesize and model our intuition about the role of community memberships in trust formation for inferring trust relations. We also formulate a novel, direct approach for modeling homophily which is complementary to and combinable with [11].

3   Inferring Trust Using Community Memberships and Novel Homophily Modeling

In this section, we propose a principled formulation for taking advantage of the community memberships of users for predicting trust relations. We also develop a method for modeling homophily which is different from and complementary to existing work. We begin by briefly laying out the problem setting and the trust model based on low-rank matrix factorization that we use. We then discuss how to encode the community membership information mathematically into that trust model. Next, we devise a new homophily modeling approach that directly connects user-user rating similarities to the pair-wise trust model.

3.1   Problem Setting

Let $\mathbf{u} = \{u_1, u_2, ..., u_n\}$ denote the set of users where n is the number of users. The matrix $G \in \mathcal{R}^{n \times n}$ represents trust relations where $G(i,j) = 1$ if we observe that u_i trusts u_j and $G(i,j) = 0$ otherwise. As G is typically very sparse and low-rank, the matrix factorization model for trust relations aims to obtain a low-rank representation of users through the following optimization [11,15].

$$\min_{U,H} \|G - UHU^T\|_F^2 + \alpha\|U\|_F^2 + \beta\|H\|_F^2 \qquad s.t., U \geq 0, H \geq 0 \qquad (1)$$

Here, $U \in \mathcal{R}^{n \times d}$ captures user representations with d being the dimension of each representation. $H \in \mathcal{R}^{d \times d}$ encodes the correlations among these latent factors with the aim of expressing $G(i,j)$ as $U(i,:)HU^T(j,:)$. $||\cdot||_F$ denotes the Frobenius norm of a matrix. Note that besides the matrix factorization term, we also have two regularization terms, whose impacts are controlled by the parameters α and β.

3.2 Encoding Community Membership Information for Trust Prediction

We offer the hypothesis that a user's tendency to develop a trust relation outside the user's community is typically lower than that of trusting a user within the community. We do not claim that this phenomenon holds true for all individuals. However, we postulate that, for *most* users, the likelihood of trusting a fellow community member is greater than (or equal to) that of trusting a user outside the community.

We now mathematically encode our hypothesis such that it can be leveraged through an optimization framework for trust prediction. The hypothesis implies that, for every user u_i, the average of the trust estimates from u_i to the users outside the community that u_i belongs to should be no higher than u_i's average within-community trust.

Let $K_w \in \mathcal{R}^{n \times n}$ be a matrix such that $K_w(i,j) = 1$ if users u_i and u_j belong to the same community, and $K_w(i,j) = 0$ otherwise. Similarly, let $K_a \in \mathcal{R}^{n \times n}$ denote the across-community counterpart of K_w. Thus, $K_a = \mathbf{1} - K_w$, where $\mathbf{1}$ is a matrix of all ones. We perform a user-wise (row-wise) normalization on the matrices K_w and K_a to get S_w and S_a respectively.

$$S_w(i,j) = \frac{K_w(i,j)}{\sum_{j'} K_w(i,j')}, \qquad S_a(i,j) = \frac{K_a(i,j)}{\sum_{j'} K_a(i,j')} \qquad (2)$$

Then, the average within-community trust for a user u_i can be expressed as follows.

$$\sum_{j=1}^{n} S_w(i,j)U(i,:)HU^T(j,:) \qquad (3)$$

Further, the difference between u_i's average across-community and within-community trust estimates can be written as follows.

$$\sum_{j=1}^{n} S_a(i,j)U(i,:)HU^T(j,:) - \sum_{j=1}^{n} S_w(i,j)U(i,:)HU^T(j,:) = \sum_{j=1}^{n} U(i,:)HU^T(j,:)(S_a(i,j) - S_w(i,j)) \quad (4)$$

The quantity that we propose to minimize to encode our community-based hypothesis for trust relations is the sum of squares of the above term over all users. We now give a relaxed version in which we minimize the term above only if the average of the estimated across-community trust values is higher than the average of the estimated within-community trust values. The final community-based factor that we propose to minimize is as follows.

$$\sum_{i=1}^{n} \left(max \left\{ 0, \sum_{j=1}^{n} U(i,:)HU^T(j,:)(S_a(i,j) - S_w(i,j)) \right\} \right)^2 \qquad (5)$$

3.3 Proposed Homophily Modeling Using Users' Item Ratings

The homophily theory suggests that the more similar the users are to each other, the more likely they are to establish trust relations [12]. In order to incorporate the notion of similarity between users into our approach, we utilize the product ratings of users. More precisely, the cosine similarity between the rating vectors of two users is taken to be the similarity value for that user pair. Unlike [11] wherein the rating similarity between user u_i and u_j is used to influence the distance between $U(i,:)$ and $U(j,:)$, our approach for modeling homophily is aimed at linking the rating similarity directly to the pair-wise trust model. Our optimization formulation to update the model parameters involves penalties that are calculated directly as per the degrees to which the user-user rating similarities and the corresponding trust estimates by the model conform to the homophily theory. And, our mathematical factor for modeling homophily is composed of terms involving both similar and dissimilar user pairs. The most basic minimization formulation that we propose for modeling homophily for trust prediction is as follows.

$$\sum_{i=1}^{n}\sum_{j=1}^{n} \left\{ w_d(1 - Z(i,j))U(i,:)HU^T(j,:) + w_s Z(i,j)(1 - U(i,:)HU^T(j,:)) \right\}^2 \qquad (6)$$

We observe that the homophily theory implies two qualities to be desired. Firstly, for dissimilar user pairs, high estimated trust values should be penalized. Secondly, for similar user pairs, low estimated trust values should be penalized. The proposed formulation encodes precisely these two desirabilities, with the left term factoring in the first desirability and the right term applying the second one. Here, Z is the homophily coefficient matrix, where $Z(i,j)$ is the rating similarity between users u_i and u_j (e.g., the cosine similarity between the two users' rating vectors). w_d and w_s are hyper-parameters to be tuned to achieve the best combination of the two terms.

Equation 6 can be expressed compactly using matrices in the following way.

$$\| \{ w_d(1 - Z) - w_s Z \} \odot (UHU^T) + w_s Z \|_F^2 \qquad (7)$$

$$= Tr \left\{ \left[Q \odot (UHU^T) + w_s Z \right]^T \left[Q \odot (UHU^T) + w_s Z \right] \right\}, \text{ where } Q = w_d(1 - Z) - w_s Z \qquad (8)$$

$$= Tr \left\{ \left[W \odot (UH^T U^T) \right] (UHU^T) + 2 \left[Q^T \odot (UH^T U^T) \right] (w_s Z) \right\}, \text{ where } W = Q^T \odot Q^T \qquad (9)$$

$$= Tr \left\{ \left[W \odot (UH^T U^T) \right] (UHU^T) + 2(UH^T U^T L) \right\}, \text{ where } L = w_s(Z \odot Q) \qquad (10)$$

We modify Q as $max\{0, w_d(1 - Z) - w_sZ\}$ in Eq. 10 based on empirical and optimization-related observations.

4   The Overall Optimization Framework

Having proposed mathematical factors that model our insight involving community memberships of users and the homophily theory in novel ways, we now present *chTrust*, our complete optimization framework for trust prediction.

4.1   Trust Inference Using the Community-Based Factor

We incorporate the proposed mathematical factor that models our community-based hypothesis for trust prediction (Eq. 5) into the standard matrix factorization optimization (Eq. 1) by writing the following optimization problem.

$$\min_{U,H}\|G-UHU^T\|_F^2 + \lambda_cTr(M \odot M) + \alpha\|U\|_F^2 + \beta\|H\|_F^2 \quad s.t., U \geq 0, H \geq 0, \quad (11)$$

where $M = \max\{0, UHU^T(S_a{}^T - S_w{}^T)\}$ is the matrix form of Eq. 5. λ_c controls the contribution of the proposed community-based term on the objective function. Before proceeding with the optimization method, we reformulate the above equation without the max function in M. Let $P \in \mathcal{R}^{n \times n}$ be a diagonal matrix defined as follows.

$$P(i,i) = \begin{cases} 1, & \text{if } (UHU^T(S_a{}^T - S_w{}^T))(i,i) > 0 \\ 0 & otherwise \end{cases} \quad (12)$$

If the strict version of the community-based factor is desired, $P(i,i)$ should always be set to 1. In this work, we use the relaxed variant. Using P, the optimization problem can be rewritten as follows.

$$\min_{U,H}\|G - UHU^T\|_F^2 + \lambda_cTr((P \odot (UHU^T(S_a{}^T - S_w{}^T)))^2) + \alpha\|U\|_F^2$$
$$+\beta\|H\|_F^2 \quad s.t., U \geq 0, H \geq 0 \quad (13)$$

Owing to P being a diagonal matrix, $P \odot (UHU^T(S_a{}^T - S_w{}^T))$ is also a diagonal matrix, allowing us to replace the $\odot$ product in Eq. 11 with a square. This reformulation makes the objective function easier for us to optimize. The constraint that the elements of U and H must be non-negative entails that the standard gradient descent method used for unconstrained optimization problems is not applicable here. We adopt a multiplicative update approach [1,4,11] to solve Eq. 13.

Let $\wedge_1$ and $\wedge_2$ denote the Lagrangian multiplier matrix variables used to incorporate the non-negativity constraints for U and H. Then, the Lagrangian function for Eq. 13, after removing constants, can be written as follows.

$$L = \lambda_cTr((M_a - M_w)^2) + Tr(-2G^TUHU^T + UH^TU^TUHU^T) \quad (14)$$
$$+\alpha Tr(U^TU) + \beta Tr(H^TH) - Tr(\wedge_1U) - Tr(\wedge_2H),$$

where $M_a = P \odot (UHU^T S_a{}^T)$ and $M_w = P \odot (UHU^T S_w{}^T)$.

To obtain the multiplicative update rules, partial derivatives of L w.r.t. U and H need to be computed. Differentiating the community-based factor part of L w.r.t U, we get the following.

$$\frac{\partial(Tr((M_a - M_w)^2))}{\partial U} = 2\left\{((M_a - M_w)^T \odot P)(S_a - S_w)UH^T + (S_a - S_w)^T((M_a - M_w) \odot P^T)UH\right\}$$

$$= 2\left\{((M_a - M_w) \odot P)(S_a - S_w)UH^T + (S_a - S_w)^T((M_a - M_w) \odot P)UH\right\}, \quad (15)$$

because, for diagonal matrices M_a, M_w and P, $M_a{}^T = M_a$, $M_w{}^T = M_w$ and $P^T = P$. Next, differentiating the community-based factor part of L w.r.t. H, we get the following.

$$\frac{\partial(Tr((M_a - M_w)^2))}{\partial H} = 2U^T\left\{(M_a - M_w)^T \odot P\right\}(S_a - S_w)U = 2U^T\left\{(M_a - M_w) \odot P\right\}(S_a - S_w)U \quad (16)$$

Using Eqs. 15 and 16, the partial derivatives of L can be written as follows.

$$\frac{\partial L}{\partial U} = 2\lambda_c\left\{((M_a - M_w) \odot P)(S_a - S_w)UH^T + (S_a - S_w)^T((M_a - M_w) \odot P)UH\right\} \quad (17)$$
$$-2G^T UH - 2GUH^T + 2UH^T U^T UH + 2UHU^T UH^T + 2\alpha U - \wedge_1^T$$
$$\frac{\partial L}{\partial H} = 2\lambda_c U^T\left\{(M_a - M_w) \odot P\right\}(S_a - S_w)U - 2U^T GU + 2U^T UHU^T U + 2\beta H - \wedge_2^T \quad (18)$$

Given partial derivatives, the multiplicative updates [1,4,11] are constructed with a term based on the negative part of the partial derivative in the numerator and a term based on the positive part of the partial derivative in the denominator. Using the above equations, the multiplicative update rules can be written as follows.

$$U(l,k) \leftarrow U(l,k)\sqrt{\frac{T_1(l,k)}{T_2(l,k)}}, \qquad H(l,k) \leftarrow H(l,k)\sqrt{\frac{T_3(l,k)}{T_4(l,k)}}, \quad (19)$$

where

$$T_1 = \lambda_c\left\{(M_a \odot P)S_w UH^T + S_w{}^T(M_a \odot P)UH + (M_w \odot P)S_a UH^T + S_a{}^T(M_w \odot P)UH\right\}$$
$$+G^T UH + GUH^T \quad (20)$$
$$T_2 = \lambda_c\left\{(M_a \odot P)S_a UH^T + S_a{}^T(M_a \odot P)UH + (M_w \odot P)S_w UH^T + S_w{}^T(M_w \odot P)UH\right\}$$
$$+UH^T U^T UH + UHU^T UH^T + \alpha U \quad (21)$$
$$T_3 = \lambda_c\left\{U^T(M_a \odot P)S_w U + U^T(M_w \odot P)S_a U\right\} + U^T GU \quad (22)$$
$$T_4 = \lambda_c\left\{U^T(M_a \odot P)S_a U + U^T(M_w \odot P)S_w U\right\} + U^T UHU^T U + \beta H \quad (23)$$

It can be easily seen that U and V remain non-negative throughout this updation process. We now prove the correctness of these update rules by showing that they satisfy the Karush-Kuhn-Tucker (KKT) complementary slackness condition like in [1]. The KKT condition for the proposed optimization problem is as follows.

$$U(l,k) \wedge_1^T (l,k) = 0, \forall l \in [1,n], k \in [1,d], \quad H(l,k) \wedge_2^T (l,k) = 0, \forall l, k \in [1,d] \tag{24}$$

To prove the correctness of the update rules, we first set $\frac{\partial L}{\partial U}$ (Eq. 17) and $\frac{\partial L}{\partial H}$ (Eq. 18) to 0, resulting in the following.

$$(T_2 - T_1) = \wedge_1^T, \quad (T_4 - T_3) = \wedge_2^T \tag{25}$$

From the update rules in Eq. 19, it is easy to observe the following.

$$U(l,k)^2(T_2(l,k) - T_1(l,k)) = 0, \quad H(l,k)^2(T_4(l,k) - T_3(l,k)) = 0 \tag{26}$$

By combining Eq. 26 with Eq. 25, we have the following.

$$U(l,k)^2 \wedge_1^T (l,k) = 0, \quad H(l,k)^2 \wedge_2^T (l,k) = 0 \tag{27}$$

These equations are equivalent to the KKT complementary slackness condition (Eq. 24), thus proving the correctness of the update rules.

4.2 Trust Inference Through Homophily Modeling Using Users' Item Ratings

Now, we combine the homophily-based factor proposed in Sect. 3.3 with our community-based factor. For tapping users' item ratings based on the homophily phenomenon, in addition to the proposed term, we also use the homophily regularization term from [11]. Building further on Eq. 13 in Sect. 4.1, we write the overall optimization as follows.

$$\min_{U,H} \|G - UHU^T\|_F^2 + \lambda_c Tr((P \odot (UHU^T(S_a{}^T - S_w{}^T)))^2) + \lambda_t Tr(U^T(D - Z)U) \tag{28}$$
$$+ \lambda_h Tr\left\{ \left[W \odot (UH^TU^T)\right](UHU^T) + 2(UH^TU^TL) \right\} + \alpha\|U\|_F^2 + \beta\|H\|_F^2 \quad s.t., U \geq 0, H \geq 0$$

As defined in [11], D is a diagonal matrix such that $D(i,i) = \sum_{j=1}^n Z(j,i)$. λ_h and λ_t control the effects of the homophily-based terms.

The update rules for this modified optimization problem can be worked out in a manner similar to the one in Sect. 4.1. They involve adding the terms $\left[W \odot (UH^TU^T)\right]UH + \left[W^T \odot (UHU^T)\right]UH^T + LUH^T + L^TUH$ and $U^T\left[W^T \odot (UHU^T)\right]U + U^TLU$ to Eqs. 21 and 23 respectively.

4.3 The chTrust Algorithm

Algorithm 1 outlines the steps involved in the proposed framework, chTrust. In line 1, we calculate matrices based on community detection. Communities can be detected either from the existing (sparse) trust network using algorithms like Infomap [10] or by clustering users based on their item ratings. Line 2 involves computing the static matrices required to model our community-based hypothesis and homophily, using the community-based matrices computed in line 1 and the input data, in accordance with the equations in Sects. 4.1 and 4.2. Between lines 4 and 12, we show how to update our model parameters U and H alternately based on the multiplicative update rules detailed in Sects. 4.1 and 4.2 using the static matrices computed in line 2. Line 13 computes the estimated user-user matrix $\hat{G}$, where $\hat{G}(i, j)$ captures the likelihood of a trust relation from user u_i to user u_j. Finally, we perform the ranking of the input list of user pairs in accordance with $\hat{G}$.

Algorithm 1. Infer Trust by Modeling Community Membership Information: chTrust

Input	Sparse user-user matrix G capturing given trust relations; Hyper parameters λ_c, λ_t, λ_h, α and β;
	Set L of ordered user pairs to be ranked based on trust estimates; User-item ratings matrix R
Output	Ranked list of ordered user pairs in L based on the corresponding estimated trust likelihoods

1: Perform community detection on G; construct K_w and K_a.
2: Compute S_w and S_a for the community-based term, and W, Q, Z and D for the homophily-based terms.
3: Initialize U and H randomly.
4: **while** below the iteration upper bound **do**
5:     Compute P, M_a and M_w.
6:     Compute $T1$, $T2$, $T3$, and $T4$.
7:     **for** $l = 1$ to n **do**
8:         **for** $k = 1$ to d **do**
9:             $U(l, k) \leftarrow U(l, k)\sqrt{\frac{T_1(l,k)}{T_2(l,k)}}$
10:     **for** $l = 1$ to d **do**
11:         **for** $k = 1$ to d **do**
12:             $H(l, k) \leftarrow H(l, k)\sqrt{\frac{T_3(l,k)}{T_4(l,k)}}$
13: Set $\hat{G} = UHU^T$.
14: Rank ordered user pairs (u_i, u_j) in L as per $\hat{G}(i, j)$ in a decreasing order.

Computational Complexity Analysis. We go over Algorithm 1 step by step to determine its computational complexity. Two of the methods explored for the community detection part in line 1 are Infomap on the network of available trust relations and k-means clustering on the item rating vectors constructed for the users. The time complexities of these two methods are $O(nlog(n))$ and $O(nm)$ respectively, considering the network to be sparse for Infomap, and the iteration count and the number of clusters to be constants for k-means, where n is the number of users and m is the number of items [9]. Computing K_w, K_a, S_w, S_a, W and L in lines 1 and 2 takes $O(n^2)$ operations. Computing Z and D from R using cosine similarity in line 2 is $O(n^2 + nm)$. The cost of updating U and H in each iteration is $O(n^2d)$, since each term in $T_1(l, k)$, $T_2(l, k)$, $T_3(l, k)$ and $T_4(l, k)$ can be computed in $O(n^2d)$ as a sequence of multiplications of two matrices. Factoring in everything, the overall computational complexity of Algorithm 1 is $O(n^2d + nm)$.

In order to lower the computational cost, we employ sparse matrices for terms involving P, M_a, M_w, S_w and G. Moreover, while computing a matrix multiplication sequence, we choose the most computationally efficient parenthesization.

5   Experiments

In this section, we present the evaluation of the proposed framework and compare it with several unsupervised trust prediction methods. We begin by describing the datasets, the evaluation metric and the experiment settings. The code is available here: https://www.dropbox.com/s/co7qds5fr4xbdsk/chTrustCode.zip

5.1   Datasets

We use two real-world datasets[1], namely Ciao and Epinions, corresponding to product review sites wherein users explicitly mark trust relations and rate items. We filter out users with one or no trustor and similarly items with less than two ratings from each dataset, as done in prior works like [11]. Table 1 gives some statistics of the pre-processed datasets.

Table 1. General statistics for Ciao and Epinions datasets

	Ciao	Epinions
Number of users	6261	8460
Number of trust relations	109524	299563
Number of items	35339	43218
Number of ratings	197162	233510
Trust network density	0.0028	0.0042

5.2   Evaluation Metric and Experiment Settings

In view of our focus on unsupervised trust prediction approaches, we employ a ranking-based metric somewhat similar to the ones used in [5,11]. Let $A = \{(u_i, u_j)|G(i,j) = 1\}$ be the set of all ordered user pairs such that u_i trusts u_j. The user pairs in A are sorted on the basis of the time stamps corresponding to when the trust relations were formed, in the ascending order. The top $x\%$ of the relations in A are chosen as old (training) trust relations O and the last $y\%$ chosen as new (test) trust relations N. Let $B = \{(u_i, u_j)|G(i,j) = 0\}$ be a set of randomly chosen ordered user pairs such that a trust relation from u_i to u_j is not specified. In our experimentation, $|B| = 5 \times |N|$ and test set fraction $y = 25\%$. In the case of Ciao, since the time stamps are not available, we assume the given set of user-user trust relations to be in the ascending order w.r.t. the time of trust formation.

Before using G as an input to a trust predictor, we eliminate new (test) trust relations N from G by setting $G(i,j) \leftarrow 0$, $\forall (u_i, u_j) \in N$. Each trust predictor ranks user pairs in $B \cup N$ in the descending order of confidence, i.e., estimated trust likelihood. The top $|N|$ pairs from the ranked list returned by the trust predictor are taken as predicted trust relations C. Using these, the evaluation metric that we use, referred to as the prediction accuracy (PA), is calculated as $PA = \frac{100 \times |N \cap C|}{|N|}$.

---

[1] http://www.cse.msu.edu/~tangjili/trust.html.

Table 2. Trust prediction accuracy of various methods across different sizes of training data

Approach\ x	All user pairs						Low-degree user pairs only					
	Ciao			Epinions			Ciao			Epinions		
	75%	65%	50%	75%	65%	50%	75%	65%	50%	75%	65%	50%
triMF	25.843	21.791	19.088	38.258	19.656	10.577	14.019	11.95	10.178	17.125	6.482	3.197
sTrust	21.385	19.115	15.684	37.445	28.662	17.28	11.16	10.056	8.644	21.972	17.43	10.315
hTrust	25.835	21.619	15.562	44.115	30.369	19.804	13.631	11.631	8.505	29.725	20.077	12.041
chTrust	**35.705**	**29.526**	**28.067**	**44.339**	**30.54**	**19.952**	**24.574**	**20.53**	**17.154**	**29.804**	**20.111**	**12.218**

5.3 Baselines

Since our work involves unsupervised trust prediction, we provide a comparison against the following unsupervised methods. (1) triMF [11]: In this approach, low-rank matrix factorization is performed on the explicit trust relations matrix (Eq. 1) for trust prediction. (2) hTrust [11]: Trust relations are predicted by tapping the social theory of homophily using the item ratings of users through an approach involving matrix factorization. (3) sTrust [15]: This approach also uses matrix factorization, but infers trust by utilizing status theory as well. We do not compare against propagation-based trust prediction methods. As discussed in Sect. 2, the success of propagation-based techniques is endangered by the sparsity of available trust relations and the power law distribution [11]. For example, the absence of a propagation path from user u_i to u_j may cause propagation-based methods to fail to estimate trust from u_i to u_j [14]. Moreover, we compare against approaches using matrix factorization, namely [11,15], which in turn have reported superior performance to [3], a propagation-based method. The baselines considered differ in terms of the types of information used. Baselines 1 (triMF) and 3 (sTrust) use only the (trust) network information. Baseline 2 (hTrust) combines the network information and item ratings.

5.4 Results

Comparison with Other Unsupervised Methods: Table 2 compares the prediction accuracy (PA) of the proposed approach against various baselines for the Ciao and Epinions datasets, for varying amounts of training data (controlled by x denoting the % of available trust relations used for estimating trust likelihoods). The test data is kept fixed. For each method, the average prediction accuracy over multiple runs is reported. We use the same initialization values for U and H for all approaches using matrix factorization to provide a fair and meaningful comparison. For the λ hyper-parameter in hTrust, the values are set to 35 and 0.01 for Epinions and Ciao respectively. The values of the λ hyper-parameter in sTrust are set to 6 and 2.5 for Epinions and Ciao respectively. These hyper-parameter values are determined experimentally by searching over various possible configurations. For the proposed approach, λ_t, λ_c, λ_h and w_s are set to 20, 0.4, 1.5 and 0.5 for Epinions, and 0, 7, 15 and 1 for Ciao. For generating the community-based matrices, we choose k-means clustering using the rating vectors of users that we construct after replacing missing ratings values with appropriate averaged estimates. We set $\alpha = 0.1$, $\beta = 0.1$ and $w_d = 1$ across datasets for all methods to which the hyper-parameters apply. Expectedly, the accuracy numbers for all the methods improve, as we increase the size of the training data (while keeping the test data fixed).

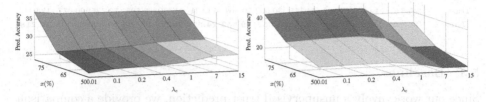

Fig. 1. Performance of the proposed approach for different λ_c values for Ciao

Fig. 2. Performance of the proposed approach for different λ_c values for Epinions

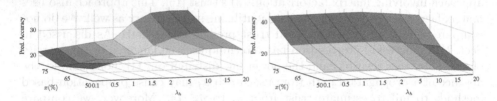

Fig. 3. Performance of the proposed approach for different λ_h values for Ciao

Fig. 4. Performance of the proposed approach for different λ_h values for Epinions

<u>Low Degree User Pairs:</u> Further, we investigate the efficacy of the proposed approach for users with relatively small numbers of trust relations. The right side of Table 2 captures the prediction accuracy numbers of all the methods for the low-degree user pairs in the test data. We define a user pair to be a low-degree one if either of the two users has a degree less than the mean degree. As seen in Table 2, our proposed approach (*chTrust*) outperforms all baselines across datasets and training data settings, for the entire set of test user pairs as well as the low-degree segment.

<u>Impact of Hyper-parameters:</u> Figures 1 and 2 show how the performance of our framework changes as λ_c is varied for the Ciao and Epinions datasets resp. Figures 3 and 4 capture *chTrust*'s accuracy variations across different λ_h values for Ciao and Epinions resp.

6   Conclusion

In this paper, we proposed a trust prediction approach by offering and modeling a hypothesis relating users' community memberships to trust relations. We also presented a new, direct way to model the homophily effect for inferring trust relations. We derived mathematical factors using low-rank matrix factorization based on these diverse entities, and combined them into our versatile unsupervised optimization framework for trust prediction.

References

1. Ding, C., Li, T., Jordan, M.I.: Nonnegative matrix factorization for combinatorial optimization: spectral clustering, graph matching, and clique finding. In: ICDM, pp. 183–192. IEEE (2008)
2. Golbeck, J.: Generating predictive movie recommendations from trust in social networks. In: Stølen, K., Winsborough, W.H., Martinelli, F., Massacci, F. (eds.) iTrust 2006. LNCS, vol. 3986, pp. 93–104. Springer, Heidelberg (2006). https://doi.org/10.1007/11755593_8
3. Guha, R., Kumar, R., Raghavan, P., Tomkins, A.: Propagation of trust and distrust. In: WWW, pp. 403–412. ACM (2004)
4. Lee, D.D., Seung, H.S.: Algorithms for non-negative matrix factorization. In: NIPS, pp. 556–562 (2001)
5. Liben-Nowell, D., Kleinberg, J.: The link prediction problem for social networks. JASIST 58(7), 1019–1031 (2007)
6. Liu, H., Lim, E., Lauw, H., Le, M., Sun, A., Srivastava, J., Kim, Y.: Predicting Trusts among users of online communities: an epinions case study. In: EC, pp. 310–319 (2008)
7. Massa, P., Avesani, P.: Controversial users demand local trust metrics: an experimental study on epinions.com community. In: AAAI, vol. 5, pp. 121–126 (2005)
8. Nguyen, V., Lim, E., Jiang, J., Sun, A.: To trust or not to trust? Predicting online trusts using trust antecedent framework. In: ICDM, pp. 896–901. IEEE (2009)
9. Papadopoulos, S., Kompatsiaris, Y., Vakali, A., Spyridonos, P.: Community detection in social media. DMKD 24(3), 515–554 (2012)
10. Rosvall, M., Bergstrom, C.T.: Maps of random walks on complex networks reveal community structure. Natl. Acad. Sci. 105(4), 1118–1123 (2008)
11. Tang, J., Gao, H., Hu, X., Liu, H.: Exploiting homophily effect for trust prediction. In: WSDM, pp. 53–62 (2013)
12. Tang, J., Chang, Y., Liu, H.: Mining social media with social theories: a survey. SIGKDD Expl. 15(2), 20–29 (2014)
13. Tang, J., Gao, H., Liu, H., Das Sarma, A.: eTrust: understanding trust evolution in an online world. In: KDD, pp. 253–261. ACM (2012)
14. Tang, J., Liu, H.: Trust in social media. Synth. Lect. Inf. Secur. Priv. Trust 10(1), 1–129 (2015)
15. Wang, Y., Wang, X., Tang, J., Zuo, W., Cai, G.: Modeling status theory in trust prediction. In: AAAI, pp. 1875–1881 (2015)
16. Zolfaghar, K., Aghaie, A.: A syntactical approach for interpersonal trust prediction in social web applications: combining contextual and structural data. Knowl.-Based Syst. 26, 93–102 (2012)

Deep Learning Theory and Applications in KDD

Text-Visualizing Neural Network Model: Understanding Online Financial Textual Data

Tomoki Ito[1]($\boxtimes$), Hiroki Sakaji[1], Kota Tsubouchi[2], Kiyoshi Izumi[1], and Tatsuo Yamashita[2]

[1] The University of Tokyo, Tokyo, Japan
{m2015titoh,sakaji}@socsim.org, izumi@sys.u-tokyo.ac.jp
[2] Yahoo Japan Corporation, Tokyo, Japan
{ktsubouc,tayamash}@yahoo-corp.jp

Abstract. This study aims to visualize financial documents to swiftly obtain market sentiment information from these documents and determine the reason for which sentiment decisions are made. This type of visualization is considered helpful for nonexperts to easily understand technical documents such as financial reports. To achieve this, we propose a novel interpretable neural network (NN) architecture called gradient interpretable NN (GINN). GINN can visualize both the market sentiment score from a whole financial document and the sentiment gradient scores in concept units. We experimentally demonstrate the validity of text visualization produced by GINN using a real textual dataset.

Keywords: Interpretable neural network · Text mining
Support system

1 Introduction

1.1 Motivation and Purpose

Understanding technical documents such as financial reports and legal documents is often difficult for nonexperts. One of the reasons is that the meaning of a word or phrase in a specific domain may differ from the general meaning. For example, the word "climb" generally has a neutral sentiment, but in the financial domain, it means a price rise and has a positive sentiment; in this context, its meaning is similar to "increase", "rise", "boost", and "boom". This research aims to present sentiments and concepts included in words and phrases that appear in specialized documents and help nonexperts understand these documents. Therefore, a keyword list containing sentiments and similarity information in specialized fields is necessary; however, manually building a

Electronic supplementary material The online version of this chapter (https://doi.org/10.1007/978-3-319-93040-4_20) contains supplementary material, which is available to authorized users.

© Springer International Publishing AG, part of Springer Nature 2018
D. Phung et al. (Eds.): PAKDD 2018, LNAI 10939, pp. 247–259, 2018.
https://doi.org/10.1007/978-3-319-93040-4_20

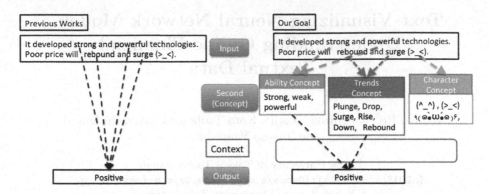

Fig. 1. Previous visualization methods (left side) vs. our visualization goal (right side)

keyword list for each specialized area requires enormous effort. Therefore, we develop a method for constructing a keyword list from specialized documents using neural networks. We then propose a method of visualizing financial texts for nonexperts.

As an example, consider the sentence "It developed strong and powerful technologies. Poor price will rebound and surge." We aim to visualize this sentence on the right side of Fig. 1 in the following steps.

Step 1. "Strong" and "Powerful" are positive in the sense of the Trend concept, and "Rebound" and "Surge" are positive in the sense of Ability concept.

Step 2. "Trends" and "Ability" concepts are important in this context.

Step 3. Therefore, this sentence is positive.

We define a set of synonyms and antonyms as *concept cluster* and sense of each concept cluster as *concept*. It would be helpful to describe some terms in each concept cluster for capturing the sense of the concept. By visualizing texts in the above manner, even nonexperts can easily capture the market sentiments of financial documents and explain the process of market sentiment analysis.

1.2 Main Approach and Problem Settings

Our aim is to develop market sentiment analysis models that can visualize documents as shown on the right side of Fig. 1. It is certain that linear models like support vector machine (SVM) [1] and methods for interpreting NNs [2,3] can be useful for text visualization. Using these previous works, the visualization as shown on the left side of Fig. 1 can be realized. However, visualizing texts as shown on the right side of Fig. 1 by simply using these previous works is difficult because they alone cannot represent concepts. To achieve our goal, we propose a novel interpretable NN architecture called *gradient interpretable neural networks* (*GINN*) as shown in Fig. 2. Layers of GINN can be interpreted as follows.

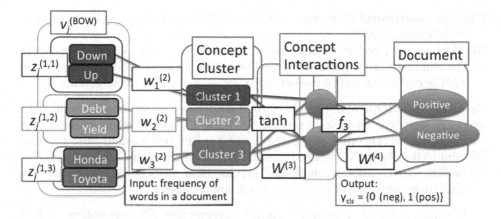

Fig. 2. GINN architecture

The input layer represents the words in a document. Each node in the input layer corresponds to a word.

The second layer (concept layer) represents the sentiment scores of concept units. Each node in the second layer corresponds to a concept.

The output layer represents an entire sentiment value of the document.

Using GINN, we can visualize text in the following steps:

Step 1: Extract word sentiment scores from the weight matrix between the input and second layers and concept sentiment scores from the second layer.

Step 2: Extract concept clusters that are important for the sentiment analysis decision using the gradient method [2].

Step 3: Extract an entire sentiment score of a document from the output layer.

To conduct the above text-visualization accurately, GINN must satisfy the following three conditions:

Condition 1: the connections between input and second layer nodes are determined by cluster analysis: if word X is in concept cluster Y, there is a link between X and Y,

Condition 2: when word X is in concept cluster Y, the value of the link between X and Y corresponds to the sentiment score of X, and

Condition 3: the output layer value is valid.

To evaluate whether Conditions 1–3 are satisfied and the validity of the text visualization by GINN, we evaluate the following *Interpretability, Cluster interpretability* and *Market mood predictability*.

Interpretability refers to the degree of accuracy with which the sentiment scores of words can be extracted from only weight matrix values between the input and second layers. Here, we consider words that frequently appear in positive (negative) documents than negative (positive) ones as positive (negative). We aim to satisfy Condition 2 by improving the interpretability.

Cluster interpretability refers to the validity of word clustering in the process of developing GINN.

Market mood predictability refers to the validity of the output layer value. We aim to satisfy Condition 3 by achieving high market mood predictability. This is equivalent to the predictability for an entire document sentiment.

By *clustering interpretability* and *interpretability*, we can evaluate the validity of Step 1 in the text visualization process by GINN. By *market mood predictability*, we can evaluate the validity of Steps 2 and 3 in the text visualization process. We aim to develop GINN whose structure satisfies Condition 1 and whose *interpretability*, *cluster interpretability* and *market mood predictability* are high.

The main contributions of our research are as follows.

- We proposed and developed a novel interpretable NN architecture called *GINN* that can visualize financial texts in the way as shown in Fig. 1 using a novel method, *Importance of infiltration (II) algorithm* (Sect. 2).
- We experimentally demonstrated validity for the text visualizations by *GINN*. (Sect. 3).

2   Importance of Infiltration (II) Algorithm

This section introduces the framework for developing *GINN*. We develop *GINN* according to the following steps.

Step 1. Prepare a dataset of documents and their positive or negative tags.
Step 2. Cluster words and construct the NN model (Subsect. 2.1).
Step 3. Initialize parameter values using *Init* and obtain parameter values from the learning process using *Update* (Subsect. 2.2).

We refer to the series of flows from Step 2 to Step 3 as the II algorithm. Conditions 1 and 2–3 in Sect. 1.1 are realized by Step 2 and Step 3, respectively. We develop the II algorithm based on the following two ideas:

1. Assigning sentiment scores from a manually created polarity dictionary to specific edges between the input and second layers, and propagating the sentiment scores to the other edge values through the learning process. Consequently, each unit in the second layer will represent its sentiment information.

2. Necessitating the addition of certain limitations for the polarity propagation process, and such limitations should not reduce the *market mood predictability* of the model.

Ideas 1 and 2 are realized by *Init* and *Update* in Step 3, respectively.

2.1   Setup of NN Model

To cluster words, we represent each word as a numerical vector using word2vec [4]. For a given number of clusters, K, we cluster similar words into the same cluster using the spherical K-means method [5] by their cosine distances. These clusters correspond to *concept clusters*. Using the results of clustering words, we construct an NN model that satisfies Condition 1 using the following layers:

Input Layer: We assign a cluster number, k $(k = 1, 2, \cdots, K)$ to each cluster and an ID number in the cluster to each word. Let $w_{k,i}$ be a word that is included in the kth cluster and whose ID number in the cluster is i, $z_{j,i}^{(1,k)}$ be the frequency of the word $w_{k,i}$ in a document j, $n(k)$ be the number of words included in the kth cluster, m be $\sum_{k=1}^{K} n(k)$, and $z_j^{(1,k)}$ be $\left[z_{j,1}^{(1,k)}, z_{j,2}^{(1,k)} \cdots, z_{j,n(k)}^{(1,k)} \right]^T$. We represent the input vector value $v_j^{(\mathrm{BOW})} \in \mathbb{R}^m$ (i.e., the frequencies of the words that appear in document j) as $v_j^{(\mathrm{BOW})} := [z_j^{(1,1)T}, z_j^{(1,2)T}, \cdots, z_j^{(1,K)T}]^T$.

Second (Concept) Layer: We set the second-layer vector, $v_j^{(\mathrm{CS})} \in \mathbb{R}^K$, as

$$v_j^{(\mathrm{CS})} := \tanh([z_j^{(1,1)} \cdot w_1^{(2)}, \cdots, z_j^{(1,K)} \cdot w_K^{(2)}]^T)$$

where $w_k^{(2)} \in \mathbb{R}^{n(k)}$ for each k. Let $w_{k,i}^{(2)}$ be the ith element of $w_k^{(2)}$ and $v_{k,j}^{(\mathrm{CS})}$ be the kth element of $v_j^{(\mathrm{CS})}$. If $w_{k,i}^{(2)}$ represents the sentiment score of word $w_{k,i}$, then $v_j^{(\mathrm{CS})}$ represent the sentiment scores of concept cluster units.

Output Layer: Let $W^{(3)} \in \mathbb{R}^{K2 \times K}$ be the weight matrix between the second and third layers, $W^{(4)} \in \mathbb{R}^{2 \times K2}$ be the weight matrix between the third and output layers, $w_i^{(l)T}$ and $w_{i,j}^{(l)}$ be the ith row and the (i, j) component of $W^{(l)}$ ($l = 3, 4$), and $b_0 \subset \mathbb{R}^2$ be the bias vector. Here $K2$ is a scalar value. We represent the output layer value as

$$y_j = \mathrm{Softmax}(W^{(4)} f_3(W^{(3)} v_j^{(\mathrm{CS})}) + b_0), y_j^{(cls)} = \mathrm{argmax}\ y_j,$$

where $y_j^{(cls)} \in \{0 \text{ (negative)}, 1 \text{ (positive)}\}$ is the output layer value that corresponds to the predicted tag for the document j. We set f_3 to be tanh.

2.2 Initialization and Learning of Parameters

Initialization. After constructing the NN model, we initialize $w_{k,i}^{(2)}$ using a manually-created polarity dictionary. Let $PS(w_{k,i})$ be the sentiment score for $w_{k,i}$ given by the polarity dictionary. We set the initial value of $w_{k,i}^{(2)}$ as

$$w_{k,i}^{(2)} = \begin{cases} PS(w_{k,i}) & (w_{k,i} \text{ is included in the polarity dictionary}) \\ 0 & \text{(otherwise)} \end{cases}.$$

This initialization strategy realizes Idea 2, and we refer to this as *Init*.

Learning. We determine the parameter values not in $\{w_k^{(2)}\}_{k=1}^K$ via the general backpropagation method with the softmax cross entropy as a loss function. However, we determine the values of $\{w_k^{(2)}\}_{k=1}^K$ by updating $\{w_k^{(2)}\}_{k=1}^K$ according to Algorithm 1 (called as *Update*) in each training iteration. In *Update*, using $H^{*(j,t)}$ instead of $H^{(j,t)}$ is specific and necessary for realizing the high *interpretability* of *GINNs*.

The values of $w_k^{(2)}$ change during the learning process by the propagation of the sentiment scores from the dictionary (Fig. 4). After the learning stage is completed, we obtain the sentiment scores of unknown words by extracting the $\boldsymbol{w}_k^{(2)}$ values. The value of $w_{k,i}^{(2)}$ corresponds to the sentiment score of word $w_{k,i}$.

Algorithm 1. Update strategy of $\{\boldsymbol{w}_k^{(2)}\}_{k=1}^K$ in the tth training iteration (*Update*)

Input: $\{\boldsymbol{w}_k^{(2)}\}_{k=1}^K$, $\boldsymbol{W}^{(4)}$, $\boldsymbol{W}^{(3)}$, minibatch dataset in the tth training iteration Ω_m;

1: **for** $j \in \Omega_m$ **do**

2: $\quad d_j := \begin{cases} (0,1)^T & (j \text{ is positive}) \\ (1,0)^T & (j \text{ is negative}) \end{cases}$, $\boldsymbol{u}_j^{(2)} := \tanh^{-1}(\boldsymbol{v}_j^{(CS)})$, $\boldsymbol{u}_j^{(3)} := \boldsymbol{W}^{(3)} \boldsymbol{v}_j^{(CS)}$;

3: $\quad \Delta_j^{(4)} := \boldsymbol{y}_j - \boldsymbol{d}_j$; $\boldsymbol{H}^{(j,t)} := \boldsymbol{W}^{(4)} \text{diag}(f_3'(\boldsymbol{u}_j^{(3)})) \boldsymbol{W}^{(3)} (\in \mathbb{R}^{2 \times K})$;

4: $\quad \boldsymbol{H}^{*(j,t)} \in \mathbb{R}^{2 \times K} \leftarrow$ zeros; Here, $(\boldsymbol{H}^{*(j,t)})_{l,k} = h^{*(j,t)}_{l,k}$ and $(\boldsymbol{H}^{(j,t)})_{l,k} = h^{(j,t)}_{l,k}$.

5: $\quad$ **for** $k \leftarrow 1$ **to** K **do**

6: $\quad\quad$ **if** $h_{1,k}^{(j,t)} < 0$ **then** $h_{1,k}^{*(j,t)} \leftarrow h_{1,k}^{(j,t)}$; **if** $h_{2,k}^{(j,t)} > 0$ **then** $h_{2,k}^{*(j,t)} \leftarrow h_{2,k}^{(j,t)}$;

7: $\quad \Delta_j^{(2)*} := (1 - \tanh^2(\boldsymbol{u}_j^{(2)})) \odot (\boldsymbol{H}^{*(j,t)T} \Delta_j^{(4)})$,

8: **for** $k \leftarrow 1$ **to** K **do**

9: $\quad \partial w_k^{(2)*} := \frac{1}{N} \sum_{j \in \Omega_m} \Delta_{k,j}^{(2)*} z_j^{(1,k)}$ where $\Delta_{k,j}^{(2)*}$ is the kth component of $\Delta_j^{(2)*}$;

10: $\quad$ Update $\boldsymbol{w}_k^{(2)}$ using $\partial w_k^{(2)*}$ instead of using the gradient value of $\boldsymbol{w}_k^{(2)}$;

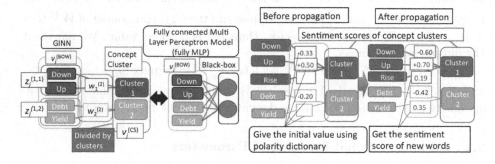

Fig. 3. GINN vs MLP (fully)   **Fig. 4.** Polarity propagation process

2.3   Proposed and Baseline Models

We introduce two types of baseline models: base multilayer perceptron (MLP), plus MLP, and our proposed model, GINN. Their structures are constructed as discussed in Subsect. 2.1, but they exhibit the following differences (Fig. 3).

In **base MLP**, neither Init nor Update is used (i.e., developed by the general backpropagation method).

In **plus MLP**, Init is not used; however, Update is used.

In **GINN**, both Init and Update are used (i.e., developed by the II algorithm).

Let t^+ and t^- be positive values, $\Omega_{pw}^{(k,t^+)}$ (*positive word set*) be a set of words that satisfy $p^+(w_{k,i}) > t^+$ and whose cluster number is k, and $\Omega_{nw}^{(k,t^-)}$ (*negative word set*) be a set of words that satisfy $p^-(w_{k,i}) > t^-$ and whose cluster number is k. We can theoretically explain that the II algorithm develops GINNs whose *interpretability* and *market mood predictability* are both high in the ideal case: the II algorithm assigns the value of $w_{k,i}^{(2)}$ to a positive value if $w_{k,i} \in \Omega_{pw}^{(k,t^+)}$, and a negative value if $w_{k,i} \in \Omega_{nw}^{(k,t^+)}$ obtaining a local optimization solution in the ideal case (from Propositions 1–3 in Appendix A).

3 Text Visualization Demonstration Using Real Data

This section applies our text-visualization method for financial textual data. First, we evaluate our method in terms of interpretability, clustering interpretability, and market mood predictability (introduced in Subsect. 1.2). Then, we present an example of text-visualization produced by GINN.

3.1 Dataset and Model Development

We used a dataset constructed from posts on the Yahoo! Finance Board[1] between September 1, 2015 and September 30, 2015 and their sentiment tags (i.e., *Yahoo! dataset*). We extracted all the posts tagged as negative (want to sell strongly) or positive (want to buy strongly), and sorted them in descending order by the date when they were posted. We then divided them into five equal parts while maintaining the order for a five fold cross-validation. After that, we prepared five train-validation and test dataset pairs by extracting each part in turn for use as the test dataset and using the remaining four parts as the train-validation dataset. We randomly extracted 10% of the train-validation data, taking equal percentages of samples from each class, for use as validation data. The remaining train-validation data were used as training data. The numbers of negative and positive posts were 15,887 and 50,843, respectively, and m was 28,261.

Using each train-validation data, we developed the following five prediction models for the evaluations: SVM, fully connected MLP (fully MLP), base MLP, plus MLP and GINN. Here, fully, plus and base MLPs and GINN had four layers, and the kernel of SVM was linear. The hyper-parameters were determined using the validation data, and we used stratified sampling [6], the Adam optimizer [7], and Dropout [8]. The number of words that were included in the manually created polarity dictionary and used in the process of *Init* was 285. See the supplementary material[2] for the details.

3.2 Interpretability Evaluation

We evaluated each model's interpretability by the following $Fw_{Sew,D}^{t^+,t^-}$ score.

---

[1] http://textream.yahoo.co.jp/category/1834773.
[2] In http://socsim.t.u-tokyo.ac.jp/wp/index.php/2017/11/15/titoh/ginn/.

Step 1: We set positive and negative word sets, $\Omega_{pw}^{(k,t+)}$ and $\Omega_{nw}^{(k,t^-)}$, for each k using a document dataset D according to Subsect. 2.3. For each word $w \in S^{ew}$, we assign a positive (negative) label for the answer label if $w \in \cup_{k=1}^{K} \Omega_{pw}^{(k,t^+)}$ ($w \in \cup_{k=1}^{K} \Omega_{nw}^{(k,t^-)}$).

Step 2: We assigned a positive or negative label for the prediction label to each word $w \in S^{ew}$ using the prediction model. For the GINN and the plus and base MLPs, we assigned word $w_{k,i}$ a positive (negative) label if $w_{k,i}^{(2)} > 0$ ($w_{k,i}^{(2)} < 0$).

Step 3: We evaluated each method by the macro F_1 score for the answer and prediction labels (defined as $Fw_{S^{ew},D}(t^+, t^-)$). We set S_D^{ew} to be a set of words that appear more than ten times in D and were not included in the manually created polarity dictionary, t^+ to be the mean value of $\{p^+(w)|w \in S_D^{ew}\}$ and t^- to be $1 - t^+$. We evaluated methods in both the case where D was a training dataset and that where D was a union set of validation and test datasets (i.e., test-valid dataset). We compared the results of plus and base MLPs and GINN.

Results. The first and second columns of Table 1 summarize the results. GINN shows significant improvement over baseline approaches: base and plus MLPs.

Table 1. Fw scores are F_1 score results for interpretability: "train" and "test-valid" mean the case where D is a training dataset and that where D is a test-valid dataset, respectively. HF scores are F_1 score results for human interpretability.

Methods	Fw score		HF
	Training dataset	Test-valid dataset	
Base MLP (baseline model)	0.488	0.493	0.465
Plus MLP (baseline model)	0.516	0.506	0.484
GINN (proposed model)	**0.739**	**0.630**	**0.742**

Discussion: These results demonstrate that the II algorithm realized the high interpretability of the GINN as intended. To measure the limit value for interpretability, we also measured how much $Fw_{S^{ew},D}^{t^+,t^-}$ scores could be produced by other high-performance methods for assigning sentiment scores to words: the gradient method with fully MLP and the SVM method. Such methods cannot achieve our goal because they cannot visualize concept cluster information. For the gradient method with fully MLP, we assigned each word $w \in S^{ew}$ a positive (negative) label if the input gradient value corresponding to the word w calculated by the gradient method [2] and the fully MLP model was positive (negative) (See the supplementary material[2] for the details). For the SVM method, we assigned each word $w \in S^{ew}$ a positive (negative) label if the support vector value corresponding to word w was positive (negative). The $Fw_{S^{ew},D}^{t^+,t^-}$

scores in the case where D was a training dataset and that where D was a valid-test dataset were 0.704 and 0.604, respectively, for the SVM method and 0.753 and 0.620, respectively, for the gradient method with fully MLP. These results show that GINN was able to produce more satisfactory results than other methods when D was a valid-test dataset, demonstrating the high interpretability of GINN.

Human Interpretability Evaluation (Additional Evaluation): We also evaluated word sentiment scores given by GINN in terms of whether they fit peoples' feelings. We randomly extracted 100 posts tagged as negative and positive from the test dataset. Three individual investors then manually extracted important words for the sentiment decision from each post and tagged them as positive or negative. We evaluated the models by their ability to accurately assign sentiment tags to these words in the same way as Step 3. We used the mean F_1 score for the three investors as the evaluate base (i.e., HF score). The right column of Table 1 summarizes the result, showing that GINN had more satisfactory results than the base and plus MLPs. Moreover, HF scores for the SVM method and the gradient method with fully MLP were 0.753 and 0.759, respectively, close to the HF score of GINN. Thus, we consider that sentiment scores given to terms by GINN sufficiently fit peoples' feelings.

3.3 Clustering Interpretability Evaluation

We briefly checked the validity of word clustering in the II algorithm. After deciding the cluster number K as 1000 and clustering words appeared in the Yahoo! dataset using the clustering method in Subsect. 2.1, we randomly extracted six clusters and 100 words in total from these six clusters. We then randomly selected one word that was not included in the extracted 100 words from each cluster in the six clusters as a base word (total six words). Two individual investors then manually reclustered the 100 words into six clusters by deciding the closest word to each word in these words from six base words. We evaluated the clustering result by measuring the proximity of the manually clustered result to the clustering result that uses the clustering method in Subsect. 2.1 in terms of macro F_1 score. The mean F_1 score between investors was **0.93($\gg$ 0.16)**. From this result, we consider that the word clustering by our approach sufficiently fits peoples' feelings, and clustering interpretability is sufficiently high.

3.4 Market Mood Predictability Evaluation

We evaluate the market mood predictability by whether each model can accurately predict sentiment tags of documents in the test dataset in terms of the mean F_1 scores for the five-fold cross-validation, and compare the results between the following methods: SVM, fully, base and plus MLPs, and GINN. The F_1 scores were 0.733, 0.737, 0.692, 0.681 and **0.743** for SVM, fully MLP, base MLP, plus MLP and **GINN**, respectively. These results show that GINN produced the more satisfactory result than the others in market mood predictability.

Algorithm 2. Extract the important clusters for the sentiment analysis

---

Input: document j, the second and output layer unit values, $v_j^{(CS)}$ and y_j

1: $loss \leftarrow \max y_j$, $H_{grad}^{(2)} \leftarrow \frac{\partial loss}{\partial v_j^{(CS)}} \odot v_j^{(CS)}$ (by the gradient method[2]);

2: $I_{grad}^{(3)} \leftarrow$ sorted indices in ascending order by the values of $H_{grad}^{(2)}$;

3: **return** the first four indices of $I_{grad}^{(3)}$;

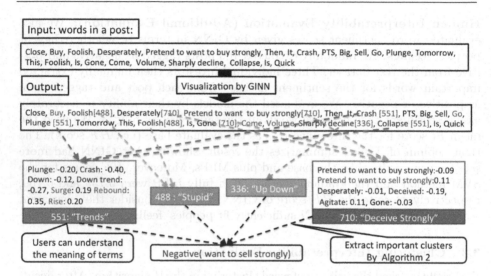

Fig. 5. Text-visualization examples from GINN and Yahoo finance board posts. The numbers in green that follow some words are their cluster numbers, and these numbers are the results of the extraction of the most four important concept clusters in Algorithm 2. This post was originally in Japanese and we manually replaced each Japanese word to the corresponding English word for this study. (Color figure online)

3.5   Text Visualization

From the evaluations for interpretability, clustering interpretability and market mood predictability of GINN, we can demonstrate both the validity for visualization by GINN and the improvement by the II algorithm.

We then present a text-visualization example produced by the GINN. We visualized an input post in the following step. We colored each word $w_{k,i}$ in a post as blue if $w_{k,i}^{(2)} < -0.05$ and red if $0.05 < w_{k,i}^{(2)}$, and displayed concept cluster information of words appeared in a post by displaying some words included in the same clusters. We then extracted the four most important concept clusters for the decision by Algorithm 2, and printed the cluster numbers after the terms included in these clusters. Figure 5 shows a text-visualization example of a document in the test dataset using the GINN ($K = 1000, K2 = 100$). By visualizing documents as above, we can quickly capture in what sense each word in a document is positive or negative and how the prediction was made.

4 Related Work

As for useful techniques for text-visualization, we can present methods using linear models such as SVM [1] and the topic models [9,10]. Regarding methods using NNs, methods for interpreting NNs [2,3,11] can be useful. Most of them try to interpret NNs by obtaining the gradient of input values for output values. Using interpretable NNs can also be helpful, and various methods for the development of interpretable NNs have been proposed. For example, a method using interpretable tree structure [12], a method for improving convolution layer [13] and methods using the attention mechanism [14–16] have been proposed. Unfortunately, these previous works could not visualize in what sense each word means positive or negative and could not be used to achieve our goal. One of the solutions to this problem is to develop NNs, where the nodes in the hidden layers have both concept and sentiment information. However, to our knowledge, there are very few techniques for the development of such interpretable NNs.

5 Conclusion

In this paper, we proposed a novel NN architecture, GINN, to aid text visualization and developed GINN using the II algorithm. GINNs allow us to easily understand in what sense each word is positive or negative and how sentiment tags were predicted. We experimentally demonstrated the validity of the text visualization produced by GINN using a real financial dataset and presented a text-visualization example produced by GINN. In future, we intend to research how many word sentiment scores need for the success of our II algorithm, and propose a method that can be applied to more complicated NNs.

Acknowledgment. This work was supported in part by JSPS Fellows Grant Number 17J04768.

A Theoretical Analysis of the II Algorithm

Let $\Omega_{dw}^{(k)}$ be a set of words included in the kth cluster and included in the polarity dictionary, $D^{(p)}$ and $D^{(n)}$ be the positive and negative document sets, $\partial w_{k,i}^{(2)*}$ be the ith component of $\partial w_k^{(2)*}$, $p^-(w_{k,i})$ be $p\left(j \in D^{(n)} | z_{j,i}^{(1,k)} > 0\right)$, $p^+(w_{k,i})$ be $1 - p^-(w_{k,i})$, and $\partial H^{(j,t)}$ be the gradient value of $H^{(j,t)}$ in *Update*. Then,

Proposition 1. *If we utilize* Update *for the parameter updates, then,*

$$
\begin{cases}
E[\partial w_{k,i}^{(2)*}] < 0 \left(\dfrac{p^+(w_{k,i})}{p^-(w_{k,i})} > \dfrac{E[|\triangle_{j,k}^{(2)*}||z_{j,i}^{(1,k)}=1 \cap j \in D^{(n)}]}{E[|\triangle_{j,k}^{(2)*}||z_{j,i}^{(1,k)}=1 \cap j \in D^{(p)}]} \right) \\[3ex]
E[\partial w_{k,i}^{(2)*}] > 0 \left(\dfrac{p^+(w_{k,i})}{p^-(w_{k,i})} < \dfrac{E[|\triangle_{j,k}^{(2)*}||z_{j,i}^{(1,k)}=1 \cap j \in D^{(n)}]}{E[|\triangle_{j,k}^{(2)*}||z_{j,i}^{(1,k)}=1 \cap j \in D^{(p)}]} \right)
\end{cases} .
\tag{1}
$$

is established. Proposition 1 indicates that if **Cond 1:** *the values of t^+ and t^- are sufficiently large,* and **Cond 2:** *for every word $w_{k,i^+} \in \Omega_{dw}^{(k)} \cap \Omega_{pw}^{(k)}$, and $w_{k,i^-} \in \Omega_{dw}^{(k)} \cap \Omega_{nw}^{(k)}$, the initial values of $w_{k,i^+}^{(2)}$ and $w_{k,i^-}^{(2)}$ given by Init are positive and sufficiently large, and negative and sufficiently small, respectively,* are met for every k, then, the II algorithm is expected to award each positive word $\in \Omega_{pw}^{(k)}$ (negative word $\in \Omega_{nw}^{(k)}$) a positive (negative) sentiment score.

Let $\boldsymbol{H}^{d(j,t)}$ be $\boldsymbol{H}^{(j,t)} - \boldsymbol{H}^{*(j,t)}$. Then, the following propositions important for explaining the *market mood predictability* of GINN are established.

Proposition 2. *If the initial values of $|\boldsymbol{W}^{(3)}|$ and $|\boldsymbol{W}^{(4)}|$ are sufficiently small* (**Cond 3**) *and for every $j \in \Omega_m^{(t)}$, the values of $z_j^{(2)}$ are $\begin{cases} positive \ (j \in D^{(p)}) \\ negative \ (j \in D^{(n)}) \end{cases}$, then, the first and second row vector values of $\partial \boldsymbol{H}^{(j,t)}$ are positive and negative respectively, and $\dfrac{\sum_{j \in \Omega_m^{(t+1)}} \|\boldsymbol{H}^{d(j,t+1)}\|_1}{\sum_{j \in \Omega_m^{(t+1)}} \|\boldsymbol{H}^{(j,t+1)}\|_1} \leq \dfrac{\sum_{j \in \Omega_m^{(t+1)}} \|\boldsymbol{H}^{d(j,t)}\|_1}{\sum_{j \in \Omega_m^{(t+1)}} \|\boldsymbol{H}^{(j,t)}\|_1}.$*

Proposition 3. *If, for every k, Cond 1–3 are established, the values $|\Omega_{pw}^{(k,t^+)}|$, $|\Omega_{nw}^{(k,t^-)}|$ and $|\Omega_m|$ are sufficiently large, then, $\lim_{t \to \infty} \dfrac{\sum_{j \in \Omega_m^{(t)}} \|\boldsymbol{H}^{d(j,t)}\|_1}{\sum_{j \in \Omega_m^{(t)}} \|\boldsymbol{H}^{(j,t)}\|_1} = 0.$*

See the supplementary material (See footnote 2) for the proofs and the details.

References

1. Ravi, K., Ravi, V.: A survey on opinion mining and sentiment analysis: tasks, approaches and applications. Knowl.-Based Syst. **89**(C), 14–46 (2015)
2. Hechtlinger, Y.: Interpretation of prediction models using the input gradient. In: NIPS 2016 Workshop on Interpretable Machine Learning in Complex Systems (2016)
3. Bach, S., Binder, A., Montavon, G., Klauschen, F., Muller, K.R., Samek, W.: On pixel-wise explanations for nonlinear classifier decisions by layer-wise relevance propagation. PLoS ONE **10**(7), 1–46 (2015)
4. Mikolov, T., Chen, K., Sutskever, I., Corrado, G., Dean, J.: Distributed representations of words and phrases and their compositionality. In: NIPS 2013, pp. 3111–3119 (2013)
5. Hornik, K., Feinerer, I., Kober, M., Buchta, C.: Spherical k-means clustering. J. Stat. Softw. **50**(10), 1–22 (2012)
6. Zhao, P., Zhang, T.: Accelerating minibatch stochastic gradient descent using stratified sampling. arXiv:1405.3080v1 (2014)
7. Kingma, D.P., Ba., J.L.: Adam: a method for stochastic optimization. In: ICLR 2015 (2015)
8. Srivastava, N., Hinton, G., Krizhevsky, A., Sutskever, I., Salakhutdinov, R.: Dropout: a simple way to prevent neural networks from overfitting. J. Mach. Learn. Res. **15**(1), 1929–1958 (2014)
9. Kjellin, P.E., Liu, Y.: A survey on interactivity in topic models. IJACSA **7**(14), 456–461 (2016)

10. Tandem anchoring: a multiword anchor approach for interactive topic modeling. In: ACL 2017, pp. 896–905 (2017)
11. Shrikumar, A., Greenside, P., Kundaje, A.: Learning important features through propagating activation differences. In: ICML 2017 (2017)
12. Xu, Q., Zhao, Q., Pei, W., Yang, L., He, Z.: Design interpretable neural network trees through self-organized learning of features. In: IJCNN 2004 (2004)
13. Zhang, Q., Wu, Y.N., Zhu, S.: Interpretable convolutional neural networks. arXiv:1710.00935 (2017)
14. Mnih, V., Heess, N., Graves, A., Kavukcuoglu, K.: Recurrent models of visual attention. In: NIPS 2014, pp. 2204–2212 (2014)
15. Xu, K., Ba, J., Kiros, R., Cho, K., Courville, A., Salakhutdinov, R., Zemel, R., Bengio, Y.: Show, attend and tell: neural image caption generation with visual attention. In: ICML 2015, pp. 77–81 (2015)
16. Dong, Y., Su, H., Zhu, J, Zhang, B.: Improving interpretability of deep neural networks with semantic information. In: CVPR 2017 (2017)

MIDA: Multiple Imputation Using Denoising Autoencoders

Lovedeep Gondara[✉] and Ke Wang

Department of Computing Science, Simon Fraser University, Burnaby, Canada
lgondara@sfu.ca, wangk@cs.sfu.ca

Abstract. Missing data is a significant problem impacting all domains. State-of-the-art framework for minimizing missing data bias is multiple imputation, for which the choice of an imputation model remains non-trivial. We propose a multiple imputation model based on overcomplete deep denoising autoencoders. Our proposed model is capable of handling different data types, missingness patterns, missingness proportions and distributions. Evaluation on several real life datasets show our proposed model significantly outperforms current state-of-the-art methods under varying conditions while simultaneously improving end of the line analytics.

1 Introduction

Missing data is an important issue, even small proportions of missing data can adversely impact the quality of learning process, leading to biased inference [4,14]. Many methods have been proposed over the past decades to minimize missing data bias [9,14] and can be divided into two categories. One that attempt to model the missing data process and use all available partial data for directly estimating model parameters, and two that attempt to fill in/impute missing values with plausible predicted values. Imputation methods are preferred for their obvious advantage. That is, providing users with a complete dataset that can be analyzed using user specified models.

Methods for imputing missing data range from replacing missing values by the column average to complex imputations based on various statistical and machine learning models. All standalone methods share a common drawback, imputing a single value for one missing observation. Which is then treated as the gold standard, same as the observed data in any subsequent analysis. This implicitly assumes that imputation model is perfect and fails to account for error/uncertainty in the imputation process. This is overcome by replacing each missing value with several slightly different imputed values, reflecting our uncertainty about the imputation process. This approach is called *multiple imputation* [10,15] and is the most widely used framework for missing data analytics. The

---

Ke Wang's work was supported by a discovery grant from the Natural Sciences and Engineering Research Council of Canada.

D. Phung et al. (Eds.): PAKDD 2018, LNAI 10939, pp. 260–272, 2018.
https://doi.org/10.1007/978-3-319-93040-4_21

biggest challenge in multiple imputation is the correct specification of an imputation model [11]. It is a nontrivial task because of the varying model capabilities and underlying assumptions. Some imputation models are incapable of handling mixed data types (categorical and continuous), some have strict distributional assumptions (multivariate normality) and/or cannot handle arbitrary missing data patterns. Existing models capable of overcoming aforementioned issues are further limited in their ability to model highly nonlinear relationships, high volume data and complex interactions while preserving inter-variable dependencies.

Recent advancements in deep learning have established state-of-the-art results in many fields [6]. Deep architectures have the capability to automatically learn latent representations and complex inter-variable associations. Which is not possible using classical models. Part of the deep learning framework, Denoising Autoencoders (DAEs) [18] are designed to recover clean output from noisy input. Missing data is a special case of noisy input, making DAEs ideal as an imputation model. But, missing data can depend on interactions/latent representations that are not observable in the input dataset space. Hence, we propose to use an *overcomplete* DAE as an imputation model. Whereby projecting our input data to a higher dimensional subspace from where we then recover missing information. We propose a multiple imputation framework with overcomplete DAE as the base model, where we simulate multiple predictions by initializing our model with a different set of random weights at each run. Details of our method are presented in Sect. 3. Our proposed method has several advantages over the current methods, some of which we outline below.

- Previous studies on imputing missing data using machine learning methods use complete observations for the training phase. We show that our model outperforms state-of-the-art methods even when users do not have the luxury of having complete observations for initial training, a common scenario in real life.
- Our model is capable of preserving attribute correlations, which are of a concern using traditional imputation methods and can significantly affect end of the line analytics.
- Our model is better equipped to deal with different missing data generation processes, such as data missing not at random, which is a performance bottleneck for other imputation methods. Experimental results using real life datasets show that our model outperforms state-of-the-art methods under varying dataset and missingness conditions and improves end of the line analytics.

The rest of the paper is organized as following. Section 2 provides preliminary background to missing data terminology and introduces denoising autoencoders. Section 3 introduces our model with Sect. 4 presenting empirical evaluation and the effect of imputation on end of the line analytics followed by our conclusions.

2   Background

Missing data is a well researched topic in statistics. Most of the early work on missing data, including definitions, multiple imputation and subsequent analysis is attributed to works of Little and Rubin [9,10,14]. From machine learning perspective, it has been shown that auto-associative neural networks are better at imputing missing data when attribute interdependencies are of concern [12]. A common scenario in real life datasets. Denoising autoencoders have been recently used in completing traffic and health records data [1,5] and collaborative filtering [8]. Below we provide some preliminary introduction to missing data mechanisms and denoising autoencoders.

2.1   Missing Data

Mechanisms: Impact of missing data depends on the underlying missing data generating mechanism. We define three missing data categories [10] with the aid of data from Table 1, representing an income questionnaire in a survey where we denote missing data with "?". Data is Missing Completely At Random ($MCAR$) if missingness does not depend on observed or unobserved data. Example: Survey participants flip a coin to decide whether to answer questions or not. Data is Missing At Random (MAR) if missingness can be explained using observed data. Example: Survey participants that live in postal code 456 and 789 refuse to fill in the questionnaire. Data is Missing Not At Random ($MNAR$) if missingness depends on an unobserved attribute or on the missing attribute itself. Example: Everyone who owns a six bedroom house refuses the questionnaire. Bigger house is an indirect indicator for greater wealth and a better paying job, but we don't have the related data. When data are MAR or MCAR, it is known as ignorable missing data as observed data can be used to account for missingness. But, given the observed data, it is impossible to distinguish between MNAR and MAR [17] and sometimes, missing data can be a combination of both.

Table 1. Data snippet for income questionnaire with missing data represented using '?'

Id	Age	Sex	Income	Postal	Job	Marital status
1	50	M	100	123	a	Single
2	45	?	?	456	?	Married
3	?	F	?	789	?	?

Multiple Imputation: In a multiple imputation scenario, we will create multiple copies of the dataset presented in Table 1 with '?' replaced by slightly different imputed values in each copy. Multiple imputation accounts for uncertainty in predicting missing data using observed data by modelling variability into the imputed values as the true values for missing data are never known. Multiple imputed datasets are then analyzed independently and the results combined. A single statistic such as classification accuracy or root mean square error (RMSE) can be simply averaged from multiple imputations.

2.2 Autoencoders and Denoising Autoencoders

An autoencoder takes an input $x \in [0,1]^d$ and maps (encodes) it to an interme-diate representation $h \in [0,1]^{d'}$ using an encoder. Where d' represents a different dimensional subspace. The assumption is, in the dataset, h captures the coor-dinates along the main factors of variation. The encoded representation is then decoded back to the original d dimensional space using a decoder. Encoder and decoder are both artificial neural networks. The two stages are represented as

$$h = s(Wx + b) \tag{1}$$

$$z = s(W'h + b') \tag{2}$$

where z is the decoded result and s is any nonlinear function. Reconstruction error between x and z is minimized during training phase.

Denoising autoencoders are a natural extension to autoencoders [18]. By cor-rupting the input data and forcing the network to reconstruct the clean output forces the hidden layers to learn robust features. Corruption can be applied in different ways, such as randomly setting some input values to zero or using dis-tributional additive noise. DAEs reconstruction capabilities can be explained by thinking of DAEs implicitly estimating the data distribution as the asymp-totic distribution of the Markov chain that alternates between corruption and denoising [2].

3 Models

This section introduces our multiple imputation model and the competitors used for comparison.

3.1 Our Model

Architecture: Our default architecture is shown in Fig. 1. We employ atypical overcomplete representation of DAEs. That is, more units in successive hidden layers during encoding phase compared to the input layer. This mapping of our input data to a higher dimensional subspace creates representations capable of adding lateral connections, aiding in data recovery. Usefulness of this approach is empirically validated in the supplemental material. We start with an initial n dimensional input, then at each successive hidden layer, we add Θ nodes, increasing the dimensionality to $n+\Theta$. For initial comparisons, we use $\Theta = 7$. We tried different values for Θ for various datasets and decided to use 7 as it provided consistent better results. It is an arbitrary choice and can be dealt with by

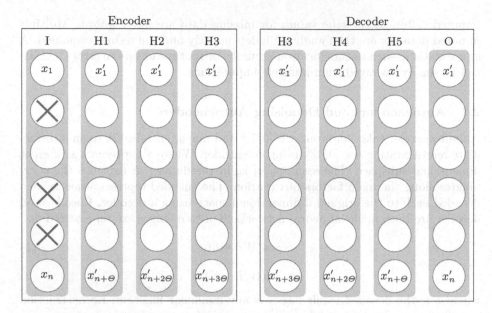

Fig. 1. Our basic architecture, encoder block increases dimensionality at every hidden layer by adding Θ units with decoder symmetrically scaling it back to original dimensions. Crossed out inputs represent stochastic corruption at input by setting random inputs to zero. H_1, H_2, H_3, H_4, H_5 are hidden layers with I and O being the input and output layers respectively. Encoder and decoder are constructed using fully connected artificial neural networks.

viewing Θ as another tuning hyperparameter. Our model inputs are standardized between 0 and 1 to facilitate faster convergence for small to moderate sample sizes. Our model is trained with 500 epochs using an adaptive learning rate with a time decay factor of 0.99 and Nesterov's accelerated gradient [13]. The input dropout ratio to induce corruption is set to 0.5. So that in a given training batch, half of the inputs are set to zero. Tanh is used as an activation function as we find it performs better than ReLU for small to moderate sample sizes, especially when some inputs are closer to zero. We use early stopping rule to terminate training if desired mean squared error (MSE) of 1e-06 is achieved or if simple moving average of length 5 of the error deviance does not improve. The training-test split of 70-30 is used with all results reported on the test set. Multiple imputation is accomplished by using multiple runs of the model with a different set of random initial weights at each run. This provides us with the variation needed for multiple imputations. Algorithm 1 explains our multiple imputation process.

Usage: We start with imitating a real life scenario. Where the user only have a dataset with pre-existing missing values. That is, the user does not have the luxury of access to the clean data and user does not know the underlying missing data generating mechanism or the distribution. In scenarios where missingness

---

Algorithm 1. Multiple imputation using DAEs

Require: k: Number of imputations needed
1: **for** $i = 1 \rightarrow k$ **do**
2:     Initialize DAE based imputation model using weights from random uniform distribution
3:     Fit the imputation model to training partition using stochastic corruption
4:     Reconstruct test set using the trained model
5: **end for**

---

is inherent to the datasets, training imputation models using complete data can bias the learner. But as DAEs require complete data at initialization, we initially use the respective column average in case of continuous variables and most frequent label in case of categorical variables as placeholders for missing data at initialization. Training phase is then initiated with a stochastic corruption process setting random inputs to zero. Where our model learns to map corrupted input to clean output. Our approach is based on one assumption, that is, we have enough complete data to train our model. So the model learns to recover *true* data using stochastic corruption on inputs, and is not learning to map placeholders as valid imputations. The results show this assumption is readily satisfied in real life scenarios. Even datasets with small sample sizes are enough for DAE based imputation model to achieve better performance compared to state-of-the-art.

3.2   Competitors and Comparison

Competitor: For multiple imputation, we need methods that can inject variation in successive imputations, providing slightly different imputation results at each iteration. Simple models such as linear/logistic regression or deterministic methods based on matrix decomposition fail to take this into account. Current state-of-the-art in multiple imputation is the Multivariate Imputation by Chained Equations (MICE) [3]. Which is a fully conditional specification approach and works better than Joint Modelling approaches where multivariate distributions cannot provide a reasonable description of the data or where no suitable multivariate distributions can be found. MICE specifies multivariate model on variable by variable basis using a set of conditional densities, one for each variable with missing data. MICE draws imputations by iterating over conditional densities. It has an added advantage of being able to model different densities for different variables. Internal imputation model in MICE is vital and a model with properties to handle different data types and distributions is essential for effective imputations. Predictive mean matching and Random Forest are the best available options within MICE framework [16]. We compared them both and found predictive mean matching to provide more consistent results with varying dataset types and sizes. Hence it is used as the internal component of our competitor MICE model.

Comparison: Imputation results are compared using sum of root mean squared error calculated per attribute on the test set, given as

$$RMSE_{sum} = \sum_{i=1}^{m} \sqrt{E(\sum_{i=1}^{n} (\hat{t}_i - t_i))} \tag{3}$$

where we have m attributes, n observations, $\hat{t}$ is the imputed value and t is the observed value. $RMSE_{sum}$ is calculated on scaled datasets to avoid disproportionate attribute contributions. $RMSE_{sum}$ provides us a measure of relative distance, that is, how far the dataset completed with imputed values is from the original complete dataset. For multiple imputation scenarios with k imputations, we have k values for $RMSE_{sum}$ per dataset. The results are then reported using average $RMSE_{sum}$ along with the range.

4   Experiments

We start our empirical evaluation for multiple imputation on several publicly available real life datasets under varying missingness conditions.

4.1   Datasets

Table 2 shows the properties of various real life publicly available datasets [7] used for model evaluation. Models based on deep architectures are known to perform well on large sample, high dimensional datasets. Here we include some extremely low dimensional and low sample size datasets to test the extremes and to prove that our model has real world applications. Most of the datasets have a combination of continuous, categorical and ordinal attributes. Which further challenges the convergence of our model using small training samples.

4.2   Inducing Missingness

To provide a wide range of comparisons. Initially for each data set, we introduce missingness in four different ways with a fixed missingness proportion of 20% using the steps detailed below.

1. Append a uniform random vector v with n observations to the dataset with values between 0 and 1, where n is number of observations in the dataset.
2. *MCAR, uniform*: Set all attributes to have missing values where $v_i \leq t$, $i \in 1 : n$, t is the missingness threshold, 20% in our case.
3. *MCAR, random*: Set randomly sampled half of the attributes to have missing values where $v_i \leq t$, $i \in 1 : n$.
4. *MNAR, uniform*: Randomly sample two attributes x_1 and x_2 from the dataset and calculate their median m_1 and m_2. Set all attributes to have missing values where $v_i \leq t$, $i \in 1 : n$ and ($x_1 \leq m_1$ or $x_2 \geq m2$).
5. *MNAR, random*: Randomly sample two attributes x_1 and x_2 from the dataset and calculate their median m_1 and m_2. Set randomly sampled half of the attributes to have missing values where $v_i \leq t$, $i \in 1 : n$ and ($x_1 \leq m_1$ or $x_2 \geq m2$).

Table 2. Datasets used for evaluation. Dataset acronyms are shown in parenthesis that we will be using in the results section.

	Observations	Attributes
Boston housing (BH)	506	14
Breast cancer (BC)	699	11
DNA (DN)	3186	180
Glass (GL)	214	10
House votes (HV)	435	17
Ionosphere (IS)	351	35
Ozone (ON)	366	13
Satellite (SL)	6435	37
Servo (SR)	167	5
Shuttle (ST)	58000	9
Sonar (SN)	208	61
Soybean (SB)	683	36
Vehicle (VC)	846	19
Vowel (VW)	990	10
Zoo (ZO)	101	17

4.3 Main Results

Table 3 shows the multiple imputation results on real life datasets, comparing five imputations by our model with five imputations by MICE. That is, each missing value is imputed five times with a slightly different value. The results show that our model outperforms MICE in 100% of cases with data MCAR and MNAR with uniform missing pattern and in > 70% of cases with random missing pattern. Our model's superior performance in this scenario using small to moderate dataset sizes with constrained dimensionality is indicative of it's utility when datasets are large and are of higher dimensionality. Which is a performance bottleneck for other multiple imputation models whereas our model is capable of handling massive data by design. Another advantage is that our model does not need a certain proportion of available data to predict missing value. As in the case of dataset VW-MNAR, MICE was unable to provide complete imputations.

Computational cost associated with our model is at par or better than imputations based on MICE for small to moderate sized datasets. This might seem counter-intuitive to some readers as our model is much more complex. But, computational gains are significant when we are modelling a complete dataset in a single attempt compared to iterative variable by variable imputation in MICE.

Table 3. Imputation results comparing our model and MICE. Results are displayed using sum of root mean square error ($RMSE_{sum}$). Providing a measure of relative distance of imputation from original data. As results are from multiple imputation (5 imputations), mean $RMSE_{sum}$ from 5 imputations is displayed outside with min and max $RMSE_{sum}$ inside parenthesis providing a range for imputation performance. Value for MNAR is NA for dataset VW as MICE was unable to impute a complete dataset.

	Data	Uniform missingness		Random missingness	
		DAE	MICE	DAE	MICE
MCAR	BH	2.9(2.9,3)	3.7(3.5,3.8)	0.9(0.9,1)	0.9(0.7,1)
	BC	2.9(2.9,2.9)	3.9(3.6,4.2)	1.2(1.2,1.3)	1.3(1.1,1.4)
	DN	25.7(25.7,25.7)	36.5(36.3,36.6)	13.1(13.1,13.2)	16.9(16.9,17)
	GL	1.1(1,1.1)	1.5(1.3,1.7)	1.3(1.2,1.4)	1.4(1.3,1.6)
	HV	2.4(2.4,2.4)	3.4(3.1,3.7)	1.1(1.1,1.2)	1.2(0.9,1.3)
	IS	13(12.9,13.1)	17.1(16.2,17.7)	5.8(5.6,6.2)	7(6.7,7.5)
	ON	2.1(2.1,2.1)	3.1(3,3.3)	0.9(0.9,1)	1(1,1.2)
	SL	3.6(3.6,3.7)	4.5(4.4,4.6)	1.8(1.7,1.8)	0.7(0.7,0.7)
	SR	1.2(1.,1.2)	1.5(1.4,1.7)	0.4(0.4,0.5)	0.4(0.4,0.5)
	ST	16.5(16.5,16.7)	27.9(27.5,28.2)	6.5(6.4,6.7)	13(12.5,13.8)
	SN	5.1(5,5.1)	7.3(7.2,7.5)	2.3(2.2,2.3)	3.2(3.2,3.3)
	SB	1.8(1.8,1.8)	2.4(2.3,2.4)	1.2(1.1,1.2)	1.1(1,1.1)
	VC	4.1(4,4.1)	5.6(5.5,5.7)	1.6(1.6,1.6)	2.2(2.1,2.3)
	VW	5.8(5.7,6.2)	7.7(7,8.1)	2.6(2.4,2.9)	3.8(3.3,4.2)
	ZO	2.1(2.1,2.1)	3.4(3.1,4.3)	1.1(1.1,1.2)	1.1(1.1,1.1)
MNAR	BH	2.3(2.2,2.4)	3.2(2.9,3.4)	0.9(0.8,1)	0.7(0.7,0.8)
	BC	2.9(2.8,3)	3.6(3.4,3.8)	1.7(1.7,1.8)	1.4(1.3,1.5)
	DN	25.3(25.2,25.3)	34.5(34.5,34.7)	5.7(5.7,5.8)	7.2(7.1,7.2)
	GL	1.3(1.3,1.4)	1.5(1.3,1.8)	0.4(0.3,0.4)	0.2(0.11,0.2)
	HV	2.6(2.6,2.6)	3.5(3.3,3.7)	1.3(1.2,1.3)	1.3(1.3,1.4)
	IS	11.7(11.5,11.8)	15.4(14.9,16.5)	4.8(4.5,5.1)	6.3(5.6,6.8)
	ON	1.5(1.5,1.5)	2.2(2,2.4)	1.2(1.1,1.2)	1.3(1.1,1.5)
	SL	3.4(3.4,3.4)	3.8(3.8,3.9)	1.6(1.6,1.6)	0.5(0.5,0.5)
	SR	1.2(1.2,1.2)	1.6(1.5,1.7)	0.4(0.3,0.4)	0.3(0.2,0.3)
	ST	11.8(11.7,11.9)	22.4(22.1,22.7)	4.5(4.3,4.7)	9.5(8.4,10.3)
	SN	4.6(4.6,4.6)	6.8(6.5,7.1)	2.3(2.3,2.4)	3.1(3,3.2)
	SB	1.7(1.7,1.7)	2.3(2.2,2.4)	0.6(0.6,0.6)	0.9(0.9,0.9)
	VC	3.5(3.4,3.7)	4.6(4.4,4.8)	1.7(1.7,1.8)	2.4(2.3,2.4)
	VW	5.9(5.9,5.9)	NA	2.3(2.1,2.5)	NA
	ZO	3.3(2.8,5.5)	3.9(3.6,4.6)	0.9(0.8,1.0)	1.1(0.7,1.7)

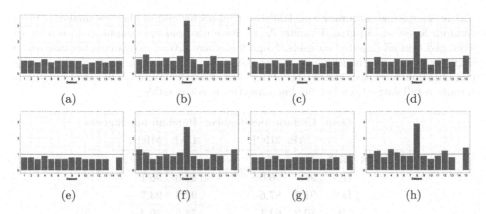

Fig. 2. Results for imputation with increased missingness proportions. Figures (a), (b), (c), (d) show imputation results with 40% missing data and figures (e), (f), (g), (h) show imputation results with 60% missing data. Red line is drawn as a reference line at y-intercept of 1 to signify superior/inferior performance of our model vs MICE. Results are displayed using E_R where values less than one signify our model performing better and values greater than one signify MICE's superior performance. X-axis show different datasets (1–15) and Y-axis display E_R. For some cases, MICE has trouble imputing dataset 14 (VW) whereas our model provides consistent imputations. (Color figure online)

4.4   Increased Missingness Proportion

Missing data proportion is known to affect imputation performance. Which deteriorates with increasing missing data proportion. To test the impact of varying missing data proportion on our model, we introduce missingness in all 15 datasets with missingness proportion set at 40% and 60% using methods described in experimental setup section. Keeping all model parameters same for our model and MICE, we multiple imputed datasets with five imputations each. For a better visual representation, we compare the imputation results between our model and MICE using mean error ratio E_R, given as

$$E_R = \frac{\frac{1}{n}\sum_{i=1}^{n} E_{Di}}{\frac{1}{n}\sum_{i=1}^{n} E_{Pi}} \tag{4}$$

where E_D is imputation error of our model, E_P is imputation error of MICE and n is number of imputations. E_R values of less than one signify average superior performance of our model over MICE, whereas values greater than one signify MICE performing better.

Figure 2 shows the results, a reference line at y-intercept of 1 is drawn to aid visual comparisons. Our model performs better on average compared to MICE, irrespective of missing data proportion. Results echo the findings of our main results, where we observe our model performing better than MICE on average of >85% cases.

Table 4. Average accuracy and RMSE estimates for end of the line analytics using random forest on imputed datasets. As we have used multiple imputation, results are averaged over all imputed datasets. * signifies where RMSE is reported because target variable is numeric, hence lower values the better. All other datasets report average classification accuracy, higher the better. For dataset VW, as MICE was unable to impute a full dataset, end of the line analytics is not possible.

	Data	Uniform missingness		Random missingness	
		DAE	MICE	DAE	MICE
MNAR	BH*	3.9	4.5	3.7	4.1
	BC	96.0	96.0	97.0	96.1
	DN	91.6	87.6	93.3	93.7
	GL	70.2	64.1	74.6	70.4
	HV	98.5	99.2	95.0	98.2
	IS	90.5	86.6	90.7	90.3
	ON*	4	3.8	3.6	4.2
	SL	90.0	80.9	89.6	89.6
	SR*	6.6	8.1	6.9	6.4
	ST	86.4	80.8	76.5	72.9
	SN	90.9	70.5	85.3	84.2
	SB	72.6	62.7	73.7	77.1
	VC	74.7	63.8	72.7	70.6
	VW	93.9	NA	77.7	NA
	ZO	99.9	98.5	99.9	99.9

4.5   Impact on Final Analysis

Main goal of imputing missing data is to generate complete datasets that can be used for analytics. While imputation accuracy provides us with a measure of how close the imputed dataset is to the complete dataset, we still do not know how well inter-variable correlations are preserved. Which severely impacts end of the line analytics. To check the imputation quality in relation to a dataset's overall structure and to quantify the impact of imputation on end of the line analytics. We use all imputed datasets as the input to classification/regression models based on random forest with 5 times 5 fold cross validation. The task is to use the target variable from all datasets and store the classification accuracy/RMSE for each dataset imputed using our model and MICE. Higher values for classification accuracy and lower RMSE will signify a better preserved predictive dataset structure. We calculate mean accuracy/RMSE from all five runs of multiple imputation. Datasets with data MNAR (uniform and random) are used as MNAR datasets pose greatest challenges for imputation.

Results in Table 4 show that multiple imputation using our model provides higher predictive power for end of the line analytics compared to MICE imputed data. The difference is even more significant when data are MNAR uniform compared to when data are MNAR random.

5 Conclusion

We have presented a new method for multiple imputation based on deep denoising autoencoders. We have shown that our proposed method outperforms current state-of-the-art using various real life datasets and missingness mechanisms. We have shown that our model performs well, even with small sample sizes, which is thought to be a hard task for deep architectures. In addition to not requiring a complete dataset for training, we have shown that our proposed model improves end of the line analytics.

References

1. Beaulieu-Jones, B.K., Moore, J.H.: The pooled resource open-access ALS, and clinical trials consortium. Missing data imputation in the electronic health record using deeply learned autoencoders. In: Pacific Symposium on Biocomputing, vol. 22, pp. 207. NIH Public Access (2016)
2. Bengio, Y., Yao, L., Alain, G., Vincent, P.: Generalized denoising auto-encoders as generative models. In: Advances in Neural Information Processing Systems, pp. 899–907 (2013)
3. Buuren, S., Groothuis-Oudshoorn, K.: MICE: multivariate imputation by chained equations in R. J. Stat. Softw. 45(3), 1–68 (2011)
4. Chen, P.: Optimization algorithms on subspaces: revisiting missing data problem in low-rank matrix. Int. J. Comput. Vis. 80(1), 125–142 (2008)
5. Duan, Y., Lv, Y., Kang, W., Zhao, Y.: A deep learning based approach for traffic data imputation. In: 2014 IEEE 17th International Conference on Intelligent Transportation Systems (ITSC), pp. 912–917. IEEE (2014)
6. LeCun, Y., Bengio, Y., Hinton, G.: Deep learning. Nature 521(7553), 436–444 (2015)
7. Leisch, F., Dimitriadou, E.: Machine learning benchmark problems (2010)
8. Li, S., Kawale, J., Fu, Y.: Deep collaborative filtering via marginalized denoising auto-encoder. In: Proceedings of the 24th ACM International on Conference on Information and Knowledge Management, pp. 811–820. ACM (2015)
9. Little, R.J.A.: Missing-data adjustments in large surveys. J. Bus. Econ. Stat. 6(3), 287–296 (1988)
10. Little, R.J.A., Rubin, D.B.: Statistical Analysis with Missing Data. Wiley, Hoboken (2014)
11. Morris, T.P., White, I.R., Royston, P.: Tuning multiple imputation by predictive mean matching and local residual draws. BMC Med. Res. Methodol. 14(1), 75 (2014)
12. Nelwamondo, F.V., Mohamed, S., Marwala, T.: Missing data: A comparison of neural network and expectation maximisation techniques. arXiv preprint arXiv:0704.3474 (2007)
13. Nesterov, Y.: A method of solving a convex programming problem with convergence rate O (1/k2) (1983)

14. Rubin, D.B.: Inference and missing data. Biometrika **63**, 581–592 (1976)
15. Schafer, J.L.: Multiple imputation: a primer. Stat. Methods Med. Res. **8**(1), 3–15 (1999)
16. Shah, A.D., Bartlett, J.W., Carpenter, J., Nicholas, O., Hemingway, H.: Comparison of random forest and parametric imputation models for imputing missing data using MICE: a CALIBER study. Am. J. Epidemiol. **179**(6), 764–774 (2014)
17. Sterne, J.A.C., White, I.R., Carlin, J.B., Spratt, M., Royston, P., Kenward, M.G., Wood, A.M., Carpenter, J.R.: Multiple imputation for missing data in epidemiological and clinical research: potential and pitfalls. BMJ **338**, b2393 (2009)
18. Vincent, P., Larochelle, H., Bengio, Y., Manzagol, P.-A.: Extracting and composing robust features with denoising autoencoders. In: Proceedings of the 25th International Conference on Machine Learning, pp. 1096–1103. ACM (2008)

Dual Control Memory Augmented Neural Networks for Treatment Recommendations

Hung Le$^{(\boxtimes)}$, Truyen Tran, and Svetha Venkatesh

Applied AI Institute, Deakin University, Geelong, Australia
{lethai,truyen.tran,svetha.venkatesh}@deakin.edu.au

Abstract. We formulate the task of treatment recommendation as a sequence-to-sequence prediction model that takes the time–ordered medical history as input, and predicts a sequence of future clinical procedures and medications. It is built on the premise that an effective treatment plan may have long–term dependencies from previous medical history. We approach the problem by using a memory–augmented neural network, in particular, by leveraging the recent differentiable neural computer that consists of a neural controller and an external memory module. Differing from the original model, we use *dual controllers*, one for encoding the history followed by another for decoding the treatment sequences. In the encoding phase, the memory is updated as new input is read; at the end of this phase, the memory holds not only the medical history but also the information about the current illness. During the decoding phase, the memory is *write–protected*. The decoding controller generates a treatment sequence, one treatment option at a time. The resulting dual controller write–protected memory–augmented neural network is demonstrated on the MIMIC-III dataset on two tasks: procedure prediction and medication prescription. The results show improved performance over both traditional bag-of-words and sequence-to-sequence methods.

1 Introduction

A core task in healthcare is to generate effective treatment plans. Machine-assisted treatment recommendations have potential to improve healthcare efficiency. We approach the task by learning from rich electronic medical records. An electronic medical record (EMR) is a digital record of patient health information over time such as details of symptoms, data from monitoring devices, and clinicians' observations. Amongst these data elements, diagnosis, clinical procedure and drug prescription codes form core information, and are temporally correlated. A medical history is a sequence of clinic visits, each of which has a set of diagnoses, treatment procedures, and discharge medications. In MIMIC-III dataset [9], diagnoses are "ordered by priority", procedures follow the order that "the procedures were performed" and the drugs follow the prescription dates[1].

---

[1] https://mimic.physionet.org/mimictables/.

© Springer International Publishing AG, part of Springer Nature 2018
D. Phung et al. (Eds.): PAKDD 2018, LNAI 10939, pp. 273–284, 2018.
https://doi.org/10.1007/978-3-319-93040-4_22

The temporal dependency in EMR clinical codes can be long-term. For example, once diagnosed with diabetes (Type I or II), the conditions (and hence its medications, if any) are persistent through the patient's life, even though it might not be coded at every visit. Since EMR data are temporally sequenced by patient medical visits, clinical codes at current admission may be related to other codes appearing in previous admissions.

These long-term dependencies pose a great challenge for prediction models. Recent efforts dealing with medical prediction have largely focused on modeling the admission's diagnoses and treatments as two sets of codes and capture sequential dependencies between codes from different admissions, i.e., sequence of sets [3,11,13–15]. This approach may expose limitations since using the admission set representation ignores internal sequential dependencies.

To tackle these issues, we propose a novel treatment recommendation model using a memory-augmented neural network (MANN) to capture the long-term dependencies from EMR data. Our model is built upon Differentiable Neural Computer (DNC) [6], a recent powerful and fully differentiable MANN. A DNC is an expressive recurrent neural network consisting of a controller augmented with a memory module. At each time step, the controller reads an input, updates the memory, and generates an output. DNC has demonstrated its efficacy in various tasks that require long chains of computation such as graph prediction and question-answering suggesting its power of solving sequence prediction problems. Despite its potential, DNC has yet to be applied to healthcare, especially in clinical treatment sequence prediction.

We adapt the DNC to the task of treatment recommendation with two key modifications. We formulate the treatment recommendations as a sequence-to-sequence prediction problem, where the entire medical history sequence stored in EMR is used to produce a sequence of treatment options. The output sequence allows modeling dependencies between current treatments, and between treatment and the distant history. We modify the DNC by using two controllers to handle dual processes: history encoding and treatment recommendations. Each controller will employ different "remembering" strategies for each process helping improvement in prediction and increasing the learning speed. In the second modification, we apply a write-protected policy for the decoding controller, that is, memory is read-only in the decoding phase.

In summary, our main contributions are: (i) handling long-term dependencies in treatment recommendations by solving the sequence prediction problem, (ii) proposing a novel memory-augmented architecture that uses dual controllers and write-protected mechanism (DCw-MANN) to suit sequence-to-sequence task, (iii) empirically evaluating our model on a real-world EMR dataset (MIMIC-III) and showing that our method outperforms existing methods in treatment recommendations. The significance of DCw-MANN lies in its versatility as our model can be applied to other sequential domains with similar data characteristics.

2   Methods

2.1   Problem Formulation

In EMR data, a hospital visit is documented as one admission record consisting of diagnosis and treatment codes for the admission. Diagnoses are coded using WHO's ICD (International Classification of Diseases) coding schemes. For example, in the ICD10 scheme, E10 encodes Type 1 diabetes mellitus, E11 encodes Type 2 diabetes mellitus and F32 indicates depressive episode. The treatment can be procedure or drug. The procedures are typically coded in CPT (Current Procedural Terminology) or ICHI (International Classification of Health Interventions) schemes. The drugs are often coded in ATC (Anatomical Therapeutic Chemical) or NDC (National Drug Code). Once diagnoses are confirmed, we want to predict the output sequence (treatment codes of the current visit) given the input sequence (all diagnoses followed by treatments from the first visit to the previous visit plus the diagnoses of the current visit). More formally, we denote all the unique medical codes (diagnosis, procedures and drugs) from the EMR data as $c_1, c_2, ..c_{|C|} \in C$, where $|C|$ is the number of unique medical codes. A patient's n-th admission's input is represented by a sequence of codes:

$$\left[c_{d_1}^1, c_{d_2}^1, ..., \text{‰}, c_{p_1}^1, c_{p_2}^1 ..., \varnothing, ..., c_{d_1}^{n-1}, c_{d_2}^{n-1}, ..., \text{‰}, c_{p_1}^{n-1}, c_{p_2}^{n-1}, ..., \varnothing, c_{d_1}^n, c_{d_2}^n, ..., \text{‰} \right] \tag{1}$$

Here, $c_{d_j}^k$ and $c_{p_j}^k$ are the j-th diagnosis and treatment code of the k-th admission, respectively. ‰, $\varnothing$ are special characters that informs the model about the change from diagnosis to treatment codes and the end of an admission respectively. This reflects the natural structure of a medical history, which is a sequence of clinical visits, each of which typically includes a subsequence of diagnoses, and a subset of treatments. A diagnosis subsequence usually started with the primary condition followed by secondary conditions. In a subset of treatments, the order is not strictly enforced, but it may reflect the coding practice. The output of the patient's n-th admission is : $\left[c_{p_1}^n, c_{p_2}^n, ..., c_{p_{L_{out}}}^n, \varnothing \right]$, in which L_{out} is the length of the treatment sequence we want to predict and $\varnothing$ is used to inform the model to stop predicting. Finally, each code is represented by one-hot vector $v_c \in [0,1]^{\|C\|}$, where $v_c = [0, ..., 0, 1, 0.., 0]$ ($v_c[i] = 1$ if and only if v_c represents c_i). Unlike set encoding of each admission, representing the data in this way preserves the admission's internal order information allowing sequence-based methods to demonstrate their power of capturing sequential events.

2.2   DNC Overview

In this subsection, we briefly review DNC [6]. A DNC consists of a controller, which accesses and modifies an external memory module using a number of read and write heads. In DNC, the memory module is more powerful and can "remember" a longer sequence than recurrent neural nets such as LSTM [8] or other MANNs such as NTM [5]. Given some input x_t, and a set of R read values from memory $r_{t-1} = \left[r_{t-1}^1, r_{t-1}^2, ..., r_{t-1}^R \right]$, the controller produces the output:

$o_t \in \mathbb{R}^{|C_p|}$, where $|C_p|$ is the number of possible output and the key $k_t \in \mathbb{R}^D$, where D is the word size in memory. This key will be used for locating the read/write slots in memory matrix M_t using cosine similarity:

$$D\left(M_t(i), k_t\right) = \frac{k_t \cdot M_t(i)}{||k_t|| \cdot ||M_t(i)||} \tag{2}$$

This is used to produce a content-based read-weight and write-weight vector $w_t^{cr}, w_t^{cw} \in \mathbb{R}^N$ whose elements are computed according to a softmax over memory's locations. N is the number of memory locations.

Dynamic Memory Allocation and Write Weightings: DNC maintains a memory usage vector $u_t \in [0,1]^N$ to define the allocation write-weight:

$$a_t\left[\Phi_t[j]\right] = (1 - u_t\left[\Phi_t[j]\right]) \prod_{i=1}^{j-1} u_t\left[\Phi_t[i]\right] \tag{3}$$

in which, Φ_t contains elements from u_t sorted by ascending order from least to most used. Given the write gate g_t^w and allocation gate g_t^a, the final write-weight then can be computed by the following interpolation:

$$w_t^w = g_t^w\left[g_t^a a_t + (1 - g_t^a) w_t^{cw}\right] \tag{4}$$

Temporal Memory Linkage and Read Weightings: DNC uses a temporal link matrix $L_t \in [0,1]^{N \times N}$ to keep track of consecutively modified memory locations, and $L_t[i,j]$ represents the degree to which location i was the location written to after location j. Each time a memory location is modified, the link matrix is updated to remove old links to and from that location, and add new links from the last-written location. The final read-weight is given as follow:

$$w_t^{rk} = \pi_t^k[1] L_t^\top w_{t-1}^{rk} + \pi_t^k[2] w_t^{crk} + \pi_t^k[3] L_t w_{t-1}^{rk} \tag{5}$$

The read mode weight π_t^k is used to balance between the content-based read-weight and the forward $L_t w_{t-1}^{rk}$ and backward $L_t^\top w_{t-1}^{rk}$ of the previous read. Then, the k-th read value r_t^k is retrieved using the final read-weight vector:

$$r_t^k = \sum_i^N w_t^{rk}(i) M_t(i) \tag{6}$$

2.3 Proposed Model

We now present our main contribution to solve the task of treatment recommendations – a deep neural architecture called Dual Controller Write-Protected Memory Augmented Neural Network (DCw-MANN) (see Fig. 1). Our DCw-MANN introduces two simple but crucial modifications to the original DNC:

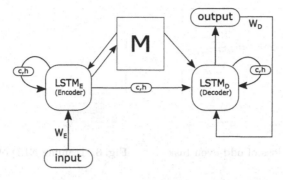

Fig. 1. Dual controller write-protected memory augmented neural network. $LSTM_E$ is the encoding controller. Both are implemented as a LSTM. $LSTM_D$ is the decoding controller.

(i) using two controllers to handle dual processes of encoding and decoding, respectively; and (ii) applying a write-protected policy in the decoding phase.

In the encoding phase, after going through embedding layer W_E, the input sequence is fed to the first controller (encoder) $LSTM_E$. At each time step, the controller reads from and writes to the memory information necessary for the later decoding process. In the decoding phase, the states of the first controller is passed to the second controller (decoder) $LSTM_D$. The use of two controllers instead of one is important in our setting because it is harder for a single controller to learn many strategies at the same time. Using two controllers will make the learning easier and more focused. Also different from the encoder, the decoder can make use of its previous prediction (after embedding layer W_D) as the input together with the read values from the memory. Another important feature of DCw-MANN is its write-protected mechanism in the decoding phase. This has an advantage over the writing strategy used in the original DNC since at decoding step, there is no new input that is fed into the system. Of course, there remains dependencies among codes in the output sequence. However, as long as the dependencies among output codes are not too long, they can be well-captured by the cell memory c_t inside the decoder's LSTM. Therefore, the decoder in our design is prohibited from writing to the memory. To be specific, at time step $t + 1$ we have the hidden state and cell memory of the controllers calculated as:

$$h_{t+1}, c_{t+1} = \begin{cases} LSTM_E\left([W_E v_{d_t}, r_t], h_t, c_t\right); & t \leq L_{in} \\ LSTM_D\left([W_D v_{p_t}, r_t], h_t, c_t\right); & t > L_{in} \end{cases} \qquad (7)$$

where v_{d_t} is the one-hot vector representing the input sequence's code at time $t \leq L_{in}$ and v_{p_t} is the predicted one-hot vector output of the decoder at time $t > L_{in}$, defined as $v_{p_t} = onehot\left(o_t\right)$, i.e.,:

$$v_{p_t}[i] = \begin{cases} 1 & ; i = \underset{1 \leqslant j \leqslant |C_p|}{argmax}(o_t[j]) \\ 0 & ; otherwise \end{cases}. \qquad (8)$$

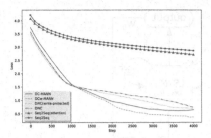

Fig. 2. Training loss of odd-even task   **Fig. 3.** Training NLD of odd-even task

We propose a new memory update rule to enable the write-protected mechanism:

$$M_t = \begin{cases} M_{t-1} \circ \left(E - w_t^w e_t^\top\right) + w_t^w v_t^\top; & t \leq L_{in} \\ M_{t-1}; & t > L_{in} \end{cases} \tag{9}$$

where E is an $N \times D$ matrix of ones, $w_t^w \in [0,1]^N$ is the write-weight, $e_t \in [0,1]^D$ is an erase vector, $v_t \in \mathbb{R}^D$ is a write vector, $\circ$ is point-wise multiplication, and L_{in} is the length of input sequence.

3   Results

In this section, we perform experiments both on real-world data and synthetic tasks. The purpose of the synthetic task is to study the incremental impact of modifications we propose.

3.1   Synthetic Task: Odd-Even Sequence Prediction

In this task, the input is sequence of random odd numbers chosen without replacement from the set $S_o = \{1, 3, 5, ..., 49\}$ and the output is sequence of even numbers from the set $S_e = \{2, 4, 6, ..98\}$. The n-th number y_n in the output sequence is computed as: $y_n = \begin{cases} 2x_n & n \leq \lfloor \frac{L}{2} \rfloor \\ y_{n-1} + 2 & n > \lfloor \frac{L}{2} \rfloor \end{cases}$. x_n is the n-th number in the input sequence and L is the length of both input and output sequence chosen randomly from the range $[1, 20]$. The formula is designed to reflect health-care situations where treatment options depend both on diagnoses in the input sequence and other treatments in the same output sequence. Here is an example of an input-output sequence pair with $L = 7$: *input* := $[11, 7, 25, 39, 31, 1, 13]$ and *output* := $[22, 14, 50, 52, 54, 56, 58]$. We want to predict the even numbers in the output sequence given odd numbers in the input sequence, hence we name it odd-even prediction task. In this task, the model has to "remember" the first half of the input sequence to compute the first half of the output sequence, then it should switch from using input to using previous output at the middle of the output sequence to predict the second half.

Table 1. Test results on odd-even task (lower is better)

Model	NLD
Seq2Seq	0.679
Seq2Seq with attention	0.637
DNC	0.267
DNC (write-protected)	0.250
DC-MANN	0.161
DCw-MANN	**0.082**

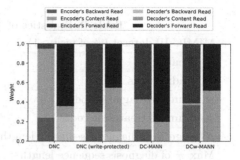

Fig. 4. Read modes of MANNs on odd-even task

Evaluations: Our baselines are Seq2Seq [20], its attention version [1] and the original DNC [6]. Since we want to analyze the impact of new modifications, in this task, we explore two other models: DNC with write-protected mechanism in the decoding phase and dual controller MANN without write-protected mechanism (DC-MANN). We use the Levenshtein distance (edit distance) to measure the model's performance. To account for variable sequence lengths, we normalize this distance over the length of the longer sequence (between 2 sequences). The predicted sequence is good if its Normalized Levenshtein Distance (NLD) to the target sequence is small.

Implementation Details: For all experiments, deep learning models are implemented in Tensorflow 1.3.0. Optimizer is Adam [10] with learning rate of 0.001 and other default parameters. The hidden dimensions for LSTM and the embedding sizes for all models are set to 256 and 64, respectively. Memory's parameters including number of memory slots and the size of each slot are set to 128 and 128, respectively.

Results: After training with 4000 input-output pair of sequences, the models will be tested for the next 1000 pairs. The learning curves of the models are plotted in Figs. 2 and 3. The average NLD of the predictions is summarized in Table 1. As is clearly shown, the proposed model outperforms other methods. Seq2Seq-based methods fail to capture the data pattern and underperform other methods. The introduction of two controllers helps boost the performance of DNC significantly. Additional DNC-variant with write-protected also performs better than the original one, which suggests the benefit of decoding without writing.

Figure 4 plots read mode weights for three reading strategies employed in encoding and decoding phases. We can observe the differences in the way the models prefer reading strategies. The biggest failure of DNC is to keep using backward read in the decoding process. This is redundant because in this problem, it is the forward of the previous read location (if the memory location that corresponds to x_{n-1} is the previous read, then its forward is the memory location that corresponds to x_n) that defines the current output (y_n). On the other hand,

Table 2. Statistics of MIMIC-III sub-datasets

MIMIC-III dataset (# of visit >1)	Procedure as output	Drug as output
# of patients	6,314	5,620
# of admissions	16,317	14,656
# of unique diagnosis codes	4,669	4,563
# of unique treatment codes	1,439	2,446
Average # of diagnosis sequence length	13.3	13.8
Max # of diagnosis sequence length	39	39
Average # of treatment sequence length	4.7	11.4
Max # of treatment sequence length	40	186
Average # of visits per patient	2.5	2.6
Max # of visits per patient	29	29

dual controllers with write-protected mechanism seems help the model avoid bad strategies and focus more on learning reasonable strategies. For example, using dual controllers tends to lessen the usage of content-based read in the encoding phase. This strategy is reasonable in this example since the input at each time step is not repeated. Write-protected policy helps balance the forward and content-based read in the decoding phase, which may reflect the output pattern – half-dependent on the input and half-dependent on the previous output.

3.2 Treatment Recommendation Tasks

The dataset used for this task is MIMIC-III [9], which is a publicly available dataset consisting of more than 58k EMR admissions from more than 46k patients. An admission history in this dataset can contain hundreds of medical codes, which raises a great challenge in handling long-term dependencies. In MIMIC-III, there are both procedure and drug codes for the treatment process so we consider two separate treatment recommendation tasks: procedure prediction and drug prescription. In practice, if we use all the drug codes in an EMR record, the drug sequence can be very long since, each day in hospital, the doctor can prescribe several types of drugs for the patient. Hence, we only pick the first drug used in a day during the admission as the representative drug for that day. We also follow the previous practice that only focuses on patients who have more than one visit [13–15]. The statistics of the two sub-datasets is detailed in Table 2.

Evaluations: For comprehensiveness, beside direct competitors, we also compare our methods with classical for healthcare predictions, which are Logistic Regression and Random Forests. Because traditional methods are not designed for sequence predictions, we simply pick the top outputs (ignoring ordering information). In treatment recommendation tasks, we use precision, which is defined as the number of correct predicted treatment codes (ignoring the order) divided

Table 3. Results on MIMIC-III dataset for procedure prediction and drug prescription (higher is better).

Model	Procedure output		Drug output	
	Precision	Jaccard	Precision	Jaccard
Logistic regression	0.256	0.185	0.412	0.311
Random forest	0.276	0.199	0.491	0.405
Seq2Seq	0.263	0.196	0.220	0.138
Seq2Seq with attention	0.272	0.204	0.224	0.142
DNC	0.285	0.214	0.577	0.529
DCw-MANN	**0.292**	**0.221**	**0.598**	**0.556**

by the number of predict treatment codes. More formally, let S_p^n be the set of ground truth treatments for the n-th admission, S_q^n be the set of treatments that the model outputs. Then the precision is: $\frac{1}{N} \sum_{n=1}^{N} \frac{|S_p^n \cap S_q^n|}{|S_q^n|}$, where N is total number of test patients. To measure how closely the generated treatment compares against the real treatment, we use Mean Jaccard Coefficient[2], which is defined as the size of the intersection divided by the size of the union of ground truth treatment set and predicted treatment set: $\frac{1}{N} \sum_{n=1}^{N} \frac{|S_p^n \cap S_q^n|}{|S_p^n \cup S_q^n|}$.

Implementation Details: We randomly divide the dataset into the training, validation and testing set in a $0.7 : 0.1 : 0.2$ ratio, where the validation set is used to tune model's hyper-parameters. For the classical Random Forests and Logistic Classifier, the input is bag-of-words. Also, we apply One-vs-Rest strategy [17] to enable these classifiers to handle multi-label output and the hyper-parameters are found by grid-searching.

Results: Table 3 reports the prediction results on two tasks (procedure prediction and drug prescription). The performance of the proposed DCw-MANN is higher than that of baselines on the testing data for both tasks, validating the use of dual controllers with write-protected mechanism. Without memory, Seq2Seq methods seem unable to outperform classical methods, possibly because the evaluations are set-based, not sequence-based. In the drug prescription task, there is a huge drop in performance of the Seq2Seq-based approaches. It should be noted that, in drug prescription, the drug codes are given day by day; hence, the average length of output sequence are much longer than the procedure's one. This could be a very challenging task for Seq2Seq. Memory-augmented models, on the other hand, have an external memory to store information, so it can cope with long-term dependencies. Figures 5 and 6 show that compared to DNC, DCw-MANN is the faster learner. This case study demonstrates that

---

[2] The metrics actually are at disadvantage to the proposed sequence-to-sequence model, but we use to make them easy to compare against non-sequential methods.

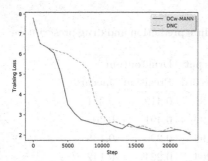

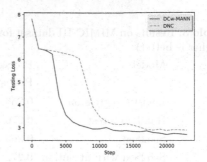

Fig. 5. Training loss of drug prescription task

Fig. 6. Testing loss of drug prescription task

a MANN with dual controller and write-protected mechanism can significantly improve the performance of the sequence prediction task in healthcare.

4   Related Works

The recent success of deep learning has drawn board interest in building AI systems to improve healthcare. Several studies have used deep learning methods to better categorize diseases and patients: denoising autoencoders, an unsupervised approach, can be used to cluster breast cancer patients [21], and convolutional neural networks (CNNs) can help count mitotic divisions, a feature that is highly correlated with disease outcome in histological images [4]. Another branch of deep learning in healthcare is to solve biological problems such as using deep RNN to predict gene targets of microRNAs [24]. Despite these advances, a number of challenges exist in this area of research, most notably how to make use of other disparate types of data such as electronic medical records (EMRs). Recently, more efforts have been made to utilize EMR data in disease prediction [15], unplanned admission and risk prediction [14] problems. Other works apply LSTMs, both with and without attention to clinical time series for heart failure prediction [3] or diagnoses prediction [11]. Treatment recommendation is also an active research field with recent deep learning works that model EMR codes as sequence such as [2] using sequence of billing codes for medicine suggestions or [23] using set of diagnoses for medicine sequence prediction. Differing from these approaches, our work focuses on modeling both the admission data and the treatment output as two sequences to capture order information from input codes and ensure dependencies among output codes at the same time.

Memory augmented neural networks (MANN) have emerged as a new promising research topic in deep learning. Memory Networks (MemNNs) [22] and Neural Turing Machines (NTMs) [5] are the two classes of MANNs that have been applied to many problems such as meta learning [18] and question answering [19]. In healthcare, there is limited work applying MemNN-based models to handle medical-related problems such as clinical textual QA [7] or diagnosis inference

[16]. However, these works have been using clinical documents as input, rather than just using medical codes stored in EMRs. Our work, on the other hand, learns end-to-end from raw medical codes in EMRs by leveraging Differentiable Neural Computer (DNC) [6], the latest improvement over the NTM. In practice, DNC and other NTM variants have been used for various domains such as visual question answering [12], and one-shot learning [18], yet it is the first time DNC is adapted for healthcare tasks.

5 Conclusion

We have introduced a dual controller write-protected MANN designed for healthcare treatment recommendations. Under our design, the order dependencies for each admission and between admissions are preserved allowing memory-based methods to make use of this sequential information for better performance. Differing from other approaches, our work is one of the first attempts to apply MANN to healthcare domain and promising results on MIMIC-III dataset have shown that modifications such as using two controllers and write-protected mechanism are necessary to make MANN work for real-world problems like treatment prediction. In additions, our method can be generalized to other sequence prediction tasks that require special handling of long-term dependencies. Future work will focus on extending the model to handle multiple healthcare tasks, and developing new capabilities for medical question answering.

Acknowledgments. The paper is partly supported by the Telstra-Deakin CoE in Big Data and Machine Learning.

References

1. Bahdanau, D., Cho, K., Bengio, Y.: Neural machine translation by jointly learning to align and translate. In: ICLR (2015)
2. Bajor, J.M., Lasko, T.A.: Predicting medications from diagnostic codes with recurrent neural networks. In: ICLR (2017)
3. Choi, E., Bahadori, M.T., Sun, J., Kulas, J., Schuetz, A., Stewart, W.: RETAIN: an interpretable predictive model for healthcare using reverse time attention mechanism. In: Advances in Neural Information Processing Systems, pp. 3504–3512 (2016)
4. Cireşan, D.C., Giusti, A., Gambardella, L.M., Schmidhuber, J.: Mitosis detection in breast cancer histology images with deep neural networks. In: Mori, K., Sakuma, I., Sato, Y., Barillot, C., Navab, N. (eds.) MICCAI 2013. LNCS, vol. 8150, pp. 411–418. Springer, Heidelberg (2013). https://doi.org/10.1007/978-3-642-40763-5_51
5. Graves, A., Wayne, G., Danihelka, I.: Neural turing machines. arXiv e-prints, October 2014
6. Graves, A., Wayne, G., Reynolds, M., Harley, T., Danihelka, I., Grabska-Barwińska, A., Colmenarejo, S.G., Grefenstette, E., Ramalho, T., Agapiou, J., et al.: Hybrid computing using a neural network with dynamic external memory. Nature **538**(7626), 471–476 (2016)

7. Hasan, S.A., Zhao, S., Datla, V.V., Liu, J., Lee, K., Qadir, A., Prakash, A., Farri, O.: Clinical question answering using key-value memory networks and knowledge graph. In: TREC (2016)
8. Hochreiter, S., Schmidhuber, J.: Long short-term memory. Neural Comput. **9**(8), 1735–1780 (1997)
9. Johnson, A.E.W., Pollard, T.J., Shen, L., Lehman, L.H., Feng, M., Ghassemi, M., Moody, B., Szolovits, P., Celi, L.A., Mark, R.G.: MIMIC-III, a freely accessible critical care database. Sci. Data **3**, 160035 (2016)
10. Kingma, D., Ba, J.: Adam: a method for stochastic optimization. arXiv preprint arXiv:1412.6980 (2014)
11. Lipton, Z., Kale, D., Elkan, C., Wetzel, R.: Learning to diagnose with LSTM recurrent neural networks. In: International Conference on Learning Representations (ICLR 2016) (2016)
12. Ma, C., Shen, C., Dick, A., van den Hengel, A.: Visual question answering with memory-augmented networks. arXiv preprint arXiv:1707.04968 (2017)
13. Ma, F., Chitta, R., Zhou, J., You, Q., Sun, T., Gao, J.: Dipole: diagnosis prediction in healthcare via attention-based bidirectional recurrent neural networks. In: Proceedings of the 23rd ACM SIGKDD International Conference on Knowledge Discovery and Data Mining, pp. 1903–1911. ACM (2017)
14. Nguyen, P., Tran, T., Wickramasinghe, N., Venkatesh, S.: Deepr: a convolutional net for medical records. J. Biomed. Health Inform. **21**(1), 22–30 (2017)
15. Pham, T., Tran, T., Phung, D., Venkatesh, S.: Predicting healthcare trajectories from medical records: a deep learning approach. J. Biomed. Inform. **69**, 218–229 (2017)
16. Prakash, A., Zhao, S., Hasan, S.A., Datla, V.V., Lee, K., Qadir, A., Liu, J., Farri, O.: Condensed memory networks for clinical diagnostic inferencing. In: AAAI, pp. 3274–3280 (2017)
17. Rifkin, R., Klautau, A.: In defense of one-vs-all classification. J. Mach. Learn. Res. **5**(Jan), 101–141 (2004)
18. Santoro, A., Bartunov, S., Botvinick, M., Wierstra, D., Lillicrap, T.: Meta-learning with memory-augmented neural networks. In: International Conference on Machine Learning, pp. 1842–1850 (2016)
19. Sukhbaatar, S., Weston, J., Fergus, R., et al.: End-to-end memory networks. In: Advances in Neural Information Processing Systems, pp. 2440–2448 (2015)
20. Sutskever, I., Vinyals, O., Le, Q.V.: Sequence to sequence learning with neural networks. In: International Conference on Machine Learning (2014)
21. Tan, J., Ung, M., Cheng, C., Greene, C.S.: Unsupervised feature construction and knowledge extraction from genome-wide assays of breast cancer with denoising autoencoders. In: Pacific Symposium on Biocomputing Co-Chairs, pp. 132–143. World Scientific (2014)
22. Weston, J., Chopra, S., Bordes, A.: Memory networks. arXiv preprint arXiv:1410.3916 (2014)
23. Zhang, Y., Chen, R., Tang, J., Stewart, W.F., Sun, J.: Leap: learning to prescribe effective and safe treatment combinations for multimorbidity. In: Proceedings of the 23rd ACM SIGKDD International Conference on Knowledge Discovery and Data Mining, pp. 1315–1324. ACM (2017)
24. Zurada, J.: End effector target position learning using feedforward with error backpropagation and recurrent neural networks. In: 1994 IEEE International Conference on IEEE World Congress on Computational Intelligence Neural Networks, vol. 4, pp. 2633–2638. IEEE (1994)

Denoising Time Series Data Using Asymmetric Generative Adversarial Networks

Sunil Gandhi[1(✉)], Tim Oates[1], Tinoosh Mohsenin[1], and David Hairston[2]

[1] University of Maryland, Baltimore County, Baltimore, USA
{sunilga1,oates,tinoosh}@umbc.edu
[2] U.S. Army Research Laboratory, Adelphi, USA
william.d.hairston4.civ@mail.mil

Abstract. Denoising data is a preprocessing step for several time series mining algorithms. This step is especially important if the noise in data originates from diverse sources. Consequently, it is commonly used in biomedical applications that use Electroencephalography (EEG) data. In EEG data noise can occur due to ocular, muscular and cardiac activities. In this paper, we explicitly learn to remove noise from time series data without assuming a prior distribution of noise. We propose an online, fully automated, end-to-end system for denoising time series data. Our model for denoising time series is trained using unpaired training corpora and does not need information about the source of the noise or how it is manifested in the time series. We propose a new architecture called *AsymmetricGAN* that uses a generative adversarial network for denoising time series data. To analyze our approach, we create a synthetic dataset that is easy to visualize and interpret. We also evaluate and show the effectiveness of our approach on an existing EEG dataset.

1 Introduction

Time series data mining is an important area of research and has applications in a variety of domains including healthcare, econometrics, and speech recognition. Consequently, a large number of methods for time series classification, clustering, anomaly detection and motif discovery have been proposed. Although these methods can handle some noise, they are not effective when noise originates from different sources and has diverse characteristics.

Consider, for example, a widely used method for time series featurization called Symbolic Aggregate approXimation (SAX) [10] that assumes time series are generated from a single normal distribution. As shown in [5] this assumption does not hold in several real life time series datasets. Other techniques assume noise comes from a Gaussian distribution and estimate the parameters of that distribution [13]. This assumption does not hold for data sources like Electroencephalography (EEG), where noise can have diverse characteristics and originate from different sources [15]. Hence, in this work, we focus on learning the characteristics of noise in EEG data and removing it as a preprocessing step.

© Springer International Publishing AG, part of Springer Nature 2018
D. Phung et al. (Eds.): PAKDD 2018, LNAI 10939, pp. 285–296, 2018.
https://doi.org/10.1007/978-3-319-93040-4_23

Electroencephalography (EEG) is a technique that records the electrical activity of the brain by placing electrodes on the scalp. As it is noninvasive and cost-effective, it is widely used in brain-computer interfaces (BCI), determining cognitive states, seizure detection, and monitoring neurological disorders [11,14]. Unfortunately, EEG data is often contaminated by different forms of noise called "artifacts". Artifacts are undesired signals in EEG data originating from sources other than brain activity. Artifacts can occur from diverse sources like ocular, muscular, and cardiac activities, or external sources like electrodes and line noise. The occurrence of external artifacts can be reduced by proper placement of electrodes, but it is impossible to avoid artifacts of biological origin. Not only do artifacts increase the chance of false alarms in seizure detection [12], they can also alter the shape of neurological events [16]. Therefore, artifact detection and removal is an important preprocessing step for EEG data.

Given the importance of artifact removal from EEG data, a large number of denoising techniques have been proposed in the neuroimaging literature. This includes techniques like artifact rejection that focus on detecting artifacts and removing segments where they are present. Although simple, artifact rejection methods can lead to excessive loss of information. Regression of denoised signals using a reference channel is another commonly used technique. Regression techniques need reference signals and cannot be trained using unpaired training data. Thus the approach is specific to EEG data and not broadly applicable to time series data. Another popular technique is to use independent component analysis (ICA) to decompose the signal into independent source signals and identify sources corresponding to noise. The identification of source signals has to be done manually or using supervised classifiers. This requires a human in the loop or additional annotation. Also, ICA has high computational complexity and large memory requirements, making it unsuitable for real-time applications.

We solve these problems by proposing an online, fully automated, end-to-end system for denoising time series trained using unpaired training corpora. An online and fully automated system makes it useful in real-time applications. Unlike [8], our system is trained end-to-end and has fewer hyperparameters to be optimized and is easy to deploy. Being able to train on unpaired training corpora allows our method to be useful in a wide variety of applications. For training of our network, we only need a set of clean signals and set of noisy signals. We do not need paired training data, i.e., we do not need clean versions of the noisy data. This is particularly useful for applications like artifact removal in EEG data as we cannot record clean versions of noisy EEG.

We create this method by leveraging recent advances in unpaired image to image translation [19] using generative adversarial networks. We modify an existing generative adversarial architecture [19] for denoising time series data. We analyze our approach on a synthetic dataset and examine the effectiveness of our approach in removing artifacts from EEG data. The remainder of this paper is organized as follows. Section 2 discusses related work. Section 3 defines the problem. Section 4 describes our system for learning a model for denoising time

series from unpaired training data. In Sect. 5, we evaluate our model on synthetic
and EEG data. Section 6 concludes and discusses future work.

2   Related Work

Time series data mining algorithms can broadly be classified as model-based
and model-free. Model-based algorithms focus on creating features and models
that are robust to noise in the data by assuming priors on the distribution of
the noise. Model-free algorithms do not assume a prior on the distribution of
noise and explicitly learn to remove noise as a preprocessing step [6]. Model-
based algorithms often make assumptions about the distribution of noise in the
data. For example, [13] assumes that noise comes from a Gaussian distribution
and estimates the parameters of the distribution. This makes these techniques
ineffective when noise has different characteristics and originates from diverse
sources [6]. For such applications, model-free denoising is often used. In this
work, we focus on model-free denoising because of its applicability for a wide
range of applications.

Model-free denoising techniques are often used in biomedical applications
for processing EEG or functional Magnetic Resonance Imaging (fMRI) data.
A simple technique for denoising EEG data is artifact rejection that focuses
on detecting artifacts and removing the segments of data where artifacts are
present. Although simple, artifact rejection methods can lead to excessive loss
of information [15]. In this work, we propose a method for denoising the time
series instead of removing entire segments of noisy data.

Regression of a denoised signal using a reference channel like Electrocardiog-
raphy (ECG), Electrooculography (EOG) or Electromyography (EMG) is a common approach to removing artifacts from EEG data [15]. These techniques fail
if a reference signal is not available. The need for a reference signal makes these
approaches EEG-specific. Wavelet transforms are commonly used to decompose
a signal into a set of coefficients at various scales, which represent the similarity
of the signal to the wavelet at that scale. The artifacts found by this method
are often removed by thresholding the coefficients and reconstructing the signal
from the filtered representation. Artifact removal based on the wavelet transform
relies on the artifacts being decomposable in a wavelet basis. Thus, the mother
wavelet, the shrinkage rule, and the threshold are important to the design of the
noise removal method [2]. Also, these techniques process each channel separately
and could miss important clues for removing the artifact.

Blind source separation methods that use Independent Component Analysis (ICA) are the most popular techniques for artifact removal from EEG data
[3,15]. ICA is used to recover independent source signals called components and
then the components corresponding to artifacts are identified. These components
are either manually identified or a classifier is trained to identify components
corresponding to noise. This requires a human in the loop or the annotation
of components corresponding to noise. Such annotation is expensive and may
not always be available. Recently, a technique was proposed that used unpaired

training data to identify components corresponding to noise by formulating it as multi-instance learning problem [8]. Although this method was trained using unpaired training data, it performed ICA which has high computational complexity and large memory requirements, making it unsuitable for real-time applications. Also, their architecture consists of several subsystems like ICA, SAX and multi-instance learning. Each subsystem has its own hyperparameters and tuning them jointly is a challenging task.

Recently, several architectures for time series classification using deep neural networks have been proposed. Wang and Oates proposed a method for encoding time series data as images and using them to classify the time series [17]. Other methods that use raw data as input to a convolutional network for time series classification have been proposed [18]. Convolutional neural networks have also been used for classification of EEG data [1]. These methods demonstrate usefulness of a deep neural network for processing time series data. However, none of these methods have used convnets to denoise time series data.

Generative Adversarial Networks (GANs) [7] have achieved impressive results in computer vision tasks like image generation, image editing, and representation learning. GANs have also been used for removing compression artifacts from images [4]. Recently, CycleGAN, a method for performing unpaired image to image translation using generative adversarial networks, was proposed [19]. The key idea was introduction of an objective function that translates an image from the source domain to the target domain and reconstructs the original image. We modify this network for denoising time series data. To the best of our knowledge, ours is the only work that uses GANs for denoising time series data.

3   Problem Definition

Given a noisy time series A, our goal is to generate time series B that is a denoised version of A. If N is the noise in time series A, then we assume $A = B + N$. We want to learn to characteristics of noise from training data and remove noise from the original signal.

The main challenge in learning a mapping from noisy signal A to clean signal B is a lack of availability of paired training examples. We generally do not have a clean version of a noisy signal. This is because manually removing noise is an expensive task and needs domain expertise. But, it is much easier to collect signals with artifacts and signals without artifacts. Thus, we want to learn a mapping from a noisy signal to a clean signal using only a set of unpaired noisy signals and a set of clean signals.

Recently, a cycle generative adversarial network (cycleGAN) [19] was proposed to perform unpaired image to image translation. They presented an approach for translating an image from source domain A to target domain B, trained using unpaired data. This translation was performed by two networks, network F that translates the image from domain A to domain B and network G that translates the image from domain B to domain A. For training the network, cycleGAN introduced a cycle-consistency loss to enforce $G(F(A)) \approx A$. In this architecture network G does not use any information from the original image to reconstruct the signal. This makes the cycleGAN architecture unsuitable for denoising of time series data. For reconstruction of the original signal from a noisy signal, the network does not have access to the predicted noise component. Thus it does not know the nature or location of the noise. We solve this problem by creating an asymmetric variation of this architecture. In this architecture, we preserve the noise in the signal and use it for reconstruction of the original signal. We explain the architecture of our generative adversarial network in the next section.

4   Asymmetric Generative Adversarial Network

Our goal is to train an end-to-end system to remove artifacts from time series data using generative adversarial networks. We want to learn a mapping function from noisy signal A to denoised signal B given training signals $\{a_i\}_{i=1}^N$ and $\{b_j\}_{j=1}^M$ where $a_i \in A$ and $b_j \in B$. We denote the data distribution of these signals as $a \sim p_{data}(a)$ and $b \sim p_{data}(b)$.

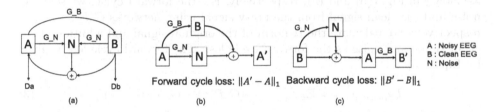

Forward cycle loss: $\|A' - A\|_1$ Backward cycle loss: $\|B' - B\|_1$

A : Noisy EEG
B : Clean EEG
N : Noise

(a)   (b)   (c)

Fig. 1. Asymmetric GAN architecture

Figure 1(a) gives the architecture of our network. In this figure, boxes represent the signals and arrows represent neural networks or operations performed on input signals. As illustrated in the figure, G_B represents the function that maps noisy time series to clean time series. G_N is function that extracts noise if the input is a noisy time series A. It generates noise if the input is a clean time series B. The noise generated from G_N and clean time series B are added to get a noisy time series. The functions G_B and G_N are realized by convolutional autoencoders that are described in Sect. 4.1.

D_a and D_b are two adversarial discriminators. D_a aims to distinguish between noisy time series A and time series generated by adding noise $B + G\_N(B)$. D_b aims to distinguish between clean time series B and denoised time series $G\_B(A)$. We describe the architecture of the discriminators in Sect. 4.1.

To train this architecture, just like in cycleGAN, four losses are used, two adversarial losses and two cycle consistency loss. The adversarial losses are used for training the two mapping functions. For mapping function $G\_B : A \to B$, discriminator D_b matches the distribution of time series denoised by the generator and the distribution of clean time series. This loss function is given by Eq. 1.

$$L_{gan}(G_B, D_b, A, B) = \mathbb{E}_{b \sim p_{data(b)}}[log D_b(b)] + \mathbb{E}_{a \sim p_{data(a)}}[log(1 - D_b(G\_B(a)))] \tag{1}$$

Similarly, for mapping function $B \rightarrow A$, discriminator D_a matches the distribution of time series with noise generated by G_N and the distribution of noisy time series. This loss function is given by Eq. 2.

$$L_{gan}(G_A, D_a, B, A) = \mathbb{E}_{a \sim p_{data(a)}}[log D_a(a)] + \mathbb{E}_{b \sim p_{data(b)}}[log(1 - D_a(b + G\_N(b)))] \tag{2}$$

Adversarial losses ensure that the distribution of generated signals A and B match the target distribution. But generator networks can map input signals to any random permutation of signals in the target domain. Adversarial losses cannot guarantee that $G\_B(A)$ is a denoised version of noisy input time series A. To enforce this relation between noisy input and clean output by the generator we use a cycle consistency loss.

There are two cycle consistency losses, forward cycle loss and backward loss as shown in Fig. 1(b) and (c), respectively. For the forward cycle, we separate noise and the clean signal from the noisy signal using networks G_B and G_N, respectively. We calculate the l1-norm of the original signal and addition of the clean signal and noise as the forward cycle loss. The forward cycle loss is given by Eq. 3.

$$L_{forward\_cyc} = \mathbb{E}_{a \sim p_{data(a)}}||(G\_B(a) + G\_N(a)) - a||_1 \tag{3}$$

For the backward cycle, we add noise generated by generator G_N and use network G_B to clean the signal. The l1-norm of the original clean signal and the signal denoised by G_B is the backward cycle loss. This loss is given by Eq. 4.

$$L_{backward\_cyc} = \mathbb{E}_{b \sim p_{data(b)}}||G\_B(b + G\_N(b)) - b||_1 \tag{4}$$

Notice that the forward and backward cycles are not symmetric. Because of this, we call our GAN architecture asymmetricGAN. Asymmetry is introduced because in the forward cycle we use the noise extracted from the noisy signal to reconstruct the original signal. The usage of noise is necessary to ensure that the network is not penalized for adding noise at the wrong location. It also reduces the burden of generating the exact noise signal for correctly reconstructing the original signal. Forward and backward cycle losses are multiplied with hyperparameters λ_A and λ_B before adding to the final loss of the network.

4.1   Generator and Discriminator Network Architecture

Figure 2(a) shows the architecture of the generator. It is a convolutional autoencoder used for implementing functions G_B and G_N in asymmetricGAN. In

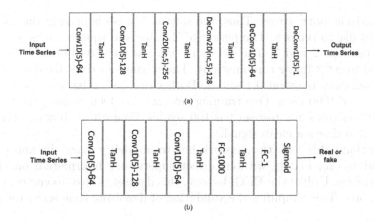

Fig. 2. (a) Generator architecture (b) Discriminator architecture

the figure, layer conv1D(5)-64 represents a layer of 64 features of $1D$ convolution with filter size 5. Figure 2(b) shows the architecture of the discriminator used for predicting if the input signal is real or generated. This architecture was used to implement networks D_a and D_b. In our architecture, we use $1D$ convolution with filter size K to capture temporal patterns in the time series and $2D$ convolution with filter size $N \times k$ to capture patterns across different channels, where N is the number of channels in time series. To create a network that is independent of the relative ordering of time series channels, we do not use $2D$ convolution with a spatial dimension of filter size less than N. This is important for the EEG data as the channel indices do not correspond to the spatial locations of the electrodes.

To train the asymmetricGAN and reduce model oscillation, we use the strategy of saving the history of generated time series [19]. We update the discriminators using a history of generated time series rather than the ones produced by the latest generative networks. We keep a buffer that stores the 50 previously generated time series for training. In the next section, we show the effectiveness of our method on synthetic and EEG dataset.

5 Experimental Setup and Results

5.1 Synthetic Dataset

Artifact removal from EEG data is a problem where ground truth does not exist because we do not have a clean version of the noisy signal. This makes the evaluation of artifact removal methods difficult. Also, visualizing and understanding EEG data is a time consuming task. We solve this problem by creating a simpler synthetic dataset.

In our synthetic dataset, we create a clean signal as a linear combination of a sine and a square wave. We create a noisy signal as a linear combination of sine, square and sawtooth waves. Thus, the sawtooth wave is playing the role of the noise we'd like to remove. The period of the sine and square waves is randomly selected between 2 and 5. The sawtooth wave has a period of 6. These signals are mixed to get 3 linear combinations. The mixing matrix is fixed and contains random numbers between 0.1 and 2. Both clean and noisy time series have a sample size of 1000 each. Our training set contains 4000 signals, the validation set has 1000 signals and test set has 100 signals. Figure 3(a) shows a noisy signal and Fig. 3(b) shows a clean signal.

To perform this task the network has to implicitly learn the nature of the noise and mixing matrix without direct supervisory information on either of these variables. Unlike the EEG dataset, this dataset is easy to create, visualize and interpret. This simplifies the validation of denoising time series data.

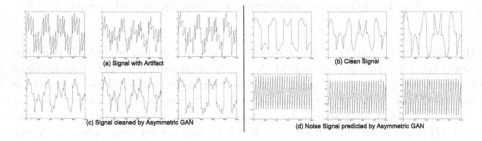

Fig. 3. Artificial data result

We train our network on the synthetic dataset using the adam optimizer and a learning rate of 0.0002. Both weights on cycle loss, λ_A and λ_B are 0.5. All weights were initialized from a Gaussian distribution with mean 0 and standard deviation 0.02. We test our network by measuring mean squared error (MSE) between the generated signal and ground truth clean signal. Figure 3 visualizes an example result on the test data. Figure 3(a) shows the noisy signal, Fig. 3(b) shows the ground truth clean signal, Fig. 3(c) shows the signal denoised by our network, and Fig. 3(d) is the noise detected by the network. We can observe from Fig. 3 that the ground truth clean signal and denoised signal are similar, and noise predicted by the network is similar to the sawtooth wave. The MSE between the ground truth clean signal and denoised signal for the example in Fig. 3 is 0.4065. The average MSE error for entire test set is 0.4180 and standard deviation is 0.011. This demonstrates the effectiveness of our network on the synthetic dataset.

5.2   EEG Dataset

We test our network for artifact removal on a dataset generated by the US Army Research Laboratory that has been previously discussed in [9]. We briefly describe the dataset here. Readers can refer to [9] for more details. The dataset was recorded using a 64-channel Biosemi ActiveTwo System. Participants in the study performed a block of artifact-inducing facial and head movements. The exact details of each movement were not controlled by the experimenter but rather left up to the participant to perform in their most natural manner. The seven movements performed included clenching the jaw, moving the jaw vertically, blinking both eyes, moving the eyes leftward then back to center, moving the eyes upwards then back to center, raising and lowering eyebrows, and rotating the head side-to-side. Each type of movement was performed in a separate run 20 times. At the beginning of each run, participants were told which movement to perform. For each run, a male voice initially counted down from 3 at a rate of every 2 s, followed by a tone every 2 s, and participants performed the movement in time with the tone. The participants were told to make the movement for the first second of the 2 s period, and then to return to a relaxed state for the remaining 1 s. A baseline dataset was recorded for each participant. Participants were told to not move and look straight at the computer screen for the baseline. We use this part of the dataset as "clean" data. Analyses here focuses on eyebrow movements in a single channel frontal electrode, with the intention of extrapolating results to other more complicated movements in the future.

Despite the fact that during collection of clean data patients were instructed to not move and look straight at the computer screen, we noticed that there were artifacts even in "clean" data. Manually annotating all artifacts from all channels is a time consuming task. So in this work, we focus on ocular artifacts in the Fp1 electrode of the frontal region. We manually remove all patients that have more than two ocular artifacts in the clean data and do not have artifacts in the region of eyebrow raising. In the resulting dataset, we have 4 patients with clean data and 10 patients with noisy data. Each patient's clean data has 4836 samples and noisy data has 420354 samples. We use this manually annotated data in all experiments below.

We train an asymmetricGAN on EEG data where the noisy signal contains artifacts corresponding to eyebrow raising and the clean signal does not contain any ocular artifacts. We use a sliding window of size 1000 over clean and noisy data as input to the network. Our network does not need to know the exact location of the artifact. Any window that contains an entire or partial artifact is considered noisy. The sliding window approach makes our model invariant to artifact location. Also, as the network can remove artifacts from a window and does not need to process the entire time series, it can be used in online real-time applications.

The DC component in EEG data is different for each recording. We normalize every window of clean and noisy data to remove the DC offset from the data. We remove the DC offset by subtracting the median of the data in the window. Normalizing by subtracting the median is more robust to outliers compared to subtraction of the mean. This preprocessing step ensures that the amplitude of clean segments of data in clean and noisy signals is centered around zero.

We use preprocessed windows of clean and noisy EEG data to train an asymmetricGAN. The parameters used for the network, the optimizer, and the initialization are similar to the one used for the synthetic dataset. Except in this case the number of input and output channels is one and λ_A and λ_B are 0.05.

Evaluation of EEG data is challenging as the ground truth noiseless signals are not known. Multiple approaches to evaluation have been proposed in recent years, however, authors do not agree on a single mechanism for evaluating artifact removal [15]. In this work, we give qualitative results and use an artifact detector to evaluate our method.

Qualitative Results. We have clean EEG for 4 patients and noisy EEG for 10 patients. To generate qualitative results, we train an asymmetricGAN using clean EEG from all 4 patients and noisy EEG from 7 randomly selected patients. We use noisy EEG from the remaining 3 patients to test our network. We generate qualitative results by performing a forward pass through network G_B over non-overlapping windows of 1000 samples on noisy data. We concatenate the output of the network over non-overlapping windows to get denoised EEG data. Figure 4 shows the result of denoising the noisy EEG of a patient from the test set. Visualization of the other two patients is shown in the appendix due to space constraints.

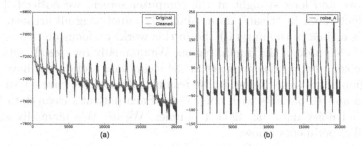

Fig. 4. EEG data result

Figure 4(a) shows the original noisy EEG signal and the signal after artifact removal. Figure 4(b) shows the artifact predicted by the network in EEG data. For collection of the original noisy EEG, the patient was instructed to raise their eyebrow every 2 s. For every eyebrow raise, there was spike in amplitude at the Fp1 electrode. As shown in the figure, the network learns that the spike in amplitude is because of the artifact and removes it. Artifacts extracted by our network as shown in Fig. 4(b) are similar to the artifacts that occur when an eyebrow is raised according to the existing literature [15].

Evaluation by Detection. In this section, we use artifact detection as a way of measuring the performance of artifact removal. This is less subjective, automatic and provides a quantitative measure of performance of artifact removal. We first train artifact detection and artifact removal on different datasets. We use the artifact detector to classify every window in EEG data denoised by the artifact removal algorithm. The error is given by the percent of windows where an artifact was detected in denoised EEG data.

For this experiment, we split the datasets into two parts. The first split contains clean EEG for 2 patients and noisy EEG for 6 patients. The first split is further divided into clean EEG of 2 patients and noisy EEG of 4 patients as a training set for the asymmetricGAN. The noisy EEG of the remaining 2 patients from the first split is used as test data for the asymmetricGAN.

The second split contains clean EEG of 2 patients and noisy EEG of 4 patients. It is used to train artifact detection network. The second split is further divided into training and test sets for artifact detection, each containing clean EEG for one patient and noisy EEG for two patients.

We use the network similar to the discriminator network explained in Sect. 4.1 to train an artifact detection algorithm. The preprocessing method of sliding a window over clean and noisy EEG and median normalization is also the same for the artifact detection network. The network is trained using the Adam optimizer and a learning rate of 0.0002. On training, we get an accuracy of 97.39% on test data for artifact detection.

To get an error metric, we use trained artifact removal to denoise noisy EEG in the test data. Then we classify every window in the denoised signal using artifact detector. The total number of windows in the test data were 38002. All these windows originally had either entire or partial artifact. The total number of windows having artifacts based on classification of the artifact detector after denoising are 10499. This shows that our artifact removal algorithm was able to change the classification of artifact detector from noisy to clean 72.37% of the time.

6   Conclusion

This paper presents an online, fully automated, end-to-end system for denoising time series data. Our system for denoising is trained using unpaired training corpus. It does not need any information about the source of the noise or how it is manifested in the time series data. We created a synthetic dataset and used it to evaluate our network. We also used model to remove artifacts from existing EEG dataset. In future, we intend to use our architecture for removing artifacts originating from other sources like muscular or cardiac activities from EEG data.

References

1. Bashivan, P., Rish, I., Yeasin, M., Codella, N.: Learning representations from EEG with deep recurrent-convolutional neural networks. arXiv preprint arXiv:1511.06448 (2015)
2. Daly, I., Nicolaou, N., Nasuto, S.J., Warwick, K.: Automated artifact removal from the electroencephalogram: a comparative study. Clin. EEG Neurosci. **44**(4), 291–306 (2013)

3. Fitzgibbon, S.P., Powers, D.M., Pope, K.J., Clark, C.R.: Removal of EEG noise and artifact using blind source separation. J. Clin. Neurophysiol. **24**(3), 232–243 (2007)
4. Galteri, L., Seidenari, L., Bertini, M., Del Bimbo, A.: Deep generative adversarial compression artifact removal. arXiv preprint arXiv:1704.02518 (2017)
5. Gandhi, S., Oates, T., Boedihardjo, A., Chen, C., Lin, J., Senin, P., Frankenstein, S., Wang, X.: A generative model for time series discretization based on multiple normal distributions. In: Proceedings of the 8th Workshop on Ph.D. Workshop in Information and Knowledge Management, pp. 19–25. ACM (2015)
6. Gao, J., Sultan, H., Hu, J., Tung, W.W.: Denoising nonlinear time series by adaptive filtering and wavelet shrinkage: a comparison. IEEE Sig. Process. Lett. **17**(3), 237–240 (2010)
7. Goodfellow, I., Pouget-Abadie, J., Mirza, M., Xu, B., Warde-Farley, D., Ozair, S., Courville, A., Bengio, Y.: Generative adversarial nets. In: Advances in Neural Information Processing Systems, pp. 2672–2680 (2014)
8. Jafari, A., Gandhi, S., Konuru, S.H., Hairston, W.D., Oates, T., Mohsenin, T.: An EEG artifact identification embedded system using ICA and multi-instance learning. In: 2017 IEEE International Symposium on Circuits and Systems (ISCAS), pp. 1–4. IEEE (2017)
9. Lawhern, V., Hairston, W.D., McDowell, K., Westerfield, M., Robbins, K.: Detection and classification of subject-generated artifacts in EEG signals using autoregressive models. J. Neurosci. Methods **208**(2), 181–189 (2012)
10. Lin, J., Keogh, E., Wei, L., Lonardi, S.: Experiencing SAX: a novel symbolic representation of time series. Data Min. Knowl. Disc. **15**(2), 107–144 (2007)
11. Ocbagabir, H.T., Aboalayon, K.A., Faezipour, M.: Efficient EEG analysis for seizure monitoring in epileptic patients. In: 2013 IEEE Long Island Systems, Applications and Technology Conference (LISAT), pp. 1–6. IEEE (2013)
12. Seneviratne, U., Mohamed, A., Cook, M., D'Souza, W.: The utility of ambulatory electroencephalography in routine clinical practice: a critical review. Epilepsy Res. **105**(1), 1–12 (2013)
13. Tracey, B.H., Miller, E.L.: Nonlocal means denoising of ECG signals. IEEE Trans. Biomed. Eng. **59**(9), 2383–2386 (2012)
14. Turner, J., Page, A., Mohsenin, T., Oates, T.: Deep belief networks used on high resolution multichannel electroencephalography data for seizure detection. In: 2014 AAAI Spring Symposium Series (2014)
15. Urigüen, J.A., Garcia-Zapirain, B.: EEG artifact removal state-of-the-art and guidelines. J. Neural Eng. **12**(3), 031001 (2015)
16. Vaughan, T.M., Heetderks, W., Trejo, L., Rymer, W., Weinrich, M., Moore, M., Kübler, A., Dobkin, B., Birbaumer, N., Donchin, E., et al.: Brain-computer interface technology: a review of the second international meeting (2003)
17. Wang, Z., Oates, T.: Imaging time-series to improve classification and imputation. arXiv preprint arXiv:1506.00327 (2015)
18. Wang, Z., Yan, W., Oates, T.: Time series classification from scratch with deep neural networks: a strong baseline. In: 2017 International Joint Conference on Neural Networks (IJCNN), pp. 1578–1585. IEEE (2017)
19. Zhu, J.Y., Park, T., Isola, P., Efros, A.A.: Unpaired image-to-image translation using cycle-consistent adversarial networks. arXiv preprint arXiv:1703.10593 (2017)

Shared Deep Kernel Learning
for Dimensionality Reduction

Xinwei Jiang[1]([✉]), Junbin Gao[2], Xiaobo Liu[1], Zhihua Cai[1], Dongmei Zhang[1],
and Yuanxing Liu[1]

[1] School of Computer Science, China University of Geosciences,
Wuhan 430074, China
ysjxw@hotmail.com, {xbliu,zhcai,yxliu}@cug.edu.cn, cugzdm@foxmail.com
[2] Discipline of Business Analytics, The University of Sydney Business School,
The University of Sydney, Sydney, NSW 2006, Australia
junbin.gao@sydney.edu.au

Abstract. Deep Kernel Learning (DKL) has been proven to be an effective method to learn complex feature representation by combining the structural properties of deep learning with the nonparametric flexibility of kernel methods, which can be naturally used for supervised dimensionality reduction. However, if limited training data are available its performance could be compromised because parameters of the deep structure embedded into the model are large and difficult to be efficiently optimized. In order to address this issue, we propose the Shared Deep Kernel Learning model by combining DKL with shared Gaussian Process Latent Variable Model. The novel method could not only bring the improved performance without increasing model complexity but also learn the hierarchical features by sharing the deep kernel. The comparison with some supervised dimensionality reduction methods and deep learning approach verify the advantages of the proposed model.

Keywords: Dimensionality reduction · Gaussian processes
Deep learning

1 Introduction

In the big data era, there are enormous data with high-dimensional features/variables, and the major concern becomes how to efficiently discover unknown patterns embedded in the observed data with high dimensionality which poses serious problems for data storage and analysis. Therefore, Dimensionality Reduction (DR) techniques have been widely used for data pre-processing, data analysis and visualization by reducing the number of features and simultaneously removing the noises and redundancies embedded in the high-dimensional observations. Applications in computer vision, gene data and remote sensing images analysis [7,9,15] demonstrate that DR techniques are essential for high-dimensional data analysis which could facilitate the final data analysis

© Springer International Publishing AG, part of Springer Nature 2018
D. Phung et al. (Eds.): PAKDD 2018, LNAI 10939, pp. 297–308, 2018.
https://doi.org/10.1007/978-3-319-93040-4_24

tasks such as regression and classification models. Typically DR can be divided into feature selection and extraction. Feature selection tries to find subsets of the original features while feature extraction transforms high-dimensional features to low-dimensional feature space. In this paper, we focus on the typical feature extraction techniques.

In the past two decades many DR algorithms based on feature extraction have been developed. Roughly these approaches can be classified into the unsupervised and supervised techniques according to whether or not the extra labels (response variables) are utilized. For the unsupervised DR models, the classic Principal Component Analysis (PCA) could be the most representative method, which tries to linearly project the high-dimensional observation to a lower-dimensional space in such a way that the variance of the data in the dimensionality-reduced space is maximized. However, the linear assumption in PCA could be violated because there exists nonlinear structures in many tasks, resulting into unsatisfactory performance. Thus, many nonlinear extensions of PCA have been introduced, such as Kernel PCA (KPCA) [26] based on the kernel tricks and Auto-Encoder (AE) [8] based on neural network, etc. Besides, the manifold learning based methods such as ISOMAP and Locally Linear Embedding (LLE) [17] are capable of finding meaningful and nonlinear relationships embedded in high-dimensional observations, and latent variable models (LVMs) including Gaussian Process Latent Variable Model (GPLVM) [13], Thin Plate Spline Latent Variable Model (TPSLVM) [10], and Relevance Units Latent Variable Model (RULVM) [5] try to explicitly represent the unknown mappings from low-dimensional latent space to high-dimensional observation space with nonlinear models like Gaussian Process Regression (GPR) [18], Thin Plate Spline (TPS) [10] and Relevance Units Machine (RUM) [5], respectively. However, it becomes difficult to obtain the low-dimensional latent representation for a new testing sample due to the lack of mapping from high-dimensional input to latent space which is typically called the out-of-sample problem. In order to address this issue the back constraint extensions of LVMs are developed by introducing the extra mapping, leading to the back constraint GPLVM (BC-GPLVM) [13]. To further extend GPLVM to multiview data, the shared GPLVM [3] is proposed by using multiple GPLVMs to represent the relationship between the shared latent variables and multiple observations.

However, it would be unwise to perform unsupervised DR when the additional labels are available. Various works based on supervised DR have been proposed to extract discriminative features. For example, Linear discriminant analysis (LDA) tries to carry on linear DR by maximizing the between-class separation and minimizing the within-class separation, which shows better performance than PCA. Also, many unsupervised manifold learning algorithms have been extended to the supervised settings, such as supervised LLE [12]. More interestingly, LVMs based methods can naturally handle the extra labels giving rise to various supervised LVMs.

The first type of supervised LVMs could be roughly termed as *the flat extensions*, where LVMs are extended to the multiview settings. For example, Supervised Probabilistic PCA (SPPCA) [27] tries to use two Probabilistic PCA

(PPCA) [13] models to linearly learn the shared latent variables with low dimensionality from two views: the relationship between the latent variables and the high-dimensional observations and the relationship between the latent variables and the output labels. Supervised GPLVM (SGPLVM) [6] further employs nonlinear GPLVM rather than the linear PPCA in SPPCA to learn the shared latent variables. Besides, there are some LVMs extensions with various discriminative priors, which can also be viewed as the flat extensions. Discriminative GPLVM (DGPLVM) [22] adds typical LDA prior to GPLVM, and Gaussian Markov Random Field (GMRF) [19] further makes use of the discriminative graph Laplacian prior which can be seen a more general extension of LDA prior. Recently, Discriminative Shared GPLVM (DS-GPLVM) [4] and Shared Autoencoder Gaussian Process (SAGP) [14] try to combine shared GPLVM with various discriminative priors to directly handle the multiview data. The flat extensions may be efficient when limited training data are available, but the performance of these models could be compromised if the LVMs can not represent the complex relationship between the observations and the latent variables.

Alternatively, there are some supervised LVMs trying to go deep, which can be named as *the deep extensions*. Deep learning has attracted many researchers' attention and made breakthroughs in many tasks because good features can be learnt automatically, which is highly related to feature extraction or DR. For instance, Auto-Encoder (AE) based deep learning techniques try to stack AE to learn hierarchical feature representation from massive unlabelled data and then a few data with labels are utilized to fine-tune the hierarchical models [23]. However, large amounts of training data are needed for training deep learning algorithms because there are many parameters to be optimized. Thus, many nonparametric methods have been adopted to reduce the number of parameters of typical deep learning models. For example, Non-Parametrically Guided Autoencoder (NPGA) [21] makes use of the nonparametric GP to relate the latent variables in AE to the additional labels. GP is further extended to Deep Gaussian Process (DGP) [2] by stacking the nonlinear GPR or GPLVM. However, applying DGP in large-scale data could be challenge due to the high complexity. Another line could be the semiparametric models, which have been typically used for performing supervised DR and regression/classification simultaneously. Supervised Latent Linear GPLVM (SLLGPLVM) [11] uses the semiparametric model to perform supervised DR with linear model mapping the high-dimensional features to the latent variables and Gaussian Process Classification (GPC) [18] model transforming the latent variables to the discrete labels. Furthermore, by combining the structural properties of deep learning with the nonparametric flexibility of kernel methods Deep Kernel Learning (DKL) [24] is proposed, which can be regarded as the deep extension of semiparametric model. By using some approximation techniques, DKL can easily scale linearly with the number of training data. DKL can outperform scalable GPs with expressive kernels, stand-alone DNNs, and GPs applied to the outputs of trained DNNs if large amount of training data could be provided, but the performance may be unsatisfactory if there are limited training data because the number of parameters to be optimized in deep structures is large.

In this paper, we introduce the Shared Deep Kernel Learning (SDKL) by combining the flat and the deep extensions of LVMs for supervised DR. Specifically, a shared deep kernel is integrated into the shared GPLVM to represent the two relationships: the mapping from the latent variables to the high-dimensional observations, and the mapping from the latent variables to the output labels. It is expected that the novel method could gain the complementary advantages of the two approaches, and particularly be adaptive for the dataset with a small number of training samples because the high-dimensional observations and the output labels are fully utilized. Different from original DKL, we make use of shared latent variables to connect the high-dimensional inputs and the output labels based on shared GPLVM. Also, compared to recently proposed Shared Autoencoder Gaussian Process (SAGP), parametric deep models instead of non-parametric GPs are adopted to represent the back constraint mapping which transforms the high-dimensional observation to latent variables, which could learn complex feature representation. The novel algorithm can not only bring the improved performance without increasing model complexity but also learn the hierarchical features by sharing the deep kernel. The proposed SDKL is verified in terms of data visualization in 2D latent space and classification accuracy based on the simple K-Nearest Neighbors (KNN) approach in the learnt latent space.

The paper is organized as follows. In the next section, some related works are briefly reviewed, and then we will introduce the proposed model in Sect. 3, followed by the experiments section. Finally, we summarize the paper in Sect. 5 with conclusions.

2   Related Works

2.1   GP and GPLVM

Denote a dataset by $\mathcal{D} = \{(\boldsymbol{x}_1, \boldsymbol{y}_1), \cdots, (\boldsymbol{x}_N, \boldsymbol{y}_N)\}$, where $X = [\boldsymbol{x}_1, ..., \boldsymbol{x}_N]^T$ are inputs data in a high-dimensional space $\mathcal{R}^D$, i.e., $\boldsymbol{x}_n \in \mathcal{R}^D$, and $Y = [\boldsymbol{y}_1, ..., \boldsymbol{y}_N]^T$ are the corresponding outputs data with each $\boldsymbol{y}_n \in \mathcal{R}^M$ (real and discrete values for regression and classification tasks, respectively). For the sake of simplicity, the output is set to be scalar ($M = 1$) for regression tasks, where the aim is to learn the distribution of prediction $p(y|\boldsymbol{x}^*)$ for any test sample $\boldsymbol{x}^*$. In GPR [18], each point y_n is assumed to be generated from the unknown functional variable f with independent Gaussian noise

$$y = f(\boldsymbol{x}) + \epsilon$$

where f is the (zero-mean) GP with the covariance/kernel function $k(\cdot, \cdot)$ defined on input space and ϵ is the additive Gaussian noise with zero mean and covariance σ^2.

For a new test sample $\boldsymbol{x}^*$, the predictive distribution conditioned on the given observation can be derived by

$$g^*|\boldsymbol{x}^*, X, Y \sim \mathcal{N}(K_{\boldsymbol{x}^*X}(K_{XX} + \sigma^2 \mathbf{I})^{-1}Y,$$
$$K_{\boldsymbol{x}^*\boldsymbol{x}^*} - K_{\boldsymbol{x}^*X}(K_{XX} + \sigma^2 \mathbf{I})^{-1}K_{X\boldsymbol{x}^*}) \tag{1}$$

where Ks are the matrices of the covariance/kernel function such as the radial basis function (RBF).

If the output Y are not given, and we similarly assume that there is an unknown mapping modelled by GPR which projects unknown latent variables $Z = \{z_1, z_2, \cdots, z_N\} \subset \mathcal{R}^p$ to high-dimensional observation X, resulting into the unsupervised DR algorithm GPLVM. In detail, the observation can be formulated by $x = f(z) + \epsilon$ with the unknown zero-mean GP f and independent Gaussian noise ϵ similarly defined as GPR. Then, with Gaussian distribution assumption the data likelihood can be written by

$$P(X|f) = \prod_{n=1}^{N} \mathcal{N}(x_n|f(z_n), \sigma^2) \tag{2}$$

and the GP prior over $f = (f(x_1), \cdots, f(x_N))^T$ is $P(f|\theta) = \mathcal{N}(f|0, K_{Z,Z})$ with the covariance/kernel matrix over the latent variables $K_{Z,Z}$ and the parameters of kernel θ. With Bayesian theory, marginalizing the mapping function f leads to the marginal likelihood as follows

$$P(X|Z, \theta) = \mathcal{N}(X|0, K_{Z,Z} + \sigma^2 I) \tag{3}$$

Furthermore, the posterior distribution of the latent variables Z can be obtained by

$$P(Z|X, \theta) \propto P(X|Z, \theta)P(Z) \tag{4}$$

where $P(Z)$ is a prior like the spherical Gaussian prior. Finally, the latent variables Z and the hyperparameters θ can be learnt by maximizing the log-posterior $L = \log P(Z|X, \theta)$ with gradients based optimization methods. However, as there is not mapping from the observation space to the latent space, it becomes difficult to project a new testing observation to latent space. Thus the back constrain mapping typically modelled by linear function or kernel regression are introduced, resulting into BC-GPLVM [13].

To further extend the unsupervised GPLVM to supervised settings, discriminative priors instead of the simple spherical Gaussian prior $P(Z)$ can be added, giving rise to the discriminative GPLVMs [4,22]. Another line is the supervised extensions based on shared GPLVM such as Supervised GPLVM (SGPLVM) [6]. As can be seen from the graphical representation in Fig. 1(a) the mapping from latent variables Z to high-dimensional inputs X and the transformation from latent variables Z to output labels Y are modelled by GPLVM, resulting into the posterior distribution of the latent variables Z by

$$P(Z|X, Y) \propto P(X|Z)P(Y|Z)P(Z) \tag{5}$$

Based on the idea of shared GPLVM, Discriminative Shared GPLVM (DS-GPLVM) extends shared GPLVM to multiview data with the discriminative graph Laplacian prior. Shared Autoencoder Gaussian Process (SAGP) adds back constraint mapping also modelled by nonparametric GPR to shared GPLVM, leading to the general extension of multiview Auto-Encoder.

2.2 DKL

DKL [24] combines the deep neural network and kernel learning by introducing the deep kernel as

$$k(\boldsymbol{x}_i, \boldsymbol{x}_j | \theta) \rightarrow k(dnn(\boldsymbol{x}_i, W), dnn(\boldsymbol{x}_j, W) | \theta, W) \tag{6}$$

where $dnn(\boldsymbol{x}, W)$ is a nonlinear mapping represented by a deep model. To add flexibility, the Spectral Mixture (SM) [25] kernel instead of typical RBF is proposed. Based on the aforementioned GPR formulation, the DKL can be similarly derived. Thus, DKL can be seen as a GP with the deep kernel which firstly transforms high-dimensional inputs X to hierarchical feature representation. The framework of DKL is also shown in Fig. 1(b) where the function g indicates GP.

With the recently proposed local kernel interpolation, inducing points and structure exploiting algebra techniques, DKL can scale linearly with the number of training data [25], and it has shown better performance than the stand-alone deep learning architectures and GP on many datasets [25].

From the prospective of GPLVM, DKL can be also regarded as the BC-GPLVM with back constraint mapping represented by a deep architecture with the difference that only high-dimensional observations X are available in BC-GPLVM while there are observations X and output labels Y in DKL. Throughout this paper, we call this kind of unsupervised model as BC-GPLVM with Deep Kernel (BC-GPLVM-DK) with the graphical representation in Fig. 1(c).

3 The Proposed Model

In this section, we introduce the proposed Shared Deep Kernel Learning (SDKL) for supervised dimensionality reduction along with the model inference and optimization.

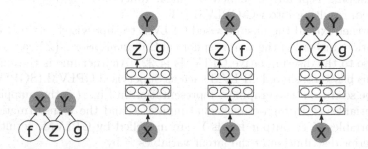

Fig. 1. Graphical representations of (a) Supervised GPLVM (SGPLVM); (b) Deep Kernel Learning (DKL); (c) BC-GPLVM with Deep Kernel (BC-GPLVM-DK); (d) the proposed Shared Deep Kernel Learning (SDKL).

The graphical representation of SDKL is shown in Fig. 1(d) where the white nodes mean unknown variables/functions, the grey nodes indicate the observed variables and the nodes f and g are GPs. Given a dataset similarly represented by $\mathcal{D} = \{X, Y\}$ where $X = [\boldsymbol{x}_1, ..., \boldsymbol{x}_N]^T$ and $Y = [\boldsymbol{y}_1, ..., \boldsymbol{y}_N]^T$ are high-dimensional observations in $\mathcal{R}^D$ space, and the corresponding output labels with each $\boldsymbol{y}_n \in \mathcal{R}^M$. We aim to discover the latent variables $Z = \{\boldsymbol{z}_1, \boldsymbol{z}_2, \cdots, \boldsymbol{z}_N\} \subset \mathcal{R}^p$ corresponding to the observations. Throughout this paper, we focus on the classification tasks with discrete labels encoded by the "one-hot" scheme because it is easy for data visualization.

Firstly, we assume that the mapping from the latent variables to the high-dimensional observations and the projection from the latent variables to the output labels are modelled by shared GPLVM, which can be formulated by

$$P(X|Z, \theta_x) = \mathcal{N}(X|0, K_{Z,Z}) = \frac{1}{(2\pi)^{DN/2}|K|^{1/2}} \exp \left\{ -\frac{1}{2} \text{tr}(K^{-1}XX^T) \right\} \quad (7)$$

$$P(Y|Z, \theta_y) = \mathcal{N}(Y|0, K_{Z,Z}) = \frac{1}{(2\pi)^{MN/2}|K|^{1/2}} \exp \left\{ -\frac{1}{2} \text{tr}(K^{-1}YY^T) \right\}$$

where the shared kernel matrices K are based on the same latent variables Z with different hyperparameters θ_x, θ_y.

Furthermore, by utilizing the notation of deep kernel in DKL in Eq. (6), which explicitly transforms the high-dimensional observation $\boldsymbol{x}_n$ to hierarchical features by deep neural networks with the last layer to be the corresponding latent variable $\boldsymbol{z}_n = dnn(\boldsymbol{x}_n, W)$, followed by the aforementioned shared GPLVM.

Based on the conditional independence property, the joint distribution can be written by

$$P(X, Y, Z|\theta, W) = P(X|Z, \theta_x, W)P(Y|Z, \theta_y, W)P(Z) \quad (8)$$

with the hyperparameter $\theta = \{\theta_x, \theta_y\}$ and the parameters of deep structure in deep kernel W.

By minimizing the negative log joint distribution $P(X, Y, Z|\theta, W)$ with respect to the hyperparameter θ and the parameters of deep kernel W, the objective function can be written by

$$L = -\log P(X, Y, Z|\theta, W) \quad (9)$$

$$= \frac{1}{2}\log|K| + \frac{1}{2}\text{tr}(K^{-1}XX^T) + \frac{1}{2}\log|K| + \frac{1}{2}\text{tr}(K^{-1}YY^T) - \log P(Z)$$

where the kernel matrices K in the first two items and the second two items are based on the same latent variables with different hyperparameters θ_x, θ_y, thus they could not be combined together.

In this paper, the scaled conjugate gradient (SCG) is employed to optimized the model, and we notice that the derivative of $\frac{\partial L}{\partial K}$ can be directly computed by (for the sake of simplification, we ignore the prior $P(Z)$)

$$\frac{\partial L}{\partial K} = K^{-1}XX^TK^{-1} - K^{-1} + K^{-1}YY^TK^{-1} - K^{-1} \quad (10)$$

Then according to the chain rule, the derivative of $\dfrac{\partial L}{\partial W}$ can be evaluated by

$$\frac{\partial L}{\partial W} = \frac{\partial L}{\partial K}\frac{\partial K}{\partial Z}\frac{\partial Z}{\partial W} \tag{11}$$

where the computation of the derivative of the kernel matrix K with respect to the latent variable Z depends on the specific kernel function, and the derivative of latent variables Z with respect to the parameters of deep kernel W can be easily evaluated by the chain rule again.

Similarly the derivative of $\dfrac{\partial L}{\partial \theta}$ can be written by

$$\frac{\partial L}{\partial \theta} = \frac{\partial L}{\partial K}\frac{\partial K}{\partial \theta} \tag{12}$$

Once the two derivatives are calculated, the SDKL can be trained by using the SCG, and the testing step is simple. Given any testing sample x^*, the corresponding latent variable z^* can be evaluated by the deep neural network $dnn(x^*, W)$ embedded in the deep kernel, followed by a simple KNN classifier to make classification in the learnt latent space to objectively evaluate the supervised DR method.

For the sake of convenience, we simply use RBF kernel in the paper, and it would be easy to adopt the Spectral Mixture (SM) kernel and the Kernel Interpolation for Scalable Structured Gaussian Processes (KISS-GP) introduced in [24] to add model flexibility and reduce model complexity from $\mathcal{O}(N^3)$ to $\mathcal{O}(N)$. Compared to original DKL, we have to highlight that the proposed SDKL does not increase the complexity, and could also scale linearly like DKL.

4   Experiments

In order to verify the novel model, we compare the proposed SDKL to original DKL, SGPLVM based on shared GPLVM, the stand-alone deep learning method Stacked AE (SAE), and unsupervised BC-GPLVM with the back constrain represented by deep neural network (BC-GPLVM-DK) in five datasets in terms of data visualization and classification accuracy. It can be seen from the experiments that the proposed SDKL outperforms other four models especially when the number of training data is small.

For the fair comparison, we run 200 iterations with RBF kernel function for all the models. The number of K in KNN is 5. All deep structure are pretrained by SAE, and the network architecture of the deep structure in deep kernel is chosen from the architecture D-$(2 \times D$-$N1$-$N2$-$N3)$-2 with the maximum and minimum numbers of layers being 6 and 3 respectively where the number of hidden units w.r.t. $N1, N2, N3$ ranges from 10 to 100. Only the latent variables corresponding to the testing data are visualized. The first dataset is the oil flow data [20] consisting of 1000 samples with the dimensionality of 12. There are 3 classes associated with the data. For the five models, we use 150 samples (50

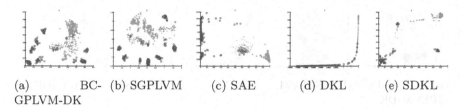

(a)   BC- (b) SGPLVM   (c) SAE   (d) DKL   (e) SDKL
GPLVM-DK

Fig. 2. Oil flow data visualized by BC-GPLVM-DK, SGPLVM, SAE, DKL and the proposed SDKL.

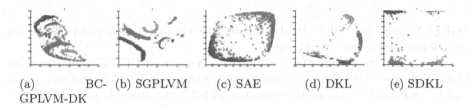

(a)   BC- (b) SGPLVM   (c) SAE   (d) DKL   (e) SDKL
GPLVM-DK

Fig. 3. Swiss roll data visualized by BC-GPLVM-DK, SGPLVM, SAE, DKL and the proposed SDKL.

points randomly selected from each class) to learn the corresponding 2D latent representation for data visualization. The data visualization for the testing data (850) is displayed in Fig. 2 with the best network architectures being 12-50-30-2, 12-50-30-2, 12-30-50-20-2, and 12-50-30-2 for BC-GPLVM-DK, SAE, DKL and SDKL, respectively.

The second dataset is the swiss roll data typically used to validate the DR algorithms. According to the introduction for this dataset[1], we similarly generate 2000 samples in 3D space where there are 500 points in each of the four classes. 400 samples are randomly selected to be the training data, and the remaining 1600 testing data are reduced to 2D space by the five methods with illustration in Fig. 3, where the best network architectures are 3-50-20-2, 3-50-10-2, 3-70-30-2, and 3-50-30-2 for BC-GPLVM-DK, SAE, DKL and SDKL, respectively. The single speaker vowel dataset [1] is also used to illustrate the performance of SDKL which consists of 9 classes and the dimensionality of features is 24. 900 points (100 points randomly selected from each class) are picked from 2700 data. Figure 4 shows the learnt latent variables in 2D space w.r.t. the remaining 1800 testing data with the best network architectures being 24-48-70-40-2, 24-48-30-2, 24-48-50-20-10-2, and 24-48-50-30-10-2 for BC-GPLVM-DK, SAE, DKL and SDKL, respectively.

As can been seen from Figs. 1, 2, 3 and 4 that the points in the latent space learnt by the proposed SDKL which originally belong to the same class in high-dimensional observation space could stay more closer than BC-GPLVM-DK,

---

[1] http://people.cs.uchicago.edu/~dinoj/manifold/swissroll.html.

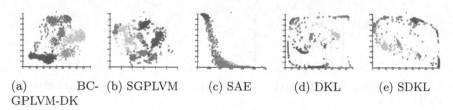

(a)   BC-   (b) SGPLVM   (c) SAE   (d) DKL   (e) SDKL
GPLVM-DK

Fig. 4. Vowel data visualized by BC-GPLVM-DK, SGPLVM, SAE, DKL and the proposed SDKL.

SGPLVM, SAE and DKL. It also means that the novel model can handle the additional labels more effectively.

The final datasets are the iris and wine data from UCI Machine Learning Data Repository [16]. The iris data consist of three classes of 50 instances each, and the wine data contains three classes with 178 instances in $\mathcal{R}^{13}$. We use the two small data to objectively testify the performance of the proposed SDKL when few training data are available. In this experiments, only 10 samples are randomly picked from each class meaning that the total number of training data for the two data is 30, and the remaining data become the testing samples. We compare the five models in terms of the best classification accuracy based on KNN in the learnt 2D latent space along with the corresponding optimal network architecture. It can be seen from Table 1 that the proposed SDKL outperforms other four algorithms, which could prove that the new model is efficient and particularly adaptive for the tasks with small numbers of training data because SDKL can fully make use of the high-dimensional observations and output labels.

Table 1. The classification accuracy comparison of the five methods in iris and wine datasets with the best deep architectures and best accuracy reported in experiments.

Models	Architecture (Iris)	Acc (Iris)	Architecture (Wine)	Acc (Wine)
BC-GPLVM-DK	4-8-50-30-2	88.33%	13-26-50-20-2	87.16%
SGPLVM	4-2	90.00%	13-2	85.90%
SAE	4-8-30-50-10-2	97.50%	13-26-50-2	93.92%
DKL	4-8-50-2	97.50%	13-26-50-20-2	86.49%
SDKL	4-8-30-2	98.33%	13-26-50-30-2	95.27%

5   Conclusion

In this paper, we introduce the Shared Deep Kernel Learning (SDKL) for dimensionality reduction by combining DKL with shared GPLVM. The new method can learn hierarchical features by the shared deep kernel structure in nonparametric manner which brings more flexibility, and also makes use of the labels information more effectively. The new model could easily be applied to the typical classification and regression tasks.

For the future works, firstly it can be extended to multiview tasks motivated by the recently proposed SAGP and the semisupervised learning framework by making use of the unlabeled data. Also, extra discriminative priors over the latent variables and the parameters of the deep structure in SDKL can be added to further regularize the model. We believe the combination of deep neural networks and kernel methods could be promising.

Acknowledgments. This work is supported by the National Natural Science Foundation of China under Grants 61402424, 61603355, 61773355, the National Science and Technology Major Project under Grant 2016ZX05014003-003, and the Fundamental Research Funds for the Central Universities, China University of Geosciences (Wuhan).

References

1. Bilmes, J.A., Malkin, J., Li, X., Harada, S., Kilanski, K., Kirchhoff, K., Wright, R.: The vocal joystick. In: IEEE Conference on Acoustics, Speech and Signal Processing (2006)
2. Damianou, A.C., Lawrence, N.D.: Deep Gaussian processes. In: Artificial Intelligence and Statistics (AISTATS) (2013)
3. Ek, C.H.: Shared Gaussian process latent variables models. Ph.D. thesis, Oxford Brookes University (2009)
4. Eleftheriadis, S., Rudovic, O., Pantic, M.: Discriminative shared Gaussian processes for multiview and view-invariant facial expression recognition. IEEE Trans. Image Process. 24(1), 189–204 (2015)
5. Gao, J., Zhang, J., Tien, D.: Relevance units latent variable model and nonlinear dimensionality reduction. IEEE Trans. Neural Netw. 21, 123–135 (2010)
6. Gao, X., Wang, X., Tao, D., Li, X.: Supervised Gaussian process latent variable model for dimensionality reduction. IEEE Trans. Syst. Man Cybern. Part B Cybern. 41(99), 425–434 (2011)
7. He, X., Yan, S., Hu, Y., Niyogi, P., Zhang, H.J.: Face recognition using laplacianfaces. IEEE Trans. Pattern Anal. Mach. Intell. 27(3), 328–340 (2005)
8. Hinton, G.E., Salakhutdinov, R.R.: Reducing the dimensionality of data with neural networks. Science 313(5786), 504–507 (2006)
9. Jiang, X., Fang, X., Chen, Z., Gao, J., Jiang, J., Cai, Z.: Supervised Gaussian process latent variable model for hyperspectral image classification. IEEE Geosci. Remote Sens. Lett. 14(10), 1760–1764 (2017)
10. Jiang, X., Gao, J., Wang, T., Shi, D.: TPSLVM: a dimensionality reduction algorithm based on thin plate splines. IEEE Trans. Cybern. 44(10), 1795–1807 (2014)
11. Jiang, X., Gao, J., Wang, T., Zheng, L.: Supervised latent linear Gaussian process latent variable model for dimensionality reduction. IEEE Trans. Syst. Man Cybern. Part B Cybern. 42(6), 1620–1632 (2012)
12. Kouropteva, O., Okun, O., Pietikäinen, M.: Supervised locally linear embedding algorithm for pattern recognition. Pattern Recogn. Image Anal. 2652, 386–394 (2003)
13. Lawrence, N.: Probabilistic non-linear principal component analysis with gaussian process latent variable models. J. Mach. Learn. Res. 6, 1783–1816 (2005)

14. Li, J., Zhang, B., Zhang, D.: Shared autoencoder Gaussian process latent variable model for visual classificatio. IEEE Trans. Neural Netw. Learn. Syst. (2017, in Press)
15. Li, X., Shu, L.: Kernel based nonlinear dimensionality reduction for microarray gene expression data analysis. Expert Syst. Appl. **36**, 7644–7650 (2009)
16. Lichman, M.: UCI machine learning repository (2013). http://archive.ics.uci.edu/ml
17. van der Maaten, L., Postma, E.O., van den Herik, H.J.: Dimensionality reduction: a comparative review. Technical report, Tilburg University (2008)
18. Rasmussen, C.E., Williams, C.K.I.: Gaussian Processes for Machine Learning. The MIT Press, Cambridge (2006)
19. Rue, H., Held, L.: Gaussian Markov Random Fields: Theory and Applications. Chapman and Hall/CRC, Boca Raton (2005)
20. Scholkopf, B., Smola, A., Muller, K.R.: Nonlinear component analysis as a kernel eigenvalue problem. Neural Comput. **10**(5), 1299–1319 (1998)
21. Snoek, J., Adams, R.P., Larochelle, H.: Nonparametric guidance of autoencoder representations using label information. J. Mach. Learn. Res. **13**, 2567–2588 (2012)
22. Urtasun, R., Darrell, T.: Discriminative Gaussian process latent variable model for classification. In: International Conference on Machine learning (ICML), pp. 927–934. ACM (2007)
23. Wang, W., Huang, Y., Wang, Y., Wang, L.: Generalized autoencoder: a neural network framework for dimensionality reduction. In: IEEE Conference on Computer Vision and Pattern Recognition Workshops (CVPR Workshops), pp. 496–503 (2014)
24. Wilson, A.G., Hu, Z., Salakhutdinov, R., Xing, E.P.: Deep kernel learning. In: The 19th International Conference on Artificial Intelligence and Statistics (AISTATS) (2016)
25. Wilson, A.G., Nickisch, H.: Kernel interpolation for scalable structured Gaussian processes (KISS-GP). In: International Conference on Machine Learning (2015)
26. Yang, J., Jin, Z., Yang, J.Y., Zhang, D., Frangi, A.F.: Essence of kernel fisher discriminant: KPCA plus LDA. Pattern Recogn. **37**, 2097–2100 (2004)
27. Yu, S., Yu, K., Tresp, V., Kriegel, H.P., Wu, M.: Supervised probabilistic principal component analysis. In: International Conference on Knowledge Discovery and Data Mining (KDD), pp. 464–473. ACM Press (2006)

CDSSD: Refreshing Single Shot Object Detection Using a Conv-Deconv Network

Vijay Gabale[1(⊠)] and Uma Sawant[2]

[1] Huew, Bangalore, India
vijay@huew.co
[2] IIT Bombay, Mumbai, India
uma.sawant@gmail.com

Abstract. Single shot multi-box object detectors [13] have been recently shown to achieve state-of-the-art performance on object detection tasks. We extend the single shot detection (SSD) framework in [13] and propose a generic architecture using a deep convolution-deconvolution network. Our architecture does not rely on any pretrained network, and can be pretrained in an unsupervised manner for a given image dataset. Furthermore, we propose a novel approach to combine feature maps from both convolution and deconvolution layers to predict bounding boxes and labels with improved accuracy. Our framework, Conv-Deconv SSD (CDSSD), with its two key contributions – unsupervised pretraining and multi-layer confluence of convolution-deconvolution feature maps – results in state-of-the-art performance while utilizing significantly less number of bounding boxes and improved identification of small objects. On 300×300 image inputs, we achieve 80.7% mAP on VOC07 and 78.1% mAP on VOC07+12 (1.7% to 2.8% improvement over StairNet [21], DSSD [5], SSD [13]). CDSSD achieves 30.2% mAP on COCO performing at-par with R-FCN [3] and faster-R-FCN [18], while working on smaller size input images. Furthermore, CDSSD matches SSD performance while utilizing 82% of data, and reduces the prediction time per image by 10%.

Keywords: Single shot detection · Unsupervised learning
Feature map confluence

1 Introduction

Image object detection involves identifying bounding boxes encapsulating objects and classifying each bounding box to recognize the underlying object category. Recently there has been mounting interest in the research community to detect multiple objects in an image using Single Shot Detection techniques [13,16]. These techniques effectively combine region proposal and classification into a single step by foregoing the candidate box proposal (or region proposal) module employed by several two-step detection techniques [1,6,7,11,18]. Not only this results in much faster object detection but it also improves

© Springer International Publishing AG, part of Springer Nature 2018
D. Phung et al. (Eds.): PAKDD 2018, LNAI 10939, pp. 309–321, 2018.
https://doi.org/10.1007/978-3-319-93040-4_25

Fig. 1. Detection output comparison of (a) SSD [13], (b) Stairnet [21], and (c) CDSSD. CDSSD results in superior performance in detecting small as well as large objects

accuracy [5,13,16,21]. One of the two prominent works, You Only Look Once (YOLO) [16], considers the global feature map of an image and utilizes a fully-connected layer to output object detections with a fixed set of regions. The other prominent work, Single Shot MultiBox Detector (SSD, henceforth) [13], considers a set of layers (or feature maps) and a set of boxes at various scales, and employs convolutional filters to predict objects inside each box. Owing to its design choice to consider multiple feature maps from different layers in a deep network (multi-scale representation), SSD performs significantly better than YOLO (Fig. 1).

While SSD [13] has achieved state-of-the-art results, it has three fundamental drawbacks. (a) When applying default bounding boxes, SSD considers each feature map in isolation (see Fig. 2). Thus it can not exploit the semantic information of later layers for better object detection on initial layers. Consequently, SSD does not perform well on smaller size object detection which is attempted by initial layers. (b) SSD architecture relies on features maps pretrained on the classical Imagenet dataset [9,20] without attempting to learn robust feature maps from the vast collection of unlabeled datasets. (c) SSD needs to evaluate several thousands of bounding boxes to detect only a few objects in an image.

Several follow-up works attempt to eliminate limitation (a) by combining feature maps at different layers of convolution networks, or inserting additional context by extending the base convolution block with a deconvolution block [1,2,5,10,11,13,16,17,21]. However, none of the prior approaches explore unsupervised pretraining to learn robust features; but use either VGG-16 [20] or ResNet-101 [9] to bootstrap the object detection training. [5,11,13,16,17,21] partially exhibit some scope to improve the performance on objects of different sizes and scales by combining information from different feature maps. However, they rely on features computed only from convolution networks, or result in considerably slower speed detection [5], or are not end-to-end trainable. In contrast to this prior work, we draw inspirations from convolution-deconvolution techniques used in semantic segmentation tasks [15,22], and base our design on convolution auto-encoders. Specifically, our contributions are as follows:

– We design an end-to-end trainable convolution-deconvolution based single shot detection framework to detect multiple objects in an image. This framework enables unsupervised pretraining of the underlying network.

– We design a refined SSD technique that carefully combines feature maps from both convolution and deconvolution blocks. Fusing of generic features from initial layers close to the input with semantically rich features of later layers close to the output detection from *both convolution and deconvolution blocks* helps us significantly reduce the required number of default bounding boxes.
– On input image size of 300 × 300, we achieve state-of-the-art accuracy on several object detection tasks with 80.7% mAP on VOC07, 78.1% mAP on VOC07+12 (1.7% improvement over StairNet [5,21], 2.8% improvement over [13]), and 30.2% mAP on COCO. We improve detection performance of both small as well as large objects, as well as visually impoverished objects while reducing the prediction time per image by 10%.

2 Limitations of Related Work

As compared to SSD, some recent approaches [6,7,18] first learn a separate bounding box (or region) proposal network, followed by learning a separate classification network on top of the proposal network. However, such two-stage object detectors suffer from high memory usage and poor inference time. In comparison, SSD networks [13,16,19] have been shown to perform better and faster. Furthermore, most of the object detection techniques, including Overfeat [19], SPPnet [8], Fast R-CNN [6], Faster R-CNN [18], and YOLO [16], utilize only a single layer (typically the top-most layer) of a convolution network to detect objects. This approach does not exploit different feature sets learned by different feature maps at different scales [5,13,21], and therefore is severely limited in identifying objects of different sizes and scales. In comparision, the state-of-the-art SSD networks [13,17] utilize feature maps from different layers in order to focus on objects that appear in certain sizes. However, they operate on each feature map independently without combining them in a meaningful manner. Hence, these SSD networks [13,17] do not particularly perform well towards identification of smaller size objects [1,5,11,21].

In order to consider feature maps from different layers in a combined fashion, [1] concatenates features of different layers before applying box proposals to detect objects. Taking a step further, [2] applies deconvolution on multiple layers of the underlying convolution network to increase feature map resolution. However, it results in significant memory and prediction time requirement. [11] too leverages the pyramidal shape of the convolution network and attempt to utilize semantics at different scales of feature maps by inserting nearest neighbor upsampling. In another work, instead of focusing only on the convolution block, [5] adds a deconvolution context layer to address the problem of shrinking resolution of feature maps in the convolution block. [21] further exploit the deconvolution context and design a top-down feature combining module that progressively encodes semantic information with low level features.

Our approach is partially inspired by [5,21] in terms of adding deconvolution context and utilizing feature maps at different layers in a network. However, as shown in Fig. 2 neither [5,21] nor any of the prior approaches explore unsupervised learning to improve SSD [13] performance. Moreover, none of the prior

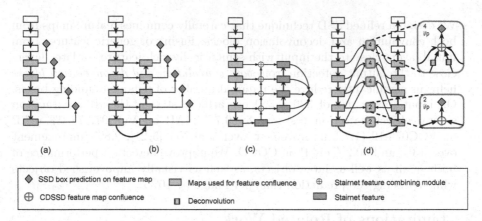

Fig. 2. Difference in SSD architectures in using deconvolution and feature map confluence (a) SSD [13], (b) DSSD [5], (c) Stairnet [21], (d) CDSSD (this work)

work exploits the difference in features learned by different layers in both convolution and deconvolution blocks. By refreshing SSD with unsupervised learning and confluence of feature maps from convolution and deconvolution blocks, we show that our approach results in state-of-the-art performance on benchmark datasets [4,12].

3   CDSSD Architecture

In this section, we first give a primer on SSD architecture. We then progressively introduce unsupervised learning and feature map confluence in the SSD architecture. Finally we showcase a method to reduce the requirement of default bounding boxes, and then explain our methodology of training and testing.

3.1   SSD

The SSD network is a convolutional architecture that utilizes different layers to predict presence of multiple objects in an image. To recognize objects at different scales, SSD utilizes predictions on different feature maps, each from a different layer, of a single network. These feature maps are processed by a fixed-size collection of bounding boxes customized for each layer. For feature map f of size $m \times n$ with p channels, K default-sized bounding boxes are applied on each of $m \times n$ cells. Subsequently, C filters of size $3 \times 3 \times p$ are applied for each cell and for a given bounding box to produce individual scores to predict each of C classes, and 4 additional filters are applied to produce offsets (center co-ordinate, height, width) to position the box on the underlying cell in order to encapsulate the object (as shown in Fig. 3(c)). Note that, for a given feature map f, the default boxes are scaled with a scaling factor f^{scale} with respect m and n and thus, they are customized to have different aspect ratios. Hence, bounding boxes

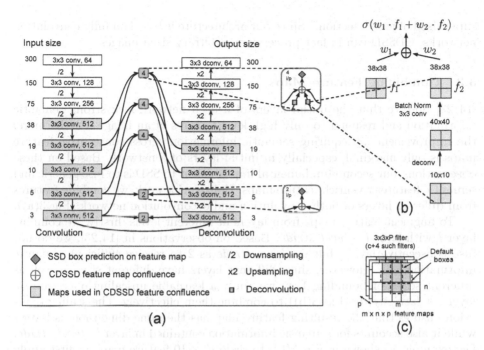

Fig. 3. CDSSD combines information from convolution and deconvolution feature maps

on initial stage feature maps cover a smaller receptive field to identify objects at a smaller scale, whereas bounding boxes on later stage feature maps cover larger receptive fields to identify objects with larger scale. By utilizing predictions for all the default boxes with different scales and aspect ratios from all locations of many feature maps, SSD attempts a diverse set of predictions, covering various input object sizes and shapes.

3.2   Unsupervised Pretraining

Our first fundamental improvement to SSD is to facilitate unsupervised training of the underlying network architecture. As we show in Sect. 4, this results in significant performance improvement. We use ResNet 101 architecture [9] and construct a convolution-deconvolution based auto-encoder (shown in Fig. 3(a)). Previously [5] have shown that ResNet 101 architecture results in more than 1.4% mAP gain in SSD as compared to VGG16 [20]. For the deconvolution block, we use learned upsampling and learned deconvolution, instead of bi-linear upsampling. The deconvolution block produces an image of the same dimension as input. We use an input image of $300 \times 300 \times 3$, with 7 meta-layers of convolution and pooling and 7 meta-layers of deconvolution with learned upsampling. Given an image dataset, we first pretrain the architecture before applying supervised object detection.[1] Since our architecture is based on fully convolution networks, CDSSD can in fact process any arbitrary sized images.

3.3 Combining Feature Maps

[14,23] observe that the initial layers of a deep network lack strong semantic information and respond to only high-level features of an image. Furthermore, the improvement in acquiring semantic information across consecutive feature maps is only marginal, especially in initial layers of a network. Based on these observations, our second fundamental improvement to SSD is to fuse generic and semantic features to enrich feature maps. Unlike prior work, we combine features from different layers of both convolution and deconvolution network (Fig. 3(a)).

To augment feature maps from layers at different levels, firstly, we combine layer l with layer $l + level\_stride$. Based on observations in [14,23], we do not fuse consecutive layers, but set $level\_stride$ as 2 to receive sufficient semantic information gain. However, since different layers have different sizes as well as different scales of bounding box, we apply a learnable upscaling operation on layer $l + level\_stride$ (Fig. 3(b)) to combine them effectively. The scaling operation ensures that the resulting feature map has the same dimension as layer l while it also accounts for semantic information contained in layer $l+level\_stride$. For example, as shown in Fig. 3(b), to scale 10×10 feature map, we first apply $4 \times 4 \times 512$ deconvolution operation and then apply a $3 \times 3 \times 512$ convolution operation to reduce the feature map size to 38×38. This is followed by a batch normalization layer to receive the final 38×38 feature map. Note that, we apply similar operation on both convolution and deconvolution blocks to process different layers. Addition of context from deconvolution block only improves the performance as we show in Sect. 4, without affecting the detection speed.

Secondly, for a given level of a meta-layer, we combine all the four feature maps; two from the convolution block and two from deconvolution block, as shown in Fig. 3(a), into a final feature map by taking element-wise learnable ReLU operation. Based on observations in [5], we further apply 3×3 filter on this feature map to extract another layer of features. Similar to SSD, we then apply a set of K default-boxes and $(C + 4) \times m \times n \times K$ filters on the resulting feature map to predict detection of objects. We apply this set of operations on meta-layer 3 to meta-layer 5 as shown in Fig. 3(a). Since there are no feature maps to pair with the last $level\_stride$ of convolution and initial $level\_strides$ of deconvolution feature maps (6th and 7th meta-layer), we combine them in element-wise learnable ReLU and process the resulting feature map. Since 6th and 7th meta-layers have higher reception field and contain richer semantic information, they are quite capable of detecting bigger size and scale objects [5].

---

[1] Our network is not symmetric. During deconvolution, we simply apply learned upsampling and learned deconvolution without residual blocks.

3.4   Box Pooling: Reducing the Number of Default Boxes

In the original SSD implementation, the authors apply default bounding boxes to every cell of $m \times n$ feature map with p channels. We consider a *box-pooling approach* where we pick the dominant cell in $l \times l$ window, with a stride of 1 on $m \times n$ feature map, and apply a set of default boxes on the dominant cell. This reduces the number default boxes by l^2 per feature map. This design choice is governed by two phenomena observed during our ablation study: (1) Unsupervised pre-training helps in learning significantly better feature maps (2) Given that we combine feature maps from different layers of both convolution and deconvolution blocks, there is no need to exhaustively search for objects for every cell of every feature map. We show in Sect. 4 that box-pooling does not affect precision and recall of object detection.

Similar to SSD [13], we tile the default boxes of different scales on different features maps so that specific feature maps learn to be responsive to particular scales of the objects. To compute different aspect ratios for each cell, we take a statistical approach and compute a cumulative distribution of aspect ratios of the ground truth boxes in a given dataset. We then divide the distribution into B bins and pick the average value of a bin as one of the aspect ratio, thus resulting in B aspect ratios. For each $b_i \in B$, for a feature map with size $m \times n$ and scale of f^{scale}, we then set height to be $m \times b_i \times f^{scale}$ and width to be $n \times b_i \times f^{scale}$. With optimized aspect ratios that fit the underlying dataset and different scales for different layers, we apply appropriate default boxes at box-pooled locations in each feature map, covering different object sizes and shapes.

4   Results

Our experiments are governed to answer the following key question: *can we achieve state-of-the-art results on object detection benchmarks by employing unsupervised learning and confluence of feature maps from convolution and deconvolution blocks?* Towards answering this question, we compare our approach with prior work on two benchmark datasets: PASCAL VOC and MS COCO. We compare our approach with the original SSD [13] that employs only convolution block, DSSD [5] that uses deconvolution blocks as additional context for convolution blocks, and Stairnet [21] that progressively merges feature maps close to traditional classification layers with feature maps close to input layers. SSD, DSSD and Stairnet do not employ unsupervised learning and do not consider confluence contextual and semantic features from convolution and deconvolution blocks. We also do an extensive ablation study to quantify improvement by each of the modules that we have contributed to extend SSD framework. We develop CDSSD as a Tensorflow module.

4.1   Training

The configuration of our network architecture is shown in Fig. 3. We keep the dropout layers during unsupervised training and remove them while training for object detection. We train our models on Azure GPU instances that have NVIDIA K80 GPUs with 12 GB of memory. We use batch size of 16, momentum as 0.9 and weight decay 0.0005. Similar to SSD [13], we match a default box to target ground truth boxes, if Jaccard overlap is larger than a threshold (e.g. 0.5). We compute the target ground truth box for each layer of the network by scaling it with respect to the feature map and original image sizes. We minimize the joint localization loss (i.e., smooth L1) and confidence loss (i.e., softmax-cross-entropy). To avoid the imbalance between the positive and negative training examples, we sort the negative boxes using the joint loss for each default box and then pick the top ones to maintain a 2:1 negative to positive ratio. We found 2:1 ratio leads to faster optimization as compared to the ratio of 3:1 as mentioned in the original SSD paper.

We further make the model robust to different input object sizes and shapes by invoking extensive augmentation. Specifically, we sample a patch from a ground truth box so that the minimum Jaccard overlap with the objects is 0.5, 0.7, or 0.9. Furthermore, we randomly sample a patch between [0.5, 1] of the original image size, and the aspect ratio is between [1, 2]. Also, we randomly flip each patch horizontally with probability of 0.5, apply different transformations such as gaussian blur, emboss, edge prominence, random black-out of 20% of pixels, and color (hue, saturation, contrast) distortions. We apply 3×3 box pooling for layer 3 and 4, 2×2 box pooling for layer 5, and no box pooling for layer 6 and 7. We apply non-maximum suppression (NMS) to post-process the predictions to get final detection results.

4.2 PASCAL VOC

When training on VOC07+12 trainval, we train the entire network with learning rate at 10^{-3} for 45K batches, and then with learning rate of 10^{-4} for 60K batches to execute unsupervised pretraining on the underlying train dataset[2]. During

Table 1. Comparison of single-shot detection techniques trained on VOC07+12 trainval and evaluated on VOC2007 test dataset. CDSSD outperforms other state-of-the-art methods while maintaining high speed of detection.

Method	Network	mAP	Boxes	fps	lib
YOLOv2_352 [16]	DarkNet-19	73.7	98	81	DarkNet
SSD300 [13]	VGGNet	77.5	8732	62	Caffe
DSSD321 [5]	ResNet-101	78.6	43688	9.5	Caffe
Stairnet [21]	VGGNet	78.8	8732	30	PyTorch
CDSSD300	ResNet-101	**80.7**	1182	51	TF
CDSSD300 (82% data)	ResNet-101	77.9	1182	51	TF

---

[2] Due to reduced batch size, the number of batches or iterations are increased as compared to the original SSD work.

Table 2. mAP comparison of single-shot detection techniques trained on VOC07 train-valtest, VOC12 trainval and evaluated on VOC12 test dataset. CDSSD results in state-of-the-art performance for several object categories.

Method	Aero	Bike	Bird	Boat	Bottle	Bus	Car	Cat	Chair	Cow
SSD300 [13]	**88.1**	82.9	74.4	61.9	47.6	82.7	**78.8**	91.5	58.1	80.0
DSSD321 [5]	87.3	83.3	75.4	64.6	46.8	82.7	76.5	**92.9**	59.5	78.3
StairNet [21]	87.7	83.1	74.6	64.2	51.3	**83.6**	78.0	92.0	58.9	**81.8**
CDSSD224	85.2	79.5	71.4	60.1	44.5	79.1	74.8	84.3	57.9	79.2
CDSSD300	87.4	**83.9**	**78.3**	**69.5**	**54.7**	80.2	76.3	88.7	**63.4**	79.9
CDSSD300 (82%)	85.8	82.7	75.3	64.5	50.5	80.1	75.2	85.8	60.0	78.4

Method	Table	Dog	Horse	mbike	Person	Plant	Sheep	Sofa	Train	Tv
SSD300 [13]	64.1	89.4	85.7	85.5	82.6	50.2	79.8	73.6	86.6	72.1
DSSD321 [5]	64.3	**91.5**	86.6	**86.6**	82.1	53.3	79.6	75.7	85.2	73.9
StairNet [21]	66.2	89.6	86.0	84.0	**82.6**	50.9	**80.5**	71.8	**86.2**	73.5
CDSSD224	63.8	85.1	84.3	84.3	82.9	52.4	77.2	72.8	83.6	72.8
CDSSD300	**69.2**	89.3	**87.8**	85.6	82.3	**56.8**	76.9	**76.2**	84.3	**77.4**
CDSSD300 (82%)	66.4	83.4	82.1	84.7	80.3	53.7	75.8	71.9	80.6	74.5

object detection training, we again fine-tune the entire network with learning rate of 2×10^{-3} for 40K iterations, and 60K iterations with learning rate of 10^{-4}. Results over VOC07 test dataset are shown in Table 1. To evaluate on VOC12 test dataset, as shown in Table 2, we use VOC07 trainvaltest, VOC12 trainval for training. We train CDSSD model for 65K iterations with 10^{-3} learning rate and 2×10^{-4} learning rate for 80k iterations for unsupervised pretraining, and 10^{-3} and 10^{-4} learning rate for supervised training for 40K and 65K iterations respectively.

We see that by adding unsupervised pretraining and confluence of feature maps, CDSSD consistently outperforms SSD, DSSD, Stairnet by 1% to 5% points for several object categories. CDSSD especially shows significant improvement for small objects such as bird, tv and bottle. Furthermore, CDSSD also shows significant improvement for objects such as boat and horse that have definite backgrounds. CDSSD detects majority objects with high confidence with less localization error and less confusion for similar object categories[3]. Recall of CDSSD is 93.5% for "strong" criteria of jaccard of overlap of 0.5, about 10% better than SSD. Finally, CDSSD achieves high-precision at high-recall range and outperforms SSD and Stairnet (Table 3).

4.3 Ablation Study

To further quantify the benefits of CDSSD, we do an ablation study to progressively add its features and measure mAP on VOC12 test dataset. To quantify the performance over different sized objects, we consider objects of three different

---

[3] Details omitted due to lack of space.

Table 3. VOC 2012 test dataset to observe mAP at recall greater than 0.7

Method	Data	Recall			
		0.5	0.7	0.9	mAP@70%
SSD300	07+12	91.9	79.7	34.4	44.9
Stairnet	07+12	94.3	83.5	38.8	48.1
CDSSD	07+12	**96.1**	**87.0**	**44.2**	**52.6**

Table 4. Effects of progressively adding confluence of feature maps on convolution block, deconvolution block, unsupervised learning, and box pooling. Box pooling does not hamper the performance while drastically reducing the box requirement.

Conv-feat confluence	Deconv-feat confluence	Box pooling	Unsup pretraining	Total boxes	Overall mAP	Small-O mAP	Medium-O mAP	Large-O mAP
No	No	No	No	17464	74.5	42.6	76.9	80.6
No	No	Yes	No	1182	70.4	35.1	71.5	75.3
Yes	No	No	No	17464	74.9	46.5	77.1	80.9
No	Yes	No	No	17464	75.4	47.9	77.8	81.8
No	Yes	Yes	No	1182	74.5	45.2	76.6	78.9
Yes	Yes	No	No	8752	76.2	56.5	80.2	83.7
Yes	Yes	No	Yes	8752	**78.3**	**59.0**	**81.6**	**85.0**
Yes	Yes	Yes	Yes	1182	78.1	57.4	81.2	84.7

sizes. Following the methodology in [21], we order the ground truth bounding boxes on test set for each class by area. We further divide the boxes into three part: small: less than 25%, medium: between 25% to 75%, and large: above 75% of image size. Furthermore, when evaluating objects of each size, we ignore the ground truth labels for other sizes. As shown in Table 4, CDSSD shows significant improvement using confluence of feature maps, on individual convolution and deconvolution blocks as well as combination of convolution and deconvolution feature maps. CDSSD especially shows considerable improvement on small size objects; it performs about 9% to 14% mAP better than prior work.

To quantify the performance of unsupervised pretraining when not pretrained on the underlying dataset, we train our convolution and deconvolution network on imagenet dataset to initialize the weights of the network (similar to SSD [13], DSSD [5], Stairnet [21]). From the table, we also observe that unsupervised learning gives a 2.1% jump in overall mAP. Furthermore, after applying box pooling, i.e, after reducing the number of boxes from 8732 to 1182, we observe that CDSSD sees only marginal reduction in mAP. Note that, box pooling is not effective without unsupervised learning and confluence of feature maps as shown in Table 4. Thus, combining unsupervised learning with feature map confluence and box pooling, CDSSD results in state-of-the-art results on object detection datasets while reducing the number of default bounding boxes.

The original version of SSD [13] uses 8732 boxes, DSSD uses substantially more (17080 to 43688 boxes), whereas CDSSD uses only 1183 boxes. As a result, SSD takes 46 FPS and DSSD takes 9.5 FPS where CDSSD clocks 51 FPS on

Table 5. Evaluation of CDSSD on MSCOCO dataset

Method	Avg. precision, IoU 0.5:0.95/0.5/0.75	Avg. precision, area S/M/L	Avg. recall, #Dets 1/10/100	Avg. recall, area S/M/L
SSD300	25.1/43.1/25.8	6.6/25.9/41.4	23.7/35.1/37.2	11.2/40.4/58.4
DSSD321	28.0/46.1/29.2	7.4/28.1/47.6	25.5/37.1/39.4	12.7/42.0/62.6
CDSSD300	29.2/48.2/29.9	8.8/31.2/49.3	26.1/39.2/42.3	13.6/44.3/63.7

Titan X GPU with a batch size of 1. While Residual-101 network is slower than VGGNet used in SSD, the reduction in default boxes not only decreases the prediction time but also time spent in non maximal suppression. Furthermore, the extra deconvolution layers do not incur an overhead since the confluence operation is light weight, and CDSSD operates on the same number of feature maps as the original SSD. Thus, CDSSD achieves improved accuracy while maintaining one of the fastest detection performance.

4.4 MSCOCO

To evaluate CDSSD on MSCOCO dataset, we first optimize the sizes of default bounding boxes as per the dataset (as explained in Sect. 3.4) to train and test prediction of classes and offsets. We train the network in an unsupervised manner for 260K iterations with learning rate of 10^{-3}. We use the trainval35k dataset and train the network in a supervised fashion for 210K iterations with learning rate of 10^{-3} and 120K iterations with learning rate of 2×10^{-4}. We show the results on test-dev2015. As shown in Table 5, CDSSD performs consistently better than SSD and DSSD even at higher Jaccard overlap threshold (0.75), and for different sized objects. Improvement in detection of large objects shows that CDSSD is able to learn better and robust features. These results corroborate the

Fig. 4. CDSSD out-performs in capturing objects of different size and scale in comparison to SSD [13]

benefits of CDSSD on generic object detection datasets towards a better single-shot detection framework. Figure 4 shows object detections on COCO test set images. Our model shows improvements on several fronts such as small objects like donuts; dense objects e.g. airplanes; objects with distinct context such as clocks; and objects that have specific relationships with the background.

5   Conclusion

We design an end-to-end framework using convolution-deconvolution deep networks to improve the state-of-the-art of single shot object detection techniques. Using a combination of unsupervised learning and confluence of feature maps with different receptive fields, we demonstrate substantial improvement in mAP for different objects in PASCAL VOC and MS COCO datasets while reducing the bounding box requirement by 8 times, thus improving inference time by 10%. As a future work, our approach can be used to improve region proposal based detection techniques as well. We also believe that our work can inspire several extensions to find more effective and efficient ways to combine feature maps of convolution and deconvolution blocks to improve image classification, object detection and semantic segmentation approaches.

References

1. Bell, S., Zitnick, C.L., Bala, K., Girshick, R.: Inside-outside net: detecting objects in context with skip pooling and recurrent neural networks. In: CVPR (2016)
2. Cai, Z., Fan, Q., Feris, R.S., Vasconcelos, N.: A unified multi-scale deep convolutional neural network for fast object detection. In: Leibe, B., Matas, J., Sebe, N., Welling, M. (eds.) ECCV 2016. LNCS, vol. 9908, pp. 354–370. Springer, Cham (2016). https://doi.org/10.1007/978-3-319-46493-0_22
3. Dai, J., Li, Y., He, K., Sun, J.: R-FCN: object detection via region-based fully convolutional networks. CoRR abs/1605.06409 (2016)
4. Everingham, M., Van Gool, L., Williams, C.K.I., Winn, J., Zisserman, A.: The pascal visual object classes (VOC) challenge. IJCV **88**(2), 303–338 (2010)
5. Fu, C., Liu, W., Ranga, A., Tyagi, A., Berg, A.C.: DSSD: deconvolutional single shot detector. CoRR abs/1701.06659 (2017)
6. Girshick, R.: Fast R-CNN. In: ICCV (2015)
7. Girshick, R., Donahue, J., Darrell, T., Malik, J.: Rich feature hierarchies for accurate object detection and semantic segmentation. In: CVPR (2014)
8. He, K., Zhang, X., Ren, S., Sun, J.: Spatial pyramid pooling in deep convolutional networks for visual recognition. CoRR abs/1406.4729 (2014)
9. He, K., Zhang, X., Ren, S., Sun, J.: Deep residual learning for image recognition. CoRR abs/1512.03385 (2015)
10. Jeong, J., Park, H., Kwak, N.: Enhancement of SSD by concatenating feature maps for object detection. CoRR abs/1705.09587 (2017)
11. Lin, T.Y., Dollár, P., Girshick, R., He, K., Hariharan, B., Belongie, S.: Feature pyramid networks for object detection. In: CVPR (2017)
12. Lin, T., Maire, M., Belongie, S.J., Bourdev, L.D., Girshick, R.B., Hays, J., Perona, P., Ramanan, D., Dollár, P., Zitnick, C.L.: Microsoft COCO: common objects in context. CoRR abs/1405.0312 (2014)

13. Liu, W., Anguelov, D., Erhan, D., Szegedy, C., Reed, S., Fu, C.-Y., Berg, A.C.: SSD: single shot MultiBox detector. In: Leibe, B., Matas, J., Sebe, N., Welling, M. (eds.) ECCV 2016. LNCS, vol. 9905, pp. 21–37. Springer, Cham (2016). https://doi.org/10.1007/978-3-319-46448-0_2
14. Luo, W., Li, Y., Urtasun, R., Zemel, R.S.: Understanding the effective receptive field in deep convolutional neural networks. CoRR abs/1701.04128 (2017)
15. Noh, H., Hong, S., Han, B.: Learning deconvolution network for semantic segmentation. CoRR abs/1505.04366 (2015)
16. Redmon, J., Divvala, S., Girshick, R., Farhadi, A.: You only look once: Unified, real-time object detection. In: CVPR (2016)
17. Ren, J.S.J., Chen, X., Liu, J., Sun, W., Pang, J., Yan, Q., Tai, Y.W., Xu, L.: Accurate single stage detector using recurrent rolling convolution. In: CVPR (2017)
18. Ren, S., He, K., Girshick, R., Sun, J.: Faster R-CNN: towards real-time object detection with region proposal networks. In: NIPS (2015)
19. Sermanet, P., Eigen, D., Zhang, X., Mathieu, M., Fergus, R., Lecun, Y.: OverFeat: integrated recognition, localization and detection using convolutional networks. In: ICLR (2014)
20. Simonyan, K., Zisserman, A.: Very deep convolutional networks for large-scale image recognition. CoRR abs/1409.1556 (2014)
21. Woo, S., Hwang, S., Kweon, I.S.: StairNet: top-down semantic aggregation for accurate one shot detection. CoRR abs/1709.05788 (2017)
22. Yu, F., Koltun, V.: Multi-scale context aggregation by dilated convolutions. CoRR abs/1511.07122 (2015)
23. Zhou, B., Khosla, A., Lapedriza, À., Oliva, A., Torralba, A.: Object detectors emerge in deep scene CNNs. CoRR abs/1412.6856 (2014)

Binary Classification of Sequences Possessing Unilateral Common Factor with AMS and APR

Yujin Tang$^{(\boxtimes)}$, Kei Yonekawa, Mori Kurokawa, Shinya Wada, and Kiyohito Yoshihara

KDDI Research Inc., 2-1-15 Ohara, Fujimino, Saitama 356-8502, Japan
{yu-tang,ke-yonekawa,mo-kurokawa,sh-wada,yosshy}@kddi-research.jp

Abstract. Most real-world sequence data for binary classification tasks appear to possess unilateral common factor. That is, samples from one of the classes occur because of common underlying causes while those from the other class may not. We are interested in resolving these tasks using convolutional neural networks (CNN). However, due to both the technical specification and the nature of the data, learning a classifier is generally associated with two problems: (1) defining a segmentation window size to sub-sequence for sufficient data augmentation and avoiding serious multiple-instance learning issue is non-trivial; (2) samples from one of the classes have common underlying causes and thus present similar features, while those from the other class can have various latent characteristics which can distract CNN in the learning process. We mitigate the first problem by introducing a random variable on sample scaling parameters, whose distribution's parameters are jointly learnt with CNN and leads to what we call adaptive multi-scale sampling (AMS). To address the second problem, we propose activation pattern regularization (APR) on only samples with the common causes such that CNN focuses on learning representations pertaining to the common factor. We demonstrate the effectiveness of both proposals in extensive experiments on real-world datasets.

Keywords: Supervised learning · Deep learning · Sequence mining
Adaptive multi-scale sampling · Activation pattern regularization

1 Introduction

Binary sequence classification has attracted a remarkable amount of interest in both academic and industry communities, particularly in the fields of failure prediction [1–3] and anomaly detection [4–6]. We observe that for a majority of these applications, the datasets appear to possess unilateral common factor. That is, samples from one of the classes occur because of common underlying causes while those from the other class may not. Take the task of predicting seizure from electroencephalograph (EEG) data for example, EEG waves can be

© Springer International Publishing AG, part of Springer Nature 2018
D. Phung et al. (Eds.): PAKDD 2018, LNAI 10939, pp. 322–334, 2018.
https://doi.org/10.1007/978-3-319-93040-4_26

different when the subject is undergoing different activities [7], but when the subject is about to suffer from epilepsy they are empirically shown to reflect the underlying pathological features. (For brievity, we call the class having common factor positive and the other class negative in the rest of the paper.) On the other hand, encouraged by the recent advance of deep learning [8], researchers have successfully demonstrated its superiority in sequence classifications as well [9–13]. In this paper, we are interested in tasks of binary classification of sequences possessing unilateral common factor using CNN.

Due to both the technical specification and the nature of the data, learning CNN for these tasks is generally associated with two problems. First, implementation design often requires chopping an original sequence into subsequences which is also an important step for data augmentation when training a deep learning model. An appropriate window size for segmentation can provide the learning process with sufficient amount of training data and avoid serious multiple-instance learning issues wherein a great portion of sub-sequences from the original sequence does not actually carry representative features of the corresponding class. Defining such a window size with sufficient high-quality data augmentation is non-trivial without domain knowledge. Secondly, while samples from the positive class present similar features due to the common underlying causes, those from the negative class can have various latent characteristics and may prevent CNN from learning discriminative representations. To address the first problem, we define a random variable on a set of sample scaling parameters. Following the random variable's distribution, we sample sub-sequences of different lengths from the original series and scale them according to the sampled scaling parameter to a common length. We fit the CNN to these scaled subsequences, at the end of every k iterations we update the scaling parameters' distribution's parameters and thus it's trained jointly with the CNN. We show that as the training progresses, the distribution we sample from converges and is going to peak on a few scales that are optimal for the task. We call this process adaptive multi-scale sampling (AMS) and we give an explanation of why it works from the perspective of reinforcement learning (RL). With the knowledge of unilateral common factor, we mitigate the second problem using activation pattern regularization (APR) which acts as an extra term to the objective function that regularizes the activation patterns of only samples from the positive class. (For example, in the previous example of predicting seizure from EEG data, we apply APR on samples collected in the onset of epilepsy.) Concretely speaking, when training the CNN we construct in each mini-batch a Gramian matrix for each positive sample that represents its activation pattern and we minimize the variances of the matrices' entries. To demonstrate the advantage of our proposals, we conducted extensive experiments on real-world datasets.

Our main contribution in this paper is the proposal of the deep learning scheme with a combination of AMS and APR. To the best of our knowledge, we are the first to give tentative solutions to the aforementioned two problems in the context of training a CNN model for binary sequence classification and have demonstrated their effectiveness on real-world datasets. The rest of this paper is organized as follows. We briefly introduce related literatures in Sect. 2, then give description of AMS and APR in Sect. 3. Experiments to prove the effectiveness of our proposals are introduced in Sect. 4 and we conclude our work in Sect. 5.

2   Related Work

A plenty of literatures ranging from heuristic methods to solutions utilizing probabilistic models on the topic of sequence classification have been published [14]. Encouraged by the huge success of deep learning applications recently, some researchers have demonstrated the effectiveness of applying deep learning models to sequence classifications as well. For instance, some approaches encode raw time series inputs into images first, and then fit CNN with 2D convolutional filters to the images, thus reducing the problem entirely to image classification and all relevant tools and parameter tuning techniques can be exploited [9,10]. In another flavour, [11,12] chopped the target time series input into equal length segments and convolutions are then performed on these transformed data. In these cases, 1D convolutional filters shared across data channels are applied along the time dimension and a fusion mechanism such as probability voting is adopted to determine the label of the original time series. To get rid of the uniform input length constraint, [13] adopted a sequence-to-sequence model that is common in natural language tasks, wherein the authors squash the original input sequences of various lengths into common length feature sequences, these feature sequences are then fed into the trailing fully-connected layers. To help learning the feature sequences, they also introduced the attention mechanism that helps the model to extract discriminative features by focusing on the most relevant part of the original sequences during feature generation.

3   Our Proposals

3.1   Network Architecture and Notations

The network we use in our experiments is of the form $input \rightarrow 3Conv \rightarrow 2FC \rightarrow output$, where the preceding numbers indicate number of layers. All layers except the output have ReLU activations. It is noteworthy that we purposefully designed the network to be simple and we did not tune its architecture in our experiments because we want to make sure all the performance advantages are from our proposals, though advanced setups such as batch normalization and skip connections are theoretically compatible.

Table 1 summarizes the notations we use frequently in this paper. The objective function that we minimize is defined as follow:

$$l(\mathbf{X}, \mathbf{y}; \theta, \phi) = \mathbb{E}[f_\theta(\mathbf{X}, \mathbf{y}) + \lambda g_\theta(\mathbf{X}) \mid \phi] \tag{1}$$

where $f_\theta(\mathbf{X}, \mathbf{y})$ is the binary cross-entropy loss and $g_\theta(\mathbf{X})$ is the term from APR, both of which are parameterized by the CNN's parameters θ. λ is a hyper-parameter that determines how much weight we should put on APR. The expectation is taken over the training dataset that depends on the parameter ϕ from

Table 1. Notations

Notation	Description
N	Mini-batch size
l	CNN input sequence length
m	Number of data channels (E.g. #senor)
$\mathbf{X}$	$\mathbf{X} \in \mathcal{R}^{N \times l \times m}$ is a mini-batch of input sequences
$\mathbf{X}_i$	$\mathbf{X}_i \in \mathcal{R}^{l \times m}$ is the i-th sample in the batch
$\mathbf{y}$	$\mathbf{y} = [y_1, \cdots, y_N]$ is the vector of binary valued ground truths
$\hat{\mathbf{y}}$	$\hat{\mathbf{y}} = [\hat{y_1}, \cdots, \hat{y_N}]$ are CNN predictions
n	Number of filters in the last convolutional layer
l'	Filter size in the last convolutional layer
z_i	$z_i \in \mathcal{R}^{l' \times n}$ is sample i's activation of the last conv layer
$\mathbf{Z}$	$\mathbf{Z} \in \mathcal{R}^{N \times n^2}$ are flattened Gramian matrices
$\mathbf{Z}_i$	$\mathbf{Z}_i \in \mathcal{R}^{1 \times n^2}$ is a flattened Gramian matrix from sample i
p	Number of positive samples in the mini-batch
$\mathbf{A}$	$\mathbf{A} \in \mathcal{R}^{N \times p}$ is the matrix that selects positive samples
S	Random variable over candidate scaling parameters
$h_S(\cdot)$	Sequence scaling operator
$\pi_\theta(\cdot)$	CNN parameterized by θ
ϕ	$\phi \in \mathcal{R}^{|S|}$ are parameters of S's distribution
η	Temperature for softmax function
λ	Weight for APR

AMS. If we remove both AMS and APR, the objective function becomes simply $f_\theta(\mathbf{X}, \mathbf{y})$ and is the typical loss for training CNN for a binary classification problem. Also notice that neither AMS nor APR adds new network parameters to CNN, the network's capacity remains the same.

3.2 Adaptive Multi-scale Sampling (AMS)

Multi-scale Sampling. In AMS, we define several sample scaling parameters $\{s_1, s_2, \cdots\}$ and assign a random variable S over them. S is categorical, and we define its distribution to be $P(S = s_i) \triangleq \frac{\exp(\phi_i/\eta)}{\sum_j \exp(\phi_j/\eta)}$ where $\phi = [\phi_1, \phi_2, \cdots, \phi_{|S|}]$ are its parameters and η is a constant that acts as the temperature for this softmax function. We initialize ϕ to be a vector of all ones and during CNN training we randomly crop a mini-batch of sub-sequences of lengths $l \cdot S$ by drawing samples from $P(S)$. We scale all sub-sequences to be of a common length l that is pre-defined as our network's input dimension and feed the mini-batch to the CNN. Mathematically, let us denote $x_{1,\cdots,lS}$ as a randomly cropped sub-sequence, $x'_{1,\cdots,l}$ as the sub-sequence after scaling, and $h_S(\cdot)$ as the scaling operator, we can have several scaling strategies. For example, we can take the mean of every S samples from $x_{1,\cdots,lS}$ and hence $h_S(x_{1,\cdots,lS}) = \{x'_{1,\cdots,l} \mid x'_i = mean(x_{(i-1)S+1,\cdots,iS})\}$. Or in a simpler form, we can just define x'_i to be the j-th element of every S consecutive samples which leads to $h_S(x_{1,\cdots,lS}) = \{x'_{1,\cdots,l} \mid x'_i = x_{(i-1)S+j}\}$.

To keep notations uncluttered, we merge some operators into $h_S(\cdot)$ and $\pi_\theta(\cdot)$. Firstly, we merge sub-sequencing operator into $h_S(\cdot)$ such that if its input is longer than $l{\cdot}S$ then sub-sequencing (random crop in training; segmentation with minimum overlapping in evaluation) is performed prior to scaling. Furthermore, because we segment the original time series, we need a fusion mechanism for predictions. E.g., if we want the prediction score of the i-th sample in a test batch, $\pi_\theta(h_S(\mathbf{X}_i))$ gives several scores each of which for the sub-sequences originated from $h_S(\mathbf{X}_i)$, we need to merge these scores to get the prediction score for $\mathbf{X}_i$. In this paper, we define the score for the original series $score(\mathbf{X}_i) \triangleq \mathbb{E}_{P(S)}[mean(\pi_\theta(h_S(\mathbf{X}_i)))]$, we assume this fusion operation is merged into $\pi_\theta(\cdot)$ and is applied whenever necessary.

Adaptive Update. At the beginning of training, a mini-batch consists of sub-sequences of multiple scales, each of which has equal sampling weight. We train the CNN with sub-sequences of mixed scales, and at the end of every k-th training iteration, we randomly sample some time series from the entire training set, apply scaling operator $h_S(\cdot)$ to this set and feed the scaled dataset to the CNN trained so far. We then calculate the first term in (1) by evaluating (2), the definition of $f_\theta(h_S(\mathbf{X}), \mathbf{y})$ is given in (3). Minimization of (2) thus becomes jointly learning the CNN's parameters θ and the parameters ϕ for $P(S)$. In our experiments, at the end of every k CNN training iterations, we perform a one-step gradient descent on ϕ using (4). The gradient of ϕ is easy to derive and is given in (5), where matrix $\mathbf{J}$ has entry $\mathbf{J}_{ij} = \frac{1}{\eta}P(S_i)(1 - P(S_i))$ if $i = j$ or $\mathbf{J}_{ij} = -\frac{1}{\eta}P(S_i)P(S_j)$ if $i{\neq}j$. Adaptive update is just a single step of gradient, whose computational cost is negligible considering its parameters' size.

$$\mathbb{E}[f_\theta(\mathbf{X}, \mathbf{y}) \mid \phi] = \mathop{\mathbb{E}}_{P(S)}[f_\theta(h_S(\mathbf{X}), \mathbf{y})] \tag{2}$$

$$f_\theta(h_S(\mathbf{X}), \mathbf{y}) = -\frac{1}{N} \sum_{i=1}^{N} [y_i \log (\pi_\theta(h_S(\mathbf{X}_i)))$$
$$+ (1 - y_i) \log (1 - \pi_\theta(h_S(\mathbf{X}_i)))] \tag{3}$$

$$\phi := \phi - \nabla\phi \tag{4}$$

$$\nabla\phi = \mathbf{J}^\mathsf{T}\mathbf{L} = \begin{bmatrix} \frac{\partial P(S_1)}{\partial \phi_1} & \cdots & \frac{\partial P(S_1)}{\partial \phi_{|S|}} \\ \vdots & \ddots & \vdots \\ \frac{\partial P(S_{|S|})}{\partial \phi_1} & \cdots & \frac{\partial P(S_{|S|})}{\partial \phi_{|S|}} \end{bmatrix}^\mathsf{T} \begin{bmatrix} f_\theta(h_{S_1}(\mathbf{X}), \mathbf{y}) \\ \vdots \\ f_\theta(h_{S_{|S|}}(\mathbf{X}), \mathbf{y}) \end{bmatrix} \tag{5}$$

Interpretation from RL. RL is a hot topic in the machine learning society recently, [15] serves as an excellent introduction material. In RL, an agent interacts with an environment by taking some actions according to a policy that usually takes into account observed state of the environment. The agent receives reward/penalty from the environment that depends on both the environment and its actions, and its goal is to maximize/minimize the accumulated reward/penalty by adjusting its policy. Policy iteration is one of the RL algorithms that searches for the best policy by iterating between two operations: policy evaluation (given a policy, estimate the expected reward/penalty in each state) and policy improvement (given the estimated reward/penalty in each state, improve the policy by taking greedy actions). If we consider the snapshot of CNN's parameters θ as a state of the environment and $P(S)$ as the agent's policy from which we sample actions (that is, to pick a sample scaling parameter), then AMS resembles policy iteration. Specifically, if we regard (2) as the expected penalty, in policy evaluation we estimate the value of (2) upon our current settings of ϕ, and in the policy improvement phase, we update our policy by taking a gradient step of ϕ.

3.3   Activation Pattern Regularization (APR)

APR takes the form as an augmented term $g_\theta(\mathbf{X})$ given by:

$$g_\theta(\mathbf{X}) = \frac{1}{n^2} \sum_{i=1}^{n^2} var(\mathbf{A}^\mathsf{T}\mathbf{Z}, i) \tag{6}$$

where $\mathbf{Z} \in \mathcal{R}^{N \times n^2}$ are flattened Gramian matrices calculated from the activations of the last convolutional layer, $\mathbf{A} \in \mathcal{R}^{N \times p}$ is a matrix that selects only the positive samples in $\mathbf{X}$. The variance operator is taken along each column i of the matrix product $\mathbf{A}^\mathsf{T}\mathbf{Z} \in \mathcal{R}^{p \times n^2}$. To be concrete, let $z_i \in \mathcal{R}^{l' \times n}$ be the activation of the last convolutional layer from the i-th sample in the batch, then each row of $\mathbf{Z}$ is given by $\mathbf{Z}_i = flat(z_i^\mathsf{T} z_i) \in \mathcal{R}^{1 \times n^2}$. It is easy to see that $\mathbf{Z}_i$ encodes the relationships between each filter's activation from the i-th sample, therefore the flattened Gramian matrices $\mathbf{Z}$ describe the activation pattern of each sample across a batch. Because we are imposing regularizations on only the positive samples, we construct a mask matrix $\mathbf{A}$ whose entry is defined as $\mathbf{A}_{i,j} = 1$ if $\mathbf{X}_i$ is the j-th positive sample in the batch and 0 otherwise for $i = 1, \cdots, N$ and $j = 1, \cdots, p$. Since the variance operator in (6) is applied along columns, $g_\theta(\mathbf{X})$ is a measure of activation patterns' variance among positive samples. When this term from APR is augmented to the original objective function, the training process becomes a multiple-task learning problem, and we impose a hyper-parameter λ on it to control its strength. Following (1), when APR is applied together with AMS, the gradient in (5) should be updated by re-defining $\mathbf{L}$'s entry to be $\mathbf{L}_i = f_\theta(h_{S_i}(\mathbf{X}), \mathbf{y}) + \lambda g_\theta(h_{S_i}(\mathbf{X}))$.

Table 2. NAB statistics (the scores are averaged accuracies)

	Description	Size	#Anomaly	Baseline	AMS	APR	AMS+APR	t-test
1	NYC taxi passengers	10320	5	0.951	0.963	0.954	**0.966**	✓
2	Twitter mentions of AMZN	15832	4	0.890	0.927	0.894	**0.945**	✓
3	Twitter mentions of FB	15834	2	0.890	0.940	0.890	**0.941**	✓
4	Twitter mentions of GOOG	15843	4	0.915	0.945	0.922	**0.954**	✓
5	Twitter mentions of KO	15852	3	0.902	0.920	0.902	**0.930**	✓
6	Twitter mentions of CVS	15854	4	0.927	0.962	0.934	**0.968**	✓
7	Twitter mentions of PFE	15859	4	0.904	**0.911**	0.899	0.909	✗
8	Twitter mentions of UPS	15867	5	0.898	0.895	0.903	**0.908**	✗
9	Twitter mentions of IBM	15894	2	0.916	0.929	0.930	**0.932**	✗
10	Twitter mentions of AAPL	15903	4	0.925	0.968	0.920	**0.973**	✓
11	Twitter mentions of CRM	15903	3	0.955	1.000	0.955	**1.000**	✓
12	Machine temperatures	22696	4	0.943	0.924	**0.947**	0.945	✗

4 Experiments

4.1 Experimental Setup

We conducted extensive experiments on two datasets, each of which consists of data from multiple tasks. For each task, we compare the cross validation results of 4 methods: baseline (vanilla CNN training scheme), AMS, APR and AMS+APR. We used the same network architecture as described in Sect. 3.1 through all tasks. Although parameters such as kernel sizes and strides differ for each task, we used the same set of parameters for all methods for each task to ensure fair comparisons. At training time we randomly sample sub-sequences from the training split (rebalance samples to make the positive to negative ratio 1 : 1), and at test time we chop each validation segment with minimum overlap. To merge sub-sequences' scores for the original series, we take their averaged value. In the case when AMS is involved, we use the simple strategy $h_S(x_{1,\cdots,lS}) = \{x'_{1,\cdots,l} \mid x'_i = x_{(i-1)S+j}\}$ where $j = 1$, and we take the expected value of scores from all scaling parameters as the final score for the original series.

4.2 Experiments on Dataset 1

Dataset Description. Numenta Anomaly Benchmark (NAB) [16] is a dataset consisting over 50 labeled real-world and artificial time series data files. The type of data included are for example, Amazon Web Services (AWS) server metrics, Freeway traffic, and Tweets volume. From the real-world datasets, we choose the subset whose size is larger than 10K and has at least 2 anomalies. The resulting selections are summarized in Table 2, due to the space restriction we refer the readers to [16] for detailed description of NAB[1]. The default segmentation window sizes are a quarter of the sequence lengths and $S = \{1, 2, 3, 4\}$.

---

[1] https://github.com/numenta/NAB/tree/master/data.

Table 3. Dataset statistics

	Dog 1	Dog 2	Dog 3	Dog 4	Dog 5	Patient 1	Patient 2
# Preictal segments	24	42	72	97	30	18	18
# Interictal segments	480	500	1440	804	450	50	42
# Test segments	502	1000	907	990	191	195	150
# Electrodes	16	16	16	16	15	15	24
Sampling frequency (Hz)	400	400	400	400	400	5000	5000
Segment length (min)	10	10	10	10	10	10	10

Results and Discussions. For each dataset, we conducted 11 trials each of which is a 2-fold cross validation, and we report the averaged accuracies in Table 2, the winning scores are in boldface, and the last column indicates whether the winning method passes the t-test. From a macro view, our proposals beat the baseline on every dataset and demonstrated the generalisability of our methods. Zoom in for a micro analysis, we notice AMS alone presents a strong improvement (10 improvements out of 12 datasets). Contrary to that, APR alone does not deliver satisfying results (though winning 7 out of 12, few passed the t-test), one possibility is that the default definition of segmentation window size is inappropriate and prevents APR from finding common factors from such a setting. This explanation is supported by the results from the combination of AMS and APR that show significant improvement on almost all datasets.

4.3   Experiments on Dataset 2

Dataset Description. This is a dataset from a past Kaggle data analysis contest[2]. The dataset contains EEG recordings from 5 dogs and 2 humans and the task is to distinguish between ten minute long data clips covering an hour prior to a seizure (preictal segments), and ten minute clips of interictal activity. Preictal segments (positive class) are provided covering one hour prior to seizure with a five minute seizure horizon, and interictal segments (negative class) were chosen randomly from the full data record, with the restriction that interictal segments be as far from any seizure as can be practically achieved. Test segments without ground-truth labels are provided for final evaluation. Table 3 summarizes the statistics of the datasets. We down-sample the sample rate of patients to 500 Hz and define the segmentation window size to be 4000. We conducted 3-fold cross validation on each subject.

Results and Discussions. In this Kaggle contest, AUROC was set to be the evaluation metric and 504 teams in total made submissions. We report the scores from 2 result merging schemes: mean ensemble (predictions are the averaged scores from the models of cross validation) and max ensemble (predictions are the scores from the cross validation model with best validation scores). The

---

[2] https://www.kaggle.com/c/seizure-prediction.

Table 4. Kaggle submission (scores are AUROCs)

(a) Mean Ensemble

	Private	Public	Rank
Baseline	0.6525	0.7097	131st
APR	0.6649	0.7272	102nd
AMS	0.6918	0.7542	71st
AMS+APR	**0.7054**	**0.7792**	**58th**

(b) Max Ensemble

	Private	Public	Rank
Baseline	0.6864	0.7513	77th
APR	0.6866	0.7789	77th
AMS	0.7049	0.7764	58th
AMS+APR	**0.7762**	**0.8098**	**14th**

scores and ranks are summarized in Table 4, because this contest is over, we are able to receive a (Public, Private) scores pair from each submission. For APR and AMS + APR, the best scores from the grid search on λ is reported. We notice first that even the baseline method could achieve high ranks (top 26% for mean ensemble and top 15% for max ensemble), this is consistent with the recent reports on successful applications of deep learning to general datasets besides images and audios. Secondly we notice the max ensemble gives higher scores than mean ensemble, but the trend in either group is consistent. When AMS is activated, we observe obvious improvements on both public and private scores. Compared to the baseline, the combination of AMS and APR gives an average relative improvement of near 10%, and sent us to as high as the 14-th place on the leaderboard. The observations here are consistent with the results from Experiment 1 and emphasizes the stability of our proposals.

4.4 Analyzing AMS and APR

We give some analysis of AMS and APR in this section. Because the effect and trend of our analysis are similar for both experiments we show the statistics from Experiment 2 only for the sake of brevity.

Figure 1 gives the distribution of scaling parameters for each subject after AMS application. Our initial guess of input length (4000, equivalent to 10sec/8sec of time frame for Dogs/Humans) is not optimal and AMS learnt to find the combinations of sub-sequence lengths better suited for the task. Although AMS puts most weight on the largest scaling parameter for dogs 1, 2, 3 and 5, it considers combinations of sub-sequence lengths for dog 4 and both human patients. This suggests AMS does not always prefer longer sequences but is indeed looking for patterns residing in sub-sequences of different lengths that is optimal for the task. Figure 2 gives the evolution of training loss, validation accuracy and validation f-score for both the baseline and AMS. For almost all subjects and on any of the three criteria, we find AMS learns faster than the baseline. One may consider the faster training loss convergence is due to that in these trials AMS kept selecting longer sequences and thus led to smaller sample spaces, and a large network can easily fit to a smaller sample space, leading to faster convergence. This might be partially responsible, but it is also well known that with the same network capacity and smaller sample spaces, a flexible model like CNN can easily overfit the training dataset. However, we see strong faster

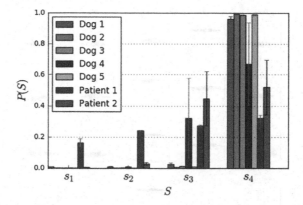

Fig. 1. Scaling parameters distribution ($S_1 = 1, S_2 = 2, S_3 = 3, S_4 = 4$).

rising accuracy and f-score lines, and do not observe any sign of overfitting in Fig. 2. We hence argue the faster learning speed (in terms of both training loss and validation scores) is indeed due to the involvement of AMS.

To analyze the difference between the baseline and APR, we select a random batch of validation data for both models and we check the differences of their activation patterns. Concretely speaking, from the models we learnt in cross validation, we pick a baseline model and a series of APR models with different λ settings from the same fold. We input the validation batch of preictal segments into the baseline model and the series of APR models, and we record the Gramian matrices **Z** that encodes their activation patterns. For each model, we calculated the variances of each entry of the Gramian matrices and analyze the difference

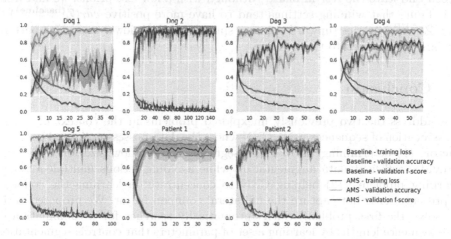

Fig. 2. Learning curves of baseline and AMS (depending on the line the Y-axes can stand for binary cross-entropy loss, accuracy or f-score, the X-axes are training iterations in hundred; shorter lines are due to early stopping).

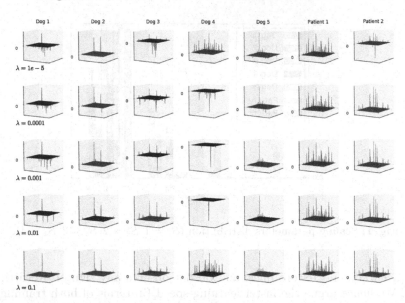

Fig. 3. Difference of positive samples' activations variances (settings with higher or equal (valid_accuracy, valid_f_score) scores are in green). (Color figure online)

$var(\mathbf{Z}^{(\mathbf{baseline})}) - var(\mathbf{Z}^{(\mathbf{APR})})$ where the variance is taken on each entry of the matrices. Figure 3 gives an image of what we described. We have scaled the values to have unit standard deviation and have labelled the 0-level in each diagram for visual convenience. In order to find patterns, we have put the settings that have higher or equal (valid_accuracy, valid_f_score) scores than the baseline in green and leave the rest in black. Although with a few exceptions, we find from this figure that winning settings tend to have more positive $var(\mathbf{Z}^{(\mathbf{baseline})}) - var(\mathbf{Z}^{(\mathbf{APR})})$ entries (upward pointing spikes), meaning lower activation pattern variances compared to the baseline.

5   Conclusion

We address the two previously unexplored problems in the context of binary classification of sequences using CNN where the data possess unilateral common factor: (1) determining the optimal segmentation window size for CNN that provides sufficient data augmentation while avoiding serious multiple-instance learning problems; (2) helping CNN to concentrate on learning the common representations that capture the unilateral common factor. We proposed AMS to solve the first problem which automatically searches for a combination of sub-sequence lengths by learning a set of parameters that controls segmentation window size jointly with the learning of CNN parameters. And we use APR to shift the CNN's attention to positive samples by minimizing the variances of the Gramian matrices' entries formed from the last convolutional layer's activations.

From our experiments, AMS alone is able to give performance boosts and when APR is augmented, the improvements are significant. Our extensive experiments on multiple real-world tasks demonstrate the effectiveness, the generalisability and stability of AMS and APR.

References

1. Murray, J.F., Hughes, G.F., Kreutz-Delgado, K.: Machine learning methods for predicting failures in hard drives: a multiple-instance application. J. Mach. Learn. Res. **6**, 783–816 (2005)
2. Liu, Y., Yao, K.T., Liu, S., Raghavendra, C.S., Balogun, O., Olabinjo, L.: Semi-supervised failure prediction for oil production wells. In: 2011 IEEE 11th International Conference on Data Mining Workshops, pp. 434–441, December 2011
3. Chalermarrewong, T., Achalakul, T., See, S.C.W.: Failure prediction of data centers using time series and fault tree analysis. In: 2012 IEEE 18th International Conference on Parallel and Distributed Systems, pp. 794–799, December 2012
4. Steinwart, I., Hush, D., Scovel, C.: A classification framework for anomaly detection. J. Mach. Learn. Res. **6**, 211–232 (2005)
5. Chitrakar, R., Chuanhe, H.: Anomaly detection using support vector machine classification with k-Medoids clustering. In: 2012 Third Asian Himalayas International Conference on Internet, pp. 1–5, November 2012
6. Bhattacharyya, D.K., Kalita, J.K.: Network Anomaly Detection: A Machine Learning Perspective. Chapman & Hall/CRC, Boca Raton (2013)
7. Niedermeyer, E., da Silva, F.L.: Electroencephalography: Basic Principles, Clinical Applications, and Related Fields, 5th edn. Lippincott Williams & Wilkins, Philadelphia (2004)
8. Goodfellow, I., Bengio, Y., Courville, A.: Deep Learning. MIT Press, Cambridge (2016). http://www.deeplearningbook.org
9. Wang, Z., Oates, T.: Encoding time series as images for visual inspection and classification using tiled convolutional neural networks. In: Workshops at the Twenty-Ninth AAAI Conference on Artificial Intelligence, AAAI (2015)
10. Zhiguang wang, T.O.: Imaging time-series to improve classification and imputation. In: Proceedings of the 24th International Joint Conference on Artificial Intelligence (IJCAI), pp. 3939–3945. AAAI (2015)
11. Zheng, Y., Liu, Q., Chen, E., Ge, Y., Zhao, J.L.: Time series classification using multi-channels deep convolutional neural networks. In: Li, F., Li, G., Hwang, S., Yao, B., Zhang, Z. (eds.) WAIM 2014. LNCS, vol. 8485, pp. 298–310. Springer, Cham (2014). https://doi.org/10.1007/978-3-319-08010-9_33
12. Yang, J.B., Nguyen, M.N., San, P.P., Li, X.L., Krishnaswamy, S.: Deep convolutional neural networks on multichannel time series for human activity recognition. In: Proceedings of the 24th International Conference on Artificial Intelligence, IJCAI 2015, pp. 3995–4001. AAAI Press (2015)
13. Tang, Y., Xu, J., Matsumoto, K., Ono, C.: Sequence-to-sequence model with attention for time series classification. In: 2016 IEEE 16th International Conference on Data Mining Workshops (ICDMW), pp. 503–510, December 2016
14. Xing, Z., Pei, J., Keogh, E.: A brief survey on sequence classification. SIGKDD Explor. Newsl. **12**(1), 40–48 (2010)

15. Sutton, R.S., Barto, A.G.: Introduction to Reinforcement Learning, 1st edn. MIT Press, Cambridge (1998)
16. Lavin, A., Ahmad, S.: Evaluating real-time anomaly detection algorithms - the numenta anomaly benchmark. In: 2015 IEEE 14th International Conference on Machine Learning and Applications (ICMLA), pp. 38–44, December 2015

Automating Reading Comprehension by Generating Question and Answer Pairs

Vishwajeet Kumar[1,2(✉)], Kireeti Boorla[1], Yogesh Meena[1],
Ganesh Ramakrishnan[1], and Yuan-Fang Li[3]

[1] Indian Institute of Technology Bombay, Mumbai, India
kireeti.boorla@gmail.com, yogesiitbcse@gmail.com, ganesh@cse.iitb.ac.in
[2] IITB-Monash Research Academy, Mumbai, India
vishwajeet@cse.iitb.ac.in
[3] Faculty of Information Technology, Monash University, Melbourne, Australia
yuanfang.li@monash.edu

Abstract. Neural network-based methods represent the state-of-the-art in question generation from text. Existing work focuses on generating only questions from text without concerning itself with answer generation. Moreover, our analysis shows that handling rare words and generating the most appropriate question given a candidate answer are still challenges facing existing approaches. We present a novel two-stage process to generate question-answer pairs from the text. For the first stage, we present alternatives for encoding the span of the pivotal answer in the sentence using Pointer Networks. In our second stage, we employ sequence to sequence models for question generation, enhanced with rich linguistic features. Finally, global attention and answer encoding are used for generating the question most relevant to the answer. We motivate and linguistically analyze the role of each component in our framework and consider compositions of these. This analysis is supported by extensive experimental evaluations. Using standard evaluation metrics as well as human evaluations, our experimental results validate the significant improvement in the quality of questions generated by our framework over the state-of-the-art. The technique presented here represents another step towards more automated reading comprehension assessment. We also present a live system (Demo of the system is available at https://www.cse.iitb.ac.in/~vishwajeet/autoqg.html.) to demonstrate the effectiveness of our approach.

Keywords: Pointer network · Sequence to sequence modeling
Question generation

1 Introduction

Asking relevant and intelligent questions has always been an integral part of human learning, as it can help assess the user's understanding of a piece of text

© Springer International Publishing AG, part of Springer Nature 2018
D. Phung et al. (Eds.): PAKDD 2018, LNAI 10939, pp. 335–348, 2018.
https://doi.org/10.1007/978-3-319-93040-4_27

(an article, an essay *etc.*). However, forming questions manually can be sometimes arduous. Automated question generation (QG) systems can help alleviate this problem by learning to generate questions on a large scale and in lesser time. Such a system has applications in a myriad of areas such as FAQ generation, intelligent tutoring systems, and virtual assistants.

The task for a QG system is to generate meaningful, syntactically correct, semantically sound and natural questions from text. Additionally, to further automate the assessment of human users, it is highly desirable that the questions are relevant to the text and have supporting answers present in the text.

Figure 1 below shows a sample of questions generated by our approach using a variety of configurations (vanilla sentence, feature tagged sentence and answer encoded sentence) that will be described later in this paper.

Sentence: It was adopted into an Oscar-winning film in 1962 by director Robert Mulligan, with a screenplay by Horton Foote.
Feature Tagged Sentence: It|PRP|O|nsubjpass was|VBD|O|auxpass adapted|VBN|O|ROOT into|IN|O|case
.... Horton|N|N|O|Compound Foote|N|N|O|Compound

With Features:	Who was the director of the film ?
With Features and Answer as "Horton Foote":	Who wrote the movie ?
With Features and Answer as "Robert Mulligan":	Who was the director of the Oscar-winning movie ?

Fig. 1. Example: sample questions generation from text by our models.

Initial attempts at automated question generation were heavily dependent on a limited, ad-hoc, hand-crafted set of rules [7,18]. These rules focus mainly on the syntactic structure of the text and are limited in their application, only to sentences of simple structures. Recently, the success of sequence to sequence learning models [16] opened up possibilities of looking beyond a fixed set of rules for the task of question generation [5,15]. When we encode ground truth answers into the sentence along with other linguistic features, we get improvement of upto 4 BLEU points along with improvement in the quality of questions generated. A recent deep learning approach to question generation [15] investigates a simpler task of generating questions only from a triplet of subject, relation and object. In contrast, we build upon recent works that train sequence to sequence models for generating questions from natural language text.

Our work significantly improves the latest work of sequence to sequence learning based question generation using deep networks [5] by making use of (i) an additional module to predict span of best answer candidate on which to generate the question (ii) several additional rich set of linguistic features to help model generalize better (iii) suitably modified decoder to generate questions more relevant to the sentence.

The rest of the paper is organized as follows. In Section 2 we formally describe our question generation problem, followed by a discussion on related work in Sect. 3. In Sect. 4 we describe our approach and methodology and summarize our main contributions. In Sects. 5 and 6 we describe the two main components of our framework. Implementation details of the models are described in Sect. 7, followed by experimental results in Sect. 8 and conclusion in Sect. 9.

2   Problem Formulation

Given a sentence **S**, viewed as a sequence of words, our goal is to generate a question **Q**, which is syntactically and semantically correct, meaningful and natural. More formally, given a sentence **S**, our model's main objective is to learn the underlying conditional probability distribution $P(\mathbf{Q}|\mathbf{S}; \theta)$ parameterized by θ to generate the most appropriate question that is closest to the human generated question(s). Our model learns θ during training using sentence/question pairs such that the probability $P(\mathbf{Q}|\mathbf{S}; \theta)$ is maximized over the given training dataset.

Let the sentence **S** be a sequence of M words $(w_1, w_2, w_3, ...w_M)$, and question **Q** a sequence of N words $(y_1, y_2, y_3, ...y_N)$. Mathematically, the model is meant to generate $\mathbf{Q}^*$ such that:

$$\mathbf{Q}^* = \underset{\mathbf{Q}}{\operatorname{argmax}}\ P(\mathbf{Q}|\mathbf{S}; \theta) \tag{1}$$

$$= \underset{y_1, ..y_n}{\operatorname{argmax}} \prod_{i=1}^{N} P(y_i|y_1, ..y_{i-1}, w_1..w_M; \theta) \tag{2}$$

Equation (2) is to be realized using a RNN-based architecture, which is described in detail in Sect. 6.1.

3   Related Work

Heilman and Smith [7] use a set of hand-crafted syntax-based rules to generate questions from simple declarative sentences. The system identifies multiple possible answer phrases from all declarative sentences using the constituency parse tree structure of each sentence. The system then over-generates questions and ranks them statistically by assigning scores using logistic regression.

[18] use semantics of the text by converting it into the Minimal Recursion Semantics notation [3]. Rules specific to the summarized semantics are applied to generate questions. Most of the approaches proposed for the QGSTEC challenge [10] are also rule-based systems, some of which put to use sentence features such as part of speech (POS) tags and named entity relations (NER) tags.

All approaches mentioned so far are heavily dependent on rules whose design requires deep linguistic knowledge and yet are not exhaustive enough. Recent successes in neural machine translation [2,16] have helped address this problem by letting deep neural nets learn the implicit rules through data. This approach has inspired application of sequence to sequence learning to automated question generation. [15] propose an attention-based [1,9] approach to question generation from a pre-defined template of knowledge base triples (subject, relation, object). Additionally, recent studies suggest that the sharp learning capability of neural networks does not make linguistic features redundant in machine translation. [14] suggest augmenting each word with its linguistic features such as POS, NER. [6] suggest a tree-based encoder to incorporate features, although for a different application.

We build on the recent sequence to sequence learning-based method of question generation by [5], but with significant differences and improvements from all previous works in the following ways. (i) Unlike [5] our question generation technique is *pivoted* on identification of the best candidate answer (span) around which the question should be generated. (ii) Our approach is enhanced with the use of several syntactic and linguistic features that help in learning models that generalize well. (iii) We propose a modified decoder to generate questions relevant to the text.

4   Approach and Contributions

Our approach to generating question-answer pairs from text is a two-stage process: in the first stage we select the most relevant and appropriate candidate answer, i.e., the *pivotal answer*, using an answer selection module, and in the second stage we encode the answer span in the sentence and use a sequence to sequence model with a rich set of linguistic features to generate questions for the pivotal answer.

Our sentence encoder transforms the input sentence into a list of fixed-length continuous vector word representation, each input symbol being represented as a vector. The question decoder takes in the output from the sentence encoder and produces one symbol at a time and stops at the EOS (end of sentence) marker. To focus on certain important words while generating questions (decoding) we use a global attention mechanism. The attention module is connected to both the sentence encoder as well as the question decoder, thus allowing the question decoder to focus on appropriate segments of the sentence while generating the next word of the question. We include linguistic features for words so that the model can learn more generalized syntactic transformations. We provide a detailed description of these modules in the following sections. Here is a summary of our three main contributions: (1) a versatile neural network-based answer selection and Question Generation (QG) approach (2) incorporation of rich set of linguistic features that help generalize the learning to syntactic and semantic transformations of the input, and (3) a modified decoder to generate the question most relevant to the text.

5   Answer Selection and Encoding

In applications such as reading comprehension, it is natural for a question to be generated keeping the answer in mind (hereafter referred to as the 'pivotal' answer). Identifying the most appropriate pivotal answer will allow comprehension be tested more easily and with even higher automation. We propose a novel named entity selection model and answer selection model based on Pointer Networks [17]. These models give us the span of pivotal answer in the sentence, which we encode using the BIO notation while generating the questions.

5.1   Named Entity Selection

In our first approach, we restrict our pivotal answer to be one of the named entities in the sentence, extracted using the Stanford CoreNLP toolkit. To choose the most appropriate pivotal answer for QG from a set of candidate entities present in the sentence we propose a named entity selection model. We train a multi-layer perceptron on the sentence, named entities present in the sentence and the ground truth answer. The model learns to predict the pivotal answer given the sentence and a set of candidate entities. The sentence $S = (w_1, w_2, ..., w_n)$ is first encoded using a 2 layered unidirectional LSTM encoder into hidden activations $H = (h_1^s, h_2^s, ..., h_n^s)$. For a named entity $NE = (w_i, ..., w_j)$, a vector representation ($\mathbf{R}$) is created as $< h_n^s; h_{mean}^s; h_{mean}^{ne} >$, where h_n^s is the final state of the hidden activations, h_{mean}^s is the mean of all the activations and h_{mean}^{ne} is the mean of hidden activations $(h_i^s, ..., h_j^s)$ between the span of the named entity. This representation vector $\mathbf{R}$ is fed into a multi-layer perceptron, which predicts the probability of a named entity being a pivotal answer. Then we select the entity with the highest probability as the answer entity. More formally,

$$P(NE_i|S) = softmax(\mathbf{R}_i.W + B) \tag{3}$$

where W is weight, B is bias, and $P(NE_i|S)$ is the probability of named entity being the pivotal answer.

5.2   Answer Selection Using Pointer Networks

We propose a novel Pointer Network [17] based approach to find the span of pivotal answer given a sentence. Using the attention mechanism, a *boundary* Pointer Network output start and end positions from the input sequence. More formally, the problem can be formulated as follows: given a sentence $\mathbf{S}$, we want to predict the start index a_k^{start} and the end index a_k^{end} of the pivotal answer. The main motivation in using a boundary pointer network is to predict the span from the input sequence as output. While we adapt the boundary pointer network to predict the start and end index positions of the pivotal answer in the sentence, we also present results using a sequence pointer network instead. **Answer sequence pointer network** produces a sequence of pointers as output. Each pointer in the sequence is word index of some token in the input. It only ensures that output is contained in the sentence but isn't necessarily a substring. Let the encoder's hidden states be $H = (h_1, h_2, \ldots, h_n)$ for a sentence the probability of generating output sequence $O = (o_1, o_2, \ldots, o_m)$ is defined as,

$$P(O|S) = \prod P(o_i|o_1, o_2, o_3, \ldots, o_{i-1}, H) \tag{4}$$

We model the probability distribution as:

$$u^i = v^T tanh(W^e \hat{H} + W^d D_i) \tag{5}$$

$$P(o_i|o_1, o_2, \ldots, o_{i-1}, H) = softmax(u^i) \tag{6}$$

Here, $W^e \in R^{d \times 2d}$, $W^D \in R^{d \times d}$, $v \in R^d$ are the model parameters to be learned. $\hat{H}$ is $<H; 0>$, where a 0 vector is concatenated with LSTM encoder hidden states to produce an end pointer token. D_i is produced by taking the last state of the LSTM decoder with inputs $<softmax(u^i)\hat{H}; D_{i-1}>$. D_0 is a zero vector denoting the start state of the decoder. **Answer boundary pointer network** produces two tokens corresponding to the start and end index of the answer span. The probability distribution model remains exactly the same as answer sequence pointer network. The boundary pointer network is depicted in Fig. 2.

We take sentence S $= (w_1, w_2, \ldots, w_M)$ and generate the hidden activations H by using embedding lookup and an LSTM encoder. As the pointers are not conditioned over a second sentence, the decoder is fed with just a start state.

Example: For the Sentence: "other past residents include composer journalist and newspaper editor william henry wills , ron goodwin, and journalist angela rippon and comedian dawn french", the answer pointers produced are:
Pointer(s) by answer sequence: [6,11,20] → journalist henry rippon
Pointer(s) by answer boundary: [10,12] → william henry wills

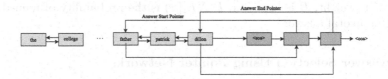

Fig. 2. Answer selection using boundary pointer network.

6   Question Generation

After encoding the pivotal answer (prediction of the answer selection module) in a sentence, we train a sequence to sequence model augmented with a rich set of linguistic features to generate the question. In sections below we describe our linguistic features as well as our sequence to sequence model.

6.1   Sequence to Sequence Model

Sequence to sequence models [16] learn to map input sequence (sentence) to an intermediate fixed length vector representation using an encoder RNN along with the mapping for translating this vector representation to the output sequence (question) using another decoder RNN. Encoder of the sequence to sequence model first conceptualizes the sentence as a single fixed length vector before passing this along to the decoder which uses this vector and attention weights to generate the output.

Sentence Encoder: The sentence encoder is realized using a bi-directional LSTM. In the forward pass, the given sentence along with the linguistic features is fed through a recurrent activation function recursively till the whole sentence is processed. Using one LSTM as encoder will capture only the left side sentence dependencies of the current word being fed. To alleviate this and thus to also capture the right side dependencies of the sentence for the current word while predicting in the decoder stage, another LSTM is fed with the sentence in the reverse order. The combination of both is used as the encoding of the given sentence.

$$\overrightarrow{\hat{h}_t} = f(\overrightarrow{W} w_t + \overrightarrow{V} \overrightarrow{\hat{h}_{t-1}} + \overrightarrow{b}) \tag{7}$$

$$\overleftarrow{\hat{h}_t} = f(\overleftarrow{W} w_t + \overleftarrow{V} \overleftarrow{\hat{h}_{t+1}} + \overleftarrow{b}) \tag{8}$$

$$\hat{h}_t = g(U h_t + c) = g(U[\overrightarrow{\hat{h}_t}, \overleftarrow{\hat{h}_t}] + c) \tag{9}$$

The hidden state $\hat{h}_t$ of the sentence encoder is used as the intermediate representation of the source sentence at time step t whereas $W, V, U \in R^{n \times m}$ are weights, where m is the word embedding dimensionality, n is the number of hidden units, and $w_t \in R^{p \times q \times r}$ is the weight vector corresponding to feature encoded input at time step t.

Attention Mechanism: In the commonly used sequence to sequence model [16], the decoder is directly initialized with intermediate source representation ($\hat{h}_t$). Whereas the attention mechanism proposed in [9] suggests using a subset of source hidden states, giving more emphasis to a, possibly, more relevant part of the context in the source sentence while predicting a new word in the target sequence. In our method we specifically use the global attention mechanism. In this mechanism a context vector c_t is generated by capturing relevant source side information for predicting the current target word y_t in the decoding phase at time t. Relevance between the current decoder hidden state h_t and each of the source hidden states ($\hat{h}_1, \hat{h}_2 ... \hat{h}_N$) is realized through a dot similarity metric: $score(h_t, \hat{h}_i) = h_t^T \cdot \hat{h}_i$.

A softmax layer (10) is applied over these scores to get the variable length alignment vector α_t which in turn is used to compute the weighted sum over all the source hidden states ($\hat{h}_1, \hat{h}_2, \ldots, \hat{h}_N$) to generate the context vector c_t (11) at time t.

$$\alpha_t(i) = align(h_t, \hat{h}_i) = \frac{\exp(score(h_t, \hat{h}_i))}{\sum_{i'} \exp(score(h_t, \hat{h}_{i'}))} \tag{10}$$

$$c_t = \sum_i \alpha_{ti} \hat{h}_i \tag{11}$$

Question Decoder: Question decoder is a two layer LSTM network. It takes output of sentence encoder and decodes it to generate question. The question decoder is designed to maximize our objective in Eq. 2. More formally decoder computes probability $P(Q|S; \theta)$ as:

$$P(Q|S; \theta) = softmax(W_s(tanh(W_r[h_t, c_t] + b))) \tag{12}$$

where W_s and W_r are weight vectors and *tanh* is the activation function. The hidden state of the decoder along with the context vector c_t is used to predict the target word y_t. It is a known fact that decoder may output words which are not even present in the source sentence as it learns a probability distribution over the words in the vocabulary. To generate questions relevant to the text we suitably modified decoder and integrated an attention mechanism (described in Sect. 6.1) with the decoder to attend to words in source sentence while generating questions. This modification to the decoder increases the relevance of question generated for a particular sentence.

6.2 Linguistic Features

We propose using a set of linguistic features so that the model can learn better generalized transformation rules, rather than learning a transformation rule per sentence. We describe our features below:

POS Tag: Parts of speech tag of the word. Words having same POS tag have similar grammatical properties and demonstrate similar syntactic behavior. We use the Stanford ConeNLP -pos annotator to get POS Tag of words in the sentence.

Named Entity Tag: Name entity tag represent coarse grained category of a word for example PERSON, PLACE, ORGANIZATION, DATE, etc. In order to help the model identify named entities present in the sentence, named entity tag of each word is provided as a feature. This ensures that the model learns to pose a question about the entities present in the sentence. We use the Stanford CoreNLP -ner annotator to assign named entity tag to each word.

Dependency Label: Dependency label of a word is the edge label connecting each word with the parent in the dependency parse tree. Root node of the tree is assigned label 'ROOT'. Dependency label help models to learn inter-word relations. It helps in understanding the semantic structure of the sentence while generating question. Dependency structure also helps in learning syntactic transformations between sentence and question pair. Verbs and adverbs present in the sentence signify the type of the question (which, who .. etc.) that would be posed for the subject it refers to. We use dependency parse trees generated using the Stanford CoreNLP parser to obtain the dependency labels.

Linguistic features are added by the conventional feature concatenation of tokens using the delimiter '|'. We create separate vocabularies for words (encoded using glove's pre-trained word embedding) and features (using one-hot encoding) respectively.

7 Implementation Details

We implement our answer selection and question generation models in Torch[1]. The sentence encoder of **QG** is a 3 layer bi-directional LSTM stack and the

---

[1] http://torch.ch/.

question decoder is a 3 layer LSTM stack. Each LSTM has a hidden unit of size 600 units. we use pre-trained glove embeddings[2] [12] of 300 dimensions for both the encoder and the decoder. All model parameters are optimized using Adam optimizer with a learning rate of 1.0 and we decay the learning rate by 0.5 after 10th epoch of training. The dropout probability is set to 0.3. We train our model in each experiment for 30 epochs, we select the model with the lowest perplexity on validation set.

The linguistic features for each word such as POS, named entity tag *etc.*, are incorporated along with word embeddings through concatenation.

8   Experiments and Results

We evaluate performance of our models on the SQUAD [13] dataset (denoted S). We use the same split as that of [5], where a random subset of 70,484 instances from S are used for training (S^{tr}), 10,570 instances for validation (S^{val}), and 11,877 instances for testing (S^{te}).

We performed both human-based evaluation as well as automatic evaluation to assess the quality of the questions generated. For automatic evaluation, we report results using a metric widely used to evaluate machine translation systems, called BLEU [11].

We first list the different systems (models) that we evaluate and compare in our experiments. A note about abbreviations: Whereas components in blue are different alternatives for encoding the pivotal answer, the brown color coded component represents the set of linguistic features that can be optionally added to any model.

Baseline System (QG): Our baseline system is a sequence-to-sequence LSTM model (see Sect. 6) trained only on raw sentence-question pairs without using features or answer encoding. This model is the same as [5].

System with Feature Tagged Input (QG+F): We encoded linguistic features (see Sect. 6.2) for each sentence-question pair to augment the basic **QG** model. This was achieved by appending features to each word using the "|" delimiter. This model helps us analyze the isolated effect of incorporating syntactic and semantic properties of the sentence (and words in the sentence) on the outcome of question generation.

Features + NE Encoding (QG+F+NE) : We also augmented the feature-enriched sequence-to-sequence QG+F model by encoding each named entity predicted by the named entity selection module (see Sect. 5.1) as a pivotal answer. This model helps us analyze the effect of (indiscriminate) use of named entity as potential (pivotal) answer, when used in conjunction with features.

Ground Truth Answer Encoding (QG+GAE) : In this setting we use the encoding of ground truth answers from sentences to augment the training of the basic QG model (see Sect. 5). For encoding answers into the sentence we employ

---

[2] http://nlp.stanford.edu/data/glove.840B.300d.zip.

the BIO notation. We append "**B**" as a feature using the delimiter "|" to the first word of the answer and "**I**" as a feature for the rest of the answer words. We used this model to analyze the effect of answer encoding on question generation, independent of features and named entity alignment.

We would like to point out that any direct comparison of a generated question with the question in the ground truth using any machine translation-like metric (such as the BLEU metric discussed in Sect. 8.1) makes sense only when both the questions are associated with the same pivotal answer. This specific experimental setup and the ones that follow are therefore more amenable for evaluation using standard metrics used in machine translation.

Features + Sequence Pointer Network Predicted Answer Encoding (QG+F+AES) : In this setting, we encoded the pivotal answer in the sentence as predicted by the sequence pointer network (see Sect. 5.2) to augment the linguistic feature based **QG+F** model. In this and in the following setting, we expect the prediction of the pivotal answer in the sentence to closely approximate the ground truth answer.

Features + Boundary Pointer Network Predicted Answer Encoding (QG+F+AEB) : In this setting, we encoded the pivotal answer in the sentence as predicted by the boundary pointer network (see Sect. 5.2) to augment the linguistic feature based **QG+F** model.

Features + Ground Truth Answer Encoding (QG+F+GAE): In this experimental setup, building upon the previous model (QG+F), we encoded ground truth answers to augment the QG model.

8.1   Results and Analysis

We compare the performance of the 7 systems QG, QG+F, QG+F+NE, QG+GAE, QG+F+AES, QG+F+AEB and QG+F+GAE described in the previous sections on (the train-val-test splits of) $\mathcal{S}$ and report results using both human and automated evaluation metrics. We first describe experimental results using human evaluation followed by evaluation on other metrics.

Human Evaluation: We randomly selected 100 sentences from the test set ($\mathcal{S}^{te}$) and generated one question using each of the 7 systems for each of these 100 sentences and asked three human experts for feedback on the quality of questions generated. Our human evaluators are professional English language experts. They were asked to provide feedback about a randomly sampled sentence along with the corresponding questions from each competing system, presented in an anonymised random order. This was to avoid creating any bias in the evaluator towards any particular system. They were not at all primed about the different models and the hypothesis.

We asked the following binary (yes/no) questions to each of the experts: (a) is this question syntactically correct?, (b) is this question semantically correct?, and (c) is this question relevant to this sentence?. Responses from all three experts were collected and averaged. For example, suppose the cumulative scores of the 100

Table 1. Human evaluation results on $\mathcal{S}^{te}$. Parameters are, **p1**: percentage of syntactically correct questions, **p2**: percentage of semantically correct questions, **p3**: percentage of relevant questions.

System	Syntactically correct (%)	Semantically correct (%)	Relevant (%)
QG [5]	51.6	48	52.3
QG+F	59.6	57	64.6
QG+F+NE	57	52.6	67
QG+GAE	44	35.3	50.6
QG+F+AES	51	47.3	55.3
QG+F+AEB	61	60.6	**71.3**
QG+F+GAE	**63**	**61**	67

Table 2. Automatic evaluation results on $\mathcal{S}^{te}$. BLEU, METEOR and ROUGE-L scores vary between 0 and 100, with the upper bound of 100 attainable on the ground truth. QG [5]:Result obtained using latest version of Torch.

Model	BLEU-1	BLEU-2	BLEU-3	BLEU-4	METEOR	ROUGE-L
QG [5]	39.97	22.39	14.39	9.64	14.34	37.04
QG+F	41.89	24.37	15.92	10.74	15.854	37.762
QG+F+NE	41.54	23.77	15.32	10.24	15.906	36.465
QG+GAE	43.35	24.06	14.85	9.40	15.65	37.84
QG+F+AES	43.54	25.69	17.07	11.83	16.71	38.22
QG+F+AEB	42.98	25.65	17.19	12.07	16.72	38.50
QG+F+GAE	46.32	28.81	19.67	13.85	18.51	41.75

binary judgements for syntactic correctness by the 3 evaluators were $(80, 79, 73)$. Then the average response would be 77.33. In Table 1 we present these results on the test set $\mathcal{S}^{te}$.

Evaluation on Other Metrics: We also evaluated our system on other standard metrics to enable comparison with other systems. However, as explained earlier, the standard metrics used in machine translation such as BLEU [11], METEOR [4], and ROUGE-L [8], might not be appropriate measures to evaluate the task of question generation. To appreciate this, consider the candidate question "who was the widow of mcdonald's owner ?" against the ground truth "to whom was john b. kroc married ?" for the sentence "it was founded in 1986 through the donations of joan b. kroc, the widow of mcdonald 's owner ray kroc.". It is easy to see that the candidate is a valid question and makes perfect sense. However its BLEU-4 score is almost zero. Thus, it may be the case that the human generated question against which we evaluate the system generated questions may be completely different in structure and semantics, but still be perfectly valid,

Sentence 1: Manhattan was on track to have an estimated 90,000 hotel rooms at the end of 2014, a 10% increase from 2013.
Answer Predicted: 90000
Question Generated: How many hotel rooms did Manhattan have ?

Sentence 2: American idol premiered in June 2002 and became the surprise summer hit show of 2002.
Answer Predicted: June 2002
Question Generated: When did American idol begin ?

Sentence 3: Shuman then constructed a full-scale steam engine powered by low-pressure water, enabling him to patent the entire solar engine system by 1912.
Answer Predicted: 1912
Question Generated: When was the solar engine system invented ?

Fig. 3. Sample output: the pivotal answer predicted and the question generated about the answer using model QG+F+AEB - that is comparable to the best performing system that also used ground truth answers.

as seen previously. While we find human evaluation to be more appropriate, for the sake of completeness, we also report the BLEU, METEOR and ROUGE-L scores in each setting. In Table 2, we observe that our models, QG+F+AEB, QG+F+AES and QG+F+GAE outperform the state-of-the art question generation system QG [5] significantly on all standard metrics.

Our model QG+F+GAE, which encodes ground truth answers and uses a rich set of linguistic features, performs the best as per every metric. And in Table 1, we observe that adding the rich set of linguistic features to the baseline model (QG) further improves performance. Specifically, addition of features increases syntactic correctness of questions by 2%, semantic correctness by 9% and relevance of questions with respect to sentence by 12.3% in comparison with the baseline model QG [5].

In Fig. 3 we present some sample answers predicted and corresponding questions generated by our model QG+F+AEB. Though not better, the performance of models QG+F+AES and QG+F+AEB is comparable to the best model (that is QG+F+GAE, which additionally uses ground truth answers). This is because the ground truth answer might not be the best and most relevant pivotal answer for question generation, particularly since each question in the SQUAD dataset was generated by looking at an entire paragraph and not any single sentence. Consider the sentence *"manhattan was on track to have an estimated 90,000 hotel rooms at the end of 2014, a 10 % increase from 2013."*. On encoding the ground truth answer, *"90,000"*, the question generated using model QG+GAE is *"what was manhattan estimated hotel rooms in 2014 ?"* and additionally, with linguistic features (QG+F+GAE), we get *"how many hotel rooms did manhattan have at the end of 2014 ?"*. This is indicative of how a rich set of linguistic features help in shaping the correct question type as well generating syntactically and semantically correct question. Further when we do not encode any answer (either pivotal answer predicted by sequence/boundary pointer network or ground truth answer) and just augment the linguistic features (QG+F) the question generated is *"what was manhattan 's hotel increase in 2013 ?"*, which is clearly a poor quality question. Thus, both answer encoding and augmenting rich set of linguistic features are important for generating high quality (syntactically correct, semantically correct and relevant) questions. When we select pivotal answer from amongst the set of named entities present in the sentence (*i.e.*, model QG+F+NE), the question generated on encoding the named entity *"manhattan"* is *"what was the 10 of hotel 's city rooms ?"*, which is clearly a poor quality question. The poor performance of QG+F+NE can be attributed to the fact that only 50% of the answers in SQUAD dataset are named entities.

9 Conclusion

We introduce a novel two-stage process to generate question-answer pairs from text. We combine and enhance a number of techniques including sequence to sequence models, Pointer Networks, named entity alignment, as well as rich linguistic features to identify potential answers from text, handle rare words, and generate questions most relevant to the answer. To the best of our knowledge this is the first attempt in generating question-answer pairs. Our comprehensive evaluation shows that our approach significantly outperforms current state-of-the-art question generation techniques on both human evaluation and evaluation on common metrics such as BLEU, METEOR, and ROUGE-L.

References

1. Bahdanau, D., Cho, K., Bengio, Y.: Neural machine translation by jointly learning to align and translate. arXiv preprint arXiv:1409.0473 (2014)
2. Cho, K., Van Merriënboer, B., Bahdanau, D., Bengio, Y.: On the properties of neural machine translation: encoder-decoder approaches. arXiv preprint arXiv:1409.1259 (2014)
3. Copestake, A., Flickinger, D., Sag, I.A., Pollard, C.: Minimal recursion semantics: an introduction (1999). http://www-csli.stanford.edu/~sag/sag.html, draft
4. Denkowski, M., Lavie, A.: Meteor universal: Language specific translation evaluation for any target language. In: Proceedings of the EACL 2014 Workshop on Statistical Machine Translation (2014)
5. Du, X., Shao, J., Cardie, C.: Learning to ask: neural question generation for reading comprehension. In: Proceedings of the 55th Annual Meeting of the ACL (Volume 1: Long Papers), vol. 1, pp. 1342–1352 (2017)
6. Eriguchi, A., Hashimoto, K., Tsuruoka, Y.: Tree-to-sequence attentional neural machine translation. CoRR abs/1603.06075 (2016). http://arxiv.org/abs/1603.06075
7. Heilman, M., Smith, N.A.: Good question! statistical ranking for question generation. In: Human Language Technologies: the 2010 Annual Conference of the North American Chapter of the Association for Computational Linguistics, pp. 609–617. Association for Computational Linguistics (2010)
8. Lin, C.Y.: Rouge: a package for automatic evaluation of summaries. In: Marie-Francine Moens, S.S. (ed.) Text Summarization Branches Out: Proceedings of the ACL-04 Workshop, pp. 74–81. Association for Computational Linguistics, Barcelona, July 2004. http://www.aclweb.org/anthology/W04-1013
9. Luong, M., Pham, H., Manning, C.D.: Effective approaches to attention-based neural machine translation. CoRR abs/1508.04025 (2015). http://arxiv.org/abs/1508.04025

10. Mannem, P., Prasad, R., Joshi, A.: Question generation from paragraphs at UPenn: QGSTEC system description. In: Proceedings of QG2010: The Third Workshop on Question Generation, pp. 84–91 (2010)

11. Papineni, K., Roukos, S., Ward, T., Zhu, W.J.: BLEU: a method for automatic evaluation of machine translation. In: Proceedings of the 40th Annual Meeting on Association for Computational Linguistics, pp. 311–318. Association for Computational Linguistics (2002)

12. Pennington, J., Socher, R., Manning, C.D.: Glove: global vectors for word representation. In: Empirical Methods in Natural Language Processing (EMNLP), pp. 1532–1543 (2014). http://www.aclweb.org/anthology/D14-1162

13. Rajpurkar, P., Zhang, J., Lopyrev, K., Liang, P.: Squad: 100,000+ questions for machine comprehension of text. arXiv preprint arXiv:1606.05250 (2016)

14. Sennrich, R., Haddow, B.: Linguistic input features improve neural machine translation. CoRR abs/1606.02892 (2016). http://arxiv.org/abs/1606.02892

15. Serban, I.V., García-Durán, A., Gulcehre, C., Ahn, S., Chandar, S., Courville, A., Bengio, Y.: Generating factoid questions with recurrent neural networks: the 30m factoid question-answer corpus. arXiv preprint arXiv:1603.06807 (2016)

16. Sutskever, I., Vinyals, O., Le, Q.V.: Sequence to sequence learning with neural networks. In: Advances in Neural Information Processing Systems, pp. 3104–3112 (2014)

17. Vinyals, O., Fortunato, M., Jaitly, N.: Pointer networks. In: NIPS, pp. 2692–2700 (2015)

18. Yao, X., Bouma, G., Zhang, Y.: Semantics-based question generation and implementation. Dialogue Discourse Spec. Issue Question Gener. 3(2), 11–42 (2012)

Emotion Classification with Data Augmentation Using Generative Adversarial Networks

Xinyue Zhu[1,2], Yifan Liu[1], Jiahong Li[3], Tao Wan[4(✉)], and Zengchang Qin[1(✉)]

[1] Intelligent Computing and Machine Learning Lab, School of ASEE,
Beihang University, Beijing, China
{yifan_liu,zcqin}@buaa.edu.cn
[2] School of Electronic Engineering,
Bejing University of Posts and Telecommunications, Beijing, China
zxysee@bupt.edu.cn
[3] Beijing San Kuai Yun Technology Co., Ltd., Beijing, China
[4] School of Biological Science and Medical Engineering,
Beijing Advanced Innovation Centre for Biomedical Engineering,
Beihang University, Beijing, China
taowan@buaa.edu.cn

Abstract. It is a difficult task to classify images with multiple class labels using only a small number of labeled examples, especially when the label (class) distribution is imbalanced. Emotion classification is such an example of imbalanced label distribution, because some classes of emotions like *disgusted* are relatively rare comparing to other labels like *happy or sad*. In this paper, we propose a data augmentation method using generative adversarial networks (GAN). It can complement and complete the data manifold and find better margins between neighboring classes. Specifically, we design a framework using a CNN model as the classifier and a cycle-consistent adversarial networks (CycleGAN) as the generator. In order to avoid gradient vanishing problem, we employ the least-squared loss as adversarial loss. We also propose several evaluation methods on three benchmark datasets to validate GAN's performance. Empirical results show that we can obtain 5% 10% increase in the classification accuracy after employing the GAN-based data augmentation techniques.

Keywords: Data augmentation · Emotion classification
Imbalanced data processing · GAN · CycleGAN

1 Introduction

In recent development of deep learning, neural networks with more and more layers are proposed [10,22]. Such neural network models have much larger capacity that needs a larger training set. However, it is always expensive to obtain adequate and balanced dataset with manual labels. This has been a general problem

© Springer International Publishing AG, part of Springer Nature 2018
D. Phung et al. (Eds.): PAKDD 2018, LNAI 10939, pp. 349–360, 2018.
https://doi.org/10.1007/978-3-319-93040-4_28

in machine learning as well as in computer vision. An effective way of synthesizing images to supplement training set may help boost accuracy in image classification. Using data augmentation for enlarging training set in image classification has reported in various literatures [2,12,21]. Model performance can be improved because data augmentation can overcome the problem of inadequate data and imbalanced label distribution. However, it is an unsolved problem of how to generate (sample) data from the 'true distribution' of given limited training data. In this research, based on Generative Adversarial Networks (GANs), we propose a new method for data augmentation in order to generate new samples via adversarial training, thus to supplement the data manifold to approximate the 'true distribution' and that may lead to better margins between different categories of data.

Since the invention of GAN [7], it has been well used in different machine learning applications [3,25], especially in computer vision and image processing [11,14]. In this paper, we explore how to use GAN to generate images helping enlarge original dataset effectively and balance label distribution from data augmentation. We focus on the emotion (facial expression images) classification task because it is a typical classification task with inadequate data and imbalanced label distribution. On one hand, the training dataset obtained from the laboratory are limited to diversity and quantity. On the other hand, some classes of emotion images (such as *disgust*) have few samples in the training set from the real world than other classes, also some images can be very nuanced, making it difficult even for humans to agree on their correct labeling [18]. In our work, we build a classical convolutional neural network (CNN) classifier for emotion image classification and train the CycleGAN model [29] with least-squared loss [17] to achieve image-to-image transformation, which can synthesis images for the unusual classes of emotion from a related image source. Given a classical CNN classifier, our effort is on constructing GAN model for improving performance in generating images given a specific class. The CycleGAN model is used for image translation between two unpaired domains.

The main contributions of this paper can be summarized as follows. (1) We propose a framework for data augmentation by using GAN to generate supplementary data in emotion classification task. (2) Through empirical studies on three benchmark datasets, we found that performance of the new model is significantly improved compared to the baselines. (3) We combine least-squared loss from LSGAN and adversarial loss in CycleGAN to avoid the problem of vanishing gradients, this is verified to be effective in training process.

2   Related Work

Facial expression recognition (or emotion classification) has attracted much attention in computer vision in past few decades. Current techniques related to facial expression mainly focus on recognizing seven prototypical emotions (*neutral, happy, surprised, fear, angry, sad,* and *disgusted*), which are considered basic and universal emotions for human. Such recognition is sometimes very difficult since there is only a slight difference between different emotions, which requires an efficient and subtle feature extractor to be trained. Moreover, Ng *et al.* [18] pointed out that the imbalanced distribution among emotion classes may lead to low accuracy in classes with fewer samples. To deal with imbalanced datasets, many methods were proposed, such as undersampling [13], creating 'box' around minorities [6] and etc. Different from these works, we aim to resolve this problem by generating data of minority classes from low-dimensional manifold, which improves the data distribution from feature level.

2.1 Generative Adversarial Networks

Generative adversarial networks (GANs) can be used to generate images from an adversarial training. The generator attempts to produce a realistic image to fool the discriminator, which tries to distinguish whether its input image is from the training set or the generated set. Since the invention of GAN [7] in 2014, variant models based on GAN were proposed [17,19,29]. Generative adversarial nets are now widely used in many image tasks such as cartoon image coloring [14], image manipulation [28], synthesis [3] and image-to-image translation [11].

Zhu *et al.* [29] proposed the CycleGAN. It can do image-to-image transition between two unpaired image domain, which is helpful to build our framework. The main idea of the cycleGAN is "If we translate from one domain to another and back again we must arrive where we start" [29]. The LSGAN [17] used a least square distance to evaluate the difference between the distribution, which is more stable than the Jensen-Shannon (JS) divergence used in [7] and convergence more quickly than the Wasserstein loss [1]. In our research, we choose CycleGAN [29] and the techniques in LSGAN [17] to generate labeled emotion images and show that these images are helpful in final image classification task.

2.2 Data Augmentation

In the field of deep learning, where the scale of dataset has a great influence on the final outcome, data augmentation is often used to expand the training corpus. As for the existing techniques of data augmentation, they can be grouped into two main types: (a) geometric transformation which is relatively generic and computationally cheap and (b) task-specific or guided-augmentation methods which are able to generate synthetic samples given specific labels [5]. In the case of image classification, the first group of data augmentation methods always focus on generating image data through label-preserving linear transformations (translation, rotation, scaling, horizontal shearing) such as Affine [2], elastic deformations [21], patches extraction and RGB channels intensities alteration [12]. However, if we look deeper into these methods, they only lead to an image-level transformation through depth and scale and actually not helpful for dividing a clear boundary between data manifolds. Such data augmentation does not improve data distribution which is determined by higher-level features. For the second group, more complex manually-specified augmentation schemes are proposed. For instance, authors in [9] proposed an approach to learn multivariate normal distribution of each class in the whole mean manifold. In [5], an attribute-guided augmentation in feature space is designed. In the field of 3D motion capture, 2D images are used for generating 3D ones [20]. Our approach aims to solve similar task in [9] but is very different from all these methods above. In this paper, new training corpus is generated from CycleGAN, which remain high-level features extracted from original images. Although in the research related to GANs, data augmentation is sometimes mentioned as an important application of GAN, such as generating realistic license plate in [24], synthesizing photorealistic facial expressions in [27], few of them lay emphasis on exploring GAN's role of doing data augmentation in addressing imbalanced datasets. Moreover, few comparisons are made to support a seemingly obvious idea that GAN-synthesized images are qualified enough to supplement original training corpus as augmented data.

3   Data Augmentation Using CycleGAN

Figure 1 shows our framework of GAN-based data augmentation. Both reference images and target images are collected from the original data and flow into the CycleGAN as domains R and T, respectively. G and F are two generators, transferring $R \rightarrow T$ and $T \rightarrow R$, respectively. Supplementary data is generated through generator G. A CNN classifier is trained using original data and supplementary data as input. In CycleGAN model, L_R is the LSGAN loss relative to reference domains and L_T is the LSGAN loss relative to target domains. Besides, a cycle loss, namely L_{cyc}, is calculated to keep cycle consistency of the whole model.

3.1   Cycle-Consistent Adversarial Networks

In this work, CycleGAN [29] is used to realize unpaired image-to-image translation, learning mapping functions between images of reference class (R) and target class (T). We use generators G and F to achieve domain transfer G: $R \rightarrow T$ and F: $T \rightarrow R$. Discriminators are denoted by D_R and D_T, where D_R aims to distinguish between real images in R and translated fake images $F(T)$ in reference domain. D_T is the discriminator in the target domain. We not only want to make the generated images $G(R)$ look like the target images in T, but also the reconstructed images $F(G(R)) \approx R$ to guarantee the cycle-consistency in addition to the adversarial loss. As for adversarial loss, G tries to generate $G(r)$ which is so similar to t that can fool the discriminator D_T. Therefore, the loss related to G and D_T is:

$$L(G, D_T, R, T) = E_{t \sim p_{data}(t)}[\log D_T(t)] + E_{r \sim p_{data}(r)}[\log(1 - D_T(G(r)))] \quad (1)$$

However, this logarithm form makes training and convergence difficult since it is likely to cause gradient vanishing problem [1]. Here we apply a least-squared loss proposed in LSGAN [17] to avoid this phenomenon and maintain the same

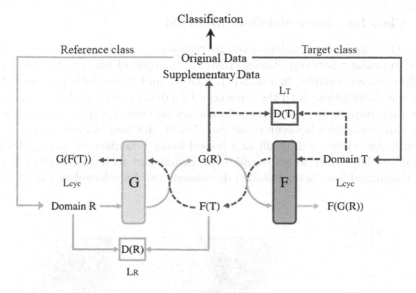

Fig. 1. An illustration of the proposed framework of using CycleGAN for data augmentation and classification using a CNN classifier.

function as adversarial loss in original CycleGAN. For the reference domain R, the loss is defined by:

$$L_{LSGAN}(G, D_R, T, R) = E_{r \sim p_{data}(r)}[(D_R(r) - 1)^2] + E_{t \sim p_{data}(t)}[D_R(G(t))^2] \quad (2)$$

For the target domain T, the loss is:

$$L_{LSGAN}(G, D_T, R, T) = E_{t \sim p_{data}(t)}[(D_T(t) - 1)^2] + E_{r \sim p_{data}(r)}[D_T(G(r))^2] \quad (3)$$

We can then define the final loss by:

$$\begin{aligned} L(G, F, D_S, D_R) &= L_R + L_T + L_{cyc} \\ &= L_{LSGAN}(G, D_R, S, R) + L_{LSGAN}(F, D_S, R, S) + \lambda L_{cyc}(G, F) \end{aligned} \quad (4)$$

where cycle consistency loss (L_{cyc}) is defined by:

$$\begin{aligned} L_{cyc}(G, F) &= E_{r \sim p_{data}(r)}[\|F(G(r)) - r\|_1] \\ &+ E_{t \sim p_{data}(t)}[\|G(F(t)) - t\|_1] \end{aligned} \quad (5)$$

where $\|\cdot\|_1$ is the L_1 norm. With these loss functions, the final functions we aim to solve is:

$$G^*, F^* = arg \min_{F,G} \max_{D_T, D_R} L(G, F, D_T, D_R) \quad (6)$$

other details of CycleGAN can be referred to [29].

3.2   Class Imbalance and Data Manifold

When the classes have imbalanced distributions, the classifier prone to learn biased boundary between classes. Take an example of binary classification in one-dimensional sample. In Fig. 2-(a), Class 1 and 2 are both generated from Gaussian distributions with the same standard deviation 1, and has the means of μ_1 and μ_2, respectively. Ideally, the boundary function $x = (\mu_1 + \mu_2)/2$, denoted by S_i can distinguish between these two classes. However, an imbalanced distribution in two classes will result in a biased linear boundary S_r moving towards the minor class, since given samples are insufficient to form a correct margin with minimized loss. Some detailed discussions can be referred to [23].

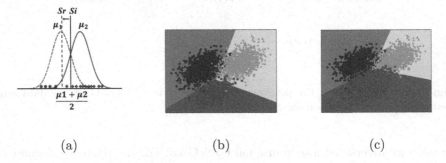

(a)   (b)   (c)

Fig. 2. (a) A binary classification problem with one-dimensional data. (b) Data manifolds with learned boundaries with imbalanced distributions. (c) The manifolds of balanced distribution. Data are generated with 2-dimensional Gaussians and using SVMs with linear kernel as classifier.

Now back to our emotion classification task. Under the assumption that image samples lie on several sub-manifolds in a high-dimensional space, where images of the same emotion lie in the same sub-manifold, image classification task is actually a task to explore the underlying geometric structure of data distribution, thus to find best-split hyper-planes between different categories in this space. These hyper-planes divide the space into several parts according to margins, each represents a clustering of a specific class (Fig. 2-(c)). When the dataset is imbalanced, it is very likely to form an incomplete manifold. Since in the same space, minorities are distributed more sparsely in their regions. In this case, biased margins or hyper-planes are learned, making it a difficult task for classifier to predict correct labels for given instances (Fig. 2-(b)). Although some data augmentation techniques can alleviate this problem from several aspects, the most essential solution is to further complement and complete the data manifold.

The reason we choose cycleGAN in stead of the classical GAN is because the original GAN [7] learns a mapping from low-dimensional manifold (directly determined by noise z) to high-dimensional data spaces, while the CycleGAN, as a tool for translation between two domains of both high-dimensional data, need to learn a low-dimensional manifold and also the parameters to map it back to high-dimensional space. Here we use $\mathbb{P}_r$ and $\mathbb{P}_t$ to represent real distributions of domain R and T, respectively, and $\mathbb{M}$ is the low-dimensional manifold. As domain R has only a small number of training samples, when it is projected into a low dimension space, sparse distribution cannot form a complete $\mathbb{M}$ with efficient feature information. By using CycleGAN, sufficient samples in $\mathbb{P}_r$ may lead to a more complete $\mathbb{M}$. G_θ can be learned through minimizing distance between $G(r)$ and T (e.g, $||G(r)-T||^2$) to 'pull' $\mathbb{M}$ to $\mathbb{P}_t$. If we again project $\mathbb{P}_t$ into manifold $\mathbb{M}$, they will form a meaningful feature-level manifold thanks to the generated samples.

4   Experimental Studies

Before doing experiments on emotion datasets, we first validate GAN's role in completing data manifold on a toy dataset. We use three two-dimensional Gaussian distributions $(x_1, x_2) \sim N_m(\mu_m, \sigma_m)$, $m \in \{1, 2, 3\}$ to simulate the distribution of three classes of data, where $\mu_1 = [0, 6]$, $\mu_2 = [6.5, 7]$, $\mu_3 = [2, 2]$ and the covariance matrix is $([2, 1]; [1, 2])$ for all three distributions. Imbalanced dataset is artificially created by randomly sampling 1000, 1000 and 100 points from each class for training, and for each class, we have 100 data points for test. Support vector machines (SVM) with linear kernel function is employed for classification task. After that, we train a CycleGAN to generate 900 target class (minority class) from reference class (one of majority classes) and these supplementary samples are added to the original dataset and trained on the same SVM classifier.

We draw two figures (Fig. 2-(b) and (c)) to show the data distribution and learned margins before and after adding the CycleGAN-based augmentation. The original biased margins in imbalanced dataset Fig. 2-(b) show a clear change to more correct ones in Fig. 2-(b). Moreover, the classification accuracy has increased from 93.3% to 98.0%. Although the distribution and dimension of data in this toy experiment is much simpler than real image data, the results can validate the effectiveness of the new approach by improving data manifold in imbalanced datasets by doing data augmentation.

4.1   Benchmark Datasets

In our experiments, three benchmark datasets are tested: Facial Expression Recognition Database (FER2013) [8], Static Facial Expressions in the Wild (SFEW) [4] and Japanese Female Facial Expression (JAFFE) Database [15]. All these datasets contain 7 types of face emotion including *angry, disgust, fear, happy, sad, surprise,* and *neutral* (labeled by 0–6 during training and test process). Samples from FER2013 database are shown in Fig. 3 (left) as an example. The distribution of this dataset is imbalanced. In order to verify the effectiveness of the data augmentation, we sample the images by 20% for each class in FER2013. We also test our data augmentation model on other two small datasets

SFEW and JAFFE. During the training process of CycleGAN, we choose 'neutral' class as our reference class and the other six are regarded as target ones, since it is natural to generate faces with emotion from non-emotional ones.

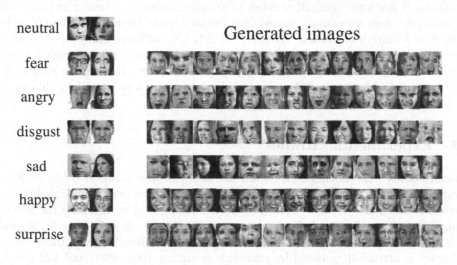

Fig. 3. The original samples and generated samples for each classes. The left two columns are original data and the rest ones are generated by CycleGAN. The neutral class, as reference class, has no generated samples in our experiment.

4.2   Experimental Results

We first train a classical CNN model based on original FER2013 datasets (20% sampled) as our baseline and the result is shown in Table 1. We use 7% and 14% samples in FER2013 for test. Our baseline, though is fairly simple, is sufficient to extract features from emotion samples and works as a qualified classifier. In order to get the most intuitive results, we choose class *disgust* and *sad* from FER2013 as our target classes, which are much smaller than the other classes, as a result, it cannot obtain sufficient learning and optimizing, thus reach a relatively low accuracy when trained on the baseline (see Table 1). In this case, two CycleGANs are trained to generate *disgust* and *sad* images respectively with class *neutral* as reference class (see Fig. 3), and then are filled into the original datasets (denoted by +*disgust* and +*sad*) to balance the distribution and complete the data manifold (Table 1).

From Table 1, we can have the following observations: (a) the overall test accuracy is improved and (b) accuracy of target class raise significantly and it is worth mentioning that (c) the accuracy of reference class *neutral* also increases. Therefore, we can intuitively verify the capability of CycleGAN in generating reliable images, which is helpful in enlarging minorities. Furthermore, this data augmentation of one class also improves accuracy of other classes, since by generating new samples, the data manifold is further supplemented and becomes more completed, thus make more clearly the margins between classes.

Table 1. Accuracy of the baseline model (CNN) and our proposed model (CNN+CycleGAN) on FER2013. The best results are highlighted.

Class	Accuracy-(7%)			Accuracy-(14%)		
	baseline	+disgust	+sad	Baseline	+disgust	+sad
angry	93.70	**93.71**	93.05	93.47	93.36	92.89
disgust	73.91	91.30	**95.65**	79.62	88.89	94.44
fear	90.88	92.18	94.46	90.38	91.43	**94.58**
happy	91.87	96.34	93.70	91.75	**96.37**	94.21
sad	87.86	93.61	**97.44**	89.22	93.26	94.61
surprise	94.27	**99.12**	96.48	93.46	97.09	96.85
neutral	89.55	91.94	**94.63**	88.24	93.06	94.48
All	91.04	94.25	**94.65**	90.77	93.82	94.32

In order to provide more powerful verification that this data augmentation indeed contributes to the shape of data manifold, we apply a t-distributed stochastic neighbor embedding (t-SNE) algorithm [16] to visualize the distribution of training samples by reducing high-dimensional data (48×48) to 2D plane (Fig. 4). Compared to the baseline (Fig. 4-(a)), where sample size of *disgust* and *sad* is too small to form a clear margin with other classes, (b) and (c) in Fig. 4 shows great improvement in enlarging the sample size, supplementing the data manifold and completing data distribution. Figure 4-(d) is a much stronger validation where both two classes stand out to improve data manifold.

After generating specific classes to validate GAN's positive role in data augmentation, we make further experiments on our framework based on all three datasets mentioned in Sect. 4.2. During this process, a baseline model and a model using our data augmentation framework (pre-train+fine-tune) is trained respectively. In our framework, all classes except *neutral* are generated from CycleGAN and then added as supplementary training corpus for training classification task. (See Fig. 3 for generated images in FER2013 database) and then the model is fine-tuned based on original datasets. Because there is a small number of examples in datasets SFEW and JAFFE, we set the FER2013 database as our pre-trained model and fine-tune it using these two datasets. Similar experiments were reported in [26]. In order to reduce the inference of complex background in SFEW, we apply a simple cropping method to extract faces from original images. For testing, we use 7% and 14% samples from FER2013, the given testing corpus of SFEW and 20% samples from JAFFE, respectively. Results are shown in Table 2. In the column DAG: Pre-train + Fine-tune, Pre-train represents the first 10K steps training on generated images from all six classes and Fine-tune represents another 10k fine-tuning steps training on original datasets. SFEW and JAFFE datasets are trained based on the FER2013 model. After applying our framework of data augmentation using GAN, accuracy of all the three datasets has visibly improved. As for FER2013 database which has obvious imbalanced

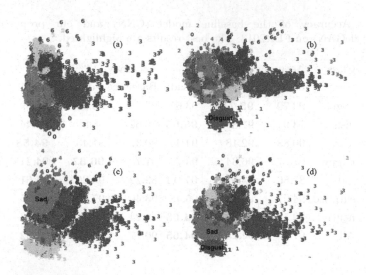

Fig. 4. Data manifold of four types of training samples using t-SNE algorithm: (a) baseline model, (b) adding generated *disgust* samples or (c) *sad* samples, and (d) samples of both two classes.

distribution among classes, our data augmentation technique is able to complete data manifold, especially for those which have much smaller samples. And for small datasets like SFEW and JAFFE, our technique can generate feature-level synthetic images from existing samples to enlarge the original datasets and form clear margins or hyper-planes between neighboring classes.

Table 2. Test accuracy of baseline and our framework (DAG).

Datasets	Accuracy			
	baseline		DAG: pre-train+fine-tune	
FER2013	91.04(7%)	90.77(14%)	**94.71**(7%)	94.35(14%)
FER+SFEW	31.92		**39.07**	
FER+JAFFE	93.87		**95.80**	

5 Conclusions and Discussions

In this paper, we explored using GAN for data augmentation in task of emotion classification task. We propose a framework for data augmentation by using CycleGAN to generate auxiliary data to minority classes in training. During the process of training CycleGAN model, a least-squared loss is combined with original adversarial loss to avoid gradient vanishing problem. Besides, we show the GAN's ability of supplementing low-dimensional data manifold. Because of possessing a more complete data manifold, the classifier can be better learned to find margins or hyper-planes of neighboring classes. Experiments on three benchmark datasets show that our GAN-based data augmentation techniques can lead to improvement in distribution integrity and margin clarity between classes, and can obtain 5%–10% increase in the accuracy of emotion classification task.

The work still has some limitations. For instance, the datasets we select are limited to emotions and only CycleGAN is used in our model. Therefore, we consider our future work to apply our model for the general image classification problems, and try other GAN models to evaluate data augmentation method with stronger evidence.

References

1. Arjovsky, M., Chintala, S., Bottou, L.: Wasserstein GAN. arXiv preprint arXiv:1701.07875 (2017)
2. Cireşan, D.C., Meier, U., Masci, J., Gambardella, L.M., Schmidhuber, J.: High-performance neural networks for visual object classification. arXiv preprint arXiv:1102.0183 (2011)
3. Denton, E.L., Chintala, S., Fergus, R., et al.: Deep generative image models using a Laplacian pyramid of adversarial networks. In: NIPS, pp. 1486–1494 (2015)
4. Dhall, A., Goecke, R., Lucey, S., Gedeon, T.: Static facial expression analysis in tough conditions: data, evaluation protocol and benchmark. In: ICCV Workshops, pp. 2106–2112. IEEE (2011)
5. Dixit, M., Kwitt, R., Niethammer, M., Vasconcelos, N.: AGA: attribute guided augmentation. arXiv preprint arXiv:1612.02559 (2016)
6. Goh, S.T., Rudin, C.: Box drawings for learning with imbalanced data. In: Proceedings of the 20th ACM SIGKDD, pp. 333–342. ACM (2014)
7. Goodfellow, I., Pouget-Abadie, J., Mirza, M., Xu, B., Warde-Farley, D., Ozair, S., Courville, A., Bengio, Y.: Generative adversarial nets. In: NIPS, pp. 2672–2680 (2014)
8. Goodfellow, I.J., et al.: Challenges in representation learning: a report on three machine learning contests. In: Lee, M., Hirose, A., Hou, Z.-G., Kil, R.M. (eds.) ICONIP 2013. LNCS, vol. 8228, pp. 117–124. Springer, Heidelberg (2013). https://doi.org/10.1007/978-3-642-42051-1_16
9. Hauberg, S., Freifeld, O., Larsen, A.B.L., Fisher, J., Hansen, L.: Dreaming more data: class-dependent distributions over diffeomorphisms for learned data augmentation. In: Artificial Intelligence and Statistics, pp. 342–350 (2016)
10. He, K., Zhang, X., Ren, S., Sun, J.: Deep residual learning for image recognition. In: Proceedings of the IEEE Conference on CVPR, pp. 770–778 (2016)
11. Isola, P., Zhu, J.Y., Zhou, T., Efros, A.A.: Image-to-image translation with conditional adversarial networks. arXiv preprint arXiv:1611.07004 (2016)
12. Krizhevsky, A., Sutskever, I., Hinton, G.E.: Imagenet classification with deep convolutional neural networks. In: NIPS, pp. 1097–1105 (2012)
13. Liu, X.Y., Wu, J., Zhou, Z.H.: Exploratory undersampling for class-imbalance learning. IEEE Trans. Syst. Man Cybern. Part B (Cybern.) **39**(2), 539–550 (2009)

14. Liu, Y., Qin, Z., Luo, Z., Wang, H.: Auto-painter: cartoon image generation from sketch by using conditional generative adversarial networks. arXiv preprint arXiv:1705.01908 (2017)
15. Lyons, M.J., Akamatsu, S., Kamachi, M., Gyoba, J., Budynek, J.: The Japanese female facial expression (JAFFE) database. In: Proceedings of Third International Conference on Automatic Face and Gesture Recognition, pp. 14–16 (1998)
16. van der Maaten, L., Hinton, G.: Visualizing data using t-SNE. J. Mach. Learn. Res. **9**(Nov), 2579–2605 (2008)
17. Mao, X., Li, Q., Xie, H., Lau, R.Y., Wang, Z., Smolley, S.P.: Least squares generative adversarial networks. arXiv preprint arXiv:1611.04076 (2016)
18. Ng, H.W., Nguyen, V.D., Vonikakis, V., Winkler, S.: Deep learning for emotion recognition on small datasets using transfer learning. In: Proceedings of the 2015 ACM on International Conference on Multimodal Interaction, pp. 443–449. ACM (2015)
19. Radford, A., Metz, L., Chintala, S.: Unsupervised representation learning with deep convolutional generative adversarial networks. arXiv preprint arXiv:1511.06434 (2015)
20. Rogez, G., Schmid, C.: Mocap-guided data augmentation for 3D pose estimation in the wild. In: NIPS, pp. 3108–3116 (2016)
21. Simard, P.Y., Steinkraus, D., Platt, J.C., et al.: Best practices for convolutional neural networks applied to visual document analysis. In: ICDAR, vol. 3, pp. 958–962 (2003)
22. Simonyan, K., Zisserman, A.: Very deep convolutional networks for large-scale image recognition. arXiv preprint arXiv:1409.1556 (2014)
23. Wallace, B.C., Small, K., Brodley, C.E., Trikalinos, T.A.: Class imbalance, redux. In: 2011 IEEE 11th International Conference on Data Mining (ICDM), pp. 754–763. IEEE (2011)
24. Wang, X., You, M., Shen, C.: Adversarial generation of training examples for vehicle license plate recognition. arXiv preprint arXiv:1707.03124 (2017)
25. Yu, L., Zhang, W., Wang, J., Yu, Y.: Seqgan: sequence generative adversarial nets with policy gradient. In: National Conference on Artificial Intelligence, pp. 2852–2858 (2016)
26. Yu, Z., Zhang, C.: Image based static facial expression recognition with multiple deep network learning. In: Proceedings of the 2015 ACM on International Conference on Multimodal Interaction, pp. 435–442. ACM (2015)
27. Zhou, Y., Shi, B.E.: Photorealistic facial expression synthesis by the conditional difference adversarial autoencoder. arXiv preprint arXiv:1708.09126 (2017)
28. Zhu, J.-Y., Krähenbühl, P., Shechtman, E., Efros, A.A.: Generative visual manipulation on the natural image manifold. In: Leibe, B., Matas, J., Sebe, N., Welling, M. (eds.) ECCV 2016. LNCS, vol. 9909, pp. 597–613. Springer, Cham (2016). https://doi.org/10.1007/978-3-319-46454-1_36
29. Zhu, J.Y., Park, T., Isola, P., Efros, A.A.: Unpaired image-to-image translation using cycle-consistent adversarial networks. arXiv preprint arXiv:1703.10593 (2017)

Trans2Vec: Learning Transaction Embedding via Items and Frequent Itemsets

Dang Nguyen$^{(\boxtimes)}$, Tu Dinh Nguyen, Wei Luo, and Svetha Venkatesh

Center for Pattern Recognition and Data Analytics,
School of Information Technology, Deakin University, Geelong, Australia
{d.nguyen,tu.nguyen,wei.luo,svetha.venkatesh}@deakin.edu.au

Abstract. Learning meaningful and effective representations for transaction data is a crucial prerequisite for transaction classification and clustering tasks. Traditional methods which use frequent itemsets (FIs) as features often suffer from the data sparsity and high-dimensionality problems. Several supervised methods based on discriminative FIs have been proposed to address these disadvantages, but they require transaction labels, thus rendering them inapplicable to real-world applications where labels are not given. In this paper, we propose an *unsupervised* method which learns *low-dimensional* continuous vectors for transactions based on information of both singleton items and FIs. We demonstrate the superior performance of our proposed method in classifying transactions on four datasets compared with several state-of-the-art baselines.

1 Introduction

A transaction dataset consists of multiple transactions, each of which is a set of discrete and distinct items. It can be found in many different domains such as the products purchased in a supermarket basket and the symptoms diagnosed in a patient's admission. Turning such data into useful information and knowledge requires the applications of machine learning methods such as Support Vector Machine (SVM) or K-means. This task, however, is challenging because machine learning methods typically require inputs as fixed-length vectors, which are not applicable to transactions.

A common solution in data mining is to use frequent itemsets (FIs) as features [5]. This method first mines FIs (i.e., itemsets whose *supports* (or frequencies) are not less than a minimum support threshold δ [6]) from the dataset. It then represents a transaction as a vector with binary components indicating whether this transaction contains a particular frequent itemset. Given a dataset $\mathcal{D}$ and the set of FIs discovered from $\mathcal{D}$, $\mathcal{F}(\mathcal{D}) = \{X_1, X_2, ..., X_F\}$, the feature vector of a transaction T is defined as $f(T) = [x_1, x_2, ..., x_F]$, where $x_i = 1$ if $X_i \subseteq T$ otherwise $x_i = 0$. We can see that the dimension of the feature space is huge since the number of FIs is often very large. For example, on some datasets, the number of FIs is more than 10^5 with $\delta < 5\%$. Consequently, this leads to the high-dimensionality and data sparsity problems.

© Springer International Publishing AG, part of Springer Nature 2018
D. Phung et al. (Eds.): PAKDD 2018, LNAI 10939, pp. 361–372, 2018.
https://doi.org/10.1007/978-3-319-93040-4_29

To tackle these two disadvantages, many researchers have attempted to extract only significant FIs using discriminative measures such as support difference [8], support ratio [9], or information gain [5]. However, all these measures require labels of transactions, making the mining process *supervised*. Due to the supervised nature of these methods, they have two limitations. First, their transaction representations are constructed for a particular mining task (e.g., transaction classification), thus the representations cannot be directly transferred to another task (e.g., transaction clustering). Second, the success of these methods relies on an enormous availability of labels for all training examples, a condition often not met in real applications.

Our Approach. To overcome the weaknesses of FI-based methods and supervised FI-based methods, we propose a novel method for learning *low-dimensional* representations (also called *embedding method*) for transactions in a fully *unsupervised* fashion. In particular, our embedding method (named **Trans2Vec**) first represents a transaction using two different sets: a set of singleton items and a set of FIs. It then proposes two models to learn transaction embeddings: one learns embeddings from these two sets separately (*individual-training* model) and another learns embeddings from these two sets simultaneously (*joint-training* model). **Trans2Vec** owns two advantages. First, it is fully unsupervised. Compared to supervised FI-based methods, it can be directly used for learning transaction embeddings in domains where labeled examples are difficult to obtain. Moreover, the low-dimensional representations learned are well-generalized to many different tasks such as transaction classification and transaction clustering. Second, it leverages not only the information of singleton items but also that of FIs which have many benefits. Regarding [5], FIs are useful for constructing transaction features since (1) They can capture the associations among individual items; and (2) They can capture the relationships among transactions.

In short, we make the following contributions:

1. We propose **Trans2Vec**, an *unsupervised* method, to learn *low-dimensional* continuous representations for transaction data.
2. We propose two models in **Trans2Vec**, which *learn* transaction embeddings from information of both singleton items and FIs. The embeddings learned are meaningful and discriminative.
3. We demonstrate **Trans2Vec** in transaction classification where it achieves significant improvements on several benchmark datasets.

2   Related Work

Our method is related to FI-based approaches. FIs have been used to construct feature vectors for transactions [5], which are essential inputs for many machine learning tasks such as transaction classification and clustering. However, this traditional approach suffers from the data sparsity and high-dimensionality problems due to the huge number of FIs discovered. To solve these two disadvantages, recently proposed methods have tried to extract significant and discriminative

FIs only. For example, Cheng et al. [5] developed an approach which first mined FIs and then selected the most discriminative ones based on their information gain. Following the same procedure, discriminative FIs were discovered based on their support difference [8] and support ratio [9]. Although discriminative FIs can help to reduce the feature space and are useful for classification, they require transaction labels, making the mining process *supervised*. Related to transaction classification, FIs have been also used to build rule-based classifiers, often called *associative classification*. These classifiers are constructed from high-confidence and high-support association rules which represent the strong associations between FIs and labels. A testing example is then predicted using one single rule [1] or multiple rules [11].

Our method is also related to embedding methods. Embedding learning has become a hot trend since 2013 when Mikolov introduced Word2Vec [12] to learn embedding vectors for words in text. In recent years, embedding methods have been developed to learn low-dimensional vectors for nodes in network [7], symptoms in healthcare [13], and documents in text [4,10]. As far as we know, learning embedding vectors for transactions has not been studied yet. In this paper, we propose the first method to learn transaction embeddings. Different from supervised FI-based methods and associative classification, our approach is fully *unsupervised* and leverages information of both items and FIs to learn transaction embeddings.

3   Framework

3.1   Problem Definition

We follow the notations in [6]. Given a set of items $\mathcal{I} = \{i_1, i_2, ..., i_M\}$, a *transaction dataset* $\mathcal{D} = \{T_1, T_2, ..., T_N\}$ is a set of transactions where each transaction T_i is a set of distinct items (i.e., $T_i \subseteq \mathcal{I}$).

Our goal is to learn a mapping function $f : \mathcal{D} \to \mathbb{R}^d$ such that every transaction $T_i \in \mathcal{D}$ is mapped to a d-dimensional continuous vector. The mapping needs to capture the similarity among the transactions in $\mathcal{D}$, in the sense that T_i and T_j are similar if $f(T_i)$ and $f(T_j)$ are close to each other on the vector space, and vice versa. The matrix $\mathbf{X} = [f(T_1), f(T_2), ..., f(T_N)]$ then contains feature vectors of transactions, which can be direct inputs for many traditional machine learning and data mining tasks, particularly classification.

3.2   Learning Transaction Embeddings Based on Items

We adapt the Paragraph Vector-Distributed Bag-of-Words (PV-DBOW) model introduced in [10] to learn embedding vectors for transactions, where each transaction is treated as a document and items are treated as words. Given a target transaction T_t whose representation needs to be learned, and a set of items $\mathcal{I}(T_t) = \{i_1, i_2, ..., i_k\}$ contained in T_t, our goal is to maximize the log probability of predicting the items $i_1, i_2, ..., i_k$ which appear in T_t:

$$\max \sum_{j=1}^{k} \log \Pr(i_j \mid T_t) \tag{1}$$

Furthermore, $\Pr(i_j \mid T_t)$ is defined by a softmax function:

$$\Pr(i_j \mid T_t) = \frac{\exp(g(i_j) \cdot f(T_t))}{\sum_{i' \in \mathcal{I}} \exp(g(i') \cdot f(T_t))}, \tag{2}$$

where $g(i_j) \in \mathbb{R}^d$ and $f(T_t) \in \mathbb{R}^d$ are embedding vectors of the item i_j and the transaction T_t respectively, and $\mathcal{I}$ is the set of all singleton items.

Computing the summation $\sum_{i' \in \mathcal{I}} \exp(g(i') \cdot f(T_t))$ in Eq. 2 is very expensive since the number of items in $\mathcal{I}$ is often very large. To solve this problem, we approximate it using the negative sampling technique proposed in Word2Vec [12]. The idea is that instead of iterating over all items in $\mathcal{I}$, we randomly select a relatively small number of items which are not contained in the target transaction T_t (these items are called *negative items*). We then try to distinguish the items contained in T_t from the negative items by minimizing the following binary objective function of logistic regression:

$$\mathcal{O}_1 = -\left[\log \sigma(g(i_j) \cdot f(T_t)) + \sum_{n=1}^{K} \mathbb{E}_{i^n \sim \mathcal{P}(i)} \log \sigma(-g(i^n) \cdot f(T_t))\right], \tag{3}$$

where $\sigma(x) = \frac{1}{1+e^{-x}}$ is a sigmoid function, $\mathcal{P}(i)$ is the negative item collection, i^n is a negative item draw from $\mathcal{P}(i)$ for K times, and $g(i^n) \in \mathbb{R}^d$ is the embedding vector of i^n.

We minimize $\mathcal{O}_1$ in Eq. 3 using stochastic gradient descent (SGD) where the gradients are derived as follows:

$$\frac{\partial \mathcal{O}_1}{\partial g(i^n)} = -\sigma(g(i^n) \cdot f(T_t) - \mathbb{I}_{i_j}[i^n]) \cdot f(T_t)$$

$$\frac{\partial \mathcal{O}_1}{\partial f(T_t)} = -\sum_{n=0}^{K} \sigma(g(i^n) \cdot f(T_t) - \mathbb{I}_{i_j}[i^n]) \cdot g(i^n), \tag{4}$$

where $\mathbb{I}_{i_j}[i^n]$ is an indicator function to indicate whether i^n is an item i_j (i.e., the negative item is contained in the target transaction T_t) and when $n = 0$, then $i^n = i_j$.

3.3   Learning Transaction Embeddings Based on Frequent Itemsets

As discussed in Sect. 1, FIs are more advantageous than singleton items since they can capture more information in transactions. We believe that if we learn transaction embeddings based on FIs instead of items, then the transaction representations learned are more meaningful and discriminative.

Following the notations in [6], we define a frequent itemset as follows. Given a set of items $\mathcal{I} = \{i_1, i_2, ..., i_M\}$ and a transaction dataset $\mathcal{D} = \{T_1, T_2, ..., T_N\}$, an *itemset* X is a set of distinct items (i.e., $X \subseteq \mathcal{I}$). The *support* of X is defined as $\sup(X) = \frac{|\{T_i \in \mathcal{D} | X \subseteq T_i\}|}{|\mathcal{D}|}$, i.e., the fraction of transactions in $\mathcal{D}$, which contain X. Given a minimum support threshold $\delta \in [0,1]$, X is called a *frequent itemset* if $\sup(X) \geq \delta$.

Example 1. Consider an example transaction dataset with five transactions, as shown in Fig. 1(a). Let $\delta = 0.6$. The itemset $\{b, c\}$ (or bc for short) is contained in three transactions T_1, T_2, and T_4; thus, its support is $\sup(bc) = 3/5 = 0.6$. We say that bc is a frequent itemset since $\sup(bc) \geq \delta$. With $\delta = 0.6$, there are in total six FIs discovered from the dataset, as shown in Fig. 1(b), and each transaction now can be represented by a set of FIs, as shown in Fig. 1(c).

Trans	Items
T_1	{a, b, c}
T_2	{b, c, d}
T_3	{a, d}
T_4	{b, c, d, e}
T_5	{a, c, d}

(a)

FI	Items	sup
X_1	{a}	0.6
X_2	{b}	0.6
X_3	{c}	0.8
X_4	{d}	0.8
X_5	{b, c}	0.6
X_6	{c, d}	0.6

(b)

Trans	FIs
T_1	$\{X_1, X_2, X_3, X_5\}$
T_2	$\{X_2, X_3, X_4, X_5, X_6\}$
T_3	$\{X_1, X_4\}$
T_4	$\{X_2, X_3, X_4, X_5, X_6\}$
T_5	$\{X_1, X_3, X_4, X_6\}$

(c)

Fig. 1. Two forms of a transaction: a set of single items and a set of FIs. Table (a) shows a transaction dataset with five transactions where each of them is a set of items. Table (b) shows six FIs discovered from the dataset (here, $\delta = 0.6$). Table (c) shows each transaction represented by a set of FIs.

Following the same procedure in Sect. 3.2, given the set of FIs $\mathcal{F}(T_t) = \{X_1, X_2, ..., X_l\}$ contained in the target transaction T_t, the objective function to learn the embedding vector for T_t based on its FIs is defined as follows:

$$\mathcal{O}_2 = - \left[\log \sigma(h(X_j) \cdot f(T_t)) + \sum_{n=1}^{K} \mathbb{E}_{X^n \sim \mathcal{P}(X)} \log \sigma(-h(X^n) \cdot f(T_t)) \right], \quad (5)$$

where $h(X_j) \in \mathbb{R}^d$ is the embedding vector of the frequent itemset $X_j \in \mathcal{F}(T_t)$, $\mathcal{P}(X)$ is the *negative frequent itemset* collection (i.e., a small set of random FIs which are not contained in T_t), X^n is a negative frequent itemset drawn from $\mathcal{P}(X)$ for K times, and $h(X^n) \in \mathbb{R}^d$ is the embedding vector of X^n. We minimize $\mathcal{O}_2$ in Eq. 5 using SGD.

3.4 Trans2Vec Method for Learning Transaction Embeddings

When learning an embedding vector for a transaction T_t based on its FIs, there is a possible situation that T_t does not contain any FIs. In this case, we cannot learn a useful embedding vector; instead, we simply use a zero vector with the size of d (i.e., $f(T_t) = [0, 0, ..., 0]$). To avoid this problem, we propose two models which combine information of both items and FIs to learn embedding vectors for transactions. These two models named *individual-training* and *joint-training* are presented next.

Individual-Training Model to Learn Transaction Embeddings. The basic idea, as illustrated in Fig. 2, is that given a transaction T_t, we learn an embedding vector $f_1(T_t)$ for T_t based on its items (see Sect. 3.2) and an embedding vector $f_2(T_t)$ for T_t based on its FIs (see Sect. 3.3). We then take the average of two embedding vectors to obtain the final embedding vector $f(T_t) = \frac{f_1(T_t) + f_2(T_t)}{2}$ for that transaction.

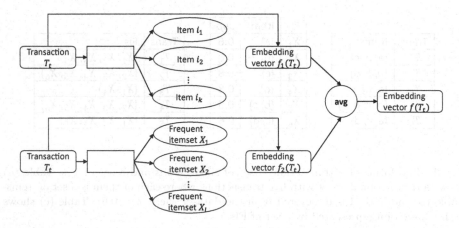

Fig. 2. *Individual-training* model. Given a transaction T_t, we learn the embedding vectors $f_1(T_t)$ and $f_2(T_t)$ based on its items and FIs, respectively. We then take the average of $f_1(T_t)$ and $f_2(T_t)$ to obtain the final embedding vector $f(T_t)$.

Joint-Training Model to Learn Transaction Embeddings. In the *individual-training* model, the relationships between items and FIs are not considered since they are used independently. Consequently, the transaction embeddings only capture the latent relationships between transactions and items and those between transactions and FIs separately. To tackle this weakness, we further propose the *joint-training* model which uses information of both items and FIs of a transaction simultaneously. The overview of this model is shown in Fig. 3. Specifically, given a transaction T_t, our goal is to minimize the following objective function:

$$\mathcal{O} = -\left[\sum_{i_j \in \mathcal{I}(T_t)} \log \Pr(i_j \mid T_t) + \sum_{X_j \in \mathcal{F}(T_t)} \log \Pr(X_j \mid T_t) \right], \qquad (6)$$

where $\mathcal{I}(T_t)$ is the set of singleton items contained in T_t and $\mathcal{F}(T_t)$ is the set of FIs contained in T_t.

Equation 6 can be simplified to:

$$\mathcal{O} = -\sum_{p_j \in \mathcal{I}(T_t) \cup \mathcal{F}(T_t)} \log \Pr(p_j \mid T_t), \tag{7}$$

where $p_j \subseteq T_t$ is an item or a frequent itemset (in general, we call p_j a *pattern*).

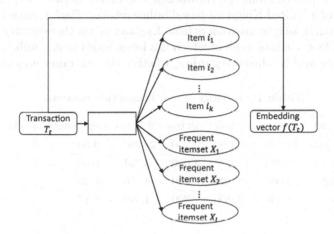

Fig. 3. *Joint-training* model. Given a transaction T_t, we learn the embedding vector $f(T_t)$ for T_t based on both its items and FIs.

Following the same procedure in Sect. 3.2, we minimize the following objective function:

$$\mathcal{O} = -\left[\log \sigma(q(p_j) \cdot f(T_t)) + \sum_{n=1}^{K} \mathbb{E}_{p^n \sim \mathcal{P}(p)} \log \sigma(-q(p^n) \cdot f(T_t)) \right], \tag{8}$$

where $q(p_j) \in \mathbb{R}^d$ is the embedding vector of the pattern $p_j \in \mathcal{I}(T_t) \cup \mathcal{F}(T_t)$, $\mathcal{P}(p)$ is the *negative pattern* collection (i.e., some random patterns which are not contained in T_t), p^n is a negative pattern drawn from $\mathcal{P}(p)$ for K times, and $q(p^n) \in \mathbb{R}^d$ is the embedding vector of p^n.

We minimize Eq. 8 using SGD. After the learning process is completed, the embedding vector $f(T_t)$ is learned for the transaction T_t, and the embedding vectors of two transactions T_i and T_j are close to each other if they have similar items and FIs.

4   Experiments

We conduct extensive experiments on real-world transaction datasets to quantitatively evaluate the performance of **Trans2Vec** in transaction classification.

4.1 Datasets

We use four benchmark datasets whose characteristics are summarized in Table 1. *Snippets* [14] consists of web search transactions where each of them is a set of keywords (e.g., "supplier", "export") and is classified into one of eight categories (e.g., "business"). *Cancer* [13] is a dataset of patient admissions where each admission is a list of diagnosed symptoms (e.g., "cough", "headache") and is labeled regarding the re-admission status of a patient. *Retail* [3] is a transaction dataset which contains the transactions occurring between 01/12/2010 and 09/12/2011 of a United Kingdom-based online retailer. Each transaction is a set of products purchased by customers from England or another country. *Food*[1] is a collection of food baskets, each of which is a list of foods (e.g., "milk") purchased by a customer and is labeled regarding whether the customer uses coupon.

Table 1. Statistics of four transaction datasets.

Dataset	# trans	# train	# test	# items	avg. length	# classes
Snippets	12,340	10,060	2,280	23,686	13.00	8
Cancer	15,000	12,000	3,000	3,234	6.00	3
Retail	3,000	2,400	600	3,376	26.93	2
Food	4,000	3,200	800	1,559	25.87	2

4.2 Baselines

For a comprehensive comparison, we employ six state-of-the-art up-to-date baselines[2] which can be categorized into three main groups:

– **Natural Language Processing (NLP)-based methods:** By treating a transaction as a document and items as words, we can apply methods in NLP to represent transactions. We select two well-known methods, namely Bag-of-Words (BOW) and Term Frequency-Inverse Document Frequency (TF-IDF).
– **FI-based methods:** Given a dataset $\mathcal{D}$ and the set of FIs discovered from $\mathcal{D}$, $\mathcal{F}(\mathcal{D}) = \{X_1, X_2, ..., X_F\}$, we employ two methods to represent transactions based on FIs. Given a transaction T, the first method (named FI-BIN) constructs the feature vector for T as $f(T) = [x_1, x_2, ..., x_F]$ where $x_i = 1$ if $X_i \subseteq T$ otherwise $x_i = 0$ while the second method (named FI-SUP) constructs the feature vector for T as $f(T) = [x_1, x_2, ..., x_F]$ where $x_i = \sup(X_i)$ if $X_i \subseteq T$ otherwise $x_i = 0$.
– **Embedding methods:** We learn embedding vectors for transactions using two simple ways. The first method is based on items (see Sect. 3.2), which we name TRANS-IT. The second method is based on FIs (see Sect. 3.3), which we name TRANS-FI.

---

[1] Available at https://github.com/neo4j-examples/neo4j-foodmart-dataset.
[2] Since our method is unsupervised, we only compare it with unsupervised baselines.

Our proposed method **Trans2Vec** has two different models which use different combinations of items and FIs. We denote **Trans2Vec-IND** for the model which learns transaction embeddings from items and FIs separately and then takes the average (see Sect. 3.4) and denote **Trans2Vec-JOI** for the model which learns transaction embeddings from items and FIs simultaneously (see Sect. 3.4).

4.3 Evaluation Metrics

Once the vector representations of transactions are constructed or learned, we feed them to an SVM with linear kernel [2] to classify the transaction labels. We use the linear-kernel SVM (a simple classifier) and do not tune the parameter C of SVM (here, we fix $C = 1$) since our focus is on the transaction embedding learning, not on a classifier. Each dataset is randomly shuffled and split into the training and test sets as shown in Table 1. All methods are applied to the same training and test sets. We repeat the classification process on each dataset 10 times and report the average classification accuracy and the average F1-macro score. We do not report the standard deviation since all methods are very stable (their standard deviations are less than 10^{-2}).

4.4 Parameter Settings

Our method **Trans2Vec** has two important parameters: the minimum support threshold δ for extracting FIs and the embedding dimension d for learning transaction embeddings. Since we develop **Trans2Vec** in a fully unsupervised learning fashion, the values for δ and d are assigned without using transaction labels. We set $d = 128$ (a common value used in embedding methods [7]) and set δ following the *elbow method* in [15]. Figure 4 illustrates the elbow method. From the figure, we can see when the δ value decreases, the number of FIs slightly increases until a δ value where it significantly increases. This δ value, highlighted in red in the figure and chosen by the elbow method without considering the transaction

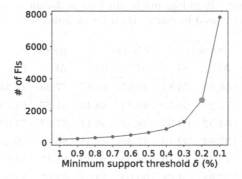

Fig. 4. The number of FIs discovered from the training set of *Snippets* dataset per δ. The δ value selected via the elbow method is indicated by the red dot. (Color figure online)

labels, is used in our experiments. In Sect. 4.6, we analyze the potential impact of selecting two parameters δ and d on the classification performance.

For a fair comparison, we use the same δ for **Trans2Vec** and three baselines FI-BIN, FI-SUP, and TRANS-FI. We also set $d = 128$ for two baselines TRANS-IT and TRANS-FI.

4.5   Results and Discussion

From Table 2, we can see two models in our method **Trans2Vec** clearly results in better classification on all datasets compared with other baselines. Compared with NLP-based methods, **Trans2Vec-JOI** achieves 4–19% and 2–13% improvements in accuracy over BOW and TF-IDF, respectively. Similar improvements can be also observed when comparing with FI-based methods. On three datasets *Snippets*, *Retail*, and *Food*, **Trans2Vec-JOI** outperforms FI-BIN and FI-SUP by large margins (achieving 7–23% and 2–108% gains over FI-BIN and FI-SUP).

For most cases, embedding baselines (TRANS-IT and TRANS-FI) are better than NLP- and FI-based methods. Moreover, TRANS-FI always outperforms TRANS-IT. This demonstrates that learning transaction embeddings from FIs is more effective than learning transaction embeddings from items, as discussed in Sect. 3.3. Two our models (**Trans2Vec-IND** and **Trans2Vec-JOI**) are always superior than two embedding baselines. This proves that our proposal to incorporate information of both singleton items and FIs into the transaction embedding learning is a better strategy than learning transaction embeddings from items or FIs only.

We also observe **Trans2Vec-JOI** produces better results than **Trans2Vec-IND** on all datasets. This verifies our intuition in Sect. 3.4 that the transaction embeddings learned from items and FIs simultaneously are more meaningful and discriminative since they can capture different latent relationships of transactions simultaneously.

Table 2. Accuracy (AC) and F1-macro (F1) of our **Trans2Vec** and six baselines on four transaction datasets. Bold font marks the best performance in a column. The last row denotes the δ values used by our method for each dataset.

Method	Snippets		Cancer		Retail		Food	
	AC	F1	AC	F1	AC	F1	AC	F1
BOW	66.32	65.83	48.57	48.57	77.33	77.31	63.12	63.12
TF-IDF	70.26	69.52	49.43	49.43	81.67	81.56	64.50	64.49
FI-BIN	64.52	63.96	48.52	48.44	77.67	77.62	61.75	61.74
FI-SUP	37.94	32.19	47.30	46.54	80.33	79.20	71.00	70.03
TRANS-IT	75.23	74.79	49.35	49.32	81.17	81.07	65.95	65.81
TRANS-FI	77.88	77.41	50.03	49.92	82.07	81.90	69.14	68.95
Trans2Vec-IND	78.80	77.92	50.10	50.02	82.23	82.12	69.22	69.12
Trans2Vec-JOI	**79.05**	**78.31**	**50.34**	**50.28**	**83.43**	**83.36**	**72.51**	**72.47**
δ (%)	0.2%		0.2%		0.7%		0.2%	

4.6 Parameter Sensitivity

We examine how the different choices of two parameters δ and d affect the classification performance of **Trans2Vec-JOI** on three datasets *Snippets*, *Cancer*, and *Food*. Figure 5 shows the classification results as a function of one chosen parameter when another is set to its default value. From Fig. 5(a), we can see the values for δ selected by the elbow method always lead to the best accuracy. This demonstrates that the elbow method is an effective way to choose δ for methods which use frequent patterns, the same finding was also mentioned in [15]. Another observation is that on *Cancer*, δ is gain of relatively little relevant to the predictive task where our classification performance just slightly changes with different values for δ.

From Fig. 5(b), we observe a first-increasing and then-decreasing accuracy line on two datasets *Snippets* and *Food* when d is increased whereas the classification performance shows an increasing trend with an increasing d on *Cancer*. This finding differs from those in document embedding methods, where the embedding dimension mostly shows a positive effect on document classification [4].

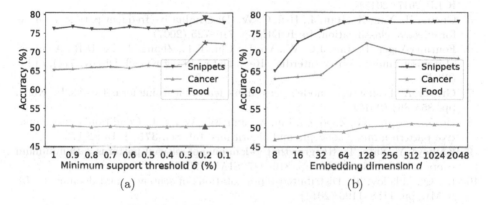

(a) (b)

Fig. 5. Parameter sensitivity in transaction classification on the *Snippets*, *Cancer*, and *Food* datasets. The minimum support δ values selected via the elbow method and used in our experiments are indicated by red markers. (Color figure online)

5 Conclusion

We have presented **Trans2Vec**, an unsupervised method for learning transaction embeddings from information of both singleton items and FIs. Our comprehensive experiments on four transaction datasets demonstrated the meaningful and discriminative representations learned by our method in the transaction classification task. In particularly, **Trans2Vec** significantly outperforms several state-of-the-art baselines in both accuracy and F1-macro scores. One of our future work is to investigate the quality of our embeddings in the transaction clustering task. Another possible extension is to utilize other information of items, e.g., quantity or weight, when learning transaction embeddings.

Acknowledgment. This work is partially supported by the Telstra-Deakin Centre of Excellence in Big Data and Machine Learning. Tu Dinh Nguyen gratefully acknowledges the partial support from the Australian Research Council (ARC).

References

1. Liu, B., Hsu, W., Ma, Y.: Integrating classification and association rule mining. In: KDD, pp. 80–86 (1998)
2. Chang, C.-C., Lin, C.-J.: LIBSVM: a library for support vector machines. ACM Trans. Intell. Syst. Technol. **2**(3), 1–27 (2011)
3. Chen, D., Sain, S.L., Guo, K.: Data mining for the online retail industry: a case study of RFM model-based customer segmentation using data mining. J. Database Market. Customer Strategy Manag. **19**(3), 197–208 (2012)
4. Chen, M.: Efficient vector representation for documents through corruption. In: ICLR 2017 (2017)
5. Cheng, H., Yan, X., Han, J., Hsu, C.-W.: Discriminative frequent pattern analysis for effective classification. In: ICDE, pp. 716–725 (2007)
6. Fournier-Viger, P., Lin, J.C.-W., Vo, B., Chi, T.T., Zhang, J., Le, H.B.: A survey of itemset mining. Wiley Interdisc. Rev.: Data Mining Knowl. Discov. **7**(4), e1207 (2017)
7. Grover, A., Leskovec, J.: node2vec: scalable feature learning for networks. In: KDD, pp. 855–864 (2016)
8. He, Z., Feiyang, G., Zhao, C., Liu, X., Jun, W., Wang, J.: Conditional discriminative pattern mining: concepts and algorithms. Inf. Sci. **375**, 1–15 (2017)
9. Kameya, Y., Sato, T.: RP-growth: top-k mining of relevant patterns with minimum support raising. In: SDM, pp. 816–827. SIAM (2012)
10. Le, Q., Mikolov, T.: Distributed representations of sentences and documents. In: ICML, pp. 1188–1196 (2014)
11. Li, W., Han, J., Pei, J.: CMAR: accurate and efficient classification based on multiple class-association rules. In: ICDM, pp. 369–376. IEEE (2001)
12. Mikolov, T., Sutskever, I., Chen, K., Corrado, G.S., Dean, J.: Distributed representations of words and phrases and their compositionality. In: NIPS, pp. 3111–3119 (2013)
13. Nguyen, D., Luo, W., Phung, D., Venkatesh, S.: Control matching via discharge code sequences. In: NIPS 2016 Workshop on Machine Learning for Health (2016)
14. Phan, X.-H., Nguyen, L.-M., Horiguchi, S.: Learning to classify short and sparse text & web with hidden topics from large-scale data collections. In: WWW, pp. 91–100 (2008)
15. Rousseau, F., Kiagias, E., Vazirgiannis, M.: Text categorization as a graph classification problem. In: ACL, pp. 1702–1712 (2015)

Detecting Complex Sensitive Information via Phrase Structure in Recursive Neural Networks

Jan Neerbek[1,2(✉)], Ira Assent[1], and Peter Dolog[3]

[1] Department of Computer Science, Aarhus University, Aarhus, Denmark
{jan.neerbek,ira}@cs.au.dk
[2] Alexandra Institute, Aarhus, Denmark
[3] Department of Computer Science, Aalborg University, Aalborg, Denmark
dolog@cs.aau.dk

Abstract. State-of-the-art sensitive information detection in unstructured data relies on the frequency of co-occurrence of keywords with sensitive seed words. In practice, however, this may fail to detect more complex patterns of sensitive information. In this work, we propose learning phrase structures that separate sensitive from non-sensitive documents in recursive neural networks. Our evaluation on real data with human labeled sensitive content shows that our new approach outperforms existing keyword based strategies.

Keywords: Sensitive information · Recursive neural networks
Data leak prevention · Natural text understanding

1 Introduction

Detecting sensitive information in unstructured data is crucial for data leak prevention. State-of-the-art approaches are based on *defining keywords* [1,2,5,6,10], i.e., assume that the sensitive topic is described in full by a small set of keywords. While effective for simple sensitive topics as in named entity recognition [2] and personal identifiable information [10], they ignore context, i.e., the way in which people describe sensitive topics in natural language phrases. As a result, they may fail to report *complex sensitive information* or report false positives.

Concretely, complex sensitive information is characterized by the fact that words are sensitive or not sensitive depending on their context. For example, describing sensitive financial transactions might use the same vocabulary as in the non-sensitive case, but using different expressions in natural language.

In this work, we therefore propose to extract the phrase structure from sensitive information to learn these expressions, and create a recursive neural network model (RNN) [7,11] that uses phrases to predict the sensitivity of documents. We suggest a training approach for the RNN that requires only document level sensitivity information and thereby does not require labeling individual sentences, phrases, or even words. Such fine-grained labels are required for existing

© Springer International Publishing AG, part of Springer Nature 2018
D. Phung et al. (Eds.): PAKDD 2018, LNAI 10939, pp. 373–385, 2018.
https://doi.org/10.1007/978-3-319-93040-4_30

backpropagation-through-structure training, but are generally not available in practice. Our evaluation on real sensitive content with humanly curated labels demonstrates superior detection accuracy compared to state-of-the-art keyword-based approaches. We furthermore show that by boosting relative importance of incorrectly predicted samples we can increase sensitive detection accuracy - in the extreme at the expense of an increase in false prediction rate. This adds flexibility to our model and allows for domain-based adjustment between prediction accuracy and end-user confidence in detected samples.

Our contributions include:

- Introducing and analyzing complex sensitive information detection
- A new RNN model based on representations of multiword structured phrases
- Training of our RNN model on document labels alone.

2   Complex Sensitive Information Detection

We assume a corpus of documents $D = \{d_1, d_2, \ldots, d_m\}$, where each document is a sequence of words $d = (w_1, w_2, \ldots, w_{n_d})$ such that $w_i \in V$ and $d \in D$, and training labels $L : D \to \{0, 1\}$, where 0 means non-sensitive, and 1 means there is (some) sensitive information.

Note that the problem is asymmetric in that non-sensitive documents are known to be completely non-sensitive, whereas sensitive documents may only contain very little sensitive information. Thus, the problem is recall-oriented, focusing on finding the pieces of sensitive information [1].

In existing work, each word w is assigned a sensitivity score $sen(w)$, without considering its context of use. In this paper, however, we differentiate based on how the word is used, i.e., the sensitivity of a word is conditional on its context d, the sequence of words in which it occurs: $sen(w|d)$.

Definition 1. *If for all words w and pairs of documents, we have d, d' $sen(w|d) = sen(w|d')$, then sensitivity is* context-less.

Conversely, if there exists a word w' and a pair of documents d, d' such that $sen(w'|d) \neq sen(w'|d')$, then sensitivity is context-based.

Please note that the definition reflects the asymmetry in sensitive information detection as discussed above.

3   SPR - Sensitive Phrase Based RNN Model

Existing sensitive information detection approaches count co-occurrence of a keyword, or small set of keywords, with other words in the text. Co-occurrence is then taken as an indication of sensitivity. This works well e.g. for topics like HIV where co-occurrence with terms such as AIDS is easily detected.

However, as we argue here, more complex topics, as in intricate financial transactions, require models that can capture context. Sentences can have arbitrary length and structure, which our model should be able to process. We therefore encode the context of *phrase structures*, which are semantic substructures in the text extracted through *constituency parse-trees*. Constituency parse-trees structure a text into constituents, i.e., compositionality through sub-phrases [14]. Creating phrase structure embeddings in a recursive manner, we generate an encoding of the entire context. Varying sizes of context are encoded by iterating in a large structure. Our SPR (Sensitive Phrase RNN) model thereby captures the complexity in sensitive information detection in natural language.

3.1   Phrase Structure

While sentences obviously can be interpreted as sequences, this does not reflect the way in which humans understand them. Consider the sentence

`We may have to move to cash margining if necessary.`

It begins with a sequence of common words, "We may have to move to", that could appear in many different contexts. The particular context for this sentence becomes clearer from the words "cash margining". In sequence order, the state after processing "We may have to move to" would be a general state, whereas given "cash margining" first, we expect a more specific hidden state. The latter order also reflects the grammatical structure in natural language.

We therefore propose to learn from sentences following their grammatical structure instead. In Natural Language Processing, this structure is captured in constituency parse-trees. Their leaves correspond to words, and nodes to sub-phrases, as illustrated in Fig. 1. A depth-first descent from the root node visits the words of the sentence in order. Constituency parse-trees can be automatically generated in high quality for most languages [13]. Using phrase structures as features we can successfully capture the context of words in the way they are used in natural language. On the other hand, however, phrase trees provide features which can be of arbitrary size and structure. We therefore propose to build a model that can handle this variable input by recursively taking in parts of the input. Concretely, we build a recursive neural network structure.

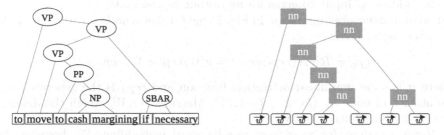

Fig. 1. Constituency parse-tree (left). Node labels are syntactic tags, e.g., *NP* Noun Phrase, *VP* Verb Phrase; cf. Penn Treebank [15]; note *SPR* relies on phrase structure alone, not the particular label; corresponding RNN (right).

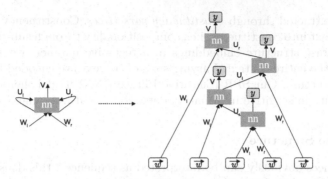

Fig. 2. RNN; on the right unfolded in structure. A node has 4 possible inputs of which 2 are active at a time. For readability, not all output edges are labeled V

3.2   Recursive Neural Networks with Phrase Structure

Our approach is to create embeddings of the complex structure in phrase trees using neural networks. To handle input of arbitrary size we propose building a *recursive neural network (RNN)* [12]. In a nutshell, a RNN recursively takes in a new part of the input from the input structure. It includes output from the previous step in the current step by applying the same architecture repeatedly. In this manner, recursive neural networks are capable of transferring knowledge between the steps, and of taking in complex input structures.

Given a constituency parse-tree we process sentences in grammatical order using the RNN recursively in a bottom-up fashion, where in each step we process a new node in the parse-tree, ending at the root node. Processing phrase trees through RNNs, we automatically learn what the relevant structures in language are for the complex sensitive information detection problem. We do not need to define the size or structure of the context a-priori, which makes this a flexible and easily applicable model.

As illustrated in Fig. 2 (left), the recursive neural network is a function that for each node in the constituency parse-tree is evaluated with the representations of its children as input to generate its output representation (as shown in the unfolded in structure illustration in Fig. 2 (right)). For a node n with child nodes n_ℓ (left) and n_r (right)

$$rep_n = RNN(rep_\ell; rep_r) = \sigma[W_\ell rep_\ell + W_r rep_r + b] \qquad (1)$$

where σ is some non-linear activation function and rep_x is the generated representation of node n_x (for $x \in \{n, \ell, r\}$). Matrices W_ℓ, W_r weigh the different inputs, b is an additive bias vector.

Representations for words are usually word embeddings [7], however, (1) assumes that word representations (leaves) have same number of dimensions as structure representations (nodes). Thus, it might be beneficial to consider words and structure as two different learning problems. Fortunately, this can be done with a minor adjustment to the equation above. We define a transformation function t for child nodes n_x as

$$t(rep_x) = \begin{cases} U_x rep_x + b_U & n_x \text{ is node} \\ W_x rep_x + b_W & n_x \text{ is leaf} \end{cases} \tag{2}$$

with $n_x \in \{n_\ell, n_r\}$. We can then rewrite (1) as

$$rep_n = RNN(rep_\ell; rep_r) = \sigma[(t(rep_\ell) + t(rep_r)]$$

The decoupling in (2) gives us freedom to select optimal word embedding size and hidden representation size [7].

The output of a node in the tree is

$$o_n = \sigma(V rep_n + b_p),$$

a prediction (sensitive/non-sensitive) based on the current node representation. Thus for internal nodes in the constituency parse-tree, we obtain our prediction having only seen part of the structure so far.

3.3 Training SPR

For training, we use the *unfolded* network, i.e., copies of the RNN for each node in the constituency parse-tree as illustrated to the right in Fig. 2. We may backpropagate errors through the tree in a top-down order and aggregate the errors across all copies of the neural network. This can be viewed as a *backpropagation-through-structure (BPTS)* [4] approach.

Our SPR makes use of the well established *softmax* activation function $\sigma(x)[i] = e^{x[i]}/(\sum_j e^{x[j]})$ which provides a differentiable *soft* max of the output. We expect only one answer to be true and therefore maximize (softly) the best probability as the output. Using the softmax function allows us to interpret the outcome as probabilities.

The prediction y_n of SPR is given as $y_n = \text{argmax}_{x \in \{0,1\}} o_n[x]$ where $o_n[x]$ can be seen as probability that the input is x, with $x \in \{0, 1\}$.

Different sensitive domains require different focus on the model's ability to correctly predict sensitive vs. non-sensitive information. We model this using relative weighing of the two types of information in the loss function (Different weights are studied in the experiments). Adding everything together and using cross-entropy for error function, we obtain our loss function

$$L = -\sum_{n=1}^{N} [wt_n \log(p_n) + (1 - t_n) \log(1 - p_n)] \tag{3}$$

where w is the weighting hyper-parameter. If $w > 1$ then we weigh sensitive loss higher than non-sensitive loss.

Using the ground truth labels and BPTS we can learn our parameters θ for maximal likelihood given our data. However, supervised BPTS as defined in [4], requires a label for each node in the tree, which is not available and would be difficult to obtain, as this would require assigning sensitivity scores to phrases of increasing complexity. Still we can solve this challenge by propagating labels from the root to the internal nodes of the tree. We can then view SPR as learning to assign probabilities to each node wrt. being sensitive or not. That is, if we have the same sentence s occurring in, say, 3 different documents with label assignment $l_1 = 0$, $l_2 = 1$, $l_3 = 0$, SPR will minimize the loss function by assigning probability $SPR(s) = 1/3$. A similar argument can be made for internal nodes in the constituency parse-trees. Thus sentences n that contain actual sensitive information have MLE of $SPR(n) = 1$ because they never occur in a non-sensitive document. On the other hand, if there are sentences that never occur in a sensitive document, their nodes have MLE of 0. We find that this approach tends to provide a self-regularizing effect and we found no improvement of the accuracy of the models when adding further L1 or L2 regularizing terms to the loss function, i.e. (3).

4   Evaluation

4.1   Evaluation Methodology and Data

Complex sensitive information detection is not only a challenging task, but also challenging to evaluate. Existing work has created evaluation ground truth for sensitivity using three main strategies. The first strategy uses word co-occurrence with seed words to semi-automatically label sensitive information as ground truth [2,5]. This ground truth uses the same assumption as keyword-based detection that sensitive information co-occurs with other sensitive terms, but does not reveal performance on more complex sensitive information. Another strategy use sources of actual sensitive information, such as WikiLeaks, and insensitive information from other sources. The major disadvantage of this strategy is that insensitive and sensitive information exhibit major differences in structure and content [6]. Thus, evaluation may actually measure how well the differences in structure and content are learned and not necessarily how well sensitive content is detected. The final strategy uses human labeled data. However, the sensitive information in these approaches is typically simple like named entities (e.g. names of cities) [1], or in general Personal Identifiable Information detection, such as person names, sicknesses [10].

In this work we propose evaluation on the Enron dataset [8], an actual private dataset which contains both sensitive and non-sensitive data, and which is labeled by human experts. Our dataset contains examples of complex sensitive information that cannot be characterized by a few keywords. It shows the varied structure as seen in internal corporate communications and there are different sensitive issues.

While this dataset has been used before for sensitive information detection, e.g. [2], there were no complex sensitive information labels in the ground truth. We here propose to exploit the human expert labels given to the Enron corpus as part of a competition of the 2010 TREC conference, legal track [3,16]. Specifically, we study the case of *prepay transactions*.

The Enron dataset has $1.2M+$ documents [8], and there are 2720 documents labeled by human experts.

For each sentence in any document, we obtain constituency parse-trees, 11390 in total, which we split into training, validation and test sets, see Table 1. All word vectors come from the Stanford Glove word vector set [9]. To evaluate performance, we study accuracy of detection of complex sensitive information, with a particular focus on finding sensitive data. Formally, if ground truth label $l_s = 1$ on sentence s and prediction of our model $SPR(s) = 1$ then s is *true positive*, if $l_s = 0$ and $SPR(s) = 1$ it is *false positive*. Similarly for negatives. $C_{tp}, C_{fp}, C_{tn}, C_{fn}$ denote the counts of true positives, false positives, true negatives and false negatives respectively. Accuracy of a data set D is then $acc_D = \frac{C_{tp} + C_{tn}}{|D|}$.

Table 1. Overview over our labeled extract of the ENRON/TREC data: number of sentences and constituency parse-trees for each split of the dataset.

Set	Sensitive	Non-sen	Total	Non-sen/total
Train	2985	6015	9000	0.6683
Validation	462	968	1430	0.6769
Test	322	638	960	0.6646
Total	3769	7621	11390	0.6691

4.2 Performance Evaluation for Complex Sensitive Information

We begin by studying the impact of input dimension of the word embeddings, varying from 10 to 300. The hidden state internally is fixed at size 200. The results are shown in Table 2, and as we can see, there is a clear increase in accuracy as the dimension of the word embeddings is increased, but this trend diminishes. Going from 100 to 200 the increase in accuracy score is 0.028, whereas going from 200 to 300 the increase is negligible at only tiny 0.007.

Next, we investigate the impact of the size of the hidden state, here we fixed word embedding size to 300. Similarly, to input word embedding size which positively affects performance up to a certain point, we observe similar performance increase when increasing internal hidden representation size, as shown Table 3. There is a drop between using 200 neurons with another increase for 300 neurons, after which the performance decreases again at 500 neurons. This suggests that we reach a level which provides a good size for the internal hidden representation after which we most likely see effects of worse performance due to overfitting as the model complexity exceeds what is useful for a well generalizing model. We also evaluate the use of several layers at each node (Table 4). In all experiments the sum of all neurons in the models is fixed to allow for the same potential expressive power, and the number of layers varies between 1 and 3. Here we find that the best hidden representation is a single, wide layer rather than stacked thinner layers. In [7], the best performance for sentiment analysis is observed for

Table 2. *SPR*-input; performance for varying size of input word embeddings

Vector size	Acc_{val}
10	0.7224
50	0.7531
100	0.7580
200	0.7608
300	**0.7615**

Table 3. *SPR*-hidden; varying the size of the hidden state

Hidden state	Acc_{val}
10	0.7406
50	0.7497
100	0.7650
200	0.7615
300	**0.7678**
500	0.7643

a stacked architecture with 3 layers. While a throughout examination of these differences falls outside the scope of this work, we make the following observation; Sentiment analysis has a more *Boolean* behavior than sensitive information detection. In sentiment a single "not" in the sentence can flip the expected label as in "good" vs. "not good". This Boolean property means that the model can make localized decisions which suggest thin and high models, whereas complex sensitive information can be viewed as a measure on the complete context. I.e., sensitive information is a property of what is communicated and is not as easily flipped by a single word, which then suggests that a wide layered model will perform better.

Table 4. Comparison of *SPR*-layer models as the number of layers is varied, keeping the total number of neurons fixed

Hidden state	Layers	Acc_{val}
300	**1**	**0.7678**
150	2	0.7554
100	3	0.7594

We compare our final model with state-of-the-art in the field of sensitive information detection both in terms of overall accuracy, and in terms of the performance when focusing mostly on the sensitive information, as opposed to correctly predicting non-sensitive documents. As discussed in more detail in Related Work in Sect. 5, state-of-the-art is based on word counting, i.e., word co-occurrences (*n*-gram), inference rules and mutual information models are all keyword-based. In Table 5, Assoc Rules denotes [2] with default parameters for "Email" corpus, C-sanitized [10] with α values as reported in their experiments. Keyword-Based denotes a generic keyword based approach, given optimal keyword set and is an upper bound for Assoc Rules and C-sanitized. *SPR*-input: best model of experiments in Table 2, *SPR*-hidden: best model of experiments in Table 3, *SPR*-layer: best model of experiments in Table 4.

As we can see in Table 5, all approaches improve upon the base line accuracy value of 0.6646. Our models, as studied in the experiments above, outperform

Table 5. Overall accuracy comparison; Assoc Rules [2], C-sanitized [10], Keyword-Based generic keyword based, given optimal keyword set, SPR-input best of Table 2, SPR-hidden best of Table 3, SPR-layer best of Table 4

Approach	Parameter	Acc
Baseline		0.6646
Assoc Rules	supp $= 2$, conf $= 0.6$	0.7104
C-sanitized	$p_{sen} = 0.3354$, $\alpha = 2.0$	0.6479
C-sanitized	$p_{sen} = 0.3354$, $\alpha = 1.5$	0.6479
C-sanitized	$p_{sen} = 0.3354$, $\alpha = 1.0$	0.7240
Keyword-based		0.7476
SPR-input	300	0.7615
SPR-hidden	300	**0.7678**
SPR-layer	1	**0.7678**

this baseline, and also existing keyword based approaches with respect to overall accuracy. However, as discussed before, in sensitive information detection, we are typically much more interested in successfully identifying sensitive information than we are in correctly predicting non-sensitive information. We therefore conduct an experiment that investigates this case in depth. Overall, SPR successfully identifies complex sensitive information and outperforms state-of-the-art particularly when focusing on documents containing sensitive information as opposed to identifying non-sensitive information. SPR shows value in capturing the phrase structure used to describe complex sensitive information that might go unnoticed when relying on keywords alone.

Table 6. Weighing sensitive examples in loss function; accuracy for sensitive information; $F1$ measure

Approach	Acc_{Sen}	$F1$
Baseline	0	N/A
Keyword-based	0.2795	0.2004
$SPR_{w=1}$	0.3540	0.2360
$SPR_{w=2}$	0.3540	0.2400
$SPR_{w=3}$	0.7236	**0.2572**
$SPR_{w=4}$	**0.9224**	0.2536

In Table 6 we weigh errors on false negatives (unidentified sensitive examples) higher in the loss function. We report class-based accuracy Acc_{Sen} and also the $F1$ score which weights false negatives and false positives in a single score according to $F1 = \frac{C_{tp}}{C_{tp}+C_{fn}+C_{fp}}$. We obtain close to 100% accuracy on sensitive example detection by increasing this weight, but for weights greater than 3 the $F1$ score starts decreasing, reflecting that the number of false positives now is so high that overall performance across both sensitive and non-sensitive information degrades unreasonably. We observe that all of the SPR models in Table 6 have higher performance than the best result we can obtain using previous state-of-the-art algorithms. The weight parameter allows our SPR model to be adjusted to the domain in a natural manner; for some domains false negatives may be associated with high cost in which case a corresponding high weight can be used to ensure a relatively low number of false negatives. In other domains, on the other hand, too many false positives may be undesirable and thus here it would intuitively make sense to have a lower weight for false negatives.

4.3   Qualitative Analysis

For our qualitative analysis (see Table 7) we investigate the semantics that SPR learns, i.e., how close selected groups of phrases are in terms of Euclidean distance in SPR embedding space. Table 7 lists single words such as "May-02" as being close to similar other dates. Please note that we did not do any preprocessing for dates and "May-02" is not in our input vocabulary. We can see that all our phrases in Fig. 3 seem far away from unknown input words, which suggests that our model learns semantic meaning for all phrases. The SPR model extracts these semantics automatically, as exemplified also for *Names* and *Goodbyes*. We observe that the space in Fig. 3 also contains structure where dates seem almost localized to a distinct point, where-as *Prepay* sentences form a curly line-like structure. The final listing in Table 7, which we termed *Oil & Gas* due to its apparent semantics, is from a region of the space completely devoid of sensitive

Table 7. SPR semantics in related single words (top) and sentences (bottom)

Label	Close SPR embeddings
Dates	May-02, Feb-02, Oct-03, Apr-01, Jun-99
Names	Stacy Dickson, Martha Braddy, James Westgate, GEA Rainey, Citibank ISDA
Good-byes	"Yours sincerely, EnronEntityName"
	"Signature of Company Officer"
Prepay	"However, he did not attempt to calculate a VaR statistic for the daily cash requirements for the exchange traded positions"
	"Li identified several of these and they are given on the flowchart (gas settlements, merchant assets, etc.)"
Oil & Gas	"The oil flows through the orifice and into the bearings and forms a film that cools and lubricates the journal"
	"In accordance with NFPA, the fire and gas detection controls will be powered by a dedicated 24V DC battery system"

phrases. *Prepay* on the other hand, has 99.2% sensitive sentences, and is thereby a strong indicator of how sensitive information is captured by *SPR*.

Our qualitative analysis shows that the *SPR* model indeed learns semantic meaning from the documents. This suggests that the successful identification of sensitive information in *SPR* is based on its ability to identify distinct compositional semantic information.

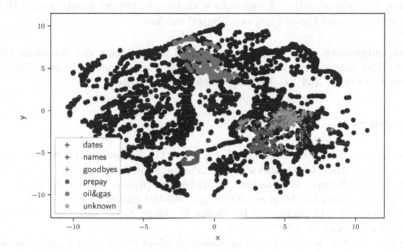

Fig. 3. 2-dimensional view of *SPR* phrase model with colors highlighting 200 closest phrases of label groups listed in Table 7

5   Related Work

Sensitive information detection in [2] uses a seed set of sensitive words, and creates inference rules based on word co-occurrences. Given a word, confidence is defined as the probability of the document containing sensitive information if it contains the word. Text containing highly sensitive words is considered sensitive. [5] applies the inference rule approach to sensitive information in software code. An ontology such as WordNet provides synonyms for actually redacting the sensitive information. In [6], word-to-word (bi-grams) co-occurrence is used to infer sensitive words. The focus is on a small number of false positives, using in particular the *false-discovery-rate (FDR)*, i.e., the inverse of precision. [1] presents a semi-automatic approach based on *total utility*, which measures the model's probability that the word is sensitive together with the *gain* associated with learning the word's true label. [10] uses *pointwise mutual information*, based on co-occurrence of bi-grams, where one is a sensitive seed word, and a threshold on *information content* which puts the occurrence of sensitive words in relation to the size of the corpus.

All approaches above use keyword definitions for sensitive content and word pair co-occurrence counting. Our proposed *SPR* model additionally learns relevant structure extracted from natural language to characterize sensitivity.

6 Conclusion

We introduce the *complex sensitive information* detection problem where context and structure in language has to be taken into account. Our *SPR* model extracts phrase structure to learn a recursive neural network model. Our experimental evaluation, which is the first to use a real document corpus containing both sensitive and non-sensitive documents with human expert labels, shows that we outperform state-of-the-art keyword based models.

Acknowledgments. This project has received funding from the European Union's Horizon 2020 research and innovation programme under grant agreement No. 645198 (Organicity Project) and No. 732240 (Synchronicity Project).

References

1. Berardi, G., Esuli, A., Macdonald, C., Ounis, I., Sebastiani, F.: Semi-automated text classification for sensitivity identification. In: CIKM, pp. 1711–1714 (2015)
2. Chow, R., Philippe, G., Staddon, J.: Detecting privacy leaks using corpus-based association rules. In: ACM SIGKDD, pp. 893–901 (2008)
3. Cormack, G.V., Grossman, M.R., Hedin, B., Oard, D.W.: Overview of the TREC 2010 legal track. In: TREC (2010)
4. Goller, C., Kuchler, A.: Learning task-dependent distributed representations by backpropagation through structure. In: IEEE ICNN, pp. 347–352 (1996)
5. Grechanik, M., McMillan, C., Dasgupta, T., Poshyvanyk, D., Gethers, M.: Redacting sensitive information in software artifacts. In: ICPC, pp. 314–325 (2014)
6. Hart, M., Manadhata, P., Johnson, R.: Text classification for data loss prevention. In: Fischer-Hübner, S., Hopper, N. (eds.) PETS 2011. LNCS, vol. 6794, pp. 18–37. Springer, Heidelberg (2011). https://doi.org/10.1007/978-3-642-22263-4_2
7. Irsoy, O., Cardie, C.: Deep recursive neural networks for compositionality in language. In: NIPS, pp. 2096–2104 (2014)
8. Klimt, B., Yang, Y.: The enron corpus: a new dataset for email classification research. In: Boulicaut, J.-F., Esposito, F., Giannotti, F., Pedreschi, D. (eds.) ECML 2004. LNCS (LNAI), vol. 3201, pp. 217–226. Springer, Heidelberg (2004). https://doi.org/10.1007/978-3-540-30115-8_22
9. Pennington, J., Socher, R., Manning, C.D.: Glove: global vectors for word representation. In: EMNLP, pp. 1532–1543 (2014)
10. Sánchez, D., Batet, M.: C-sanitized: a privacy model for document redaction and sanitization. JASIST **67**, 148–163 (2016)
11. Socher, R., Huang, E.H., Pennin, J., Manning, C.D., Ng, A.Y.: Dynamic pooling and unfolding recursive autoencoders for paraphrase detection. In: NIPS (2011)
12. Socher, R., Lin, C.C., Manning, C., Ng, A.Y.: Parsing natural scenes and natural language with recursive neural networks. In: ICML, pp. 129–136 (2011)
13. Socher, R., Manning, C.D., Ng, A.Y.: Learning continuous phrase representations and syntactic parsing with recursive neural networks. In: NIPS WS Deep Learning and Unsupervised Feature Learning, pp. 1–9 (2010)

14. Socher, R., Perelygin, A., Wu, J.Y., Chuang, J., Manning, C.D., Ng, A.Y., Potts, C.: Recursive deep models for semantic compositionality over a sentiment treebank. In: EMNLP, pp. 1631–1642 (2013)
15. Taylor, A., Marcus, M., Santorini, B.: The Penn treebank: an overview. In: Abeillé, A. (ed.) Treebanks. Text, Speech and Language Technology, vol. 20. Springer, Dordrecht (2003). https://doi.org/10.1007/978-94-010-0201-1_1
16. Tomlinson, S.: Learning task experiments in the TREC 2010 legal track. In: TREC (2010)

Clustering and Unsupervised Learning

A Distance Scaling Method to Improve Density-Based Clustering

Ye Zhu[1]($\boxtimes$), Kai Ming Ting[2], and Maia Angelova[1]

[1] School of Information Technology, Deakin University,
Burwood, VIC 3125, Australia
ye.zhu@ieee.org, maia.a@deakin.edu.au
[2] Faculty of Science and Technology, Federation University,
Churchill, VIC 3842, Australia
kaiming.ting@federation.edu.au

Abstract. Density-based clustering is able to find clusters of arbitrary sizes and shapes while effectively separating noise. Despite its advantage over other types of clustering, it is well-known that most density-based algorithms face the same challenge of finding clusters with varied densities. Recently, ReScale, a principled density-ratio preprocessing technique, enables a density-based clustering algorithm to identify clusters with varied densities. However, because the technique is based on one-dimensional scaling, it does not do well in datasets which require multi-dimensional scaling. In this paper, we propose a multi-dimensional scaling method, named `DScale`, which rescales based on the computed distance. It overcomes the key weakness of ReScale and requires one less parameter while maintaining the simplicity of the implementation. Our empirical evaluation shows that `DScale` has better clustering performance than ReScale for three existing density-based algorithms, i.e., DBSCAN, OPTICS and DP, on synthetic and real-world datasets.

Keywords: Density-ratio · Density-based clustering
Varied densities · Scaling

1 Introduction

Clustering, as the most common unsupervised knowledge discovery technique, has become one of the most popular automatic data-labelling techniques. It has been widely studied for data mining and knowledge discovery [8]. The goal of clustering is to partition a set of data points into a set of homogeneous groups based on their similarity [6].

There are different kinds of clustering algorithms depending on the specific assumption and model used. Density-based clustering algorithms find clusters in regions of high density which are separated by regions of low density. The clusters are typically identified by grouping points which are above a global density threshold [5]. In contrast to traditional partitioning methods, which can only discover globular clusters, density-based clustering finds clusters of arbitrary sizes

© Springer International Publishing AG, part of Springer Nature 2018
D. Phung et al. (Eds.): PAKDD 2018, LNAI 10939, pp. 389–400, 2018.
https://doi.org/10.1007/978-3-319-93040-4_31

and shapes while effectively separating noise and outliers. Therefore, density-based clustering has received substantial attention in theory and practice.

Despite its popularity, it is well-known that most density-based algorithms face the same challenge of finding clusters with differing densities [4]. Zhu et al. [12] have analysed the cause of this weakness and determined the kind of data distribution in which most density-based clustering will fail. Armed with the analytical result, they introduce a less restrictive assumption that finds clusters in regions of *locally* high density separated by regions of *locally* low density; and propose a density-ratio based approach to overcome this weakness.

ReScale [12], a density-ratio based approach, has been shown to success-fully overcome the weakness of density-based clustering in finding clusters of varied densities. It is an adaptive scaling approach which operates as a pre-processing step to rescale a given dataset, and then apply the rescaled dataset to an existing density-based clustering algorithm. This enables the clustering algorithm to detect all clusters with varied densities. However, ReScale is an one-dimensional scaling method which is applied to each individual attribute inde-pendently. Therefore, it becomes less effective when a data distribution demands multi-dimensional scaling. Such a dataset produces a significant overlap between clusters if the one-dimension projection is conducted [12].

This paper makes the following contributions:

1. Introducing a new distance scaling method and proving that it is equivalent to a density-ratio estimation. It is a multi-dimensional scaling method which considers all dimensions simultaneously.
2. Demonstrating DScale's effectiveness by applying it to existing density-based algorithms: DBSCAN [5], OPTICS [2] and DP [11].

The advantages of DScale over ReScale are: it overcomes the aforementioned weakness of ReScale and requires only one parameter rather than two; while maintaining the simplicity of the implementation. As a result, DScale enables an existing density-based algorithm to find clusters of varied densities in a more general context and has less parameter tuning than ReScale.

The rest of the paper is organised as follows: we provide an overview of density-based clustering algorithms and related work in Sect. 2. Section 3 describes the varied density problem faced by existing density-based clustering algorithms. Section 4 presents the principle and the proposed DScale method. We provide the empirical evaluation results in Sect. 5, followed by conclusion in the last section.

2   Related Work

The classic density based clustering algorithms, such as DBSCAN [5] and DEN-CLUE [7], model the data distribution of a given dataset using a density esti-mator and then apply a threshold to identify "core" points which have densities higher than the threshold. A linking method is employed to link all neighbour-ing core points to form a cluster. If a point is neither a core point nor in the neighbourhood of a core point, it is considered to be noise. Since both DBSCAN and DENCLUE use a global density threshold to identify clusters, they face the same challenge of finding clusters with differing densities [1].

Many variants of DBSCAN have attempted to overcome the issue of detecting clusters with varied densities. OPTICS [2] produces a "reachability" plot such that all points are ordered in a special linear order where spatially adjacent points follow close to each other in the x-axis, and the reachability distances are shown in the y-axis. Cluster centres normally have the higher density or lower "reachability distance" than the cluster boundaries. Thus, a hierarchical method can be employed to extract "valleys" from this plot as clusters. The clustering performance of OPTICS depends on the hierarchical method employed.

Density peaks-based clustering (DP) algorithm [11] identifies clusters with density maxima. For each point, DP calculates its density value (ρ) using an ϵ-neighbourhood density estimator, and the minimum distance (δ) between it and another point with a higher density value. Then, it selects k cluster centres which have locally maximum density and have a relatively large distance from any points with higher local densities, i.e., the k points with the highest $\rho \times \delta$. Each remaining point is assigned to its nearest neighbour of higher density; and the points, which are connected or transitively connected to the same cluster centre, are grouped into the same cluster. Finally, points with low densities at border regions are classified as noise.

Recently, density-ratio based clustering [12] is proposed to detect clusters as regions of *local high densities* that are separated by regions of *local low densities*. Instead of identifying a "core" point based on its density, it advises to use density-ratio which is a ratio of the density of an instance and the density of its η-neighbourhood (η is a large radius). Points located at locally maximum density regions have higher density-ratio values than that located at locally minimum density regions. Thus it allows for a single threshold to be used to separate all clusters with varied densities. Figure 1a illustrates a density distribution of an one-dimensional data, where the x-axis is the point attribute value and the y-axis is the density value. A single density threshold cannot separate all clusters. However, the density-ratio distribution allows a single threshold to be used to separate these clusters, as shown in Fig. 1b.

ReScale [12] is a preprocessing technique based on density-ratio and designed for a density-based clustering algorithm which uses a global density threshold to identify with varied densities. ReScale first rescales each attribute of a dataset and then applies an existing density-based clustering algorithm directly to the rescaled dataset. This converts the density-based clustering algorithm to perform density-ratio clustering because the estimated density of each rescaled point approximates the estimated density-ratio of that point in the original space. For example, Fig. 1c is the density distribution of rescaled points in Fig. 1a, where the density value of each point approximates the density-ratio value in Fig. 1b. ReScale requires two parameters η and ψ, used to define the local neighbourhood and control the precision of η-neighbourhood density estimation, respectively. A key weakness of ReScale is that the rescale is conducted on each individual

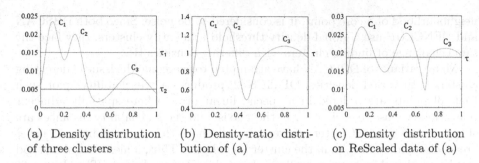

(a) Density distribution of three clusters   (b) Density-ratio distribution of (a)   (c) Density distribution on ReScaled data of (a)

Fig. 1. (a) A mixture of three Gaussian distributions that cannot be separated using a single density threshold; (b) Density-ratio distribution of (a) which allows for a single threshold to be used to separate all three clusters; (c) Density distribution on ReScaled data of (a), where the density value of each point approximates the density-ratio value in (b).

attributes independently. If there is a significant overlap between clusters on some attributes, ReScale may become less effective [12].

In this paper, we focus on the density-ratio approach because it is a principled method and can be used to resolve the varied density issue in existing density-based clustering algorithms.

3   The Problem of Varied Densities

We first provide notations used throughout this paper.

Let $D = \{x_1, x_2, \ldots, x_n\}$, $x_i \in R^d$, $x_i \sim F$ denote a dataset of n points, each sampled independently from a distribution F. Let $\widehat{pdf}(x)$ denote the density estimate of point x which approximates the true density $pdf(x)$.

Let $\mathcal{N}_\epsilon(x)$ be the ϵ-neighbourhood of x, $\mathcal{N}_\epsilon(x) = \{y \in D | s(x, y) \leqslant \epsilon\}$, where $s(\cdot, \cdot)$ is the distance function ($s : R^d \times R^d \to R$).

In general, $pdf(x)$ can be estimated via a small ϵ-neighbourhood (as used by density-based clustering algorithm DBSCAN [5]) as follows:

$$\widehat{pdf}_\epsilon(x) = \frac{1}{nV_\epsilon}|\mathcal{N}_\epsilon(x)| = \frac{|\{y \in D | s(x, y) \leqslant \epsilon\}|}{nV_\epsilon} \tag{1}$$

where $V_\epsilon \propto \epsilon^d$ is the volume of a d-dimensional ball of radius ϵ.

A set of clusters $\{C_1, \ldots, C_\varsigma\}$ is defined as non-empty and non-intersecting subsets: $C_i \subset D, C_i \neq \emptyset, \forall_{i \neq j} C_i \cap C_j = \emptyset$. Let $c_i = \arg\max_{x \in C_i} \widehat{pdf}(x)$ denote the mode (point of the highest estimated density) for cluster C_i; and $p_i = \widehat{pdf}(c_i)$ denote the corresponding peak density value.

A commonly used density-based clustering method uses a global density threshold to identify core points (which have densities higher than the threshold); and then it links neighbouring core points together to form clusters.

For this kind of algorithm to find all clusters in a dataset, the data distribution must have the following necessary condition: the peak density of any cluster is greater than the maximum of the minimum density along any path linking any two modes. It is formally described by Zhu et al. [12] as follows:

$$\min_{k \in \{1,...,\varsigma\}} p_k > \max_{i \neq j \in \{1,...,\varsigma\}} g_{ij} \tag{2}$$

where p_k is the peak density of cluster C_k from a total of ς clusters; and g_{ij} is the largest of the minimum density along any path linking clusters C_i and C_j.

This condition implies that there must exist a threshold τ that can be used to break all paths between the modes by assigning regions with density less than τ to noise. Otherwise, if the mode of some cluster has a density lower than that of a low-density region between some clusters, then this kind of clustering algorithm will fail to find all clusters, i.e., either some high-density clusters are merged together when a lower density threshold is used, or some low-density clusters are designated as noise when a higher density threshold is used. To illustrate, Fig. 1a shows that using a high threshold τ_1 will treat all points in Cluster C_3 as noise but use a low threshold τ_2 will assign points in C_1 and C_2 to the same cluster.

Density-ratio based clustering is a principled approach to overcome this weakness of density-based clustering. It identifies clusters as regions of locally high density.

The density-ratio of a point is the ratio of its density and the density of its η-neighbourhood, which can be estimated as

$$\widehat{rpdf}_{\epsilon,\eta}(x) = \frac{\widehat{pdf}_\epsilon(x)}{\widehat{pdf}_\eta(x)} \tag{3}$$

where $\widehat{pdf}_\eta(x) = \frac{1}{nV_\eta}|\mathcal{N}_\eta(x)|$ denotes the average density value over the η-neighbourhood of x, providing $\eta > \epsilon$.

There is a lemma about the density-ratio value [12], restated as follows:

Lemma 1. *If x is at a local maximum density of $\mathcal{N}_\eta(x)$, then $\widehat{rpdf}_{\epsilon,\eta}(x) \geqslant 1$; if x is at a local minimum density of $\mathcal{N}_\eta(x)$, then $\widehat{rpdf}_{\epsilon,\eta}(x) \leqslant 1$.*

A global density-ratio threshold around unity can be used to identify all cluster peaks and break all paths between different clusters when the following conditions are met: (i) the peak density of each cluster is higher than average density over the η-neighbourhood around the peak; and (ii) points along every path linking between any two clusters have lower density than the average density over the η-neighbourhood of these points.

ReScale [12] is a density-ratio based scaling approach which operates as a pre-processing step to rescale a given dataset D to D'. For an one-dimensional dataset, ReScale uses a η-neighbourhood density estimation to rescale the data according to the cumulative distribution function of the radius (η) estimation using the following mapping function:

$$\forall_{x \in D, y \in D'} \; y = \widehat{cdf}_\eta(x) \tag{4}$$

where $\widehat{cdf}_\eta(x) = \int_{-\infty}^x \widehat{pdf}_\eta(x')dx'$ is the cumulative distribution function. Note that this is a probability integral transform [10] such that $\forall_{x \in D, y \in D'} \widehat{pdf}_\eta(y) = \widehat{pdf}_\eta(\widehat{cdf}_\eta(x)) = 1/n$, which is a uniform distribution.

The aim of this scaling procedure is to make the data approximately uniform (on each dimension). After rescaling, the density-based clustering with a bandwidth ($\epsilon < \eta$) will approximate the density-ratio based clustering, i.e., the estimated density of each rescaled point is approximately the estimated density-ratio of that point in the original space [12]. In a nutshell, ReScale enables clusters with varied densities to be identified using a single global threshold that would otherwise be impossible on the original dataset.

For a multi-dimensional dataset, ReScale uses the one-dimensional mapping method mentioned above to scale each attribute independently. However, if a data distribution has a significant overlap between cluster peaks on one-dimension projection, i.e., not all cluster peaks are located at locally maximum density areas on the projection, the effectiveness of ReScale would be weakened [12]. In order to overcome this weakness, it is important to propose a multi-dimensional scaling. A new distance scaling method called DScale is provided to meet this demand.

4   A Distance Scaling Method for Density-Ratio Estimation

Rather than rescaling on each individual original attribute, we propose a new density-ratio method which rescales the pairwise distance as a multi-dimensional scaling, such that points located at locally high-density areas have higher densities than points located at locally low-density areas.

Given a point $x \in D$, we can rescale the distance between x and its η-neighbourhood point $y \in \mathcal{N}_\eta(x, s)$ using a mapping function:

$$\forall_{x, y \in D, y \in \mathcal{N}_\eta(x,s)} \ s'(x, y) = s(x, y) \times r(x) \tag{5}$$

where $r(x)$ is the scaling function and $s'(\cdot, \cdot)$ is the scaled distance.

Similar to ReScale, the scaling function should make the data approximately uniformly distributed in the scaled η-neighbourhood. Then we have:

$$\widehat{pdf}_{\eta'}(x, s') = \frac{|\mathcal{N}_{\eta'}(x, s')|}{nV_{\eta'}} = \frac{|\mathcal{N}_\eta(x, s)|}{nV_{\eta'}} = \frac{1}{V_m} \tag{6}$$

where $\eta' = \eta \times r(x)$; and V_m is the total volume of a d-dimensional ball of radius m in the original space; and $m = \max_{x,y \in D} s(x, y)$. Note that Eq. 5 is a linear scaling, thus the number of points in $\mathcal{N}_\eta(x, s)$ remains the same as in $\mathcal{N}_{\eta'}(x, s')$, i.e., $|\mathcal{N}_{\eta'}(x, s')| = |\mathcal{N}_\eta(x, s)|$.

Since $V_{\eta'} = V_{\eta \times r(x)} = r(x)^d \times V_\eta \propto r(x)^d \times \eta^d$ and $V_m \propto m^d$, substituting them in Eq. 6 gives:

$$r(x) = \left(\frac{|\mathcal{N}_\eta(x, s)| \times V_m}{n \times V_\eta} \right)^{\frac{1}{d}} = \left(\frac{|\mathcal{N}_\eta(x, s)| \times m^d}{n \times \eta^d} \right)^{\frac{1}{d}} \tag{7}$$

Finally, the distance between x and any point $y \in D \setminus \mathcal{N}_\eta(x, s)$ is normalised as if the maximum distance is m. Here we use a simple *min-max* normalisation:

$$\forall_{y \in D \setminus \mathcal{N}_\eta(x,s)} \; s'(x, y) = (s(x, y) - \eta) \times \frac{m - \eta'}{m - \eta} + \eta' \tag{8}$$

The reason we use the *min-max* normalisation is to keep the same instance rank. Other similar normalisation methods can also be used in Eq. 8.

When using a small radius ϵ' to estimate the new density such that $(\epsilon' < \eta' < m) \wedge (\epsilon' = \epsilon \times r(x)) \wedge (\eta' = \eta \times r(x))$, then we can provide a theorem about the ϵ'-neighbourhood density estimator on the first part of rescaled distance:

Theorem 1. *The estimated density $\widehat{pdf}_{\epsilon'}(x, s')$ in terms of the rescaled distance s' is proportional to the density-ratio $\widehat{rpdf}_{\epsilon,\eta}(x, s)$, where the density-ratio is estimated based on two radii ϵ and η in terms of the original distance s.*

Proof. After scaling the distance between x and other points with Eqs. 5 and 8, the newly estimated density of x using an ϵ'-neighbourhood density estimator is:

$$\widehat{pdf}_{\epsilon'}(x, s') = \frac{1}{nV_{\epsilon'}} |\mathcal{N}_{\epsilon'}(x, s')| = \frac{1}{nV_{\epsilon'}} |\mathcal{N}_\epsilon(x, s)|$$

$$= \frac{|\mathcal{N}_\epsilon(x, s)|}{nV_\epsilon \times r(x)^d} = \frac{|\mathcal{N}_\epsilon(x, s)|}{nV_\epsilon \times \frac{|\mathcal{N}_\eta(x,s)| \times V_m}{n \times V_\eta}} = \frac{V_\eta}{V_m \times V_\epsilon} \frac{|\mathcal{N}_\epsilon(x, s)|}{|\mathcal{N}_\eta(x, s)|}$$

$$= \frac{\widehat{pdf}_\epsilon(x, s)}{V_m \times \widehat{pdf}_\eta(x, s)} = \frac{1}{V_m} \widehat{rpdf}_{\epsilon,\eta}(x, s) \propto \widehat{rpdf}_{\epsilon,\eta}(x, s) \tag{9}$$

$\square$

Based on Lemma 1, using the rescaled distance, a single density threshold τ around $\frac{1}{V_m}$ can be used to identify all points which are located at locally high-density areas in a dataset with varied densities.

Usually, ϵ is set less than η, i.e., $\epsilon' < \eta'$. However, it is possible to set $\epsilon > \eta$ in practice. Even in this case, Eq. 8 still can transform the original density distribution to one which is more uniform in two conditions: (i) When both $\widehat{pdf}_\epsilon(x, s)$ and $\widehat{pdf}_\eta(x, s)$ are higher than the uniformly distributed density $\frac{1}{V_m}$, we have $r(x) > 1$, and then get $\epsilon' - \epsilon = (\epsilon - \eta) \times \frac{m - \eta'}{m - \eta} + \eta' - \epsilon = \frac{\epsilon\eta - \epsilon\eta' + m\eta' - m\eta}{m - \eta} = \frac{(\eta - \eta')(\epsilon - m)}{m - \eta} > 0$ and $V_{\epsilon'} > V_\epsilon$, since $\eta' = \eta \times r(x) > \eta$ and $m > \epsilon > \eta$. As $|\mathcal{N}_\epsilon(x, s)| = |\mathcal{N}_{\epsilon'}(x, s')|$, we have $\widehat{pdf}_{\epsilon'}(x, s') < \widehat{pdf}_\epsilon(x, s)$. (ii) If both $\widehat{pdf}_\epsilon(x, s)$ and $\widehat{pdf}_\eta(x, s)$ are lower than the uniformly distributed density, we have $\widehat{pdf}_{\epsilon'}(x, s') > \widehat{pdf}_\epsilon(x, s)$ by a similar derivation. Thus, it can reduce the density gaps between clusters with varied densities and enable a density-based clustering algorithm to use a density threshold to detect clusters with varied densities.

We name the above new distance scaling method as `DScale`. Note that `DScale` has one parameter η only. ϵ is a parameter in the density-based clustering algorithm. The parameters ϵ' and η' are not real parameters that a user needs to set; they are used above for the proof and explanation only.

After applying DScale for each and every $x \in D$ in the distance matrix, it allows an existing density-based clustering algorithm to use a single threshold to find all clusters with varied densities, that would otherwise be impossible. This is because the density distribution in the rescaled space meets the necessary condition for a density-based clustering algorithm to detect all clusters, as shown in Eq. 2.

It is worth mentioning that the rescaled distance is asymmetric. Based on Eqs. 5 and 7, $\forall_{x,y \in D}$, if $|N_\eta(x,s)| \neq |N_\eta(y,s)|$, then $r(x) \neq r(y)$; therefore $s'(x,y) \neq s'(y,x)$. This has no effect on most of the existing density-based algorithms because density estimation is point based.

The implementation of DScale based on distance s is shown in Algorithm 1. It requires one parameter η only. Both the time complexity and space complexity of DScale are $O(n^2)$. Because many existing density-based clustering algorithms have time and space complexities $O(n^2)$, DScale as a preprocessing step does not increase their overall complexities.

Note that DScale is not a typical distance normalisation as used in some distance measures such as cosine distance because DScale aims to achieve the required density-ratio estimation but the typical normalisation does not. It is also different from multidimensional scaling [3] that preserves as well as possible the original pairwise distances between instances.

5   Empirical Evaluation

This section presents experiments designed to evaluate the effectiveness of DScale. We compare DScale with ReScale using three existing density-based clustering algorithms (DBSCAN, OPTICS and DP) in terms of best F-measure: given a clustering result, we calculate the precision score P_i and the recall score R_i for each cluster C_i based on the confusion matrix, and then the F-measure score of C_i is the harmonic mean of P_i and R_i. The overall F-measure score is the unweighted average over all clusters: F-measure $= \frac{1}{\varsigma} \sum_{i=1}^{\varsigma} \frac{2P_i R_i}{P_i + R_i}$.

We used 2 artificial datasets and 9 real-world datasets with different data sizes and dimensions from UCI Machine Learning Repository [9] to ascertain the ability of density-ratio in handling datasets with varied densities. Table 1 presents the data properties of the datasets.

3L is a 2-dimensional data containing three elongated clusters with different densities, as shown in Fig. 2a. 4C is a 2-dimensional data containing four clusters with different densities (three Gaussian clusters and one elongated cluster), as shown in Fig. 2b. Note that DBSCAN is unable to correctly identify all clusters in both datasets because they do not satisfy the condition specified in Eq. 2. Furthermore, clusters in 3L are significantly overlapped on individual attribute projection, which violates the requirement of ReScale such that the one-dimensional projections cannot identify the density peaks of any clusters.

All algorithms used in our experiments were implemented in Matlab (the source code can be obtained at https://sourceforge.net/p/distance-scaling/). All datasets were normalised using the *min-max* normalisation to yield each attribute to be in [0, 1] before the experiments began.

Algorithm 1. DScale(S, η, d)

Input: S - input distance matrix ($n \times n$ matrix); η - radius of the neighbourhood; d - dimensionality of the dataset.

Output: S' - distance matrix after scaling.

1: $m \leftarrow$ the maximum distance in S
2: Initialising $n \times n$ matrix S'
3: **for** $i = 1$ to n **do**
4:     $\mathcal{N}_\eta(x_i) \leftarrow \{x_j \in D \mid S[x_i, x_j] \leqslant \eta\}$ /* Identify the η-neighbourhood of x_i */
5:     $r(x_i) = (\frac{|\mathcal{N}_\eta(x_i)| \times m^d}{n \times \eta^d})^{\frac{1}{d}}$ /* Calculate the scaling factor based on Equation 7 */
6:     $\forall_{x_j \in \mathcal{N}_\eta(x_i)}$ $S'[x_i, x_j] = S[x_i, x_j] \times r(x_i)$ /* Scale the distances between x_i and instances in the η-neighbourhood of x_i based on Equation 5 */
7:     $\forall_{x_j \in D \backslash \mathcal{N}_\eta(x_i)}$ $S'[x_i, x_j] = (S[x_i, x_j] - \eta) \times \frac{m - \eta \times r(x_i)}{m - \eta} + \eta \times r(x_i)$ /* Scale the distances to instances outside the η-neighbourhood of x_i based on Equation 8 */
8: **end for**
9: **return** S'

Table 1. Data properties

Dataset	Data size	#Dimensions	#Clusters
Iris	150	4	3
GPS	163	6	2
Thyroid	215	5	3
Ecoli	336	7	8
Libras	360	90	15
Wilt	500	5	2
Breast	699	9	2
Pima	768	8	2
Segment	2310	19	7
3L	560	2	3
4C	1250	2	4

For DP, we normalised both ρ and δ to be in $[0, 1]$ before selecting k cluster centres so that these two variables have the same weight in their product $\rho \times \delta$. We report the best clustering performance within a reasonable range of parameter search for each algorithm. Table 2 lists the parameters and their search ranges for each algorithm. Note that the parameter ξ in OPTICS is used to identify downward and upward areas of the reachability plot in order to extract all clusters using a hierarchical method [2]. ψ in ReScale controls the precision of $\widehat{cdf}_\eta(x)$, i.e., the number of intervals used for estimating $\widehat{cdf}_\eta(x)$.

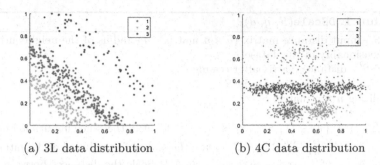

(a) 3L data distribution   (b) 4C data distribution

Fig. 2. (a) A two-dimensional data containing three line-shaped clusters. (b) A two-dimensional data containing four clusters.

Table 2. Parameters and their search ranges for each algorithm. The search ranges of ψ and η are as used by Zhu et al. [12].

Algorithm	Parameter with search range
DBSCAN	$Minpts \in \{2, 3, ..., 10\};\ \epsilon \in [0, 1]$
OPTICS	$Minpts \in \{2, 3, ..., 10\};\ \xi \in \{0.01, 0.02, ..., 0.99\}$
DP	$k \in \{2, 3, ..., 20\};\ \epsilon \in [0, 1]$
ReScale	$\psi = 100;\ \eta \in \{0.1, 0.2, ..., 0.5\}$
DScale	$\eta \in \{0.1, 0.2, ..., 0.5\}$

5.1   Clustering Performance

Table 3 shows the best F-measures of DBSCAN, OPTICS, DP, and their ReScale and DScale versions. The average F-measures, showed in the second last row, reveal that DScale improves the clustering performance of every existing clustering algorithm in more datasets than those of ReScale. The extent of improvement is most pronounced for DBSCAN, i.e., from 0.59 to 0.74. The gap decreases for OPTICS and DP because they are more powerful algorithms which do not rely on a single density threshold to identify cluster modes.

Regarding the number of top 1 performers (showed in the last row), DScale-DBSCAN is the top 1 performer on 10 out of 11 datasets. DScale-OPTICS and DScale-DP are the top 1 performers on 7 and 8 out of 11 datasets, respectively.

For the 4C dataset, both ReScale and DScale significantly improved the clustering performances of DBSCAN, OPTICS and DP. However, ReScale cannot improve clustering performances on the 3L dataset because of using one-dimensional projections. In contrast, DScale significantly improved the F-measure of DBSCAN and OPTICS from 0.59 and 0.83 to 0.90 and 0.95, respectively.

DScale has only one parameter η to define the η-neighbourhood. The density-ratio based on a small η will approximate 1 and provides no information, while on a large η will approximate the true density and show no advantage.

Table 3. Best F-measure of DBSCAN, OPTICS, DP, and their ReScale and `DScale` versions on 11 datasets. For each clustering algorithm, the best performer in each dataset is boldfaced. Orig, ReS and DS represent Original, ReScale and `DScale`, respectively.

Data	DBSCAN			OPTICS			DP		
	Orig	ReS	DS	Orig	ReS	DS	Orig	ReS	DS
Iris	0.85	0.90	**0.93**	0.85	0.84	**0.88**	**0.97**	**0.97**	**0.97**
GPS	0.75	0.75	**0.80**	0.76	0.76	**0.763**	0.81	**0.821**	0.82
Thyroid	0.58	0.79	**0.83**	0.59	0.85	**0.90**	0.87	**0.93**	0.87
Ecoli	0.37	0.40	**0.54**	0.44	**0.57**	0.50	0.48	0.55	**0.63**
Libras	0.40	0.44	**0.46**	0.50	**0.52**	0.49	0.52	0.38	**0.523**
Wilt	0.38	0.39	**0.54**	0.677	0.58	**0.68**	0.54	**0.68**	0.54
Breast	0.82	0.95	**0.96**	0.84	**0.96**	0.95	0.97	0.97	**0.972**
Pima	0.43	0.48	**0.64**	0.65	0.65	**0.66**	0.62	0.66	**0.67**
Segment	0.59	**0.62**	0.61	0.69	0.67	**0.70**	0.78	0.77	**0.80**
3L	0.59	0.63	**0.90**	0.83	0.83	**0.95**	0.82	0.81	**0.86**
4C	0.71	0.90	**0.92**	0.87	**0.95**	0.94	0.87	0.92	**0.95**
Average	0.59	0.66	**0.74**	0.70	0.74	**0.76**	0.75	0.77	**0.78**
#Top 1	0	1	**10**	0	4	**7**	1	4	**8**

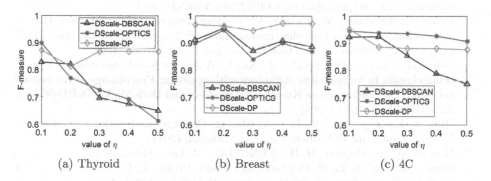

(a) Thyroid   (b) Breast   (c) 4C

Fig. 3. F-measure on 3 datasets with different η values.

Figure 3 shows the average F-measure on 3 datasets when `DScale` uses η from 0.1 to 0.5. It shows that this scaling effect is different for different datasets in terms of getting the best clustering results. Generally, $\eta \in [0.1, 0.2]$ and ϵ is set slightly smaller than η.

6 Conclusion

We introduce a new density-ratio method `DScale` to enable existing density-based clustering algorithms to find clusters of varied densities. It is a multi-dimensional scaling method which operates as a pre-processing step to rescale the distance for a given dataset. Applying the rescaled distance to an existing density-based clustering algorithm enables the algorithm to detect clusters with varied densities that would otherwise be impossible.

DScale has two advantages over the existing method ReScale. First, DScale overcomes the key weakness of ReScale which relies on one-dimension projections. Second, DScale requires one less parameter than ReScale, and retains the same simplicity of ReScale. As a result, DScale enables an existing density-based algorithm to find clusters of varied densities in a more general context and has less parameter tuning than ReScale. Our empirical evaluation shows that DScale has better improvement on clustering performance than ReScale for three existing density-based algorithms: DBSCAN, OPTICS and DP.

References

1. Aggarwal, C.C., Reddy, C.K.: Data Clustering: Algorithms and Applications. Chapman and Hall/CRC Press, Boca Raton (2013)
2. Ankerst, M., Breunig, M.M., Kriegel, H.P., Sander, J.: OPTICS: ordering points to identify the clustering structure. In: Proceedings of the 1999 ACM SIGMOD International Conference on Management of Data. SIGMOD 1999, pp. 49–60. ACM, New York (1999)
3. Borg, I., Groenen, P.J., Mair, P.: Applied Multidimensional Scaling. Springer, Heidelberg (2012). https://doi.org/10.1007/978-3-642-31848-1
4. Ertöz, L., Steinbach, M., Kumar, V.: Finding clusters of different sizes, shapes, and densities in noisy, high dimensional data. In: Proceedings of the 2003 SIAM International Conference on Data Mining, pp. 47–58. SIAM (2003)
5. Ester, M., Kriegel, H.P., Sander, J., Xu, X.: A density-based algorithm for discovering clusters in large spatial databases with noise. In: Proceedings of the Second International Conference on Knowledge Discovery and Data Mining (KDD-96), pp. 226–231. AAAI Press (1996)
6. Han, J., Kamber, M., Pei, J.: Data Mining: Concepts and Techniques, 3rd edn. Morgan Kaufmann Publishers Inc., San Francisco (2011)
7. Hinneburg, A., Gabriel, H.-H.: DENCLUE 2.0: fast clustering based on kernel density estimation. In: R. Berthold, M., Shawe-Taylor, J., Lavrač, N. (eds.) IDA 2007. LNCS, vol. 4723, pp. 70–80. Springer, Heidelberg (2007). https://doi.org/10.1007/978-3-540-74825-0_7
8. Kaufman, L., Rousseeuw, P.J.: Finding Groups in Data: An Introduction to Cluster Analysis. Wiley, Hoboken (1990)
9. Lichman, M.: UCI machine learning repository (2013). http://archive.ics.uci.edu/ml
10. Pearson, E.S.: The probability integral transformation for testing goodness of fit and combining independent tests of significance. Biometrika $30(1/2)$, 134–148 (1938)
11. Rodriguez, A., Laio, A.: Clustering by fast search and find of density peaks. Science $344(6191)$, 1492–1496 (2014)
12. Zhu, Y., Ting, K.M., Carman, M.J.: Density-ratio based clustering for discovering clusters with varying densities. Pattern Recogn. 60, 983–997 (2016)

Neighbourhood Contrast: A Better Means to Detect Clusters Than Density

Bo Chen[1($\boxtimes$)] and Kai Ming Ting[1,2]

[1] Monash University, Clayton, VIC 3168, Australia
bo.chen@monash.edu
[2] Federation University Australia, Churchill, VIC 3842, Australia
kaiming.ting@federation.edu.au

Abstract. Most density-based clustering algorithms suffer from large density variations among clusters. This paper proposes a new measure called Neighbourhood Contrast (NC) as a better alternative to density in detecting clusters. The proposed NC admits all local density maxima, regardless of their densities, to have similar NC values. Due to this unique property, NC is a better means to detect clusters in a dataset with large density variations among clusters. We provide two applications of NC. First, replacing density with NC in the current state-of-the-art clustering procedure DP leads to significantly improved clustering performance. Second, we devise a new clustering algorithm called Neighbourhood Contrast Clustering (NCC) which does not require density or distance calculations, and therefore has a linear time complexity in terms of dataset size. Our empirical evaluation shows that both NC-based methods outperform density-based methods including the current state-of-the-art.

Keywords: Neighbourhood Contrast · Clustering

1 Introduction

Density-based clustering methods rely on the estimated density distribution to detect clusters in a dataset. High density regions are recognized as a cluster and low density areas are regarded as separations between clusters [6]. However, most density-based methods are known to have difficulties clustering datasets with hugely varying densities [2,3,10]. For example, the density-based spatial clustering of applications with noise (DBSCAN) [4], which employs a single cut-off threshold to identify high density points, often can not detect all the clusters with hugely varying densities. Another example is the recently proposed clustering by fast search and find of density peaks (DP) [11]. It uses a novel way of detecting clusters by finding local density peaks in the first step. The local density peaks are points that have high densities and relatively distant from other peaks. Because it combines density and distance rather than using density alone in detecting density peaks, DP has a much improved capability

© Springer International Publishing AG, part of Springer Nature 2018
D. Phung et al. (Eds.): PAKDD 2018, LNAI 10939, pp. 401–412, 2018.
https://doi.org/10.1007/978-3-319-93040-4_32

than DBSCAN in detecting clusters. However, it still suffers from large density variations among clusters in some data distributions.

The unaddressed issue of using density to identify clusters is that low density clusters are often overlooked in a dataset having large density variations. For a measure which addresses this issue, the necessary property is to admit all cluster centers, regardless of their densities, to have approximately the same highest value of the measure. Density, by definition, does not posses this property.

To address this issue from its root cause, we propose a new measure which has the above-mentioned property. The key contributions of this paper are:

1. Introducing a new measure called Neighbourhood Contrast (NC) with a unique property, i.e., all local density maxima have similar NC values, regardless of their densities. This property makes NC a better means to detect clusters than density.
2. Using NC in clustering. This is done in two ways: First, NC is incorporated in a state-of-the-art density-based clustering algorithm DP [11]. By replacing density with NC in the procedure of DP, we show that NC-DP, i.e., the NC version of DP, significantly improves DP's clustering performance. Second, we devise a new clustering algorithm called Neighbourhood Contrast Clustering (NCC) which does not require pairwise distance calculations or nearest neighbour search and hence has a linear time complexity.
3. Conducting experiments to examine the effectiveness of NC. In our experiments, both NC-DP and NCC outperform two major density-based algorithms including the state-of-the-art method DP.

The rest of this paper is organised as follows. Section 2 introduces Neighbourhood Contrast. Section 3 describes NC-DP. Section 4 proposes NCC. The experiments are reported in Sect. 5, followed by the conclusion of the paper.

2   Neighbourhood Contrast

For $\mathbf{x} \in R^d$, let T be a pair of neighbouring non-overlapping and symmetric regions which is generated from a random process, and one of two regions must cover $\mathbf{x}$. Let $T(\mathbf{x})$ denote the region covering $\mathbf{x}$ and $T'(\mathbf{x})$ denote the other region.

Definition 1. *Given a dataset D, Neighbourhood Contrast of $\mathbf{x}$ is the probability that $T(\mathbf{x})$ has larger probability mass than $T'(\mathbf{x})$, i.e.,*

$$NC(\mathbf{x}) = P(|T(\mathbf{x})| > |T'(\mathbf{x})|),$$

where $|T(\mathbf{x})| = |\{\mathbf{y} \in D : \mathbf{y} \in T(\mathbf{x})\}|$ is the number of instances in $T(\mathbf{x})$.

2.1   Property of Neighbourhood Contrast

Theorem 1. *If a local density distribution is isotropic in an adjacent region of a density maximum $\mathbf{x}^*$, i.e., the density decreases at the same rates while moving away from $\mathbf{x}^*$ along any direction, then $NC(\mathbf{x}^*) = 1$.*

Proof. Let $\mathbf{x}^*$ be an isotropic density maximum, as shown in Fig. 1. For any point $\mathbf{x}$ near $\mathbf{x}^*$, the larger the distance $d(\mathbf{x}, \mathbf{x}^*)$, the smaller the density of $\mathbf{x}$. Suppose a random pair of regions $T(\mathbf{x}^*)$ and $T'(\mathbf{x}^*)$ is generated as shown in Fig. 1(b) and (c). For an arbitrary point $\mathbf{x}$ in $T(\mathbf{x}^*)$, let $\mathbf{x}'$ be its mirror counterpart in $T'(\mathbf{x}^*)$. Because $d(\mathbf{x}, \mathbf{x}^*) < d(\mathbf{x}', \mathbf{x}^*)$, hence $f(\mathbf{x}) > f(\mathbf{x}')$, for all $\mathbf{x} \in T(\mathbf{x}^*)$. Therefore, $\int_{T(\mathbf{x}^*)} f(\mathbf{x})d\mathbf{x} > \int_{T'(\mathbf{x}^*)} f(\mathbf{x}')d\mathbf{x}'$. In other words, the probability mass in $T(\mathbf{x}^*)$ is always larger than that in $T'(\mathbf{x}^*)$, which leads to $NC(\mathbf{x}^*) = 1$. □

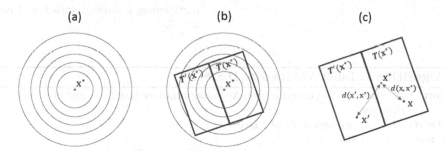

Fig. 1. (a) A local density maximum $\mathbf{x}^*$ where the density of its nearing points decreases isotropically, with concentric contours centered at $\mathbf{x}^*$. (b) A random region $T(\mathbf{x}^*)$ and its sister region $T'(\mathbf{x}^*)$. (c) An arbitrary point $\mathbf{x}$ in $T(\mathbf{x}^*)$ and its mirror counterpart $\mathbf{x}'$ in $T'(\mathbf{x}^*)$: $\mathbf{x}'$ is always further away from $\mathbf{x}^*$ than $\mathbf{x}$.

Although the estimated density contours near a density peak may not be strictly isotropic, the region $T(\mathbf{x}^*)$ which covers the density peak is likely to have larger mass than $T'(\mathbf{x}^*)$. Hence based on Theorem 1, we provide the key property of Neighbourhood Contrast $NC(\mathbf{x})$ as follows:

Property 1. *For any local density maximum* $\mathbf{x}^*$, *its Neighbourhood Contrast* $NC(\mathbf{x}^*)$ *approximates 1, regardless of its density.*

A comparison of density and NC distributions of a synthetic dataset is shown in Fig. 2. In Fig. 2(a) the sparse cluster in the middle exhibits significantly lower density than the other three clusters. In contrast, Fig. 2(b) shows that core regions of all 4 clusters have similar NC, by virtue of Property 1.

2.2 Estimating Neighbourhood Contrast

To estimate NC given a dataset D, random pairs of regions need to be generated. Binary trees are used to partition the data space and produce such regions. Each tree partitions a randomly oriented initial hyper-rectangular space S, that covers the whole dataset, into small cells. A cell is a region corresponding to a leaf node of the tree. We use Algorithm 1 to build an ensemble of trees. The two functions it calls are given in Algorithms 2 and 3. A demonstration of an ensemble of two trees is given in Fig. 3.

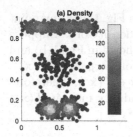 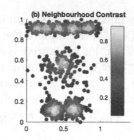 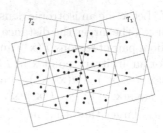

Fig. 2. Density vs NC distribution.

Fig. 3. A demonstration of two trees partitioning a dataset with $h = 4$ and $\mathcal{L} = 3$.

Algorithm 1. Build_NC_Regions($D, t, h, \mathcal{L}$)

---

input : D - dataset; t - ensemble size; h - maximum tree level; $\mathcal{L}$ - leaf node mass threshold

output: $\{T_j\}_{j=1}^{t}$ - an ensemble of t trees

1 **for** $j = 1, ..., t$ **do**
2     $\mathbf{u} \leftarrow$ a randomly orientated orthonormal basis of $\Re^d$
3     $D' \leftarrow D\mathbf{u}$
4     $q \leftarrow$ a randomly selected value in $\{1, ..., d\}$
5     $S \leftarrow$ Initial_Space(D')
6     $T_j \leftarrow$ Build_Tree($D', h, 1, S, q, \mathcal{L}$)
7 **end**

Given a dataset $D \in \Re^d$, a random rotation of D is applied before each tree is built. That is, we randomly rotate the coordinate system by multiplying D with a randomly orientated orthonormal basis $\mathbf{u}$. Let $D' = D\mathbf{u}$ denote the projection of D in the new coordinate system. The initial space S is then generated via Algorithm 2 and it is axis-aligned with the basis $\mathbf{u}$.

Let T denote a binary tree. The root node of the tree represents the initial region S. At each level, a feature $q \in \{1, ..., d\}$ is selected in a round-robin manner; and each branch node is split into two child nodes at the middle point of feature q of the node space. A node becomes leaf when either it reaches level h or its mass is no larger larger than threshold $\mathcal{L}$. The region corresponding to a leaf node is called a cell. The tree building procedure is given in Algorithm 3.

An ensemble of trees $\{T_j\}_{j=1}^{t}$ is built independently to estimate $NC(\mathbf{x})$. Let $T(\mathbf{x})$ denote the leaf node of tree T in which $\mathbf{x}$ falls. Let $T'(\mathbf{x})$ denote the sister node of $T(\mathbf{x})$. Note that $T'(\mathbf{x})$ can be either a branch node or a leaf node. The Neighbourhood Contrast of an instance $\mathbf{x} \in D$ is then estimated by

$$NC(\mathbf{x}) = \frac{1}{t} \sum_{j=1}^{t} I_{\{|T_j(\mathbf{x})| > |T'_j(\mathbf{x})|\}},$$

Algorithm 2. Initial_Space(D)

---

input : D - dataset;
output: S - an axis-aligned hyper-rectangular region such that $D \subset S$
  1 **for** $q = 1, ..., d$ **do**
  2 | $min_q \leftarrow \min\{x_q : \mathbf{x} \in D\}$
  3 | $max_q \leftarrow \max\{x_q : \mathbf{x} \in D\}$
  4 | $z_q \leftarrow$ uniformly random value in $[min_q, max_q]$
  5 | $r_q \leftarrow max_q - min_q$
  6 | $S_q^l \leftarrow z_q - r_q$, the lower bound of S on q
  7 | $S_q^u \leftarrow z_q + r_q$, the upper bound of S on q
  8 **end**

Algorithm 3. Build_Tree($D, h, l, S, q, \mathcal{L}$)

---

input : D - dataset; h - maximum tree level; l - current tree level; S - current
           space; q - current attribute; $\mathcal{L}$ - leaf node mass threshold
output: T - a binary tree that partitions S
  1 **if** $l > h$ **then**
  2 | Terminate and return S as a leaf node region
  3 **else**
  4 | **if** $|D| \leq \mathcal{L}$ **then**
  5 | | Terminate and return S as a leaf node region
  6 | **else**
  7 | | $q \leftarrow q + 1$
  8 | | **if** $q > d$ **then**
  9 | | | $q \leftarrow q - d$
 10 | | **end**
 11 | | $s_q \leftarrow (S_q^l + S_q^u)/2$
 12 | | $D_{(l)} \leftarrow \{\mathbf{x} \in D : w_q < s_q\}$
 13 | | $D_{(r)} \leftarrow \{\mathbf{x} \in D : x_q \geq s_q\}$
 14 | | Split S at s_q into $S_{(l)}$ and $S_{(r)}$
 15 | | $left \leftarrow$ Build_Tree($D_{(l)}, h, l + 1, S_{(l)}, q, \mathcal{L}$)
 16 | | $right \leftarrow$ Build_Tree($D_{(r)}, h, l + 1, S_{(r)}, q, \mathcal{L}$)
 17 | **end**
 18 **end**

---

where $I_{\{.\}}$ is an indicator. For notation brevity, we use $NC_i = NC(\mathbf{x}_i)$ to denote
the Neighbourhood Contrast of instance $\mathbf{x}_i$.

3   Improving DP with Neighbourhood Contrast

It is easy to utilize Neighbourhood Contrast in existing clustering procedures to
improve their performance. By simply replacing density with NC in the procedure of DP [11], we create NC-DP, a version that better handles density variation.
The procedure of NC-DP consists of following three steps which is exactly the
same as DP, except density is replaced with NC.

The first step is to estimate NC. Given a dataset D, $NC(\mathbf{x})$ for all $\mathbf{x} \in D$ are estimated as described in Sect. 2.

The second step is to find K points that have the largest $NC(\mathbf{x}) \times \delta(\mathbf{x})$ values as cluster centers, where K is a parameter deciding the number of clusters and δ is defined as follows,

$$
\delta(\mathbf{x}) = \begin{cases} \min\limits_{NC(\mathbf{y})>NC(\mathbf{x})} d(\mathbf{x}, \mathbf{y}), \forall \mathbf{x} \in D \setminus \{\mathbf{x}^\omega\} \\ \max\limits_{\mathbf{y} \in D} d(\mathbf{x}, \mathbf{y}), \text{ if } \mathbf{x} = \mathbf{x}^\omega \end{cases},
$$

where $d(\cdot, \cdot)$ is a distance measure and $\mathbf{x}^\omega$ is the point having the maximal NC.

The last step is to assign every unassigned point to one of the K cluster centers. All points are sorted in descending order of NC, then one by one from top down, each unassigned point is assigned to the same cluster as its nearest neighbour with a higher NC.

4 Neighbourhood Contrast Clustering

The NC-DP described above improves the capability of DP in detecting clusters of varying densities, which will be shown in Sect. 5. However, it requires pairwise distance calculations and nearest neighbour search which hinder its scalability.

In this section, we present a new clustering algorithm named Neighbourhood Contrast Clustering (NCC). Reusing the trees built for estimating NC, NCC performs clustering without any distance calculation or nearest neighbour search—it is hence highly scalable. It consists of following key steps:

1. Estimate NC for each point in the given dataset.
2. Cluster nexuses are identified and the number of clusters is detected.
3. For each point, a membership score w.r.t. each cluster is calculated; and each point is assigned to the cluster in which it has the highest membership score.

The key algorithmic differences from DP are: (i) NCC employs cluster nexuses instead of cluster centres; and (ii) DP assigns points based on nearest neighbour having a higher density, and NCC assigns points based on membership scores which are computed without distance calculations.

The top layer procedure of NCC is provided in Algorithm 4. The implementation of step 1 has been provided in Sect. 2. In the following subsections we provide the details of steps 2 and 3.

4.1 Core Points and Cluster Nexuses

After obtaining $\{NC_i\}_{i=1}^N$ in step 1, points that have higher NCs than threshold γ are selected as core points. If two core points are covered by the same cell which reaches the maximum level h, then these two core points are linked. A group of transitively linked core points is called a cluster nexus, denoted by M_k. This process is given in Algorithm 5.

A demonstration of forming cluster nexuses is given in Fig. 4(a), (b) and (c). Note that the four groups of points at the top in Fig. 4(c) belong to a single cluster nexus because they are transitively linked.

Algorithm 4. NCC($D, t, h, \mathcal{L}, \gamma$)

---

input : D - dataset; t - ensemble size; h - maximum tree level; $\mathcal{L}$ - leaf node mass
threshold; γ - core point threshold
output: $\{G_k\}_{k=1}^{K}$ - K groups of points
1   $\{T_j\}_{j=1}^{t} \leftarrow$ Build_NC_Regions($D, t, h, \mathcal{L}$)
    $NC_i \leftarrow \frac{1}{t} \sum_{j=1}^{t} I_{\{|T_j(\mathbf{x}_i)| > |T_j'(\mathbf{x}_i)|\}}$, for $i = 1, ..., N$
    Let $D, \{T_j\}, \{NC_i\}$ be global variables accessible by all functions
2   $\{M_k\}_{k=1}^{K} \leftarrow$ Form_Cluster_Nexuses($\gamma, \mathcal{L}, h$)
3   $\{\bar{\eta}_k(\mathbf{x}_i)\}_{k=1,i=1}^{K,N} \leftarrow$ Membership_Score($\{M_k\}$)
    $Y_i \leftarrow \arg\max_k (\bar{\eta}_k(\mathbf{x}_i)), \forall i$
    $G_k \leftarrow \{\mathbf{x}_i \in D : Y_i = k\}, \forall k$

Algorithm 5. Form_Cluster_Nexuses($\gamma, \mathcal{L}, h$)

---

input : γ - core point threshold; $\mathcal{L}$ - leaf node mass threshold; h - maximum tree
level
output: $\{M_k\}_{k=1}^{K}$ - K cluster nexuses
1   Set of core points $Z \leftarrow \{\mathbf{x}_i : NC_i > \gamma\}$
2   **for** *each tree* T_j *in* $\{T_j\}_{j=1}^{t}$ **do**
3   **for** *each level-h cell in tree* T_j **do**
4   **if** *at least 2 core points in* Z *are in this cell* **then**
5   link these core points together
6   **end**
7   **end**
8   **end**
9   $K \leftarrow$ number of groups of transitively linked core points
10   $\{M_k\}_{k=1}^{K} \leftarrow$ the K groups of transitively linked core points

---

4.2   Assigning Non-core Points

The intuition for assigning non-core points is based on a membership function
which is a function of the masses of the cells. Starting from a nexus, where it has
the highest membership score, non-core points have non-increasing membership
scores as they are farther away from the nexus.

For each cluster nexus M_k, membership scores $\eta_k(\mathbf{x})$ are computed for all
points. To be efficient, $\eta_k(\cdot)$ is computed via a nexus expansion process which
is done for all nexuses in one go. The procedure is given in Algorithm 6, where
C_m^j denote the m-th cell in tree T_j.

A brief description is provided as follows. All points are initialized to have
$\eta_k(\mathbf{x}) = 1$. Then, $\eta_k(\mathbf{x})$ is updated for all nexuses. An illustration of this nexuses
expansion process is in Fig. 5. At the end of this process, for every point $\mathbf{y}$ which
is not reached by M_k, $\eta_k(\mathbf{y})$ is set to 0. Example distributions of the membership
scores are given in Fig. 4(d), (e), (f) and (g).

Note that the order in which the trees are examined may affect the expansion
path of M_k and hence, the values of $\eta_k(\cdot)$. To address this issue, an averaged

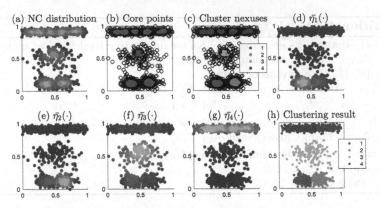

Fig. 4. A demonstration of NCC procedure on the synthetic dataset. The four cluster nexuses identified are shown in (c). The membership score distribution for each of the four clusters is shown in (d), (e), (f) and (g), respectively.

Algorithm 6. Membership_Score($\{M_k\}$)

---

input : $\{M_k\}_{k=1}^K$ - cluster nexuses
output: $\bar{\eta}_k(\mathbf{x}_i), \forall i, k$

1   Initialize $\eta_k(\mathbf{x}_i) \leftarrow 1, \forall i, k$
2   **for** $j = 1, ..., t$ **do**
3   **if** $M_k = D, \forall k$ **then**
4   Exit for-loop
5   **end**
6   **for** $m = 1, ..., \# \ of \ cells \ in \ T_j$ **do**
7   **for** $k = 1, ..., K$ **do**
8   **if** $C_m^j \cap M_k \neq \emptyset$ and $C_m^j \setminus M_k \neq \emptyset$ **then**
9           $\eta_k(\mathbf{x}_i) \leftarrow \min(\frac{|C_m^j|}{N}, \min\{\eta_k(\mathbf{x}_o) : \mathbf{x}_o \in C_m^j\}), \forall i \in \{o : \mathbf{x}_o \in C_m^j \setminus M_k\}$
10           $M_k \leftarrow M_k \cup C_m^j$
11   **end**
12   **end**
13   **end**
14 **end**
15 $\eta_k(\mathbf{x}_i) \leftarrow 0, \forall i \in \{o : \mathbf{x}_o \in D \setminus M_k\}, \forall k$
16 Repeat W times steps 1-15, with $\{T_j\}$ shuffled, producing $\eta_k^w(\mathbf{x}_i), w = 1, ..., W$
17 $\bar{\eta}_k(\mathbf{x}_i) = \frac{1}{W} \sum_{w=1}^W \eta_k^w(\mathbf{x}_i), \forall i, k$

$\bar{\eta}_k(\cdot)$ is produced by calculating $\eta_k(\cdot)$ multiple times, each time with a randomly shuffled order of trees.

After the membership score calculation, for each non-core point $\mathbf{x}_i$, its cluster label is assigned as $Y_i = \arg\max_k(\bar{\eta}_k(\mathbf{x}_i))$ (stated in step 3 in Algorithm 4).

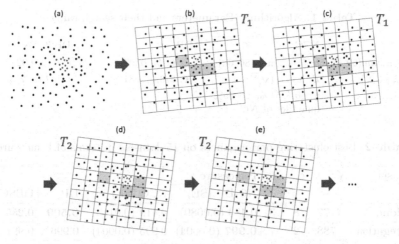

Fig. 5. An illustration of the expansion process of a cluster nexus M_k, described in Algorithm 6. Red points are members of M_k while black ones denote non-members of M_k. (a) The initial M_k. (b) For tree T_1, cells that cover both member and non-member points of M_k are identified (shaded cells). (c) Non-members in these cells become members of M_k, and their $\eta_k()$ get updated to be the smaller quantity of the following two: the normalized mass of the cell and the minimum of current $\eta_k(\cdot)$ of all points in the cell. (d) (e) When tree T_1 is done, another tree T_2 is used and the process continues until all trees are exhausted, or all points are already members of M_k. (Color figure online)

5 Experiments

We compare NC-DP and NCC to the state-of-the-art method DP and the commonly used method DBSCAN on 15 datasets used in the literature [5,7 9,12][1]. The clustering performance is measured in terms of F-measure[2].

For all algorithms, their parameters are searched as shown in Table 1 and the best F-measure is reported. Because NC-DP and NCC are randomized methods, for each dataset, we report the average result of 10 runs and its standard error. For DP and DBSCAN, they are executed only once. The ensemble size t of all NC estimations is set to 1000, except for the three smallest datasets "iris", "shape" and "seeds", where t is set to 5000 for better stability.

The clustering performances of NC-DP, NCC, DP and DBSCAN are given in Table 2. NC-DP outperforms DP with 11 wins 4 losses; and NCC outperforms

---

[1] Dataset "jain" is from [7]. "d31" is from [12]. "aggregation" is from [5]. "shape" is from [9]. The other datasets are from the UCI repository [8].

[2] The F-measure is calculated as follows, $F = \sum_{k=1}^{K} \frac{|G_k|}{N} \times \frac{2p_k r_k}{p_k + r_k}$, where p_k and r_k are the precision and recall, respectively, based on the confusion matrix. Note that noise points will lower the recall and hence dampen the F-measure. This is one of the reasons we choose F-measure over other external evaluation measures such as Purity or Normalized Mutual Information [1] which ignore noise points in their evaluations. These measures favour clustering results that label more points as noise.

Table 1. Algorithms: Parameters and their search ranges

NC-DP	NCC	DP	DBSCAN
h: $5, 6, .., \min(80, 6d)$	h: $5, 6, .., \min(80, 6d)$	d_C: $0.1\%, 0.2\%, \ldots, 10\%$	ϵ: $0.01, 0.02, .., 2$
$\mathcal{L}$: $3, .., \lceil \sqrt{N} \rceil$	$\mathcal{L}$: $3, \ldots, \lceil \sqrt{N} \rceil$	K: $2, 3, \ldots, 31$	$minPts$: $2, 3, \ldots, 50$
K: $2, 3, .., 31$	γ: $50\%, 51\%, \ldots, 99\%$ quantiles of NC		

Table 2. Best clustering performances on 15 datasets in terms of F measure.

Dataset	N	d	K	F measure			
				NC-DP (SE)	NCC (SE)	DP	DBSCAN
abalone	4177	8	3	0.483 (0.0089)	0.419 (0.0113)	**0.509**	0.255
aggregation	788	2	7	**0.997** (0.0004)	0.995 (0.0004)	0.996	0.991
breast	699	9	2	**0.965** (0.0021)	0.962 (0.0016)	0.917	0.867
d31	3100	2	31	0.970 (0.0003)	**0.974** (0.0003)	0.970	0.914
diabetes	768	8	2	**0.622** (0.0096)	0.618 (0.0142)	0.602	0.538
haberman	306	3	2	**0.643** (0.0026)	0.634 (0.0040)	0.616	0.630
htru2	17898	8	2	**0.972** (0.0004)	0.957 (0.0023)	0.944	0.889
iris	150	4	3	0.952 (0.0084)	0.954 (0.0049)	**0.967**	0.880
jain	373	2	2	**0.989** (0.0078)	0.957 (0.0023)	0.972	0.964
seeds	210	7	3	0.896 (0.0022)	**0.910** (0.0018)	0.909	0.750
shape	160	17	9	**0.743** (0.0033)	0.548 (0.0082)	0.699	0.642
thyroid	215	5	3	0.854 (0.0015)	**0.863** (0.0087)	0.707	0.584
wdbc	569	30	2	0.875 (0.0158)	**0.890** (0.0144)	0.830	0.547
wilt	4339	5	2	0.974 (0.0000)	**0.975** (0.0000)	0.974	**0.975**
yeast	1484	8	10	**0.399** (0.0045)	0.377 (0.0053)	0.359	0.220
win/draw/loss wrt NC-DP					6/0/9	4/0/11	1/0/14
win/draw/loss wrt NCC				9/0/6		5/0/10	2/1/12
average rank				1.67	2.13	2.53	3.60

DP with 10 wins 5 losses. NC-DP is the best performer in terms of average rank, followed by NCC. P-values of pairwise Friedman tests are reported in Table 3. NC-DP is significantly better than DP at 10% significance level; both NC-DP and NCC is significantly better than DBSCAN at 1% significance level.

NCC performed poorly on some datasets, e.g., shape and jain. This is due to a weakness in the assignation process in step 3: when clusters are not separated by a low enough density region, some low density points might receive similar membership scores for different clusters. Note that this is not the same issue as in the varying densities problem which has prevented existing density-based

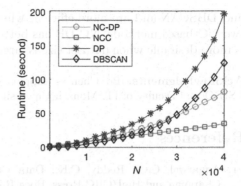

Table 3. Pairwise Friedman tests: p-values.

	NCC	DP	DBSCAN
NC-DP	0.4386	0.0707	0.0008
NCC		0.1967	0.0075
DP			0.0045

Fig. 6. Runtimes of the four methods as the dataset size N increases.

clustering from identifying all clusters. This weakness in NCC causes low density points to be incorrectly assigned, rather than high density points.

A scalability test[3] with respect to dataset size is provided in Fig. 6. It shows that NCC, having a linear time complexity $O(N)$, is much more scalable than the other three methods which all have complexity $O(N^2)$. Note that when N is small, both NC-based methods take longer time due to the overhead computation of building the ensemble of trees. However, when N grows large, NC-based methods are more efficient than density-based ones. This is even the case for NC-DP which has complexity $O(N^2)$, where the gap between NC-DP and DP increases as the data size increases. This is because when N is large, NC estimation is more efficient than density estimation.

6 Conclusions

It is common knowledge that density-based clustering methods fail to detect all clusters in datasets that have hugely varying densities. However, many existing improvements still rely on density to detect clusters. Our proposal of Neighbourhood Contrast (NC) addresses this issue from its root cause by providing an alternative means for detecting clusters. We show that NC is a better means than density for clustering procedures, especially in the presence of hugely varying densities. This is because of the unique property of NC, i.e., it admits all local density maxima, regardless of their densities, to have similar NC values.

We provide two ways of applying NC. We show that NC can be easily incorporated in an existing procedure to replace density by proposing NC-DP. We also devise a new procedure Neighbourhood Contrast Clustering (NCC), which is based on space partitioning and hence has a linear time complexity.

We evaluate the clustering performance of four methods: NC-DP, NCC, DP and DBSCAN. The results show that both NC-based methods outperform DP

---

[3] The datasets are draw randomly from a mixture of bivariate Gaussian distributions with increasing sample size. The four methods achieve similar F-measures.

and DBSCAN and are more efficient when dataset size is large. Comparing the two NC-based methods, NC-DP has better performance than NCC; while NCC is more desirable when dataset size is large because of its linear time complexity.

Acknowledgments. Bo Chen is supported by scholarships provided by Data61, CSIRO and Faculty of IT, Monash University.

References

1. Aggarwal, C.C., Reddy, C.K.: Data Clustering: Algorithms and Applications. Chapman and Hall/CRC Press, Boca Raton (2013)
2. Borah, B., Bhattacharyya, D.: DDSC: a density differentiated spatial clustering technique. J. Comput. **3**(2), 72–79 (2008)
3. Ertöz, L., Steinbach, M., Kumar, V.: Finding clusters of different sizes, shapes, and densities in noisy, high dimensional data. In: Proceedings of the SIAM Conference on Data Mining, pp. 47–58. SIAM (2003)
4. Ester, M., Kriegel, H.P., Sander, J., Xu, X.: A density-based algorithm for discovering clusters in large spatial databases with noise. In: Proceedings of the 2nd International Conference on Knowledge Discovery and Data Mining, pp. 226–231 (1996)
5. Gionis, A., Mannila, H., Tsaparas, P.: Clustering aggregation. ACM Trans. Knowl. Discov. Data (TKDD) **1**(1), 4 (2007)
6. Han, J., Kamber, M.: Data Mining: Concepts and Techniques, 3rd edn. Morgan Kaufmann, Burlington (2011)
7. Jain, A.K., Law, M.H.C.: Data clustering: a user's dilemma. In: Pal, S.K., Bandyopadhyay, S., Biswas, S. (eds.) PReMI 2005. LNCS, vol. 3776, pp. 1–10. Springer, Heidelberg (2005). https://doi.org/10.1007/11590316_1
8. Lichman, M.: UCI machine learning repository (2013). http://archive.ics.uci.edu/ml
9. Müller, E., Günnemann, S., Assent, I., Seidl, T.: Evaluating clustering in subspace projections of high dimensional data. In: Proceedings of the VLDB Endowment, vol. 2, no. 1, pp. 1270–1281 (2009)
10. Ram, A., Sharma, A., Jalal, A.S., Agrawal, A., Singh, R.: An enhanced density based spatial clustering of applications with noise. In: Proceedings of the IEEE International Advance Computing Conference, pp. 1475–1478. IEEE (2009)
11. Rodriguez, A., Laio, A.: Clustering by fast search and find of density peaks. Science **344**(6191), 1492–1496 (2014)
12. Veenman, C.J., Reinders, M.J.T., Backer, E.: A maximum variance cluster algorithm. IEEE Trans. Pattern Anal. Mach. Intell. **24**(9), 1273–1280 (2002)

Clustering of Multiple Density Peaks

Borui Cai[1]([✉]), Guangyan Huang[1], Yong Xiang[1], Jing He[2], Guang-Li Huang[3],
Ke Deng[3], and Xiangmin Zhou[3]

[1] School of Information Technology, Deakin University, Melbourne, Australia
{bcai,guangyan.huang,yong.xiang}@deakin.edu.au
[2] Data Science Research Institute, Swinburne University of Technology,
Melbourne, Australia
lotusjing@gmail.com
[3] School of Science, RMIT University, Melbourne, Australia
guangli.huang@student.rmit.edu.au, {ke.deng,xiangmin.zhou}@rmit.edu.au

Abstract. Density-based clustering, such as Density Peak Clustering
(DPC) and DBSCAN, can find clusters with arbitrary shapes and have
wide applications such as image processing, spatial data mining and text
mining. In DBSCAN, a core point has density greater than a threshold, and can spread its cluster ID to its neighbours. However, the core
points selected by one cut/threshold are too coarse to segment fine clusters that are sensitive to densities. DPC resolves this problem by finding
a data point with the peak density as centre to develop a fine cluster.
Unfortunately, a DPC cluster that comprises only one centre may be too
fine to form a natural cluster. In this paper, we provide a novel clustering of multiple density peaks (MDPC) to find clusters with arbitrary
number of regional centres with local peak densities through extending
DPC. In MDPC, we generate fine seed clusters containing single density peaks, and form clusters with multiple density peaks by merging
those clusters that are close to each other and have similar density distributions. Comprehensive experiments have been conducted on both
synthetic and real-world datasets to demonstrate the accuracy and effectiveness of MDPC compared with DPC, DBSCAN and other base-line
clustering algorithms.

Keywords: Clustering · Density peaks · Cluster merge

1 Introduction

Clustering can discover the relationship of points by grouping similar points into
the same cluster; this capability makes it attractive in many data mining tasks.
K-means finds the best k centres to minimize the overall distance between the
points and their centres [13]. Affinity Propagation (AP) [7] finds the best point
to represent the whole cluster. However, both of them are not effective in finding
non-spherical clusters.

Mean-shift [9] is able to find non-spherical clusters, but it highly relies on
the significance of density gradients among data points. Density-based clustering,

© Springer International Publishing AG, part of Springer Nature 2018
D. Phung et al. (Eds.): PAKDD 2018, LNAI 10939, pp. 413–425, 2018.
https://doi.org/10.1007/978-3-319-93040-4_33

such as Density Peak Clustering (DPC) [14] and DBSCAN [6], use critical data points to form clusters. DBSCAN finds natural shape clusters by finding core data points which spread cluster IDs to their neighbours; and a core point is a data point that has density greater than a threshold. However, core data points that are selected based on one cut/threshold in DBSCAN are too coarse to segment fine clusters that are sensitive to density, as shown in the third dataset in Fig. 4(c), where we can only accurately find the top left clusters (two sparse clusters) or the bottom left clusters (two dense clusters), but not both. Also, in practice it is hard to find the optimal values of DBSCAN's two parameters.

DPC [14] resolves the problem of DBSCAN by finding a data point with the peak density as a centre to develop a fine cluster, and it only needs one parameter d_c, the cut-off distance. DPC converts n-dimensional features into two features: *density* and *delta* (the distance to the point which spreads cluster ID to it), chooses one point with the peak density as the centre for each cluster, and assigns the rest points to the relative cluster centres. DPC can find clusters with more fine densities than DBSCAN. This advantage makes DPC a potential solution to many data mining tasks [15,17]. Unfortunately, a DPC cluster that comprises only one centre (density peak) may be too fine to form a natural cluster, since this is too strict for a natural cluster that comprises multiple regional centres with local peak densities. DPC is unable to achieve a satisfying result due to two problems. First, it is difficult to choose correct cluster centres from candidate density peaks (with "anomalously" large *density* and *delta*); for example, in Fig. 1(c), although we know there are five cluster centres, it is challenging to pick them from those candidate density peaks (large red dots) in Fig. 1(c') and requires exhaustively searching. Second, the assumption in DPC that each cluster only comprises one centre is not always true thus DPC cannot find natural shape clusters accurately; three examples are shown in Fig. 4(b).

Hierarchical clustering inspires us to merge DPC clusters and form the correct clusters that comprise multiple density peaks. We initially generate seed clusters by DPC and merge seed clusters according to their distances. Thus, a good cluster distance is the key for the accuracy. We have tested four existing cluster distances ("single", "average", "complete" and Hausdorff linkage [1]), but their accuracy is not good in merging DPC seed clusters, as shown in Fig. 3.

In this paper, we propose a novel Multiple Density Peaks Clustering (MDPC) method to flexibly find clusters with arbitrary number of regional centres that have local peak densities (as they exist in the real world) through extending DPC. In MDPC, we generate fine seed clusters, which are simpler clusters since each seed cluster has only one density peak, and discover a natural shape cluster with multiple density peaks by merging those seed clusters that are close to each other and have similar density distributions. We conduct comprehensive experiments on both synthetic and real-world datasets to demonstrate the accuracy and effectiveness of MDPC. Therefore, this paper has three contributions:

- We provide a novel MDPC method, which improves DPC in two aspects: to find natural shape clusters with arbitrary number of local density peaks and to form seed clusters by automatic selection of cluster centres.

- We define a new distance between two seed clusters, based on which seed clusters are merged more accurately in MDPC than four counterpart cluster distances ("single", "average", "complete" and Hausdorf [1]).
- We conduct experiments on both synthetic and real-world datasets to prove that MDPC achieves more accurate results than DPC, DBSCAN and other baseline clustering algorithms.

The rest of this paper is organized as follows. Section 2 presents the preliminary knowledge and problem definition. Section 3 details the proposed MDPC method. Section 4 conducts experimental studies to evaluate MDPC. Finally, Sect. 5 concludes the paper.

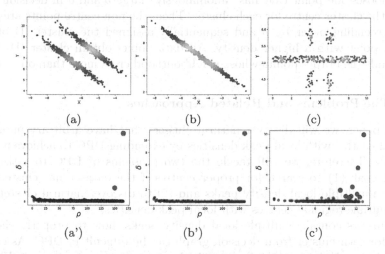

Fig. 1. (a)–(c) are three datasets with color indicates density, and (a')–(c') are corresponding decision graphs with large red dots as candidate cluster centers. (Color figure online)

2   Preliminary Knowledge and Problem Definition

In this section, we present the DPC algorithm, point out its two problems and discuss the state of the art methods that approach these problems.

2.1   The Algorithm of DPC

As we mentioned in Sect. 1, DPC improves DBSCAN to find more fine clusters with only one density peak as a centre. To help understand this, we introduce the main idea of DPC as follows. In DPC, the first step is to determine cluster centres using a 2-dimensional "decision graph" as follows. The n-dimensional feature space of a point is mapped into the 2-dimensional feature space: ρ and δ. The density ρ_i of data point i is given by:

$$\rho_i = \Sigma_j \chi \left(d_{ij} - d_c \right), \tag{1}$$

where $\chi(x) = 1$ if $x < 0$, and $\chi(x) = 0$ otherwise. d_c is a threshold distance and d_{ij} is the Euclidean distance between data points i and j. Then, these local density peaks are those points that have the greatest ρ in its d_c region. Delta δ_i is the minimum distance between point i (with density ρ_i) and any other point, j, with higher density ρ_j:

$$\delta_i = \min_{j:\rho_j > \rho_i} (d_{ij}). \tag{2}$$

In DPC, the two-dimensional features (ρ and δ) form a "decision graph"; an example is shown in Fig. 1(a'), which is transferred from raw data in Fig. 1(a). User chooses one point that has "anomalously" large ρ and δ in decision graph [14] as the cluster centre for each cluster. Then, the non-centre points are sorted in a descending order by ρ and sequentially assigned the cluster ID of their nearest point with a higher density. All data points obtain cluster IDs before DPC finds the border points whose neighbourhood cross more than one clusters.

2.2   The Problems and Related Approaches

In this paper, we will discover natural clusters that have arbitrary number of regional centres with local peak densities by extending DPC to achieve a better accuracy. Therefore, we will tackle the two problems of DPC to satisfy this paper's goal: (1) to search for proper centres if the dataset has clusters that comprise multiple local density peaks and (2) to discover natural clusters that have multiple regional centres with local peak densities.

If clusters contain multiple local density peaks, how to properly choose a centre for each cluster from decision graph can be difficult in DPC. As demonstrated in the dataset containing two single centre clusters as shown in Fig. 1(a) it is intuitive to pick up two large red dots as centres in its decision graph (Fig. 1(a')). However, the cluster with two regional centres with local peak densities in Fig. 1(b) generates nearly identical decision graph (Fig. 1(b')), thus the cluster may incorrectly be divided into two in DPC. Meanwhile in some cases, even though the number of clusters is given, it is still difficult to distinguish centres since those candidates are close to each other in the decision graph. One example is shown in Fig. 1(c), where the large cluster in the middle contains several local density peaks. Although we know there are five clusters, it is challenging to select five centres for five clusters from the candidate large red dots in Fig. 1(c') and requires exhaustively searching. A recursive dividing DPC (3DC) is proposed in [12], but it still uses the heuristic in [14] to pick two greatest $\rho_i \cdot \delta_i$ cluster centres, which is not true in the clusters with multiple density peaks. By applying graph kernel to data vectors before inferring DPC, the candidate centres are differentiated from the common points in [18], however, it still needs to choose cluster centres manually.

Even if we luckily select the right cluster centres, the non-centre local density peaks still can severely decrease the clustering accuracy. The cluster ID propagation rule of DPC is that one non-centre point obtains a cluster ID from its nearest point that has a higher density. If the nearest point with a higher density of one non-centre local density peak belongs to another cluster, this local density peak will wrongly be assigned to that cluster, and this mistake is propagated to other data points to form incorrect clusters, as shown in the three datasets in Fig. 4(b). In the first dataset of Fig. 4(b), the ring shape cluster is incorrectly divided into three parts and the left and right parts are assigned to the other two spherical clusters, respectively; in the second dataset, the bottom arch-shape cluster is also divided into two with the top part group with the top arch; in the third dataset, the dot shape cluster is wrongly grouped with the ring shape cluster surrounding it. To resolve the second problem of DPC, a two-step DPC is developed in [19] using the notion of core-reachable of DBSCAN. However, the strategy is neither well explained nor proved with convincing experiments.

3   The MDPC Approach

In this section, we present our MDPC approach that effectively finds clusters with multiple local density peaks as regional centres by two steps:

- Find seed clusters. We discover the seed clusters that have single density peaks by allowing automatic selection of cluster centres.
- Merge seed clusters. We define a new distance between seed clusters and hierarchically merge these seed clusters using this distance.

3.1   Find Seed Clusters

Different from DPC, natural shape clusters are regarded as the combination of several seed clusters which contain single density peaks in the scope of MDPC. That is, all of the density peak points are centres of seed clusters. Accordingly, we define a cluster centre/density peak as follows:

$$c : \rho_c \geq \rho_j, dist_{ij} \leq d_c, \forall j \in D. \tag{3}$$

According to Eq. (2), the cluster centres who have peak densities in their d_c radius have δ larger than d_c. Accordingly, we write the cluster centres, C, as:

$$C = \{c | \delta_c > d_c, c \in D\}. \tag{4}$$

With the cluster centres C, we find relevant seed clusters, which comprises only one density peak and satisfies the characteristic of a DPC cluster (data points in a cluster coarsely following a density descending order from centre to border).

In our implementation, the DPC's original definition of density in Eq. (1) performs not good: it either finds the unnecessary centres in a small region or incorrectly labels the non-centralized points as the cluster centres, because it regards each neighbour in d_c region of one point the same weight without considering the distance to the centre. As a result, we modify the fuzzy density defined in [5] as:

$$\rho_i = \Sigma_j \left(1 - \frac{dist_{ij}^2}{d_c^2} \right), dist_{ij} \leq d_c, \forall j \in D. \tag{5}$$

In Fig. 2, the cluster centres found by fuzzy density metric (Eq. (5)) are more significant than those found by DPC density (Eq. (1)). When d_c is fixed, cluster centres (cross) found by the fuzzy density (Fig. 2(c)-(d)) are sparser than those found by the DPC density (Fig. 2(a)-(b)). Meanwhile, the fuzzy density is also more robust to d_c; the number of the cluster centres in Fig. 2(a) significantly decreases when d_c increases to 0.04, while the cluster centres found using fuzzy density keep stable when increasing d_c from 0.03 (Fig. 2(c)) to 0.04 (Fig. 2(d)). The pseudocode of finding seed clusters is shown in Algorithm 1. We calculate ρ and δ of each data point at lines 1–7 and discover those cluster centres at line 6. Then we find seed cluster centres and assign cluster IDs to the non-centre data points at lines 8–13.

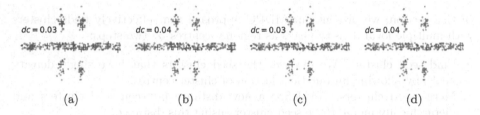

Fig. 2. Centres (colored cross) discovered by original density ((a)-(b)), and fuzzy density ((c)-(d)), with d_c as the cut-off distance. (Color figure online)

Algorithm 1. Find Seed Clusters (n_i: nearest point with a higher density)

Input: Dataset D, Cutoff distance d_c
1: **for** each $i \in D$ **do**
2:     Calculate ρ_i Eq. (5)
3: **end for**
4: **for** each $i \in D$ **do**
5:     Calculate δ_i by Eq. (2), get n_i
6:     Add i into C if $\delta_i > d_c$
7: **end for**

8: Sort D in descending order of ρ
9: **for** each $i \in D$ **do**
10:     **if** i in C **then** $clusterID_i = $ new id
11:     **else** $clusterID_i = clusterID_{n_i}$
12:     **end if**
13: **end for**
Output: points $clusterID$

3.2   Merge Seed Clusters

Borrowing the idea from hierarchical clustering, we iteratively merge those seed clusters. In our MDPC, merging is conducted only on two clusters which have border regions (mutually have border points whose neighbourhood cross the two clusters). We need to define a new distance metric to achieve an effective merging since the wide-used linkage distances are too coarse for these fine seed clusters. First, two rules are defined to determine whether the seed clusters should be merged based on the characteristic of these seed clusters:

- Rule 1: Two clusters should be merged together if they have comparable scales and spatially close border points, as exhibited in Fig. 1(b).
- Rule 2: One small seed cluster should be absorbed by a larger cluster, if its scale is similar to the border region of the larger cluster and have spatially close border regions.

A seed cluster in MDPC has a coarsely monotonic density distribution in a descending order from its centre to border. Based on this characteristic, we define a new cluster distance by both their density distributions and their spatial distance. As a density-based metric, our cluster distance should define the comparable scale in Rule 1 by measuring their difference of density distributions. If the density is monotonic, this difference can be defined by two aspects: the average density and the density-descending rate. Given two seed clusters A and B with average density ρ_{avg}^A and ρ_{avg}^B, the distance between the average densities of A and B is:

$$d_{merge} = \frac{\|\rho_{avg}^A - \rho_{avg}^B\|}{max\left(\rho_{avg}^A, \rho_{avg}^B\right)}. \tag{6}$$

Similarly, in Rule 2, we define the cluster distance to be the difference of the density distributions of the smaller cluster and the border region of the larger one. Assuming A absorbs B ($\rho_{avg}^A > \rho_{avg}^B$), and the border region of A, $border^A$ is defined as $\{i | dist(i, j) < d_c, i \in A, j \in B\}$. Thus ρb_{avg}^A, the average density of $border^A$, is calculated as the average density of points in $border^A$ together with their neighbours. In this way, the distance between ρb_{avg}^A and ρ_{avg}^B is defined as:

$$d_{absorb} = \frac{\|\rho b_{avg}^A - \rho_{avg}^B\|}{max\left(\rho b_{avg}^A, \rho_{avg}^B\right)}. \tag{7}$$

We normalize the above distances and enable them to be combined together and get the overall measurement as $d_{density}$:

$$d_{density} = min\left(d_{merge}, d_{absorb}\right). \tag{8}$$

The distance of the density descending rate and the spatial closeness of the clusters are measured by d_{border}, with the density of cluster centre as ρ_c and the density descending rate from the cluster centre to point i as $r_i = \frac{\rho_c}{\rho_i}$.

$$d_{border} = \inf_{i \in border^A, j \in border^B} \frac{d_{ij}}{2d_c} \times \left(r_i^A + r_j^B\right) \|r_i^A - r_j^B\|. \tag{9}$$

We calculate d_{border} to satisfy (1) A and B is spatial close if one point in B falls in the d_c region of A's centre; (2) when one point of B falls in the d_c region of A's border point, note that the border has smaller density and thus their distance must be proportionally smaller to compensate the density difference. Also, the larger the density of a pair of border points (one from A and the other from B), the more likely these two border points are changed into inner points by merging the two clusters. We explain how the new metric can measure the difference of the density descending rates and clusters' spatial closeness as follows. If two clusters are not close to each other, that means the distance of the border points pair d_{ij} is great then makes d_{border} great. A greater value of $\|r_i^A - r_j^B\|$ means their density descending rates are more different. In addition, we prefer to merge those clusters whose border points have greater densities (i.e., the value of $(r_i^A + r_j^B)$ is small). Combining the above density distance $S_{density}$ with border distance S_{border} the total distance of two seed clusters is:

$$d_{AB} = d_{border} \times exp(d_{density}), \tag{10}$$

where d_{AB} is non-negative and symmetric, and exponential (exp) is used to ensure that d_{AB} is sensitive to small density variation and prevent merging of two clusters if their densities are globally distinctive.

Using the d_{AB} we develop a hierarchical merging algorithm (Algorithm 2) to form the final clusters with arbitrary numbers of density peaks. We calculate the pair-wise distances of seed clusters at lines 1–6. Then, we merge those with distances under a threshold and re-assign new cluster IDs to all the data points at lines 7–12. Results of each iteration are stored at line 13.

Algorithm 2. Hierarchical Merging of Seed Clusters (*dists*: cluster distance list)

Input: Seed clusters C_{seed}	8: for seed cluster $a, b \in C_{seed}$ **do**
1: **for** seed clusters pair $a, b \in C_{seed}$ **do**	9: **if** $d^{ij} \leq d$ **then** merge a b
2: **if** a, b has border points **then**	10: **end if**
3: calculate d_{AB} with (10)	11: **end for**
4: add d_{AB} into *dists*	12: $clusterID_{merge}$ = merged IDs;
5: **end if**	13: add $clusterID_{merge}$ into L;
6: **end for**	14: **end for**
7: **for** each $d \in dists$ **do**	**Output:** Points cluster IDs list L

3.3   Algorithm Complexity

We denote the number of neighbours of each point as N_{dc} regarding to d_c, the size of border point as N_b, and N_s as the number of seed cluster pairs that mutually have border points, in practice, N_{dc}, N_b and N_s are far less than the dataset size, N. The time complexity of finding seed clusters (Algorithm 1) and border points is $O(2N \log N + N \times N_{dc} + N)$. The time complexity of calculating the average densities of all seed clusters is $O(N)$, and the complexity of calculating all average border densities is $O(N_b)$. $O(N_s + N_s \log N_s)$ is spent to compute and sort distances of seed cluster pairs, and $O(N_s)$ is spent on hierarchical merging. Thus, the overall time complexity of merging seed clusters (Algorithm 2) is $O(N + N_b + 2N_s + N_s \log N_s)$.

4   Experiments

In this section, we use various datasets including both synthetic datasets (four shape datasets and two density datasets) and one real-world dataset to evaluate the accuracy of our MDPC. Accuracy is measured using Normalized Mutual Information (NMI) [16]. All algorithms, including both the proposed MDPC and the counterpart methods (DPC, DBSCAN, K-means, AP and mean-shift), are compared using the best performance under their optimal parameters. We use the heuristic $\rho \times \delta$ in [14] to select the optimal cluster centres for DPC and we automatically search the best NMI of our MDPC from its output list. All the algorithms are implemented in Python 3.2 (with packages scikit-learn, numpy) and experiments are run on Windows 10 with 3.4 GHz CPU and 16 GB RAM.

4.1   Synthetic Datasets

We use six different synthetic datasets with true labels. Four shape datasets with natural shape clusters (shape datasets) are Pathbased [3], Jain [11], Flame [8] and Spiral [3]. Two datasets with uneven density-pattern clusters (density datasets) are Compound [20] and Aggregation [10]. We evaluate the proposed cluster distance by comparing with four counterpart cluster distances ("single", "average", "complete" and Hausdorff [1]). We set the same experimental environment by replacing our proposed cluster distance with each of the four in MDPC. In Fig. 3, we can see that our proposed cluster distance achieves the best accuracy compared to the other four distances in all of the six datasets. The best performance improvement of the proposed distance for MDPC is in Flame, where the accuracy of the proposed method wins the best of the other four distances (Single) by 1.000 to 0.521.

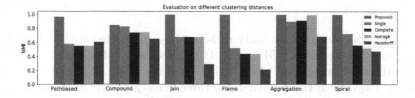

Fig. 3. Evaluation of the cluster distance of MDPC with four cluster distances.

We compare our MDPC with two other density-based methods (DPC and DBSCAN) using three challenging datasets containing natural shape clusters with multiple local density peaks (Pathbased, Jain and Compound) (Fig. 4). We can see from the first row that MDPC achieves the best accuracy in the Pathbased dataset, where both DPC and DBSCAN incorrectly split the ring shape cluster into three and two parts, respectively. In Jain dataset as shown in the second row, MDPC correctly separates the two arch-shaped clusters, while DPC incorrectly divides the bottom arch-shaped cluster into two clusters and wrongly assigns the upper cluster to the other arch-shaped cluster and DBSCAN wrongly splits the top arch-shaped cluster into two parts. In Compound dataset at the last row, only MDPC successfully finds both the two clusters (a ring and a dot the middle of the ring) at the bottom left and two spatial close round-shaped clusters at the top left, while DBSCAN only can correctly identify the two dense clusters at the bottom left or the two close sparse round-shaped clusters but not both since its one cut/threshold strategy cannot adapt itself to the clusters with different density patterns.

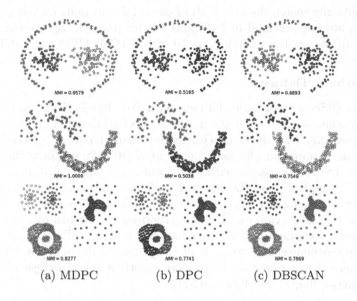

| (a) MDPC | (b) DPC | (c) DBSCAN |

Fig. 4. (a)–(c) are the clustering results of MDPC, DPC and DBSCAN.

We also evaluate the scalability of MDPC by comparing with both density-based clustering methods (DPC and DBSCAN) and other popular ones (K-means, Affinity Propagation (AP) and mean-shift) using all of the six datasets (Table 1). The overall trend is that the three density-based clustering methods, MDPC, DPC and DBSCAN, perform better than the three non-density-based clustering methods. In Table 1 MDPC excels both DPC and DBSCAN in all of the six datasets, while DPC performs better than DBSCAN in Flame and Aggregation and DBSCAN performs better than DPC in Pathbased, Jain and Compound. This validates that our MDPC is more adaptive to density sensitive datasets than too coarse clustering in DBSCAN and too fine clustering in DPC.

4.2 Real Datasets

We use the iris dataset (with four features and three labels) from the real-world dataset UCI [2] to evaluate the accuracy of the three density-based clustering

Table 1. NMI on synthetic datasets, with the best in bold and negative as '−'.

Type	Dataset	MDPC	DPC	DBSCAN	K-means	AP	Mean-shift
Shape	Pathbased	**0.9579**	0.5165	0.6893	0.5102	0.3530	0.5431
	Jain	**1.0000**	0.5038	0.7549	0.3362	0.2073	0.3282
	Flame	**1.0000**	**1.0000**	0.8654	0.4478	0.3011	0.4442
	Spiral	**1.0000**	**1.0000**	**1.0000**	-	0.3142	0.2767
Density	Compound	**0.8277**	0.7482	0.7869	0.7421	0.5289	0.8110
	Aggregation	**1.0000**	**1.0000**	0.9787	0.8376	0.5035	0.8983

methods (MDPC, DPC and DBSAN). We use PCA to map the original four features into three features for visualization. From the result in Fig. 5, we can see that the number of mislabelled items (marked in red color) by MDPC in Fig. 5(a) is significantly less than both DPC in Fig. 5(b) and DBSCAN in Fig. 5(c) where DBSCAN wrongly groups the three categories into only two clusters.

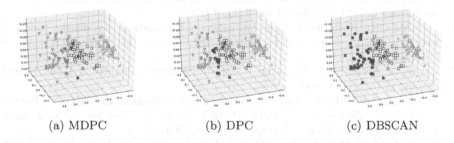

(a) MDPC   (b) DPC   (c) DBSCAN

Fig. 5. Clustering results for the iris dataset, (a)–(c) correspond to MDPC, DPC and DBSCAN. Incorrectly clustered points are in red color. (Color figure online)

5   Conclusion

In this paper, we provide the MDPC clustering method for shape and density sensitive datasets. MDPC overcomes the two problems of DPC by automatically selecting cluster centres and finding natural shape clusters with multiple local density peaks. Extensive experiments based on both synthetic and real-world datasets have demonstrated that our MDPC is a more adaptive clustering method for density sensitive datasets, compared with too coarse clustering in DBSCAN and too fine clustering in DPC, and thus achieves the best accuracy and effectiveness.

Acknowledgement. This work was partially supported by Australia Research Council (ARC) DECRA Project (DE140100387).

References

1. Basalto, N., Bellotti, R., De Carlo, F., Facchi, P., Pantaleo, E., Pascazio, S.: Hausdorff clustering. Phys. Rev. E **78**(4), 046112 (2008)
2. Blake, C.L., Merz, C.J.: UCI repository of machine learning databases. Department of Information and Computer Science, University of California, Irvine, vol. 55 (1998). http://www.ics.uci.edu/~mlearn/mlrepository.html
3. Chang, H., Yeung, D.Y.: Robust path-based spectral clustering. Pattern Recogn. **41**(1), 191–203 (2008)
4. Cho, M., MuLee, K.: Authority-shift clustering: hierarchical clustering by authority seeking on graphs. In: CVPR, pp. 3193–3200. IEEE (2010)
5. Du, M., Ding, S., Xue, Y.: A robust density peaks clustering algorithm using fuzzy neighborhood. Int. J. Mach. Learn, Cyb (2017). https://doi.org/10.1007/s13042-017-0636-1
6. Ester, M., Kriegel, H.P., Sander, J., Xu, X., et al.: A density-based algorithm for discovering clusters in large spatial databases with noise. In: SIGKDD, pp. 226–231 (1996)
7. Frey, B.J., Dueck, D.: Clustering by passing messages between data points. Science **315**(5814), 972–976 (2007)
8. Fu, L., Medico, E.: Flame, a novel fuzzy clustering method for the analysis of DNA microarray data. BMC Bioinform. **8**(1), 3 (2007)
9. Fukunaga, K., Hostetler, L.: The estimation of the gradient of a density function, with applications in pattern recognition. IEEE Trans. Inf. Theory **21**(1), 32–40 (1975)
10. Gionis, A., Mannila, H., Tsaparas, P.: Clustering aggregation. ACM Trans. Knowl. Discov. Data **1**(1), 4 (2007)
11. Jain, A.K., Law, M.H.C.: Data clustering: a user's dilemma. In: Pal, S.K., Bandyopadhyay, S., Biswas, S. (eds.) PReMI 2005. LNCS, vol. 3776, pp. 1–10. Springer, Heidelberg (2005). https://doi.org/10.1007/11590316_1
12. Liang, Z., Chen, P.: Delta-density based clustering with a divide-and-conquer strategy: 3DC clustering. Pattern Recogn. Lett. **73**, 52–59 (2016)
13. Ray, S., Turi, R.H.: Determination of number of clusters in k-means clustering and application in colour image segmentation. In: ICAPRDT, Calcutta, India, pp. 137–143 (1999)
14. Rodriguez, A., Laio, A.: Clustering by fast search and find of density peaks. Science **344**(6191), 1492–1496 (2014)
15. Shi, Y., Chen, Z., Qi, Z., Meng, F., Cui, L.: A novel clustering-based image segmentation via density peaks algorithm with mid-level feature. Neural Comput. Appl. **28**(1), 29–39 (2017)
16. Strehl, A., Ghosh, J.: Cluster ensembles-a knowledge reuse framework for combining multiple partitions. J. Mach. Learn. Res. **3**(Dec), 583–617 (2002)
17. Wang, P., Xu, B., Xu, J., Tian, G., Liu, C.L., Hao, H.: Semantic expansion using word embedding clustering and convolutional neural network for improving short text classification. Neurocomputing **174**, 806–814 (2016)
18. Yang, H., Zhao, D., Cao, L., Sun, F.: A precise and robust clustering approach using homophilic degrees of graph kernel. In: Bailey, J., Khan, L., Washio, T., Dobbie, G., Huang, J.Z., Wang, R. (eds.) PAKDD 2016. LNCS (LNAI), vol. 9652, pp. 257–270. Springer, Cham (2016). https://doi.org/10.1007/978-3-319-31750-2_21

19. Yaohui, L., Zhengming, M., Fang, Y.: Adaptive density peak clustering based on K-nearest neighbors with aggregating strategy. Knowl.-Based Syst. **133**, 208–220 (2017)
20. Zahn, C.T.: Graph-theoretical methods for detecting and describing gestalt clusters. IEEE Trans. Comput. **100**(1), 68–86 (1971)

A New Local Density for Density Peak Clustering

Zhishuai Guo[1,2], Tianyi Huang[1], Zhiling Cai[1], and William Zhu[1(✉)]

[1] Lab of GRC and AI, Institute of Fundamental and Frontier Sciences,
University of Electronic Science and Technology of China, Chengdu, China
wfzhu@uestc.edu.cn
[2] Yingcai Honors College, University of Electronic Science and Technology of China,
Chengdu, China

Abstract. Density peak clustering is able to recognize clusters of arbitrary shapes, so it has attracted attention in academic community. However, existing density peak clustering algorithms prefer to select cluster centers from dense regions and thus easily ignore clusters from sparse regions. To solve this problem, we redefine the local density of a point as the number of points whose neighbors contain this point. This idea is based on our following finding: whether in dense clusters or in sparse clusters, a cluster center would have a relatively high local density calculated by our new measure. Even in a sparse region, there may be some points with high local densities in our definition, thus one of these points can be selected to be the center of this region in subsequent steps and this region is then detected as a cluster. We apply our new definition to both density peak clustering and the combination of density peak clustering with agglomerative clustering. Experiments on benchmark datasets show the effectiveness of our methods.

Keywords: Local density · K-nearest neighbors
Density peak clustering · Agglomerative clustering

1 Introduction

Clustering has been widely used in knowledge discovery, computer vision, pattern recognition and so on [2,3,9,10,14,24]. It tries to partition data into clusters, where similar points are clustered together and dissimilar points are separated to different clusters [13,18,28].

Many clustering algorithms, such as K-means [11], often fail when dealing with clusters with nonspherical shapes [8]. As an important way to solve this problem, density peak clustering (DPC) [22] is able to recognize clusters regardless of their shape and of the dimensionality. It is based on the assumptions that a cluster center is characterized by a higher local density than its neighbors and by a relatively large distance from any point with a higher local density. After cluster centers are selected, each remaining point is assigned to the same cluster as its nearest neighbor of higher local density.

© Springer International Publishing AG, part of Springer Nature 2018
D. Phung et al. (Eds.): PAKDD 2018, LNAI 10939, pp. 426–438, 2018.
https://doi.org/10.1007/978-3-319-93040-4_34

Based on DPC, many researches have been conducted [4,5,7,16,17,26,27,30]. Xu et al. developed an efficient hierarchical clustering algorithm based on DPC and granular computing [27,31,32]. Some researchers constructed DPC-AC to combine DPC with agglomerative clustering [16,17,30]. DPC-AC uses DPC to generate initial clusters and then merges them according to some affinity measure between clusters. DPC-AC behaves better than DPC on datasets where a cluster has more than one density peak.

However, existing clustering algorithms based on DPC tend to select cluster centers from points in dense regions and easily ignore points in other regions. Therefore, clusters from sparse regions cannot be detected and points in these clusters are often assigned to wrong clusters or are mistakenly treated as noises. Although a sparse region with a few points is likely to be a noisy region, a sparse region with a significantly large number of points should be treated as a cluster. Even though several different definitions of local density have been proposed [5,7,16,26], DPC and DPC-AC still do not perform well when different clusters have big density differences.

To address this issue, we redefine the local density of a point as the number of points whose K-nearest neighbors contain this point. Even in a sparse region, there may be some points with high local densities in our new definition, thus one of these points can be selected to be the center of this region in subsequent steps and this region is then detected as a cluster. Therefore, in density peak clustering with our new local density (NDPC), sparse clusters can be correctly detected instead of being merged into other clusters or be treated as noises as in DPC. Moreover, outliers in sparse regions with a few points would have small local densities in our definition, so they are unlikely to be selected to be cluster centers in our algorithms.

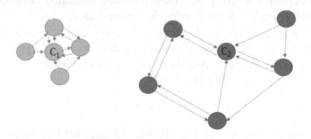

Fig. 1. A directed 2-nearest-neighbor graph. There are two directed edges from every point, directing to its 2-nearest neighbors.

Figure 1 is an illustration of our new local density. There are two clusters in Fig. 1. The points in the left cluster are distributed densely, thus every point has many neighbors within a small distance. The points in the right cluster are distributed sparsely, thus every point has fewer neighbors within a small distance or even no neighbor. Intuitively, C_1 and C_2 should be selected as cluster centers. However, DPC easily ignores C_2 because C_2 would have a very low local density in DPC. In our new definition, both C_1 and C_2 have high local densities because they are contained in the 2-nearest neighbors of many points. Thus, in NDPC, both C_1 and C_2 can be selected to be cluster centers and then both clusters can be easily detected.

Further, we apply our new local density to DPC-AC to construct NDPC-AC. Similar to DPC, DPC-AC does not perform well when different clusters have very different densities because DPC-AC also tends to ignore sparse clusters. With our new local density, NDPC-AC would detect sparse clusters better.

In experiments on benchmark datasets, NDPC behaves much better than DPC. NDPC-AC behaves better than not only DPC-AC but also some state-of-the-art clustering algorithms.

2   Related Work

Density peak clustering (DPC) is able to recognize clusters regardless of their shape and of the dimensionality of the space in which they are embedded [22]. It is based on the assumptions that a cluster center is characterized by a higher local density than its neighbors and by a relatively large distance from any point with a higher local density.

Given a set of data points $X = \{x_1, x_2, \ldots, x_n\}$, DPC defines the local density of a point x_i as

$$\rho_i = \sum_j \chi(dist(i,j) - d_c), \tag{1}$$

where $\chi(d) = \begin{cases} 1, & if\ d < 0 \\ 0, & otherwise \end{cases}$, d_c is a cutoff distance and $dist(i,j)$ is the Euclidean distance between x_i and x_j. Basically, ρ_i is equal to the number of points within distance d_c from x_i. The minimum distance between x_i and any other point with a higher local density is denoted as δ_i.

To select cluster centers, DPC plots a decision graph, which has ρ and δ as its axes. In this graph, points with reasonably high ρ_i and large δ_i are manually selected to be cluster centers. Another way to select cluster centers is to select points whose γ_i are reasonably large, where

$$\gamma_i = \rho_i \delta_i. \tag{2}$$

After cluster centers are selected, each remaining point is assigned to the same cluster as its nearest neighbor of higher local density.

Even though DPC provides a new view for clustering, existing clustering algorithms based on DPC do not perform well on datasets where different clusters have very different densities, while such datasets are common in the real world [7].

To solve this problem, several diffierent definitions of local density were proposed. In DPC-KNN-PCA [7], the local density of a point x_i is defined as

$$\rho_i = exp(-\frac{1}{K} \sum_{x_j \in \mathcal{N}_i^K} dist(i,j)^2), \tag{3}$$

where $\mathcal{N}_i^K$ is the set of K-nearest neighbors of x_i.

In FKNN-DPC [26], the local density of x_i is defined as

$$\rho_i = \sum_{x_j \in \mathcal{N}_i^K} exp(-dist(i,j)). \tag{4}$$

In APDC-KNN [16], the local density of x_i is defined as

$$\rho_i = \sum_{x_j \in \mathcal{N}_i^K} exp(-dist(i,j)^2/d_c^2). \tag{5}$$

However, the above three local densities still have not solved the problem well. The distances from a point in a sparse region to its K-nearest neighbors are relatively large, therefore, its local density is relatively low in any above method.

Another problem of DPC is that it does not perform well when a cluster has more than one density peak [30]. To solve this problem, some researchers constructed DPC-AC to combine DPC with agglomerative clustering [17,30]. Agglomerative clustering, e.g., Chameleon [15] and GDL [29], begins with a large number of initial clusters and then merges them according to the affinity between clusters [10]. Initial clusters in agglomerative clustering cannot capture the local structure of data well, thus harming the subsequent steps. On the other hand, too many initial clusters will imped the efficiency. DPC-AC generates initial clusters by DPC and then merges them according to the affinity between clusters. Compared with agglomerative clustering, DPC-AC has fewer initial clusters but captures the local structure of data better because it takes the local density into consideration. However, the local density in DPC-AC is the same as that in DPC. Thus, DPC-AC still does not perform well when different clusters have very different densities in DPC.

3   Methodology

In this section, we propose a new local density to improve DPC on datasets where different clusters have very different densities in DPC. Furthermore, we apply our new local density to DPC-AC.

3.1   A New Local Density

We find that, whether in dense clusters or in sparse clusters, a cluster center would be contained in K-nearest neighbors of more points than others. Therefore, we define the local density of a point as the number of points whose K-nearest neighbors contain this point.

Our new local density, denoted as ρ', is calculated as follows. We construct a K-nearest-neighbor graph G. For any point x_i in G, if x_j is one of K-nearest neighbors of x_i, then there are a directed edge from point x_i to point x_j and a weight W_{ij} for this edge. Basically, W_{ij} is calculated by

$$W_{ij} = \begin{cases} 1, if\ x_j \in \mathcal{N}_i^K, \\ 0, otherwise. \end{cases} \tag{6}$$

In our measure, the local density of x_i is

$$\rho_i' = \sum_j W_{ji}. \tag{7}$$

3.2 NDPC Algorithm

We substitute the local density in DPC with our new local density to construct NDPC. Accordingly, γ_i in Formula (2) is as follows,

$$\gamma_i = \rho_i' \delta_i. \tag{8}$$

Points whose γ_i are reasonably large are selected to be cluster centers.

The NDPC algorithm is summarized in Algorithm 1.

Algorithm 1. NDPC

Input: A set of data points $X = \{x_1, x_2, ..., x_n\}$; n_T, the number of final clusters.
Output: A set of n_T clusters $V^c = \{C_1, C_2, ..., C_{n_T}\}$, where C_p is a cluster.
1: Calculate ρ_i' of each point x_i according to Formula (7).
2: Calculate δ_i of each point x_i.
3: Calculate γ_i of each point x_i according to Formula (8).
4: Sort all the points by $\gamma = \{\gamma_1, \gamma_2, ..., \gamma_n\}$ in descending order and then select the first n_T points in the sorted list to be cluster centers.
5: Assign each point x_i, except those selected as cluster centers, to the same cluster as its nearest neighbor x_j of higher local density ρ_j', thus forming the set of n_T clusters $V^c = \{C_1, C_2, ..., C_{n_T}\}$.
6: **Return** V^c.

In NDPC, dense clusters will lose superiority over sparse clusters. A point in a dense region has more neighbors nearby, but this does not necessarily mean that there are more points that contain it in their K-nearest neighbors. On the other hand, even in a sparse region, there may be some points that have high local densities in our definition, thus one of these points can be selected to be a center and then this region is detected as a cluster. Therefore, in NDPC, sparse clusters can be correctly detected instead of being merged into other clusters or be treated as noises as in DPC.

In the following part, we analyze DPC, DPC-KNN-PCA and NDPC on a synthetic dataset Unbalance [21] and a real-world dataset Lung [25].

The synthetic dataset Unbalance in Fig. 2(a), Fig. 2(b) and Fig. 2(c) is composed of eigth Gaussian clusters. Points in the left three clusters are distributed densely, while points in the right five clusters are distributed sparsely. We use each of the above three algorithms to select eight cluster centers and then to

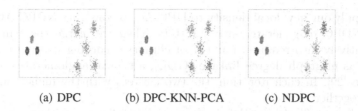

(a) DPC   (b) DPC-KNN-PCA   (c) NDPC

Fig. 2. Analysis of DPC, DPC-KNN-PCA and NDPC on synthetic dataset Unbalance. Pentagrams indicate the points that are selected as cluster centers.

generate eight clusters. As we can see from Fig. 2(a) and Fig. 2(b), DPC selects five cluster centers from the three dense clusters and DPC-KNN-PCA selects six cluster centers from the three dense clusters. This suggests that both DPC and DPC-KNN-PCA are suffering from the density difference between clusters. In contrast, as shown in Fig. 2(c), NDPC selects one center from every cluster and gets the correct clustering result.

Lung dataset is a real-world dataset about lung cancer. Intuitively, there are two sparse clusters in the two rings in Fig. 3(a). This is coincident with the ground truth. As shown in Fig. 3(b), DPC selects all the cluster centers from the dense region, so it does not detect the two clusters in the rings. From Fig. 3(c), we can see that DPC-KNN-PCA succssfully detects one of those two sparse clusters, but fails to detect the other one. In contrast, as shown in Fig. 3(d), our NDPC selects one center in each of those two sparse clusters and thus successfully detects both of those two sparse clusters.

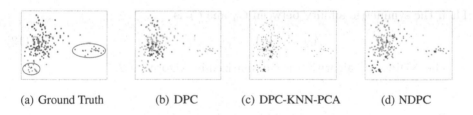

(a) Ground Truth   (b) DPC   (c) DPC-KNN-PCA   (d) NDPC

Fig. 3. Analysis of DPC, DPC-KNN-PCA and NDPC on Lung dataset.

3.3 NDPC-AC Algorithm

Since DPC does not perform well on datasets where a cluster has more than one density peak, some researchers combined DPC with AC to develop a new algorithm, DPC-AC. Though DPC-AC can deal with datasets where a cluster has more than one density peak better than DPC, it still does not perform well on datasets where different clusters have very different densities.

We apply our new local density to DPC-AC to construct NDPC-AC. Firstly, we use NDPC to generate initial clusters. Then we merge these initial clusters iteratively into required number of clusters according to the same affinity measure as in graph degree linkage (GDL), a typical agglomerative clustering algorithm [29]. In each iteration, the two clusters with the highest affinity are merged together.

The cluster affinity in GDL is computed as follows. Firstly, we construct a K-nearest-neighbor graph G'. For any point x_i in G', if x_j is one of K-nearest neighbors of x_i, then there are a directed edge from point x_i to point x_j and a weight W'_{ij} for this edge. W'_{ij} is calculated by

$$W'_{ij} = \begin{cases} exp(-\frac{dist(i,j)^2}{\sigma^2}), & if\ x_j \in \mathcal{N}_i^K, \\ 0, & otherwise, \end{cases} \quad (9)$$

where $\sigma^2 = \frac{1}{n \times K}[\sum_{i=1}^n \sum_{x_j \in \mathcal{N}_i^K} dist(i,j)^2]$.

Given a point x_i, the average indegree from and the average outdegree to a cluster C are defined as $deg_i^-(C) = \frac{1}{|C|} \sum_{j \in C} W'_{ji}$ and $deg_i^+(C) = \frac{1}{|C|} \sum_{j \in C} W'_{ij}$, respectively, where $|C|$ is the cardinality of set C. The affinity between a point and a cluster is defined as the product of the average indegree and average outdegree,

$$A_{i \to C} = deg_i^-(C)deg_i^+(C). \quad (10)$$

The affinity from cluster C_b to C_a is

$$A_{C_b \to C_a} = \sum_{x_i \in C_b} A_{i \to C_a} = \sum_{x_i \in C_b} deg_i^-(C_a)deg_i^+(C_a). \quad (11)$$

Then the symmetric affinity between C_a and C_b is

$$A_{C_a, C_b} = A_{C_b \to C_a} + A_{C_a \to C_b}. \quad (12)$$

The NDPC-AC algorithm is summarized in Algorithm 2.

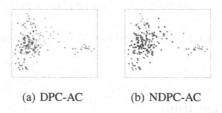

(a) DPC-AC   (b) NDPC-AC

Fig. 4. Analysis of DPC-AC and NDPC-AC on Lung dataset.

Figure 4 is an illustration of clustering results of DPC-AC and NDPC-AC on Lung dataset. For both algorithms, we generate ten initial clusters and then merge them into five clusters. Compared with the ground truth shown in Fig. 3 (a), DPC-AC fails, while our NDPC-AC successfully detects all the five clusters.

Algorithm 2. NDPC-AC

Input: A set of data points $X = \{x_1, x_2, ..., x_n\}$; n_c, the number of initial clusters; n_T, the number of final clusters.

Output: A set of n_T clusters $V^c = \{C_1, C_2, ..., C_{n_T}\}$.

1: Generate a set of n_c initial clusters $V^c = \{C_1, C_2, ..., C_{n_c}\}$ by NDPC algorithm.

2: **while** $n_c > n_T$ **do**

3:   Search two clusters C_p and C_q, such that $\{C_p, C_q\} = argmax_{\{C_a, C_b | a \neq b\}} A_{C_a, C_b}$, where A_{C_a, C_b} is the affinity measure between C_a and C_b, calculated by Formula (12).

4:   $V^c \longleftarrow V^c \cup \{C_p \cup C_q\} \setminus \{C_p, C_q\}$ and $n_c = n_c - 1$.

5: **end while**

6: **Return** V^c.

---

4 Experiments

In this section, we demonstrate the effectiveness of our NDPC and NDPC-AC on benchmark datasets.

4.1 Comparison Scheme

In this subsection, we describe datasets that we use, comparison algorithms and some details in the experiments.

Seven benchmark datasets are used in our experiments: Coil20 [19], Mnist [6], USPS [12], Isolet [1], Lung [25], Vote [1] and ORL [23]. The basic information of benchmark datasets are shown in Table 1.

Table 1. Description of the datasets

Datasets	# Instances	# Attributes	# Clusters	Category
Coil20	1440	1024	20	Object images
Mnist	10000	784	10	Handwritten digit images
USPS	11000	256	10	Handwritten text images
Isolet	1560	617	26	Spoken letters
Lung	203	3312	5	Lung cancer
Vote	435	16	2	Voting records
ORL	400	10304	40	Human faces

We compare our NDPC and NDPC-AC with K-means [11], N-Cut [24], CLR [20], Chameleon [15], GDL [29], DPC [22], DPC-KNN-PCA [7], and DPC-AC [17,30]. K-means iteratively updates the cluster centers with points in corresponding clusters and assigns all the points to the nearest center until some condition is reached. N-Cut works by dividing the data graph according to inter-group dissimilarity and intra-group similarity. CLR is also a graph-based clustering algorithm. Its data similarity graph is adaptive in iterations of clustering procedure and it learns a graph with exactly n_T connected components, where n_T is the number of clusters in ground truth. Chameleon is a classical agglomerative clustering algorithm. It generates initial clusters by dividing the K-nearest-neighbor graph of the data set into several sub-graphs such that the edge cut is minimized. Then it merges these initial clusters according to cluster affinity which is based on relative interconnectivity and relative closeness. GDL is an effective agglomerative clustering algorithm. It constructs initial clusters as weakly connected components of a 1-nearest-neighbor graph. Then it merges these initial clusters according to cluster affinity which is based on indegree and outdegree. DPC-KNN-PCA is an algorithm based on DPC. It uses K-nearest neighbors to calculate local density and uses PCA to preprocess high-dimensional data. DPC-AC uses DPC to generate initial clusters and merges them according to the affinity between clusters. For the sake of fairness in comparison, the affinity measure used in DPC-AC is the same as that in NDPC-AC.

In all the algorithms based on DPC, d_c is calculated by

$$d_c = D_{\lceil N \times p/100 \rceil}, \tag{13}$$

where $N = n^2$, $\lceil \cdot \rceil$ is a ceiling function and $D_{\lceil N \times p/100 \rceil} \in D = [D_1, D_2, \ldots, D_N]$. D is the set of the Euclidean distances between every two points and is sorted in ascending order.

For NDPC and NDPC-AC, we use a Gaussian kernel to calculate W_{ij} in graph G and the virance is set to be d_c. The K in graph G is set to be $\lceil n \times p/1000 \rceil$, then p is the only parameter for both DPC and NDPC. For NDPC and NDPC-AC, p is selected from 1, 3 and 10. For NDPC, n_c, the number of initial clusters, is set to be $e \times n_T$ and e is selected from 5, 10, 100 and 200. For GDL, DPC-AC and NDPC-AC, the K in graph G' is selected from 5 and 20. For all the algorithms, the number of clusters remained after clustering is set to be n_T.

We evaluate the experiment results based on two widely used clustering criterions: normalized mutual information (NMI) and clustering accuracy (ACC), both of which measure clustering results referring to ground truth information.

4.2 Experimental Results on Benchmark Datasets

Figure 5 shows NMI and ACC of DPC and NDPC on three datasets, Coil20, Mnist, and USPS, with p varying. As we can see, NDPC behaves better than DPC stably when p varies. This demonstrates the rationality of our new local density, since the measure of local density is the only difference between DPC and NDPC.

Table 2 shows the NMI and ACC of all the algorithms on benchmark datasets. We can see that NDPC behaves much better than not only DPC, but also DPC-KNN-PCA. NDPC-AC behaves better than not only DPC-AC, but also excellent state-of-the-art clustering algorithms such as GDL and CLR. These results have proved the advantages of our new local density and the combination of our NDPC with AC.

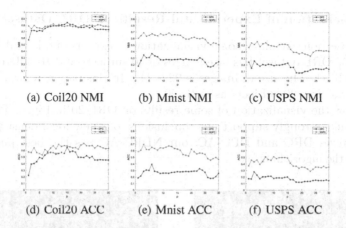

(a) Coil20 NMI   (b) Mnist NMI   (c) USPS NMI

(d) Coil20 ACC   (e) Mnist ACC   (f) USPS ACC

Fig. 5. Performances of DPC and NDPC when parameter p varies.

Table 2. Experimental results on benchmark datasets

		Coil20	Mnist	USPS	Isolet	Lung	Vote	ORL
NMI	K-means	0.749	0.502	0.461	0.764	0.577	0.348	0.801
	N-Cuts	0.836	0.519	0.384	0.770	0.320	0.357	0.802
	CLR	0.964	0.702	0.721	0.815	0.685	0.072	0.913
	Chameleon	0.715	0.442	0.377	0.729	0.627	0.291	0.811
	GDL	0.969	0.843	0.814	0.817	0.678	0.076	0.899
	DPC	0.841	0.308	0.260	0.566	0.330	0.414	0.816
	DPC-KNN-PCA	0.796	0.384	0.389	0.700	0.230	0.427	0.836
	DPC-AC	0.934	0.648	0.593	0.783	0.730	0.427	0.901
	NDPC	0.806	0.688	0.487	0.816	0.526	0.514	0.878
	NDPC-AC	0.958	0.822	0.837	0.816	0.779	0.514	0.925
		Coil20	Mnist	USPS	Isolet	Lung	Vote	ORL
ACC	K-means	0.643	0.515	0.454	0.547	0.670	0.821	0.585
	N-Cuts	0.674	0.578	0.352	0.546	0.424	0.828	0.595
	CLR	0.874	0.539	0.607	0.601	0.872	0.570	0.783
	Chameleon	0.594	0.495	0.436	0.577	0.645	0.798	0.625
	GDL	0.897	0.850	0.760	0.603	0.852	0.568	0.708
	DPC	0.647	0.301	0.279	0.310	0.714	0.844	0.550
	DPC-KNN-PCA	0.674	0.309	0.300	0.557	0.704	0.853	0.585
	DPC-AC	0.861	0.626	0.614	0.597	0.905	0.853	0.778
	NDPC	0.708	0.695	0.507	0.611	0.734	0.894	0.740
	NDPC-AC	0.935	0.906	0.907	0.687	0.941	0.894	0.830

4.3   Visualization of Experimental Results on ORL Dataset

In this subsection, we show some visualization of experimental results on ORL dataset [23]. ORL dataset has face images of 40 subjects and 10 different images for each subject. The size of images is 92×112. In this paper, we select the 200 images of the first 20 subjects as ORL-20.

We show the visualization of some results on ORL-20 in Fig. 6. The experimental results strongly support our conclusions: our new local density substantially improves DPC and DPC-AC; our NDPC-AC has the best performance among all the algorithms.

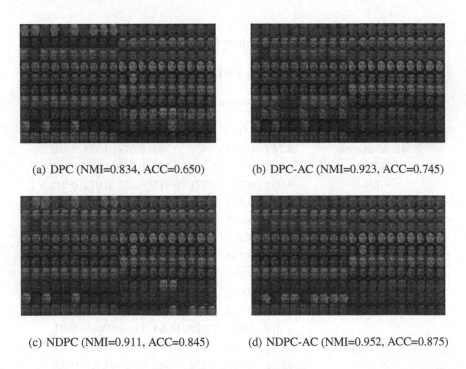

(a) DPC (NMI=0.834, ACC=0.650)   (b) DPC-AC (NMI=0.923, ACC=0.745)

(c) NDPC (NMI=0.911, ACC=0.845)   (d) NDPC-AC (NMI=0.952, ACC=0.875)

Fig. 6. Analysis on ORL-20.

5   Conclusion

In this paper, we redefine the local density and apply our new local density to DPC and DPC-AC. The analysis and experimental results show that our new local density has improved both DPC and DPC-AC substantially. NDPC is better than DPC, and NDPC-AC is better than not only DPC-AC but also some excellent state-of-the-art clustering algorithms.

Acknowledgements. This work was partially supported by the National Natural Science Foundation of China under Grant No. 61772120.

References

1. Asuncion, A., Newman, D.: UCI machine learning repository (2007)
2. Baraldi, A., Blonda, P.: A survey of fuzzy clustering algorithms for pattern recognition I. IEEE Trans. Syst. Man Cybern. Part B **29**(6), 778–785 (1999)
3. Brecheisen, S., Kriegel, H.-P., Pfeifle, M.: Parallel density-based clustering of complex objects. In: Ng, W.-K., Kitsuregawa, M., Li, J., Chang, K. (eds.) PAKDD 2006. LNCS (LNAI), vol. 3918, pp. 179–188. Springer, Heidelberg (2006). https://doi.org/10.1007/11731139_22
4. Chen, G., Zhang, X., Wang, Z.J., Li, F.: Robust support vector data description for outlier detection with noise or uncertain data. Knowl.-Based Syst. **90**, 129–137 (2015)
5. Cheng, D., Zhu, Q., Huang, J., Yang, L., Wu, Q.: Natural neighbor-based clustering algorithm with local representatives. Knowl.-Based Syst. **123**, 238–253 (2017)
6. Deng, L.: The MNIST database of handwritten digit images for machine learning research [best of the web]. IEEE Signal Process. Mag. **29**(6), 141–142 (2012)
7. Du, M., Ding, S., Jia, H.: Study on density peaks clustering based on k-nearest neighbors and principal component analysis. Knowl.-Based Syst. **99**, 135–145 (2016)
8. Ertöz, L., Steinbach, M., Kumar, V.: Finding clusters of different sizes, shapes, and densities in noisy, high dimensional data. In: ICDM 2003, pp. 47–58. SIAM (2003)
9. Ester, M., Kriegel, H.P., Sander, J., Xu, X., et al.: A density-based algorithm for discovering clusters in large spatial databases with noise. In: KDD 1996. vol. 96, pp. 226–231 (1996)
10. Frey, B.J., Dueck, D.: Clustering by passing messages between data points. Science **315**(5814), 972–976 (2007)
11. Hastie, T., Tibshirani, R., Friedman, J.: The Elements of Statistical Learning. SSS. Springer, New York (2009). https://doi.org/10.1007/978-0-387-84858-7
12. Hull, J.J.: A database for handwritten text recognition research. IEEE Trans. Pattern Anal. Mach. Intell. **16**(5), 550–554 (1994)
13. Jain, A.K., Murty, M.N., Flynn, P.J.: Data clustering: a review. ACM Comput. Surv. **31**(3), 264–323 (1999)
14. Jing, L., Ng, M.K., Xu, J., Huang, J.Z.: Subspace clustering of text documents with feature weighting K-means algorithm. In: Ho, T.B., Cheung, D., Liu, H. (eds.) PAKDD 2005. LNCS (LNAI), vol. 3518, pp. 802–812. Springer, Heidelberg (2005). https://doi.org/10.1007/11430919_94
15. Karypis, G., Han, E.H., Kumar, V.: Chameleon: hierarchical clustering using dynamic modeling. IEEE Comput. **32**(8), 68–75 (1999)
16. Liu, Y., Ma, Z., Yu, F.: Adaptive density peak clustering based on K-nearest neighbors with aggregating strategy. Knowl.-Based Syst. **133**, 208–220 (2017)
17. Mehmood, R., El-Ashram, S., Bie, R., Dawood, H., Kos, A.: Clustering by fast search and merge of local density peaks for gene expression microarray data. Sci. Rep. **7** (2017)
18. Meila, M., Shi, J.: A random walks view of spectral segmentation. In: AISTATS 2001 (2001)
19. Nene, S.A., Nayar, S.K., Murase, H., et al.: Columbia object image library (coil-20) (1996)
20. Nie, F., Wang, X., Jordan, M.I., Huang, H.: The constrained Laplacian rank algorithm for graph-based clustering. In: AAAI 2016, pp. 1969–1976 (2016)

21. Rezaei, M., Fränti, P.: Set-matching methods for external cluster validity. IEEE Trans. Knowl. Data Eng. **28**(8), 2173–2186 (2016)
22. Rodriguez, A., Laio, A.: Clustering by fast search and find of density peaks. Science **344**(6191), 1492–1496 (2014)
23. Samaria, F., Harter, A.: Parameterisation of a stochastic model for human face identification. In: WACV 1994, pp. 138–142. IEEE (1994)
24. Shi, J., Malik, J.: Normalized cuts and image segmentation. IEEE Trans. Pattern Anal. Mach. Intell. **22**(8), 888–905 (2000)
25. Singh, D., Febbo, P.G., Ross, K., Jackson, D.G., Manola, J., Ladd, C., Tamayo, P., Renshaw, A.A., D'Amico, A.V., Richie, J.P., et al.: Gene expression correlates of clinical prostate cancer behavior. Cancer Cell **1**(2), 203–209 (2002)
26. Xie, J., Gao, H., Xie, W., Liu, X., Grant, P.W.: Robust clustering by detecting density peaks and assigning points based on fuzzy weighted K-nearest neighbors. Inf. Sci. **354**, 19–40 (2016)
27. Xu, J., Wang, G., Deng, W.: Denpehc: density peak based efficient hierarchical clustering. Inf. Sci. **373**, 200–218 (2016)
28. Zaïane, O.R., Foss, A., Lee, C.-H., Wang, W.: On data clustering analysis: scalability, constraints, and validation. In: Chen, M.-S., Yu, P.S., Liu, B. (eds.) PAKDD 2002. LNCS (LNAI), vol. 2336, pp. 28–39. Springer, Heidelberg (2002). https://doi.org/10.1007/3-540-47887-6_4
29. Zhang, W., Wang, X., Zhao, D., Tang, X.: Graph degree linkage: agglomerative clustering on a directed graph. In: Fitzgibbon, A., Lazebnik, S., Perona, P., Sato, Y., Schmid, C. (eds.) ECCV 2012. LNCS, vol. 7572, pp. 428–441. Springer, Heidelberg (2012). https://doi.org/10.1007/978-3-642-33718-5_31
30. Zhang, W., Li, J.: Extended fast search clustering algorithm: widely density clusters, no density peaks. arXiv preprint arXiv:1505.05610 (2015)
31. Zhu, W., Wang, F.Y.: Reduction and axiomization of covering generalized rough sets. Inf. Sci. **152**, 217–230 (2003)
32. Zhu, W., Wang, F.Y.: On three types of covering-based rough sets. IEEE Trans. Knowl. Data Eng. **19**(8), 1131–1144 (2007)

An Efficient Ranking-Centered Density-Based Document Clustering Method

Wathsala Anupama Mohotti[(⊠)] and Richi Nayak

Queensland University of Technology (QUT), Brisbane, Australia
wathsalaanupama.mohotti@hdr.qut.edu.au,
r.nayak@qut.edu.au

Abstract. Document clustering is a popular method for discovering useful information from text data. This paper proposes an innovative hybrid document clustering method based on the novel concepts of ranking, density and shared neighborhood. We utilize ranked documents generated from a search engine to effectively build a graph of shared relevant documents. The high density regions in the graph are processed to form initial clusters. The clustering decisions are further refined using the shared neighborhood information. Empirical analysis shows that the proposed method is able to produce accurate and efficient solution as compared to relevant benchmarking methods.

Keywords: Density estimation · Ranking function · Graph-based clustering

1 Introduction

Document clustering is a popular method to discover useful information from the text corpuses [1]. It has been used to organize the data based on similarity in many applications such as social media analytics, opinion mining and recommendation systems. A myriad of clustering methods exist that can be classified into the popular categories of partitional, hierarchical, matrix factorization, and density based clustering [1, 2]. The centroid based partitional methods such as k-means are known to suffer from the data concentration problem when dimensionality is high and the data distribution is sparse. Specifically, the difference between data points becomes negligible [3]. Hierarchical clustering suffers from the same problem due to the requirement of multiple pairwise computation at each step of decision making [1]. Matrix factorization, a dimension reduction method for high dimensional text, is also commonly used in finding clusters in low dimension data. In these methods, information loss is inevitable [1] as well as the required time for low rank approximation of a large text data through optimization increases with the size of the datasets.

Density-based methods such as DBSCAN and OPTICS have been found highly efficient in traditional data [4]. They generate diverse shapes of clusters without taking the number of clusters as an input – the desired requirements for document datasets [2]. Moreover, text data has shown to experience the Hub phenomena, i.e., "the number of times some points appear among k nearest neighbors of other points is highly skewed" [3]. A density based clustering method should be ideal to identify these naturally spread dense sub regions made of frequent nearest neighbors that assist in estimating density.

D. Phung et al. (Eds.): PAKDD 2018, LNAI 10939, pp. 439–451, 2018.
https://doi.org/10.1007/978-3-319-93040-4_35

However, this approach is hardly explored in document clustering due to manifold reasons [5].

Firstly, density based methods become stagnated in high dimensional data clustering as the document datasets exhibit varying densities due to sparse text representation and the density definition cannot identify core points to form clusters [1]. Secondly, techniques employed for efficient neighborhood inquiry to expand clusters do not scale well to high dimensional feature space [1]. A handful of solutions have been proposed by using different shapes, sizes, density functions and applying constraints in the high dimensional data [2, 5]. Semi supervised and active learning approaches have been used in density document clustering with DBSCAN to obtain improved clustering performance [2].

Majority of density based methods utilize the concept of Shared Nearest Neighbor (SNN) [6] whereby the similarity between points is defined based on the number of neighbors they share [5, 7]. The SNN concept facilitates the relatively uniform regions to form a graph and to identify clusters by differentiating varying densities. In document clustering setting where data representation is naturally sparse this is an ideal solution to identify dense regions. However, the computation of a SNN graph is expensive due to the high number of pairwise comparisons required.

In this paper, we propose a novel and effective method called as Ranking centered Density based Document Clustering (RDDC). It first builds the SNN graph based on the concepts of Inverted index and Ranking and, then, iteratively form clusters by finding density regions within the shared boundary of documents in the SNN graph.

Information Retrieval (IR) is an established field that uses the document similarity concept to provide the ranked results in response to the user query [10]. An IR system is able to process queries per second on collections of millions of documents using efficient inverted index data structure on a traditional desktop computer [11]. Given a query and the documents organized in the form of inverted index on a standard desktop machine, a search engine will efficiently retrieve the related documents ranked by the relevancy order to the query. We conjecture that a document neighborhood can be generated using this relevant documents set found by an IR system without the expensive pairwise documents comparisons. In RDDC, we propose to explore this neighborhood of relevant documents to build the SNN graph effectively that, in turn, reveals the core dense points and form clusters.

The conventional density clustering methods are known for not covering all data points in clusters and leaving the higher number of documents un-clustered [7]. To deal with this problem, we identify multiple hubs in the shared neighborhoods sets and reassign these un-clustered documents to the closest hub based on prior calculated relevancy scores. Empirical analysis using several document corpuses reveals that RDDC is able to cluster high percentage of documents accurately and efficiently compared to other state-of-the-art methods.

More specifically, in this paper we propose a novel density based clustering method RDDC for sparse text data. RDDC explores the dense patches in high dimensional setting using a shared nearest neighbor graph built with ranked results of an IR system. RDDC further enhances the clustering decision using these shared nearest neighbors as hubs in higher dimensionality. It efficiently calculates the similarity for hubs using relevancy scores provided by the IR system. These approaches of cluster allocation enable RDDC obtaining improved accuracy and efficiency for document clustering.

To our best of knowledge, RDDC is the first such method that extends the IR concepts of Inverted index and Ranking to density document clustering. Recently, a couple of researchers have used the ranking concept to partitional document clustering, to produce relevant clusters instead of all clusters in semi-supervised clustering [8] and to select centroids using ranked retrieval in k-means [9]. However, the approach employed in RDDC is entirely different from these two works. RDDC does not need a user-defined cluster number k and the expensive steps of centroid updates in these methods. RDDC finds the density regions in the SNN graph which is built efficiently using the document ranking scores obtained from the text data through an IR system.

2 Ranking-Centered Density Document Clustering (RDDC)

Let $D = \{d_1, d_2, d_3, \ldots, d_N\}$ be a document corpus and d_i be a document represents with set of M distinct terms $\{t_1, t_2, t_3, \ldots, t_M\}$. RDDC uses an IR system to index all documents in D based on their terms and frequencies. The indexed documents become input to the clustering process that includes three main steps. (1) Firstly, the nearest neighbor sets which possess common documents, $D^{SNN} \subseteq D$ are identified using the document ranking scores obtained from the IR system in order to build the SNN graph. (2) Secondly, the graph G^{SNN} is built using documents in D^{SNN} as vertices and the corresponding number of shared relevant documents as edge weight. Dense regions are found in the graph and a distinct cluster label in $\mathbf{C} = \{c_1, c_2, c_3, \ldots, c_l\}$ is assigned to documents in high dense regions. Another set of documents D^{O2} that appear in low density regions is separated out. (3) Lastly, RDDC assigns cluster labels to $d_i \in D^{O2}$ according to their maximum affinity to a hub residing within a cluster that is identified in previous step.

2.1 Obtaining Nearest Neighbors as Relevant Documents

Document Querying. Given a document $d_i \in D$ as a query and D organized as inverted index, an IR system generates the most relevant documents ranked in the order of relevancy to d_i. A query representing the document, $q = \{t_1, t_2, t_3, \ldots, t_s\} \in d_i$ should be generated such that the most accurate nearest neighbors are obtained. RDDC represents the document as a query using the top-s terms ranked in the order of term frequency according to the length of the document. A set of s distinct terms with $0 \leq s \leq M$ is obtained as:

$$s = (|d_i|)/k : i = 1, 2, \ldots, N \tag{1}$$

If the length of the document is less than s, all the terms in the document is used as the query. A factor k controls the query length in various sized documents. A smaller k value (e.g., $k = 15$) is used in large text data yielding larger queries while a larger value (e.g., $k = 35$) is used in short text data yielding smaller queries. The value of k is empirically learned ranging from 3–50 for each corpus.

Document Ranking. Given the document query q, a set of m most relevant documents with their ranking score r vector is obtained using the ranking function R_f as in Eq. (2). A number of ranking functions such as Term Frequency-Inverse Document Frequency ($tf * idf$), Okapi Best Matching 25 ($BM25$) can be used to calculate the relevancy score [10]. RDDC uses the $tf * idf$ ranking function given in Eq. (3) where tf represents how often a term appears in the document, idf represents how often the term appears in the document collection and field length normalization depicts how the length of the field which contains the term is important.

$$R_f : q \rightarrow D_q = \left\{ \left(d_j^q, r_j \right) \right\} : j = 1, 2, \ldots, m \tag{2}$$

$$r_{d_j} = score(q, d_j) = \sum_{t \, in \, q} \left(\sqrt{tf_{t,d_j}} \times idf_t^2 \times norm(t, d_j) \right) \tag{3}$$

Claim 1 shows that the ranking results obtained by an IR system using a ranking function provides the relevant neighbors to d_i with a reduced computational time and high accuracy, in comparison to the pairwise document comparison.

Claim 1. Let $N(d_i)$ be the neighborhood documents calculated from the pairwise document comparisons of document d_i with rest of the documents in the collection D of size N obtained with δ_1 time and ∂_1 level of accuracy. Let $R(d_i)$ be the IR ranked result of document d_i obtained with δ_2 time and ∂_2 level of accuracy. $R(d_i) \subset N(d_i)$ will be built with $\delta_2(< \delta_1)$ time and $\partial_2(> \partial_1)$ level of accuracy.

Proof:

- In order to obtain $N(d_i)$, (cosine) similarity has to be obtained by comparing d_i with every document in D. This process consists of $N - 1$ steps which takes δ_1 time and allows to obtain the top-k neighbours where $k \geq 1$ according to similarity values with ∂_1 level of accuracy.
- $R(d_i)$ is obtained using inverted indexed documents in D and a ranking function such as $tf * idf$ [10]. The $tf * idf$ ranking function only computes similarity scores for documents containing high $tf * idf$ weights for query terms. This is a one step process which takes δ_2 time and gives most relevant neighbour documents with ∂_2 level of accuracy.
- The cluster hypothesis [12] states that "associated documents appear in a returned result set of a query". The reversed cluster hypothesis in the optimum clustering framework [10] further states that "the returned documents in response to a query will be in the same cluster" and can be considered as nearest neighbours.
- Since, $R(d_i) \subset N(d_i) \subset D$, $R(d_i)$ will contain neighbours with $\partial_2(> \partial_1)$ level of accuracy obtained with $\delta_2(< \delta_1)$ time.

2.2 Graph Based Clustering

The IR ranked results should contain the relevant neighborhood documents to the query document. RDDC uses the top-10 ranked documents as the nearest neighborhood set as they possess sufficient information richness [13]. A $D^{SNN} \subseteq D$ is identified by calculating common documents for each $d_i \in D$ with its top-10 retrieved documents. Let retrieved results set of d_i, d_j be $R(d_i)$ and $R(d_j)$ respectively where $d_j \in R(d_i)$. If $R(d_i) \cap R(d_j) > 2$, documents d_i and d_j $(d_i, d_j \in D^{SNN})$ become vertices in the graph G^{SNN} and the corresponding number of shared relevant documents $((|R(d_i) \cap R(d_j)|)$ be the edge weight. G^{SNN} construction leaves out a set of orphan documents D^{O1} $(D = D^{SNN} \cup D^{O1})$ that do not appear in D^{SNN}.

$$D^{O1} = \left\{ (d_i \in D) \cap (d_i \notin D^{SNN}) \right\} : i = 1, 2, \ldots, N \tag{4}$$

The next task is to identify dense nodes in G^{SNN}. A dense node contains the number of documents (higher than a threshold) connected in the region with the edge weight higher than a threshold. These nodes are defined as core points in G^{SNN} that become initial cluster representatives. Each cluster boundary is then expanded to include documents with the same edge weight. This process gives us a set of documents with cluster labels C, as well as it identifies documents D^{O2} that do not fit into any cluster boundaries.

$$D^{O2} = \left\{ (d_i \in D) \cap (d_i \in D^{SNN}) \cap (d_i \notin C) \right\} : i = 1, 2, \ldots, N \tag{5}$$

Algorithm 1 details the process of obtaining density based clusters. Claim 2 shows that the SNN graph can be built accurately using ranking results.

Claim 2. Let the SNN graph created with k nearest neighbourhoods $N_k(D)$ in the document collection D be $G^{N_k(D)}$ and the SNN graph created with IR ranked results $R(D)$ be $G^{R(D)}$. If $R(d_i) \subset N(d_i)$ for $d_i \in D$, then graph $G^{R(D)} \subseteq G^{N_k(D)}$ and $G^{R(D)}$ contains $\partial_2 (> \partial_1)$ level of accuracy where ∂_1 is accuracy level of the $G^{N_k(D)}$.

Proof:

- Let $V(R(D))$, $V(N_k(D))$ be the vertices of two graphs $G^{R(D)}$ and $G^{N_k(D)}$ represented by the documents in $R(D)$ and $N_k(D)$ respectively and $E(R(D))$, $E(N_k(D))$ be the edges represented by the number of shared documents within document pairs in $R(D)$ and $N_k(D)$ respectively.
- For document d_i to obtain k relevant neighbourhoods $N_k(d_i)$ we have to prune meaningless neighbourhood levels. Hence, $N_k(d_i) \subset N(d_i)$.
- If $N_k(d_i) \subset N(d_i)$ and $R(d_i) \subset N(d_i)$ then $R(d_i) \subseteq N_k(d_i)$ as $R(d_i)$ contains only the most relevant neighbours according to the optimum clustering framework [10]. Thus $R(D) \subseteq N_k(D)$.
- Therefore $V(R(D)) \subseteq V(N_k(D))$ and $E(R(D)) \subseteq E(N_k(D))$.   It   proves $G^{R(D)} \subseteq G^{N_k(D)}$.

- The IR ranked results $R(d_i)$ of d_i contains the relevant documents to d_i with $\partial_2(> \partial_1)$ level of accuracy as shown in Claim1. Hence, $G^{R(D)}$ contains all required document information to represent SNN graph with $\partial_2(> \partial_1)$ level of accuracy.

In this phase, a repository $H = \{H_1, H_2, H_3, \ldots, H_\emptyset\}$ is also built to store the shared relevant documents where each node $H_j \in H$ contains a set of shared documents { $d_i, d_j, \ldots d_k\}$ and $\emptyset$ is the total number of sets of shared documents. Usually, $|H| > |C|$ and a node H_j contains documents from the same cluster. The set of relevant nodes within a cluster is comparable to the concept of Hubs in high dimensionality [3]. These hubs actually represent the sub dense regions within clusters. RDDC accurately cluster higher percentage of documents using affinity calculation for these hub nodes (Fig. 1) and avoid the problem of higher number of un-clustered documents in many other density based methods.

ALGORITHM 1: Density based document clustering via graph

Input: Set of indexed documents in D^{SNN}, a threshold α (set to 3 for the minimum number of documents and edge weight)

Output: Distinct set of cluster labels $C = \{c_1, c_2, ., ., \ldots c_k\}$ assigned to $d_i \in D^{SNN}$, set of orphan documents D^{O2}

$k = 0$
for each $d_i \in D^{SNN}$ do
  $D^{RD} = ReleavantDocs(d_i) = graphQuery(d_i, \alpha)$
  If $sizeof(D^{RD}) < \alpha$ then
    $D^{O2} = D^{O2} \cup \{d_i\}$
  else
    $++k; c_k = c_k \cup \{d_i\}; C \leftarrow c_k$
    for each $d_r \in D^{RD}$ do
      $D^{RRD} = ReleavantDocs(d_r) = graphQuery(d_r, \alpha)$
      If $sizeof(D^{RRD}) \geq \alpha$ then
        $D^{RD} = D^{RD} \cup D^{RRD}$
        If $d_r \notin C$ then
          $c_k = c_k \cup \{d_r\}$
    end
  end
end

ALGORITHM 2: Relevancy based clustering

Input: Set of indexed documents D^{O2} where $D^{O2} \in D^{SNN}$, Set of shared relevant documents $H = \{H_1, ., ., \ldots H_\emptyset\}$ with their cluster labels $C = \{c_1, ., ., \ldots c_k\}$

Output: Distinct set of cluster labels C for documents, set of noise documents

for each $d_i \in D^{O2}$ do
  $V_{d_i} \leftarrow \{AS(d_i)_j : j = 1,2, \ldots, \emptyset\}$
  $H_j \leftarrow max\{V_{d_i}\}$
  $c_k \leftarrow H_j$
  $c_k = c_k \cup \{d_i\}$
end

Fig. 1. Algorithm for RDDC

2.3 Relevancy Based Clustering

Algorithm 2 details the process of clustering documents D^{O2} that remain un-clustered in the first phase, based on the maximum relevancy to the set of documents in the repository H. In the high-dimensional data such as text, a cluster is shown to contain multiple hubs of documents instead of a uniform spread across the cluster [3]. In RDDC, the sets of shared relevant documents present in H are considered as hubs within a cluster. We envisage that the hubs of documents stored in the repository H will share higher affinity to $d_i \in D^{O2}$ instead of a cluster represented as mean (centroid) vectors. For each $d_i \in D^{O2}$, we calculate its affinity with each node in H as follows.

$$AS(d_i) = \left\{ \left(\frac{\sum_{u=1}^{sizeof(H_j)} score(d_i, d_u)}{sizeof(H_j)} \right), j : 1, 2, ., \emptyset, \right. \tag{6}$$

$$\left. score(d_i, d_u) \; calculated \; as \; per \; Eq.(3) \right\}$$

The affinity score of hub node H_j is calculated using the ranking score of each document that it contains, when the orphan document d_i was posed as a query. Usually, the hub calculation in existing clustering methods is found very expensive due to the need of pairwise computation between all documents within a cluster [8, 14]. However, RDDC uses (already calculated) relevancy scores in IR ranked results to measure hub affinity and makes the process computationally efficient. Document d_i is then assigned with the cluster label of the maximum relevant node $H_j \in H$ that yields the largest affinity score to d_i.

3 Empirical Analysis

We used multiple datasets with varying dimensionality such as 20 Newsgroups, Reuters 21578, Media Eval Social Event Detection (SED) 2013 and SED 2014, and the TDT5 English corpus, as reported in Table 1. We created smaller subsets of 20 newsgroups datasets as given in Table 1 to compare the RDDC performance with the density-based document clustering method by Zhao et al. [2]. Additionally, several other density-based clustering methods including SNN based DBSCAN [7], SNN based clustering for coherent topic clustering [5] and DBSCAN [4] as well as the well-known matrix factorization method, NMF [15] were used for benchmarking.

Table 1. Summary of the datasets in the experiment.

Datasets	# Docs	#Terms	Avg. #Terms per doc	Std. Dev. terms per corpus	#Clusters (ground truth)
20 Newsgroups-20ng_DS1	300	6595	104	88	3
20 Newsgroups-20ng_DS2	2000	22841	100	119	20
20 Newsgroups-20ng_DS3	7528	43946	97	104	20
Reuters 21578 - R52_DS	9100	19479	46	41	52
SED 13 - SED13_DS	99989	61806	17	23	3711
SED 14 - SED14_DS	120000	64056	16	16	3875
TDT 5 - TDT5_DS	3905	38631	172	124	40

Each dataset was pre-processed to remove stop words and words are stemmed. Document term frequency was selected as the weighting schema for the query representation after extensive experiments. In all the experiments, query length was set to optimum query length according to Eq. (1). The minimum number of documents and minimum weight for graph α was set to 3 based on experiments and prior research [16] for all the datasets. Experiments were done using python 3.5 on 1.2 GHz – 64 bit processor with 264 GB Memory. The Elasticsearch with fast bulk indexing was used as the search engine to obtain relevant documents. Standard pairwise F1-score and Normalized Mutual Information (NMI) were used as cluster evaluation measures [2].

3.1 Accuracy Analysis

Results in Table 2 and Fig. 2(a) show the comparative performance of RDDC with benchmarking methods. As shown by the average performance of all datasets in Table 2, RDDC has produced much higher accuracy as compared to benchmarking methods. Results in Fig. 2(a) ascertain that RDDC forms tight natural clusters. It is able to identify sub clusters within the specified clusters as shown by finding the higher number of clusters (Fig. 2(a)), but still produce higher NMI (Table 2). Sometimes, this leads to producing low F1- score. Density based clustering is known not to cover every data point in clusters, due to the requirement of fitting the clustered objects into a density region [4]. Figure 2(a) shows that RDDC is able to assign a large share of documents to clusters with high accuracy due to the inclusion of relevancy based clustering in the third step. RDDC shows two-fold increase in the percentage of documents clustered using the graph-based clustering to the relevance clustering with 52% and 17% increase in NMI and F1-score respectively. In some datasets, DBSCAN has shown to cover more documents than RDDC, however, a closer investigation reveals that it produces a few larger clusters only that will hold a large number of documents, yielding poor clustering solution.

Table 2. Performance comparison of different datasets, methods, and metrics

Datasets	NMI					F1-score				
	RD	SD	ST	DB	MF	RD	SD	ST	DB	MF
20ng_DS3	0.28	-	-	0.00	0.18	0.28	-	-	0.09	0.14
R52_DS	0.36	0.04	-	0.07	0.38	0.41	0.38	-	0.43	0.26
SED13_DS	0.87	-	-	0.00	-	0.66	-	-	0.00	-
SED14_DS	0.87	-	-	0.00	-	0.65	-	-	0.00	-
TDT5_DS	0.61	0.36	0.21	0.22	0.70	0.35	0.32	0.25	0.22	0.54
Average	**0.60**	0.20	0.21	0.06	0.42	**0.47**	0.35	0.25	0.15	0.31

Methods: RDDC (RD), SNN based DBSCAN (SD), SNN based topic clustering (ST), DBSCAN (DB) and NMF (MF)
Note: "-" denotes out of run-time or memory

Table 3. Performance comparison with semi supervised clustering [2]

Datasets	RDDC			Semi-supervised DBSCAN		
	NMI	F1-score	# constraints	NMI	F1-score	# constraints
20ng_DS1	0.66	0.75	0	0.62	0.62	25
20ng_DS2	0.40	0.32	0	0.22	0.42	50
TDT5_DS	0.61	0.35	0	0.22	0.31	75

Zhao et al. [2] used DBSCAN in semi-supervised setting for document clustering. We have created 20ng_DS1, 20ng_DS2, and TDT5_DS according to the explanation given in their paper as we are unable to find the method implementation. RDDC is an unsupervised method and can be considered equivalent to zero constraint level of the method in [2]. As shown in Table 3, results produced by unsupervised RDDC are mostly superior to semi-supervised DBSCAN [2]. These results show the effectiveness of using relevancy scores obtained with the concepts of ranking and inverted index, in building SNN graph, finding dense regions and forming clusters.

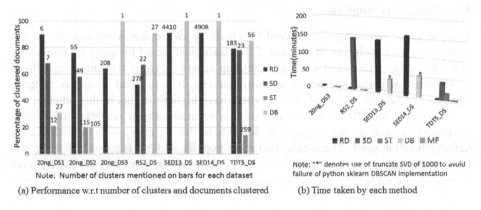

(a) Performance w.r.t number of clusters and documents clustered

(b) Time taken by each method

Fig. 2. Performance comparison with percentage of clustered documents and time taken

3.2 Scalability and Complexity Analysis

Figure 2(b) shows that the traditional SNN based methods failed to scale with large datasets due to the computational complexity introduced by the number of comparisons made for k NN search. It is $O(nk + nd)$ where n is the number of instances in dataset, d is feature dimensionality and k is the number of nearest neighbors. Whereas, the relevant document calculation of RDDC has computational complexity of $O(m + n)$ where m is the query length to obtain relevant neighbors. RDDC consumes more time than DBSCAN as in Fig. 2(b) due to additional graph construction and maximum relevancy calculation. However, as shown in Table 2 the tradeoff by achieving 0.54 and 0.32 increase on average accuracy in terms of NMI and F1-score respectively in RDDC is well justified. Incremental sampling on the SED13 collection is used to demonstrate the scalability of RDDC. Figure 3(a) shows that RDDC exhibits near linear increase in time with the size of the corpus, whereas traditional SNN based methods are not scalable as shown by the runtime in Table 2. Further, performance of RDDC with the increased feature dimensionality in Fig. 3(b) shows that the RDDC performance comes to stabilize after a linear increase in runtime with dimensionality.

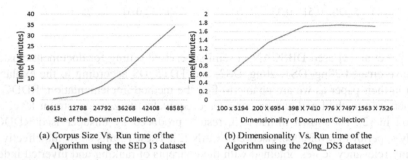

(a) Corpus Size Vs. Run time of the   (b) Dimensionality Vs. Run time of the
Algorithm using the SED 13 dataset   Algorithm using the 20ng_DS3 dataset

Fig. 3. Scalability performance of RDDC using the SED 13 and 20ng_DS3 dataset

3.3 Sensitivity Analysis

The parameter sensitivity is analyzed by obtaining two independent samples of different sizes from each dataset in Table 2. The document model for query formation was evaluated using tf, idf and $tf - idf$ weighting schema. Figure 4(a) shows that the tf presentation outperformed others in many datasets. It is justified as important terms which determine the theme of the document have higher weights in this scheme.

Success of the RDDC relies on obtaining accurate nearest neighbors of a given document to build the SNN graph. It depends on how a query is presented and the ranking function used. There exist two most popular ranking functions, $tf * idf$ and $BM25$ [10]. Figure 4(b) shows that a higher performance in terms of NMI and F1 score is obtained by using $tf * idf$, so it is used in all experiments and can be set as default.

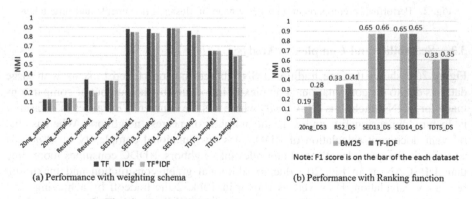

(a) Performance with weighting schema   (b) Performance with Ranking function

Fig. 4. Term weighting schema and ranking functions

Next we explored the relationship between query length and the text size in document corpuses. Figure 5(a) shows that there is a linear relationship between query length and document size. Smaller document corpuses need smaller queries while larger documents sized corpuses need lager queries. In RDDC factor k is included to control the size of query length for documents in a corpus. We analyze the parameter k against size of the datasets as in Fig. 5(b). The factor k shows inverse linear relationship, that is, a large k value should be set for short text data and a smaller k value should be set for large text data. The parameter k adjusts the query size as per length of the document.

The threshold α in RDDC denotes the minimum number of documents to be shared between two documents to define as similar, and the number of documents to be considered as dense. Prior research has shown this value to be set as 3 [16]. As shown in Fig. 6, the best performance (i.e. maximum number of documents clustered) is obtained when α is set to 3. For future use, the default value can bet set to 3.

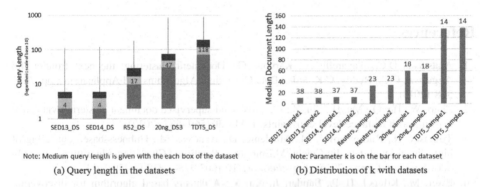

(a) Query length in the datasets   (b) Distribution of k with datasets

Fig. 5. Query length and Parameter k

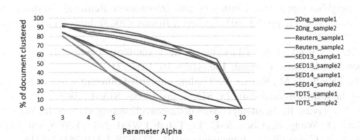

Fig. 6. Parameter Alpha vs. Clustered documents

4   Conclusion

This paper was inspired by the conjecture that text documents have sparse data representation so we should leverage the techniques that suit to those representation. We proposed a novel ranking-centered density-based document clustering method *RDDC* based on the concepts of density estimation, inverted indexing, ranking and hubs. RDDC introduces the innovative concept of finding nearest neighbors using the document relevancy ranking scores to construct a SNN graph and finds the dense regions to form the clusters. We showed that the use of document ranking score is more effective compared to calculating the pairwise similarity between data points in text data by reducing the computational complexity and improving accuracy. We also introduce a refinement phase to increase the percentage of clustered documents by assigning orphan documents to hubs within clusters, rather than to cluster itself. The hubness affinity calculation utilizes the prior calculated relevancy ranking scores, thus, not incurring any overheads. We proved that closeness to shared relevant neighbors can improve the performance of text clustering due to the existence of multiple hubs in a text cluster. Empirical results conducted on several datasets, benchmarked with several clustering methods, show that RDDC overcomes the issues attach with sparse vectors and cluster text data with considerably higher performance, including accuracy and scalability.

References

1. Anastasiu, D.C., Tagarelli, A., Karypis, G.: Document clustering: the next frontier. In: Aggarwal, C.C., Reddy, C.K. (eds.) Data Clustering: Algorithms and Applications, pp. 305–328 (2013)
2. Zhao, W., He, Q., Ma, H., Shi, Z.: Effective semi-supervised document clustering via active learning with instance-level constraints. KAIS **30**, 569–587 (2012)
3. Tomašev, N., Radovanović, M., Mladenić, D., Ivanović, M.: Hubness-based clustering of high-dimensional data. In: Celebi, M.Emre (ed.) Partitional Clustering Algorithms, pp. 353–386. Springer, Cham (2015). https://doi.org/10.1007/978-3-319-09259-1_11
4. Ester, M., Kriegel, H.-P., Sander, J., Xu, X.: A density-based algorithm for discovering clusters in large spatial databases with noise. In: KDD, pp. 226–231 (1996)
5. Ertöz, L., Steinbach, M., Kumar, V.: Finding topics in collections of documents: a shared nearest neighbor approach. Clustering and Information Retrieval. Network Theory and Applications, vol. 11, pp. 83–103. Springer, Boston (2003). https://doi.org/10.1007/978-1-4613-0227-8_3
6. Jarvis, R.A., Patrick, E.A.: Clustering using a similarity measure based on shared near neighbors. IEEE Trans. Comput. **100**, 1025–1034 (1973)
7. Ertöz, L., Steinbach, M., Kumar, V.: Finding clusters of different sizes, shapes, and densities in noisy, high dimensional data. In: SIAM, pp. 47–58. SIAM (2003)
8. Sutanto, T., Nayak, R.: Semi-supervised document clustering via loci. In: Wang, J., Cellary, W., Wang, D., Wang, H., Chen, S.-C., Li, T. (eds.) WISE 2015. LNCS, vol. 9419, pp. 208–215. Springer, Cham (2015). https://doi.org/10.1007/978-3-319-26187-4_16
9. Broder, A., Garcia-Pueyo, L., Josifovski, V., Vassilvitskii, S., Venkatesan, S.: Scalable k-means by ranked retrieval. In: 7th WSDM, pp. 233–242. ACM (2014)
10. Fuhr, N., Lechtenfeld, M., Stein, B., Gollub, T.: The optimum clustering framework: implementing the cluster hypothesis. Inf. Retr. **15**, 93–115 (2012)
11. Zhang, J., Long, X., Suel, T.: Performance of compressed inverted list caching in search engines. In: 17th WWW, pp. 387–396. ACM (2008)
12. Jardine, N., van Rijsbergen, C.J.: The use of hierarchic clustering in information retrieval. Inf. Storage Retr. **7**, 217–240 (1971)

13. Zhang, B., Li, H., Liu, Y., Ji, L., Xi, W., Fan, W., Chen, Z., Ma, W.-Y.: Improving web search results using affinity graph. In: 28th ACM SIGIR, pp. 504–511. ACM (2005)
14. Hou, J., Nayak, R.: The heterogeneous cluster ensemble method using hubness for clustering text documents. In: Lin, X., Manolopoulos, Y., Srivastava, D. (eds.) WISE 2013. LNCS, vol. 8180, pp. 102–110. Springer, Heidelberg (2013). https://doi.org/10.1007/978-3-642-41230-1_9
15. Lin, C.J.: Projected gradient methods for nonnegative matrix factorization. Neural Comput. **19**, 2756–2779 (2007)
16. Hajek, B.: Adaptive transmission strategies and routing in mobile radio networks. Urbana **51**, 61801 (1983)

Fast Manifold Landmarking Using Locality-Sensitive Hashing

Zay Maung Maung Aye[(✉)], Benjamin I. P. Rubinstein,
and Kotagiri Ramamohanarao

School of Computing and Information Systems, The University of Melbourne,
Parkville, VIC 3052, Australia
zaye@student.unimelb.edu.au,
{benjamin.rubinstein,kotagiri}@unimelb.edu.au

Abstract. Manifold landmarks can approximately represent the low-dimensional nonlinear manifold structure embedded in high-dimensional ambient feature space. Due to the quadratic complexity of many learning algorithms in the number of training samples, selecting a sample subset as manifold landmarks has become an important issue for scalable learning. Unfortunately, state-of-the-art Gaussian process methods for selecting manifold landmarks themselves are not scalable to large datasets. In an attempt to speed up learning manifold landmarks, uniformly selected minibatch stochastic gradient descent is used by the state-of-the-art approach. Unfortunately, this approach only goes part way to making manifold learning tractable. We propose two adaptive sample selection approaches for gradient-descent optimization, which can lead to better performance in accuracy and computational time. Our methods exploit the compatibility of locality-sensitive hashing (via LSH and DBH) and the manifold assumption, thereby limiting expensive optimization to relevant regions of the data. Landmarks selected by our methods achieve superior accuracy than training the state-of-the-art learner with randomly selected minibatch. We also demonstrate that our methods can be used to find manifold landmarks without learning Gaussian processes at all, which leads to orders-of-magnitude speed up with only minimal decrease in accuracy.

1 Introduction

A common approach to high-dimensional data analysis is an assumption that observations lie on a low-dimensional manifold embedded in the ambient space. Leveraging the underlying manifold can improve computational tractability of learning and generalization of the learned model. In order to estimate the underlying manifold structure, many manifold learning methods have been proposed in the literature, such as Isometric Mapping [13], Locally Linear Embedding [12], and t-distributed Stochastic Neighbor Embedding [11]. These early manifold learning techniques primarily focus on dimensionality reduction. Recently, manifold structures have been successfully exploited in many machine learning

© Springer International Publishing AG, part of Springer Nature 2018
D. Phung et al. (Eds.): PAKDD 2018, LNAI 10939, pp. 452–464, 2018.
https://doi.org/10.1007/978-3-319-93040-4_36

methods such as semi-supervised learning [7], clustering [5], matrix factorization [3], topic modeling [8] and hashing cross-modal similarities [14].

Manifold landmarking algorithms provide a small set of locations along the manifold that capture the low-dimensional nonlinear structure of the data embedding in high-dimensional space. Potential applications of manifold landmarks include: (1) fast manifold learning and dimensionality reduction; (2) sampling procedures for supervised learning; (3) semi-supervised learning (e.g., anchor graph regularization and finding important unlabeled data in active learning) and (4) unsupervised learning (e.g., spectral clustering). A state-of-the-art manifold landmarking algorithm based on active learning with Gaussian processes has been recently proposed [10]. This method forms a packing of landmarks by finding new landmarks that are repelled by current landmarks in a greedy manner. It requires computing pairwise distances between samples in landmark optimization and as a result, it is prohibitively expensive when the number of observations is large. This drawback limits the usefulness of this method on many applications where dataset size is non-trivial. We show in this paper how Locality-Sensitive Hashing can be used to overcome this computational complexity.

Locality-Sensitive Hashing (LSH) is a randomized algorithm for approximate nearest neighbor search [9]. It partitions high-dimensional data in one efficient linear pass. The samples in each partition have a higher chance of similarity than samples across other partitions. Assuming (1) nearby data points are likely to have the same labels, and (2) points on the same structure are likely to have the same label, often called the cluster or manifold assumption [15], we can utilize the local sensitivity of LSH to approximately select landmarks. This locality property ensures that data in each partition are similar, while an additional heuristic to incorporate label information or manifold substructure separates the data from different structures that are projected into the same bucket. Thus, we mitigate the complexity associated with the expensive optimization step usually required in manifold landmarking. Experimental results demonstrate that our method achieves similar accuracy as expensive state-of-the-art landmark optimization methods at a significantly reduced computational cost.

Our contributions include: (1) proposing two supervised LSH and DBH-based sample compression methods for efficient manifold landmark learning using Gaussian processes; (2) illustrating how our supervised methods are superior to unsupervised hashing-based sample selection or compression; (3) demonstrating that our approaches can also be used to find manifold landmarks without Gaussian process optimization; (4) extensive evaluation on a synthetic manifold dataset and real-life image and text datasets.

2   Background

In this section, we describe the main baseline method used for comparing manifold landmark quality and hashing methods utilized for data-dependent sampling.

2.1   Landmarking Manifolds with Gaussian Processes

Liang and Paisley [10] proposed a Gaussian process-based model to learn the location of the landmarks on a latent manifold sequentially. Let X be a training set with examples $\{x_1, x_2, \ldots, x_N\}$ from $\mathbb{R}^d$ and S be a set of samples $\{s_1, s_2, \ldots, s_m\}$ selected randomly from X. The procedure for selecting landmark l_{n+1} given n previously selected landmarks corresponds to maximising objective function f given as:

$$\delta_{x_i}(l_{n+1}) = \exp(-\|l_{n+1} - x_i\|^2/\eta),$$
$$\phi_S(l_{n+1}) = [\delta_{s_1}(l_{n+1}), \ldots, \delta_{s_m}(l_{n+1})]^\top,$$
$$\Phi_{S,n} = [\phi_S(l_1), \ldots, \phi_S(l_n)],$$
$$W = I - \Phi_{S,n}(\Phi_{S,n}^\top \Phi_{S,n})^{-1}\Phi_{S,n}^\top,$$
$$f(l_{n+1}, X) = \sum_{i=1}^{N}\sum_{j=1}^{N} W[i,j]\delta_{x_i}(l_{n+1})\delta_{x_j}(l_{n+1}), \qquad (1)$$

where η is the kernel width (total variance of all features is used in this paper), S is a minibatch randomly sampled from the training set X without replacement and $W[i,j]$ is the row i and column j of the matrix W. A local minimum of f can be found using gradient descent.

Computing full gradients using the whole training data set, i.e. $S = X$, is prohibitively expensive on large datasets. Therefore, minibatch stochastic gradient descent is used in order to speed up the learning time [10]. Training with smaller minibatch is faster; however, it can significantly reduce the accuracy. With randomly sampled small minibatch S, it can take many iterations of the learning algorithm to explore sufficient regions of the landscape, leading to slower convergence. In this paper, we propose a data-dependent sampling approach to select minibatches that can lead to faster convergence utilizing LSH and DBH.

2.2   Locality-Sensitive Hashing

Locality-Sensitive Hashing (LSH) is a randomized algorithm for quickly finding similar items to a query point, with high probability. It ensures that closer points have a higher chance to be mapped into the same partitions on a random projection. A hash function that approximately preserves Euclidean distance [4] is defined as:

$$h_{r,b}(x) = \left\lfloor \frac{x^\top r + b}{w} \right\rfloor \qquad (2)$$

where x is a data point vector, r is a random vector with components drawn i.i.d. from the Gaussian distribution $\mathcal{N}(0, 1)$, $w > 0$ is the bucket width and b is a uniformly distributed random variable between 0 and w. In each hash function, the random projection $x'r$ is quantized into a hash value. Hash buckets are represented by concatenating respective hash values from multiple hash functions so that neighboring points in the original space will fall into the same bin with

---

Algorithm 1. LSH-SC(X, Y, p, w)

Input: Training set $X = \{x\}_{i=1}^{N}$ and $Y = \{y\}_{i=1}^{N}$, hash length p, bucket width w
Output: Landmarks Set $Z_{p,w}$

1: create hash table T
2: generate iid random vectors r_j with components from $\mathcal{N}(0,1)$ where $j = 1, 2, .., p$
3: generate bias b_j uniformly from $[0, w]$ where $j = 1, 2, .., p$
4: $Z_{p,w} = \phi$
5: **for** i:=1 to $|X|$ **do**
6:    **for** k:=1 to p **do**
7:        $H_j \leftarrow \left\lfloor \frac{x_i^T r_j + b_j}{w} \right\rfloor$
8:        $H \leftarrow H_1 \parallel H_2 \parallel ... \parallel H_p$, for $\parallel$ the concatenation operator
9:        $T[H].\text{append}(x_i)$
10: **for** H in T **do**
11:    **for** c in C **do**
12:        $x_c \leftarrow center(\{x_i\})$, where $\{x_i\} \subseteq T[H], y_i = c$
13:        $Z_{p,w}.\text{append}(x_c)$
14: **return** $Z_{p,w}$

---

high probability. Each LSH is designed for a particular distance function, and it is not trivial to design LSH for desired distance (e.g. manifold distance). DBH can be used to overcome this limitation as it can incorporate desired distance (radial basis function), however, it is slightly more expensive.

2.3 Distance-Based Hashing

The FastMap [6] line projection function, which is the foundation of DBH [2] is defined as:

$$H_{x_1, x_2}(x) = \frac{D(x, x_1)^2 + D(x_1, x_2)^2 - d(x, x_2)^2}{2\,D(x_1, x_2)} \tag{3}$$

where x is a query point, x_1 and x_2 are random data sampled from uniform distribution and D can be an arbitrary distance (we use RBF as it is the distance used by the Gaussian processes for manifold landmarking). This function can approximately preserve arbitrary distance measures by projecting the query point to the randomly selected line defined by x_1 and x_2. The output of this function is binarized by DBH for indexing the buckets as explained in Algorithm 2.

3 LSH for Finding Manifold Landmarks

Gaussian processes for manifold landmarks compute gradients using the whole training data. This procedure is prohibitively slow. Accordingly, stochastic mini-batch gradients are computed from randomly selected small samples sets (mini-batches). While this yields speedups, randomly selected samples may not cover all relevant regions of the data. Therefore, we proposed data-dependent adaptive sampling. Based on the nature of manifold landmarking using Gaussian processes, our data-dependent approach leverages the following improvements.

Algorithm 2. DBH-SC(X, Y, p)

Input: Training set $X = \{x\}_{i=1}^N$ and $Y = \{y\}_{i=1}^N$, hash length p
Output: Landmarks Set Z_p
1: create hash table T, matrix $G \in \mathbb{R}^{p \times N}$
2: **for** k:=1 to p **do**
3:     Uniformly sample two samples x_1 and x_2 without replacement
4:     **for** i:=1 to $|X|$ **do**
5:         $G_{k,i} \leftarrow \frac{D(x_i, x_1)^2 + D(x_1, x_2)^2 - D(x_i, x_2)^2}{2\, D(x_1, x_2)}$
6:     $t_k \leftarrow \text{median}(G_k)$
7: **for** k:=1 to p **do**
8:     **for** i:=1 to $|X|$ **do**
9:         $H_k \leftarrow 1$ if $G_{k,i} \geq T_k$, 0 otherwise
10:     $H \leftarrow H_1 \parallel H_2 \parallel ... \parallel H_p$
11:     $T[F]$.append(x_i)
12: Line 10-14 of Algorithm 1

Improvement 1: samples should cover diverse regions. Gaussian processes should observe all regions of the data space, in order to learn accurate manifold landmarks. Observing all regions takes many iterations if the randomly selected minibatch size is small. We assume that the locality-sensitive property (via LSH or DBH) ensures that each of these partitions approximately represents the diverse regions of the data. We can then randomly select a sample per partition to form a diverse sample set for computing gradients within a Gaussian process.

Improvement 2: centroids represent the regions well. This can be further sped up by selecting the centroids of the data in partitions instead of picking individual samples. The assumption here is that average distance between all pairs of samples between two partitions can be approximated by their centroids. By exploiting this, we can reduce the variations caused by sample positions inside partitions, achieving near-same accuracy in fewer gradient updates.

Improvement 3: incorporating label information leads to more accurate landmarks. Centroids computed by improvement 2 can misrepresent the manifold if multiple manifold substructures are mapped into a partition. We can mitigate this by computing a centroid per manifold substructure by leveraging label information.

3.1   Manifold Assumption and Locality-Sensitive Hashing

We focus on designing adaptive sample selection to avoid high computational complexity in learning manifold landmarks. Hashing methods with complexity linear in the number of samples present an ideal method for our purposes. Owing to the locality-sensitive property, these methods approximately partition the samples by projecting similar samples in Euclidean space into the same partition with high probability. However, similar samples in Euclidean space may not be similar

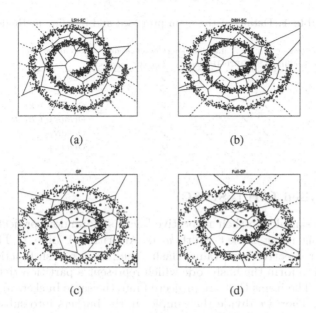

(a)   (b)

(c)   (d)

Fig. 1. Representation quality of fifferent methods on swiss-roll manifold

according to the underlying manifold distance. Randomized hashing methods can project false positive samples into partitions. For our method we need to handle two types of false positives: (1) false positives in Euclidean space which are the actual false positives of the hashing method and (2) false positives in manifold space which are true positives in Euclidean space. We can handle the first type of false positives by tuning the parameters of LSH as it works directly in Euclidean space. However, for the second type, we need to rely on a manifold assumption.

The manifold assumption states that (1) nearby data points are likely to have the same labels; and (2) points on the same structure are likely to have the same label. LSH projects nearby points to the same buckets with high probability, and therefore, these data points should have the same label and so clustering them is a good approach to represent them, achieving improvements 1 and 2 described previously. LSH can also project false positives (far away points), which either have the same or different labels. If false positive points have different labels, we do not cluster them together with the other points as they are from different manifold substructures, achieving improvement 3. Further splitting the partitions based on label information is useful for handling the second type of false positives.

If false positives have the same label, our supervised approaches in Algorithms 1 and 2 will cluster them together and generate an incorrect cluster centroid. However, this will only happen if the data manifold is complex with different substructures having the same label and substructures being similar in Euclidean space. Therefore, we assume that these incorrect centroids rarely occur and do not have a significant impact on the performance of the learning algorithms.

Table 1. Details of the main proposed and baseline methods

Method	Preprocessing				Landmarking		Classification	Computational complexity
	None	Sampling	LSH	DBH	GP	Identity	Logistic regression	
Full-GP	✓				✓		✓	$O(E(M^2 N + M^3 + NMd))$
GP		✓			✓		✓	$O(E(M^2 m + M^3 + mMd))$
LSH-SC-GP		✓	✓		✓		✓	$O(RdNp + E(M^2 m + M^3 + mMd))$
DBH-SC-GP		✓		✓	✓		✓	$O(RdNp + E(M^2 m + M^3 + mMd))$
LSH-SC			✓			✓	✓	$O(RdNp)$
DBH-SC				✓		✓	✓	$O(RdNp)$

3.2 Supervised LSH (LSH-SC)

First we map samples to their respective buckets by random projection followed by partitioning using bucket width w in Algorithm 1 Line 7. This maps an item into its respective hash value. Hash values from p hash functions are then concatenated to form the hash code which represent a partition that the item is mapped into. The items that are projected into the same buckets are more likely to be similar. Then we divide the samples in the buckets into subsets based on their class label in Algorithm 1 Lines 10–12 to handle the false positives caused by random projection and false positives caused by the underlying manifold structure. The means of each subset are selected as landmarks.

3.3 Supervised DBH (DBH-SC)

DBH is designed to work with arbitrary distance functions, therefore we can plug in radial basis function (RBF). As the Gaussian process computes gradients for landmarks using RBF distance, it is suitable for approximating using DBH. The details of this algorithm are given in Algorithm 2. While tuning LSH-SC is more flexible with two parameters (hash length p and bucket size w), DBH-SC produces more consistent outputs across different runs as the number of buckets in DBH-SC can be estimated ahead of time, based on the number of hash functions.

As LSH-SC and DBH-SC are randomized algorithms we compress the dataset into approximately several times the desired number of landmarks. And pick a small minibatch (100 samples in this paper) from the compressed set for gradient computation of Gaussian processes in LSH-SC-GP and DHB-SC-GP.

3.4 Landmark Selection Methods

Landmark selection approaches that we evaluated in our experiments are briefly described below. Our first two baselines are preexisting approaches:

- **(Full-GP) Landmarking Manifold with Gaussian Processes:** State-of-the-art method [10] computing full gradients based on the whole training dataset X.

- **(GP) Landmarking Manifold with Gaussian Processes:** Minibatches are uniformly randomly selected from X during optimization.

The next two are only used in Fig. 2 for evaluating the effect of supervision (improvement 3).

- **(DBH-U-GP) Unsupervised DBH Sample Selection for Manifold Landmarking:** One sample per unsupervised DBH bucket is randomly collected to form a minibatch use in optimization steps.
- **(DBH-UC-GP) Unsupervised DBH Compressed Sample selection for Manifold Landmarking:** Centroids of unsupervised DBH buckets are collected as minibatch on which GP is applied.

The final four methods are our main approaches, the first two are adaptive sampling for Gaussian processes landmarking while the last two omit the Gaussian processes learning.

- **(LSH-SC-GP) Supervised LSH Manifold Landmarking with GP:** Algorithm 1 followed by Gaussian process learning.
- **(DBH-SC-GP) Supervised DBH Manifold Landmarking with GP:** Algorithm 2 followed by Gaussian process learning.
- **(LSH-SC) Supervised LSH Manifold Landmarking:** Stand-alone LSH-based landmarking as described in Algorithm 1 without additional Gaussian process learning.
- **(DBH-SC) Supervised DBH Manifold Landmarking:** Stand-alone DBH-based landmarking (Algorithm 2) without additional Gaussian process learning.

3.5 Complexity Analysis

In this section, we derive complexity analyses for Gaussian process manifold landmarking and analyze the effect of our approach, summarized in Table 1. The notations used are: d for data dimensionality, N for the number of instances in training set, m is the number of samples in each mini-batch, M is the number of landmarks, E is the number of epochs, R for number of repeated trials to obtain the desired buckets in LSH and p is the number of hash functions.

The computational complexity of Gaussian process manifold landmarking with computing full gradients is $O(E(M^2N + M^3 + NMd))$ and minibatch gradients is $O(E(M^2m + M^3 + mMd))$. We can reduce the complexity by selecting smaller minibatch of size $m \ll N$, however, it may take a large number of gradient epochs E when M is small. Our goal is to choose a better minibatch of small M that better represents the whole data space and as a result, keeping E relatively small.

The complexity of LSH-SC (Algorithm 1) and DBH-SC (Algorithm 2) are both $O(RdNp)$. However, in practice, LSH-SC is faster as DBH-SC requires additional processing with constant time complexity (e.g. each function needs to compute the distance twice, with distances to x_1 and x_2). LSH-SC-GP and DBH-SC-GP complexities are $O(RdNp + E(M^2m + M^3 + mMd))$ the first term becomes smaller as the number of landmarks M or minibatch size m increases.

4   Experimental Results

In this section, we describe our datasets, experimental setup, and our comprehensive comparative study. The datasets used in our experiments are described in Table 2. We run 10-fold cross validation on the datasets summarized. All experiments are performed on a PC with Intel i7-3770 with 3.40 GHz and 8 GB RAM.

Table 2. Experimental datasets and their statistics. $|C|$ is number of classes in the dataset. d is data dimensionality.

Name	N	$|C|$	d
Yale-B	2,414	38	1,024
CIFAR-10	60,000	10	3072
MNIST	60,000	10	784
Reuters	7,285	10	18,933
RCV1	9,625	4	29,992

4.1   Classification Using Landmark-Based Transformed Features

Our focus is on landmark selection as applied to logistic regression: linear learners are known to scale well [1]. Following the evaluation method used by [10], we first generate manifold landmarks. The landmarks are used as anchors to transform the training set into low-dimensional space. Given a set of M landmarks, $L = \{l_1, \ldots, l_M\}$, we map each sample $x_i \in X$ to M-dimensional landmark-based features as $v_i = [\delta_{l_1}(x_i), \ldots, \delta_{l_M}(x_i)] \in V$, where $\delta_{l_j}(x_i) = \exp(-\|l_j - x_i\|^2/\eta)$ and η is the total variance of all features. We set $M = 100$ for image datasets and $M = 500$ for text datasets as we would like to evaluate them in low-dimensional latent space in comparison to their ambient high-dimensional feature space (with dimensions given by d in Table 2).

A logistic regression classifier is applied to these new feature vectors V to obtain an accuracy measurement which is reported in the experimental results section.

4.2   Impact of Approach Improvements

Figure 2 depicts how our approach achieves progressively better accuracy as we incorporate improvements 1 to 3. Unsupervised centroids (Improvement 2) are learned faster than unsupervised sampling (Improvement 1) as DBH-UC-GP outperforms DBH-U-GP. DBH-SC-GP (Improvement 3) achieves superior accuracy over previous improvements.

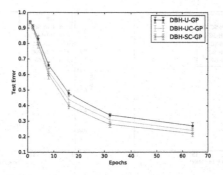

Fig. 2. Comparison of supervised vs. unsupervised DBH methods on Yale-B.

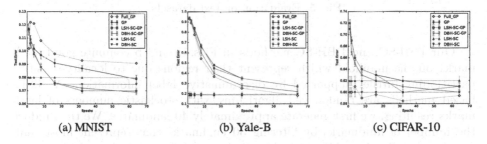

(a) MNIST　　　　　　　(b) Yale-B　　　　　　　(c) CIFAR-10

Fig. 3. Evaluation of Test Error vs. Epochs.

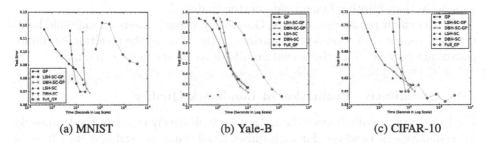

(a) MNIST　　　　　　　(b) Yale-B　　　　　　　(c) CIFAR-10

Fig. 4. Evaluation on image datasets. Note LSH-SC and DBH-SC do not have any learning component, and hense only one data point in the graph.

4.3　Qualitative Evaluation

In Fig. 1, we illustrate the representative quality of different landmarking methods on the classical swiss roll dataset with 10 different classes. In this figure, the Voronoi diagrams are created based on the landmarks and they represent the Euclidean region covered by each landmark (e.g., the landmark is the nearest to all samples in its Voronoi cell). If a cell contains samples from multiple manifolds (i.e. samples from different classes) the landmark of that cell does not represent the data space well.

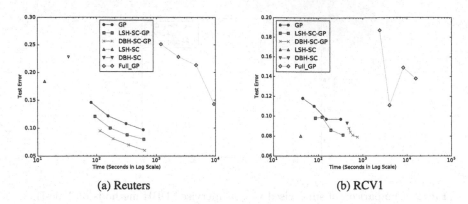

(a) Reuters   (b) RCV1

Fig. 5. Evaluation on text datasets.

Our LSH-SC and DBH-SC methods in Figs. 1a and b compute good landmarks on the manifold which represent their regions due to locality-sensitive partitioning further supported by discriminative label information. As LSH-based methods are randomized algorithms, with stochastic numbers of landmarks resulting, we first generate approximately 40 landmarks. We then reduce the number of landmarks by filtering 30 landmarks that represent most samples for visualization. Alternatively, we can apply k-means clustering over the compressed dataset (40 landmarks in this case). This would be slightly more expensive than filtering but provide better landmarks.

With default parameter settings, Gaussian process learners for manifold landmarking cannot find good representations of the data, as the resulting landmark points are very far from the actual manifolds as we see in Figs. 1c and d.

4.4   Quantitative Evaluation of Landmark Quality

For low-dimensional manifold data, we can qualitatively evaluate the landmarks by visualization. However, for high-dimensional data, we evaluate classification accuracy using landmarks as feature anchors. First, we compressed the data to approximately three times the number of landmarks using Algorithms 1 and 2. For LSH-SC and DBH-SC, we select the M landmarks ($M = 100$ for image data and 500 for text data) that cover most points in their neighborhood region. For LSH-SC-GP or DBH-SC-GP, we train a Gaussian process on the compressed training set. We set the minibatch size of the Gaussian process to 100 samples. We then report the test error achieves using the landmarks produced at particular epoch (2^{i-1} where i is the position of the point from the left margin in the figures). Figure 3 confirms our hypothesis that adaptive sampling makes better use of each epoch.

On smaller image datasets in Fig. 4, overhead costs for hashing is significant especially on the early epochs. However accuracy achieved by our methods outweighs the overhead cost as the number epochs increases. On the larger text dataset in Figs. 5a–b, our approaches are significantly more accurate and much more efficient. Full-GP is impractically slow when training with larger datasets. It also has the problem of overshooting gradient updates near local minimum and requires carefully tuning the learning rate. In Fig. 5b DBH-SC is significantly slower than on other datasets because partitioning the data is harder on this dataset and as the result, many hash functions are required to generate the desired number of landmarks.

5   Conclusion

We have presented two locality-sensitive hashing-based supervised adaptive sampling methods for finding manifold landmarks with Gaussian processes. Our new supervised approaches are efficient and achieve better accuracy than training full gradient or minibatch gradient-based Gaussian processes. Our approaches can also be used to find landmarks without having to learn Gaussian processes. This variation of our approaches makes learning extremely fast and also achieves comparable accuracy in practice as confirmed in our experiments. In both quantitative and qualitative comparisons, we find that our DBH-SC approach achieves the overall best result. For future work, we will extend our method to efficient and accurate unsupervised and semi-supervised manifold landmarking approaches. We would also like to explore how our work can be extended to spatio-temporal data where observations are independent.

Acknowledgments. We acknowledge partial support from ARC DP150103710.

References

1. Agarwal, A., Chapelle, O., Dudík, M., Langford, J.: A reliable effective terascale linear learning system. J. Mach. Learn. Res. **15**(1), 1111–1133 (2014)
2. Athitsos, V., Potamias, M., Papapetrou, P., Kollios, G.: Nearest neighbor retrieval using distance-based hashing. In: ICDE, pp. 327–336. IEEE (2008)
3. Cai, D., He, X., Wu, X., Han, J.: Non-negative matrix factorization on manifold. In: ICDM, pp. 63–72. IEEE (2008)
4. Datar, M., Immorlica, N., Indyk, P., Mirrokni, V.S.: Locality-sensitive hashing scheme based on p-stable distributions. In: SoCG, pp. 253–262. ACM (2004)
5. Elhamifar, E., Vidal, R.: Sparse manifold clustering and embedding. In: Advances in Neural Information Processing Systems, pp. 55–63 (2011)
6. Faloutsos, C., Lin, K.I.: FastMap: A fast algorithm for indexing, data-mining and visualization of traditional and multimedia datasets, vol. 24. ACM (1995)
7. Goldberg, A.B., Zhu, X., Singh, A., Xu, Z., Nowak, R.: Multi-manifold semi-supervised learning (2009)
8. Huh, S., Fienberg, S.E.: Discriminative topic modeling based on manifold learning. ACM Trans. Knowl. Discov. Data (TKDD) **5**(4), 20 (2012)
9. Indyk, P., Motwani, R.: Approximate nearest neighbors: towards removing the curse of dimensionality. In: STOC, pp. 604–613. ACM (1998)
10. Liang, D., Paisley, J.: Landmarking manifolds with Gaussian processes. In: ICML, pp. 466–474 (2015)

11. Maaten, L.V.D., Hinton, G.: Visualizing data using t-SNE. J. Mach. Learn. Res. 9(Nov), 2579–2605 (2008)
12. Roweis, S.T., Saul, L.K.: Nonlinear dimensionality reduction by locally linear embedding. Science 290(5500), 2323–2326 (2000)
13. Tenenbaum, J.B., De Silva, V., Langford, J.C.: A global geometric framework for nonlinear dimensionality reduction. Science 290(5500), 2319–2323 (2000)
14. Wang, Y., Lin, X., Wu, L., Zhang, W., Zhang, Q.: LBMCH: learning bridging mapping for cross-modal hashing. In: SIGIR, pp. 999–1002. ACM (2015)
15. Zhou, D., Bousquet, O., Lal, T.N., Weston, J., Schölkopf, B.: Learning with local and global consistency. In: NIPS, vol. 16(16), pp. 321–328 (2004)

Equitable Conceptual Clustering Using OWA Operator

Noureddine Aribi[1], Abdelkader Ouali[2], Yahia Lebbah[1], and Samir Loudni[2(✉)]

[1] Lab. LITIO, University of Oran 1, 31000 Oran, Algeria
{aribi.noureddine,lebbah.yahia}@univ-oran1.dz
[2] Normandie Univ, UNICAEN, ENSICAEN, CNRS, GREYC, 14000 Caen, France
{abdelkader.ouali,samir.loudni}@unicaen.fr

Abstract. We propose an equitable conceptual clustering approach based on multi-agent optimization, where each cluster is represented by an agent having its own satisfaction. The problem consists in finding the best cumulative satisfaction while emphasizing a fair compromise between all individual agents. The fairness goal is achieved using an equitable formulation of the Ordered Weighted Averages (OWA) operator. Experiments performed on UCI and ERP datasets show that our approach efficiently finds clusterings of consistently high quality.

1 Introduction

Structuring data in knowledge discovery is a fundamental task which helps to better understand the data and to define groups with regards to an a priori similarity measure. This is usually referred to clustering in unsupervised learning. In practice, users often would like to perform further actions such as interpreting the cluster semantically. Methods such as conceptual clustering address this by attempting to find descriptions of the clusters by means of formal concepts. Numerous approaches have been devised for conceptual clustering. Traditional approaches are of heuristic nature [17,18]. More recently, Constraint Programming (CP) [4] and Integer Linear Programming (ILP) [15] approaches have been proposed to address the problem of finding optimal conceptual clusterings in a declarative framework. They combine two exact techniques: in a first step, a dedicated mining tool (i.e., LCM [20]) is used to compute the set of all formal concepts and, in a second step, ILP or CP is used to select the best k clusters (i.e. concepts) that optimizes some given criterion. Most of the optimization measures used in these approaches lead to an unbalanced clustering where one cluster is more dominant than others. Ensuring that the clusters obtained be (roughly) balanced, i.e. of approximately the same number of data points helps in making the resulting clusterings more useful and actionable [2,22].

This paper deals with the concept of *equitably efficient solutions* to conceptual clustering problem in multi-agent decision making, where each agent represents a concept and has its own utility corresponding to a specific measure (e.g. the frequency). Here, *equity* refers to the idea of favoring solutions that fairly share

© Springer International Publishing AG, part of Springer Nature 2018
D. Phung et al. (Eds.): PAKDD 2018, LNAI 10939, pp. 465–477, 2018.
https://doi.org/10.1007/978-3-319-93040-4_37

happiness or dissatisfaction among agents [8]. The equity requirement has been fully studied by the multicriteria optimization community [9], and formalized through the three properties: **(i) Symmetry** meaning that all agents have the same importance. For instance, both utility vectors $(5, 3, 0)$ and $(0, 3, 5)$ are considered equivalent. **(ii) Pareto-monotony** which expresses that solution $(x_1, x_2, \ldots, x_n)$ is better than solution $(y_1, y_2, \ldots, y_n)$ if and only if $x_i \geq y_i$ for all i, with at least one strict inequality. **(iii) Transfer Principle** formalizes an important notion of equitable utility distribution [19]. The intuition is that any transfer between some two inequitable utilities x_i and x_j, which preserves the average of utilities, would improve the overall utility.

The common way to deal with the concept of equitably efficient solutions is to define aggregation functions that fulfills the above properties. This defines a family of the equitable aggregations which are *Schur-convex* [10]. In the literature there are several functions to aggregate individual agents' utilities by mean of *collective utility function* (CUF). The most used aggregations are `maxMin`, `minDev` and `maxSum`. The transfer principle is not ensured in the `maxMin` and `minDev`, on all of the utilities, thereby leading to the *drowning effect* [7]. The `maxSum` function is fully compensatory and thus does not capture the idea of equity.

Section 2 introduces the concepts used in this paper. Section 3 describes our ILP model for equitable conceptual clustering task. We discuss related work in Sect. 4 before demonstrating our technique's performance in Sect. 5. Section 6 concludes and points towards future research directions.

2   Background

2.1   Formal Concepts and Conceptual Clustering

Formal Concepts. Let $\mathcal{D}$ be a set of m transactions (numbered from 1 to m), $\mathcal{I}$ a set of n items (numbered from 1 to n), and $R \subseteq T \times \mathcal{I}$ a binary relation that lies transactions to items: $(t, i) \in R$ if the transaction t contains the item $i : i \in t$. An itemset (or *pattern*) is a non-null subset of $\mathcal{I}$. For instance, Table 1a gives a transactional dataset $\mathcal{D}$ with $m=11$ transactions $t_1, \ldots, t_{11}$ described by $n=8$ items. The *extent* of a set $I \subseteq \mathcal{I}$ of items is the set of transactions containing all items in I, i.e., $ext(I) = \{t \in \mathcal{D} | \ \forall i \in I, (t, i) \in R\}$. The *intent* of a subset $T \subseteq \mathcal{D}$ is the set of items contained by all transactions in T, i.e., $int(T) = \{i \in \mathcal{I} | \ \forall t \in T, (t, i) \in R\}$. These two operators induce a Galois connection between $2^{\mathcal{D}}$ and $2^{\mathcal{I}}$, i.e. $T \subseteq ext(I) \Leftrightarrow I \subseteq int(T)$. A pair such that $(I = int(T), T = ext(I))$ is called **formal concept**. This definition defines a **closure property** on dataset $\mathcal{D}$, $\texttt{closed}(I) \Leftrightarrow I = int(ext(I))$. An itemset I for which $\texttt{closed}(I) = true$ is called *closed pattern*. Using $ext(I)$, we can define the *frequency* of a concept: $\texttt{freq}(I) = |ext(I)|$, and its *diversity*: $\texttt{divers}(I) = \sum_{t \in ext(I)} |\{i \in \mathcal{I} | \ (i \notin I) \wedge (i \in t)\}|$. Additionally, we can refer to its *size*: $\texttt{size}(I) = |\{i \mid i \in I\}|$. We note $\mathcal{C}$ the set of all formal concepts.

Table 1. Running example.

(a) Dataset $\mathcal{T}$.

Trans.	Items
t_1	A B D
t_2	A E F
t_3	A E G
t_4	A E G
t_5	B E G
t_6	B E G
t_7	C E G
t_8	C E G
t_9	C E H
t_{10}	C E H
t_{11}	C F G H

(b) Three conceptual clusterings for $k=3$.

Sol.	P_1	P_2	P_3
s_1	{A, B, D}	{C, F, G, H}	{E}
s_2	{B}	{C}	{A, E}
s_3	{A}	{C}	{B, E, G}

(c) $(a_{t,c})$ matrix.

	1	2	3	4	5	6	7	8	9	10	11	12	13	14	15	16	17	18
t_1	1	1	0	0	0	1	0	0	0	0	0	0	0	0	0	0	0	0
t_2	1	0	1	1	0	0	0	0	0	0	0	0	0	0	1	0	1	0
t_3	1	0	1	0	1	0	0	0	0	0	0	0	0	0	1	1	0	1
t_4	1	0	1	0	1	0	0	0	0	0	0	0	0	0	1	1	0	1
t_5	0	0	0	0	0	1	1	0	0	0	0	0	0	0	1	1	0	1
t_6	0	0	0	0	0	1	1	0	0	0	0	0	0	0	1	1	0	1
t_7	0	0	0	0	0	0	0	1	1	1	0	0	1	1	0	1	1	0
t_8	0	0	0	0	0	0	0	1	1	1	0	0	1	0	1	1	0	1
t_9	0	0	0	0	0	0	1	1	0	1	0	0	1	1	0	0	0	0
t_{10}	0	0	0	0	0	0	1	1	0	1	0	0	1	1	0	0	0	0
t_{11}	0	0	0	0	0	0	1	0	0	0	1	1	1	0	0	1	1

Conceptual Clustering. Clustering is the task of assigning the transactions in the data to relatively homogeneous groups. Conceptual clustering aims to also provide a distinct description for each cluster - the concept characterizing the transactions contained in it. This problem can be formalized as: "find a set of k clusters, each described by a closed pattern $P_1, P_2, \ldots, P_k$, covering all transactions without any overlap between clusters". An evaluation function f can be used to express the goodness of the clustering. Different optimization criteria may be considered: maximizing the *sum of frequencies* of the selected concepts; minimizing the *sum of diversities* of the selected concepts. For instance, for dataset $\mathcal{D}$ and $k = 3$, minimizing $f(P_1, \ldots, P_k) = \sum_{1 \leq i \leq k} \texttt{divers}(P_i)$ provides one clustering s_1, with optimal value 18 (see Table 1b). Solution $s_1 = (1, 1, 9)$ has one large cluster (of size 9) covering most of the transactions, and two clusters that cover only one transaction. Such a clustering may be less interesting than those in which the clusters are all of comparable size. A common way to get more balanced clusterings is to consider dedicated optimization settings:

- *maximizing the minimal frequency* (maxMin). We search for solutions in which the minimal frequency of the selected concepts is as large as possible.
- *minimizing the deviation in cluster frequency* (minDev). We enforce a small difference between cluster frequencies: Min $\max(\texttt{freq}(P_1), \ldots)$ - $\min(\texttt{freq}(P_1), \ldots)$.

However, these two settings suffer from the so called *drowning effect* [7]. In fact, concerning maxMin (resp. minDev), the transfer principle is ensured only on the min (resp. min and max) utility, and thus intermediate utilities are not necessarily equitable. To address equity requirement, we consider, in the next section, a sophisticated operator that focuses on the whole utilities.

2.2 Equitable Multiagent Optimization

Let $N = \{1, \ldots, n\}$ be a set of n agents. A solution of a multiagent optimization problem is characterized by a utility vector $x = (x_1, \ldots, x_n) \in \mathbb{R}^n_+$, where x_i represent the utility (or a degree of satisfaction) of the ith agent. Utility vectors are commonly compared using the Pareto dominance relation (P-dominance).

The weak-P-dominance $\succsim_P$ between two utility vectors x, x' is defined as: $x \succsim_P x' \Leftrightarrow [\forall i \in N, x_i \geq x_i']$, whereas the strict P-dominance $\succ_P$ between x and x' is given by: $x \succ_P x' \Leftrightarrow [x \succsim_P x' \wedge not(x' \succsim_P x)]$. A solution x^* is Pareto-optimal (a.k.a *efficient*) if and only if there is no solution x that dominates x^*. The P-dominance can be formulated as: $max \{(x_1, \ldots, x_n) : x \in Q\}$, where Q is the set of feasible solutions. The P-dominance may lead to a large set of incomparable solutions. To refine the P-dominance, we should specialize a dominance relation so as to favor *equitable* utility vectors. The main intuition behind the equity criterion refers to the idea of selecting solutions that fairly share satisfaction between agents [19]. Formally, an equitable dominance relation $\succsim_\parallel$ should fulfill three main properties [8,10]: (i) **Symmetry**: Consider a utility vector $x \in I\!R_+^n$. For any permutation σ on N, we have $(x_{\sigma(1)}, \ldots, x_{\sigma(n)}) \sim (x_1, \ldots, x_n)$. (ii) **P-Monotony**: For all $x, y \in I\!R_+^n$, $x \succsim_P y \Rightarrow x \succsim_\parallel y$ and $x \succ_P y \Rightarrow x \succ_\parallel y$. (iii) **Transfer principle** (a.k.a *Pigou-Dalton* transfers in Social Choice Theory): Let $x \in I\!R_+^n$ and $x_i > x_j$ for some $i, j \in N$. Let e^z be a vector such that $\forall i \neq z, e_i^z = 0$ and $e_z^z = 1$. For all ϵ where $0 < \epsilon \leq \frac{x_i - x_j}{2}$, we get $x - \epsilon e^i + \epsilon e^j \succsim_\parallel x$. Any slight improvement of x_j at the expense (reduction) of x_i, which preserves the *average of utilities*, would produce a better distribution of the utilities among agents and consequently improve the overall utility of the solution. For example, if we consider two utility vectors $x = (\mathbf{11}, 10, \mathbf{7}, 10)$ and $y = (\mathbf{9}, 10, \mathbf{9}, 10)$, then the transfer principle implies that $y \succsim_\parallel x$, because there is a transfer of size $\epsilon = 2$ (i.e. $\frac{x_1 - x_3}{2}$), which allows to have y from x. Combining P-monotony and the Transfer principle leads to the so called *Generalized Lorenz dominance* defined in [5] (for more details see [8,10]).

2.3 Equitable Aggregation Functions

A usual way to assess the quality of a utility vector is to aggregate the individual utilities with a *collective utility function* (CUF) [12] $G : I\!R_+^n \rightarrow I\!R_+$, which improves the overall welfare by $max\{G(x) : x \in Q\}$. The G function can be a linear combination of individual utilities (i.e. $G(x) \stackrel{\text{def}}{=} sum(x)$), which is not suitable to fairness context. Another way to build G is based on the min function (i.e. $G(x) \stackrel{\text{def}}{=} min(x)$), but it is sensitive to the *drowning effect* [7]. Other refinements of the min function exist (e.g. augmented min, lexmin [3]), but do not really solve the problem, since all are sensitive to *drowning effect*. In order to guarantee equitable aggregations, G should conform to the three equity properties. The most known way is to use Schur-convex function ψ, which are order preserving the three equity properties : $x \succsim_\parallel y \Leftrightarrow \psi(x) \geq \psi(y)$. Precisely, when some aggregation function G is Schur-convex [10], then it is an equitable aggregation [9]. Thus Schur-convex functions play a key role in equitable aggregations (for more details, see [9,10]). In this line of reasoning, we introduce, in the next section, an aggregation function that ensures equity.

2.4   Ordered Weighted Averages (OWA)

The Ordered Weighted Averages (OWA) [21] is defined as follows: $G^w(x) = \sum_{i=1}^{n} w_i x_{\sigma(i)}$, where $w = (w_1, \ldots, w_n) \in [0,1]^n$ and $x_{\sigma(1)} \leq x_{\sigma(2)} \leq \cdots \leq x_{\sigma(n)}$. OWA provides a family of compromises between the sum and min. Golden and Perny [8] propose coefficients for the OWA aggregation method, such that it is Schur-convex:

Theorem 1. [8] *Let be the following coefficients of the OWA aggregation:* $W(x) = \sum_{k=1}^{n} \sin(\frac{(n+1-k)\pi}{2n+1}) x_{(k)}$. *W is a Schur-convex function.*

Theorem 1 is fundamental, since Schur-convex functions ensure equity [8,9].

3   ILP Models

This section describes different ILP models for finding an equitable conceptual clustering. Our approach follows the two steps approach of [15]: (1) a dedicated closed itemset mining tool (i.e., LCM [20]) is used to compute the set $\mathcal{C}$ of all closed patterns; (2) ILP is used to select a subset of $\mathcal{C}$ that is a partition of the set $\mathcal{D}$ of transactions and that optimizes some given criterion. To enforce equitable clusterings, we enhance the second step with additional constraints enabling to ensure equitable OWA aggregation.

3.1   OWA ILP Models

We present our first ILP model, called basic OWA ILP, for computing equitable conceptual clusterings using OWA operator. Then, we show how this basic model can be improved by post-processing the OWA constraints.

Let $\mathcal{D}$ be a dataset with m transactions defined on a set of n items $\mathcal{I}$. Let $\mathcal{C}$ be the set of p closed patterns representing the candidate clusters. Let $a_{t,c}$ be an $m \times p$ binary matrix where $(a_{t,c} = 1)$ iff $c \subseteq t$, i.e., the transaction t belongs to extension of the closed pattern c. The $(a_{t,c})$ matrix associated with dataset $\mathcal{D}$ of Table 1a is outlined in Table 1c. Let v be the list of closed pattern utilities (e.g., frequency, diversity, etc.). For each closed pattern $(c \in \mathcal{C})$, a binary variable x_c is associated s.t. $(x_c = 1)$ iff the cluster c selected.

(a) Basic OWA ILP model. Figure 1a gives the ILP model for equitable conceptual clustering. It uses two types of constraints: conceptual clustering constraints and OWA constraints modeling the sorting operation required in the OWA operator:

- **Conceptual clustering constraints.** Constraints (C1) enforce the subset of selected closed patterns to define a partition of $\mathcal{D}$. Constraints (C2) specify a lower bound k_{min} and/or an upper bound k_{max} on the number of selected closed patterns.
- **OWA constraints.** The objective function and constraints (O1) and (O2) implement a known linear programming formulation [14] of the OWA operator on the conceptual clustering, where the coefficients ω are fixed by theorem 1.

$\text{Max } \sum_{c=1}^{|\mathcal{C}|} \omega_c \cdot r_c$

$$
\text{s.t.} \begin{cases}
\text{Clustering.} \begin{cases}
\text{(C1)} & \sum_{c=1}^{|\mathcal{C}|} a_{t,c} \cdot x_c = 1, \quad \forall t \in \mathcal{D} \\
\text{(C2)} & k_{min} \le \sum_{c=1}^{|\mathcal{C}|} x_c \le k_{max}
\end{cases} \\
\text{OWA sorting.} \begin{cases}
\text{(O1)} & r_c - (v_i \cdot x_i) \le M z_{c,i}, \, \forall i, c = 1, ..., |\mathcal{C}| \\
\text{(O2)} & \sum_{i=1}^{|\mathcal{C}|} z_{c,i} \le c-1, \quad \forall c = 1, ..., |\mathcal{C}|
\end{cases} \\
x_c \in \{0,1\}, \, r_c \in \mathbb{R}_+, \quad \forall c = 1, ..., |\mathcal{C}| \\
z_{c,i} \in \{0,1\}, \quad \forall i, c = 1, ..., |\mathcal{C}|
\end{cases}
$$

(a) Basic OWA ILP model.

$\text{Max } \sum_{c=1}^{|\mathcal{C}|} \omega_c \cdot (v_c^{\uparrow} \cdot x_c^{\uparrow})$

$$
\text{s.t.} \begin{cases}
\text{(C1),} \quad \text{(C2)} \\
x_c \in \{0,1\}, \\
\quad \forall c = 1, ..., |\mathcal{C}|
\end{cases}
$$

(b) Improved OWA ILP model.

Fig. 1. OWA ILP models for equitable conceptual clustering.

As explained in Sect. 2.4, OWA is a weighed sum on the sorted utilities. That is why we introduced r, which is equal to the sorted version of the utility vector v. M is a sufficiently large constant. Let z be $|\mathcal{C}|^2$ boolean matrix dedicated only to formulate the sorting constraints (O1) and (O2), which enforce that the utility vector $v \cdot x$ of the closed patterns are sorted in ascending order matching the OWA coefficients ω. These sorting constraints are fully explained in [14]. It follows that the kth smallest utility value r_k will have the kth biggest weight ω_k. The objective function maximizes the weighted sum using OWA weights ω given in Theorem 1.

(b) Improved OWA ILP model. In order to find efficiently an equitable conceptual clustering, we propose the optimize model (see Fig. 1b) as follows:

- Precisely, sorting constraints (O1) and (O2) are specifically used when the utility values are given in comprehension. Fortunately, the utility values of formal concepts are known beforehand. Thus, sorting is performed immediately after finding closed patterns. We use $v^{\uparrow}$, which is the sorted version of v in ascending order.
- We assign the weights ω of equitable OWA to the sorted utility values, so that all equal utilities will have the same weight.

For our experiments, we used the improved OWA model. Our preliminary results showed that basic OWA model performs very poorly compared to the improved OWA model in terms of CPU-times. This meanly due to the fact that (n^2) additional constraints and (n^2) additional variables are used to encode the OWA sorting constraints. This constitutes a strong limitation of the size of the databases that could be managed.

Proposition 1. *Basic and improved OWA ILP models are equivalent.*

Proof. Both OWA models use weights ω given in Theorem 1, which ensure an equitable aggregation. Improved OWA is an optimization of the basic model: (1) It uses an a priori sorting of utilities (no need to sorting constraints); (2) The same weight is assigned to equal utilities (the same satisfaction level), which preserves straightforwardly the conformity with Theorem 1. Thus, both OWA models are equivalent. □

$$\text{Max } z$$

$$\text{s.t.} \begin{cases} \text{(C1),} \quad \text{(C2)} \\ \text{(C3)} \quad z \leq v_c \cdot x_c, \; \forall \, c = 1, ..., |\mathcal{C}| \\ x_c \in \{0, 1\}, \qquad \forall \, c = 1, ..., |\mathcal{C}| \\ z \geq 0 \end{cases}$$

(a) maxMin **ILP** model.

$$\text{Max } z_{max} - z_{min}$$

$$\text{s.t.} \begin{cases} \text{(C1),} \quad \text{(C2)} \\ \text{(C4)} \quad z_{max} \geq v_c \cdot x_c, \; \forall \, c = 1, ..., |\mathcal{C}| \\ \text{(C5)} \quad z_{min} \leq v_c \cdot x_c, \; \forall \, c = 1, ..., |\mathcal{C}| \\ x_c \in \{0, 1\}, \qquad \forall \, c = 1, ..., |\mathcal{C}| \\ z_{max} \geq 0, \; z_{min} \geq 0 \end{cases}$$

(b) minDev **ILP** model.

Fig. 2. ILP models for the conceptual clustering.

(c) ILP numerical stability. The set of extracted closed patterns is mostly huge, which leads to a huge OWA vector ω in the basic model, and affect the numerical stability of the ILP solver. The optimized OWA model tackles this issue, thanks to assigning the same weight to all equal utilities. This makes it possible to solve real-world instances in our experiments reported in Sect. 5.

3.2 Other ILP Models

As described in Sect. 2.1 a linear aggregation of individual utilities $max\{sum(x) : x \in Q\}$, does not fit the equity requirement. This suggests resorting to non-linear aggregation operators, especially the maxMin and minDev. The maxMin aggregation $max\{min(x) : x \in Q\}$ tackles equity by improving the worst utility. This function can be linearized by maximizing a decision variable $z \geq 0$, that is a lower bound for the utility vector $v \cdot x$ (see Fig. 2a, inequality C3), where v is the clustering criterion to be optimized (e.g. frequency). Thus, the linear formulation for the conceptual clustering is given by the ILP model of Fig. 2a.

An alternative way of ensuring equity is by achieving maximum deviation minimization minDev between both the best and the worst utilities: Min $\{max(x) -min(x) : x \in Q\}$. It can be linearized by introducing $2 \times n$ constraints and two decision variables $z_{max} \geq 0$ and $z_{min} \geq 0$ to maintain the max and the min values of the utility vector $v \cdot x$ (see Fig. 2b, inequalities C4–C5). The resulting ILP model is given in Fig. 2b.

4   Related Work

Declarative approaches. Recently, [15,16] have developed declarative frameworks using ILP, which can find optimal conceptual clusterings, where clusters correspond to concepts. Later, Chabert *et al.* have introduced two new CP models for computing optimal conceptual clusterings. The first model (denoted FullCP2) may be seen as an improvement of [6]. The second model (denoted HybridCP) follows the two step approach of [15]: the first step is exactly the same; the second step uses CP to select formal concepts. Our work is different in that we study the setting where each clustering must fulfills equity requirements.

Distance-based clustering aims at finding homogeneous clusters only based on a dissimilarity measure between objects. Different declarative frameworks have been developed, which rely on CP [6] or ILP [1,13]. There are a few existing approaches for obtaining balanced clusters. The most prominent one is the approach proposed by [2]. Our adoption of closed patterns cuts down on redundancy compared to other ways of selecting candidate clusters. Moreover, our use of OWA gives stronger guarantees about the obtained clusterings in terms of balancing.

5   Experiments and Results

The experimental evaluation is designed to address the following questions:

1. How do the ILP models compare and scale on the considered datasets?
2. How do the resulting clusters and their description compare qualitatively?
3. How (in terms of CPU-times) does our ILP model compares to the CP models of Chabert *et al.* [4]?

Experimental protocol. All experiments were conducted on Linux cluster[1], where each node has a dual-CPU Xeon E5-2650 with 16 cores, 64 GB RAM, running at 2.00 GHz. We used LCM to extract all closed patterns and CPLEX v.12.6.1 to solve the different ILP models. For all methods, a time limit of 24 h has been used.

Test instances. We used classical ML datasets, coming from the UCI database. We have also considered the same datasets (called ERP-i, with $i \in [1,7]$) used in [4] and coming from a real application case[2], which aims at extracting setting concepts from an Enterprise Resource Planning (ERP) software corresponding to groups of parameter settings groups of parameter settings. Table 2 shows the characteristics of all datasets.

Table 2. Dataset characteristics. Each row gives the number of transactions ($\#\mathcal{D}$), the number of items ($\#\mathcal{I}$), the density and the number of closed patterns extracted ($\#\mathcal{C}$).

(a) UCI datasets.

Dataset	$\#\mathcal{D}$	$\#\mathcal{I}$	Density(%)	$\#\mathcal{C}$
Soybean	630	50	32	31,759
Primary-tumor	336	31	48	87,230
Lymph	148	68	40	154,220
Vote	435	48	33	227,031
tic-tac-toe	958	27	33	42,711
Mushroom	8124	119	18	221,524
Zoo-1	101	36	44	4,567
Hepatitis	137	68	50	3,788,341
Anneal	812	93	45	1,805,193

(b) ERP datasets.

Dataset	$\#\mathcal{D}$	$\#\mathcal{I}$	Density(%)	$\#\mathcal{C}$
ERP-1	50	27	48	1,580
ERP-2	47	47	58	8,1337
ERP-3	75	36	51	10,835
ERP-4	84	42	45	14,305
ERP-5	94	53	51	63,633
ERP-6	95	61	48	71,918
ERP-7	160	66	45	728,537

---

[1] http://www.rx-racim.cerist.dz/?page_id=26.
[2] Datasets available at https://perso.liris.cnrs.fr/christine.solnon/erp.html.

Table 3. Comparing the quality of the resulting clusterings in terms of ICS and ICD.

(a) Maximizing frequency.

$\mathcal{D}$	k	OWA		minDev		maxMin		maxSum	
		ICS	ICD	ICS	ICD	ICS	ICD	ICS	ICD
Soybean	3	0.447	0.784	0.447	0.784	1.000	0.026	1.000	0.026
	4	0.331	0.865	0.331	0.865	1.000	0.026	1.000	0.026
	5	0.259	0.895	0.284	0.905	1.000	0.026	1.000	0.026
	6	0.231	0.940	0.231	0.940	1.000	0.026	1.000	0.026
	7	0.195	0.964	0.195	0.964	0.959	0.108	0.959	0.108
	8	0.186	0.987	0.186	0.987	0.671	0.474	0.959	0.108
	9	0.166	1.000	0.166	1.000	0.671	0.474	0.959	0.108
	10	0.136	0.999	0.142	0.999	0.670	0.474	0.959	0.108

(b) Minimizing diversity.

$\mathcal{D}$	k	OWA		minDev		maxMin		maxSum	
		ICS	ICD	ICS	ICD	ICS	ICD	ICS	ICD
soybean	3	0.447	0.776	0.447	0.776	1.000	0.026	0.447	0.776
	4	0.334	0.839	0.338	0.854	1.000	0.026	0.406	0.831
	5	0.296	0.900	0.301	0.900	1.000	0.026	0.389	0.843
	6	0.257	0.929	0.265	0.934	1.000	0.026	0.398	0.851
	7	0.240	0.956	0.240	0.956	0.959	0.106	0.330	0.909
	8	0.220	0.971	0.198	0.978	0.959	0.106	0.323	0.918
	9	0.183	0.991	0.184	0.989	0.959	0.106	0.216	0.975
	10	0.170	0.999	0.157	1.000	0.959	0.106	0.213	0.980

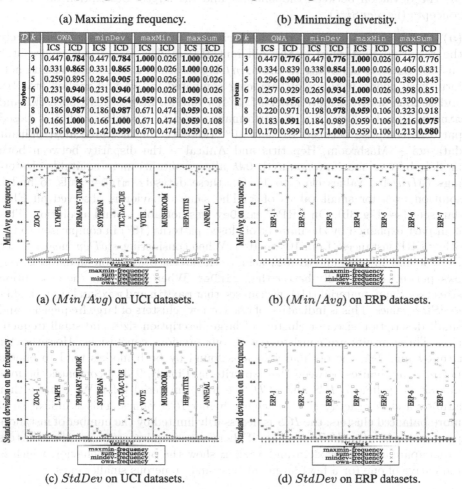

(a) (Min/Avg) on UCI datasets.

(b) (Min/Avg) on ERP datasets.

(c) $StdDev$ on UCI datasets.

(d) $StdDev$ on ERP datasets.

Fig. 3. Quality of balancing of the resulting clusterings of the ILP models.

To evaluate the quality of a clustering, we test the coherence of a clustering, measured by the intra-cluster similarity *(ICS)* and the inter-clusters dissimilarity *(ICD)*, both of which should be as large as possible. Given a similarity measure s between two transactions t and t', where $s : \mathcal{D} \times \mathcal{D} \mapsto [0,1]$, $s(t,t') = \frac{|t \cap t'|}{|t \cup t'|}$, $ICS(P_1, ..., P_k) = \frac{1}{2}\sum_{1 \leq i \leq k}(\sum_{t,t' \in P_i} s(t,t'))$ and $ICD(P_1, ..., P_k) = \sum_{1 \leq i < j \leq k}(\sum_{t \in P_i, t' \in P_j}(1 - s(t,t')))$.

To evaluate how well equitable the clusters are w.r.t frequency, we used three measures: (1) the ratio between the frequency of the smallest cluster to the average cluster frequency (i.e. Min/Avg). For m transactions put into k clusters, Avg is just (m/k); (2) the *Standard Deviation* in cluster frequencies (i.e. $StdDev$);

(3) the deviation between the smallest and the largest description of selected concepts (i.e. *devSize*).

(a) Qualitative analysis of clusterings. Figure 3a compares qualitatively the resulting clusterings of the different ILP models for various values of k on UCI datasets according to the Min/Avg measure. maxMin and maxSum perform very poorly in terms of balancing compared to OWA and minDev (maxMin and maxSum always achieve lower Min/Avg values). Interestingly, both OWA and minDev almost achieve similar performance on datasets with number of closed patterns comprise between 10^3 and 10^5. However, for the three most difficult datasets − Mushroom, Hepatitis and Anneal − the disparity between both models become more pronounced: OWA always obtains more equitable clusterings (Min/Avg values close to 1). On these datasets, minDev fails to find a solution even for small values of k. The same behavior is observed on ERP datasets (see Fig. 3b). On ERP-7, minDev was not able to find a solution. This is in part explained by the number of closed patterns (10^6) in comparison to the other ERP instances (from 10^3 to 10^5). When considering $stdDev$ measure (see Figs. 3c and 3d), OWA and minDev achieve the lowest $StdDev$ on all datasets, but OWA performs marginally better than minDev. When examining the description sizes (see Supp. material [11]), we can see that maxMin and maxSum lead to higher *devSize* values. This is indicative of one (or few) clusters of large frequencies and small description sizes, or clusters of large description sizes and small frequencies. These results are consistent with our previous conclusions. However, for minDev and OWA, the optimal solutions found by both models tend to offer a better compromises between the two criteria. Finally, Tab. 3 compares the four models according to *ICS* and *ICD* (see Supp. material [11]). We can see that minDev and OWA sacrify *ICS* to achieve higher *ICD* values. This is indicative of more balanced clusters: the *ICS* is necessarily limited by the number of instances per cluster but the *ICD* increases if there are more instances in other clusters to compare against. maxMin and maxSum show the opposite behavior, which is indicative of one (or a few) large clusters, and numerous smaller ones.

(b) Scale-up property analysis. Figures 4a and b compare the CPU-times for computing optimal clusterings for various values of k on UCI and ERP datasets when maximizing the sum of frequencies of the selected concepts. The CPU-times include the time spent by LCM to extract all closed patterns. On UCI datasets, minDev performs very poorly compared to the other ILP models. Although the qualitative results of minDev are satisfactory, this model remains hampered by long solving times: it goes beyond the timeout on 32 instances (out of 72), particularly on the three most difficult datasets Mushroom, Hepatitis and Anneal (see Fig. 4a). This probably stems from the fact that $(2 \times n)$ additional constraints are used to capture the minimal deviation. However, OWA yields quite competitive results, while achieving optimal equitable clusterings (see the qualitative analysis). It is able to solve all instances and comes in second position. Overall, maxMin gets the best performances. However, as noticed above, the optimal solutions found are far to be equitable ones; they correspond to extreme solutions (worst cases). This probably explains in part the good behaviour of maxMin

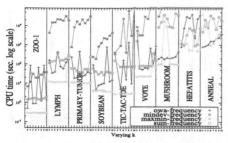

(a) UCI datasets: maximizing frequency.   (b) ERP datasets: maximizing frequency.

Instance	OWA with k not fixed $k \in [3,10]$		OWA with k fixed		OWA with k not fixed $k \in [3,|\mathcal{D}|-1]$	
	best k	Time (s.) (2)	best k	Time (s.) (2)	best k	Time (s.) (2)
Soybean	10	27.09	10	**14.82**	501	15.76
Primary-tumor	10	26.81	10	33.34	215	**14.52**
Lymph	10	77.97	10	173.00	147	**20.61**
Vote	10	89.8	10	879.22	342	**42.3**
tic-tac-toe	9	2,104.07	9	**9.95**	956	11.07
Mushroom	10	**377.21**	10	442.34	8,123	982.95
Zoo-1	10	5.47	10	1.37	59	**0.8**
Hepatitis	10	8,462.45	10	35,498.2	136	**607.51**
Anneal	10	3,674.89	10	3,666.82	459	**1,453.04**

(c) Maximizing frequency.

Instance	OWA with k not fixed $k \in [3,10]$		OWA with k fixed		OWA with k not fixed $k \in [3,|\mathcal{D}|-1]$	
	best k	Time (s.) (2)	best k	Time (s.) (2)	best k	Time (s.) (2)
Soybean	10	13.7	10	165.42	501	**9.61**
Primary-tumor	10	46.19	10	210.01	215	**18.5**
Lymph	10	123.84	10	569.63	145	**22.05**
Vote	10	146.72	10	786.84	342	**45.7**
tic-tac-toe	9	37,882.31	9	293.82	956	**7.21**
Mushroom	10	**274.62**	10	667.99	8,123	1,086.13
Zoo-1	10	0.89	10	1.82	59	**0.8**
Hepatitis	10	37,915.3	8	6,275.23	136	**630.91**
Anneal	10	6,839.68	10	25,760.25	459	**2,311.01**

(d) Minimizing diversity.

Fig. 4. CPU-times analysis.

model. The same behavior is observed for `minDev` on ERP datasets. Finally, the three ILP models – `OWA`, `maxMin` and `maxSum` – perform very similarly on all instances. We conclude that `OWA` model offers a good compromise between solution quality and computing time.

(c) ILP models vs. CP based models. Figures 5a and b compare the performance of `maxMin` ILP model with the two CP models (`FullCP2` and `HybridCP`) maximizing the minimal frequency of a cluster on UCI and ERP datasets. The CPU-times of `HybridCP` include those for the preprocessing step. `maxMin` ILP model outperforms `FullCP2` and `HybridCP` by several orders of magnitude on all datasets. None of the two CP models scales well for this objective: they fail to find a solution within the time limit for ($k \geq 4$), except for 4 datasets. Moreover, `OWA` ILP model clearly beats the two CP models. Finally, notice that `FullCP2` performs marginally better than `HybridCP`.

(d) `OWA` model with k not fixed. Our third set of experiments aims at evaluating `OWA` model capability for finding the optimal solution when k is not fixed. For this aims, we selected two settings: $k \in [3, 10]$ (`OWA-1`) and $k \in [3, |\mathcal{D}|-1]$ (`OWA-2`). Figures 4c and d compare the CPU-times when k is not fixed (Columns 3 and 7), and when k is fixed (Col. 5) on UCI datasets. Col. 4 reports the best values found for k ($3 \leq k \leq 10$) that optimize both objectives. For all datasets but two, `OWA-1` and `OWA-2` are the best performing approaches. `OWA-1` is able to solve 5 (resp. 7) instances quicker when maximising the frequency (resp. diversity). Interestingly, `OWA-1` and `OWA` (with k fixed) always agree on the best value for

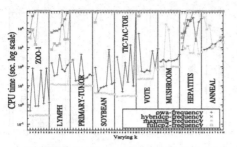

(a) UCI datasets: maximizing frequency.

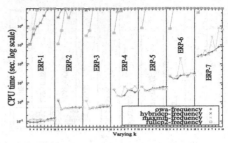

(b) ERP datasets: maximizing frequency.

Fig. 5. Comparing CPU-times of `maxMin` ILP model with the two CP models.

k. Compared to `OWA-1`, `OWA-2` scales well, particularly on the two most difficult datasets Anneal and Hepatitis (speed-up of up to 60.09). Indeed, larger values of k enable to find balanced clustering more quickly than for smaller values of k: there $|\mathcal{D}| - 1$ clusters for 3 datasets, whereas for the remaining datasets the value of k is rather high.

6   Conclusion

We have proposed an efficient approach for equitable conceptual clustering that uses closed itemset mining to discover candidates for descriptions, and ILP implementing an equitable aggregate function based on `OWA` to select the best clusters of balanced frequencies. Contrary to `maxMin` and `minDev` operators, our approach offers a good compromise between solution quality and computing time. We plan to investigate multi-criteria conceptual clustering, where the utilities are not comparable. Exploiting equity constraints within approximate approaches could become interesting to tackle very large datasets.

References

1. Babaki, B., Guns, T., Nijssen, S.: Constrained clustering using column generation. In: Simonis, H. (ed.) CPAIOR 2014. LNCS, vol. 8451, pp. 438–454. Springer, Cham (2014). https://doi.org/10.1007/978-3-319-07046-9_31
2. Banerjee, A., Ghosh, J.: Scalable clustering algorithms with balancing constraints. Data Min. Knowl. Discov. **13**(3), 365–395 (2006)
3. Bouveret, S., Lemaître, M.: Computing leximin-optimal solutions in constraint networks. Artif. Intell. **173**(2), 343–364 (2009)
4. Chabert, M., Solnon, C.: Constraint programming for multi-criteria conceptual clustering. In: Beck, J.C. (ed.) CP 2017. LNCS, vol. 10416, pp. 460–476. Springer, Cham (2017). https://doi.org/10.1007/978-3-319-66158-2_30
5. Chong, K.M.: An induction theorem for rearrangements. Candadian J. Math. **28**, 154–160 (1976)
6. Dao, T., Duong, K., Vrain, C.: Constrained clustering by constraint programming. Artif. Intell. **244**, 70–94 (2017)

7. Dubois, D., Fortemps, P.: Computing improved optimal solutions to max-min flexible constraint satisfaction problems. EJOR **118**, 95–126 (1999)
8. Golden, B., Perny, P.: Infinite order Lorenz dominance for fair multiagent optimization. In: AAMAS, pp. 383–390 (2010)
9. Kostreva, M.M., Ogryczak, W., Wierzbicki, A.: Equitable aggregations and multiple criteria analysis. EJOR **158**(2), 362–377 (2004)
10. Marshall, W., Olkin, I.: Inequalities: Theory of Majorization and its Applications. Academic Press, London (1979)
11. Material Science. https://loudni.users.greyc.fr/cclustering.html
12. Moulin, H.: Axioms of Cooperative Decision Making. Cambridge University Press, Cambridge (1989). Cambridge Books
13. Mueller, M., Kramer, S.: Integer linear programming models for constrained clustering. In: Pfahringer, B., Holmes, G., Hoffmann, A. (eds.) DS 2010. LNCS (LNAI), vol. 6332, pp. 159–173. Springer, Heidelberg (2010). https://doi.org/10.1007/978-3-642-16184-1_12
14. Ogryczak, W., Sliwinski, T.: On solving linear programs with the ordered weighted averaging objective. EJOR **148**(1), 80–91 (2003)
15. Ouali, A., Loudni, S., Lebbah, Y., Boizumault, P., Zimmermann, A., Loukil, L.: Efficiently finding conceptual clustering models with integer linear programming. IJCAI **2016**, 647–654 (2016)
16. Ouali, A., Zimmermann, A., Loudni, S., Lebbah, Y., Cremilleux, B., Boizumault, P., Loukil, L.: Integer linear programming for pattern set mining; with an application to tiling. In: Kim, J., Shim, K., Cao, L., Lee, J.-G., Lin, X., Moon, Y.-S. (eds.) PAKDD 2017. LNCS (LNAI), vol. 10235, pp. 286–299. Springer, Cham (2017). https://doi.org/10.1007/978-3-319-57529-2_23
17. Pensa, R.G., Robardet, C., Boulicaut, J.-F.: A bi-clustering framework for categorical data. In: Jorge, A.M., Torgo, L., Brazdil, P., Camacho, R., Gama, J. (eds.) PKDD 2005. LNCS (LNAI), vol. 3721, pp. 643–650. Springer, Heidelberg (2005). https://doi.org/10.1007/11564126_68
18. Perkowitz, M., Etzioni, O.: Adaptive web sites: conceptual cluster mining. In: IJCAI, vol. 99, pp. 264–269 (1999)
19. Sen, A., Foster, J.: On Economic Inequality. Clarendon Press, Oxford (1997)
20. Uno, T., Asai, T., Uchida, Y., Arimura, H.: An efficient algorithm for enumerating closed patterns in transaction databases. In: Suzuki, E., Arikawa, S. (eds.) DS 2004. LNCS (LNAI), vol. 3245, pp. 16–31. Springer, Heidelberg (2004). https://doi.org/10.1007/978-3-540-30214-8_2
21. Yager, R.R.: On ordered weighted averaging aggregation operators in multicriteria decisionmaking. IEEE Trans. Syst. Man Cybern. **18**(1), 183–190 (1988)
22. Yang, Y., Padmanabhan, B.: Segmenting customer transactions using a pattern-based clustering approach. In: ICDM, Vol. 2003, pp. 411–418 (2003)

Unsupervised Extremely Randomized Trees

Kevin Dalleau$^{(\boxtimes)}$, Miguel Couceiro, and Malika Smail-Tabbone

Universite de Lorraine, CNRS, Inria, LORIA, 54000 Nancy, France
{kevin.dalleau,miguel.couceiro,malika.smail}@loria.fr

Abstract. In this paper we present a method to compute dissimilarities on unlabeled data, based on extremely randomized trees. This method, Unsupervised Extremely Randomized Trees, is used jointly with a novel randomized labeling scheme we describe here, and that we call *AddCl3*. Unlike existing methods such as *AddCl1* and *AddCl2*, no synthetic instances are generated, thus avoiding an increase in the size of the dataset. The empirical study of this method shows that Unsupervised Extremely Randomized Trees with *AddCl3* provides competitive results regarding the quality of resulting clusterings, while clearly outperforming previous similar methods in terms of running time.

Keywords: Clustering · Unsupervised classification · Decision tree
Extremely randomized trees · Similarity measure · Distance

1 Introduction and Preliminaries

Many unsupervised learning algorithms rely on a metric to evaluate the pairwise distance between samples. Despite the large number of metrics already described in the literature [3], in many applications, the set of available metrics is reduced by intrinsic characteristics of the data and of the chosen algorithm. The choice of a metric may strongly impact the quality of the resulting clustering, thus making this choice rather critical.

Shi and Horvath [19] proposed a method to compute distances between instances in unsupervised settings using Random Forest (RF). RF [2] is a popular algorithm for supervised learning tasks, and has been used in various settings ([13,16]). It is an ensemble method, combining decision trees in order to obtain better results in supervised learning tasks. Let $L = \{(x_1, y_1), \ldots, (x_n, y_n)\}$ be a training set, where $X = \{x_1, \ldots, x_n\}$ is a list of *samples* (*i.e.*, feature vectors) and $Y = \{y_1, \ldots, y_n\}$ is the list of corresponding class labels. The algorithm begins by creating several new training sets, each one being a bootstrap sample of elements from X. A decision tree is built on each training set, using a sample of m_{try} features at each split. The prediction task is performed by performing a majority vote or by averaging the results of each tree, according to the problem at hand (classification or regression). This approach leads to better accuracy

© Springer International Publishing AG, part of Springer Nature 2018
D. Phung et al. (Eds.): PAKDD 2018, LNAI 10939, pp. 478–489, 2018.
https://doi.org/10.1007/978-3-319-93040-4_38

and generalization capacity of the model, and it reduces the variance of single decision trees [7].

The method proposed by Shi and Horvath, called Unsupervised RF (URF), derives from the common RF algorithm. Once the forest has been constructed, the training data can be run down each tree. Since each leaf only contains a small number of objects, and all objects of the same leaf can be considered similar, it is possible to define a similarity measure from these trees: if two objects i and j are in the same leaf of a tree, the overall similarity between the two objects is increased by one. This similarity is then normalized by dividing by the number of trees in the forests. In doing so, the similarities lie in the interval $[0, 1]$. The use of this RF is made possible in the unsupervised case thanks to the generation of synthetic instances, enabling binary classification between the latter and the observed instances. Two methods for data generation are presented in [19], namely, *addCl1* and *addCl2*.

In *addCl1*, the synthetic instances are obtained by a random sampling from the observed distributions of variables, whereas in *addCl2* they are obtained by a random sampling in the hyper-rectangle containing the observed instances. The authors found that *addCl1* usually leads to better results in practice. URF as a method for measuring dissimilarity presents several advantages. For instance, objects described by mixed types of variables as well as missing values can be handled. The method has already been successfully used in fields such as biology ([1, 10, 18]) and image processing [15].

However, the method, albeit its appealing character, suffers from some drawbacks. Firstly, the generation step is not computationally efficient. Since the obtained trees highly depend on the generated instances, it is necessary to construct many forests with different synthetic instances and average their results, leading to a computational burden. Secondly, the synthetic instances may bias the model being constructed to discriminate objects on specific features. For example, *addCl1* leads to forests that focus on correlated features.

Geurts et al. [8] presented a novel type of ensemble of trees method that they called Extremely Randomized Trees (or ExtraTrees, for short). This algorithm is very similar to RF in many ways. In RF, both the instance and feature samplings are performed during the construction of each tree. In ExtraTrees (ET) another layer of randomization is added. Indeed, whereas in RF the threshold of a feature split is selected according to some purity measure (the most popular ones being the entropy and the Gini impurity), in ET these thresholds are obtained totally or partially at random. In addition, instead of growing the trees from bootstrapped samples of the data, ET uses the whole training set. At each node, K attributes are randomly selected and a random split is performed on each one of them. The best split is kept and used to grow the tree. The ET algorithm is described in Fig. 1.

Two parameters are of importance in this algorithm: K, that we already defined, and n_{min}, that is the minimum sample size for a node to be split. Interestingly, the parameter K, that takes values in $\{1, \ldots, n_{features}\}$, influences the randomness of the trees. Indeed, for small values of K, the dependence of

---

Build_an_extra_tree_ensemble(S).
Input: a training set S.
Output: a tree ensemble $\mathcal{T} = \{t_1, \ldots, t_M\}$.
– For $i=1$ to M
   • Generate a tree: $t_i=$ **Build_an_extra_tree**(S);
– Return $\mathcal{T}$.

Build_an_extra_tree(S).
Input: a training set S.
Output: a tree t.
– Return a leaf labeled by class frequencies (or average output, in regression) in S if
   (i) $|S| < n_{min}$, or
   (ii) all candidate attributes are constant in S, or
   (iii) the output variable is constant in S
– Otherwise:
   1. Select randomly K attributes, $\{a_1,...,a_K\}$, without replacement, among all (non constant in S)
      candidate attributes;
   2. Generate K splits $\{s_1, \ldots, s_K\}$, where $s_i = $ **Pick_a_random_split**(S, a_i), $\forall i = 1, \ldots, K$;
   3. Select a split s_* such that Score(s_*, S) = $\max_{i=1,...,K}$ Score(s_i, S);
   4. Split S into subsets S_l and S_r according to the test s_*;
   5. Build $t_l = $ **Build_an_extra_tree**(S_l) and $t_r = $ **Build_an_extra_tree**(S_r) from these subsets;
   6. Create a node with the split s_*, attach t_l and t_r as left and right subtrees of this node and return the
      resulting tree t.

Pick_a_random_split(S,a)
Input: a training set S and an attribute a.
Output: a split.
– If the attribute a is numerical:
   • Compute the maximal and minimal value of a in S, denoted respectively by a_{min}^S and a_{max}^S;
   • Draw a cut-point a_c uniformly in $[a_{min}^S, a_{max}^S]$;
   • Return the split $[a < a_c]$.
– If the attribute a is categorical (denote by $\mathcal{A}$ its set of possible values):
   • Compute $\mathcal{A}_S$ the subset of $\mathcal{A}$ of values of a that appear in S;
   • Randomly draw a proper non empty subset $\mathcal{A}_1$ of $\mathcal{A}_S$ and a subset $\mathcal{A}_2$ of $\mathcal{A} \backslash \mathcal{A}_S$;
   • Return the split $[a \in \mathcal{A}_1 \cup \mathcal{A}_2]$.

Fig. 1. The figure presets the ExtraTrees algorithm as extracted from [8]

the constructed trees on the output variables gets weak. In the extreme case where K is set to 1 (*i.e.*, only one feature is selected and randomly split), the dependence of the trees on the observed label is removed.

Following the tracks of [19] on URF, we propose to use ET with a novel approach where synthetic case generation is no longer necessary. This approach, that we call *addCl3*, consists of a random labelling of each instance. Using properties of ET that we will discuss below, it is possible to compute a good similarity measure from a dataset where *addCl3* is applied. The method outperforms URF in running time, while giving similar or better clusters.

This paper is organized as follows. After a description of the method in Sect. 2, we focus on the empirical evaluation on real-world datasets in Sect. 3, before reviewing of the method and giving some perspectives in Sect. 4.

2 Unsupervised Extremely Randomized Trees

Two methods are used for the generation of synthetic data: *addCl1* and *addCl2*. In these methods, the generation consists in sampling in the observed data. The synthetic data is assigned a label, while the observed data is assigned another one, enabling binary classification between observed and synthetic examples. The novel method we propose and evaluate in this work that we refer to as *addCl3*, does not focus on the generation of synthetic instances, but on the generation of labels instead. In *addCl3*, the label generation runs as follows:

1. Let n_{obs} be the number of instances in the dataset. A list containing $\lfloor \frac{n_{obs}}{2} \rfloor$ times the label 0 and $n_{obs} - \lfloor \frac{n_{obs}}{2} \rfloor$ times the label 1 is generated.
2. For each instance in the dataset, a label is randomly sampled without replacement from the aforementioned list.

This procedure ensures that the label distribution is balanced in the dataset. However, this leads to the same problem arising with *addCl1* and *addCl2*: the results are highly dependent on the generation step, as different realizations of the instance-label association or of the synthetic data may lead to completely different forests. To circumvent this issue, one solution is to run multiple forests on multiple generated datasets, and to average the results. Shi and Horvath found out that averaging the results from 5 forests, with a total of 5000 trees leads to robust results. Moreover, instead of running multiple forests on many generated datasets, it may be possible - and computationally more efficient - to run a single forest with a large amount of generated data, if some care is taken regarding the reweighting of each class. This workaround, proposed by a reviewer in [19], is easier to implement when our generation scheme is used. Indeed, since we do not add new instances, it is not necessary to reweight each class. Instead, we propose to duplicate the original dataset multiple times and apply *addCl3* to obtain a balanced dataset. This approach is evaluated in Sect. 3.

With *addCl3*, the construction of the trees no longer depends on the structure of the data. Indeed, when *addCl1* or *addCl2* are used, the forests are trained to distinguish between observed and synthetic instances. In *addCl3*, the labels being assigned randomly, two similar instances may be labeled differently and may fall in different leaves. However, using ET with the number of features randomly selected at each node $K = 1$, the construction of the trees no longer depends on the class label, as described in the previous section. Hence, ET seems to be a suitable algorithm to use with *addCl3*. Algorithm 1 describes the Unsupervised Extremely Randomized Trees (UET) method.

Algorithm 1. Unsupervised Extremely Randomized Trees

Data: Observations O
Result: Similarity matrix S
$D \longleftarrow addCl3(O)$;
$T \longleftarrow Build\_an\_extra\_tree\_ensemble(D, K)$ // *Here $K = 1$*;
$S = 0_{n_{obs},n_{obs}}$ // *Initialization of a zero matrix of size n_{obs}* ;
for $d_i \in D$ **do**
$\quad$ **for** $d_j \in D$ **do**
$\quad\quad$ $S_{i,j}$ = number of times the samples d_i and d_j fall in the same leaf
$\quad\quad$ node in each tree of T = $\{t_1, t_2, ..., t_M\}$;
$\quad$ **end**
end
$S_{i,j} = \frac{S_{i,j}}{M}$;

The algorithm $Build\_an\_extra\_tree\_ensemble(D)$ is given in Fig. 1. A few parameters can influence the results of UET:

1. The number of copies of the original dataset n_{copies} before applying *addCl3*.
2. The number of trees n_{trees}.
3. The minimum number of samples for a node to be split n_{min}.

3   Empirical Evaluation

In this section, we investigate the influence of the parameters introduced at the end of Sect. 2, as well as the performance of the method.

3.1   Optimization of Parameters

For each evaluation presented in this subsection, the following process is repeated 10 times:

1. A similarity matrix is constructed using UET.
2. This similarity matrix is transformed into a distance matrix using the relation $DIS_{ij} = \sqrt{1 - SIM_{ij}}$, used in [19].
3. An agglomerative clustering (with average linkage) is performed using this distance matrix, with the relevant number of clusters for the dataset.

For each clustering, Adjusted Rand Indices (ARI) are computed. This measure quantifies the agreement between two partitions of a dataset, adjusted for chance [9,17]. ARI takes values in $[-1, 1]$, where a value of 1 indicates perfect agreement up to a permutation, while a value of 0 indicates a result no better than a random label assignment.

Three datasets are used for this evaluation process: Iris [5], Wine [6] and Wisconsin breast cancer [12]. These datasets are described Table 1.

Table 1. Properties of used datasets

Dataset	# samples	# features	# labels
Iris	150	4	3
Wine	178	13	3
Wisconsin	699	9	2

Influence of the number of copies of the dataset

The use of *addCl3* on a dataset leads to a balanced distribution of the labels. Instead of running k forests on as many datasets where *addCl3* is applied k times, it is possible to run one forest on a dataset duplicated k times. We evaluate here the influence of this duplication process. The results are presented Fig. 2.

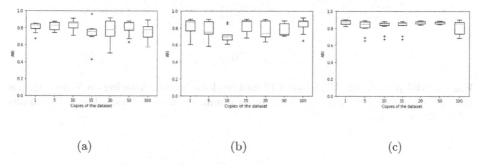

(a)   (b)   (c)

Fig. 2. ARI performing UET+*addCl3* and agglomerative clustering on Wine (a), Iris (b) and Wisconsin (c) datasets when the number of copies of the dataset increases.

The ARI are compared using the Kruskal-Wallis test [11]. The results show that the ARI does not differ significantly in Wine, Iris and Wisconsin datasets ($p = 0.26$, $p = 0.09$ and $p = 0.23$, respectively).

Intuitively, as UET grows the tree without any consideration of the labels and without bootstrapping the samples, the results should stay relatively constant when number of duplications grows. This replication was needed in URF as the generation scheme could lead to significant differences in the output similarity. This intuition is confirmed with this experiment. Here, the randomness induced by the labelling step does not induce a difference in the construction of the trees. Indeed, since we set $K = 1$, trees are constructed totally at random. Any difference in the similarity matrix is rather related to the randomness induced by the choice of features to split at each node.

Influence of the number of trees

The influence of the number of trees n_{trees} as also been studied in [8], where this parameter is referred to as the averaging strength M. For randomized method such as RF and ET used in a supervised learning setting, the average error is a monotonically decreasing function of M [2]. In our experiments, we observed no substantial gain for $n_{trees} > 50$. The difference in ARI are not significant ($p > 0.1$) for all three datasets. This observation confirms the one from Geurts *et al.* where values of $n_{trees} > 40$ outperforms Tree Bagging. However, as the time to construct the ensemble grows linearly with the number of trees, it is a good option to choose small a value of n_{trees}. We chose the value $n_{trees} = 50$ by default. We noticed that this value is way below the overall number of trees recommended for URF, 5000. The results are presented Fig. 3.

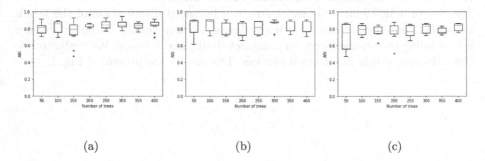

(a)   (b)   (c)

Fig. 3. ARI performing UET+*addCl3* and agglomerative clustering on Wine (a), Iris (b) and Wisconsin (c) datasets when the total number of trees varies. The ARI remains relatively constant.

Influence of the minimum number of samples to split
ET tend to produce trees having 3 to 4 times the number of leaves than those of RF. As UET computes similarities by counting the number of times objects fall into the same leaf, the results are impacted by this increase in the number of leaves. It might be useful to stop the growth of the trees, in order to group similar instances in the same leaves more often. The minimum number of objects to split a node n_{min} can control this growth. This parameter n_{min}, also called the *smoothing strength*, has an impact on the bias and the variance. As stated by Geurts *et al.* [8], the optimal value for this parameter depends on the level of noise in the dataset. They showed in [8] that larger values of n_{min} are needed when ET is applied to noisy data. In UET, the noise is maximal, as the labels are assigned randomly. The results of the evaluations performed varying n_{min} are presented below. For $n_{min} = 2$, we observe that the method fails to compute a similarity matrix leading to a good clustering. Values of n_{min} between 20% and 30% of the data seem to give better results. The ARI variations for the three datasets according to n_{min} are presented Fig. 4.

3.2   Comparative Evaluation of UET

In this section we first evaluate the relevance of clusterings obtained using UET by comparing the Normalized Mutual Information [20] (NMI) scores with the

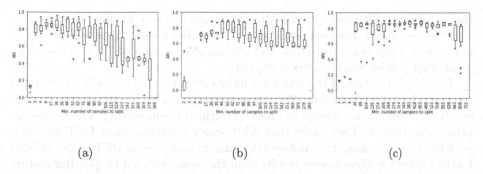

(a)   (b)   (c)

Fig. 4. ARI performing UET+*addCl3* and agglomerative clustering on Wine (a), Iris (b) and Wisconsin (c) datasets when the min. number of samples to split increases. Last value corresponds to 110% of the samples in a dataset.

values presented in [4]. This reference was chosen because results were provided for many well-known datasets. Then, we compare UET and URF, using another quality score presented previously, ARI. The ten datasets used in this section are available on the UCI website[1] and presented Table 2. UET are computed with $n_{trees} = 50$ and $n_{min} = \lceil \frac{n_{samples}}{3} \rceil$.

Table 2. Datasets used for benchmarking

Dataset	# samples	# features	# labels
Iris	150	4	3
Wine	178	13	3
Wisconsin	699	9	2
Lung	32	56	3
Breast tissue	106	9	6
Isolet	1559	617	26
Pima	768	8	2
Parkinson	195	22	2
Ionosphere	351	34	2
Segmentation	2310	19	7

Comparative evaluation with results from the literature

For each dataset, UET was run 10 times, and the similarity matrices were averaged. The obtained matrix was then transformed into a distance matrix using the equation $DIS_{ij} = \sqrt{1 - SIM_{ij}}$, and an agglomerative clustering with the relevant number of clusters was performed. The quality of the clustering was

---

[1] https://archive.ics.uci.edu/ml/index.php.

then evaluated with respect to NMI. This process is run 20 times, and we provide the mean and standard deviation of the quality metric. This evaluation was performed using *scikit-learn* [14] and our implementation of UET. This implementation will be available upon request.

In [4], NMI obtained by running k-means 20 times and averaging the results are provided for each dataset. We compare our results to the ones obtained without feature selection, as none has been performed in our setting. The results are presented Table 3. They show that NMI scores obtained using UET are competitive in most cases. It is noteworthy that in some cases, UET alone without feature selection gives better results than the ones obtained by [4] after feature selection. For instance, this is the case for *Breast tissue* dataset.

Table 3. Comparative evaluation with the results from [4]. Best obtained values are indicated in boldface. In case of a tie, both values are in boldface. Time comparison was not performed in this case.

Dataset	UET - NMI	Literature - NMI
Wisconsin	**72.95** ± 4.94	**73.61** ± 0.00
Lung	**28.89** ± 5.76	22.51 ± 5.58
Breast tissue	**59.59** ± 1.03	51.18 ± 1.38
Isolet	**69.95** ± 1.20	69.83 ± 1.74
Parkinson	21.06 ± 5.33	**23.35** ± 0.19
Ionosphere	**13.48** ± 3.25	12.62 ± 2.37
Segmentation	**69.31** ± 1.51	60.73 ± 1.71

Comparison with URF

To compare UET and URF, we used the R implementation provided by Shi and Horvath[2], and compared the ARI obtained after running the partitioning around medoids (PAM) algorithm on the distance matrices obtained by both methods. 2000 trees and 100 forests are used for URF, with a value of $m_{try} = \lfloor \sqrt{n_{features}} \rfloor$[3]. We set UET parameters to $n_{trees} = 50$ and $n_{min} = \lceil \frac{n_{samples}}{3} \rceil$, and averaged the similarity matrices of 20 runs. These experiments were run on a computer with an Intel i7-6600U (2.6 Ghz) and 16 Go of 2133 MHz DDR4 RAM.

We compared both ARI and time (in seconds) for each method. The results are presented Table 4. UET outperforms URF time-wise, while giving similar or better clusterings. Regarding the Isolet dataset, we manually terminated URF's computation as we weren't able to obtain results in an acceptable amount of time on our machine. However, we performed the computation on a more powerful machine, and were able to obtain an ARI of 28.39.

---

[2] https://labs.genetics.ucla.edu/horvath/RFclustering/RFclustering.htm.

[3] m_{try} is the number of variables used at each node when a tree is grown in RF.

Table 4. Comparative evaluation between URF and UET

Dataset	UET (ARI - Time (s))	URF (ARI - Time (s))
Wisconsin	79.30 - **823.36 s**	**81.36** - 1267.82 s
Lung	**10.81 - 7.45 s**	8.16 - 89.32 s
Breast tissue	**40.35 - 25.25 s**	39.05 - 94.55 s
Isolet	**33.44 - 4589.36 s**	* - * s
Parkinson	**17.37 - 66.91 s**	13.44 - 252.12 s
Ionosphere	**8.54 - 184.97 s**	7.59 - 722.92 s

4   Conclusion and Perspectives

In this preliminary work, we presented a novel method to perform unsupervised clustering using decision trees. This approach extends the unsupervised random forest method, by using extremely randomized trees as a base estimator. In the former method, the generation of synthetic instances was needed. This generation can be performed by two different approaches, *AddCl1* or *AddCl2*. With the approach we proposed here, the generation of instances is no longer necessary. Indeed, for some parameter choices, extremely randomized trees can be made independent of the labels. We therefore present a way to bypass the need for instance generation, *AddCl3*, where a label is randomly associated with each observation, which results in a significant reduction in running time.

A performance evaluation of our method showed that essentially one parameter influenced the results, the *smoothing parameter* n_{min}. This is explained by the fact that higher values of n_{min} give better results in the presence of noise. In our case, the data is highly noisy, as the labeling is a random process. We found that a value of $\frac{n_{samples}}{4} \leq n_{min} \leq \frac{n_{samples}}{3}$ gives good clusterings. Other parameters, such as the number of trees per forest n_{trees} did not influence much the results of the procedure for values of $n_{trees} > 50$, while increasing the time to perform the procedure. An interesting finding is that it is no longer necessary to duplicate the dataset multiple times to improve the results. However, due to the randomness of the procedure, it is still necessary to average the results of multiple UET to decrease the variance.

We compared the quality of the clustering between our method and (i) results found in the literature and (ii) results obtained by URF on multiple datasets. The quality is measured by normalized mutual information or adjusted rand index, according to the metric available in the literature. This empirical evaluation gave promising results, with overall similar or better NMI and ARI. The advantages of our method over URF are twofold. First, the generation of synthetic data is no longer necessary. Second, the method is 1.5 to more than 10 times faster than URF.

However, there is still room for improvement. We are aware that we only have tested our method on a few small datasets so far. A comparison with other metrics on large synthetic and real-world datasets would be interesting.

Moreover, one of the major advantages of using a decision tree-based method to compute a distance is that (i) it enables the use of mixed-type variables and (ii) missing data can be handled. In fact, the latter were our original motivation and they constitute topics for future work.

Acknowledgements. Kevin Dalleau's PhD is funded by the RHU FIGHT-HF (ANR-15-RHUS-0004) and the Region Grand Est (France).

References

1. Abba, M.C., et al.: Breast cancer molecular signatures as determined by sage: correlation with lymph node status. Mol. Cancer Res. **5**(9), 881–890 (2007)
2. Breiman, L.: Random forests. Mach. Learn. **45**(1), 5–32 (2001)
3. Deza, M.M., Deza, E.: Encyclopedia of distances. Encyclopedia of Distances, pp. 1–583. Springer, Heidelberg (2009). https://doi.org/10.1007/978-3-642-00234-2_1
4. Elghazel, H., Aussem, A.: Feature selection for unsupervised learning using random cluster ensembles. In: 2010 IEEE 10th International Conference on Data Mining (ICDM), pp. 168–175. IEEE (2010)
5. Fisher, R., Marshall, M.: Iris data set. RA Fisher, UC Irvine Machine Learning Repository (1936)
6. Forina, M., et al.: An extendible package for data exploration, classification and correlation. Institute of Pharmaceutical and Food Analysis and Technologies 16147 (1991)
7. Friedman, J., Hastie, T., Tibshirani, R.: The Elements of Statistical Learning. SSS, vol. 1. Springer, New York (2001)
8. Geurts, P., Ernst, D., Wehenkel, L.: Extremely randomized trees. Mach. Learn. **63**(1), 3–42 (2006)
9. Hubert, L., Arabie, P.: Comparing partitions. J. Classif. **2**(1), 193–218 (1985)
10. Kim, H.L., Seligson, D., Liu, X., Janzen, N., Bui, M., Yu, H., Shi, T., Belldegrun, A.S., Horvath, S., Figlin, R.: Using tumor markers to predict the survival of patients with metastatic renal cell carcinoma. J. Urol. **173**(5), 1496–1501 (2005)
11. Kruskal, W., Wallis, W.: Use of ranks in one-criterion variance analysis. J. Am. Stat. Assoc. **47**(260), 583–621 (1952)
12. Mangasarian, O., Wolberg, W.: Cancer diagnosis via linear programming. University of Wisconsin-Madison, Computer Sciences Department (1990)
13. Pal, M.: Random forest classifier for remote sensing classification. Int. J. Remote Sens. **26**(1), 217–222 (2005)
14. Pedregosa, F., et al.: Scikit-learn: machine learning in Python. J. Mach. Learn. Res. **12**, 2825–2830 (2011)
15. Peerbhay, K., Mutanga, O., Ismail, R.: Random forests unsupervised classification: the detection and mapping of solanum mauritianum infestations in plantation forestry using hyperspectral data. IEEE J. Sel. Top. Appl. Earth Obs. Remote Sens. **8**(6), 3107–3122 (2015)
16. Percha, B., Garten, Y., Altman, R.B.: Discovery and explanation of drug-drug interactions via text mining. In: Pacific Symposium on Biocomputing. pp. 410–421 (2012). http://psb.stanford.edu/psb-online/proceedings/psb2012/percha.pdf
17. Rand, W.M.: Objective criteria for the evaluation of clustering methods. J. Am. Stat. Assoc. **66**(336), 846–850 (1971)

18. Rennard, S.I., et al.: Identification of five chronic obstructive pulmonary disease subgroups with different prognoses in the ECLIPSE cohort using cluster analysis. Ann. Am. Thorac. Soc. **12**(3), 303–312 (2015)
19. Shi, T., Horvath, S.: Unsupervised learning with random forest predictors. J. Comput. Graph. Stat. **15**(1), 118–138 (2006)
20. Strehl, A., Ghosh, J.: Cluster ensembles-a knowledge reuse framework for combining multiple partitions. J. Mach. Learn. Res. **3**, 583–617 (2002)

Local Graph Clustering by Multi-network Random Walk with Restart

Yaowei Yan[1]([✉]), Dongsheng Luo[1], Jingchao Ni[1], Hongliang Fei[2], Wei Fan[3], Xiong Yu[4], John Yen[1], and Xiang Zhang[1]

[1] College of Information Sciences and Technology,
The Pennsylvania State University, State College, USA
{yxy230,dul262,jzn47,jyen,xzhang}@ist.psu.edu
[2] Baidu Research Big Data Lab, Sunnyvale, USA
hongliangfei@baidu.com
[3] Tencent Medical AI Lab, Palo Alto, USA
davidwfan@tencent.com
[4] Case Western Reserve University, Cleveland, USA
xiong.yu@case.edu

Abstract. Searching local graph clusters is an important problem in big network analysis. Given a query node in a graph, local clustering aims at finding a subgraph around the query node, which consists of nodes highly relevant to the query node. Existing local clustering methods are based on single networks that contain limited information. In contrast, the real data are always comprehensive and can be represented better by multiple connected networks (multi-network). To take the advantage of heterogeneity of multi-network and improve the clustering accuracy, we advance a strategy for local graph clustering based on Multi-network Random Walk with Restart (MRWR), which discovers local clusters on a target network in association with additional networks. For the proposed local clustering method, we develop a localized approximate algorithm (AMRWR) on solid theoretical basis to speed up the searching process. To the best of our knowledge, this is the first elaboration of local clustering on a target network by integrating multiple networks. Empirical evaluations show that the proposed method improves clustering accuracy by more than 10% on average with competently short running time, compared with the alternative state-of-the-art graph clustering approaches.

1 Introduction

Networks (or graphs) are natural representations of real-world relationships. Network clustering is a fundamental problem and is the basis of many applications such as online recommendation, medical diagnosis, social networks and biological networks analysis [1–4]. The clustering of a network aims to find groups of nodes that are closely related to each other. Unlike global clustering which retrieves all clusters from a network, local clustering focuses on the subgraph within neighborhood of the query node and is less demanding in computation [5]. With the

© Springer International Publishing AG, part of Springer Nature 2018
D. Phung et al. (Eds.): PAKDD 2018, LNAI 10939, pp. 490–501, 2018.
https://doi.org/10.1007/978-3-319-93040-4_39

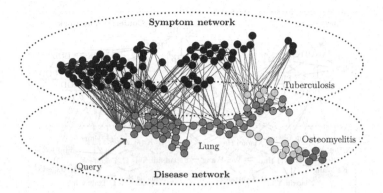

Fig. 1. An example of the Disease-Symptom network. Black nodes are symptoms. Green nodes are lung diseases. Yellow nodes and orange nodes are of tuberculosis and osteomyelitis, respectively. (Color figure online)

increasing size of networks, recently local clustering has attracted lots of research interest [6–9].

Local clustering takes a cohesive group of nodes as the target cluster. The cohesiveness is either evaluated by goodness metrics such as conductance and density [6], or the proximity between nodes according to the network topology [10]. While existing local clustering approaches are based on single networks, cohesiveness of single network does not always reveal the real clustering structure, since the single network is often noisy and incomplete. On the contrary, complete information is usually available through multiple connected networks.

For example, Fig. 1 is a corner of a disease network and an associated symptom network. Because of the incompleteness of the disease network, the lung disease cluster is divided into subgroups with relatively sparse connections in-between. When querying a lung disease from one subgroup, the lung diseases in other groups are hard to be enclosed into the desired local cluster due to the sparse inter-group connections. However, if the symptom network is taken into account, the symptom nodes serve as bridges between lung diseases groups, which integrate the lung diseases into a whole cluster. Moreover, the symptom nodes help to distinguish the lung diseases from tuberculosis and osteomyelitis, since the lung diseases share lots of common symptoms while the tuberculosis and osteomyelitis nodes are left aside.

Figure 2 provides another example of 16 scholars from research areas of data mining, database and information retrieval. In the coauthor network, the data mining researchers are isolated into two groups. Associating the authors to their top-3 most published conferences, it is clear that the data mining researchers are linked with each other by various conferences while the authors considered as database or information retrieval researchers are linked to conferences of their specialized domains. For example, among all the authors only Xiaofei He frequently publishes papers in SIGIR. A similar case concerns Flip Korn and H.

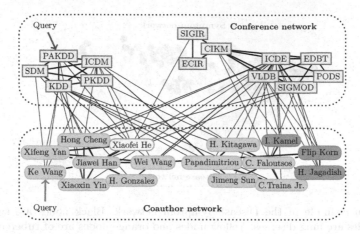

Fig. 2. An example of the Author-Conference network. Blue nodes are data mining researchers. Green nodes are researchers more relevant to information retrieval and red nodes are researchers closer to the database area. (Color figure online)

Jagadish et al., who are intensively relevant to database conferences such as SIGMOD, VLDB and ICDE.

In this paper, we propose a local clustering method which integrates multiple networks. The method targets on a specific network and uses other networks to improve the clustering accuracy. To use the links from various networks, we introduce a Multi-network Random Walk with Restart (MRWR) model, which allows the random surfer goes into different networks with differentiated cross-network transition probabilities. Our theoretical analysis shows that MRWR can measure nodes proximity across multiple networks by capturing the multi-network heterogeneity.

As in the local clustering task, only proximity of nodes close to the query node is necessary, we propose a localized approximate MRWR (AMRWR) algorithm for the multi-network node-proximity calculation based on solid theoretical basis. By the AMRWR algorithm, running time of the proposed method is limited and the method is scalable to large networks. To the best of our knowledge, the proposed method is the first local clustering approach that considers multiple networks. The effectiveness and efficiency are validated by both theoretical and empirical studies in the following sections.

2   Related Work

As all current multi-network clustering approaches are global clustering [1,2,11–15], our method is more relevant to single network local clustering [6–9,16].

Single network local clustering methods can be classified into three categories: local search [5,17], dense subgraph [6,8,16] and node proximity based methods [7,9,10]. Conventional local search algorithms examine nodes around query node to improve a goodness function [6] by greedy or combinatorial optimization methods [5,17]. The searching efficiency of such approaches is limited [7]. The dense subgraph methods try to find a dense subgraph around the query node, which can be a k-core [8], k-truss [18], k-plex [19], etc. The limitation of these methods is that many relevant nodes are not in a dense subgraph and may not be enclosed into the local clusters. Recently random walk and node proximity based methods are successful in local community detection [7,9]. These approaches rank nodes with respect to the query node and cut a part of nodes with high ranking scores as the local community. All these methods are created for single networks and ignore the heterogeneity of multi-networks.

3   Problem Definitions

In this paper, the multi-network or N-network $\mathcal{G}$ consists of N interconnected single networks. The ith single network is represented by the weighted graph $G^{(i)} = (V^{(i)}, E^{(i)})$ with adjacency matrix $\mathbf{A}^{(i)} \in \mathbb{R}_+^{n_i \times n_i}$, where $n_i = |V^{(i)}|$, and each entry $\mathbf{A}_{xy}^{(i)}$ of matrix $\mathbf{A}^{(i)}$ is the weight of edge between node $V_x^{(i)}$ and node $V_y^{(i)}$ in network $G^{(i)}$. In this multi-network model, the connection between single networks $G^{(i)}$ and $G^{(j)}$ is represented by bipartite graph $G^{(ij)} = (V^{(i)}, V^{(j)}, E^{(ij)})$ with adjacency matrix $\mathbf{C}^{(ij)} \in \mathbb{R}_+^{n_i \times n_j}$, where $n_i = |V^{(i)}|$ and $n_j = |V^{(j)}|$. The entry $\mathbf{C}_{xy}^{(ij)}$ is the weight of link between node $V_x^{(i)}$ and $V_y^{(j)}$.

In this work, we focus on the clustering of a specific network, while other networks provide extra information to improve the clustering accuracy. Without loss of generality, we assume the clustering targets on the network $G^{(1)}$ with node set $V^{(1)}$. Given a query node $q \in V^{(1)}$, the goal of local clustering on multi-network $\mathcal{G}$ is to find a local cluster $S \subseteq V^{(1)}$ such that $q \in S$. The nodes in S should be cohesive not only through $G^{(1)}$, but also through other networks connected with $G^{(1)}$.

In the following section, we will discuss the cross-graph cohesiveness measure and the consequent local clustering method. We adopt the widely used random-walk-based proximity score as the cohesiveness criterion since it has been shown to be most effective in capturing local clustering structures [9,10].

4   Methods

Random-walk-based approaches such as Random Walk with Restart (a.k.a. Personalized PageRank) are commonly used to evaluate the node proximity in the single network [20,21]. Compared with other methods, random walk takes advantage of local neighborhood structure of the network and thus has higher performance [7]. In this work, we generalize the single-network random walk to multi-network and propose local clustering methods based on the multi-network random walk.

4.1   Random Walk on Single Network

In Random Walk with Restart (RWR), a random surfer starts from the query node q, and randomly walks to a neighbor node according to the edge weights. At each time point $(t + 1)$, the surfer goes on with probability α or returns to the query node q with probability $(1 - \alpha)$. The proximity score of node p with respect to node q is defined as the converged probability that the surfer visits node p [21]:

$$\mathbf{r}^{(t+1)} = \alpha \cdot \mathbf{r}^{(t)} \cdot \mathbf{P} + (1 - \alpha) \cdot \mathbf{s}, \tag{1}$$

where $\mathbf{s}$ is the row vector of the initial distribution with q^{th} entry as 1 and all other entries 0, and $\mathbf{r}$ is the row vector whose p^{th} entry is the visiting probability of node p. $\mathbf{P}$ is the row-stochastic transition matrix with entries $\mathbf{P}_{xy} = \frac{\mathbf{A}_{xy}}{\sum_y \mathbf{A}_{xy}}$.

An alternative perspective of Random Walk with Restart is the walk-view [10]. Repeatedly insert the right-hand side of Eq. (1) into $\mathbf{r}$, then

$$\mathbf{r} = (1 - \alpha) \sum_{k=0}^{\infty} \alpha^k \mathbf{s} \mathbf{P}^k \tag{2}$$

Since $\mathbf{P}^k$ contains probabilities of all possible length-k walks on the graph, Eq. (2) can be interpreted as that the converged vector $\mathbf{r}$ contains accumulated probabilities of all possible walks for q to each node. As the probabilities of length-k walks are discounted by factor α^k, the proximity is large when there are many short walks between a certain node and the query node.

4.2   Random Walk on Multi-network

The single-network RWR has no knowledge of the heterogeneity of networks, where we can not control the surfer's behavior towards different networks. In this paper, we propose a Multi-network Random Walk with Restart (MRWR) model, in which the surfer knows the environments and chooses different transition probabilities to distinct networks.

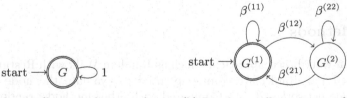

(a) Random walk on single network   (b) Random walk on 2-network

Fig. 3. Cross-network transition probabilities of random walk on single network and multi-network that contains two networks $G^{(1)}$ and $G^{(2)}$. $\beta^{(ij)}$ is the transition probability between network $G^{(i)}$ and $G^{(j)}$.

In single-network random walk, the random surfer stays in the network with probability $\beta = 1$ (Fig. 3a). However, for a multi-network with N networks, there are $N \times N$ different transition probability β. Figure 3b gives an example of 2-network, where $\beta^{(11)}$ is the chance of the random surfer staying in $G^{(1)}$, $\beta^{(12)}$ is the chance that the random surfer goes from $G^{(1)}$ to $G^{(2)}$, etc. All these transition probabilities form a matrix $\mathbf{B} = \begin{bmatrix} \beta^{(11)} & \cdots & \beta^{(1N)} \\ \vdots & \ddots & \vdots \\ \beta^{(N1)} & \cdots & \beta^{(NN)} \end{bmatrix}$.

The matrix $\mathbf{B}$ is also row-stochastic since the sum of probabilities to all networks must be 1. The cross-network decision of random surfer is made by these probabilities. There exist two kinds of MRWR with special matrix $\mathbf{B}$: every row is identical (rank-1), or the matrix is symmetric. We name these two special MRWR as Biased MRWR and Symmetric MRWR, respectively. In Biased MRWR, the random surfer is likely to walk into the networks with large entry value in $\mathbf{B}$ at any time. In Symmetric MRWR, however, the surfer walks between an arbitrary pair of network $G^{(i)}$ and $G^{(j)}$ back and forth with equal probabilities, as $\beta^{(ij)} = \beta^{(ji)}$. In this work we focus on the Biased MRWR since we bias the target network.

Comparing with single-network random walk, we may formulate the walk-view of multi-network by the following equation.

$$
\mathbf{r} = \sum_{\sigma}(1-\alpha)\alpha^k \beta_\sigma \mathbf{s}\mathbf{P}_\sigma = \sum_{\sigma}(1-\alpha)\alpha^k \prod_{i=1}^{k}\beta^{(\sigma_i\sigma_{i+1})}\mathbf{s}\prod_{i=1}^{k}\mathbf{P}^{(\sigma_i\sigma_{i+1})}
$$
$$
= \sum_{\sigma}(1-\alpha)\alpha^k\mathbf{s}\prod_{i=1}^{k}\beta^{(\sigma_i\sigma_{i+1})}\mathbf{P}^{(\sigma_i\sigma_{i+1})}, \tag{3}
$$

where $\sigma = \langle \sigma_1, \sigma_2, \ldots \sigma_{k+1}\rangle$ is an arbitrary walk with length $k \geq 0$, whose i^{th} node is in network $G^{(\sigma_i)}$, $1 \leq \sigma_i \leq N$. Similar with single-network RWR, α acts as the discount factor for length of walks. $\mathbf{P}_\sigma = \prod_{i=1}^{k}\mathbf{P}^{(\sigma_i\sigma_{i+1})}$ is the transition matrix of walk σ, and $\beta_\sigma = \prod_{i=1}^{k}\beta^{(\sigma_i\sigma_{i+1})}$ is the heterogeneous discount factor of walk σ. For the transition matrix on multi-network, $\mathbf{P}^{(ij)}_{xy} = \frac{\mathbf{A}^{(i)}_{xy}}{\sum_y \mathbf{A}^{(i)}_{xy}}$ if $i = j$, otherwise $\mathbf{P}^{(ij)}_{xy} = \frac{\mathbf{C}^{(ij)}_{xy}}{\sum_y \mathbf{C}^{(ij)}_{xy}}$.

4.3   Localized Algorithm for MRWR

Due to the nature of local clustering, only proximity of nodes close to the query node is important. The exact RWR vector is not necessary and a localized approximation vector is enough in practice. In this section, we propose an Approximate MRWR calculation method (AMRWR), motivated by the approximate Personalized PageRank (APPR) algorithm [10]. The APPR algorithm simulates ε-approximation of the RWR score vector by a slightly different starting vector $\mathbf{s}'$, where $\mathbf{s}_x - \mathbf{s}'_x \leq \varepsilon d(x)$ for every node x, ε is the pair-wise error bound and $d(x)$ is the degree of node x. Instead of calculating RWR score directly, APPR uses a local network diffusion strategy where values of the initial vector **s** diffuse towards both the target score vector **p** and itself through the network. Once each entry x of **s** is less than $\varepsilon d(x)$, the target vector **p** is the approximation of RWR score **r**.

Our method is different with APPR for the following aspects. First, instead of getting approximated RWR/PPR score on unweighted graph, we extend the calculation to weighted graph. Second, the original APPR is not aware of different networks during the diffuse operations. We solve this problem by pushing different values from vector **s** to nodes of different networks, according to the cross-network transition matrix **B**. Moreover, at each diffuse step, we push as much value as possible from the initial vector to reduce the running time.

Algorithm 1. Approximate Multi-network RWR (AMRWR)

    Input   : Multi-network $\mathcal{G}$, query node q, forward probability α, cross-network
                transition matrix **B**, error bound ε
    Output: Approximate RWR vector $\hat{\mathbf{r}}^{(1)}$ for network $G^{(1)}$
1 /* Initialization */
2 $\mathbf{p}_x \leftarrow 0$ for all nodes x of $\mathcal{G}$;
3 $\mathbf{s}_x \leftarrow 0$ for all nodes $x \neq q$; $\mathbf{s}_q \leftarrow 1$;
4 /* Multi-network diffuse operation */
5 **while** $\mathbf{s}_x \geq \varepsilon d(x)$ *for node* $x \in V^{(i)}$ **do**
6   /* $d(x)$ is degree of x on multi-network $\mathcal{G}$ */
7       $\delta \leftarrow \mathbf{s}_x - \varepsilon d(x)/2$; $\mathbf{p}_x \leftarrow \mathbf{p}_x + (1-\alpha)\delta$; $\mathbf{s}_x = \mathbf{s}_x - \delta$;
8   **for** *each neighbor* $y \in V^{(j)}$ *of* x **do**
9   /* Differentiated push */
10          $\mathbf{s}_y \leftarrow \mathbf{s}_y + \mathbf{P}_{xy}^{(ij)}\beta^{(ij)}\alpha\delta/d(x)$;
11   **end**
12 **end**
13 $\hat{\mathbf{r}}_x^{(1)} \leftarrow \mathbf{p}_x$ for all nodes $x \in V^{(1)}$;
14 **return** $\hat{\mathbf{r}}^{(1)}$;

The overall MRWR algorithm is demonstrated in Algorithm 1. In the algorithm, Line 2 to 3 are the initialization steps. The while loop from Line 5 to 12 process multi-network diffusion. Finally Line 13 and 14 output the RWR score of nodes in $G^{(1)}$. Here we prove the correctness of Algorithm 1 by Lemma 1 and Theorem 1. All the proofs in this paper can be found in the full version [22].

Lemma 1. *The multi-network diffuse operations in Algorithm 1 construct an ε-approximate RWR vector from initial vector* **s** *through graph*

$$\widetilde{\mathbf{P}} = \begin{bmatrix} \beta^{(11)}\mathbf{P}^{(11)} & \cdots & \beta^{(1N)}\mathbf{P}^{(1N)} \\ \vdots & \ddots & \vdots \\ \beta^{(N1)}\mathbf{P}^{(N1)} & \cdots & \beta^{(NN)}\mathbf{P}^{(NN)} \end{bmatrix}.$$

Proof. See Sect. A.1 in the full version [22].

Theorem 1 (Effectiveness of the AMRWR Algorithm). *The Algorithm 1 generates an ε-approximate vector for MRWR scores* $\mathbf{r}$ *in Eq. (3).*

Proof. See Sect. A.2 in the full version [22].

Theorem 2 (Time Complexity of AMRWR). *The local diffusion process in Algorithm 1 runs in* $O(\frac{1}{\varepsilon(1-\alpha)})$ *time.*

Proof. See Sect. A.3 in the full version [22].

After calculating the proximity score of each node with respect to the query node, the clusters can be generated in two ways, the cut process (MRWR-cut) or the sweep process (MRWR-cond). Specifically, MRWR-cut chooses the top-k nodes with largest proximity scores as the target local cluster, while MRWR-cond sorts the node by the proximity score in descending order, and then scans from the top node to get a cluster with minimal conductance.

5   Experiment

We perform extensive experiments to evaluate the effectiveness and efficiency of the proposed method on a variety of real networks and synthetic graphs. The experiments are performed on a PC with 16GB memory, Intel Core i7-6700 CPU at 3.40 GHz frequency, and the Windows 10 operating system. The core functions are implemented by C++.

5.1   Datasets and Baseline Methods

We have 6 multi-networks from 3 real data sources for our experiments.

Disease-Symptom Networks [1]. This dataset contains a disease similarity network with 9,721 diseases and a symptom similarity network with 5,093 symptoms. In this dataset, there are two sets of clustering ground truth for the disease network: level-1 and level-2, where the clusters in level-1 are larger than those of level-2. We use both sets of ground truth, and denote the experiments by DS1 and DS2, accordingly.

Gene-Disease Networks [23]. The gene network represents the functional relationship between 8,503 genes. The disease network is a phenotype similarity network with 5,080 diseases. By flipping these two networks, we have two multi-networks: the Disease-Gene network (DG) and the Gene-Disease network (GD).

Author-Conference Networks [24]. The dataset comprises 20,111 authors and 20 conferences from 4 research areas. Since we have labels of both authors and conferences, we use this dataset as two multi-networks: the Author-Conference network (AC) and the Conference-Author network (CA).

We compare our approaches with six state-of-the-art methods, including densest subgraph methods Query-biased Densest Connected Subgraph (QDC) [6] and k-core clustering [8]; attributed graph clustering methods Attributed Community Query (ACQ) [25] and Focused clustering (FOC) [26]; node proximity measures Random Walk with Restart (RWR) [7] and Heat Kernel Diffusion (HKD) [9]. To be fair, all the baseline methods run on multi-networks and are tuned to their best performances. For attributed network clustering method, nodes of additional network are taken as the attributes of node in the target network. All results are averaged on 10,000 queries if not specifically stated.

5.2   Effectiveness Evaluation

To evaluate effectiveness of the selected methods, we use precision and F1-score as metrics of clustering accuracy. Given the cluster node set U and the ground truth node set V, the F1-score integrates both precision and recall, and is defined as $F(U,V) = 2 \cdot \frac{p \times r}{p+r}$, where $p = \frac{|U \cap V|}{|U|}$ is the precision and $r = \frac{|U \cap V|}{|V|}$ is the recall. The experiments are deployed on real networks. For each multi-network, we randomly pick up query nodes from the labeled ground truth. The β of biased cross-network transition matrix $\mathbf{B} = \begin{bmatrix} \beta & 1 - \beta \\ \beta & 1 - \beta \end{bmatrix}$ is 0.6 by default, and the forward probability α is 0.99. We set $\beta > 0.5$ as we emphasize the target network more than the additional network. Expected cluster sizes of both MRWR-cut and RWR are set to be the average cluster size of ground truth. For sensitivity of parameter β please see section B.1 in the full version [22].

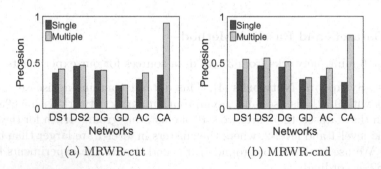

| (a) MRWR-cut | (b) MRWR-cnd |

Fig. 4. Improvement of MRWR precision, from clustering on single target network, to the multi-network clustering.

Firstly we evaluate the precision improvement of MRWR on the multi-network, comparing with MRWR on the target network only. In Fig. 4 we can see that the precision increases on every dataset, for both MRWR-cut and MRWR-cnd. The biggest performance gain is achieved by the Conference-Author (CA) network, meaning that the conference network is very noisy and much information is from the additional network. Actually the conference network is a full graph simply created by cosine similarity of keyword vectors, as shown partially in Fig. 2. The precision increment of MRWR-cnd is more obvious for most datasets (Fig. 4b). It is worth reporting that, on multi-networks, the cluster size of MRWR-cnd is 33% less than on the single networks and is closer to the ground truth cluster size. In other words, the clusters generated by MRWR-cnd on multi-network are more dense and reasonable than their single-network counterparts.

Table 1. Accuracy comparison of all selected methods on real multi-networks.

F1-scores	CUT	CND	k-core	QDC	ACQ	FOC	RWR	HKD
DS1	**0.40**	0.36	0.15	0.15	0.09	0.08	0.31	0.22
DS2	0.47	**0.50**	0.07	0.33	0.03	0.22	0.40	0.41
DG	**0.35**	0.33	0.14	0.18	0.05	0.12	0.29	0.28
GD	0.16	**0.17**	0.04	0.06	0.11	0.12	0.12	0.06
AC	**0.40**	0.22	0.32	0.02	0.05	0.04	0.35	0.30
CA	**0.93**	0.75	0.40	0.40	–	0.58	0.47	0.58

For the accuracy of selected approaches, we use F1-score as it is fair in comparing performance of different methods. Table 1 shows that our methods are overall the best on all networks. For networks DS2 and GD, MRWR-cnd is better than MRWR-cut. For all other networks, MRWR-cut has the best performance. ACQ hits the timeout on network CA and has no result in the table. The MRWR methods are balanced between precision and recall than the baselines. For example, ACQ and FOC have high precision but very low recall, as the methods can hardly enclose more nodes other than the query node due to the sparse links to the additional networks. QDC tends to detect small dense clusters, which decreases its performance on large cluster structures. On the contrary, k-core clustering inclines to a large number of nodes unrelated to the query node. Consequently, its recall is high but the precision is quite low, which also results in low F1-score.

5.3 Efficiency Evaluation

Figure 5 illustrates the average running time of all selected algorithms. For all datasets, the speed of our method outperforms all other approaches except k-core. On the DG and GD networks, MRWR is faster by 1 to 3 orders of magnitude in comparison with other algorithms except for the k-core clustering. Though k-core method is fast in clustering, its accuracy is not competent with other approaches (Table 1). The FOC takes long running time because its global searching strategy. ACQ reaches timeout on the AC network, so we omit its running time in the figure. We also measure the efficiency of MRWR on large synthetic datasets and report the result in section B.2 in the full version [22]. It shows that MRWR runs within seconds on networks with millions of nodes.

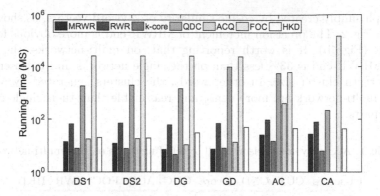

Fig. 5. Average running time of a single query in milli-seconds, comparing all methods on all networks.

6   Conclusion

In this paper, we have introduced a local graph clustering method on multi-networks. The clustering is based on the node proximity measurement by multi-network random walk. Compared with the single network local clustering algorithms, our method takes advantage of both additional network connection and the local network structure. Empirical studies show that our method is both accurate and fast in comparison with alternative approaches. The future work is to make the random walk on multi-network more smart by learning the network heterogeneity online.

Acknowledgement. This work was partially supported by the National Science Foundation grants IIS-1664629, SES-1638320, CAREER, and the National Institute of Health grant R01GM115833. We also thank the anonymous reviewers for their valuable comments and suggestions.

References

1. Ni, J., Fei, H., Fan, W., Zhang, X.: Cross-network clustering and cluster ranking for medical diagnosis. In: ICDE (2017)
2. Ni, J., Koyuturk, M., Tong, H., Haines, J., Rong, X., Zhang, X.: Disease gene prioritization by integrating tissue-specific molecular networks using a robust multi-network model. BMC Bioinform. **17**(1), 453 (2016)
3. Liu, R., Cheng, W., Tong, H., Wang, W., Zhang, X.: Robust multi-network clustering via joint cross-domain cluster alignment. In: ICDM (2015)
4. Sozio, M., Gionis, A.: The community-search problem and how to plan a successful cocktail party. In: KDD (2010)
5. Schaeer, S.E.: Graph clustering. Comput. Sci. Rev. **1**(1), 27–64 (2007)
6. Yubao, W., Jin, R., Li, J., Zhang, X.: Robust local community detection: on free rider effect and its elimination. Proc. VLDB Endow. **8**(7), 798–809 (2015)

7. Kloumann, I.M., Kleinberg, J.M.: Community membership identification from small seed sets. In: KDD (2014)
8. Cui, W., Xiao, Y., Wang, H., Wang, W.: Local search of communities in large graphs. In: SIGMOD (2014)
9. Kloster, K., Gleich, D.F.: Heat kernel based community detection. In: SIGKDD (2014)
10. Andersen, R., Chung, F., Lang, K.: Local graph partitioning using pagerank vectors. In: FOCS (2006)
11. Zhou, D., Burges, C.J.C: Spectral clustering and transductive learning with multiple views. In: ICML (2007)
12. Kumar, A., Rai, P., Daume, H.: Co-regularized multi-view spectral clustering. In: Advances in neural information processing systems (2011)
13. Kumar, A., Daumé, H.: A co-training approach for multi-view spectral clustering. In: ICML (2011)
14. Cheng, W., Zhang, X., Guo, Z., Yubao, W., Sullivan, P.F., Wang, W.: Flexible and robust co-regularized multi-domain graph clustering. In: KDD (2013)
15. Ni, J., Tong, H., Fan, W., Zhang, X.: Flexible and robust multi-network clustering. In: KDD (2015)
16. Yubao, W., Bian, Y., Zhang, X.: Remember where you came from: on the second-order random walk based proximity measures. Proc. VLDB Endow. 10(1), 13–24 (2016)
17. Schaeffer, S.E.: Stochastic local clustering for massive graphs. In: Ho, T.B., Cheung, D., Liu, H. (eds.) PAKDD 2005. LNCS (LNAI), vol. 3518, pp. 354–360. Springer, Heidelberg (2005). https://doi.org/10.1007/11430919_42
18. Huang, X., Cheng, H., Qin, L., Tian, W., Yu, J.X.: Querying k-truss community in large and dynamic graphs. In: SIGMOD (2014)
19. Martins, P.: Modeling the maximum edge-weight k-plex partitioning problem (2016). arXiv preprint arXiv:1612.06243
20. Tong, H., Faloutsos, C., Gallagher, B., Eliassi-Rad, T.: Fast best-effort pattern matching in large attributed graphs. In: KDD (2007)
21. Tong, H., Faloutsos, C., Pan, J.Y.: Fast random walk with restart and its applications (2006)
22. Yan, Y., et al.: Local Graph Clustering by Multi-network Random Walk with Restart, Technical report. https://sites.google.com/site/yanyaw00/pakdd
23. Van Driel, M.A., Bruggeman, J., Vriend, G., Brunner, H.G., Leunissen, J.A.M.: A text-mining analysis of the human phenome. Eur. J. Hum. Genet. 14(5), 535–542 (2006)
24. Ji, M., Sun, Y., Danilevsky, M., Han, J., Gao, J.: Graph regularized transductive classification on heterogeneous information networks. In: Balcázar, J.L., Bonchi, F., Gionis, A., Sebag, M. (eds.) ECML PKDD 2010. LNCS (LNAI), vol. 6321, pp. 570–586. Springer, Heidelberg (2010). https://doi.org/10.1007/978-3-642-15880-3_42
25. Fang, Y., Cheng, R., Luo, S., Jiafeng, H.: Effective community search for large attributed graphs. Proc. VLDB Endow. 9(12), 1233–1244 (2016)
26. Perozzi, B., Akoglu, L., Iglesias Sánchez, P., Müller, E.: Focused clustering and outlier detection in large attributed graphs. In: KDD (2014)

Scalable Approximation Algorithm
for Graph Summarization

Maham Anwar Beg, Muhammad Ahmad, Arif Zaman,
and Imdadullah Khan$^{(\boxtimes)}$

Department of Computer Science, School of Science and Engineering,
Lahore University of Management Sciences, Lahore, Pakistan
{14030016,17030056,arifz,imdad.khan}@lums.edu.pk

Abstract. Massive sizes of real-world graphs, such as social networks
and web graph, impose serious challenges to process and perform analytics on them. These issues can be resolved by working on a small summary
of the graph instead. A summary is a compressed version of the graph
that removes several details, yet preserves it's essential structure. Generally, some predefined quality measure of the summary is optimized to
bound the approximation error incurred by working on the summary
instead of the whole graph. All known summarization algorithms are
computationally prohibitive and do not scale to large graphs. In this
paper we present an efficient randomized algorithm to compute graph
summaries with the goal to minimize *reconstruction error*. We propose a
novel weighted sampling scheme to sample vertices for merging that will
result in the least reconstruction error. We provide analytical bounds on
the running time of the algorithm and prove approximation guarantee for
our score computation. Efficiency of our algorithm makes it scalable to
very large graphs on which known algorithms cannot be applied. We test
our algorithm on several real world graphs to empirically demonstrate
the quality of summaries produced and compare to state of the art algorithms. We use the summaries to answer several structural queries about
original graph and report their accuracies.

1 Introduction

Analysis of large graphs is a fundamental task in data mining, with applications
in diverse fields such as social networks, e-commerce, sensor networks and bioinformatics. Generally graphs in these domains have very large sizes - millions of
nodes and billions of edges are not uncommon. Massive sizes of graphs make
processing, storing and performing analytics on them very challenging. These
issues can be tackled by working instead on a compact version (summary) of the
graph, which removes certain details yet preserves it's essential structure.

Summary of a graph is represented by a 'supergraph' with weights both
on edges and vertices. Each supernode of the summary, represents a subset of
original vertices while it's weight represents the density of subgraph induced by
that subset. Weights on edges, represent density of the bipartite graph induced

© Springer International Publishing AG, part of Springer Nature 2018
D. Phung et al. (Eds.): PAKDD 2018, LNAI 10939, pp. 502–514, 2018.
https://doi.org/10.1007/978-3-319-93040-4_40

by the two subsets. Quality of a summary is measured by the 'reconstruction error', a norm of the difference of actual and reconstructed adjacency matrices. Another parameter adopted to assess summaries is the accuracy in answer to queries about original graph computed from summaries only.

Note that, since there are exponentially many possible summaries (number of partitions of vertex set), finding the best summary is a computationally challenging task. GraSS [1] uses an agglomerative approach, where in each iteration a pair of nodes is merged until the desired number of nodes is reached. Since the size of search space at iteration t is $O(\binom{n(t)}{2})$, where $n(t)$ is number of supernodes at iteration t. GraSS randomly samples $O(n(t))$ pairs and merges the best pair (pair with the least score) among them. With the data structures of GraSS, merging and evaluating score of a pair can be done in $O(\Delta(t))$ (maximum degree of the summary in iteration t). This results in the overall worst case complexity of $O(n^2\Delta)$ to compute a summary with $O(n)$ nodes. $S2L$ [2] on the other hand uses a clustering technique for Euclidean space by considering each vertex as an N-dimensional vector. The complexity of this algorithm is $O(n^2t)$ to produce a summary of a fixed size $k = O(n)$, where t is the number of iterations before convergence.

In this paper we take the agglomerative approach to compute summary of any desired size. In every iteration a pair is chosen for merging from a randomly chosen sample. We derive a closed form formula for reconstruction error of the graph resulting after merging a pair. Exact computation of this score takes $O(\Delta)$ time but with constant extra space per node, this can be approximated in constant time with bounded error. Furthermore, we define weight of each node that can be updated in constant time and closely estimate the contribution of a node to score of pairs containing it. We select a random sample of pairs by selecting nodes with probability proportional to their weights, resulting in samples of much better quality. We establish that with these weights, logarithmic sized sample yields comparable results. The overall complexity of our algorithm comes down to $O(n(\log n + \Delta))$. Our approach of sampling vertices according to their weights form a dynamic graph (where weights are changing) may be of independent interest. We evaluate our algorithm on several benchmark real world networks and demonstrate that we significantly outperform GraSS [1] and $S2L$[2] both in terms of running time and quality of summaries.

The remaining paper is organized as follows. Section 2 discusses previous work on graph summarization and related problems. In Sect. 3 we formally define the problem with it's background. We present our algorithm along with it's analysis in Sect. 4. In Sect. 5 we report results of experimental evaluation of our algorithm on several graphs. We also provide comparisons with existing solutions both in terms of runtime and quality.

2   Related Work

Graph summarization and compression is a widely studied problem and has applications in diverse domains. There are broadly two types of graph summaries (represented as supergraphs described above): *lossless* and *lossy*. The original graph can be exactly reconstructed from a lossless summary, hence the goal is to optimize the space complexity of a summary [3]. The guiding principle here is that of *minimum description length* MDL [4]. It states that minimum extra information should be kept to describe the summarized data. In lossless summary [5,6], *edge-corrections* are stored along with each supernode and super edges to identify missing edges. [7] stores information about structures like *cliques, stars, and chains* formed by subgraphs as lossless summary of a graph.

Lossy compression, on the other hand, compromises some detailed information to reduce the space complexity. There is a trade off between quality and size of the summary. Quality of a summary is measured by a norm of difference between original adjacency matrix and the adjacency matrix reconstructed from the summary, known as reconstruction error. [1] adopted an agglomerative approach to greedily merge pairs of nodes to minimize the l_1-reconstruction error. Runtime of their algorithm amounts $O(n^3)$ in the worst case.

In [2], each node is considered a vector in $\mathbb{R}^n$ (it's row in the adjacency matrix) and point-assignment clustering methods (such as k-means) are employed. Each cluster is considered a supernode and the goal is to minimize the l_2-reconstruction error. The authors suggest to use dimensionality reduction techniques for points in $\mathbb{R}^n$. This technique does not use any structural information of the graph. In [8] social contexts and characteristics are used to summarize social networks. Summarization of edge-weighted graphs is studied in [9]. Graph compression techniques relative to a certain class of queries on labeled graphs is studied in [10]. [11] uses entropy based unified model to make a homogeneous summary for labeled graphs. Compression of web graphs and social networks is studied in [12,13] and [14], respectively. See [15] for detailed overview of graph summarization techniques.

A closely related area is that of finding clusters and communities in a graph using iterative algorithms [16], agglomerative algorithms [17] and spectral techniques [18]. Identification of web communities in web graphs using maximum flow/minimum cut problem is discussed in [19].

3   Problem Definition

Given a graph $G = (V, E)$ on n vertices, let A be it's adjacency matrix. For $k \in \mathbb{Z}$, a summary of G, $S = (V_S, E_S)$ is a weighted graph on k vertices. Let $V_S = \{V_1, \ldots, V_k\}$, each $V_i \in V_S$ is referred to a supernode and represents a subset of V. More precisely, V_S is a partition of V, i.e. $V_i \subset V$ for $1 \leq i \leq k$, $V_i \cap V_j = \emptyset$ for $i \neq j$ and $\bigcup_{i=1}^{k} V_i = V$. Each supernode V_i is associated with two integers $n_i = |V_i|$ and $e_i = |\{(u, v)|u, v \in V_i, (u, v) \in E\}|$. For an edge $(V_i, V_j) \in E_S$ (known as superedge), let e_{ij} be the number of edges in the bipartite subgraph induced between V_i and V_j, i.e. $e_{ij} = |\{(u, v)|u \in V_i, v \in V_j, (u, v) \in E\}|$. Given a summary S, the graph G is approximately reconstructed by the expected adjacency matrix, $\bar{A}$, where $\bar{A}$ is a $n \times n$ matrix with

$$\bar{A}(u,v) = \begin{cases} 0 & \text{if } u = v \\ \frac{e_i}{\binom{n_i}{2}} & \text{if } u,v \in V_i \\ \frac{e_{ij}}{n_i n_j} & \text{if } u \in V_i, v \in V_j \end{cases}$$

The quality of a summary S is assessed by l_p-norm of element-wise difference between $\bar{A}$ and A.

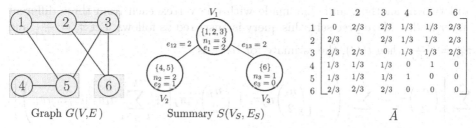

	1	2	3	4	5	6
1	0	2/3	2/3	1/3	1/3	2/3
2	2/3	0	2/3	1/3	1/3	2/3
3	2/3	2/3	0	1/3	1/3	2/3
4	1/3	1/3	1/3	0	1	0
5	1/3	1/3	1/3	1	0	0
6	2/3	2/3	2/3	0	0	0

Graph $G(V,E)$ Summary $S(V_S, E_S)$ $\bar{A}$

Definition 1 *(l_p-Reconstruction Error (RE_p)): The (unnormalized) l_p reconstruction error of a summary S of a graph G is*

$$RE_p(G|S) = RE_p(A|\bar{A}) = \left(\sum_{i=1}^{|V|} \sum_{j=1}^{|V|} |\bar{A}(i,j) - A(i,j)|^p \right)^{1/p} \tag{1}$$

Note that the case $p = 1$ considered in [1] and $p = 2$ considered in [2] are closely related to each other. In this paper we use $p = 1$ and refer to $RE_1(G|S)$ as $RE(G|S)$. A simple calculation shows that $RE(G|S)$ can be computed in the following closed form.

$$RE(G|S) = RE(A|\bar{A}) = \sum_{i=1}^{k} 4c_i - \frac{4e_i^2}{\binom{n_i}{2}} + \sum_{i=1}^{k} \sum_{j=1, j \neq i}^{k} 2e_{ij} - \frac{2e_{ij}^2}{n_i n_j} \tag{2}$$

Formally, we address the following problem.

Problem 2. Given a graph $G(V, E)$ and a positive integer $k \leq |V|$, find a summary S for G with k super nodes such that $RE(G|S)$ is minimized.

Another measure to assess quality of a summary S of G is by the accuracy of answers of queries about structure of G based on S only. In the following we list how certain queries used in the literature are answered from S.

Adjacency Queries: Given two vertices $u, v \in V$, the query whether $(u, v) \in E$ is answered with $\bar{A}(u, v)$. This can either be interpreted as the expected value of an edge being present between u and v or as returning a 'yes' answer based on the outcome of a biased coin.

Degree Queries: Given a vertex $v \in V$, the query about degree of v is answered as $\bar{d}(v) = \sum_{j=1}^{n} \bar{A}(v, j)$.

Eigenvector-Centrality Queries: Eigenvector-centrality of a vertex v, $p(v)$ measures the relative importance of v [20]. For a vertex $v \in V$, this query is answered as $\bar{p}(v) = \frac{\bar{d}(v)}{2|E|}$.

Triangle Density Queries: Let $t(G)$ be the number of triangles in G. $t(G)$ is estimated from S by counting the expected number of triangles within each super node, the expected number of triangles made with one vertex in one supernode and two in another, and that made with one vertex each from three different super nodes. More precisely, this query is answered as follows. Let $\pi_i = \frac{e_i}{\binom{n_i}{2}}$ and $\pi_{ij} = \frac{e_{ij}}{n_i n_j}$, then $\bar{t}(G)$, the estimate for $t(G)$, is

$$\sum_{i=1}^{k} \left[\binom{n_i}{3} \pi_i^3 + \sum_{j=i+1}^{k} \left(\pi_{ij}^2 \left[\binom{n_i}{2} n_j \pi_i + \binom{n_j}{2} n_i \pi_j \right] + \sum_{l=j+1}^{k} n_i n_j n_l \pi_{ij} \pi_{jl} \pi_{il} \right) \right].$$

4   Algorithm

Given a graph G and an integer k our algorithm produces a summary S on k super nodes as follows. Let S_{t-1} be the summary before iteration t with $n(t-1)$ super nodes, i.e. $S_0 = G$, and let $\bar{A}_t$ be the expected adjacency matrix of S_t. For $1 \leq t \leq n - k$, we select a pair of supernodes (u, v) and merge it to get S_t. To select an approximately optimal pair we define weight of each node v that closely estimate the contribution of this node to score of pairs $(v, *)$. We randomly sample nodes for each pair with probability proportional to their weights and evaluate score of the pairs. We derive a closed form formula to evaluate score of a pair. Furthermore, in this form these scores can be approximately computed very efficiently. Based on approximate score we select the best pair in the sample and merge it to get S_t. In what follows, we discuss implementation of each of these subroutines and their analyses.

Lemma 3. *A pair (u, v) of nodes in S_t, can be merged to get S_{t+1} in time $O(deg(u) + deg(v))$.*

Proof. In the adjacency list format, one needs to iterate over neighbors of each u and v and record their information in a new list of the merged node. However, updating the adjacency information at each neighbor of u and v could potentially lead to traversal of all the edges. To this end, as a preprocessing step, for each (x, y), in the adjacency list of x at node y, we store a pointer to the corresponding entry in the adjacency list of y. With this constant (per edge) extra book keeping we can update the merging information at each neighbor in constant time by traversing just the list of u and v. It is easy to see that this preprocessing can be done in time $O(|E|)$ once at the initialization.   □

The next important step is to determine the *goodness* of a pair (a, b). This can be done by temporarily merging a and b and then evaluating (1) or (2) respectively taking $O(n^2)$ and $O(n(t))$. For a pair of nodes (a, b) in S_{t-1}, let $S_t^{a,b}$ be the graph obtained after merging a and b. We define score of a pair (a, b) to be

$$score_t(a, b) = RE(G|S_{t-1}) - RE(G|S_t^{a,b})$$

$$= -\frac{4e_a^2}{\binom{n_a}{2}} - \sum_{\substack{i=1 \\ i \neq a}}^{n(t)} \frac{4e_{ai}^2}{n_a n_i} + \frac{4c_{ab}^2}{n_a n_b} - \frac{4e_b^2}{\binom{n_b}{2}} - \sum_{\substack{i=1 \\ i \neq b}}^{n(t)} \frac{4e_{bi}^2}{n_b n_i}$$

$$+ \frac{4(e_a + e_b + e_{ab})^2}{\binom{n_a+n_b}{2}} + \frac{4}{(n_a + n_b)} \sum_{\substack{i=1 \\ i \neq a,b}}^{n(t)} \left(\frac{e_{ai}^2}{n_i} + \frac{e_{bi}^2}{n_i} + \frac{2e_{ai}e_{bi}}{n_i} \right) \quad (3)$$

Fact 4. *Since S_{t-1} is fixed, minimizing $RE(G|S_t^{a,b})$ is equivalent to maximizing $score_t(a, b)$.*

Remark 5. Except for the last summation in (3) all other terms of $score_t(a, b)$ can be computed in constant time. Since n_a, n_b, e_a, and e_b are already stored at a and b, this can be achieved by storing an extra real number D_a at each super node a such that, $D_a = \sum_{\substack{i=1 \\ i \neq a}}^{n(t)} \frac{e_{ai}^2}{n_i}$. Note that D_a can be updated in constant time after merging of any two vertices $x, y \neq a$, i.e. after merging x, y, while traversing their neighbors for a we subtract e_{xa}/n_x and e_{ya}/n_y from D_a and add back $(e_x + e_y)/(n_x + n_y)$ to it. This value can be similarly updated at the merged node too.

The last summation in (3), $\sum_{\substack{i=1 \\ i \neq a,b}}^{n(t)} \frac{2e_{ai}e_{bi}}{n_i}$, in essence is the inner product of two $n(t)$ dimensional vectors $\mathcal{A}$ and $\mathcal{B}$, where the ith coordinate of $\mathcal{A}$ is $\frac{e_{ai}}{\sqrt{n_i}}$ ($\mathcal{B}$ is similarly defined). Storing these vectors will take $O(n(t))$, moreover computing score will take time $O(n(t))$. However, $\langle \mathcal{A}, \mathcal{B} \rangle = \mathcal{A} \cdot \mathcal{B}$ can be very closely approximated with a standard application of *count-min sketch* [21].

Theorem 6 *(c.f [21] Theorem 2). For $0 < \epsilon, \delta < 1$, let $\langle \widehat{\mathcal{A}, \mathcal{B}} \rangle$ be the estimate for $\langle \mathcal{A}, \mathcal{B} \rangle$ using the count-min sketch. Then*

$- \langle \widehat{\mathcal{A}, \mathcal{B}} \rangle \geq \langle \mathcal{A}, \mathcal{B} \rangle$

$- Pr[\langle \widehat{\mathcal{A}, \mathcal{B}} \rangle < \langle \mathcal{A}, \mathcal{B} \rangle + \epsilon ||\mathcal{A}||_1 ||\mathcal{B}||_1] \geq 1 - \delta$. $\mathcal{A}, \mathcal{B}$

Furthermore, the space and time complexity of computing $\langle \widehat{\mathcal{A}, \mathcal{B}} \rangle$ is $O(\frac{1}{\epsilon} \log \frac{1}{\delta})$. While after a merge, the sketch can be updated in time $O(\log \frac{1}{\delta})$.

Hence, for a pair of nodes (a, b) in S_{t-1}, $score_t(a, b)$ can be closely approximated in constant time. Note that the bounds on time and space complexity, though constants are quite loose in practice.

The next important issue the quadratic size of search space. This is a major hurdle to scalability to large graphs. We define weight of a node a as

$$f(a) = -\frac{4e_a^2}{\binom{n_a}{2}} - \sum_{\substack{i=1 \\ i \neq a}}^{n(t)} \frac{4e_{ai}^2}{n_a n_i} \qquad w(a) = \begin{cases} \dfrac{-1}{f(a)} & \text{if } f(a) \neq 0 \\ 0 & \text{otherwise} \end{cases} \qquad (4)$$

We select pairs by sampling nodes according to their weights so as the pairs selected will likely have higher scores. With this weighted sampling a sample of size $O(\log n)$ outperforms a random sample of size $O(n)$. Let $W = \sum_{i=1}^{n(t)} w(i)$ be the sum of weights, we select a vertex w with probability $w(a)/W$. Weighted sampling though can be done in linear time at a given iteration. In our case it is very challenging since the population varies in each iteration; two vertices are merged into one and weights of some nodes also change. To overcome this challenge, we design special data structure $\mathcal{D}$ that has the following properties.

Claim. $\mathcal{D}$ can be implemented as a binary tree such that

  i. it can be initially populated in $O(n)$,
 ii. a node can be sampled with probability proportional to it's weight in $O(\log n)$;
iii. inserting, deleting or updating a weight in $\mathcal{D}$ takes time $O(\log n)$.

Remark 7. We designed this data structure independently, but found out that it has been known to the statistics community since 1980 [22]. We note that this technique could have many applications in sampling from dynamic graphs.

Algorithm 1 is our main summarization algorithm that takes as input G, integers k (target summary size), s (sample size), w and d (where $w = \frac{1}{\epsilon}$ and $d = \log \frac{1}{\delta}$ are parameters for count-min sketch).

Algorithm 1. ScalableSumarization($G = (V, E)$, k, w, d)

1: $\mathcal{D} \leftarrow$ BUILDSAMPLINGTREE($V, W, 1, n$)   ▷ $W[1 \ldots n]$ is initialize as $W[i] = w(v_i)$
2: **while** G has more than k vertices **do**
3:     $samplePairs \leftarrow$ GETSAMPLE($\mathcal{D}, s$)   ▷ s calls to Algorithm 3
4:     $scores \leftarrow$ COMPUTEAPPROXSCORE($samplePairs$)   ▷ Uses (3) and Theorem 6
5:     $bestPair \leftarrow$ MAX($scores$)
6:     MERGE($bestPair$)   ▷ Lemma 3
7:     **for** each neighbor x of $u, v \in bestPair$ **do**
8:         UPDATEWEIGHT($x, \mathcal{D}$)

For each vertex a we maintain a variable D_a (Remark 5). Hence the weight array can be initialized in $O(n)$ time using (4). By Claim 4, $\mathcal{D}$ can be populated in $O(n)$ time. By Claim 4, Line 3 takes $O(s \log n)$ time, by Theorem 6 and (3) Line 4 takes constant time per pair, and by Lemma 3 merging can be performed in

$O(\Delta)$ time. Since delete and update in $\mathcal{D}$ takes time $O(\log n)$ and the while loop is executed $n - k + 1$ times, total runtime of Algorithm 1 is $O((n - k + 1)(s \log n + \Delta \log n)$. Generally k is $O(n)$ (typically a fraction of n) and in our experiments we take s to be $O(\log n)$ and $O(\log^2 n)$. Furthermore, since many real world graphs are very sparse, (Δ which is worst case upper bound is constant), we get that overall complexity of our algorithm is $O(n \log^2 n)$ or $O(n \log^3 n)$.

Data Structure for Sampling: We implement $\mathcal{D}$ as a balanced binary tree, where leaf corresponds to (super) node in the graph and stores weight and id of the node. Each internal node stores the sum of values of the two children. The value of the root is equal to $\sum_{i=1}^{n(t)} w(i)$. Furthermore, at each node in the graph we store a pointer to the corresponding leaf. We give pseudocode to construct this tree in Algorithm 2 along with the structure of a tree node. By construction, it is clear that hight of the tree is $\lceil \log n \rceil$ and running time of building the tree and space requirement of $\mathcal{D}$ is $O(n)$.

Algorithm 2. BuildSamplingTree(A, W, st, end)

1: **if** $A[st] = A[end]$ **then**
2:     $leaf \leftarrow$ CREATENODE()
3:     $leaf.weight \leftarrow W[st]$
4:     $leaf.vertexID \leftarrow A[st]$
5:     **return** $leaf$
6: **else**
7:     $mid = \frac{end + st}{2}$
8:     $left \leftarrow$ CREATENODE()
9:     $left \leftarrow$ BUILDSAMPLINGTREE(A, W, st, mid)
10:     $right \leftarrow$ CREATENODE()
11:     $right \leftarrow$ BUILDSAMPLINGTREE($A, W, mid + 1, end$)
12:     $parent \leftarrow$ CREATENODE()
13:     $parent.weight \leftarrow left.weight + right.weight$
14:     $left.parent \leftarrow parent$
15:     $right.parent \leftarrow parent$
16:     **return** $parent$

Structure TreeNode

int *vertexID*
double *weight*
TreeNode *∗left*
TreeNode *∗right*
TreeNode *∗parent*

The procedure to sample a vertex with probability proportional to its weight using $\mathcal{D}$ is given in Algorithm 3. This takes as input a uniform random number $r \in [0, \sum_{i=1}^{n(t)} w(i)]$. Since it traverses a single path from root to leaf, the runtime of this algorithm is $O(\log n)$. The update procedure is very similar, whenever weight of a node changes, we start from the corresponding leaf (using the stored pointer to leaf) and change weight of that leaf. Following the parent pointers, we update weights of internal nodes to the new sum of weights of children. Deleting a node is very similar, it amounts to updating weight of the corresponding leaf to 0. Inserting a node (the super node representing the merged nodes) is achieved by changing the weight of the first empty leaf in $\mathcal{D}$. A reference to first empty node is maintained as a global variable.

Algorithm 3. GetLeaf(r,*node*)

1: **if** *node.left* = **NULL** ∧*node.right* = **NULL then**
2:     **return** *node.vertexID*
3: **if** $r <$ *node.left.weight* **then**
4:     **return** GETLEAF(r, *node.left*)
5: **else**
6:     **return** GETLEAF($r -$ *node.weight*, *node.right*)

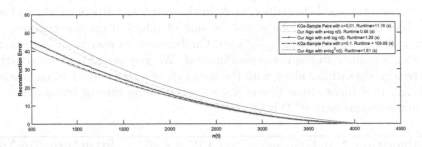

Fig. 1. Comparison between k-Gs-SamplePairs on ego-Facebook. Runtimes reported are at k = 500.

5   Evaluation

We evaluate performance of our algorithm in terms of runtime, reconstruction error and accuracies of answers to queries on standard benchmark graphs[1]. We demonstrate that our algorithm substantially outperforms existing solutions, GraSS [1] and $S2L$ [2] in terms of quality while achieving order of magnitude speed-up over them. Our Java Implementation is available at[2]. We also report the accuracies in query answered based on summaries only and show that error is very small and we save a lot of time. Errors reported are normalized by $|V|$. All runtimes are in seconds.

From Fig. 1, it is clear that the quality of our summaries compares well with that of k-Gs-SamplePairs but with much smaller sample size. We report results for $s \in \{\log n(t), 5 \log n(t), \log^2 n(t)\}$, with exact score computation. Indeed with sublinear sample size $O(\log n(t))$ and $O(\log^2 n(t))$, our reconstruction error is less than k-Gs-SamplePairs with sample size $0.01n(t)$. Although for $n(t)$ significantly smaller than $|V|$, there is a small difference in size of logarithmic and linear sample, but we benefit from our logarithmic sample size for large $n(t)$.

In Table 1, we present reconstruction errors on moderately large sized graphs. Even though $S2L$ is suitable for Euclidean errors, our algorithm still outperforms $S2L$ because by minimizing RE we preserve the original structure of the graph. We use $s = \log n(t)$, $w \in \{50, 100\}$ and $d = 2$. We also report results without approximation in score computation. Note that we are able to generate

---

[1] http://snap.stanford.edu/.
[2] https://bitbucket.org/M_AnwarBeg/scalablesumm/.

Table 1. Comparison of $S2L$ with our algorithm using different count min sketch widths. The numbers reported here for $S2L$ are as given by authors in [2].

Graph	k	w	Error		RunTime		$S2L$
			RE	$(l_2)^2$	Avg	Std	$(l_2)^2$
cgo-Faccbook	1000	50	69.98	35.00	0.83	0.01	581
$|V| = 4,039$		100	57.27	28.66	0.94	0.05	
$|E| = 88,234$		-[a]	38.98	19.51	1.01	0.08	
	1500	50	58.17	29.09	0.82	0.05	501
		100	40.05	20.04	0.83	0.03	
		-	27.14	13.58	0.87	0.04	
email-Enron	10000	50	6.03	3.01	1.95	0.05	72
$|V| = 36,692$		100	5.84	2.92	2.14	0.02	
$|E| = 183,831$		-	5.82	2.91	2.21	0.07	
	14000	50	4.25	2.13	1.69	0.05	54
		100	4.16	2.08	1.77	0.02	
		-	4.15	2.08	1.81	0.01	
web-Stanford	2000	50	24.11	12.06	65.97	0.02	48
$|V| = 281,903$		100	24.01	12.01	64.41	0.82	
$|E| = 1,992,636$		-	24.01	12.05	73.48	0.28	
	10000	50	23.65	11.83	59.87	0.15	38
		100	23.61	11.81	66.21	0.67	
		-	21.13	10.57	59.34	0.28	
amazon0601	2000	50	24.11	12.06	147.3	3.72	53
$|V| = 403,394$		100	24.10	12.05	158.03	1.59	
$|E| = 2,443,408$		-	24.10	12.05	175.29	5.81	
	8000	50	23.77	11.89	143.83	3.58	51
		100	23.74	11.87	154.34	1.92	
		-	23.73	11.87	171.00	5.63	

[a]Represents score computation without approximation using Eq. 3.

summaries with much smaller runtime on a less powerful machine, (Intel(R) Core i5 with 8.00 GB RAM and 64-bit OS) compared to one reported in [2].

Table 2 contains quality and runtime for very large graphs, on which none of the previously proposed solutions were applicable. We use $s = \log n(t)$, $w \in \{50, 100, 200\}$ and $d = 2$. We get some reduction in running times by approximating the score while the quality of summaries remains comparable. Note that large values of w result in increased runtime without any improvement in quality. This is so because the complexity of exact score evaluation depends on maximum degree, which in real world graphs is small.

In Table 3, we tabulate answers to queries that are computed from summaries only. We report mean absolute errors in estimated degrees and eigenvector-centrality scores. For triangle density we report relative error, calculated as $\frac{t(\bar{G})-t(G)}{t(G)}$. In all cases query answers are very close to the true values, signifying the fact that our summaries do preserve the essential structure of the graph.

Table 2. Quality in terms of RE of summary produced by our algorithm on large sized graph.

Parameters		as-Skitte $|E| = 1,696,415$ $|V| = 11,095,298$			wiki-Talk $|E| = 2,394,385$ $|V| = 4,659,565$			com-Youtube $|E| = 1,157,828$ $|V| = 2,987,624$		
		Time			Time			Time		
$k\times(10^3)$	w	RE	Avg	Std	RE	Avg	Std	RE	Avg	Std
10	50	25.42	521.43	12.93	7.28	311.10	5.68	9.98	207.38	4.66
	100	24.81	516.03	22.97	7.07	328.19	2.06	9.64	222.22	7.17
	200	24.35	559.91	13.07	6.82	363.37	8.63	9.30	251.94	8.28
	-	23.81	649.82	20.44	6.72	319.95	28.91	9.26	242.58	13.85
50	50	23.49	481.40	14.11	5.77	285.89	5.56	7.49	184.67	3.86
	100	21.78	480.85	23.54	5.65	299.98	1.97	7.13	195.85	6.41
	200	20.90	524.94	12.88	5.63	329.39	8.95	7.09	215.87	7.23
	-	20.77	591.35	19.36	5.62	273.24	24.05	7.08	199.48	10.85
100	50	21.28	436.84	11.52	5.12	266.15	4.83	5.90	160.11	3.51
	100	18.90	445.27	23.18	5.09	276.32	2.77	5.82	167.67	5.22
	200	18.48	486.88	13.14	5.08	303.44	8.60	5.81	183.64	5.81
	-	18.42	535.02	18.90	5.08	248.73	20.44	5.81	164.80	9.87
250	50	15.34	332.27	9.58	4.23	223.65	3.85	3.91	103.25	3.97
	100	13.79	350.47	21.77	4.22	232.39	1.81	3.90	107.79	1.03
	200	13.68	376.89	11.86	4.21	256.05	6.74	3.89	118.73	4.60
	-	13.65	392.58	13.48	4.21	203.93	18.21	3.89	98.70	4.89

Table 3. Error in queries computed by summaries generated by our algorithm. Absolute average error is reported for degree and centrality query. For triangle density, relative error is reported.

		ego-Facebook							email-Enron				
		Degree		Eigenvector-Centrality $(\times 10^{-5})$					Degree		Eigenvector-Centrality $(\times 10^{-5})$		
k	w	avg	stdev	avg	stdev	Triangle Density		k	w	avg	stdev	avg	stdev	Triangle Density
	50	22.50	29.70	6.37	8.42	-0.89			50	2.49	8.60	0.34	1.17	-0.37
500	100	14.95	26.58	4.24	7.53	-0.76		4000	100	1.94	4.26	0.26	0.58	-0.19
	-	12.01	12.70	3.40	3.60	-0.28			-	1.91	3.09	0.26	0.42	-0.16
	50	17.66	25.68	5.00	7.28	-0.84			50	1.56	4.70	0.21	0.64	-0.18
1000	100	10.67	24.19	3.02	6.85	-0.58		6000	100	1.37	2.33	0.19	0.32	-0.12
	-	7.62	9.41	2.16	2.67	-0.15			-	1.38	2.24	0.19	0.31	-0.11
	50	13.22	22.19	3.75	6.29	-0.72			50	1.15	3.10	0.16	0.42	-0.11
1500	100	7.04	18.32	2.00	5.19	-0.41		8000	100	1.06	1.77	0.14	0.24	-0.08
	-	4.50	5.67	1.28	1.61	-0.08			-	1.04	1.52	0.14	0.21	-0.08

6 Conclusion

In this work we devise a sampling based efficient approximation algorithm for graph summarization. We derive a closed form for measuring suitability of a pair of vertex for merging. We approximate this score with theoretical guarantees on error. Another major contribution of this work is the efficient weighted sampling scheme to improve the quality of samples. This enables us to work with substantially smaller sample sizes without compromising summary quality. Our algorithm is scalable to large graphs on which previous algorithms are not applicable. Extensive evaluation on a variety of real world graphs show that our algorithm significantly outperforms existing solutions both in quality and time complexity.

References

1. LeFevre, K., Terzi, E.: GraSS: graph structure summarization. In: SIAM International Conference on Data Mining SDM, pp. 454–465 (2010)
2. Riondato, M., García-Soriano, D., Bonchi, F.: Graph summarization with quality guarantees. In: IEEE International Conference on Data Mining ICDM, pp. 947–952 (2014)
3. Storer, J.: Data compression. Elsevier, Amsterdam (1988)
4. Rissanen, J.: Modeling by shortest data description. Automatica **14**(5), 465–471 (1978)
5. Navlakha, S., Rastogi, R., Shrivastava, N.: Graph summarization with bounded error. In: ACM International Conference on Management of Data SIGMOD, pp. 419–432 (2008)
6. Khan, K., Nawaz, W., Lee, Y.: Set-based approximate approach for lossless graph summarization. Computing **97**(12), 1185–1207 (2015)
7. Koutra, D., Kang, U., Vreeken, J., Faloutsos, C.: VOG: summarizing and understanding large graphs. In: SIAM International Conference on Data Mining SDM, pp. 91–99 (2014)
8. Zhuang, H., Rahman, R., Hu, X., Guo, T., Hui, P., Aberer, K.: Data summarization with social contexts. In: ACM International Conference on Information and Knowledge Management CIKM, pp. 397–406 (2016)
9. Toivonen, H., Zhou, F., Hartikainen, A., Hinkka, A.: Compression of weighted graphs. In: ACM International Conference on Knowledge Discovery and Data Mining SIGKDD, pp. 965–973 (2011)
10. Fan, W., Li, J., Wang, X., Wu, Y.: Query preserving graph compression. In: ACM International Conference on Management of Data SIGMOD, pp. 157–168 (2012)
11. Liu, Z., Yu, J.X., Cheng, H.: Approximate homogeneous graph summarization. J. Inf. Process. **20**(1), 77–88 (2012)
12. Boldi, P., Vigna, S.: The webgraph framework I: compression techniques. In: International Conference on World Wide Web WWW, pp. 595–602 (2004)
13. Adler, M., Mitzenmacher, M.: Towards compressing web graphs. In: Data Compression Conference DCC, pp. 203–212 (2001)
14. Chierichetti, F., Kumar, R., Lattanzi, S., Mitzenmacher, M., Panconesi, A., Raghavan, P.: On compressing social networks. In: ACM International Conference on Knowledge Discovery and Data Mining SIGKDD, pp. 219–228 (2009)

15. Liu, Y., Dighe, A., Safavi, T., Koutra, D.: A graph summarization: a survey (2016). arXiv preprint arXiv:1612.04883
16. Newman, M.E., Girvan, M.: Finding and evaluating community structure in networks. Phys. Rev. E **69**(2), 026113 (2004)
17. Clauset, A., Newman, M.E., Moore, C.: Finding community structure in very large networks. Phys. Rev. E **70**(6), 066111 (2004)
18. White, S., Smyth, P.: A spectral clustering approach to finding communities in graphs. In: SIAM International Conference on Data Mining SDM, pp. 274–285 (2005)
19. Flake, G.W., Lawrence, S., Giles, C.L.: Efficient identification of web communities. In: ACM International Conference on Knowledge Discovery and Data Mining SIGKDD, pp. 150–160 (2000)
20. Motwani, R., Raghavan, P.: Randomized Algorithms. Chapman & Hall/CRC, Boca Raton (2010)
21. Cormode, G., Muthukrishnan, S.: An improved data stream summary: the count-min sketch and its applications. J. Algorithms **55**(1), 58–75 (2005)
22. Wong, C.K., Easton, M.C.: An efficient method for weighted sampling without replacement. SIAM J. Comput. **9**(1), 111–113 (1980)

Privacy-Preserving and Security

RIPEx: Extracting Malicious IP Addresses from Security Forums Using Cross-Forum Learning

Joobin Gharibshah[(✉)], Evangelos E. Papalexakis, and Michalis Faloutsos

Department of Computer Science, University of California - Riverside,
900 University Ave, Riverside, CA 92521, USA
{jghar002,epapalex,michalis}@cs.ucr.edu

Abstract. Is it possible to extract malicious IP addresses reported in security forums in an automatic way? This is the question at the heart of our work. We focus on security forums, where security professionals and hackers share knowledge and information, and often report misbehaving IP addresses. So far, there have only been a few efforts to extract information from such security forums. We propose RIPEx, a systematic approach to identify and label IP addresses in security forums by utilizing a cross-forum learning method. In more detail, the challenge is twofold: (a) identifying IP addresses from other numerical entities, such as software version numbers, and (b) classifying the IP address as benign or malicious. We propose an integrated solution that tackles both these problems. A novelty of our approach is that it does not require training data for each new forum. Our approach does knowledge transfer across forums: we use a classifier from our source forums to identify seed information for training a classifier on the target forum. We evaluate our method using data collected from five security forums with a total of 31 K users and 542 K posts. First, RIPEx can distinguish IP address from other numeric expressions with 95% precision and above 93% recall on average. Second, RIPEx identifies malicious IP addresses with an average precision of 88% and over 78% recall, using our cross-forum learning. Our work is a first step towards harnessing the wealth of useful information that can be found in security forums.

Keywords: Security · Online communities mining

1 Introduction

The overarching goal of this work is to harness the user generated content in forums, especially security forums. More specifically, we focus here on collecting malicious IP addresses, which are often reported at such forums. We use the term security forums to refer to discussion forums with a focus on security, system administration, and in general systems-related discussions. In these forums,

© Springer International Publishing AG, part of Springer Nature 2018
D. Phung et al. (Eds.): PAKDD 2018, LNAI 10939, pp. 517–529, 2018.
https://doi.org/10.1007/978-3-319-93040-4_41

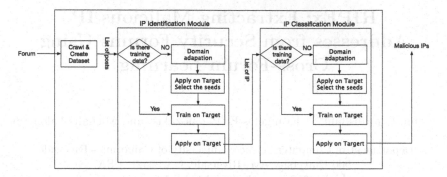

Fig. 1. The overview of key modules of our approach (RIPEx): (a) collecting data, (b) IP Identification, and (c) IP Characterization. In both classification stages, we use our Cross-Seeding approach that in order to generate seed information for training a classifier for a new forum.

security professionals, hobbyists, and hackers identify issues, discuss solutions, and in general exchange information.

We provide a few examples of the types of discussions that take place in these forums that could involve IP addresses, which is our focus. Posts could talk about a benign IP address, say in configuration files, as in the post: *"[T]his thing in my hosts file: 64.91.255.87 ... [is] it correct?"*. At the same time, posts could also report compromised or malicious IP addresses, as in the post: *"My browser homepage has been hijacked to http://69.50.191.51/2484/"*. Our goal is to automatically distinguish between the two and provide a new source of information for malicious IP addresses directly from the affected individuals.

The problem that we address here is to find all the IP addresses that are being reported as malicious in a forum. In other words, the input is all the posts in a forum and the expected output is a list of malicious IP addresses. As with any classification problem, one would like to achieve both high precision and recall. Precision represents the percentage of the correctly labeled over all addresses labeled malicious. Recall is the percentage of malicious addresses that we find among all malicious addresses reported in forums. It turns out that this is a two-step problem. First, we need to solve the **IP Identification** problem: distinguishing IP addresses from other numerical entities, such as a software version. Second, we need to solve the **IP Characterization** problem: characterizing IP address as malicious or benign. The extent of the Identification problem caught us by surprise: we find 1820 non-address dot-decimals, as we show in Table 1.

There is limited work on extracting information from security forums, and even less work on extracting malicious IP addresses. We can group prior work in the following categories. First, recent works study the number of malicious IP addresses in forums, but without providing the comprehensive and systematic solution that we propose here [7]. Second, there are recent efforts that extract other types of information from security forums, related to the black market of hacking services and tools [13], or the behavior and roles of their users [1,9]. Third, other works focus on analyzing structured sources, such as security reports and vulnerability databases [2,10]. We discuss related work in Sect. 5.

There is a wealth of information that can be extracted from security forums, which motivates this research direction. Earlier work suggests that there is close to four times more malicious IP addresses in forums compared to established databases of such IP addresses [8]. At the same time, there are tens of thousands of IP addresses in the forums, as we will see later. Interestingly, not all of the reported IP addresses are malicious, which makes the classification necessary.

We propose RIPEx[1], a comprehensive, automated solution that can detect malicious IP addresses reported in security forums. As its key novelty, our approach minimizes the need for human intervention. First, once initialized with a small number of security forums, it does not require additional training data to mine new forums. Second, it addresses both the Identification and Characterization problems. Third, our approach is systematic and readily deployable. We are not aware of prior work claiming these three properties, as we discuss in Sect. 5. The overview of our approach is shown in Fig. 1.

The key technical novelty is that we propose **Cross-Seeding**, a method to conduct a multi-step knowledge transfer across forums. We use this approach for both classification problems, when we have no training data for a new forum. With Cross-Seeding, we create training data for the new forum in the process depicted in Fig. 1. We use a classifier based on the current forums to identify seed information in the new forum. We then use this seed information to train a classifier for the new forum. This forum-specific classifier performs much better than if we have used the classifier of the current forums on the new forum. We refer to this latter knowledge transfer approach as **Basic**.

We evaluate our approach using five security forums with a total of 31 K users and 542 K posts spanning a period of roughly six years. Our results can be summarized into the following points.

a. **Identification: 98% Precision with Training Data Per Forum.** We develop a supervised learning algorithm for solving the Identification problem in the case where we have training data for the target forum. Our approach exhibits 98% precision and 96% recall on average across all our sites.

b. **Identification: 95% Precision with Cross-Seeding.** We show that our Cross-Seeding approach is effective in transferring the knowledge between forums. Using the WildersSecurity forum as source, we observe an average of 95% precision and 93% recall in the other forums.

c. **Characterization: 93% Precision with Training Data Per Forum.** We develop a supervised learning algorithm for solving the Characterization problem assuming we have training data for the target forum. Our classifier achieves 93% precision and 92% recall on average across our forums.

d. **Characterization: 88% Precision on Average with Cross-Seeding Data.** We show that our Cross-Seeding approach by using OffensiveCommunity forum as source can provide 88% precision and 82% recall on average.

---

[1] RIPEx stands for **R**iverside's **IP Ex**tractor.

e. **Cross-Seeding Outperforms Basic.** We show that Cross-Seeding is important, as it increases the precision by 28% and recall by 16% on average in the Characterization problem, and the precision by 8% and recall by 7% on average in the Identification problem.

f. **Using More Source Forums Improves the Cross-Seeding Performance.** We show that, by adding a second source forum, we can improve the precision by 13% on average over the remaining three forums.

Our work suggests that there is a wealth of information that we find in security forums and offers a systematic approach to do so.

2   Our Forums and Datasets

We have collected data from five different forums, which cover a wide spectrum of interests and intended audiences. We present basic statistics of our forums in Table 1 and we highlight the differences of their respective communities.

Our Semi-automated Crawling Tool. We have developed an efficient and customizable python-based crawler, which can be used to crawl online forums, and it could be of independent interest. To crawl a new forum, our tool requires a configuration file that describes the structure of the forum. Leveraging our current configuration files, the task of crawling a new forum is simplified significantly. Due to space limitations, we do not provide further details. Following are the descriptions of collected forums.

- **WildersSecurity (WS)** seems to attract system administrator types and focuses on defensive security: how one can manage and protect one's system.
- **OffensiveCommunity (OC)** seems to be on the fringes of legality. As the name suggests, the forum focuses on breaking into systems: it provides step by step instructions, and advertises hacking tools and services.
- **HackThisSite (HT)** seems to be in between these extremes represented by the first two forums. There are discussions on hacking challenges, but it does not act as openly as black market services compared to OffensiveCommunity.
- **EthicalHackers (EH)** seems to consist mostly of "white hat"hackers, as its name suggests. The users discuss hacking techniques, but they seem to have a strict moral code.
- **Darkode (DK)** is a forum on the dark web that has been taken down by the FBI in July 2015. The site was a black market for malicious tools and services similar to OffensiveCommunity.

Our goal is to identify and report IP addresses that the forum readers report as malicious. We currently do not assess whether the author of the post is right, though the partial overlap with blacklisted IPs indicates so. We leave for future work to detect misguided reports of IP addresses.

Determining the Ground-Truth. For both of the problems we address here, there are no well-established benchmarks and labeled datasets. To train and

Table 1. The basic statistics of our forums

	WildersSec.	OffensiveComm.	HackThisSite	EthicalHackers	Darkode
Posts	302710	25538	84125	54176	75491
Threads	28661	3542	8504	8745	7563
Users	14836	5549	5904	2970	2400
Dot-decimal	4325	7850	1486	1591	1097
IP found	3891	6734	1231	1330	1082

validate our approach, we had to rely on external databases and some manual labelling. For the Identification problem, we could not find any external sources of information and benchmarks. To establish our ground-truth, we selected dot-decimal expressions uniformly randomly, and we used four different individuals for the labelling. To ensure testing fairness, we opted for balanced datasets, which led us to a corpus of 3200 labeled entries across all our forums.

For the Characterization problem, we make use of the VirusTotal site which maintains a database of malicious IP addresses by aggregating information from many other such databases. We also provide a second level of validation via manual inspection.

We create the ground truth by uniformly randomly selecting and assessing IP addresses from our forums. If VirusTotal and the manual inspection give it the same label, we add the addresses into our ground-truth. Finally, we again ensure that we create balanced sets for training and testing to ensure proper training and testing.

3 Overview of RIPEx

We represent the key components of our approach in addressing the Identification and Characterization problems. To avoid repetitions, we present at the end the Cross-Seeding approach, which we use in our solution to both problems.

3.1 The IP Identification Module

We describe our proposed method to identify IP addresses in the forum.

The IP Address Format. The vast majority of IP addresses in the forums follow the *IPv4* dot-decimal format, which consists of 4 decimal numbers in the range [0-255] separated by dots. We can formally represent the dot-decimal notation as follows: *IPv4* $[x_1.x_2.x_3.x_4]$ with $x_i \in [0-225]$, for $i = 1, 2, 3, 4$. Note that the newer *IPv6* addresses consists of eight groups of four hexadecimal digits, and our algorithms could easily extend to this format as well. Interestingly, we found a negligible number of *IPv6* addresses, and we opted to not focus on *IPv6* addresses here. For example, in WildersSecurity forum, we find 3891 *IPv4* addresses and only 56 *IPv6* addresses. At such small numbers, it is difficult to train and test a classifier. Thus, for the rest of this paper, IP address refers to *IPv4* addresses.

The Challenge: The Dot-Decimal Format is Not Enough. If IP addresses were the only numerical expressions in the forums with this format, the Identification problem could have been easily solved with straightforward text processing and Named-Entity Recognition (NER) tools, such as the Stanford NER models [6]. However, there is a non-trivial number of other numerical expressions, which can be misclassified as addresses. For example, we quote a real post: *"factory reset brings me to the Clockworkmod 2.25.100.15 recovery menu"*. where the structure *2.25.100.15* refers to the version of Android app *"Clockworkmod"*.

To this end, we propose a method to solve the IP Identification problem, a supervised learning algorithm. We first identify the features of interest as we discuss below. We then train a classifier using the Logistic Regression method gives the best results among the several methods using 10-fold cross validation on our ground-truth as we decribed in the previous section.

Feature Selection. We use three sets of features in our classification.

a. **Contextual Information: *TextInfo.*** Inspired by how a human would determine the answer, we focus on the words surrounding the dot-decimal structure. For example, the words *"server"* or *"address"* suggests that the dot-decimal is an address, while the words *"version"* or a software name, like *"Firefox"* suggests the opposite. At the same time, we wanted to focus on words close to the dot-decimal structure. Therefore, we introduce **Word-Range**, *W,* to determine the number of surrounding words before and after the dot-decimal structure that we want to consider in our classification. We use TF-IDF [14] to normalize the frequency of a word to better estimate its discriminatory value.

b. **The Numerical Values of the Dot-Decimal: *DecimalVal.*** We use the numerical value of the four numbers in the dot-decimal structure as features. The rationale is that non-addresses, such as software versions, tend to have lower numerical values. This insight was based on our close interaction with the data.

c. **The Combined Set: *Mixed.*** We combine the two feature sets to create in order to leverage their discriminating power.

Determining the Right Number of Context Words, Word-Range. We wanted to identify the best value of parameter Word-Range for our classification. In Fig. 2, we plot the classification accuracy, precision and recall, as we vary Word-Range, $W = 1, 2, 5 \ and \ 10$, for the WildersSecurity forum and using only the *TextInfo*. We see that using one to two words gives better results compared to using five and ten words. The explanation to this counter-intuitive result is that considering more words includes text that is not relevant for inferring the nature of a dot-decimal, which we verified manually.

Using Numerical Values *DecimalVal* Improves the Performance Significantly. In Fig. 3, we plot the classification accuracy of different features sets.

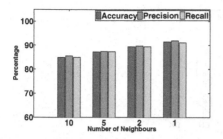

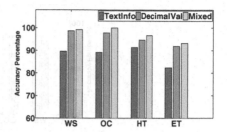

Fig. 2. Classification performance versus the number of words WordRange, W, in WildersSecurity.

Fig. 3. Classification accuracy for different features sets in 10-fold cross validation in four forums.

Recall that we are not able to include Darkode forum due to its limited number of non-IP dot-decimal expressions, as we saw in 3.1. We see that using *DecimalVal* features alone, we can get 94% overall accuracy and using both *DecimalVal* and *TextInfo*, we get 98% overall accuracy across our forums. Focusing on the IP address class, we see a an average precision of 95% using only *DecimalVal* and, 98% using both *DecimalVal* and *TextInfo*.

3.2 The IP Characterization Module

We develop a supervised learning algorithm to characterize IP addresses. Here, we assume that we have labeled data, and we discuss how we handle the absence of ground truth in Sect. 3.3. We first identify the appropriate set of features which we discuss below. We then train a classifier and find that the Logistic Regression method gives the best results among several methods that we evaluated. Due to space limitations, we show a subset of our results.

Features Sets for the Characterization Problem. We consider and evaluate three sets of features in our classification.

a. **Text Information of the Post: *PostText*.** We use the words and their frequency of appearance in the post. Here, we use the TF-IDF technique [14] again to better estimate the discriminatory value of a word by considering its overall frequency. In the future, we intend to experiment with sophisticated Natural Language Processing models for analyzing the intent of a post.

b. **The Contextual Information Set: *ContextInfo*.** We consider an extended feature set that includes both the *PostText* features, but also features of the author of the post. These features capture the behaviour of the author, including frequency of posting, average post length etc. These features were introduced by earlier work [8], with the rationale that profiling the author of a post can help us infer their intention and role and thus, improve the classification.

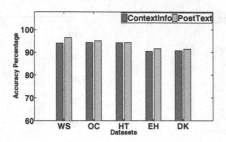

Fig. 4. Characterization: The effect of the features set on the classification accuracy with balanced testing data.

Characterization: 93% Precision with Training Data. We assess the performance of the Characterization classifier using the set of features above and by using the labeled data of each forum. We evaluate the performance using 10-fold cross validation. In Fig. 4, we show the accuracy of classification.

We can achieve 93% precision and 92% recall on average across all the forums. The results are shown in Fig. 4, where we report the results using the accuracy across both classes, given that we have balanced training datasets.

Selecting the *PostText* Feature Set. We see that, by using *PostText* features on their own, we obtain slightly better results. *PostText* feature achieves 94% accuracy on average, while using the *ContextInfo* results in 92% accuracy on average across all forums. Furthermore, text-based only features have one more key advantage: they can transfer between domains in a straightforward way. Therefore, we use the *PostText* features in the rest of the paper.

3.3 Transfer Learning with Cross-Seeding

In both classification problems, we face the following conundrum:

a. the classification efficiency is better when the classifier is trained with forum-specific ground-truth, but,
b. requiring ground-truth for a new forum will introduce manual intervention, which will limit the practical value of the approach.

We propose to do cross-forum learning by leveraging transfer learning approaches [5,12]. We use the terms *source* and *target* domain to indicate the two forums with the target forum not having ground-truth available. For both classification problems, we consider two solutions for classifying the target forum:

a. **Basic:** We use the classifier from the source forum on the target forum.
b. **Cross-Seeding:** We propose an algorithm that will help us develop a new classifier for the target forum by using the old classifier to create training data as we explain below.

Algorithm 1: Cross-Seeding. transfer learning between forums

---

1 CrossForum $(\mathcal{X}, \mathcal{Y})$:
2   Take the union of the features in forum $\mathcal{X}$ and $\mathcal{Y}$
3   Apply classifier from $\mathcal{X}$ on $\mathcal{Y}$
4   Select the high-confidence instances to create seed for $\mathcal{Y}$
5   Train a new classifier on $\mathcal{Y}$ based on the new seeds.
6   Apply the new classifier on $\mathcal{Y}$

Our Cross-Seeding Approach. We propose to create training data for the target forum following the four steps below, which are illustrated in Fig. 1 and outlined in Algorithm 1.

a. **Domain Adaptation.** The main role of this step is to ensure that the source classifier can be applied to the target forum. The main issue in our case is that the feature sets can vary among forums. Recall that, for both classification problems, we use the frequency of words and these words can vary among forums. We adopt an established approach that works well for text classification [5]: we take the union of the feature sets of the source and target forums. The approach seems to work sufficiently well in our case, as we see later.

b. **Creating Seed Information for the Target Forum.** Having resolved any potential feature disparities, we can now apply the classifier from the source forum to the target forum. We create the seeding data by selecting instances of the target domain, for which the classification confidence is high. Most classification methods provide a measure of confidence for each classified instance and we revisit this issue in Sect. 4.

c. **Training a New Classifier for the Target Forum.** Having the seed information, this is now a straightforward step of training a classifier.

d. **Applying the New Classifier on the Target Forum.** In this final step, we apply our newly-trained forum-specific classifier on the target forum.

4 Evaluation of Our Approach

We evaluate our approach focusing on the performance of Cross-Seeding for both the Identification and the Characterization problems.

Our Classifier. We use Logistic Regression as our classification engine, which performed better than several others, including SVM, Bayesian networks, and K-nearest-neighbors. In Cross-Seeding, we use the Logistic Regression's prediction probability with a threshold of 0.85 to strike a balance between sufficient confidence level and adequate number of instances above that threshold. We found this value to provide better performance than 0.8 and 0.9, which we also considered.

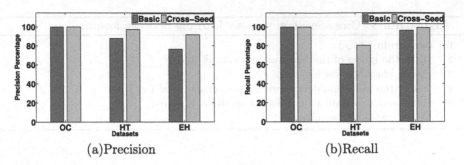

(a)Precision   (b)Recall

Fig. 5. Identification: Cross-Seeding improves both precision and recall. Using Wilder-sSecurity to classify OffensiveCommunity, HackThisSite, and EthicalHackers.

A. **The IP Identification Problem.** As we saw in Sect. 3.1, our classification approach exhibits 98% precision and 96% recall on average across all our sites, when we train with ground-truth for each forum.

    a. **Identification: 95% Precision with Cross-Seeding.** We show that our cross-training approach is effective in transferring the knowledge between domains. We use the classifier from WildersSecurity and we use it to classify three of the other forums. Note that we do not include Darkode in this part of the evaluation as it did not have sufficient data for testing (less than 15 non-address expressions in all its posts).

    In Fig. 5, we show the results for precision and recall of cross-training using Basic and Cross-Seeding. We see that Cross-Seeding improves *both* precision and recall significantly. For example, for HackThisSite, Cross-Seeding increases the precision from 57% to 79% and the recall from 60% to 78%.

    b. **Identification: Cross-Seeding Outperforms Basic.** Cross-Seeding improves the precision by 8% and recall by 7% on average for the experiment shown in Fig. 5. The average precision increased from 88% to 95% and the average recall increased from 85% to 97%.

B. **The IP Characterization Problem.** We evaluate our approach for solving the Characterization problem without per-forum training data. As we saw in Sect. 3.2, we can achieve 93% precision and 92% recall on average across all the forums, when we train with ground-truth for each forum.

    a. **Characterization: 88% Precision on Average with Cross-Seeding.** Using OffensiveCommunity as source, and we classify WildersSecurity, HackThisSite, EthicalHackers and Darkode as shown in Fig. 6. Our Cross-Seeding approach can provide 88% precision and 82% recall on average.

    b. **Characterization: Cross-Seeding Outperforms Basic.** We show that Cross-Seeding improves the classification compared to just reusing the classifier from another forum. In Fig. 6, we show the precision and recall of the two approaches. Using OffensiveCommunity as our source, we see that Cross-Seeding improves the precision by 28% and recall by

16% on average across the forums compare to the Basic approach. We also observe that the improvement is substantial: Cross-Seeding improves both precision and recall in all cases.

c. **Using More Source Forums Improves the Cross-Seeding Performance Significantly.** We quantify the effect of having more than one source forums in the classification accuracy of a new forum. We use EthicalHackers and WildersSecurity as our training forums, and we use Cross-Seeding for OffensiveCommunity, HackThisSite, and Darkode. First, we use the source forums one at a time and then both of them together. In Table 2, we show the average improvement of having two source forums over having one for each target website. Using two source forums increases the classification precision by 13% and the recall by 17% on average.

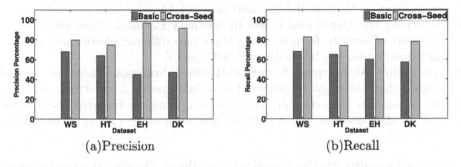

(a)Precision   (b)Recall

Fig. 6. Characterization: Cross-Seeding improves both precision and recall. Using OffensiveCommunity as source, we classify WildersSecurity, HackThisSite, EthicalHackers and Darkode.

Table 2. Average improvement of using EthicalHackers and WildersSecurity as sources together compared to each of them individually.

	OffensiveComm.	HackThisSite	Darkode	Average
Precision	3.3	20.5	17.8	13.2
Recall	8.3	6.4	38.8	17.8

Discussion: Source Forums and Training. How would we handle a new forum? Given the above observations, we would currently use all our five forums as sources for a new forum. Overall, we can argue that the more forums we have, the more we can improve our accuracy. However, we would like to point out that some forums are more "similar" and thus more suitable for cross-training. We will investigate how to best leverage a large group of source forums once we collect 20–25 more forums.

5   Related Work

We summarize related work clustered into areas of relevance.

a. **Extracting IP Addresses from Security Forums.** There two main efforts that focus on IP addresses and security forums [7,8] and neither provides the comprehensive solution that we propose here. The most relevant work [8] does not address the Identification problem, and sidesteps the problem of cross-forum training by assuming training data for each forum. The earlier work [7] focuses on the spatiotemporal properties of Canadian IP addresses in forums, but assumes that all identified addresses are suspicious.

b. **Extracting Other Information from Security Forums.** Various efforts have attempted to extract other types of information from security forums. A few recent studies identify malicious services and products in security forums by focusing on their availability and price [11,13].

c. **Studying the Users and Posts in Security Forums.** Other efforts study the users of security forums, group them into different classes, and identify their roles and social interactions [1,9,15,16,18].

d. **Analyzing Structured Security-Related Sources.** There are several studies that automate the extraction of information from structured security documents, extracting ontology and comparing the reported information, such as databases of vulnerabilities, and security reports from the industry [2,10].

Transfer Learning Methods and Applications. There is extensive literature on transfer learning [3–5] and several good surveys [12,17], which inspired our approach. However, to the best of our knowledge, we have not found any work that address the same domain-specific challenges or uses all the steps of our approach, which we described in 3.3.

6   Conclusion

We propose a comprehensive solution for mining malicious IP addresses from security forums. A novelty of our approach is it minimizes the need for human intervention. First, once it is initialized with a small number of security forums, it does not require additional training data for each new forum by using Cross-Seeding. Second, it addresses both the Identification and Characterization problems, unlike all prior work that we are aware of. We evaluate our method real data and we show that: (a) our Cross-Seeding approach works fairly well reaching precision above 85% on average for both classification problems, and (b) using more source forums increases the performance as one would expect.

Our future plans include: (a) collecting a large number of security forums, and (b) exploring additional transfer learning methods.

References

1. Abbasi, A., Li, W., Benjamin, V., Hu, S., Chen, H.: Descriptive analytics: examining expert hackers in web forums. In: ISI 2014, pp. 56–63 (2014)
2. Bridges, R.A., Jones, C.L., Iannacone, M.D., Goodall, J.R.: Automatic labeling for entity extraction in cyber security (2013) CoRR abs/1308.4941
3. Dai, W., Xue, G.R., Yang, Q., Yu, Y.: Co-clustering based classification for out-of-domain documents. In: KDD 2007, pp. 210–219, USA (2007)
4. Dai, W., Yang, Q., Xue, G.R., Yu, Y.: Boosting for transfer learning. In: ICML 2007, pp. 193–200, New York, NY, USA (2007)
5. Daume III, H.: Frustratingly easy domain adaptation. In: ACL 2007 (2007)
6. Finkel, J.R., Grenager, T., Manning, C.: Incorporating non-local information into information extraction systems by Gibbs sampling. In: ACL 2005 (2005)
7. Frank, R., Macdonald, M., Monk, B.: Location, location, location: mapping potential Canadian targets in online hacker discussion forums. In: EISIC 2016 (2016)
8. Gharibshah, J., Li, T.C., Vanrell, M.S., Castro, A., Pelechrinis, K., Papalexakis, E., Faloutsos, M.: InferIp: Extracting actionable information from security discussion forums. In: ASONAM 2017 (2017)
9. Holt, T.J., Strumsky, D., Smirnova, O., Kilger, M.: Examining the social networks of malware writers and hackers 6(1), 891–903 (2012)
10. Jones, C.L., Bridges, R.A., Huffer, K.M.T., Goodall, J.R.: Towards a relation extraction framework for cyber-security concepts (2015) CoRR abs/1504.04317
11. Motoyama, M., McCoy, D., Levchenko, K., Savage, S., Voelker, G.M.: An analysis of underground forums. In: IMC 2011, pp. 71–80, New York, NY, USA (2011)
12. Pan, S.J., Yang, Q.: A survey on transfer learning. IEEE Trans. Knowl. Data Eng. 22(10), 1345–1359 (2010)
13. Portnoff, R.S., Afroz, S., Durrett, G., Kummerfeld, J.K., Berg-Kirkpatrick, T., McCoy, D., Levchenko, K., Paxson, V.: Tools for automated analysis of cybercriminal markets. In: WWW 2017 (2017)
14. Ramos, J.: Using TF-IDF to determine word relevance in document queries. In: ICML 2003 (2003)
15. Samtani, S., Chinn, R., Chen, H.: Exploring hacker assets in underground forums. In: ISI 2015, pp. 31–36 (2015)
16. Shakarian, J., Gunn, A.T., Shakarian, P.: Exploring malicious hacker forums. In: Jajodia, S., Subrahmanian, V.S.S., Swarup, V., Wang, C. (eds.) Cyber Deception, pp. 261–284. Springer, Cham (2016). https://doi.org/10.1007/978-3-319-32699-3_11
17. Weiss, K., Khoshgoftaar, T.M., Wang, D.: A survey of transfer learning. J. Big Data 3(1), 9 (2016)
18. Zhang, X., Tsang, A., Yue, W.T., Chau, M.: The classification of hackers by knowledge exchange behaviors. Info. Syst. Front. 17(6), 1239–1251 (2015)

Pattern-Mining Based Cryptanalysis of Bloom Filters for Privacy-Preserving Record Linkage

Peter Christen[1(✉)], Anushka Vidanage[1], Thilina Ranbaduge[1],
and Rainer Schnell[2]

[1] Research School of Computer Science, The Australian National University,
Canberra, Australia
peter.christen@anu.edu.au
[2] Methodology Research Group, University Duisburg-Essen, Duisburg, Germany

Abstract. Data mining projects increasingly require records about individuals to be linked across databases to facilitate advanced analytics. The process of linking records without revealing any sensitive or confidential information about the entities represented by these records is known as privacy-preserving record linkage (PPRL). Bloom filters are a popular PPRL technique to encode sensitive information while still enabling approximate linking of records. However, Bloom filter encoding can be vulnerable to attacks that can re-identify some encoded values from sets of Bloom filters. Existing attacks exploit that certain Bloom filters can occur frequently in an encoded database, and thus likely correspond to frequent plain-text values such as common names. We present a novel attack method based on a maximal frequent itemset mining technique which identifies frequently co-occurring bit positions in a set of Bloom filters. Our attack can re-identify encoded sensitive values even when all Bloom filters in an encoded database are unique. As our experiments on a real-world data set show, our attack can successfully re-identify values from encoded Bloom filters even in scenarios where previous attacks fail.

Keywords: Privacy · Re-identification · Apriori algorithm · FPmax
Data linkage

1 Introduction

Applications in domains ranging from healthcare, business analytics and social science research all the way to fraud detection and national security increasingly require records about individual entities to be linked across several databases, and commonly across different organizations. Linked individual-level databases

This work was funded by the Australian Research Council under DP130101801 and DP160101934. Peter Christen likes to acknowledge the support of ScaDS Dresden/Leipzig (BMBF grant 01IS14014B), where parts of this work were conducted.

ⓒ Springer International Publishing AG, part of Springer Nature 2018
D. Phung et al. (Eds.): PAKDD 2018, LNAI 10939, pp. 530–542, 2018.
https://doi.org/10.1007/978-3-319-93040-4_42

allow to improve data quality and enrich data, and open novel ways of data analysis and mining not possible on a single database [4].

Due to the lack of common unique entity identifiers (such as social security numbers or patient identifiers) across databases, linking records of individuals most often requires identifying personal details such as names, addresses and dates of birth [4]. However, in many application areas growing concerns about privacy and confidentiality increasingly limit the use of sensitive personal details for linking databases across organizations [20].

Since the 1990s, the research area of *privacy-preserving record linkage* (PPRL) has aimed to develop techniques for linking records that correspond to the same entities across databases while protecting the privacy and confidentiality of these entities [20]. The general idea behind PPRL is to encode sensitive identifying attribute values (such as names, addresses, and dates of birth) and to conduct the linkage using these encoded values. At the end of the PPRL process the organizations involved only learn which of their records are matched with records from the other databases according to some decision model, but they cannot learn any sensitive information about the other databases, while any external attacker is not able to learn anything about these databases at all [20].

As surveyed by Vatsalan et al. [20], various encoding techniques have been proposed for PPRL. They can be categorized into secure multi-party computation (SMC) and perturbation based techniques, as well as hybrid techniques that combine aspects of both. While SMC techniques are accurate and provably secure, they often incur high computation and communication costs. Perturbation based techniques, on the other hand, are generally efficient and provide a trade-off between linkage quality, scalability, and privacy.

One perturbation technique that has attracted much interest is Bloom filter (BF) encoding [2,17]. As we describe in detail in the following section, a BF is a bit array (initialized to 0) into which elements of a set are mapped into by setting those positions to 1 that are selected by a set of hash functions. For PPRL, the elements of these sets are commonly character q-grams as extracted from the string values used for linking. The number of common 1-bits between two BFs approximates the number of common encoded q-grams and allows the calculation of the similarity between two BFs, as is illustrated in Fig. 1.

BF encoding has shown to allow accurate and efficient PPRL of large databases, and first practical PPRL systems based on BF encoding are now being deployed [3,16]. However, recent work has shown that BFs can be successfully attacked with the aim to re-identify the sensitive values encoded in them [5,10–13,15]. These cryptanalysis attacks assume that some bit patterns occur many times in an encoded BF database, allowing a mapping of frequent bit patterns to frequent plain-text values (such as common names) to identify the BF bit positions that possibly can correspond to certain q-grams in a plain-text value. Most existing attacks also require knowledge of certain parameters used in the BF encoding process, or they have high computational costs making them impractical for attacking large databases.

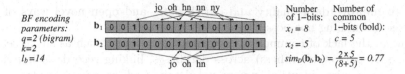

Fig. 1. An example Dice coefficient similarity calculation of the first names 'johnny' and 'john' encoded in BFs, as described in Sect. 2. Hash collisions are shown in italics.

Contributions: We present a novel attack method that employs a maximal frequent itemset based pattern-mining approach (combining Apriori [1] and FPMax [7]) to identify frequently co-occurring bit positions in a BF database that encode frequent q-grams in a plain-text database. Our attack can re-identify q-grams and plain-text values even when each BF in an encoded database is unique, and when an attacker has no knowledge of any parameters used in the BF encoding process. Our evaluation on real-world data sets shows that our attack can re-identify plain-text values even when several attributes have been encoded into a BF, a situation where previous attacks would not be successful.

2   Background and Related Work

We now describe Bloom filter (BF) encoding for PPRL in more detail, and then provide an overview of existing cryptanalysis attacks on encoded BF.

Bloom Filter Encoding for PPRL: BF encoding is a popular privacy technique for PPRL due to its ability to calculate similarities between BF encoded string [6,17,20] and numerical [9,19] values, as well as its efficiency and simplicity of implementation. As the recent study by Randall et al. [16] has shown, in real-world applications PPRL based on encoded BFs can achieve linkage quality similar to traditional linkage methods on unencoded values.

BFs were proposed by Bloom [2] to efficiently represent sets and test for set membership. A BF $\mathbf{b}$ is a bit array of length l_b initialized to 0. k independent hash functions, $h_1, \ldots, h_k$, are used to map the elements s in a set $\mathbf{s}$ into $\mathbf{b}$ by setting the bit positions $\mathbf{b}[h_j(s)] = 1$, with $1 \leq j \leq k$, $1 \leq h_j(s) \leq l_b$, and $\forall s \in \mathbf{s}$.

In PPRL, where most attributes used for linking contain strings, the set $\mathbf{s}$ can be generated from character q-grams (sub-strings of length q [4]) extracted from a string value using a sliding window approach [17]. As illustrated in Fig. 1, the Dice coefficient (which is insensitive to many matching zeros) [4] can then be used to calculate the similarity between two BFs $\mathbf{b}_1$ and $\mathbf{b}_2$, as: $sim_D(\mathbf{b}_1, \mathbf{b}_2) = 2c/(x_1 + x_2)$, where c is the number of common bit positions that are set to 1 in both BFs, and x_1 and x_2 are the number of bit positions set to 1 in $\mathbf{b}_1$ and $\mathbf{b}_2$, respectively. Note that the hashing process can lead to collisions (where several q-grams are hashed to the same position), leading to false positives [14,17].

Different methods of how to encode records into BFs have been proposed. Generating one BF per attribute, known as *attribute level* BF (ABF), is one approach which has the advantage of allowing one similarity to be calculated per attribute. However, ABFs are susceptible to attacks due to the frequency distribution of common attribute values [5]. Alternatively, two methods that hash several attribute values of a record into one combined BF are the *cryptographic long term key* (CLK) [18] and *record level* BFs (RBF) [6]. With CLK, q-grams from several attributes are hashed into one BF, while in RBF the values from different attributes are first hashed into individual ABFs, and then bits are selected from each ABF into one RBF according to weights assigned to attributes.

A double hashing scheme was initially proposed for BF encoding for PPRL [17], where the k hash functions are based on the sum of the integer representation of two independent hash functions. This approach is however vulnerable to attacks [10,15], as discussed below. An alternative to prevent against these attacks is random hashing, where an integer representation of an element s to be hashed is used as the seed of a random generator used for hashing [18].

Once the databases to be linked are encoded into BFs by their owners, they are either sent to a linkage unit to calculate the similarity between pairs of BFs and classify them as matches or non-matches, or BFs are exchanged among the database owners to distributively calculate the similarities between BFs [17,20].

Cryptanalysis Attacks on Bloom Filter Based PPRL: Because BF encoding is now being used in practical PPRL applications [3,16], it is highly important to identify the limits of this technique with regard to the privacy protection it provides. As we now describe, BFs are prone to different attacks.

The first attack method proposed by Kuzu et al. [6,11] was based on a constraint satisfaction problem (CSP) solver which assigns values to variables such that a set of constraints is satisfied. The attack is achieved by a frequency alignment of a set of BF encodings of records and sensitive plain-text values, where it requires access to a global database where the encoded records and their frequencies are drawn from. The attack was applied on a real patient database where it was successful in re-identifying four out of 20 frequent names correctly [12].

More recently, Niedermeyer et al. [15] proposed an attack on BFs based on the counts of q-grams extracted from frequent German surnames. From 7,580 unique surnames encoded into 10,000 BFs, the authors manually re-identified the 934 most frequent ones. This work was extended by Kroll and Steinmetzer [10] into a cryptanalysis of several attributes, which was able to re-identify 44% of plain-text values correctly. Both these attack methods are however only applicable with the double hashing approach used by Schnell et al. [17].

Christen et al. [5] recently proposed a novel efficient attack method that does not require any knowledge of the BF encoding function and its parameters used. The method aligns frequent BFs with frequent plain-text values and identifies sets of q-grams that could have been hashed to certain bit positions or not. The attack was successfully applied on large real databases and was able to correctly re-identify the most frequent plain-text values within a few minutes.

The most recent attack by Mitchell et al. [13] depends on the strong assumption that the adversary knows all parameters used in the BF encoding process. First, a brute-force attack is used to identify all possible q-grams encoded in a

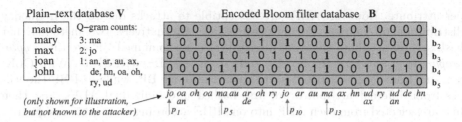

Fig. 2. Example of our proposed attack method, as described in Sect. 3. Using a frequent pattern-mining approach, we first identify that bit positions p_5 and p_{13} have co-occurring 1-bits in the *same three BFs* (b_1, b_3 and b_4) and therefore must encode 'ma' which is the only q-gram that occurs in *three plain-text values*. Next, we find that positions p_1 and p_{10} must encode 'jo' as they have co-occurring 1-bits in the *same two BFs* (b_2 and b_5) and 'jo' is the only q-gram that occurs in *two plain-text values*. Based on the identified q-grams and their bit positions, we learn that BFs b_2 and b_4 can only encode 'john' and 'joan', while b_1, b_3 and b_4 can encode 'maude', 'mary' or 'max'.

BF, then a graph is built which represents possible plain-text values that can be generated from the identified q-grams in a BF. The evaluation on real-world databases showed a 76.8% accuracy of correct one-to-one re-identifications.

The drawbacks of existing attacks on encoded BFs for PPRL are that they require knowledge about certain parameters used during BF encoding, are computationally expensive, and/or are only applicable if there are BFs and plain-text values that occur frequently such that their frequency alignments can be exploited. Our novel attack, presented next, overcomes these limitations.

3   Pattern-Mining Based Cryptanalysis Attack

We now discuss the ideas behind our attack method, as we illustrate in Fig. 2. In Sects. 3.1 and 3.2 we then describe the two main phases of our method in detail, and analyze their complexities. For notation we use bold letters for BFs, sets and lists (with upper-case bold letters for sets/lists of BFs/sets/lists) and normal type letters for integers and strings. Lists are shown with square and sets with curly brackets, where lists have an order while sets do not.

Previous attack methods exploit the bit patterns in and between BFs that occur frequently in an encoded BF database. Our new method is the first to exploit the co-occurrence between BF bit positions without requiring frequent BFs. As with existing attacks on BFs for PPRL, we assume an attacker has access to an encoded BF database, **B**, where it is unknown which plain-text value(s) are encoded in a BF; and a plain-text database, **V**, that contains values from one or several attributes. An attacker can guess which attributes are encoded in **B** based on the distribution of the number of 1-bits in BFs in **B** [18]. However, unlike with other attacks [10–13,15], the attacker does not require any information used in the BF encoding, such as the number and type of hash functions.

In the first phase of our attack (Sect. 3.1), we identify sets of frequently co-occurring bit positions (columns) in $\mathbf{B}$ that can encode q-grams that are frequent in $\mathbf{V}$. In the second phase (Sect. 3.2) we then re-identify possible plain-text values $v \in \mathbf{V}$ that can be encoded in a BF $\mathbf{b} \in \mathbf{B}$ based on these identified frequent q-grams. Our attack exploits the way BFs are constructed as follows:

Proposition 1. *Assuming a q-gram q occurs in $n_q < n$ records in a plain-text database $\mathbf{V}$ that contains $n = |\mathbf{V}|$ records, and $k \geq 1$ independent hash functions are used to encode q-grams from $\mathbf{V}$ into the encoded database $\mathbf{B}$ of n BFs, i.e. $|\mathbf{V}| = |\mathbf{B}|$. Then, (i) each BF bit position that can encode q must contain a 1-bit in at least n_q BFs in $\mathbf{B}$, and (ii) if $k > 1$ then up to k bit positions must contain a 1-bit in the same subset of BFs $\mathbf{B}_q \subseteq \mathbf{B}$, with $n_q = |\mathbf{B}_q|$, that encode q.*

Proof. For (i), based on the BF construction principle [2,17], if a BF contains a 0-bit in a bit position p where any of the k hash functions have mapped a q-gram q into, then this BF cannot encode q. Formally, $\exists\, h_j, 1 \leq j \leq k : (h_j(q) = p) \wedge (\mathbf{b}[p] = 0) \Rightarrow \mathbf{b}$ cannot encode q. For (ii), given two distinct BF bit positions, p_x and p_y (with $1 \leq p_x, p_y \leq l_b$ and $p_x \neq p_y$), for a given BF $\mathbf{b}$, if $\mathbf{b}[p_x] \neq \mathbf{b}[p_y]$ then $\mathbf{b}$ cannot encode a q-gram q because $\forall h_i, h_j, 1 \leq i \leq k, 1 \leq j \leq k, i \neq j :$ $(h_i(q) = p_x \wedge h_j(q) = p_y) \Rightarrow (\mathbf{b}[p_x] = 1) \wedge (\mathbf{b}[p_y] = 1)$.

Because of possible collisions of the k hash functions used to map q-grams into BFs [14], potentially less than k bit positions will encode a certain q-gram, and thus less than k bit positions might be co-occurring frequently. We can calculate the probability of this to happen using the birthday paradox [14], which, for example for the commonly used BF settings $l_b = 1,000$ and $k = 30$ [18], leads to a probability of 0.64 of no collision for a single q-gram, and a probability of less than 0.07 of more than one collision. For $k = 20$ the probability of no collision is 0.83 while for $k = 10$ it is 0.96. Our attack should therefore be able to clearly identify frequently co-occurring bit positions that encode the same q-gram with high accuracy, as we validate experimentally on a real data set in Sect. 4.

3.1 Identifying Co-occurring Bit Positions in Bloom Filters

The first phase of our attack, as detailed in Algorithm 1, identifies the sets of co-occurring bit positions (columns) in the BF database $\mathbf{B}$ that correspond to the frequent q-grams in the plain-text database $\mathbf{V}$.

The algorithm first converts $\mathbf{B}$ from its row storage (one BF per record) into a column-wise format $\mathbf{B}_c$ of l_b bit arrays each of length $|\mathbf{B}|$ to allow efficient access to individual bit positions. Then, in line 3, the plain-text values $v \in \mathbf{V}$ are converted into one set of q-grams per record, where $\mathbf{Q}$ is the list of q-gram sets of all records in $\mathbf{V}$. Various lists and sets are then initialized, as is the queue Q in line 8 with the first partition that contains all BFs and all bit positions. A partition is a subset of BFs (rows) and bit positions (columns) in $\mathbf{B}$.

The main loop of the algorithm starts in line 9, where the largest partition in the queue Q will be processed. In each iteration, we use a tuple containing c_i

(the column filter of which bit positions in $\mathbf{B}_c$ to consider), $\mathbf{r}_i$ (the row filter of which BFs in $\mathbf{B}_c$ to consider), and $\mathbf{m}_i$ and $\mathbf{n}_i$ (the set of must and cannot contain q-grams, respectively, that a q-gram set in $\mathbf{Q}$ must/cannot contain in order to be considered in this iteration). The function **GetTwoMostFreqQGrams()** called in line 11 returns the two q-grams, q_1 and q_2, and their respective frequencies, f_1 and f_2 (with $f_1 \geq f_2$), that occur in most q-gram sets in $\mathbf{Q}$, conditional on only considering q-gram sets in $\mathbf{Q}$ that contain all q-grams in $\mathbf{m}_i$ and no q-gram from $\mathbf{n}_i$.

The percentage difference between f_1 and f_2 is then calculated as d_i in line 12, and only if d_i is at least the user provided minimum threshold d (i.e. q-gram q_1 occurs a certain times more often than q_2) will the current partition i be processed. This ensures q_1 is clearly more frequent than q_2 in the partition for it to be assigned to the set of co-occurring bit positions $\mathbf{f}_i$ as described next.

Algorithm 1: *Identify frequent q-grams co-occurring with frequent bit positions* – Phase 1

Input:
- **B**: List of Bloom filters (BFs) from the sensitive encoded database, one BF per record
- **V**: List of plain-text values from a public database, one string value per record
- l_q: Length of sub-strings to extract from plain-text values
- d: Minimum percentage difference between two most frequent q-grams in a partition
- m: Minimum partition size (a subset of BFs and bit positions in **B**) as a number of BFs

Output:
- **F**: List of frequent q-grams and their identified sets of co-occurring BF bit positions
- $\mathbf{A}^+$, $\mathbf{A}^-$: Lists of must have and cannot have q-gram sets assigned to each BF in **B**

```
1:   Bc = ConvColWise(B)              // One bit array per BF bit position (column) for efficient access
2:   lb = |Bc|                        // Number of BF bit positions (BF length)
3:   Q = GenQGramSets(V, lq)          // Convert plain-text values into q-gram sets, one set per record
4:   F = [], A⁺ = [], A⁻ = []         // Initialize lists to be generated
5:   r = {b : 1 ≤ b ≤ |B|}            // Initialize row partition filter (so all BFs are considered)
6:   c = {p : 1 ≤ p ≤ lb}             // Initialize column filter (so all BF bit positions are considered)
7:   m = {}, n = {}                   // Initialize empty sets of must and cannot include q-grams
8:   Q = [(|B|, c, r, m, n)]          // Initialize queue with first partition (the full BF data set)
9:   while |Q| > 0 do:                // Main loop: As long as the queue is not empty
10:      (psi, ci, ri, mi, ni) = Q.pop()   // Start iteration i: Get first tuple from queue
11:      q1, f1, q2, f2 = GetTwoMostFreqQGrams(Q, mi, ni)   // Most frequent in partition
12:      di = 2(f1 − f2)/(f1 + f2) · 100   // Percentage difference between two most frequent q-grams
13:      if di ≥ d then:              // Only continue if large enough percentage difference
14:         si = |B| · (f1 + f2)/(2|V|)   // Minimum support for frequent pattern-mining
15:         fi = GetLongFreqCoOccurBitPos(Bc, si, ci, ri)   // Run pattern-mining
16:         if fi ≠ ∅ then:           // A set of frequent co-occurring bit positions was identified
17:            F[q1] = fi             // Assign bit positions to most frequent q-gram q1 (from line 11)
18:            ci = ci \ fi           // Remove identified bit positions so they are not considered anymore
19:            ri⁺, ri⁻ = UpdateRowFilter(fi, ri)   // Update row filter based on bit positions in fi
20:            ∀b ∈ r⁺ : A⁺[b] = A⁺[b] ∪ {q1}   // Assign q1 to must have q-gram sets of BFs
21:            ∀b ∈ r⁻ : A⁻[b] = A⁻[b] ∪ {q1}   // Assign q1 to cannot have q-gram sets of BFs
22:            if |ri⁺| ≥ m then:     // Partition containing q1 is large enough
23:               Q.add((|ri⁺|, ci, ri⁺, mi ∪ {q1}, ni))   // New partition of BFs containing q1
24:            if |ri⁻| ≥ m then:     // Partition not containing q1 is large enough
25:               Q.add((|ri⁻|, ci, ri⁻, mi, ni ∪ {q1}))   // New partition of BFs not containing q1
26:            Q.sort()               // Sort queue by partition size with largest first
27: return F, A⁺, A⁻
```

In line 14 the average frequency of q_1 and q_2 is converted into a support count s_i of required 1-bits in $\mathbf{B}_c$ to be used for frequent pattern-mining. This is based on Proposition 1, because any set of co-occurring bit positions in $\mathbf{B}_c$ that potentially can encode q_1 must have 1-bits in at least s_i common BFs (rows) in $\mathbf{B}_c$. The function **GetLongFreqCoOccurBitPos()** (line 15) employs the

Apriori [1] and FPmax [7] algorithms on the bit positions in $\mathbf{B}_c$, where bit positions correspond to items and sets of co-occurring bit positions to itemsets. The aim of frequent pattern-mining is to find the longest set (likely of size k) of co-occurring bit positions, $\mathbf{f}_i$, that have a 1-bit in at least s_i common BFs.

Because the length of the longest pattern is likely large (values of $15 < k < 30$ have been used [6, 17, 18]), employing only the Apriori algorithm would generate too many candidate bit position sets. On the other hand, running the FPmax algorithm would generate a very large FPtree with a number of branches in the order of $O(|\mathbf{B}|)$. To reduce the size of the tree generated by FPmax, we therefore first run Apriori to find those bit positions that occur in frequent triplets (i.e. frequent 3-itemsets) of co-occurring bit positions that have a 1-bit in at least s_i common BFs. The FPmax algorithm is then run on only those bit positions, and only the longest set of found frequent bit positions is returned in $\mathbf{f}_i$.

The parameters $\mathbf{c}_i$ and $\mathbf{r}_i$ are the column and row filters that select a subset of BF bit positions (columns) and rows (BFs) from $\mathbf{B}_c$ to be considered in this partition i (as described below). If a non-empty set of co-occurring bit positions $\mathbf{f}_i$ is returned, it is assigned in line 17 to the identified most frequent q-gram q_1 of this partition, as we assume that the bit positions in $\mathbf{f}_i$ must encode q_1.

In line 18, we remove the identified bit positions $\mathbf{f}_i$ from the set $\mathbf{c}_i$ of positions to be considered in the following iterations. In the function **UpdateRowFilter()**, based on the current set $\mathbf{r}_i$ of BFs that have been considered in this iteration, we generate the two new subsets $\mathbf{r}_i^+$ of BFs that do contain q_1 (have 1-bits in all positions in $\mathbf{f}_i$) and $\mathbf{r}_i^-$ of BFs that cannot contain q_1, where $\mathbf{r}_i^- = \mathbf{r}_i \setminus \mathbf{r}_i^+$. We then add the frequent q-gram q_1 to the sets of must have (line 20) and cannot have (line 21) q-gram sets, $\mathbf{A}^+$ and $\mathbf{A}^-$, for all BFs in $\mathbf{r}_i^+$ and $\mathbf{r}_i^-$.

Finally, we generate two new partitions (if their size is at least the minimum partition size, m), where in line 23 the new partition contains those BFs that contain the frequent q-grams q_1 (so q_1 is added to the must contain q-gram set $\mathbf{m}_i$) and in line 25 the new partition contains those BFs that do not contain q_1 (and therefore q_1 is added to the set $\mathbf{n}_i$ of not contained q-grams). In line 26 we sort the queue q such that the largest partition is first.

To estimate the complexity of Algorithm 1, we assume $n = |\mathbf{B}| = |\mathbf{V}|$ is the number of BFs and plain-text values, respectively. The initialization steps and function calls in lines 1, 3, 11 and 19 are of complexity $O(n)$ as they require linear scans over $\mathbf{B}$ or $\mathbf{V}$. Assuming in each iteration of the main loop k bit positions are assigned to a frequent q-gram in line 17, then the expected number of iterations is $O(l_b/k)$. The complexities of the Apriori and FPMax algorithms in line 15 are known to be linear in the number of transactions (in our case BFs) and quadratic in the number of items (bit positions) [8] and therefore of $O(l_b^2 n)$. In the worst case the FPmax tree has a height of l_b and contains n branches. The overall complexity of Algorithm 1 is therefore $O(l_b^3 n/k)$.

3.2   Plain-Text Value Re-identification

As detailed in Algorithm 2, the second phase of our attack aims to re-identify plain-text values from $\mathbf{V}$ that could have been encoded into BFs in $\mathbf{B}$ according to the lists of must have and cannot have q-gram sets for BFs, $\mathbf{A}^+$ and $\mathbf{A}^-$ (from Algorithm 1). The algorithm only considers plain-text values and BFs that contain at least n_m must have q-grams assigned to them, because considering a single or only a few frequent q-grams would result in too many possible plain-text values that could match a BF (for example, nearly 4,000 surnames from our experimental data set of 224,061 records contain the q-gram 'sm').

The algorithm consists of three main steps, where in the first (lines 3 to 6) we find all BFs from $\mathbf{A}^+$ that (according to phase 1 of our attack) contain at least n_m identified q-grams. We build an inverted index, $\mathbf{I}_B$ where q-gram sets $\mathbf{q}_b$ are index keys, each with a list of BFs that contain $\mathbf{q}_b$ (line 6). In the second step (lines 7 to 10), the function **GetLongQGramSetInVal()** finds for each value $v \in \mathbf{V}$ the longest q-gram set $\mathbf{q}_v$ from $\mathbf{I}_B$. If a value v contains at least n_m identified q-grams then we add it to the inverted index list of $\mathbf{q}_v$ in $\mathbf{I}_V$.

Algorithm 2: *Re-identify plain-text values in Bloom filters* – Phase 2

Input:
- **V:**   List of plain-text values from a public database, one string value per record
- **$\mathbf{A}^+$, $\mathbf{A}^-$:** Lists of must have and cannot have q-gram sets assigned to each BF in **B**
- n_m: Minimum number of identified q-grams in a BF from **B**

Output:
- **R:** List of re-identified plain-text values from **V** for BFs from **B**

```
 1: I_B = [], I_V = []           // Initialize inverted indexes of BFs and plain-text values
 2: R = []                       // Initialize list of re-identified plain-text values to be generated
 3: for b ∈ A^+ do:              // Step 1: Find all BFs in A^+ with enough identified q-gram sets
 4:    q_b = A^+[b]              // Get the set of must have q-grams for this BF
 5:    if |q_b| ≥ n_m then:      // There are enough must have q-grams for this BF
 6:       I_B[q_b].add(b)        // Add the BF identifier to the inverted index list of this q-gram set
 7: for v ∈ V then:              // Step 2: Find longest identified q-gram set for plain-text values
 8:    q_v = GetLongQGramSetInVal(I_B, v)   // Get longest known q-gram set in value v
 9:    if |q_v| ≥ n_m then:      // There are enough must have q-grams for this value
10:       I_V[q_v].add(v)        // Add the value to the list of this q-gram set in the inverted index
11: for q_b ∈ I_B do:            // Step 3: Assign identified values to BFs
12:    if q_b ∈ I_V then:        // Q-gram set occurs both in BFs and plain-text values
13:       for b ∈ I_B[q_b] do:   // All BFs that contain this q-gram tuple
14:          for v ∈ I_V[q_b] do:   // All plain-text values that contain this q-gram tuple
15:             if ∀q ∈ A^-[b] : q ∉ v then:  // If the value does not contain any cannot have q-grams
16:                R[b] = R[b] ∪ {v}     // Add to set of re-identified values for BF identifier b
17: return R
```

In the final step (lines 11 to 16), we loop over all q-gram sets $\mathbf{q}_b$ that have both BFs from $\mathbf{A}^+$ and plain-text values from $\mathbf{V}$. For each $\mathbf{q}_b$ and each of its BF identifiers b from $\mathbf{I}_B$, we find all possible corresponding values v from $\mathbf{I}_V$ that do not contain any cannot have q-grams for this BF according to $\mathbf{A}^-$ (line 15). All possible values v are added to the set $\mathbf{R}[b]$ of re-identified values for b.

The computational complexity of the first two steps of Algorithm 2 is $O(n)$ because they loop over $\mathbf{A}^+$ and $\mathbf{V}$, respectively. In the third step, we loop over the identified q-gram tuples, and for each over its associated BFs and plain-text values. The maximum number of unique q-gram tuples in the inverted indexes $\mathbf{I}_B$ and $\mathbf{I}_V$ will be n, which would be the case where every value and every BF would contain a different q-gram tuple. In this case, each list in $\mathbf{I}_B$ and $\mathbf{I}_V$ would contain one BF identifier. The minimum number of q-gram tuples would be 1, the case where all values and BFs contain the same q-gram tuple. In this case the length of the corresponding list in $\mathbf{I}_B$ and $\mathbf{I}_V$ is n BF identifiers. Based on this, the overall complexity of step 3 of Algorithm 2 is $O(n^2)$.

4 Experiments and Results

We evaluated our attack method using real data from the North Carolina Voter Registration (NCVR) database (http://dl.ncsbe.gov/data/). We used one snapshot of NCVR from April 2014 as $\mathbf{B}$ and another snapshot from June 2014 as $\mathbf{V}$. We extracted pairs of records that correspond to the same voter but had name and/or address changes over time, resulting in two files of 222, 251 and 224, 061 records, respectively. Using the CLK approach [18] discussed in Sect. 2, we encoded combinations of between two and four of the attributes *first name*, *last name*, *street address* and *city* into BFs. For combinations of three and four attributes, all plain-text values in $\mathbf{V}$ and bit patterns in $\mathbf{B}$ were unique.

We used the following BF encoding settings: $q = 2$, $l_b = 1,000$, double and random hashing [18] as described in Sect. 2, and different values for k (top row in Fig. 3). We calculate the optimal number, *opt*, of hash functions such that the average number of 1-bits in a BF is 50% to minimize the false positive rate [19]. In Algorithm 1 we set the values for the minimum percentage difference as $d = [1.0, 5.0]$ and the minimum partition size as $m = [2,000, 10,000]$ (bottom row in Fig. 3); and in Algorithm 2 we set the minimum q-gram tuple size $n_m = 3$, as these values provided good results in setup experiments.

We present the quality of the identified frequent q-grams from Algorithm 1 as the precision and recall of how many bit positions were correctly identified for a q-gram in $\mathbf{F}$ averaged over all q-grams in $\mathbf{F}$. For Algorithm 2, we evaluated the quality of re-identified values in $\mathbf{R}$ as the percentages of (1) *exact* matches of a plain-text value with the true value encoded in a BF, (2) *partial* matches where not all words matched (for example, *first* and *last name* matched but *city* was different), and (3) *wrong* matches. We only considered BFs with 10 or less plain-text values assigned to them in $\mathbf{R}$, and we separately present averaged results for *1-to-1* ($|\mathbf{R}[b]| = 1$) and *1-to-many* ($1 < |\mathbf{R}[b]| \leq 10$) re-identifications.

We compared our attack method with the recently proposed attack by Christen et al. [5] (the only other attack that does not require knowledge of the BF encoding parameters) which aligns frequent BFs and plain-text values to allow re-identification of the most frequent plain-text values. We implemented both attack methods using Python 2.7 and ran experiments on a server with 64-bit Intel Xeon 2.4 GHz CPUs, 128 GBytes of memory and running Ubuntu 14.04. The programs and data sets are available from: https://dmm.anu.edu.au/pprlattack.

Discussion: In Fig. 3 we show the results for the first phase of our attack (Algorithm 1). As can be seen, both precision and recall of the identified frequent q-grams are very high, above 0.88, for all settings of parameters. This validates that

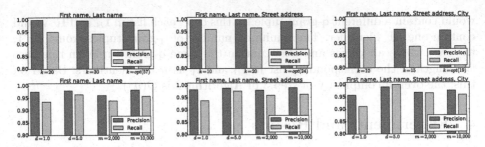

Fig. 3. Precision and recall results for the re-identified frequent q-grams from Algorithm 1 (see Sect. 3.1), with different numbers of hash functions k (top), and different minimum partition sizes m and different q-gram frequency percentage differences d (bottom).

our attack can successfully identify bit positions of q-grams with high accuracy even when no frequent BFs are available in an encoded database. Both precision and recall decrease slightly as more attributes are encoded into BFs, which is due to the increased number of unique encoded q-gram sets that make accurate frequent pattern-mining more difficult. A smaller difference d between the two most frequent q-grams lowers re-identification accuracy, because the chance of a wrong frequent q-gram being identified is increased. The minimum partition size, m, seems to have no strong effect upon the q-gram re-identification accuracy.

Table 1. Re-identification percentages of exact (E), partial (P) and wrong (W) matches of plain-text values from Algorithm 2 (Sect. 3.2), averaged over the settings used in Fig. 3.

		Two attributes			Three attributes			Four attributes		
		E	P	W	E	P	W	E	P	W
Christen et al. [5]	1-to-1	5.0	0	95.0	0	0	100	0	0	100
	1-to-many	6.0	0	94.0	0	0	100	0	0	100
Our approach	1-to-1	20.7	30.9	48.4	0.2	61.0	38.8	0.5	73.2	26.3
	1-to-many	27.5	46.5	26.0	0.4	83.5	16.1	0.5	87.9	11.6

As Table 1 shows, the re-identification results for the second phase (Algorithm 2) of our attack has led to around 51% to 74% of 1-to-1 matches to be exact or partially correct, which means in the majority of cases where only one plain-text value $v \in \mathbf{V}$ was identified to match one BF $\mathbf{b} \in \mathbf{B}$, an attacker has information that likely allows re-identification of the individual represented by v. If between 2 and 10 values v match a BF, then the over 74% of re-identifications are exact or partially correct. The small values of exact matches for three and four attributes is because less than $1,800$ of over $222,200$ ($<1\%$) combined values between the two NCVR snapshots are exact matches. In future work, we will concentrate on improving the accuracy and efficiency of this second phase of our approach.

As can also be seen from Table 1, the frequency based attack by Christen et al. [5] is only able to correctly re-identify a very small percentage of values when two attributes are encoded into BFs, because with three and four attributes no frequency information is available that could be exploited by this attack.

Our experiments show that basic BFs, even when each BF in an encoded database is unique, can successfully be attacked by identifying frequently co-occurring BF bit positions. These results highlight the need for improved BF encoding, as well as new PPRL encoding methods that do not exhibit the weaknesses of basic BFs that allow the re-identification of encoded values.

5   Conclusions and Future Work

We have presented a pattern-mining based attack method on BF encoding as used for PPRL. Our attack can successfully re-identify encoded q-grams and plain-text values even when all BFs in an encoded database are unique. Given that BF based PPRL is now employed in real-world applications [3,16], it is vital to study the limits of BF encoding. We believe our attack is important for PPRL because it allows data custodians to understand security flaws in BF encodings and ensure their encoded databases are not vulnerable to such attacks.

As future work we plan to improve the second phase of our attack by analyzing the differences in bit patterns between BFs to identify additional encoded q-grams which will allow an improved re-identification of plain-text values. We furthermore aim to formalize different attack scenarios for PPRL, and to conduct extensive experiments across a variety of parameter settings (such as different values of q, k, and l_b) and different encoding methods, and using data sets from a variety of domains to identify the limitations of our attack method. Finally, we will also explore how BF *hardening* techniques, such as balancing and XOR-folding [18], will influence the feasibility of our pattern-mining based attack.

References

1. Agrawal, R., Srikant, R.: Fast algorithms for mining association rules. In: VLDB, Santiago de Chile (1994)
2. Bloom, B.: Space/time trade-offs in hash coding with allowable errors. Commun. ACM **13**(7), 422–426 (1970)
3. Boyd, J.H., Randall, S.M., Ferrante, A.M.: Application of privacy-preserving techniques in operational record linkage centres. In: Gkoulalas-Divanis, A., Loukides, G. (eds.) Medical Data Privacy Handbook, pp. 267–287. Springer, Cham (2015). https://doi.org/10.1007/978-3-319-23633-9_11
4. Christen, P.: Data Matching - Concepts and Techniques for Record Linkage, Entity Resolution, and Duplicate Detection. Springer, Heidelberg (2012). https://doi.org/10.1007/978-3-642-31164-2

5. Christen, P., Schnell, R., Vatsalan, D., Ranbaduge, T.: Efficient cryptanalysis of bloom filters for privacy-preserving record linkage. In: Kim, J., Shim, K., Cao, L., Lee, J.-G., Lin, X., Moon, Y.-S. (eds.) PAKDD 2017. LNCS (LNAI), vol. 10234, pp. 628–640. Springer, Cham (2017). https://doi.org/10.1007/978-3-319-57454-7_49

6. Durham, E.A., Kantarcioglu, M., Xue, Y., Toth, C., Kuzu, M., Malin, B.: Composite Bloom filters for secure record linkage. IEEE TKDE **26**(12), 2956–2968 (2014)

7. Grahne, G., Zhu, J.: Fast algorithms for frequent itemset mining using FP-trees. IEEE TKDE **17**(10), 1347–1362 (2005)

8. Hegland, M.: The Apriori algorithm - a tutorial. Math. Comput. Imaging Sci. Inf. Process. **11**, 209–262 (2005)

9. Karapiperis, D., Gkoulalas-Divanis, A., Verykios, V.S.: FEDERAL: a framework for distance-aware privacy-preserving record linkage. IEEE TKDE **30**(2), 292–304 (2017)

10. Kroll, M., Steinmetzer, S.: Automated cryptanalysis of bloom filter encryptions of databases with several personal identifiers. In: BIOSTEC, Lisbon (2015)

11. Kuzu, M., Kantarcioglu, M., Durham, E., Malin, B.: A constraint satisfaction cryptanalysis of bloom filters in private record linkage. In: Fischer-Hübner, S., Hopper, N. (eds.) PETS 2011. LNCS, vol. 6794, pp. 226–245. Springer, Heidelberg (2011). https://doi.org/10.1007/978-3-642-22263-4_13

12. Kuzu, M., Kantarcioglu, M., Durham, E., Toth, C., Malin, B.: A practical approach to achieve private medical record linkage in light of public resources. JAMIA **20**(2), 285–292 (2013)

13. Mitchell, W., Dewri, R., Thurimella, R., Roschke, M.: A graph traversal attack on Bloom filter-based medical data aggregation. IJBDI **4**(4), 217–226 (2017)

14. Mitzenmacher, M., Upfal, E.: Probability and Computing: Randomized Algorithms and Probabilistic Analysis. Cambridge University Press, Cambridge (2005)

15. Niedermeyer, F., Steinmetzer, S., Kroll, M., Schnell, R.: Cryptanalysis of basic Bloom filters used for privacy preserving record linkage. JPC **6**(2), 59–79 (2014)

16. Randall, S., Ferrante, A., Boyd, J., Bauer, J., Semmens, J.: Privacy-preserving record linkage on large real world datasets. JBI **50**, 205–212 (2014)

17. Schnell, R., Bachteler, T., Reiher, J.: Privacy-preserving record linkage using Bloom filters. BMC Med. Inform. Decis. Making **9**(1), 41 (2009)

18. Schnell, R., Borgs, C.: Randomized response and balanced Bloom filters for privacy preserving record linkage. In: ICDMW DINA, Barcelona (2016)

19. Vatsalan, D., Christen, P.: Privacy-preserving matching of similar patients. JBI **59**, 285–298 (2016)

20. Vatsalan, D., Sehili, Z., Christen, P., Rahm, E.: Privacy-preserving record linkage for big data: current approaches and research challenges. In: Zomaya, A.Y., Sakr, S. (eds.) Handbook of Big Data Technologies, pp. 851–895. Springer, Cham (2017). https://doi.org/10.1007/978-3-319-49340-4_25

A Privacy Preserving Bayesian Optimization with High Efficiency

Thanh Dai Nguyen[✉], Sunil Gupta, Santu Rana, and Svetha Venkatesh

Center for Pattern Recognition and Data Analytics, Deakin University,
Waurn Ponds 3216, Australia
{thanh,sunil.gupta,santu.rana,svetha.venkatesh}@deakin.edu.au

Abstract. Bayesian optimization is a powerful machine learning technique for solving experimental design problems. With its use in industrial design optimization, time and cost of industrial processes can be reduced significantly. However, often the experimenters in industries may not have the expertise of optimization techniques and may require help from third-party optimization services. This can cause privacy concerns as the optimized design of an industrial process typically needs to be kept secret to retain its competitive advantages. To this end, we propose a novel Bayesian optimization algorithm that can allow the experimenters from an industry to utilize the expertise of a third-party optimization service in privacy preserving manner. Privacy of our proposed algorithm is guaranteed under a modern privacy preserving framework called Error Preserving Privacy, especially designed to maintain high utility even under the privacy restrictions. Using several benchmark optimization problems as well as optimization problems from real-world industrial processes, we demonstrate that the optimization efficiency of our algorithm is comparable to the non-private Bayesian optimization algorithm and significantly better than its differential privacy counterpart.

1 Introduction

Bayesian optimization is a popular machine learning technique to find the optimum of expensive black-box functions in an efficient manner. It is widely applicable for optimizing experimental design in industrial processes [7], hyperparameter tuning of machine learning algorithms [1], robotics [8], etc. In these optimization problems, the objective functions are costly to evaluate (e.g. running a physical process) and time consuming (e.g. training a highly complex model).

Bayesian optimization works in a sequential manner. It uses the available set of observations from the function to develop a surrogate model of the black-box function and uses this model to suggest a new sample where the function should be evaluated next. When making this suggestion, the algorithm addresses two goals. The first is to exploit the existing knowledge to maximize the chance of observing the function optimum (exploitation). The second goal is to acquire information about the black-box function in the best possible way (exploration).

© Springer International Publishing AG, part of Springer Nature 2018
D. Phung et al. (Eds.): PAKDD 2018, LNAI 10939, pp. 543–555, 2018.
https://doi.org/10.1007/978-3-319-93040-4_43

These two goals may not always be aligned and the algorithm needs to balance them. This is achieved by constructing an acquisition function, which is then optimized to suggest the next function evaluation point. Balancing exploitation with exploration is key to the efficiency of Bayesian optimization.

Since Bayesian optimization is an efficient method for industrial design optimization, its use to optimize the industrial process can significantly reduce the time and cost of industrial processes. However, the experimenters in industries typically may not have the expertise of optimization techniques and therefore require optimization services from a third-party. This can cause privacy concerns as the optimized design of an industrial process needs to be protected to retain its competitive advantages. Consider two parties involved in the design optimization process: an experimenter A and an optimizer B. At each iteration of the Bayesian optimization, the optimizer B asks the experimenter A to perform experiments at the suggested point. A conducts the experiment, assesses the outcomes to score the function output and then returns the function value to B. This interaction repeats until the optimum is found or the number of experiments exceeds a pre-defined budget. Since both experimenter and optimizer have access to the exact knowledge of the optimum point, this algorithm does not offer any privacy. Such industrial data is sensitive and the optimum of the objective function cannot be revealed to keep competitive advantage. As an example, consider an alloy making company that needs to design an alloy with certain target properties. The task involves optimizing the mixture proportions of constituent ingredient elements. The final composition is kept secret. In such cases, the experimenter A from the industry wants to avail the service of the optimizer B without disclosing the exact function values.

Privacy preserving machine learning has recently attracted the attention from research community. One of the most popular frameworks for private data release is Differential Privacy [5]. It protects the data privacy from various type of privacy attacks, even when the adversary has auxiliary information. For example, the privacy of a data point is guaranteed even if an adversary has access to all other data points in the database. Several differentially-private machine learning and data mining algorithms have been proposed e.g. differentially-private logistic regression [3], differentially-private random forest [9], etc. Although differential privacy offers a strong privacy guarantee, a crucial drawback is that to achieve this guarantee, algorithms need to be perturbed so significantly that their utility becomes low and often unacceptable. To address this problem, Nguyen et al. [4] developed a new privacy framework, known as Error Preserving Privacy (EPP). This privacy framework is similar to differential privacy in that it protects data even under the scenario when an adversary has access to auxiliary information. The high utility is maintained by making certain assumptions about the adversary model that may be used to extract the data point of interest.

The work on privacy preserving Bayesian optimization is quite limited. Kusner et al. [6] proposed a Differentially Private Bayesian Optimization algorithm. However, the privacy setting of this algorithm is quite different from the setting considered in this paper. In Kusner's work, data are shared freely between the experimenter and the optimizer and their interactions are non-private. The optimum point is kept private from public. Once the optimization process is finished, the optimum point is perturbed appropriately before releasing it to the public. We consider a *different setting* wherein the experimenter from an industry may not trust the optimizer and does not want to share exact data with the optimizer. In this setting, Kusner's method is not applicable. To the best of our knowledge, we are the first to identify the privacy problem for Bayesian optimization in this setting. *Therefore, the problem of developing a privacy preserving Bayesian optimization algorithm with third-party optimizer remains open.*

We propose a privacy preserving Bayesian optimization that helps to find the optimum of an expensive black-box function without revealing the best point up to any optimization iteration, ensuring the privacy of the optimum under the EPP framework [4]. EPP is chosen to maintain high optimization efficiency even under the stringent privacy requirements. The proposed algorithm follows an three-step iterative procedure (see Fig. 1). The first step requires the experimenter to evaluate the function value at the input suggested by the optimizer. The second step perturbs the result evaluated by the experimenter by a noise that helps to protect the privacy of the true optimum. The third step is at the untrusted end where the perturbed point is included in the function model by the optimizer to suggest the next evaluation point. We perform theoretical analysis and derive the amount of perturbation required to guarantee the privacy. We apply our algorithm to benchmark optimization problems as well as optimization problems from real-world industrial processes and demonstrate that the optimization efficiency of our algorithm is comparable to the non-private Bayesian optimization algorithm. We also suggest a differentially-private Bayesian optimization baseline and show that the performance of EPP based Bayesian optimization algorithm is significantly better than the differentially-private baseline.

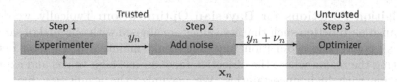

Fig. 1. Privacy preserving Bayesian optimization framework.

2 Background

2.1 Bayesian Optimization

Bayesian optimization [11] is an efficient technique to optimize expensive black-box function f. Formally, the optimization problem can be defined as:

$$\mathbf{x}^* = \text{argmax}_{\mathbf{x} \in \mathcal{X}} f(\mathbf{x}) \tag{1}$$

where $\mathcal{X}$ is domain of $\mathbf{x}$. Bayesian optimization has two parts: the first part builds a surrogate model of f using available observations and the second part uses this surrogate model to construct an acquisition function that is then used to suggest the next sample for function evaluation. The process repeats until the optimum is found or a pre-specified budget exceeds. The surrogate model uses a prior distribution about the function f and known observations to maintain the posterior of the function. The surrogate model of Bayesian optimization can be any probabilistic model that is capable of representing the uncertainty about the function, e.g. Gaussian processes [10], Bayesian neural networks [12], bootstrapped random forest [2], etc. Gaussian processes are the most popular choice and are briefly described here.

Gaussian Process: Gaussian process [10] is a probabilistic model that specifies distribution over function spaces. A Gaussian process is defined by a mean function $m(\mathbf{x})$ and a covariance function $k(\mathbf{x}, \mathbf{x}')$. Let $f(\mathbf{x})$ be a function that is drawn from a Gaussian process $f(\mathbf{x}) \sim \mathcal{GP}(m(\mathbf{x}), k(\mathbf{x}, \mathbf{x}'))$ and $\mathcal{D}_n = \{(\mathbf{x}_i, y_i)\}, i = 1, 2, \ldots, n$ be the set of observations where y_i is the noisy output value, $y_i = f(\mathbf{x}_i) + \varepsilon_i$ and $\varepsilon_i \sim \mathcal{N}(0, \sigma_\epsilon^2)$. Without loss of generality, we assume the mean function $m(\mathbf{x})$ to be zero. This makes the Gaussian process depending on the covariance function $k(\mathbf{x}, \mathbf{x}')$ only. A popular choice for covariance function is squared exponential kernel which is defined as $k(\mathbf{x}, \mathbf{x}') = \sigma_{SE}^2 \exp(-\frac{1}{2\theta^2} \|\mathbf{x} - \mathbf{x}'\|^2)$ where θ is a length scale parameter representing the smoothness of the function and σ_{SE} is a scale factor. Because any finite collection of random variables in a Gaussian process are jointly distributed as Gaussian, the joint distribution of known observations $\mathcal{D}_n$ and a new observation $(\mathbf{x}_{n+1}, y_{n+1})$ is Gaussian. The predictive distribution of the function value y_{n+1} is also Gaussian *i.e.* $p(y_{n+1}|\mathbf{y}_{1:n}, \mathbf{x}_{1:n+1}) = \mathcal{N}(\mu_n(\mathbf{x}_{n+1}), \sigma_n^2(\mathbf{x}_{n+1}))$ with mean and variance given as $\mu_n(\mathbf{x}_{n+1}) = \mathbf{k}^T(\mathbf{K} + \sigma_\epsilon^2\mathbf{I})^{-1}\mathbf{y}$, $\sigma_n^2(\mathbf{x}_{n+1}) = k(\mathbf{x}_{n+1}, \mathbf{x}_{n+1}) - \mathbf{k}^T(\mathbf{K} + \sigma_\epsilon^2\mathbf{I})^{-1}\mathbf{k}$ where $\mathbf{k} = [k(\mathbf{x}_1, \mathbf{x}_{n+1}), k(\mathbf{x}_2, \mathbf{x}_{n+1}), \ldots, k(\mathbf{x}_n, \mathbf{x}_{n+1})]^T$ and $\mathbf{K}(i, j) = k(\mathbf{x}_i, \mathbf{x}_j)$.

Acquisition Functions for Bayesian Optimization: Typically, we defined an acquisition function such that its values are aligned with our goal of sampling the next point. The acquisition function should maintain a trade-off between exploiting the region where the objective function has high values and exploring the region with high uncertainties. We maximize the acquisition function to suggest the next point that achieves the trade-off in the best way. Let $\alpha(\mathbf{x}; \mathcal{I}_n)$ denote the acquisition function where $\mathcal{I}_n$ is the Gaussian process posterior after n observations. The original optimization problem of Eq. (1) becomes maximizing $\alpha(\mathbf{x}; \mathcal{I}_n)$ as follows:

$$\mathbf{x}_{n+1}^* = \operatorname{argmax}_{\mathbf{x} \in \mathcal{X}} \alpha(\mathbf{x}; \mathcal{I}_n). \tag{2}$$

A popular choice for acquisition function is Upper Confidence Bound. For the Gaussian process surrogate model, it is defined as [13] $\alpha_{\mathrm{GPUCB}}(\mathbf{x}; \mathcal{I}_n) = \mu_n(\mathbf{x}) + \kappa_n \sigma_n(\mathbf{x})$ where κ_n is an iteration dependent positive parameter representing the exploration-exploitation trade-off. Srinivas et al. [13] provided a theoretical guarantee of GP-UCB through an upper bound on the cumulative regret $\mathcal{R}_n = \sum_{i=1}^{n}(f(\mathbf{x}^*) - f(\mathbf{x}_i))$, which grows sub-linearly. The acquisition function in (2) can be maximized using any global optimizer because it is cheap to evaluate unlike function f.

2.2 Differential Privacy

Differential privacy is a privacy preserving framework that offers a strong guarantee of privacy. The formal definition is given as:

Definition 1. A randomized function $\mathcal{A}$ gives ϵ−differential privacy if for all data sets D_1 and D_2 differing in at most one element, and all $S \subseteq Range(\mathcal{A})$,

$$\exp(-\epsilon) \leq \frac{\Pr\left[\mathcal{A}(D_1) \in S\right]}{\Pr\left[\mathcal{A}(D_2) \in S\right]} \leq \exp(\epsilon) \tag{3}$$

A popular way to make an algorithm differentially private is to add noise to the algorithm output. Sensitivity is defined as the maximum change of the algorithm output due to the change of one single datapoint. Using the sensitivity S, Laplacian mechanism adds a Laplacian noise with standard deviation S/ϵ to the algorithm output and makes the algorithm ϵ−differentially private [5].

3 The Proposed Framework

We present our proposed privacy preserving Bayesian optimization framework. We start with a brief description about the EPP framework proposed in [4]. Under this framework, we propose a novel privacy preserving Bayesian optimization algorithm to protect privacy of the objective function's maximum.

3.1 Error Preserving Privacy Framework

Let $\mathcal{D}_N = \{\mathbf{x}_1, \ldots, \mathbf{x}_N\}$ be the set of N data points and g be a quantity of interest that needs to be protected. In [4], Nguyen et al. proposed the Error Preserving Privacy (EPP) framework that provides privacy guarantees for g even in presence of auxiliary information. As in the differential privacy framework, EPP framework controls the level of privacy using a pre-specified leakage parameter ϵ. In particular, given an adversary model for estimating g, the errors in the adversary's estimates of g are guaranteed to be similar for any two datasets $\mathcal{D}_N$ and $\mathcal{D}_{N+1}$ differing by just one data point (say $\mathbf{x}_{N+1}$). Thus, the extra information gained by an adversary by knowing $\mathbf{x}_{N+1}$ brings negligible risks on the privacy of the quantity g for small ϵ. Let us assume that an adversary estimates the statistic $\hat{g}(\mathcal{D}_N)$ and $\hat{g}(\mathcal{D}_{N+1})$ using data $\mathcal{D}_N$ and $\mathcal{D}_{N+1}$ respectively. If we denote by $\mathcal{E}(\hat{g}(\mathcal{D}_N))$ the error of the adversary in estimating g using data $\mathcal{D}_N$, i.e. $\mathcal{E}(\hat{g}(\mathcal{D}_N)) = \mathbb{E}\left[(\hat{g}(\mathcal{D}_N) - g)^2\right]$ and by $\mathcal{E}(\hat{g}(\mathcal{D}_{N+1}))$ the error of the adversary in estimating g using data $\mathcal{D}_{N+1}$, i.e. $\mathcal{E}(\hat{g}(\mathcal{D}_{N+1})) = \mathbb{E}\left[(\hat{g}(\mathcal{D}_{N+1}) - g)^2\right]$, then the EPP framework ensures the following inequality:

$$\frac{\mathcal{E}(\hat{g}(\mathcal{D}_{N+1}))}{\mathcal{E}(\hat{g}(\mathcal{D}_N))} \geq \exp(-\epsilon) \tag{4}$$

where $\epsilon \geq 0$ is a pre-specified privacy leakage parameter.

3.2   The Proposed Algorithm

Let us imagine an industrial design optimization task involving two parties: "an experimenter" and "an optimizer". The optimizer is assumed to be untrusted. The "experimenter" wants to find the maximum of an objective function (the function underlying the industrial process) f and wants to utilize the services offered by the "optimizer". Let $\mathcal{D}_n = \{(\mathbf{x}_1, y_1), \ldots, (\mathbf{x}_n, y_n)\}$ be the set of observations such that $y_i = f(\mathbf{x}_i) + \varepsilon_i$ where $\varepsilon_i \sim \mathcal{N}(0, \sigma_\varepsilon^2)$ is measurement noise. Further, let $\mathcal{D}_{n+1} = \mathcal{D}_n \cup (\mathbf{x}_{n+1}, y_{n+1})$. Let $(\mathbf{x}_{n+1}^+, y_{n+1}^+)$ be the "best point so far" in $\mathcal{D}_{n+1}$ such that $\mathbf{x}_{n+1}^+ = \mathbf{x}_i$ and $y_{n+1}^+ = y_i$ and $i = \underset{j=1,\ldots,n+1}{\operatorname{argmax}} y_j$. Since the optimizer may be not trustworthy, the experimenter does not want to disclose the true optimum $(\mathbf{x}_{n+1}^+, y_{n+1}^+)$ to the untrusted optimizer. Instead, the experimenter decides to share the experimental data in a privacy preserving manner, which in this case is achieved by perturbing the function value. The quantity of interest that needs to be protected at all times is the best point at any iteration. Our *aim* is therefore to share the data between the experimenter and the optimizer in such a way so that $(\mathbf{x}_{n+1}^+, y_{n+1}^+)$ is ambiguous for the optimizer (assumed to be an adversary here) even if the optimizer has exact knowledge of data in $\mathcal{D}_n$. In the following, we refer to the *optimizer* as *adversary*.

We next develop a Bayesian optimization algorithm maintaining this privacy under the EPP framework. Let $\hat{\omega}_n = \hat{y}_{n+1}^+ \mid \mathcal{D}_n$ and $\hat{\omega}_{n+1} = \hat{y}_{n+1}^+ \mid \mathcal{D}_{n+1}$ be the estimates of the adversary about the "best point so far" using $\mathcal{D}_n$ and $\mathcal{D}_{n+1}$ respectively. The EPP framework ensures the errors of the adversary's estimates $\hat{\omega}_n$ and $\hat{\omega}_{n+1}$ are similar, which means that by acquiring $(\mathbf{x}_{n+1}, y_{n+1})$, the adversary's estimate of y_{n+1}^+ does not change significantly. This helps in hiding the true optimum y_{n+1}^+ and also provides the protection for the location of the maximum $\mathbf{x}_{n+1}^+$. Formally, we denote by $\mathcal{E}(\hat{\omega}_n)$ and $\mathcal{E}(\hat{\omega}_{n+1})$ as the errors in the adversary's estimate of y_{n+1}^+ using $\mathcal{D}_n$ and $\mathcal{D}_{n+1}$ respectively. For simplicity in notation, we refer to these quantities as $\mathcal{E}_n$ and $\mathcal{E}_{n+1}$.

In the absence of the privacy preserving scheme, the error of the adversary's estimates using $\mathcal{D}_{n+1}$ may be significantly lower than the one using $\mathcal{D}_n$ as the adversary can simply find the maximum over all the observations. To ensure the privacy, we add a Gaussian distributed perturbation noise to the function output $y_{n+1} \leftarrow y_{n+1} + \nu_{n+1}$ where $\nu_{n+1} \sim \mathcal{N}(0, q_{n+1}^2)$, q_{n+1} is the standard deviation of the noise. The following theorem characterizes the amount of noise required to guarantee the EPP framework.

Theorem 1. *The noise standard deviation q_{n+1} obtained as solution of Eq. (17) ensures Error Preserving Privacy for the Bayesian optimization algorithm.*

Proof. To prove the theorem, we need to derive the error in the adversary's estimates: $\mathcal{E}_n$ and $\mathcal{E}_{n+1}$. After deriving these errors, we plug them in Eq. (4) to obtain an equation in q_{n+1} (see Eq. (17)). The minimum value of q_{n+1} that satisfies Eq. (17) ensures the EPP privacy guarantee. □

For the adversary estimation model, we assume that given $\mathcal{D}_{n+1}$, the adversary estimates the $\hat{y}_{n+1}^+$ as $\hat{y}_{n+1}^+ = \max(y_1, \ldots, y_n, \hat{y}_{n+1} | \mathcal{D}_{n+1})$. Similarly, given

$\mathcal{D}_n$, the adversary estimates the $\hat{y}_{n+1}^+$ as $\hat{y}_{n+1}^+ = \max(y_1, \ldots, y_n, \hat{y}_{n+1}|\mathcal{D}_n)$ where $\hat{y}_{n+1}|\mathcal{D}_n$ is also estimated using a Gaussian process model.

Computation of $\mathcal{E}_n$: As per the stated adversary model, $\hat{\omega}_n$ is given as

$$\hat{\omega}_n = \max(\hat{\theta}_n, y_n^+) \tag{5}$$

where $y_n^+ = \max(y_1, \ldots, y_n)$, $\hat{\theta}_n = \hat{y}_{n+1}|\mathcal{D}_n$ is the adversary's estimate of y_{n+1} using $\mathcal{D}_n$ and is Gaussian distributed. The mean and variance of this estimate can be computed as follows:

$$\mathbb{E}\left[\hat{\theta}_n\right] = \mathbb{E}[\mathbf{k}_n^T \mathbf{K}_n^{-1} \mathbf{y}_{1:n}] = \mathbf{k}_n^T \mathbf{K}_n^{-1} \mathbf{f}_{1:n}, \; Var\left[\hat{\theta}_n\right] = \sigma_\epsilon^2 \left\| \mathbf{k}_n^T \mathbf{K}_n^{-1} \right\|_2^2 \tag{6}$$

where $\mathbf{k}_n$, $\mathbf{K}_n$ and σ_ϵ are the quantities introduced in Sect. 2. Using the mean and variance of $\hat{\theta}_n$, we can compute the distribution of $\hat{\omega}_n$. Defining $\mu_n = \mathbf{k}_n^T \mathbf{K}_n^{-1} \mathbf{f}_{1:n}$ and $\sigma_n = \sigma_\epsilon^2 \left\| \mathbf{k}_n^T \mathbf{K}_n^{-1} \right\|_2^2$, the probability density function of $\hat{\omega}_n$ can be written as:

$$p_{\hat{\omega}_n}(y) = \begin{cases} \frac{1}{\sqrt{2\pi\sigma_n^2}} \exp\left[-\frac{(y-\mu_n)^2}{2\sigma_n^2}\right] & \text{if } y > y_n^+ \\ Pr(y_{n+1} \leq y|\mathcal{D}_n) & \text{if } y = y_n^+ \\ 0 & \text{if } y < y_n^+ \end{cases} \tag{7}$$

The mean square error $\mathcal{E}_n$ can be written as follows:

$$\mathcal{E}_n = \mathbb{E}\left[(\hat{\omega}_n - y_{n+1})^2\right] = (\mathbb{E}\left[\hat{\omega}_n\right] - y_{n+1})^2 + Var\left[\hat{\omega}_n\right] \tag{8}$$

Given the distribution function $p_{\hat{\omega}_n}(y)$, we can compute the expectation and variance of $\hat{\omega}_n$ as follows:

$$\mathbb{E}\left[\hat{\omega}_n\right] = \sigma_n p + (y_n^+ - \mu_n)P + \mu_n \tag{9}$$

$$Var\left[\hat{\omega}_n\right] = \left(y_n^+ + \mu_n\right)\sigma_n p + \left((y_n^+)^2 - \sigma_n^2 - \mu_n^2\right)P + \mu_n^2 + \sigma_n^2 - (\mathbb{E}\left[\hat{\omega}_n\right])^2 \tag{10}$$

where $p = \phi\left(\frac{y_n^+ - \mu_n}{\sigma_n}\right)$ and $P = \Phi\left(\frac{y_n^+ - \mu_n}{\sigma_n}\right)$. Using Eqs. (9) and (10), we can finally compute $\mathcal{E}_n$.

Computation of $\mathcal{E}_{n+1}$: After the adversary (or optimizer) suggests $\mathbf{x}_{n+1}$, the experimenter conducts the experiment and returns a noisy value $y_{n+1} \leftarrow y_{n+1} + \nu_{n+1}$ to ensure privacy, where $\nu_{n+1} \sim \mathcal{N}(0, q_{n+1}^2)$. We assume that after receiving y_{n+1}, the adversary uses maximum statistic to estimate y_{n+1}^+ using $\mathcal{D}_{n+1}$ as:

$$\hat{\omega}_{n+1} = \max(\hat{\theta}_{n+1}, y_n^+) \tag{11}$$

where $\hat{\theta}_n = \hat{y}_{n+1}|\mathcal{D}_{n+1}$ is the adversary's estimate of y_{n+1} using $\mathcal{D}_{n+1}$. Similar to the previous derivation, the mean and variance of $\hat{\theta}_{n+1}$ can be computed as:

$$\mathbb{E}\left[\hat{\theta}_{n+1}\right] = \mathbf{k}_{n+1}^T \mathbf{K}_{n+1}^{-1} \mathbf{f}_{1:n+1} \tag{12}$$

Algorithm 1. Error Preserving Private Bayesian Optimization (EPP-BO)

1: **Input:**

2:     Initial observation set $\mathcal{D}_{n_0} = \{\mathbf{x}_{1:n_0}, y_{1:n_0}\}$, search space $\mathcal{X}$ and privacy budget ϵ.

3: **Output:** $\{\mathbf{x}_n, y_n\}_{n=1}^T$

4:     **for** $n = n_{0+1}, \ldots, T$

5:         Evaluate target function $y_n = f(\mathbf{x}_n)$

6:         Find q_n that satisfy 17 using binary search.

7:         Add a noise $\nu_n \sim \mathcal{N}(0, q_n^2)$ to y_n.

8:         Return the output to the optimizer.

9:     **end for**

$$Var\left[\hat{\theta}_{n+1}\right] = \sigma_\epsilon^2 \left\|\mathbf{k}_{n+1}^T \mathbf{K}_{n+1}^{-1}\right\|_2^2 + \gamma_{n+1}^2 \left(\sigma_\epsilon^2 + q_{n+1}^2\right) \tag{13}$$

where $\gamma = \mathbf{k}_{n+1}^T \mathbf{K}_{n+1}^{-1}$ is a vector and γ_{n+1} is the $n+1$-th element of γ. Let us define $\mu_{n+1} = \mathbb{E}\left[\hat{\theta}_{n+1}\right]$ and $\sigma_{n+1} = Var\left[\hat{\theta}_{n+1}\right]$. The error $\mathcal{E}_{n+1}$ can be derived as:

$$\mathcal{E}_{n+1} = \mathbb{E}\left[\left(\hat{\omega}_{n+1} - y_{n+1}\right)^2\right] = \left(\mathbb{E}\left[\hat{\omega}_{n+1}\right] - y_{n+1}\right)^2 + Var\left[\hat{\omega}_{n+1}\right] \tag{14}$$

where $p' = \phi\left(\frac{y_n^+ - \mu_{n+1}}{\sigma_{n+1}}\right)$, $P' = \Phi\left(\frac{y_n^+ - \mu_{n+1}}{\sigma_{n+1}}\right)$ and

$$\mathbb{E}\left[\hat{\omega}_{n+1}\right] = \sigma_{n+1} p' + (y_n^+ - \mu_{n+1})P' + \mu_{n+1} \tag{15}$$

$$Var\left[\hat{\theta}_{n+1}\right] = \left(y_n^+ + \mu_{n+1}\right)\sigma_{n+1}p' + \left((y_n^+)^2 - \sigma_{n+1}^2 - \mu_{n+1}^2\right)P'$$
$$+ \mu_{n+1}^2 + \sigma_{n+1}^2 - \left(\mathbb{E}\left[\hat{\omega}_{n+1}\right]\right)^2 \tag{16}$$

Computation of q_{n+1}: Now we have computed $\mathcal{E}_n$ and $\mathcal{E}_{n+1}$. To ensure privacy condition while maintaining high utility, we want to add a smallest noise possible that makes the following inequality satisfied:

$$\frac{\mathcal{E}_{n+1}}{\mathcal{E}_n} = \frac{\left(\mathbb{E}\left[\hat{\omega}_{n+1}\right] - y_{n+1}\right)^2 + Var\left[\hat{\omega}_{n+1}\right]}{\left(\mathbb{E}\left[\hat{\omega}_n\right] - y_{n+1}\right)^2 + Var\left[\hat{\omega}_n\right]} \geq \exp(-\epsilon) \tag{17}$$

Equation (17) can be solved by plugging Eqs. (9), (10), (15) and (16). Our objective is to find a smallest value of q_{n+1} that satisfies Eq. (17). We note that by adding more noise, the variance of $\hat{\omega}_{n+1}$ will increase and hence $\frac{\mathcal{E}_{n+1}}{\mathcal{E}_n}$ is an increasing function of q_{n+1}. We can find the smallest value α of q_{n+1} that satisfies (17) using binary search. Assigning $q_{n+1} \leftarrow \alpha$, we can then add a noise sample ν_{n+1} to y_{n+1} and keep the "best point so far" private from the adversary. Our algorithm is summarized in Algorithm 1.

3.3   Discussion of Differentially Private Bayesian Optimization

In [6], Kusner et al. proposed a Differentially Private Bayesian Optimization algorithm. This algorithm was designed to tackle the challenge in a setting that the optimizer is considered trusted and the experimenter can share all the data with the optimizer. Since our privacy setting is different, we cannot use the algorithm proposed in [6]. Instead, we suggest another differentially-private Bayesian optimization algorithm, which we'll use for comparison in this paper.

DP-BO Baseline: We suggest a privacy preserving Bayesian optimization algorithm under the differential privacy framework using Laplacian mechanism [5]. Laplacian mechanism adds a perturbation noise to the output of the algorithm to achieve the required privacy. The amount of noise depends on the sensitivity of the quantity that needs to be protected. Since by releasing y_{n+1}, the maximum possible change in y_{n+1}^+ is $S = \|y_{max} - y_{min}\|$ where y_{max} and y_{min} are the maximum and the minimum possible value of y respectively. Using the sensitivity, we iteratively add a Laplacian noise to the function output before passing it to the optimizer: $y_{n+1} \leftarrow y_{n+1} + \nu_{n+1}^{DP}$ where $\nu_{n+1}^{DP} \sim Lap(S/\epsilon_{DP})$ and ϵ_{DP} is the privacy budget for differential privacy. We refer to this algorithm as DP-BO. When the quantities y_{max} and y_{min} are not known exactly, it may be possible to estimate them using Lipschitz smoothness where possible, otherwise this algorithm may not be usable.

4   Experiments

We experiment our algorithm on several benchmark optimization problems as well as optimization problems from real-world industrial processes and demonstrate the optimization efficiency of our algorithm comparing it with various baselines. We use the following three baselines: (i) **Non-private Bayesian Optimization (Non-private BO)**: the standard non-private version of Bayesian optimization, used to show the ultimate utility of private Bayesian optimization algorithm. (ii) **Random Search**: used as a lower bound baseline. A private algorithm must achieve higher optimization efficiency than random search to justify the extra complexity. (iii) **DP-BO**: constructed under differential privacy framework as discussed in Sect. 3.3. For all our experiments, we use GP-UCB as the acquisition function and set the privacy leakage parameter $\epsilon = 0.1$.

4.1   Experiment with Benchmark Functions

To demonstrate our algorithm for a variety of functions in different number of input dimensions, we experiment with four popular benchmark functions: Branin 2D, Rosenbrock 4D, Hartmann 4D and Hartmann 6D. The optimization results are averaged over 20 different initializations. Figure 2 shows the performance of our proposed algorithm (referred to as EPP-BO) against the baselines on the benchmark functions. For all four benchmark functions, EPP-BO's performance

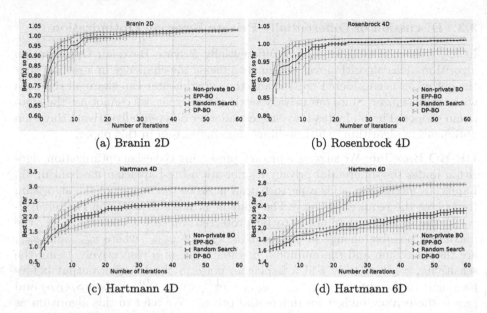

(a) Branin 2D   (b) Rosenbrock 4D

(c) Hartmann 4D   (d) Hartmann 6D

Fig. 2. Best value so far with respect to Bayesian optimization iteration.

is close to the Non-private Bayesian Optimization and clearly outperforms both Random Search and DP-BO by a significant margin.

Figure 3a illustrates the illusion our proposed EPP-BO creates for an adversary for Hartmann 4D function. We show two scenarios: high privacy scenario (using $\epsilon = 0.1$) and low privacy scenario (using $\epsilon = 0.5$). From any run of EPP-BO, two different graphs are extracted showing different views of the optimization from experimenter or optimizer (adversary) perspectives. The graph in 'magenta' color shows the best function value achieved so far from the optimizer's perspective. The graph in 'blue' color shows the best function value achieved so far from the experimenter's perspective. Between these graphs, the locations of the best points do not necessarily coincide. We also note that when the privacy decreases, the optimizer gets closer to the true optimum since less noise is added.

4.2 Experiment with Real Datasets

Alloy Heat Treatment: This dataset is a simulation model of an Al-Sc alloy heat treatment process. The strengthening process of an alloy involves nucleation and growth. During nucleation, new "phase" is formed through a self-organizing process of clusters of atoms. This process happens at low temperature. Following nucleation step, the growth step archives the requisite alloy property through diffusion. The industrial standard precipitation KWN model [14] is used for nucleation and growth. This model consists of multiple stages, each of them having two parameters: temperature and time. The output quality of the alloy heat treatment process is measured by the hardness of the alloy. Our objective

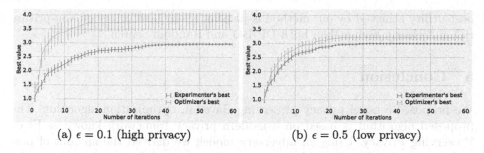

(a) $\epsilon = 0.1$ (high privacy)   (b) $\epsilon = 0.5$ (low privacy)

Fig. 3. Comparison between the best values achieved by the experimenter and the optimizer with respect to iterations on Hartman 4D function with different level of privacy: (a) $\epsilon = 0.1$ (high privacy) and (b) $\epsilon = 0.5$ (low privacy). The blue and the magenta show the best function value achieved so far from the experimenter's perspective and optimizer's perspective, respectively. (Color figure online)

is to find the best combination of time and temperature parameters that achieve the highest level of hardness in few iterations.

Figure 4a demonstrates the results of our experiments on Alloy heat treatment dataset averaged over 20 different initializations. After 30 iterations, the best hardness achieved by EPP-BO is comparable to Non-private BO and higher than the best results of both Random Search and DP-BO.

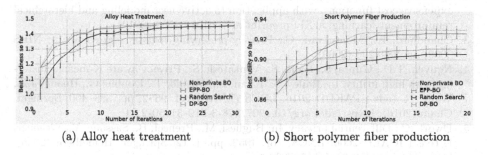

(a) Alloy heat treatment   (b) Short polymer fiber production

Fig. 4. Optimization results for two real datasets.

Short Polymer Fiber Production: This dataset was collected in a collaboration with material scientists from Deakin University. For short polymer fiber production, a particular geometric manifold is used to mix polymer rich fluid with the flow of another solvent. This manifold has 5 different parameters: device position, constriction angle, channel width, polymer flow, and coagulant speed. The objective is to find the best manifold parameter set to maximize a combine utility measured by the length and diameter of the output polymer.

Figure 4b shows the experimental results on Short polymer fiber production dataset averaged over 40 different random initializations. After 20 iterations, the best utility achieved by our method is just under the best utility of Non-private BO and clearly higher than both DP-BO and Random Search.

5   Conclusion

We proposed a novel privacy preserving Bayesian optimization algorithm. The proposed algorithm is based on a modern privacy framework known as Error Preserving Privacy. Using an adversary model, we derived the amount of perturbation required to provably guarantee the privacy. The experimental results clearly showed that our private algorithm has higher optimization efficiency than other privacy reserving counterparts.

Acknowledgment. This research was partially funded by the Australian Government through the Australian Research Council (ARC) and the Telstra-Deakin Centre of Excellence in Big Data and Machine Learning. Prof Venkatesh is the recipient of an ARC Australian Laureate Fellowship (FL170100006).

References

1. Bardenet, R., Brendel, M., Kégl, B., Sebag, M.: Collaborative hyperparameter tuning. In: International Conference on Machine Learning, pp. 199–207 (2013)
2. Brochu, E., Cora, V.M., De Freitas, N.: A tutorial on Bayesian optimization of expensive cost functions, with application to active user modeling and hierarchical reinforcement learning. arXiv preprint arXiv:1012.2599 (2010)
3. Chaudhuri, K., Monteleoni, C.: Privacy-preserving logistic regression. In: NIPS 2009, pp. 289–296 (2009)
4. Nguyen, T.D., Gupta, S., Rana, S., Venkatesh, S.: Privacy aware K-means clustering with high utility. In: Bailey, J., Khan, L., Washio, T., Dobbie, G., Huang, J.Z., Wang, R. (eds.) PAKDD 2016. LNCS (LNAI), vol. 9652, pp. 388–400. Springer, Cham (2016). https://doi.org/10.1007/978-3-319-31750-2_31
5. Dwork, C.: Differential privacy. In: Bugliesi, M., Preneel, B., Sassone, V., Wegener, I. (eds.) ICALP 2006. LNCS, vol. 4052, pp. 1–12. Springer, Heidelberg (2006). https://doi.org/10.1007/11787006_1
6. Kusner, M., Gardner, J., Garnett, R., Weinberger, K.: Differentially private Bayesian optimization. In: International Conference on Machine Learning, pp. 918–927 (2015)
7. Li, C., de Celis Leal, D.R., Rana, S., Gupta, S., Sutti, A., Greenhill, S., Slezak, T., Height, M., Venkatesh, S.: Rapid Bayesian optimisation for synthesis of short polymer fiber materials. Sci. Rep. **7**(1), 5683 (2017)
8. Lizotte, D.J., Wang, T., Bowling, M.H., Schuurmans, D.: Automatic gait optimization with Gaussian process regression. In: IJCAI, vol. 7, pp. 944–949 (2007)
9. Rana, S., Gupta, S.K., Venkatesh, S.: Differentially private random forest with high utility. In: ICDM 2015, pp. 955–960. IEEE (2015)
10. Rasmussen, C.E.: Gaussian Processes for Machine Learning. Citeseer (2006)
11. Snoek, J., Larochelle, H., Adams, R.P.: Practical Bayesian optimization of machine learning algorithms. In: NIPS, pp. 2951–2959 (2012)

12. Snoek, J., Rippel, O., Swersky, K., Kiros, R., Satish, N., Sundaram, N., Patwary, M., Prabhat, M., Adams, R.: Scalable Bayesian optimization using deep neural networks. In: ICML (2015)
13. Srinivas, N., Krause, A., Seeger, M., Kakade, S.M.: Gaussian process optimization in the bandit setting: no regret and experimental design. In: ICML (2010)
14. Wagner, R., Kampmann, R., Voorhees, P.W.: Homogeneous second-phase precipitation. In: Materials Science and Technology (1991)

Randomizing SVM Against Adversarial Attacks Under Uncertainty

Yan Chen[1], Wei Wang[2], and Xiangliang Zhang[3(✉)]

[1] Columbia University, New York, USA
yc3107@columbia.edu
[2] Beijing Jiaotong University, Beijing, China
wei.wang.email@gmail.com
[3] King Abdullah University of Science and Technology (KAUST),
Thuwal, Saudi Arabia
xiangliang.zhang@kaust.edu.sa

Abstract. Robust machine learning algorithms have been widely studied in adversarial environments where the adversary maliciously manipulates data samples to evade security systems. In this paper, we propose randomized SVMs against generalized adversarial attacks under uncertainty, through learning a classifier distribution rather than a single classifier in traditional robust SVMs. The randomized SVMs have advantages on better resistance against attacks while preserving high accuracy of classification, especially for non-separable cases. The experimental results demonstrate the effectiveness of our proposed models on defending against various attacks, including aggressive attacks with uncertainty.

Keywords: Adversarial learning · Robust SVM · Randomization

1 Introduction

Adversary machine learning is an important research track that harnesses machine learning to resolve security issues. The adversary can deliberately manipulate their data to mislead the defender of security. Machine learning is challenged by learning from poisoned training data [2,7]. Consequently, it is imperative to identify potential vulnerabilities and propose countermeasures in order to improve the robustness of machine learning algorithms against attacks [6,8,9].

Support Vector Machines (SVMs) as supervised models are among the most popular classification techniques adopted in security applications like malware detection, intrusion detection, and spam filtering [4,13]. In order to secure decision-making system against *poisoning attacks* (contaminating training data),

Y. Chen—The work was completed when the first author was visiting KAUST as an intern.

Robust SVMs as the modification to standard SVMs had been proposed by exploring robustness, kernels and dual formulations in SVMs and Bayes learning [10–13]. In general, the intuition is to make the decision boundary learned in robust SVMs not be extremely sensitive to any single training example. Recently, *Randomized SVMs* are studied for defending a classifier against *exploratory attacks*, which probe the classifier with queries in order to reveal confidential information about the training dataset [1]. Randomized SVMs aim at learning *a distribution of classifiers*, rather than a single classifier in previous study of robust SVMs, and thus make the system less vulnerable.

In this paper, we study how to design the *randomized SVMs* against *poisoning attacks with uncertainty*, which are more sophisticated than previously studied attacks, e.g., free range and restrained range attack in [13]. The idea of randomized SVM is demonstrated in Fig. 1. Standard SVM linear classifier learns a w that separates positive and negative class with maximal margin (Fig. 1(a)). Randomized SVM learns a distribution about w, for example, a Gaussian distribution $\mathcal{D}_w = \mathcal{N}(u, \Sigma)$, demonstrated in Fig. 1(c). Such a distribution can guarantee the classification accuracy of w sampled from $\mathcal{D}_w$ with a separability higher than ν (the probability that training data can be separated), i.e., $\mathbb{P}_{w\sim\mathcal{N}(u,\Sigma)}(y_i(w^T x_i) \geq 1) \geq \nu$. In well-separated cases, classifier distribution is as good as a single classifier, but provides a set of classifiers with the same performance to confuse attackers when they attempt to understand the classification system and prepare attacks. In the case where the adversary adds noise to mislead the system, the region close to decision boundary usually becomes complicated. A deterministic classifier has to separate all data with probability 1, and thus accuracy is scarified. Randomized classifier lowers the separation standard (with probability ν less than 1) and guarantees the accuracy, as shown in Fig. 1(d). Therefore, we investigate randomized SVM as a promising solution to learn robust classifiers against poisoning attacks.

The main contributions of our work are:

(1) We design randomized SVM models against different types of attacks and formulate each model into convex optimization, second order cone programming (SOCP) or semi-definite programming (SDP);
(2) The attacks we define to challenge randomized SVMs are generalized from previously studied restrained range (RR) attacks. The generalized attacks

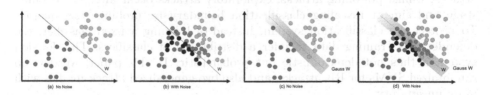

Fig. 1. Standard classifier (a, b) and randomized classifiers (c, d) when input without (a, c) and with noise (b, d) shown in black and orange. (Color figure online)

(RR with uncertainty, and distributional range attack with uncertainty) are more aggressive and complicated, and cover a wide range of attacks;

(3) We evaluate our randomized SVMs on several data sets. The experimental results show that *randomized* SVMs outperform existing *robust* SVMs on defending against attacks at different level of intensity.

2   Related Work

[4] comprehensively evaluated SVMs in adversarial environment, i.e., how SVMs cope with different types of attacks, such as poisoning (contaminating the training data), envision (circumventing the learned classifier) and privacy-breaching attacks. Our work focuses on learning optimal SVMs against poisoning attacks. Therefore, we first discuss the related work that also studies the learning of SVMs from poisoned training data, where each sample is manipulated by adding a noise Δx, i.e., $x' = x + \Delta x$.

There are many Robust SVM models modified from the standard SVM for handling noise in training data. [3] formulate SVM learning with contaminated training data by modeling the unobserved true input (uncorrupted data) as a hidden mixture component. The added noise is assumed to be bounded as $\|\Delta x_i\| \leq \delta_i$, such that a noisy sample lies within a ball of given radius w.r.t. the true non-noisy sample. [11] prove that such SVMs with norm-based regularization build in a robustness to sample noise whose probability level sets are symmetric unit balls with respect to the dual of the regularizing norm. Different but relevant work in [5,12] studies how SVMs are affected by adversarial label noise (e.g., flipping labels of certain training samples), rather than by feature noise (e.g., adding noise to training samples).

The most relevant work in [13] develops robust SVMs against two attack models: free range and restrained range, which are more realistic manipulations attackers can make, while the noise bound δ_i in [3] is fixed and known to both attackers and defenders. In this paper, we study generalized attack models of restrained range (RR), which is more advantageous than free range attacks. The generalized attacks have more flexibility on designing attacks in different forms and with more uncertainty.

The other stream of related work is Randomized SVMs. [1] investigate randomization as a suitable strategy for protecting SVMs against exploratory attacks. Unlike poisoning attacks, exploratory attacks occur after the training stage and aim at revealing classification boundary by probing with queries. To protect the classification system, instead of learning a fixed classifier, the defender uses training data to infer a distribution of classifiers. The decision system is thus not deterministic but probabilistic. In this paper, we develop randomized SVMs for RR attacks, and also two generalized attack models with more uncertainty.

3   Problem Definition

Denote a training sample x_i where $x_i \in \mathbb{R}^d$ and $y_i \in \{-1, 1\}$ as the label of x_i, $i = 1, \ldots, n$. We consider the adversary learning problem where the adversary aims to modify training data in the feature space to their desired targets to mislead the classifier learning. For example in spam detection, the x_i includes some demonstrative information from individual email while y_i is the indicator that judges if the email is malicious. For a malicious sample x_i with $y_i = 1$, the adversary can modify it to be $x_i + \delta_i$ for avoiding detection and misleading the classifier training process according to his planned goals. For the same example in spam detection, good words can be added to a spam email to defeat spam detectors. Following the assumption in [13], the adversary does not modify the innocuous data (with $y_i = -1$), e.g., the adversary has no intention to modify legitimate e-mails.

Our target is to learn a Gaussian distribution $\mathcal{D}_w = \mathcal{N}(u, \Sigma)$, where each sample w is a classifier for discriminating x_i with different labels. Given a required separable probability ν ($\mathbb{P}_{w \sim \mathcal{N}(u, \Sigma)}(y_i(w^T x_i) \geq 1) \geq \nu$), such a distribution can guarantee the classification accuracy of w sampled from $\mathcal{D}_w$. In simple words, it correctly differentiates positive and negative samples as many as possible, while allowing some samples be separable with a low probability. When training data are poisoned, our randomized SVMs are able to classify innocuous data correctly while allowing samples close to the decision boundary (that are probably manipulated samples with noise) be separable with a low probability. Therefore, randomized SVMs are expected to be more robust against poisoning attacks than deterministic SVMs.

4   Attack Model Design

[13] introduces a Restrained Range (RR) Attack model, which allows modification of x_i in a limited range, as a large modification of original x_i entails loss of malicious utility. The modification of x_i is proportional to the difference between x_i and x_i^t (the target of modifying x_i), and is usually set according to the adversary's estimate of the innocuous data. We generalize RR attack with uncertainty in different norm settings. The adversary will not only have the freedom to move data in the feature space, but also can develop attacks with different range shape. Then a most general attack model is defined by considering that the adversary probably manipulates deliberately data with uncertain distribution, unknown expectation and variance to develop an infinite dimensional attack space.

4.1   Restrained Range Attack with Uncertainty

The restrained attack in [13] is defined as

$$0 \leq (x_{ij}^t - x_{ij})\delta_{ij} \leq c_f(1 - \frac{|x_{ij}^t - x_{ij}|}{|x_{ij}| + |x_{ij}^t|})(x_{ij}^t - x_{ij})^2. \tag{1}$$

Dividing by $(x^t_{ij} - x_{ij})$, we obtain the bound of δ_{ij}

$$0 \le |\delta_{ij}| \le c_f(1 - \frac{|x^t_{ij} - x_{ij}|}{|x_{ij}| + |x^t_{ij}|})(|x^t_{ij} - x_{ij}|). \tag{2}$$

We generalize the above bound for δ_i as

$$\delta_i = \{\delta_i : P_i r, ||r|| \le c_f ||v_i||, v_{ij} = x^t_{ij} - x_{ij}\} \tag{3}$$

which provides more freedom to set the shrink matrix P. When setting P as $(1 - \frac{|x^t_{ij} - x_{ij}|}{|x_{ij}| + |x^t_{ij}|})$ and implementing L_1 norm, the generalized restrained attack is approximately reduced to prior form. The main difficult in restrained attacks is to estimate the target x^t_{ij}. The defender usually utilizes their prior knowledge to guess the most possible x^t_{ij}. In fact, the adversary is reluctant to make target data move far away from the origin, which leading to loss of maliciousness. A simple method to estimate x^t_{ij} is to calculate the mean and variance of malicious data to obtain $x^t_{ij} = x_{ij} + c_\delta \varepsilon_{ij}$, where c_δ is the standard variation of these samples and ε_{ij} is a random noise.

4.2   Distributional Range Attack with Uncertainty

To introduce an infinite-dimensional uncertain set to attack models, we define a most general attack model where the modification δ_i follows a distribution belonging to the set:

$$\Delta = \{p : supp(p) = R^n, E_p[\delta_i] = m_i, E_p[||\delta_i||] \le \sigma_i\}. \tag{4}$$

This attack model processes remarkably probability uncertainty. The parameter m is the central point that attacks may happen. According to strong law of large number, we know when the aggressive samples are sufficient large, the average results are close to the expectation of attacks. Thus, m and σ control the intensity of attacks, similar to c_f in (3). The above attack model is the most generalized result by considering all variables in the probability measure space. When the mean is properly set and the variance is sufficient large, it can be implemented to cover RR models. It is expected that by solving such a problem we would obtain an optimal classifier against all distributional attacks.

5   Randomized SVMs Learning

For the different attack strategies defined above, we develop randomized classifiers for learning against noise. Randomized SVMs introduced in [1] learn a distribution $\mathcal{D}_w = \mathcal{N}(u, \Sigma)$, from which a randomly drawn w can make a system of linear inequalities $y_i(w^T x_i) \ge 1$ satisfied with a probability that exceeds ν, where $0 \le \nu \le 1$. That is to say, the probability of data samples that are separable by D_w is at least ν,

$$\mathbb{P}_{w \sim \mathcal{N}(u, \Sigma)}(y_i(w^T x_i) \ge 1) \ge \nu. \tag{5}$$

The optimization problem of identifying the parameters of $\mathcal{D}_w$ (u and $\Sigma = s*s^T$) is defined as:

$$\min_{u,\xi,s} \frac{1}{2}\frac{u^T u}{1^T s} + C\sum_{i=1}^{n}\xi_i$$

$$s.t. \; y_i^T(u^T x_i) \geq 1 + \Phi^{-1}(\nu)\sum_{j=1}^{d} x_{ij}^2 s_j - \xi_i, \tag{6}$$

$$s_i \geq 0, i = 1, 2, \ldots, d. \qquad \xi_i \geq 0, i = 1, 2, \ldots, n.$$

We now formulate randomized SVM for the above-mentioned different types of attacks, including the Restrained Range Attack (RRA), Restrained Range Attack with Uncertainty (RRAU), and Distributional Range Attack with Uncertainty (DRAU). They all modify x_i with $y_i = 1$ to be $x_i + \delta_i$, without changing x_i with $y_i = -1$. The hinge loss of classification can be defined as

$$h(w,b,x_i) = \begin{cases} \max_{\delta_i}\max(1 - (w^T(x_i + \delta_i) + b), 0) & \text{for } y_i = 1; \\ \max(1 + w^T x_i + b, 0) & \text{for } y_i = -1. \end{cases}$$

Combing these two loss functions, the objective function in (6) becomes

$$\min_{w,b}\sum_i \max_{\delta_i}(1 - y_i(w^T x_i + b) - \frac{1}{2}(1 + y_i)w^T\delta_i)^+ + \frac{1}{2}\frac{u^T u}{1^T s}. \tag{7}$$

and can be simplified as

$$\min_{w,b}\sum_i \max_{\delta_i}(1 - y_i(w^T x_i) - \frac{1}{2}(1 + y_i)u^T\delta_i)^+ + \frac{1}{2}\frac{u^T u}{1^T s}. \tag{8}$$

The unique term in (8) relevant with δ_i is $\max_{\delta_i}(-\frac{1}{2}(1 + y_i)u^T\delta_i)$ for fixed u. Then for the following model derivation, we will focus on the sub-problem

$$\max_{\delta_i}(-\frac{1}{2}(1 + y_i)u^T\delta_i) \tag{9}$$

with different constraints in different attack models.

5.1   Randomized SVM Against RRA

RRA sets δ_i by (1). Let $e_{ij} = c_f(1 - \frac{|x_{ij}^t - x_{ij}|}{|x_{ij} + |x_{ij}^t|})(x_{ij}^t - x_{ij})^2$. The sub-problem relevant with δ_i in (9) becomes

$$\max_{\delta_i} -\frac{1}{2}(1 + y_i)u^T\delta_i$$

$$s.t. \; 0 \leq (x_{ij}^t - x_{ij})\delta_{ij} \leq e_{ij}. \tag{10}$$

Introducing $(-z_{ij} + v_{ij})(x^t_{ij} - x_{ij}) = \frac{1}{2}(1 + y_i)u_j, z_i \succeq 0, v_i \succeq 0$ into the above problem, it is transformed into $\max - \sum_j(-z_{ij}e_{ij} + v_{ij}0)$, by solving which, we can rewrite the optimization problem in (6) for Randomized SVM against RRA as follows,

$$\min_{u,s} \max \frac{u^T u}{1^T s} + C \sum_i \xi_i$$

$$s.t. \; \xi_i \geq 0, z_i \succeq 0, v_i \succeq 0, t_i \geq \sum_j(z_{ij}e_{ij}), i = 1, \ldots, n, \; s_i \geq 0, i = 1, 2, \ldots, d,$$

$$y_i^T(u^T x_i) \geq 1 + \Phi^{-1}(\nu) \sum_{j=1}^d x^2_{ij} s_j - \xi_i + t_i, \; (-z_{ij} + v_{ij})(x^t_{ij} - x_{ij}) = \frac{1}{2}(1 + y_i)u_j.$$

$$(11)$$

All constraints in the formulation are linear, we would obtain similar SDP by considering the worst situation.

5.2   Randomized SVM Against RRAU

RRAU sets δ_i by (3). Similarly, we have the sub-problem

$$\max_{\delta_i} -\frac{1}{2}(1 + y_i)u^T \delta_i$$

$$s.t. \; \delta_i = \{\delta_i : P_i r, ||r|| \leq c_f ||v_i||, v_{ij} = x^t_{ij} - x_{ij}\}. \tag{12}$$

Let $f_i = \max -\frac{1}{2}(1 + y_i)u^T P_i r$, s.t. $||r|| \leq c_f||v_i||$. Since we have $f_i \leq \frac{1}{2}(1 + y_i)||v_i||||P_i u||_*$. If the norm of v is L_1 norm, the optimization problem for Randomized SVM against RRAU as follows,

$$\min_{u,s} \max \frac{u^T u}{1^T s} + C \sum_i \xi_i$$

$$s.t. \; \xi_i \geq 0, i = 1, 2, \ldots, n, \qquad s_i \geq 0, i = 1, 2, \ldots, d,$$

$$y_i^T(u^T x_i) \geq 1 + \Phi^{-1}(\nu) \sum_{j=1}^d x^2_{ij} s_j - \xi_i + t_i, \tag{13}$$

$$t_i \geq \frac{1}{2}(1 + y_i)c_f \sum_{j=1}^d |x^t_{ij} - x_{ij}|||P_i u||_*, i = 1, 2, \ldots, n.$$

5.3   Randomized SVM Against DRAU

DRAU sets δ_i by (4). The original optimization problem described in (8) with DRAU becomes

$$\min_{w,b} \sum_i \max_{\delta_i \in \Delta} E_p(1 - y_i(w^T x_i + b) - \frac{1}{2}(1 + y_i)w^T \delta_i)^+ + \frac{1}{2}\frac{u^T u}{1^T s}. \tag{14}$$

We first consider the maximizing inner optimization problem,

$$\max_p E_p(1 - y_i(w^T x_i + b) - \frac{1}{2}(1 + y_i)w^T \delta_i)^+$$

$$s.t. \ E_p(\delta_i) = m_i, \ E_p(||\delta_i||) \leq \sigma_i, \ i = 1, 2, \ldots, n. \tag{15}$$

The *dual Lagrange function* of it can be written as $g(\lambda_1, \lambda_2) = \lambda_1 m_i + \lambda_2 \sigma_i + \max_p\{E_p(1 - y_i(w^T x_i + b) - \frac{1}{2}(1 + y_i)w^T \delta_i)^+ - \lambda_1 \delta_i - \lambda_2 ||\delta_i||\}$, which can be equivalently written as $g(\lambda_1, \lambda_2) = \lambda_1 m_i + \lambda_2 \sigma_i + \max_{z_i}\{(1 - y_i(w^T x_i + b) - \frac{1}{2}(1 + y_i)w^T z_i)^+ - \lambda_1 z_i - \lambda_2 ||z_i||\}$. Since $x^+ = max(x, 0)$, it is further written as

$$g(\lambda_1, \lambda_2) = \lambda_1 m_i + \lambda_2 \sigma_i + \max\{\max_{z_i}[(1 - y_i(w^T x_i + b)$$

$$-\frac{1}{2}(1 + y_i)(w + \lambda_1)^T z_i)^+ - \lambda_2 ||z_i||], \max_{z_i}[-\lambda_1^T z_i - \lambda_2 ||z_i||]\}.$$

By Cauchy-Schwarz inequality, note $(w + \lambda_1)^T z_i \leq ||w + \lambda_1||_* ||z_i||$ and limit the domain as a compact set. We have $\max_{||z||=\alpha}(-\frac{1}{2}(1 + y_i)(w + \lambda_1)^T z) = \alpha ||w + \lambda_1||_*$. So it yields that

$$g = \begin{cases} 1 - y_i(w^T x_i + b) & ||w + \lambda_1||_* \leq \lambda_2 \\ +\infty & \text{otherwise} \end{cases}.$$

Similarly,

$$\max_{z_i}[-\lambda_1^T z_i - \lambda_2 ||z_i||] = \begin{cases} 0 & ||\lambda_1||_* \leq \lambda_2 \\ +\infty & \text{otherwise} \end{cases}.$$

Minimize the dual function can obtain the dual problem as follows,

$$\min_{\lambda_1, \lambda_2} g(\lambda_1, \lambda_2) = \min_{\lambda_1, \lambda_2} (1 - y_i(w^T x_i + b))^+ + \lambda_1^T m_i + \lambda_2 \sigma_i$$

$$s.t. \ ||w + \lambda_1||_* \leq \lambda_2, ||\lambda_1||_* \leq \lambda_2. \tag{16}$$

The equivalent expression can be obtained by introducing constraints into objective function,

$$\min_{\lambda_1, \lambda_2} g(\lambda_1, \lambda_2) = (1 - y_i(w^T x_i + b))^+ + \min_{\lambda_1}\{\lambda_1^T m_i + \sigma_i \max[||\lambda_1 + w||_*, ||\lambda_1||_*]\}.$$

Since $||w + \lambda_1||_* \geq ||w||_* - ||\lambda_1||_*$, we have the lower bound

$$\min_{\lambda_1, \lambda_2} g(\lambda_1, \lambda_2) \geq (1 - y_i(w^T x_i + b))^+ + \min_{\lambda_1}\{\lambda_1^T m_i + \sigma_i \max[||w||_* - ||\lambda_1||_*, ||\lambda_1||_*]\}.$$

There are two cases: (1) $||w||_* - ||\lambda_1||_* \geq ||\lambda_1||_*$, it follows that $\min_{||\lambda_1||_* \leq \frac{1}{2}||w||_*}\{\lambda_1^T m_i + \sigma_i(||w||_* - ||\lambda_1||_*)\}$ has the optimal lower bound

$$\min_{\lambda_1, \lambda_2} g(\lambda_1, \lambda_2) \geq (1 - y_i(w^T x_i + b))^+ + \frac{1}{2}(\sigma_i - ||m_i||)||w||_*.$$

And, (2) $||w||_* - ||\lambda_1||_* \leq ||\lambda_1||_*$ means the optimal lower bound of $\min_{||\lambda_1||_* \geq \frac{1}{2}||w||_*}\{\lambda_1^T m_i + \sigma_i ||\lambda_1||_*\}$ is

$$\min_{\lambda_1, \lambda_2} g(\lambda_1, \lambda_2) \geq (1 - y_i(w^T x_i + b))^+ + \frac{1}{2}(\sigma_i - ||m_i||)||w||_*.$$

The lower bound is achieved when $\lambda_1 = -\frac{1}{2}w$. Thus the formulation with similar SVM form is given by

$$\min \frac{1}{2}(\sigma_i - ||m_i||)||w||_* + (1 - y_i(w^T x_i + b))^T + \frac{u^T u}{1^T s}. \tag{17}$$

Here, we take the expectation in the dual norm term for simplicity and randomized the SVM classifier in (17) to consider the approximate problem,

$$\min \frac{1}{2}\sum_i (\sigma_i - ||m_i||)||u||_* + \frac{u^T u}{1^T s} \tag{18}$$

$$s.t. \ P_{w \sim N(u, \Sigma)}(a_i^T w \leq -1 - t_i) \geq \nu, i = 1, 2, \ldots n.$$

The $a_i = -y_i x_i$ makes the above constraint the same as (5). Rewrite it by Gauss distribution and introduce slack variable, the final formulation yields,

$$\min \frac{1}{2}\sum_i (\sigma_i - ||m_i||)||u||_* + \frac{u^T u}{1^T s} + C\sum_i \xi_i$$

$$s.t. \ \xi_i \geq 0, i = 1, 2, \ldots, n, \qquad s_i \geq 0, i = 1, 2, \ldots, d, \tag{19}$$

$$y_i^T(u^T x_i) \geq 1 + \Phi^{-1}(\nu)\sum_{j=1}^d x_{ij}^2 s_j - \xi_i.$$

6   Experimental Evaluation

In this section, we evaluate our proposed randomized SVM models against different type of attacks and compare their performance with the robust SVM model proposed in [13]. When simulating different types of attacks, we should estimate x_{ij}^t, the target to which the adversary may change x_i (note that the actual modification is $x_i + \delta_i$, where δ_i depends on x_{ij}^t). We use a simple method to estimate $x_{ij}^t = x_{ij} + c_\delta \varepsilon_{ij}$, where $c_\delta \in (0, 1)$ controls the aggressiveness of attacks (c_δ is small if attackers are conservative on modifying positive samples within a small area, while c_δ is large if attackers are aggressive on poisoning a larger range of sample space), and $\varepsilon_{ij} \sim \mathcal{N}(0, 1)$.

The attack strategies are depicted by setting δ_i in different attack models defined in (1) for RRA, (3) for RRAU, and (4) for DRAU, where c_f in (1) and (3), and σ in (4) control the intensity of attacks (how much to modify x_i). We will evaluate the proposed randomized SVM models at different levels of aggressiveness and intensity. In addition, the matrix P in RRAU is set to be a diagonal matrix, whose maximum eigenvalues $eig(P) \leq 0.5$. In attack model DRAU, we set the parameter σ differently and let $m = 0$. Attacks with larger σ are intenser with higher uncertainty.

We have three different randomized SVM models designed for various attacks, RRA, RRAU and DRAU. Since they are all randomized SVM, we differentiate them by the attack model names, such as SVM-RRA, SVM-RRAU and SVM-DRAU. In SVM-RRAU, we set $||.||_*$ as L_2 norm and $||v_i||$ as L_1 norm in its optimization function (13). In SVM-DRAU, different norms L_1, L_2 and L_∞ are studied in (19).

Four binary classification data sets (SEEDS, CLIMATE, QSAR, and SPAM-BASE) from UCI repository are used as evaluation data. Linear SVMs were implemented using the LIBSVM library, while CVX package with SDPT3.0, MOSEK, Gurobi solvers is implemented to solve randomized classifier SVM models against different attacks. The probability of separation ν in (5) is set to 0.59 if not especially specified.

The experiments go through the following steps for obtaining performance measurement:

(1). Load the training data $(x_i, y_i), i = 1, \ldots, n$,

(2). Modify each x_i to obtain x_i^t,

(3). Train classifier distribution $\mathcal{N}_w(u, s * s^T)$,

(4). Obtain $y_i^t = sign(w^T x_i + b)$ to evaluate classification accuracy and failure rate.

Accuracy measures the classification correctness: Accuracy $= \frac{\sum_i \{y_i^t \neq y_i\}}{n}$. A high value of accuracy indicates the strong capability of the classification model to differentiate one class from the other. Failure rate measures how much the classification system fails to resist the attacks. Failure rate $= \frac{\sum_i \{y_i^t = -1 | y_i = 1\}}{n}$, i.e., among the manipulated malicious data samples with $y_i = 1$, how much of them are recognized as innocuous (the system fails and the attacker wins).

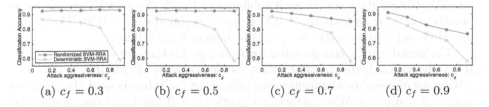

(a) $c_f = 0.3$ (b) $c_f = 0.5$ (c) $c_f = 0.7$ (d) $c_f = 0.9$

Fig. 2. Classification accuracy of randomized SVM-RRA and deterministic SVM-RRA when training data are poisoned at different levels of attack aggressiveness and intensity.

Table 1. Accuracy of SVM-RRAU when varying c_f, c_δ

	$c_f \downarrow$	$c_\delta = 0.1$	$c_\delta = 0.3$	$c_\delta = 0.5$	$c_\delta = 0.7$	$c_\delta = 0.9$
SVM-RRAU	0.1	0.92	0.92	0.91	0.90	0.90
	0.3	0.92	0.90	0.90	0.88	0.86
	0.5	0.91	0.89	0.87	0.84	0.82
	0.7	0.91	0.88	0.84	0.81	0.80
	0.9	0.90	0.87	0.81	0.80	0.80
Standard SVM		$c_\delta = 0.1$	$c_\delta = 0.3$	$c_\delta = 0.5$	$c_\delta = 0.7$	$c_\delta = 0.9$
		0.86	0.71	0.65	0.62	0.62

Table 2. Accuracy of SVM-DRAU with L_∞ when varying σ, ν

	$\sigma \downarrow$	$\nu = 0.59$	$\nu = 0.69$	$\nu = 0.79$	$\nu = 0.89$	$\nu = 0.99$
SVM-DRAU with L_∞	0.1	0.91	0.91	0.91	0.91	0.91
	0.3	0.90	0.90	0.90	0.89	0.89
	0.5	0.89	0.89	0.89	0.89	0.88
	0.7	0.88	0.88	0.88	0.87	0.87
	0.9	0.88	0.88	0.87	0.87	0.87
Standard SVM		$\sigma = 0.1$	$\sigma = 0.3$	$\sigma = 0.5$	$\sigma = 0.7$	$\sigma = 0.9$
		0.86	0.71	0.65	0.62	0.62

6.1 Comparison with Deterministic SVM-RRA

We compare our **randomized** SVM-RRA with the **deterministic** SVM-RRA in [13]. and show how randomization improves the robustness of SVM against RR attack. Figure 2 shows the comparison of classification accuracy when applying **randomized** SVM-RRA and **deterministic** SVM-RRA on poisoned training data at different levels of attack aggressiveness and intensity. The evaluation is on SPAMBASE data set. In each subfigure of Fig. 2, the attack aggressiveness (x-axis) varied from least ($c_\delta = 0.1$) to most aggressive ($c_\delta = 0.9$). The attack intensity varies from gentle ($c_f = 0.3$) to intensest ($c_f = 0.9$).

The overall conclusion we can draw from Fig. 2 is that our randomized SVM-RRA always has higher accuracy than the deterministic SVM-RRA in [13]. The advantage of using randomization for enhancing the robustness of classification system is significant. When attacks are gentle ($c_f = 0.3$ and 0.5), our randomized SVM-RRA performs well with a stable high accuracy even when attackers are aggressive on attacking a large region. However, the performance of deterministic SVM-RRA decreases significantly when c_δ increases. When attacks are intense ($c_f \geq 0.7$), the accuracy of both our randomized SVM-RRA and deterministic SVM-RRA is affected by aggressive attacks. However, our randomized SVM-RRA always performs better than the deterministic SVM-RRA.

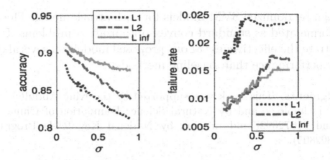

Fig. 3. Accuracy (upper part) and failure rate (lower part) of SVM-DRAU with different norms when varying σ.

6.2 Comparative Study with Standard SVMs

Attacks with uncertainty are less studied in literature. We evaluate the performance of our randomized SVM-RRAU and SVM-DRAU with L_∞ against attacks at different level of intensity and aggressiveness, compared to the performance of standard SVMs. In SVM-RRAU, the intensity and aggressiveness are controlled by c_f and c_δ respectively. In SVM-DRAU with L_∞, we vary ν in (5) to see under different separation probability how SVM-DRAU with L_∞ performs in classification against attacks at different level of aggressiveness (controlled by σ). The results given in Tables 1 and 2 are averaged on four UCI data sets.

There are several observations from these tables. First, all randomized SVM models are more robust against generalized attacks, comparing to standard SVM. Second, model robustness decreases when attacks are more aggressive and more severe. Third, in Table 2, when ν increases, the classifier is expected to separate more examples with a high probability. Then the accuracy decreases a little as it probably makes wrong separations in order to meet the requirement of ν.

6.3 SVM-DRAU with Different Norms

To further analyze the performance of SVM-DRAU with different norms, we evaluate the *classification accuracy* and *resistance failure rate* at different level of attack intensity. Here we fix the separation probability ν in (5) to be 0.59 and set attack aggressiveness $c_\delta = 0.3$. When changing the attack intensity σ from 0.1 to 0.9, the performance of SVM-DRAU with different norms is shown in Fig. 3. We can see that obviously SVM-DRAU with L_∞ is the most accurate and most robust one. It always has higher accuracy than others, while keeps resistance failure rate low. SVM-DRAU with L_2 also has low failure rate, but lower accuracy than SVM-DRAU with L_∞.

7 Conclusion

In this paper, we investigate how randomization can improve the robustness of SVMs against attack models with uncertainty. We define two general attack models and design randomized SVM models for each attack model. The randomized SVMs are formulated as standard convex optimization problems. Experimental results illustrate the effectiveness of our proposed models on several datasets and their better performance than baseline methods.

Acknowledgments. This work was supported by the King Abdullah University of Science and Technology, and by Natural Science Foundation of China, under grant U1736114 and 61672092, and in part by National Key R&D Program of China (2017YFB0802805).

References

1. Alabdulmohsin, I.M., Gao, X., Zhang, X.: Adding robustness to support vector machines against adversarial reverse engineering. In: CIKM, pp. 231–240 (2014)
2. Barreno, M., Nelson, B., Sears, R., Joseph, A.D., Tygar, J.D.: Can machine learning be secure? In: Proceedings of the 2006 ACM Symposium on Information, Computer and Communications Security, pp. 16–25 (2006)
3. Bi, J., Zhang, T.: Support vector classification with input data uncertainty. In: NIPS, pp. 161–168 (2004)
4. Biggio, B., et al.: Security evaluation of support vector machines in adversarial environments. In: Ma, Y., Guo, G. (eds.) Support Vector Machines Applications, pp. 105–153. Springer, Cham (2014). https://doi.org/10.1007/978-3-319-02300-7_4
5. Biggio, B., Nelson, B., Laskov, P.: Support vector machines under adversarial label noise. In: Proceedings of the 3rd Asian Conference on Machine Learning, pp. 97–112 (2011)
6. Brückner, M., Kanzow, C., Scheffer, T.: Static prediction games for adversarial learning problems. J. Mach. Learn. Res. **13**(1), 2617–2654 (2012)
7. Dalvi, N., Domingos, P., Mausam, Sanghai, S., Verma, D.: Adversarial classification. In: SIGKDD, pp. 99–108 (2004)
8. Dekel, O., Shamir, O., Xiao, L.: Learning to classify with missing and corrupted features. Mach. Learn. **81**(2), 149–178 (2010)
9. Globerson, A., Roweis, S.: Nightmare at test time: robust learning by feature deletion. In: ICML, pp. 353–360 (2006)
10. Großhans, M., Sawade, C., Brückner, M., Scheffer, T.: Bayesian games for adversarial regression problems. In: ICML, pp. 55–63 (2013)
11. Xu, H., Caramanis, C., Mannor, S.: Robustness and regularization of support vector machines. J. Mach. Learn. Res. **10**, 1485–1510 (2009)
12. Xu, L., Crammer, K., Schuurmans, D.: Robust support vector machine training via convex outlier ablation. In: AAAI, pp. 536–542 (2006)
13. Zhou, Y., Kantarcioglu, M., Thuraisingham, B., Xi, B.: Adversarial support vector machine learning. In: SIGKDD, pp. 1059–1067 (2012)

Recommendation and Data Factorization

One for the Road: Recommending Male Street Attire

Debopriyo Banerjee$^{(\boxtimes)}$, Niloy Ganguly, Shamik Sural,
and Krothapalli Sreenivasa Rao

Department of Computer Science and Engineering,
Indian Institute of Technology Kharagpur, Kharagpur, India
debopriyo@iitkgp.ac.in, {niloy,shamik,ksrao}@cse.iitkgp.ernet.in

Abstract. Growth of male fashion industry and escalating popularity of
affordable street fashion wear has created a demand for the intervention
of effective data analytics and recommender systems for male street wear.
This motivated us to undertake extensive image collection of male subjects in casual wear and pose; assiduously annotate and carefully select
discriminating features. We build up a classifier which predicts accurately
the attractive quotient of an outfit. Further, we build a recommendation
system - MalOutRec - which provides pointed recommendation of changing a part of the outfit in case the outfit looks unattractive (e.g. change
the existing pair of trousers with a recommended one). We employ an
innovative methodology that uses personalized pagerank in designing
MalOutRec - experimental results show that it handsomely beats the
metapath based baseline algorithm.

1 Introduction

As it is widely believed, fashion primarily deals with women wear, hence most of
the works on fashion predominantly deal with understanding and recommending
female fashion [5,6,13]; there has been a dearth of research in men's clothing [16].
Men's casual and formal attires are much more constrained and moderate compared to apparels designed for women [1,9]. It is all about technical details,
practical concepts and functional requirements that characterizes male attires.
However, the male youth of the twenty first century, have become highly fashion-
conscious. Consequently the neglected topic of male fashion is gaining momentum, and becoming a major part of the billion dollar fashion industry. This forms
a motivation towards conducting research and analysis in male street attire. An
autonomous system that helps a male consumer to choose visually appealing
street wear outfits is expected to be very useful.

To build such system, we collected street photos of men by crawling fash-
ionbeans.com, Facebook and Instagram[1]. Pictures of individuals were extracted
from each photo and their outfit consisting of a set of clothing items were manually annotated with suitable attribute and values. Initially we aimed to build a
supervised model that is capable of predicting whether an outfit looks attractive

---

[1] www.fashionbeans.com, www.facebook.com, www.instagram.com.

© Springer International Publishing AG, part of Springer Nature 2018
D. Phung et al. (Eds.): PAKDD 2018, LNAI 10939, pp. 571–582, 2018.
https://doi.org/10.1007/978-3-319-93040-4_45

(Sect. 3). To create annotated data for that purpose, each outfit is tagged either attractive or not through an extensive survey (Survey 1 - Sect. 2). Besides annotating the attributes in a picture with objective values, we categorized each outfit into three groups - conformative, individualistic and average through another survey (Survey 2 - Sect. 2). We have used this subjective attribute as an additional feature in our prediction model (Sect. 3) - the attribute turned out to be one the most important predictors. It improved the prediction accuracy of the Support Vector Machine model by a fraction of 10% and that of the Random Forest model by 8%. In order to make an efficient prediction and recommendation system, the data was analyzed thoroughly (Sect. 2.2).

Beyond the prediction model, we develop a novel recommendation model MalOutRec which suggests partial replacement of dresses in outfits that has been judged as unattractive, MalOutRec constructs a bipartite graph from the data set and shortlists recommendation through innovative usage of personalized pagerank. We further use the weights obtained in the prediction model to sanitize the recommended list even further. MalOutRec beats the recent baselines by 25%. We also make several interesting observations e.g. layering a shirt over a T-shirt followed by changing a trouser is the best way to enhance attractivity.

2   Data Collection

Our data set consists of 824 street photos; it is enriched with outfits presented on different body shapes of individuals belonging to various age groups. Every photo captured one or more people, dressed in different outfits with a street background. So in order to assess each outfit, pre-processing of the images is required followed by representation of the outfits with a set of well defined attributes. We conducted two surveys - (a) to identify the attractive outfits, and (b) to annotate each outfit as either conformative, individualistic or average.

Data Preprocessing - As mentioned earlier each captured photo may contain one or more men. So we have manually segmented out each men from an image with full frontal view and cropped it with a bounding rectangle, having dimension 400×255 pixels. Then we set the background as white (gray in some cases) and pixelated the face of each segmented image to avoid biasness due to facial features.

Description of Outfits - A set of well defined attributes with appropriate values were used to denote all the clothes and accessories visible in each image after pre-processing. Since each body part gets adorned with specific clothes, we have coded the clothes mainly according to three different body parts, viz. torso, legs and feet. Three categories of torso items are considered - torso base1 (forming the inner most layer) e.g. T-shirt, torso base2 (second layer) e.g. shirt, torso cover (first outer layer) e.g. jacket. A person may be wearing at least one, or more torso items. Each of torso base1 and torso base2 categories are associated with four attributes, and the torso cover category is associated with five attributes. Legs, feet, head and suit are the remaining dress categories. Seven attributes are associated with legs category, four attributes for feet category, and three attributes for head and suit respectively. The presence or absence of a list of accessories such as sunglass, scarf, bag, back-pack, etc. were also recorded. An item may have more than one color. We considered one or two major visible colors for each item from a set of 15 basic colors.

A set of five dress categories have been used in our recommendation model, denoted as dress_category = {*torso base1* (TB_1), *torso base2* (TB_2), *torso cover* (*TC*), *legs* (*L*), *feet* (*F*)}; the other categories like head, suit and accessories are not attempted due to lack of data. In addition to that body structure and age of each subject (person portrayed in image) were coded to form the set, subject_feature = {*age, body structure*}.

It is worth mentioning that there are items which do not have values for all the associated attributes. So we have put *na* as a value for those attributes. Table 1 shows the detailed view of the attributes, and corresponding values that have been formulated to represent each outfit.

2.1 Surveys

Survey 1 - Attractiveness Measurement - Measurement of attractiveness is relative in nature. Hence, in order to identify the set of *attractive* looking outfits in our data set, we ask the respondent to mark at least one image as attractive from a set of four different images - we have build up a mobile app for conducting the survey.

Total 10 slides were shown to every respondent with each slide having four images. The four images are carefully chosen to create each slide - at first, we grouped the images produced after pre-processing into clusters, based on similar *age group* and *body structure* and chose from each cluster. Randomly two slides were chosen and repeated arbitrarily with a view to detect whether responses are consciously given or not. It was ensured that images of each cluster got displayed uniformly across different sessions. We have created both mobile and web versions for people to take the survey.

For each outfit (image) O_i, we computed a score $S_i = \frac{v_i}{N_i}$, where v_i denotes the number of times O_i got selected as attractive, and N_i is the total number of times O_i was shown. We considered all those outfits with score greater than 0.6 as attractive and rest as unattractive. The outcome of this survey which was taken by 213 distinct individuals produced 409 attractive and 415 unattractive outfits.

Survey 2 - Categorization of Outfits - The motivation behind conducting the second survey is to add an attribute named *outfit type* with three values - conformative (CFM), individualistic (INV) and average (AVG)[2]. Such a

---

[2] Collection of dress items, which aligns a person's composite appearance with a desired social group, can be defined as a *conformative outfit*. A collection of dress items that are used to project a person's unique identity, differentiating them from their peer group members in the society, can be defined as an *individualistic outfit*. The set of clothing items that projects a balance between social acceptance and unique personal identity can be defined as an *average outfit* [15].

Table 1. List of attributes and their values

Attribute			Value
Torso	Base (1, 2)	Type	Shirt, shirt-short, shirt-short-in, shirt-short-out, shirt-long, shirt-long-in, shirt-long-out, T-shirt, polo, sweater, sweatshirt, zip-sweatshirt, jacket, waterproof-jacket, blazer, na
		Patterns	Mono, stripes, rhomb, motives, imprint, picture, checker, others, na
		Color-(1, 2)	Black, blue, white, brown, gray, khaki, green, red, yellow, beige, purple, pink, orange, teal, other, na
	Cover	Type	Jacket, jacket-suit-type, jacket-sport/casual, jacket-elegant, coat, coat-mini, coat-midi, coat-maxi, sweatshirt, zip-sweatshirt, blazer, waterproof-jacket, vest, sweater, others, na
		Patterns	Mono, stripes, rhomb, motives, imprint, picture, checker, others, na
		Material	Jeans, wool/cotton, leather, micro fleece, synthetic/mixed, na
		Color-(1, 2)	Black, blue, white, brown, gray, khaki, green, red, yellow, beige, purple, pink, orange, teal, other, na
Legs	Patterns		mono, stripes, motive, imprint, checker, others, na
	Material		Jeans, cord, wool/cotton, sport/synthetic, leather, na
	Shape		Skinny/close-fitting, normal, wide, baggy, na
	Length		Long, 7/8, half, short, na
	Age-type		Normal, washed-out, na
	Color-(1, 2)		Black, blue, white, brown, gray, khaki, green, red, yellow, beige, purple, pink, orange, teal, other, na
Feet	Type		Sport-converse, sport-athletic, sport-casual, boots-sport, boots-sport-in, boots-elegant, boots-elegant-in, boots-casual, boots-casual-in, boots-military, boots-military-in, casual shoes, elegant, sandals, moccasins, flip-flops, na
	Socks		With, without, not visible
	Color-(1, 2)		Black, blue, white, brown, gray, khaki, green, red, yellow, beige, purple, pink, orange, teal, other, na
Head	Type		Baseball cap, flat cap, hat, scarf, winter hat, beret, peaked cap, others, na
	Color-(1, 2)		Black, blue, white, brown, gray, khaki, green, red, yellow, beige, purple, pink, orange, teal, other, na
Suit	Type		With vest, without vest, na
	Color-(1, 2)		Black, blue, white, brown, gray, khaki, green, red, yellow, beige, purple, pink, orange, teal, other, na
Accessories present			Sunglass(0/1), scarf(0/1), bag(0/1), back-pack(0/1), hand-jewellery(0/1), neck-jewellery(0/1), tie(0/1), bow-tie(0/1), belt(0/1), watch(0/1), gloves(0/1)
Body structure			Thin, normal, obese
Age			≤ 25, 26–35, 36–55
Outfit type			Conformative, individualistic, average

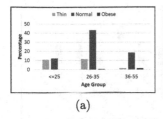 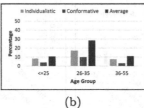 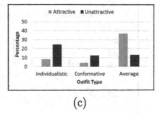

(a) (b) (c)

Fig. 1. Statistics regarding (a) Body structure, (b) Outfit preference w.r.t age, and (c) Attractive Outfits.

distinction can be used as one of the discriminative predictors while classifying an outfit as either attractive or not.

An online user interface was created where each respondent was asked to register with email-id and password. On logging into the system, a respondent was shown a response form with an image having three options below it - conformative, individualistic, average - he had to choose any one of the options. A respondent was provided with certain reference images tagged with those three categories which can be referred any time during the survey. One was allowed to log in several times and tag the set of all 824 images over a period of seven days. Five male students were chosen who took the entire survey.

An outfit may be labeled either as CFM or INV or AVG; if an outfit got 3 or more votes for a particular category, we tagged that outfit with that category. However, there may be a case where an outfit may get 2 votes each for two different category. In that case, we took another opinion (from another male student) and tagged accordingly. Finally, we got 138 outfits as conformative, 273 as individualistic, and 413 as average.

2.2 Data Statistics and Analysis

Majority of men in our data set belongs to 26–35 age group, followed by those below 25. The distribution of outfit type preference shows that average outfits are mostly preferred (50%), as majority balances between social acceptance, and individual identity. Individualistic outfits also form a sizeable group (33%), with many opting to showcase personal identity without bothering about critics. Only a very small fraction (17%) opted for conformative outfits. Most men aged below 25, and between 26–35 have normal body structure with very few being obese (Fig. 1a). Average looking outfits are mostly preferred by those between 26–35 (Fig. 1b). Specifically, men with normal body structure and of 26–55 age group prefer average outfits over conformative or individualistic (Fig. 2a). More than 50% of men with normal body structure and age between 26–35 chose average looking dress. Figure 1c shows that mostly the average outfit types are considered as attractive while conformative looking ones as unattractive.

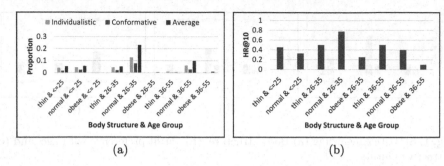

(a)   (b)

Fig. 2. (a) Preference of outfits types, and (b) Hit rate for selective recommendation of top-10 pair of items (chosen from two dress-category) across different body structures and age groups.

3 Prediction Model

Our prediction task aims to assist a male consumer to check whether a chosen outfit looks attractive or not. In the first survey, we got our data set annotated as either attractive or unattractive followed by a second survey which helped distinguish each outfit into three classes (*conformative, individualistic* and *average*). Our data set consists of 33 categorical and 11 binary predictor attributes; the target variable is binary.

In order to identify the optimal subset of discriminative features, we applied some of the standard feature selection techniques. More specifically we have used Filter [4] and Wrapper [8] methods for selecting the best set of features. Out of the 44 predictors, we got 25 discriminative predictor variables. The chosen predictors are as follows: [body structure, age, **F**[color1, type], **L**[material, shape, color2], **TB₁**[color2, type, patterns], **TB₂**[color2, type, patterns], **TC**[color1, color2, type, patterns], **head**[color2, type], **suit**[type], **accessories**[scarf, bag, back pack, tie], outfit type]. With the selected subset of features some of the classical prediction models - Random Forest (RF), and Support Vector Machine (SVM) were built.

From the trained SVM model we obtain the coefficients associated with each predictor variable. Let the predictors be denoted as $X_1, X_2, \ldots, X_n$, and the corresponding learned coefficients be denoted as $\theta_1, \theta_2, \cdots, \theta_n$ ($n = 25$ in our case). Now consider an outfit o_i from the test data set. Here we are interested in finding the positive influence of a category $j \in o_i$. It can be computed as follows:

$$I_j^{(i)} = \sum_{k \in S_j} \theta_k \cdot X_k^{(i)}, \tag{1}$$

where S_j is the set of attributes associated with the item of the category j, $X_k^{(i)}$ is the value of attribute $k \in S_j$ for outfit o_i, and θ_k is the coefficient of k as computed before. So if o_i consists of five categories, one can compute $I_j^{(i)}$ $\forall j = 1, 2, \ldots, 5$ and sort them in ascending order providing a rank of unattractiveness of the categories. As mentioned before we are considering only the five categories for recommendation.

4   Recommendation Model - MalOutRec

In this Section, we develop a method to recommend replacement of certain dress categories whenever an outfit is predicted as unattractive for a particular age group and body structure by the prediction model of Sect. 3. As per the objective of recommendation, replacement of either one or two categories of dress are initiated. Our recommendation model - MalOutRec - is inspired by the multidimensional recommendation model of [12]. [12] constructed a bipartite graph [3] by considering information across multiple dimensions. The bipartite graph consists of target and non-target nodes and through use of personalized pagerank algorithm [7] suitable target nodes are selected based on an initial input. However, beyond selecting target nodes we take advantage of the replacement concept to sieve in higher quality recommendation.

4.1   Construction of Bipartite Graph

We first elaborate the target and non-target nodes in our case.

Target Nodes: The target nodes comprise of a subset of the five dress-categories - the cardinality of the subset depends on the number of categories we want to recommend. For example, if we want to recommend $D_T = \{torso\ base1,\ feet\}$, then each target node would comprise of two categories (TB_1 and F) and its corresponding attributes.

Non Target Nodes: A non-target node $Z = \{X, Y\}$, where $X = \{$age or body structure$\}$ while Y is either one of the attribute in $\{F[type], L[material], TB_1[type], TB_2[type], TC[type]\}$ or all of them taken together.

From Picture to Node Formation: Given a picture, based on the recommendation objective we construct the bipartite graph. Suppose if we want to recommend the categories $\{torso\ base1\ (TB_1),\ feet\ (F)\}$, the target node contains the value of their attributes (here all the attributes of $TB_1\ and\ F$ are used to form the target nodes) while the corresponding non target nodes are $Z = \{X, Y\}$; the attributes associated with the target dress-category set are not used in formation of the non target nodes i.e. in this example, $\{TB_1[type], F[type]\}$ are left out. That is, given a picture and a recommendation objective, we derive a subgraph of either 10 or 8 nodes in the non-target set and one node in the target set. We form a bipartite graph by converting all the candidate pictures into corresponding subgraphs and merging them. We merge two nodes (arising from two different pictures) if the nodes have same attribute values.

Edge Formation with Weights: The edges are directed in the graph, initially when a subgraph is derived from a picture, each non-target node has a directed edge towards target node with weight 1. When two non-target nodes as well as their corresponding target nodes get merged, the weight of the edge is accordingly added. Corresponding to each incoming edge in a target node a matching outgoing edge is drawn. The sum of weight of the outgoing edges from a target node is set to 1; the corresponding weights are learned using [2]. Once the graph is formed, corresponding adjacency M, and transition probability matrix P are derived.

4.2   Traversal of the Graph

In the recommendation phase, given information of an unattractive outfit we decide to recommend either one or a subset of dress_category. With the information of all the attributes except that of the dress_category which we want to recommend, the set Y of non target nodes are formed. The set Y is then used to define a vector $\vec{q'}$ as follows:

$$q'_i = \begin{cases} 1, \text{if } v_i \in Y, \\ 0, \text{otherwise} \end{cases} \tag{2}$$

The vector $\vec{q'}$ is then normalized to $\vec{q}$, which we use as a query vector. We have used the algorithm of personalized pagerank [7] for ranking nodes in our graph. The personalized pagerank score can be expressed as

$$\vec{r} = cP^T\vec{r} + (1-c)\vec{t} \tag{3}$$

where $\vec{r}$ is the rank score vector of all the nodes, c is a constant damping factor (usually set as 0.85), P is the transition probability matrix, and $\vec{t}$ is the teleport vector or a normalized biased vector representing the interest of users. Here we have replaced the teleport vector $\vec{t}$ with our normalized query vector $\vec{q}$ to rank the nodes. So in our case, the ranking score vector can be expressed as $\vec{r} = cP^T\vec{r} + (1-c)\vec{q}$. The top-$k$ ranked target nodes are shortlisted for recommendation.

4.3   Refining Recommendation Using Positive Influence Factor

The recommendation list can be further refined by having a quick check whether the chosen target node is performing better than the one it is replacing. This is done by considering the weights obtained from the SVM in Sect. 3 and calculating the positive influence of the items recommended for replacement (Eq. 1). We check whether this influence is higher (by a value δ) than the influence of the original items and discard any recommendation which does not satisfy this constraint.

5   Experiments

We have a total of 824 example outfits; according to Survey 1, 409 outfits are attractive, and 415 are unattractive. From Survey 2, we got 138 outfits as conformative, 273 as individualistic, and 413 as average. Let AT, and UAT be the set of attractive and unattractive outfits respectively.

5.1 Prediction Accuracy

We randomly selected 270 outfits respectively from AT, and UAT to construct the training set. The test set consisted of 139 outfits from AT and 145 outfits from UAT. The 2 classification models SVM and RF were subjected to 10 fold cross validation using the training set. Optimal cutoff probability for classification has been used to compute the maximum accuracy. The SVM model have 10-fold cross validation accuracy $= 0.74$, f1-score $= 0.74$, and Area Under the Receiver Operating Characteristic Curve (ROC AUC) $= 0.8$. However, the RFC model produced a 10-fold cross accuracy of 0.73, with f1-score $= 0.73$, and ROC AUC $= 0.8$. So, we chose the slightly better SVM model over the RFC model in our prediction task. The inclusion of the attribute *outfit type* (derived from Survey 2), improved classification accuracy of the SVM model by a fraction of 0.10, and the RFC model by 0.08.

5.2 MalOutRec - Performance Assessment

Evaluation: The recommendation algorithm - MalOutRec is evaluated with the set of outfits that have been predicted as not attractive. Suppose PR_{uat} be the set used for testing the recommender. For every outfit $o_i \in PR_{uat}$, our recommendation algorithm suggests top-k set of items from one or two dress_category. The corresponding items are then replaced in o_i and the modified outfit $\hat{o}_i$ is obtained. The prediction model is then used to predict whether the newly created $\hat{o}_i$ is attractive or not. This is done for $k = 3$, 5, and 10.

We use the metric HR@k [10] for assessing the performance of our recommender. It is defined as $\frac{\#hits}{N}$, where #hits is the number of cases when at least one of the top-k recommended items transform the unattractive into an attractive outfit, and N is the cardinality of the set PR_{uat}.

Baseline Method: RecMeta In order to compare our recommendation algorithm with an existing state of the art, we adopted the metapath based recommendation algorithm in [11]. The two metapaths used are $item \rightarrow X \rightarrow item$, and $item \rightarrow X \rightarrow item \rightarrow ensemble \rightarrow item \rightarrow Y \rightarrow item$. In our set up, an item corresponds to a particular dress_category with values of all the associated attributes taken together; X, Y represents common attribute value between two items, ensemble represents an outfit from the set of all outfits. The relevance metric to measure similarity between source and target items used is AvgSim [14].

Experimentation Results Comparison: Table 2 compares MalOutRec with the baseline algorithm-RecMeta. We present results for both MalOutRec without shortlisting via positive influence factor (Sect. 4.3) (henceforth called MalOutRecPagerank) and MalOutRec. The results (averaged over all cases) are presented with top-k (k = 3, 5 or 10) recommendations having one or two items. It is seen all the three algorithms in around 12–25% of cases, can effectively recommend a single item that changes an unattractive outfit to attractive. The figures seem to be low but it is natural as outfit comprises of seven-eight items, changing one may work only in some cases. The performance improves to 38%

Table 2. Comparison of average hit rate for recommendation of items taken from all possible one (case 1) or two (case 2) dress_category, using RecMeta (M_1), MalOutRecPagerank (M_2), and MalOutRec(M_3).

	$k = 3$			$k = 5$			$k = 10$		
	M_1	M_2	M_3	M_1	M_2	M_3	M_1	M_2	M_3
*Case*1	0.1205	0.1698	0.2721	0.1507	0.2384	0.3367	0.2493	0.3466	0.4551
*Case*2	0.4017	0.3856	0.6209	0.4601	0.5117	0.7240	0.5567	0.6432	0.8237

Table 3. First row - HR@k recommendation of one(two) dress_category [($SR_{1(2)}$)], second row - HR@k recommendation of two dress_category for different outfit types (Algorithm - MalOutRec)

	$k = 3$	$k = 5$	$k = 10$	SR_1	$k = 3$	$k = 5$	$k = 10$	SR_2	$k = 3$	$k = 5$	$k = 10$
					0.60	0.63	0.67		0.5	0.61	0.71
Conformative	0.47	0.58	0.75	Individualistic	0.44	0.56	0.65	Average	1.00	1.00	1.00

when two items are recommended. MalOutRecPagerank performs better (albeit modest) than RecMeta for five of the six cases - the performance improves as we increase the value of k, However, we find that the performance of MalOutRec is way superior to RecMeta. The measure driven by positive influence helped to provide targeted recommendations according to the needs of individual users and consequently resulted in improved performance.

Performance of MalOutRec - Detailed Analysis. Figure 3 shows result of impact of recommendation of one and two items respectively. Note that through measuring the positive influence factor, we can not only refine the recommendation list for a particular item but also determine which one (two) of the five dress_category item replacement would have the best effect. This is reflected in value of $SR_{1(2)}$ in Figs. 3(a), and (b) respectively. The result is also presented in the first row of Table 3. One can now compare the result with the performance of RecMeta (from Table 2) and notice the improvement is massive (60% in certain cases). In case of recommendations having items from a single dress_category, TB_2 was found to have the maximum HR@10 measure of 69.5%, followed by TC with 54.9% and TB_1 with 42.3% (Fig. 3(a)). This indicates that in maximum cases by modifying only TB_2 - that is the second layer of clothing of torso, an unattractive outfit can be transformed into attractive one. In cases where we recommend two dress items, TB_2 and L has the highest HR@k measure (Fig. 3(b)). Thus, by putting a second layer of top-wear (e.g. shirt/sweatshirt) over the first layer (e.g. T-shirt) followed by replacing a trouser, transforms an unattractive outfit into attractive.

MalOutRecPerformance Across Different Parameters. Figure 2b shows the performance of MalOutRec across various age group and body structure. It is seen that the algorithm performs best for young and thin, normal weighted people. The performance is generally poor across obese people. Table 3 (2nd row)

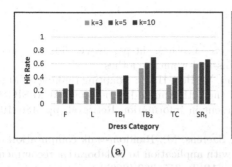 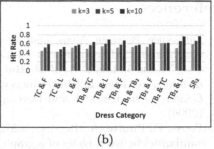

(a)   (b)

Fig. 3. Top-k performance comparison of recommendation for one (a), or two (b) items from different dress categories; $SR_{1(2)}$ denotes selective single (double) item recommendation; TB_1, TB_2, TC, L, and F correspond to the dress categories as defined before.

shows that the fraction of times recommendation given for a conformative dress, is better than individualistic. This is perhaps as the recommendation system is converting it to a more average type of dress. The performance for average dress is highest as MalOutRec just retrieves similar attractive outfits for an already average dress.

6   Conclusion

We believe this is one of the first endeavors to understand the dynamics of male street fashion; we have given a substantial effort in carefully collecting the data, annotating them and conducting a survey among more than 200 people to guage the attractiveness of an outfit. Since attractiveness is a subjective concept, predicting it is considered a hard job; our accuracy of 74% is decent in that scenario. The prediction accuracy improved substantially as we mined social science literature and understood that the concept of social conformance is an important discriminative feature. The novelty of the recommendation algorithm MalOutRec is two-fold: (a) we make an innovative use of personalized pagerank algorithm (b) we take advantage of the fact that this is a replacement recommendation task and accordingly make a quick comparison between old and new items to refine the recommendation set. The result shows remarkable improvement over metaphath based baseline. Although the method can be extended to even female fashion, concentrating on male attires provides us interesting insights, For example, it has been previously found, that most of the female dress items are tightly coupled, and as such there is very little scope to recommend separate items to enhance the attractiveness of an outfit which we found largely does not hold for male street attire. However, these are initial propositions and need to be studied in detail in the future. As a final note, we intend to make the relevant portion of the data, code and developed applications available to research community.

References

1. Difference Between Mens and Womens Clothes. https://www.slideshare.net/coughwalrus92/difference-between-mens-and-womens-clothes
2. Cao, Y., Xu, J., Liu, T.Y., Li, H., Huang, Y., Hon, H.W.: Adapting ranking SVM to document retrieval. In: Proceedings of the 29th Annual International ACM SIGIR Conference on Research and Development in Information Retrieval, pp. 186–193 (2006)
3. Fouss, F., Pirotte, A., Renders, J.M., Saerens, M.: Random-walk computation of similarities between nodes of a graph with application to collaborative recommendation. IEEE Trans. Knowl. Data Eng. 19(3), 355–369 (2007)
4. Hall, M.A.: Correlation-based feature selection for machine learning (1999)
5. Jagadeesh, V., Piramuthu, R., Bhardwaj, A., Di, W., Sundaresan, N.: Large scale visual recommendations from street fashion images. In: Proceedings of the 20th ACM SIGKDD International Conference on Knowledge Discovery and Data Mining, pp. 1925–1934. ACM (2014)
6. Kiapour, M.H., Han, X., et al.: Where to buy it: matching street clothing photos in online shops. In: International Conference on Computer Vision (2015)
7. Kleinberg, J.M.: Authoritative sources in a hyperlinked environment. J. ACM 46(5), 604–632 (1999)
8. Kohavi, R., John, G.H.: Wrappers for feature subset selection. Artif. Intell. 97(1), 273–324 (1997)
9. Leach, R.: Why men's fashion is not like women's fashion. http://www.telegraph.co.uk/men/fashion-and-style/10749463/Why-mens-fashion-is-not-like-womens-fashion.html
10. Lee, D., Park, S.E., Kahng, M., Lee, S., Lee, S.: Exploiting contextual information from event logs for personalized recommendation. In: Lee, R. (ed.) Computer and Information Science 2010. SCI, vol. 317, pp. 121–139. Springer, Heidelberg (2010). https://doi.org/10.1007/978-3-642-15405-8_11
11. Lee, H., Lee, S.: Style recommendation for fashion items using heterogeneous information network. In: Poster Proceedings of the 9th ACM Conference on Recommender Systems (2015)
12. Lee, S., Song, S.I., Kahng, M., et al.: Random walk based entity ranking on graph for multidimensional recommendation. In: 5th ACM Conference on Recommender Systems, pp. 93–100. ACM (2011)
13. Liu, S., Song, Z., Liu, G., Xu, C., Lu, H., Yan, S.: Street-to-shop: cross-scenario clothing retrieval via parts alignment and auxiliary set. In: 2012 IEEE Conference on Computer Vision and Pattern Recognition, pp. 3330–3337 (2012)
14. Meng, X., Shi, C., Li, Y., Zhang, L., Wu, B.: Relevance measure in large-scale heterogeneous networks. In: Chen, L., Jia, Y., Sellis, T., Liu, G. (eds.) APWeb 2014. LNCS, vol. 8709, pp. 636–643. Springer, Cham (2014). https://doi.org/10.1007/978-3-319-11116-2_61
15. Niinimäki, K.: Green aesthetics in clothing. Artifact 3(3), 3.1–3.13 (2014)
16. Reilly, A., Rudd, N.A.: Shopping behaviour among gay men: issues of internalized homophobia and self-esteem. Int. J. Consum. Stud. 31(4), 333–339 (2007)

Context-Aware Location Annotation on Mobility Records Through User Grouping

Yong Zhang[1], Hua Wei[2], Xuelian Lin[1(✉)], Fei Wu[2], Zhenhui Li[2],
Kaiheng Chen[1], Yuandong Wang[1], and Jie Xu[3]

[1] BDBC, School of Computer Science and Engineering, Beihang University,
Beijing, China
{zhangy11,linxl,chenkh,wangyd}@act.buaa.edu.cn

[2] College of Information Science and Technology, Pennsylvania State University,
State College, USA
{hzw77,fxw133,jessieli}@ist.psu.edu

[3] School of Computing, University of Leeds, Leeds, UK
j.xu@leeds.ac.uk

Abstract. Due to the increasing popularity of location-based services, a massive volume of human mobility records have been generated. At the same time, the growing spatial context data provides us rich semantic information. Associating the mobility records with relevant surrounding contexts, known as the location annotation, enables us to understand the semantics of the mobility records and helps further tasks like advertising. However, the location annotation problem is challenging due to the ambiguity of contexts and the sparsity of personal data. To solve this problem, we propose a Context-Aware location annotation method through User Grouping (CAUG) to annotate locations with venues. This method leverages user grouping and venue categories to alleviate the data sparsity issue and annotates locations according to multi-view information (spatial, temporal and contextual) of multiple granularities. Through extensive experiments on a real-world dataset, we demonstrate that our method significantly outperforms other baseline methods.

1 Introduction

In recent years, location-based services have been widely used in our daily lives and generated a massive volume of human mobility records (e.g., transportation records) and online spatial context data (e.g., venue database). The combination of mobility records with *relevant* contexts helps reveal the semantic of user movement and is known as the *semantic annotation of mobility records* [1]. In this paper, we use venue dataset as the context and consider the problem of mapping a user's location to a venue he might actually visit. The work can have important applications, such as user profiling, recommendation, and advertisement targeting. For example, as shown in Fig. 1, if we know a person often moves

© Springer International Publishing AG, part of Springer Nature 2018
D. Phung et al. (Eds.): PAKDD 2018, LNAI 10939, pp. 583–596, 2018.
https://doi.org/10.1007/978-3-319-93040-4_46

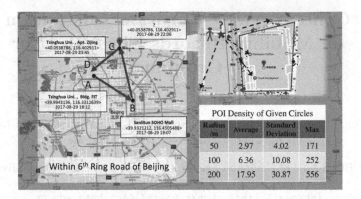

Fig. 1. An example of location annotation problem

from a university to entertainment venues at night, and go back very late, we could infer this person is a sparky college student, and recommend some recreational activities to him. Besides, smart city applications can benefit from such semantic understandings of the raw transportation data.

However, it is hard to associate right venues with mobility records. The challenges are mainly two folds: (1) Both recorded locations and surrounding contexts are *ambiguous*. For a given mobility record of a user, the observed location could have noises, and there may be many possible venues around. As Fig. 1 shows, the number of POIs in some areas of Beijing can reach 500 (according to the data from AutoNavi[1]). (2) The user data maybe *sparse*. Though the total number of mobility records is large, each user may only have a limited number of personal data. According to the data from UCAR[2], less than 10% users have their trips recorded over 3 times within one month. In addition, most POIs only have a few visit records except some popular ones.

For the annotation problem, some previous work mainly considers the distance between the context location and the location of a user [2–4], while others further consider personal preference [5,6]. However, modeling personal preference straightforwardly requires adequate records for each individual user, which conflicts with the data sparsity. Furthermore, these methods do not consider the influence of contextual mobility records (e.g., former and later records of a given record). For example, a person who has just visit a restaurant is less likely to visit a restaurant again in a short time.

To tackle these problems, we propose a Context-Aware location annotation method through User Grouping, named CAUG. In this method, the correlation between mobility contextual records is captured by contextual features. And the data sparsity issue is mainly compensated by considering information of user groups, which is based on our intuition that users who share similar mobility

---

[1] A map service provider. https://en.wikipedia.org/wiki/AutoNavi.

[2] A chauffeured car service provider in China. https://www.crunchbase.com/organization/ucar.

Table 1. Summary of notations

Notation	Terms	Description
p	Point-of-Interest (POI)	$p\langle id, name, l, c\rangle$, where $p.l$ is a location defined by longitude and latitude, and $p.c$ is categories of POI
c	Categories of POI	$c\langle c_1, c_2, c_3\rangle$, where $c.c_1$, $c.c_2$ and $c.c_3$ are p's 1st class, 2nd class and 3rd class categories, respectively
g	Grid	$g\langle row, col\rangle$, an indexed grid ($780\,\mathrm{m} \times 780\,\mathrm{m}$) in a city
tp	Period of a Day (POD)	A time period of a day, including morning, noon, afternoon, evening and late night
s	Spatio-temporal Area	$s\langle g, tp\rangle$, denoting a grid g at time period tp
z	User Activity	$z\langle v.c, tp\rangle$, where $v.c$ is a venue category and tp is a time period
x	Stop-Point	$x\langle u, l, t, p\rangle$, denoting a geographic location $x.l$ where a user $x.u$ *actually* picked up or droped off at time $x.t$ around a POI $x.p$
r	Travel Record	$r\langle x^s, x^e\rangle$, where $r.x^s$ and $r.x^e$ are the start stop-point and the end stop-point of a user, respectively
T'	Tajectory	A sequence of stop-points of a user

patterns are likely to visit similar venues under the same condition. To summarize, we make the following contributions:

1. We propose an iterative grouping method to group users based on the similarity of their mobility patterns, which are captured by a Hidden Markov Model (HMM) [7]. The user-grouping method alleviates the data sparsity to a great extent.
2. We apply a ranking model to annotate locations, with a strategy that integrates multi-view (spatial, temporal and contextual) information extracted from users and POIs' historical information at different granularities. The comprehensive consideration guarantees the effectiveness of annotation.
3. We evaluate our method with a 14-month real-world dataset from a car-hailing company. Experimental results show our method can produce effective user groups and contextual features. Meanwhile, this method outperforms other baseline methods.

2 Preliminary

This section describes some basic terms used in this paper. The notations are summarized in Table 1.

We use transportation data from UCAR for our study, including a mass of travel records. When using UCAR's online car-hailing service to book a trip, a user can search based on address or based on the name of the POI and then selects a POI for pick-up and a POI for drop-off. After the trip finished, it will be recorded as a *travel record* which consists of a start *stop-point* and an end *stop-point*. It should be noted a POI selected by a user can be either a *venue* or an *indistinct place*. A *venue* refers to a place like a restaurant or a cinema where people conduct specific activities, while an *indistinct place* is a place like a crossroad or a public parking lot which is hardly the final intention of a user. In order to make users' trips more semantic, those *stop-points* whose POIs are indistinct places are to be annotated with venues users might actually visit.

A *trajectory* represents a sequence of stop-points with contextual relations. Therefore, we actually concatenate a set of *travel records* back to a *trajectory* if the end *stop-point* of the previous trip and the start *stop-point* of the next trip has similar time or location. The notion of *trajectory* enables our model to consider contextual correlation not only between a start *stop-point* and an end *stop-point*, but also between *travel records*. Specifically, given a sequence of *stop-points* $T = x_1^s x_1^e x_2^s x_2^e \ldots x_n^s x_n^e$ of a user u, a time gap threshold $\Delta_t > 0$ and a distance threshold $\Delta_d > 0$, a subsequence $T' = x_i^s x_i^e x_{i+1}^s x_{i+1}^e \cdots x_{i+k}^s x_{i+k}^e$ is a *trajectory* of T if T' satisfies: (a) $\forall 1 < j \leq k$, $x_j^s.t - x_{j-1}^e.t \leq \Delta_t$ or $d(x_{j-1}^e.l, x_j^s.l) \leq \Delta_d$, where $d(l_a, l_b)$ stands for the distance between location l_a and l_b; and (b) there are no longer sub-sequences in T that contains T' and satisfies condition (a). Also, for a given *stop-point* x in T', its former and later points are denoted as $\overleftarrow{x}$ and $\overrightarrow{x}$, respectively.

Problem 1 (Location Annotation). *Given a trajectory T' of a user u, for each stop-point x in T' whose POI $x.p$ is an indistinct place or unknown, a location annotation method provides a list of venues u might visits.*

3   Method

3.1   Overview of CAUG

The overall framework of CAUG is shown in Fig. 2, mainly including *user grouping, feature extracting* and *venue ranking*. When extracting features from historical records, we organize records into multiple granularities.

First, from the perspective of users, we observe the personal records are usually too sparse to do the personalized annotation and the overall records are too inconsistent as different users usually have totally different mobility patterns. Thus, we are motivated to find a middle-level granularity called user group to compensate, based on our intuition that users who share similar mobility patterns may have similar visit tendencies under the same condition (e.g., a group of colleagues may live in the same housing area and often go to a specific bar near their homes after work). In this way, we organize user travel records into three levels, i.e., personal level, group level and overall level.

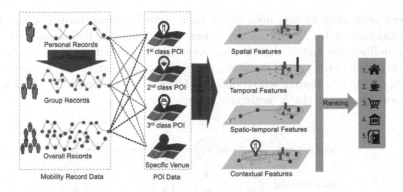

Fig. 2. An overview of CAUG

Second, from the perspective of venues, we observe though most venues are seldom visited, similar venues have similar visited tendencies. Therefore, we consider not only the visit history of venues themselves but also that of the venue categories. We adopt a classification method of venue categories defined by AutoNavi and organize a specific venue into four classes. For example, a specific Starbucks coffee shop belongs to the shop itself (4th class), Starbucks (3rd class), cafe (2nd class) and catering services (1st class). Note that the venue category can vary with different classification methods based on different POI datasets. Combining the above two kinds of granularities, i.e., three levels of user records and four classes of venues, we finally get $3 \times 4 = 12$ granularities.

For a stop-point in a trajectory of a user, CAUG first selects all venues within a distance as candidates. Then, for each candidate venue, CAUG extracts a series of spatial, temporal, spatio-temporal and contextual features for each of the twelve granularities. Finally, CAUG returns a ranked list of Top-k venues through a ranking model.

3.2   User Grouping

When people move from one place to another place, their activities are in nature sequential and have mobility patterns. The mobility patterns of users could be captured by HMM, which is widely used to model the mobility of users [4,8]. We use it for learning representations of users' mobility patterns and further use it in an iterative grouping.

HMM Formulating. We assume there are K activities $Z = \{z_1, z_2, \ldots, z_K\}$ (*i.e., hidden states*) and M spatio-temporal areas $S = \{s_1, s_2, \ldots, s_M\}$ (*i.e., observations*). As a user movement process is shown in Fig. 3, each observation s_n in the observed sequence $s_1 s_2 \ldots s_N$ corresponds to a state $z_n \in Z$, and the state seqeunce $z_1 z_2 \ldots z_N$ follows a transition regulation. Thus, we consider three factors when formulating the HMM: (1) the probabilities of activities users begin, (2) the probabilities of transition between activities, and (3) the probabilities of users appearing at one area given their activity. More formally, the HMM is parameterized by $\lambda = (\pi, A, B)$, where $\pi = (\pi_i)$ is a K-dimensional vector which defines the initial distribution over the K activities, $A = [a_{ij}]_{K \times K}$ is a matrix that defines the transition probabilities among the K activities, and $B = [b_{ij}]_{K \times M}$ is a matrix which defines the emission probabilities of M spatio-temporal areas over the K activities.

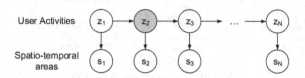

Fig. 3. The illustration of the HMM

Given R trajectories, we first generate a *observed* sequence and a corresponding *state* sequence for each trajectory by the following ways: (1) map stop-points in the trajectory to spatio-temporal areas, and consider the sequence of spatio-temporal areas as the observed sequence, (2) map stop-points in the trajectory to activities, and consider the activity sequence as the state sequence. However, because a proportion of stop-points' POIs may be indistinct places, their corresponding activities are unclear. In this case, the parameters of the model are estimated in the following way:

$$\pi_i = \frac{R_i + \alpha}{R_\pi + K\alpha}, i = 1, \ldots, K \tag{1}$$

$$a_{ij} = \frac{A_{ij} + \alpha}{\sum_{k=1}^{K} A_{ik} + K\alpha}, i = 1, \ldots, K; j = 1, \ldots, K \tag{2}$$

$$b_{ij} = \begin{cases} \frac{B_{ij} + \alpha}{\sum_{m=1}^{M} B_{im} + M^s\alpha} & , z_i.tp = s_j.tp \\ 0 & , z_i.tp \neq s_j.tp \end{cases}, i = 1, \ldots, K; j = 1, \ldots, M \tag{3}$$

where R_i is the number of state sequences begin with activity z_i and R_π is the number of state sequences whose first state is explicit. A_{ij} is the frequency of transferring from state i at time t to state j at time $t + 1$, which is counted according to state sequences, skipping indistinct activities. B_{ij} is the frequency of appearing in spatio-temporal area s_j when doing activity z_i. M^s is the number of grids and α is the smoothing parameter of the additive smoothing[3].

Iterative Grouping. Given a set of users $U = \{u_1, u_2, \ldots, u_D\}$ and their trajectories, we first initialize their groups. Then, we employ an iterative refinement framework to further group users based on their mobility patterns. During each

---

[3] https://en.wikipedia.org/wiki/Additive_smoothing.

iteration, we generate a better representation of each group's mobility pattern and then assign every user to a more appropriate group. The major steps are described as follows:

Step 1: Initialization. In this step, instead of assigning every user to a group randomly, we vectorize every user and use clustering algorithms like k-means to preliminarily cluster users according to their mobility patterns, so as to reduce the time cost in the subsequent iteration. Specifically, we first train an HMM $H_u = (\pi_u, A_u, B_u)$ for each user u by the aforementioned learning method, whose parameters reflect the mobility pattern. Then, we simply reshape (π_u, A_u, B_u) to an E-dimensional vector, where $E = K \times (1 + K + M)$. After vectorizing all users, we stack all vectors to be a matrix $I_{D \times E}$. Since E is a very large integer, we leverage PCA to reduce the dimensionality and then apply k-means to get the initial user groups $\varphi = \{g_1, g_2, \ldots, g_G\}$ where G is a user-defined group number. Next, for each $g \in \varphi$, we train an HMM H_g which represents the mobility pattern of group g. Finally, we get an initial HMM ensemble $\Phi^{(0)} = \{H_g^{(0)} \mid g \in \varphi\}$.

Step 2: Grouping. For each user u, let us denote the set of u's trajectories as J_u, in which the j-th trajectory is T_u^j. Based on the latest HMM ensemble $\Phi^{(t)}$, we assign u to a new group g^{t+1} by a way of voting, considering all trajectories in J_u. Users belong to g make up a new set $S_g^{t+1} = \{u \mid \forall j, v_u(g) \geq v_u(g_j), 1 \leq j \leq G\}$.

The voting value that u gives to g is:

$$v_u(g) = \sum_{j=1}^{|J_u|} \frac{1}{Z_u^j} p(T_u^j \mid g; \Phi^{(t)}) \tag{4}$$

where $p(T_u^j \mid g; \Phi^{(t)})$ is the probability of observing T_u^j given group g's group-level HMM $H_g^{(t)}$ and can be computed by the Forward Scoring algorithm of HMM, and $Z_u^j = \sum_{i=1}^{G} p(T_u^j \mid g_i; \Phi^{(t)})$ is the normalization term.

Step 3: Updating. For each g in φ, we utilize the trajectories belong to group g to train an HMM $H_g^{(t+1)}$ by the aforementioned learning method. Thus, we generate a new ensemble of HMMs $\Phi^{(t+1)} = \{H_g^{(t+1)} \mid g \in \varphi\}$.

Step 4: Iteration. After updating the ensemble of HMMs, we go back to step 2 for further iterations. The algorithm will stop when the number of reassigned users is lower than a preset value (e.g., 1% of total user number D).

At last, users with the similar mobility are grouped, upon which we get group-level features along with personal-level and overall-level features.

3.3 Feature Extraction

In this section, we introduce features from the point of views of multiple granularities and multiple views. Given a stop-point x belongs to a trajectory T' of a user u. We searched out venues $V = \{v_1, v_2, \ldots, v_n\}$ that are within distance d from location $x.l$ as candidates. For each candidate venue v_i, we extract a series of features, mainly based on travel history of different granularities.

Multi-granularity Features. As mentioned in Sect. 3.1, we leverage multi-granularity features to alleviate the data sparsity issue. The kinds of feature granularities can be divided into multi-user and multi-venue granularities. For multi-user granularities, we generate features for a given user u from the travel history of all users, u himself and the group g he belongs to. And for multi-venue granularities, we generate features for a given venue v from visit history of v itself (X_v) and v's category $(X_{v.c})$ which can be further sub-divided into $X_{v.c.c_1}$, $X_{v.c.c_2}$ and $X_{v.c.c_3}$. The symbol X_v stands for the set of stop-points where each stop-point x's POI $x.p$ is the same as the given venue v, and similarly $X_{v.c.c_i}$ stands for the set of stop-points where each stop-point x's POI category $x.p.c.c_i$ is the same as the given venue category $v.c.c_i$.

By combining these two kinds of granularities, we can extract history-related features from the user u's visit history to v and $v.c$, his group's visit history to v and $v.c$ and all users' visit history to v and $v.c$, respectively.

Multi-view Features. Given a stop-point x and a candidate venue v, we mainly consider four types of features: the spatial relationship - $F_s(x, v)$, the temporal relationship - $F_t(x, v)$, the spatio-temporal relationship - $F_{st}(x, v)$ and the contextual relationship - $F_c(x, T', v)$, where T' is the trajectory x belongs to. In the following, we only introduce features generated from X_v due to the limitation of space. Note that the feature generated from $X_{v.c}$ is similar.

(1) Spatial features of $F_s(x, v)$ reflect visit preference related to geographic factors, including 2 parts:

- Revised distance $dist_{rv}(x, v)$. We observe distance $d(v.l, x.l)$ sometimes mislead annotation. For example, if a user gets off a vehicle at the roadside $(x.l)$ and get into a large supermarket v represented geographically by only one point $v.l$, $d(v.l, x.l)$ may be bigger than distances from $x.l$ to many other venues. Thus, we extract typical stop-points of venues by applying Affinity Propagation Clustering to historical stop-points of venues. Then we consider the distance between $x.l$ and the closest typical stop-point as the distance feature.
- Spatial conditional frequency $freq_d(x, X_v) = |\{x_h \in X_v | d(x.l, x_h.l) < \Delta'_d\}|$: the number of historical stop-points in X_v around $x.l$. Spatial adjacent points may have similar visit preference.

(2) Temporal features of $F_t(x, v)$ consist of 6 temporal conditional frequencies, which reflect different visit preference under different temporal conditions. We first define a temporal condition set $\Lambda = \{\lambda_1, \lambda_2, \ldots, \lambda_6\}$. Given time t' of a historical stop-point and t of an unannotated stop-point, Λ contains: (1) $POD(t') = POD(t)$, where $POD(t)$ maps time t to a time period tp; (2) t' is in the weekend if t is in the weekend, otherwise t' is in the weekday; (3) $DOW(t') = DOW(t)$, where $DOW(t)$ maps time t to day of week; (4) t' and t are on the same day, which reflects the situation (e.g., a sales promotion) on that day of venue v; (5) t' is within 30 days before and after t, which reflects recent situations of venue v; and (6) t' is within 90 days before and after t, which

reflects long-term situations of venue v. Then, for each condition $\lambda(t', t) \in \Lambda$, we get a conditional frequency $freq(X_v, x, \lambda) = |\{x_h \in X_v | \lambda(x_h.t, x.t)\}|$ as a feature, which stands for the number of historical stop-points in X_v satisfying temporal condition λ.

(3) Spatio-temporal features of $F_{st}(x, v)$. Spatio-temporal conditional frequency $freq_{st}(x, X_v) = |\{x_h \in X_v \mid d(x.l, x_h.l) < \Delta'_d \wedge I(x.t, x_h.t) < \Delta'_t\}|$: the number of historical stop-points in X_v which satisfy both spatial and temporal constraints, where $I(t, t')$ computes the time interval between t and t' in the span of 24 h.

(4) Contextual features of $F_c(x, T', v)$. The relevance between x and v is related to the trajectory T'. For example, if x's former point $\overleftarrow{x}$ and another stop-point $\overleftarrow{x_h}$ have similar characters (e.g., spatial adjacent or corresponding to the same POI category), x and x_h are likely to visit the same venue if they are spatial adjacent (e.g., a transition from housing areas to a specific company followed by a group of colleagues). Contextual features contains 2 parts:

- User activity inferred by group-level HMM. Instead of using Viterbi Algorithm directly, we consider the activities already known in trajectory T'.
- Contextual conditional frequencies. We first define a contextual condition set $\Omega = \{\omega_1, \omega_2, \omega_3\}$. Given two stop-points x and x', Ω contains: (1) $d(x.l, x'.l) < \Delta'_d$; (2) $x.p = x'.p$, which is the POI limit; and (3) $x.p.c = x'.p.c$, which is the POI category limit. Then, for each $\omega(x, x') \in \Omega$, we get a frequency feature about former points $freq(X_v, x, \omega) = |\{x_h \in X_v | \omega(\overleftarrow{x_h}, \overleftarrow{x})\}|$ and a frequency feature about later points $freq(X_v, x, \omega) = |\{x_h \in X_v | \omega(\overrightarrow{x_h}, \overrightarrow{x})\}|$, both of which stand for the number of historical stop-points in X_v satisfying contextual condition ω.

3.4 Venue Ranking

Our method has three variants to rank venues by relevance: *(1) CAUG-LR.* For a stop-point, it uses Logistic Regression [9] to do binary classification for every candidate venue. Then venues are ranked by probabilities. *(2) CAUG-GB.* It uses XGBoost [10] to replace Logistic Regression, measuring the performance in tree based model. *(3) CAUG-Rank.* It uses a learning-to-rank algorithm named LambdaMart to give ranked lists, which is a boosted tree version of LambdaRank based on RankNet [11].

4 Experimental Study

4.1 Setup

Datasets. In the work, we use a real operational transportation dataset collected by UCAR within 6th Ring Road of Beijing during Jun. 1, 2015 to Aug. 31, 2016. Also, we make use of the POI dataset of AutoNavi for annotation and use the first-class category to make up the activity.

Algorithms. We implement three variants of our method, i.e. *CAUG-LR*, *CAUG-GB* and *CAUG-Rank*, which use the same features proposed in this paper. We also implement 2 straightforward methods (i.e., *Dist* and *DistR*) and 3 distinct methods (i.e., *MRF*, *DistHMM* and *LSRank*) as baselines:

- *Dist*. It directly matches the closest venue to each stop-point.
- *DistR*. It uses the revised distance $dist_{rv}$ to match the closest venue.
- *MRF*. A method based on Markov random field model [5], considering the distance factor, spatial and temporal regularity of human mobility.
- *DistHMM*. A method based on HMM [4], considering the distance and the historical consecutive transitions between POIs.
  If there is no venue within 500 m for a stop-point in the user's states, we instead use the *Dist* model.
- *LSRank*. A learning-to-rank-based local search framework [6]. It uses features including the popularity of venues, distance between stop point and venues, temporal preference to venues and personal preference to venues.

Metrics. We use Normalized Discounted Cumulative Gain $(NDCG)^4$ to measure whether the ground truth venue appears in the output ranked list weighted by the position. For each annotation, $NDCG_k = \sum_{i=1}^{k} \frac{rel_i}{log_2(i+1)}$, where $rel_i \in \{0, 1\}$, is the binary relevance of the result at position i. The higher the ground truth venue is ranked in our list, the higher the NDCG score will be, and a value of 0 indicates the ground truth is not in the Top-k ranked list. In this paper, $NDCG@k$ is the mean of the $NDCG_k$ for each annotation and $NDCG@1$ is the same as the Top-1 accuracy.

4.2   Results

In this section, we report our findings. Note that, based on the actual situation of our dataset, the percentage of the labeled data is set to 55% for an experiment of the cold-start user and 85% for feature and comparison experiments.

The Impact of Feature Views. To evaluate the impact of different feature views on our methods, we test the effectiveness of our method on one view at a time and gradually combine them all together. The result in Fig. 4(a) shows temporal features are more effective than spatial features (0.88 versus 0.68 in NDCG@1). By combining temporal, spatial and spatio-temporal features, the NDCG@1 reaches over 0.92. Furthermore, the integration of contextual features further enhances the performance apparently, demonstrating it improves annotation in a different aspect with spatio-temporal factors.

---

^{4} https://en.wikipedia.org/wiki/Discounted_cumulative_gain.

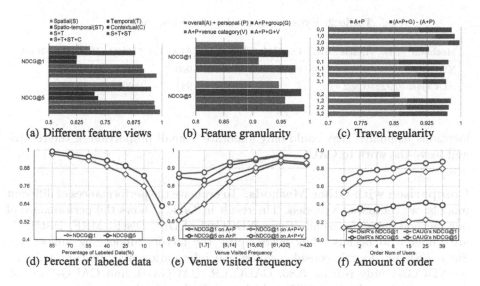

Fig. 4. Experiment results

The Impact of Feature Granularities. To evaluate the effectiveness of group and venue-category granularities, we start with features of both the overall and personal granularities without considering venue categories, then we add group and venue-category granularities. As Fig. 4(b) shows, when we only use overall and personal granularities, the NDCG@1 only reaches 0.88. By introducing the group and venue-category granularities, the NDCG@1 rises to about 0.97, which verifies the sparsity issue is alleviated. And we can find the group granularity is more effective than the venue-category granularities (0.96 versus 0.91). To further observe the impacts of group and venue-category granularities, the following two sub-experiments were conducted:

(1) Impact on travel regularity. We first define travel regularity by a 2-dimensional vector (L_{order}, L_{POI}), where L_{order} represents the level of order quantity and L_{POI} is the level of POI quantity. Intuitively, users who have visited various venues (L_{POI} is high) in his few number of travel records (L_{order} is low) are less regular in their mobilities. The result in Fig. 4(c) shows adding group granularity (G) to the model which just use overall and personal records (A+P) improves the performance, especially for those irregularly-traveling users.

(2) Impact on visited frequency. We test venues with different visited frequency. As shown in Fig. 4(e), popular venues have higher annotation accuracies. By adding venue-category granularity (V) to the model using overall and personal records (A+P), the performance improves, especially for novel and cold venues.

The Impact of Labeled Data Percentage. As shown in Fig. 4(d), the performance does not drop notably until the percentage of labeled data diminish to under 10%. This result provides an interesting message: for those who annotate

Table 2. Experiment results of different methods

Methods	Dist	DistR	MRF	DistHMM	LSRank	CAUG-LR	CAUG-GB	CAUG-Rank
NDCG@1	0.105	0.381	0.438	0.428	0.820	0.960	0.973	0.975
NDCG@5	0.222	0.562	/	/	0.913	0.985	0.990	0.991

locations manually, they could only annotate a small proportion of the records and leave the work to CAUG.

Comparison with Other Methods. To evaluate the performances of different models, we compared 8 models. The result in Table 2 shows the performance of our method outperforms all others. NDCG@1 of Dist, DistR, MRF, DistHMM are under 0.50 because they oversimplify the factors influencing annotation. Since LSRank doesn't consider contextual information and multi-granularity, its NDCG@1 only reaches 0.80. CAUG-LR, CAUG-GB, and CAUG-Rank all annotate accurately (over 0.95), which reveals the effectiveness of our features. Besides, we compare DistR and our model for cold users without enough historical data. The result in Fig. 4(f) shows as the number of personal records grows, the overall NDCG gradually improves. Specifically, for a new user who only has one travel record, the NDCG@1 of DistR only reaches 0.10, while our model is above 0.50 owing to the contextual information and multi-granularity features enrich the information for modeling user preference.

5 Related Work

Researchers proposed numerous methods [2,5,8,12,13] for semantic annotation of mobility records according to their specific tasks or data. Studies on traditional mobility data like GPS traces [2,4] mainly consider the distance between the context location and the location of the user. Without considering the history of individual's movement, they cannot provide personalized annotation. Spinsanti et al. [14] add some manually defined semantic rules to calculate the possibility of a person visiting a POI. However, rules cannot be well-rounded. Yan et al. [4,15] take transition relation of human movements into account and propose a method using HMM to annotate trajectories. Nevertheless, they ignored the temporal influences and only annotate locations with categories of POI other than specific POIs.

Due to the development of mobile Internet, massive geo-tagged social media (GeoSM) data combining texts with locations are generated. Wu et al. [3] and Zhang et al. [8] utilize noisy and sparse GeoSM data to discover proper activities or text tags of locations. Since sources of annotations are texts over the space, methods proposed by them cannot be applied to our problem. Some researchers utilize check-in data to study location annotation in [5,6,12,16], which is similar to our work. However, because check-in records are usually not continual, both of them neglect mobility transitions.

Moreover, location annotation is similar to the problem of recommending a POI to a user at a location [17,18]. Nevertheless, POI recommendation aims to rank those previously unvisited venues, while location annotation does not.

6   Conclusion

In this paper, we have proposed CAUG, an effective method to provide personalized location annotation through spatial, temporal and contextual factors, which can be generalized to many kinds of mobility data (e.g., locations collected by mobile apps). By constructing the sequence of locations, we take advantage of the transition relations among contextual mobility records to help annotate. We use HMM to model the users' mobility and group users based on their mobility patterns. With the help of user groups and venue categories, we effectively alleviate the issue of data sparsity. Experiments on a real-world dataset show that CAUG outperforms other 5 baseline models.

References

1. Parent, C., Spaccapietra, S., Renso, C., Andrienko, G., Andrienko, N., et al.: Semantic trajectories modeling and analysis. ACM Comput. Surv. (CSUR) **45**(4), 42 (2013)
2. de Graaff, V., de By, R.A., van Keulen, M.: Automated semantic trajectory annotation with indoor point-of-interest visits in urban areas. In: SAC, pp. 552–559 (2016)
3. Wu, F., Li, Z., Lee, W.C., Wang, H., Huang, Z.: Semantic annotation of mobility data using social media. In: WWW, pp. 1253–1263 (2015)
4. Yan, Z., Chakraborty, D., Parent, C., Spaccapietra, S., Aberer, K.: SeMiTri: a framework for semantic annotation of heterogeneous trajectories. In: EDBT, pp. 259–270 (2011)
5. Wu, F., Li, Z.: Where did you go: personalized annotation of mobility records. In: CIKM, pp. 589–598 (2016)
6. Lian, D., Xie, X.: Learning location naming from user check-in histories. In: SIGSPATIAL, pp. 112–121 (2011)
7. Rabiner, L., Juang, B.: An introduction to hidden markov models. IEEE ASSP Mag. **3**(1), 4–16 (1986)
8. Zhang, C., Zhang, K., Yuan, Q., Zhang, L., Hanratty, T., Han, J.: GMove: group-level mobility modeling using geo-tagged social media. In: KDD, pp. 1305–1314 (2016)
9. Berger, A.L., Pietra, V.J.D., Pietra, S.A.D.: A maximum entropy approach to natural language processing. Comput. Linguist. **22**(1), 39–71 (1996)
10. Chen, T., Guestrin, C.: XGBoost: a scalable tree boosting system. In: KDD, pp. 785–794 (2016)
11. Burges, C.J.: From RankNet to LambdaRank to LambdaMART: an overview. Learning **11**, 23–581 (2010)
12. Lian, D., Xie, X.: Mining check-in history for personalized location naming. TIST **5**(2), 1–25 (2014)

13. Nishida, K., Toda, H., Kurashima, T., Suhara, Y.: Probabilistic identification of visited point-of-interest for personalized automatic check-in. In: UbiCOMP, pp. 631–642 (2014)
14. Spinsanti, L., Celli, F., Renso, C.: Where you stop is who you are: understanding people's activities by places visited. In: BMI Workshop (2010)
15. Yan, Z., Chakraborty, D., Parent, C., Spaccapietra, S., Aberer, K.: Semantic trajectories: mobility data computation and annotation. TIST **4**(3), 49 (2013)
16. Shaw, B., Shea, J., Sinha, S., Hogue, A.: Learning to rank for spatiotemporal search. In: WSDM, pp. 717–726 (2013)
17. Liu, Y., Pham, T.A.N., Cong, G., Yuan, Q.: An experimental evaluation of point-of-interest recommendation in location-based social networks. VLDB **10**(10), 1010–1021 (2017)
18. Bao, J., Zheng, Y., Mokbel, M.F.: Location-based and preference-aware recommendation using sparse geo-social networking data. In: SIGSPATIAL, pp. 199–208 (2012)

A Joint Optimization Approach
for Personalized Recommendation
Diversification

Xiaojie Wang[1], Jianzhong Qi[1], Kotagiri Ramamohanarao[1], Yu Sun[2], Bo Li[3],
and Rui Zhang[1(✉)]

[1] The University of Melbourne, Melbourne, Australia
xiaojiew1@student.unimelb.edu.au,
{jianzhong.qi,rui.zhang,kotagiri}@unimelb.edu.au
[2] Twitter Inc., San Francisco, USA
ysun@twitter.com
[3] University of Illinois at Urbana Champaign, Champaign, USA
lxbosky@gmail.com

Abstract. In recommendation systems, items of interest are often classified into categories such as genres of movies. Existing research has shown that diversified recommendations can improve real user experience. However, most existing methods do not consider the fact that users' levels of interest (i.e., user preferences) in different categories usually vary, and such user preferences are not reflected in the diversified recommendations. We propose an algorithm that considers user preferences for different categories when recommending diversified results, and refer to this problem as personalized recommendation diversification. In the proposed algorithm, a model that captures user preferences for different categories is optimized jointly toward both relevance and diversity. To provide the proposed algorithm with informative training labels and effectively evaluate recommendation diversity, we also propose a new personalized diversity measure. The proposed measure overcomes limitations of existing measures in evaluating recommendation diversity: existing measures either cannot effectively handle user preferences for different categories, or cannot evaluate both relevance and diversity at the same time. Experiments using two real-world datasets confirm the superiority of the proposed algorithm, and show the effectiveness of the proposed measure in capturing user preferences.

1 Introduction

In most recommendation systems, items are classified by predefined categories, e.g., genres of movies or styles of musics. Recent studies show that users' interests often spread into several genres [22,23] (for ease of presentation, we will simply use genres to represent categories in the following). However, many existing algorithms (e.g., [7,8]) only try to optimize toward recommendation accuracy or item relevance, which is not optimal to cover users' diverse interests. In fact, the objectives of relevance and diversity are largely orthogonal, i.e., optimizing

© Springer International Publishing AG, part of Springer Nature 2018
D. Phung et al. (Eds.): PAKDD 2018, LNAI 10939, pp. 597–609, 2018.
https://doi.org/10.1007/978-3-319-93040-4_47

Table 1. Three lists of recommended movies in movie recommendations.

Rank	Recommendations by three different ranking measures		
	Non-diverse recomm.	Diverse without user pref.	Diverse with user pref.
1	First Shot (★)	First Shot (★)	First Shot (★)
2	Rapid Fire (★)	Snow Angels (∗)	Snow Angels (∗)
3	Black Dawn (★)	Rapid Fire (★)	Rapid Fire (★)
4	Shadow Man (★)	iss Potter (∗)	Black Dawn (★)
Count	#Action = 4 #Drama = 0	#Action = 2 #Drama = 2	#Action = 3 #Drama = 1

[1] Star (★) stands for action movies and asterisk (∗) stands for drama movies.

toward relevance may recommend very similar items, while optimizing toward diversity may present less relevant items. Recommendation diversification algorithms aim to achieve these two objectives at the same time and recommend diverse items with high relevance. Existing work in this area either separates relevance and diversity optimization [18], or does not explicitly consider the personalization in genre preferences [3,5,21] as discussed below.

Users usually have varied preferences over different genres [18]. High variances in such genre preferences require highly personalized recommendation diversification algorithms, which aim to present diverse recommendations catering to individual user's genre preference [18]. For example, Table 1 shows three lists of movies recommended to a user interested in both action and drama movies. The movies under the "non-diverse recomm." column are all action movies, which are not diverse in terms of genres. The movies under the "diverse without user pref." and "diverse with user pref." columns resolve this issue by also presenting drama movies. Suppose that the user prefers action movies. The "diverse without user pref." column treats the two genres equally (recommending two action movies and two drama movies) and does not consider the user's genre preference. The "diverse with user pref." column in this case presents a better recommendation, i.e., personalized diverse recommendations, which is the aim of this paper.

Toward this end, we propose a *personalized diversification algorithm* to jointly optimize both relevance and diversity and explicitly consider personalized genre preferences in diversification. The proposed algorithm iteratively selects the item that maximizes a function (i.e. ranking function) of two components: one models a user's rating for an item and the other models the user's genre preference for the item. The two components are collaborated by a joint optimization method to recommend items as accurately as possible (accurate rating prediction) and make an item list as personalized diverse as possible (personalized diverse ranking). The joint optimization method enables the personalized diversification algorithm to use the true ratings and pre-determined item rankings as sources of training information, where the item rankings indicate which item should be selected for personalized diverse recommendations given a selected item list.

To provide effective item rankings (i.e., training labels) to our algorithm, we need to measure the diversity of recommendations for each user, i.e., *personalized diversity*. Existing measures have limitations in evaluating personalized diversity:

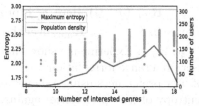
(a) The frequency-based user preference.

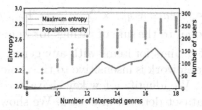
(b) The rating-based user preference.

Fig. 1. User preference analysis on a movie rating dataset (#genres = 18).

they either cannot handle the genre preferences of a user [4], or ignore the minor interests of a user [1], or cannot evaluate both relevance and diversity at the same time [18]. To overcome these limitations, we propose a new *personalized diversity measure*, which evaluates an item list based on user preferences for the covered genres of the list. This makes the item list having the highest score under our measure (i.e., the ideal list) has a desired property [18]: each genre is represented according to personalized genre preferences in the list.

The main contributions of this paper include: (1) We propose a novel recommendation diversification algorithm which can learn a ranking function by jointly optimizing the relevance and diversity. (2) We also propose a personalized diversity measure that can effectively evaluate personalized diversity of recommendations. (3) Experiments using real-world datasets of different domains show that the proposed algorithm outperforms several baseline methods and the proposed measure is more effective in capturing personalized genre preferences.

2   Problem Formulation

We assume that items to be recommended are categorized into genres. Let $\mathcal{X} = \{x_n\}_{n=1}^N$ be an item set, $\mathcal{G} = \{g_k\}_{k=1}^K$ be a genre set, $\mathbf{R} \in \mathbb{R}^{U \times N}$ be a rating matrix ($\mathbf{R}_{u,n}$ is the rating of user u for item x_n), $\mathbf{J} \in \mathbb{R}^{N \times K}$ be the genre information for items $\mathcal{X}$ ($\mathbf{J}_{n,g} = 1$ if item x_n is with genre g and $\mathbf{J}_{n,g} = 0$ otherwise). We define the problem of personalized recommendation diversification as:

Definition 1 (Personalized Recommendation Diversification). Given U users, N items, K genres, the rating matrix $\mathbf{R}$, the genre information $\mathbf{J}$, and a personalized diversity measure $\mathcal{M}$, the task is to generate the item list $\mathcal{Y}^u = [x_{y_1}, ..., x_{y_N}]$ that maximizes the measure $\mathcal{M}$ for each user u.

Intuitively, the problem is to consider **personalized genre preferences** (referred to as **user preferences** in the following for brevity) in diversification.

We consider two formulations of modeling user preferences. Let $\mathcal{X}^u$ be the item set rated by user u and $\mathcal{X}_g^u \subseteq \mathcal{X}^u$ be the subset of items with genre g. The *frequency-based user preference* is given by $p_g^u \propto |\mathcal{X}_g^u|/|\mathcal{X}^u|$ ($g \in \mathcal{G}$) [18]. Here, p_g^u is the user preference for genre g, which is proportional to the percentage of rated items with genre g. To consider the scale of ratings, we define the *rating-based user preference* as $q_g^u \propto \sum_{n:x_n \in \mathcal{X}_g^u} \mathbf{R}_{u,n} / \sum_{n:x_n \in \mathcal{X}^u} \mathbf{R}_{u,n}$ $(g \in \mathcal{G})$. Here, q_g^u is proportional to the sum of ratings for items with genre g, $\sum_{n:x_n \in \mathcal{X}_g^u} \mathbf{R}_{u,n}$, over the sum of ratings for items with any genre, $\sum_{n:x_n \in \mathcal{X}^u} \mathbf{R}_{u,n}$. Experiments show that our framework is insensitive to the choice of user preference formulation.

We identify a few key characteristics of user preferences using a movie rating dataset detailed in Sect. 5. We show the entropy of user preferences against the number of interested genres (those genres covered by $\mathcal{X}^u$) in Fig. 1. Each user is a blue dot which may be overlapped, so the population density is also drawn. For both user preference formulations, we find that: (1) Users prefer different degrees of diversity: the number of interested genres varies from 8 to 18 and the entropy of user preferences varies from 1.5 to 3.0 across users; (2) Users have varied preferences for different genres: no single user reaches the maximum entropy line where all genres are of the same interest to a user. These findings motivate us to consider user preferences in diversification.

3   Personalized Diversification Algorithms

In theory, optimizing a diversity measure is NP-hard [1], and a greedy strategy is often adopted [3]: at iteration r, $r - 1$ items $\mathcal{Y}_{r-1}$ have been selected. A marginal score function $s(x_n, \mathcal{Y}_{r-1})$ is used to select the next best item, which is then added to $\mathcal{Y}_{r-1}$. Two methods for modeling $s(x_n, \mathcal{Y}_{r-1})$ are presented as follows.

3.1   Personalized Diversification Algorithm by Greedy Re-ranking

A naive method is to use a re-ranking strategy which greedily selects next items based on predicted ratings, which is called *personalized diversification algorithm based on greedy re-ranking* (PDA-GR). It consists of: (1) A prediction phrase uses matrix factorization to predict ratings $\{\hat{\mathbf{R}}_{u,n}\}_{x_n \in \mathcal{X}}$; (2) A re-ranking phrase uses a training set to estimate user preferences $\{\hat{p}_g^u\}_{g \in \mathcal{G}}$, and a heuristic-based marginal score function to re-rank. Using the genre information $\mathbf{J}$, the marginal score function is defined as a combination of a rating component $f(\hat{\mathbf{R}}_{u,n})$, which models a user's rating for item x_n, and a genre preference component $\mathbf{J}_{n,g}(\hat{p}_g^u)^{C_g(r-1)}$, which models the user's genre preference of item x_n:

$$s(x_n, \mathcal{Y}_{r-1}) = \sum_{g \in \mathcal{G}} f(\hat{\mathbf{R}}_{u,n}) \cdot \mathbf{J}_{n,g}(\hat{p}_g^u)^{C_g(r-1)} \tag{1}$$

Here, $f(r) = 2^r$, and $C_g(r - 1)$ is the number of previous items with genre g. PDA-GR is sub-optimal because it divides optimizing accurate rating prediction and personalized diverse ranking into two separate phrases.

3.2   Personalized Diversification Algorithm by Joint Optimization

To tackle the sub-optimality, we propose a *personalized diversification algorithm based on joint optimization* (PDA-JO), which can optimize both accurate rating prediction and personalized diverse ranking simultaneously.

Algorithm 1. Personalized Diversification Algorithm by Joint Optimization

---

    Input: users $\mathcal{U}$, items $\mathcal{X}$, ratings $\mathbf{R}$, a personalized diversity measure $\mathcal{M}$

1   Pre-train $\{\mathbf{p}_u, \mathbf{b}_u\}_{u \in \mathcal{U}}, \{\mathbf{q}_n, \mathbf{b}_n\}_{x_n \in \mathcal{X}}$ based on $\mathbf{R}$

2   **while** *PDA-JO has not converge* **do**

3      $\mathcal{Z} \leftarrow \varnothing, \mathcal{B} \leftarrow \varnothing$ $\diamond$ $\mathcal{Z}$ is sampled item lists and $\mathcal{B}$ is training instances

4   **for** *each user u in $\mathcal{U}$* **do**

5   **for** *length l from 0 to $|\mathcal{X}| - 1$* **do**

6   Add the ideal list of length l under the measure $\mathcal{M}$ into $\mathcal{Z}$

7   Sample S non-ideal lists of length l and add them into $\mathcal{Z}$

8   **for** *item list $\mathcal{Y}$ in $\mathcal{Z}$* **do**

9   **for** *item pair (x_m, x_n) from $\mathcal{X} \setminus \mathcal{Y}$* **do**

10   **if** $M(\mathcal{Y} + [x_m]) > M(\mathcal{Y} + [x_n])$ **then** $L \leftarrow 1$

11   **else** $L \leftarrow 0$

12   Add $(\mathcal{Y}, x_m, x_n, y = (\mathbf{R}_{u,m}, \mathbf{R}_{u,n}, L))$ into $\mathcal{B}$

13   **for** *mini-batch b in $\mathcal{B}$* **do**

14   Update $\{\mathbf{p}_u, \mathbf{b}_u\}_{u \in \mathcal{U}}, \{\mathbf{q}_n, \mathbf{b}_n\}_{x_n \in \mathcal{X}}, \mu$ based on Equation 3

For user u, let $\mathbf{p}_u \in \mathbb{R}^F$ be the embedding and $\mathbf{b}_u \in \mathbb{R}$ be the bias. For item x_n, let $\mathbf{q}_n \in \mathbb{R}^F$ be the embedding and $\mathbf{b}_n \in \mathbb{R}$ be the bias. The rating for item x_n is predicted by $\hat{\mathbf{R}}_{u,n} = \mathbf{p}_u^{\mathsf{T}} \mathbf{q}_n + \mathbf{b}_u + \mathbf{b}_n$. We use a parameter μ to alleviate the error of rating prediction. The marginal score function is defined as:

$$s(x_n, \mathcal{Y}_{r-1}) = \sum_{g \in \mathcal{G}} f(\hat{\mathbf{R}}_{u,n} + \mu) \cdot \mathbf{J}_{n,g} (\hat{p}_g^u)^{C_g(r-1)} \tag{2}$$

Here, $f(r) = 2^r$, $\mathbf{J}_{n,g} = 1$ if item x_n is with genre g and $\mathbf{J}_{n,g} = 0$ otherwise, p_g^u is the preference of user u for genre g, and $C_y(r - 1)$ is the number of previous items with genre g. $\{\mathbf{p}_u, \mathbf{b}_u\}_{u \in \mathcal{U}}, \{\mathbf{q}_n, \mathbf{b}_n\}_{x_n \in \mathcal{X}}$, and μ are learnable parameters.

We define a training instance for user u as $(\mathcal{Y}, x_m, x_n, y)$ where $\mathcal{Y}$ is selected items, x_m and x_n are two candidate items, and $y = (\mathbf{R}_{u,m}, \mathbf{R}_{u,n}, L)$. Here, $\mathbf{R}_{u,m}$ and $\mathbf{R}_{u,n}$ are the true ratings for items x_m and x_n. Training label L indicates which item ranking is better under the measure $\mathcal{M}$: $L = 1$ if $\mathcal{M}(\mathcal{Y} + [x_m]) > \mathcal{M}(\mathcal{Y} + [x_n])$ and $L = 0$ otherwise. The probability of $L = 1$ is $P = \sigma(s(x_m, \mathcal{Y}) - s(x_n, \mathcal{Y}))$, where $\sigma(\cdot)$ is the sigmoid function. The loss function of our algorithm consists of a relevance loss $\mathcal{L}_r$ a personalized diversity loss $\mathcal{L}_d$:

$$\mathcal{L} = \underbrace{0.5[(\hat{\mathbf{R}}_{u,m} - \mathbf{R}_{u,m})^2 + (\hat{\mathbf{R}}_{u,n} - \mathbf{R}_{u,n})^2]}_{\text{The relevance loss:} \mathcal{L}_r} - \underbrace{D[L \log P + (1 - L) \log(1 - P)]}_{\text{The personalized diversity loss:} \mathcal{L}_d}$$

Here, D balances between accurate rating prediction (loss $\mathcal{L}_r$) and personalized diverse ranking (loss $\mathcal{L}_d$). We use L2 regularization to regularize the model.

Algorithm 2. Building the ideal list for p-nDCG

Input: user u, items $\mathcal{X} = \{x_n\}_{n=1}^N$, user ratings $\mathbf{R}$, genre information $\mathbf{J}$
Output: ideal list $\mathcal{Y} = \{x_{y_n}\}_{n=1}^N$
1 Estimate the user preferences $\{p_g^u\}_{g \in \mathcal{G}}$ based on the genre information $\mathbf{J}$
2 $\mathcal{Y}_0 \leftarrow \varnothing$ $\diamond$ a selected item list
3 **for** $r = 1, ..., N$ **do**
4       $x_m \leftarrow \arg\max_{x_n \in \mathcal{X} \setminus \mathcal{Y}_{r-1}} (\text{p-nDCG}(\mathcal{Y}_{r-1} + [x_n]) - \text{p-nDCG}(\mathcal{Y}_{r-1}))$
5       $\mathcal{Y}_r \leftarrow \mathcal{Y}_{r-1} + [x_m]$
6 $\mathcal{Y}^u \leftarrow \mathcal{Y}_N$

The model is trained by stochastic gradient descent with gradient given by:

$$\frac{\partial \mathcal{L}}{\partial \mathbf{p}_u} = (e_{u,m}\mathbf{q}_m - e_{u,n}\mathbf{q}_n) + DE\{\sum_{g \in \mathcal{G}} d_g f'(\hat{\mathbf{R}}_{u,m} + \mu)\mathbf{q}_m - \sum_{g \in \mathcal{G}} d_g f'(\hat{\mathbf{R}}_{u,n} + \mu)\mathbf{q}_n\}$$

$$\frac{\partial \mathcal{L}}{\partial \mathbf{q}_l} = e_{u,l}\mathbf{p}_u + DE\{\sum_{g \in \mathcal{G}} d_g f'(\hat{\mathbf{R}}_{u,l} + \mu)\mathbf{p}_u\} \, l \in \{m, n\}$$

$$\frac{\partial \mathcal{L}}{\partial \mathbf{b}_u} = (e_{u,m} - e_{u,n}) + DE\{\sum_{g \in \mathcal{G}} d_g f'(\hat{\mathbf{R}}_{u,m} + \mu) - \sum_{g \in \mathcal{G}} d_g f'(\hat{\mathbf{R}}_{u,n} + \mu)\}$$

$$\frac{\partial \mathcal{L}}{\partial \mathbf{b}_l} = e_{u,l} + DE\{\sum_{g \in \mathcal{G}} d_g f'(\hat{\mathbf{R}}_{u,l} + \mu)\} \, l \in \{m, n\}$$

$$\frac{\partial \mathcal{L}}{\partial \mu} = DE\{\sum_{g \in \mathcal{G}} d_g f'(\hat{\mathbf{R}}_{u,m} + \mu) - \sum_{g \in \mathcal{G}} d_g f'(\hat{\mathbf{R}}_{u,n} + \mu)\} \qquad (3)$$

Here, $e_{u,l} = \hat{\mathbf{R}}_{u,l} - \mathbf{R}_{u,l} \, l \in \{m, n\}$, $E = P - L$, and $d_g = (\hat{p}_g)^{C_g(r-1)}$. The total number of training instances is $\Theta(MN!)$. To speed up training, we use a sampling method similar to the negative sampling [9]: both ideal lists and a number of sampled non-ideal lists under measure $\mathcal{M}$ are used to estimate the gradient. The overall procedure of the joint optimization method is summarized in Algorithm 1. The model is first pre-trained by training ratings. Then, we sample S non-ideal lists with a certain length $l \in [0, N-1]$ for each user and update the parameters with the gradient given by Eq. 3.

Time Complexity. The training time complexity is $\Theta(E \cdot M \cdot S \cdot N^2 \cdot T)$, where E is the number of epoches, S is the number of sampled non-ideal item lists. $T = \max\{F, K\}$ is the time complexity of computing the marginal score function. The test time complexity is $\Theta(N^2 \cdot T)$ for each user.

4   Personalized Diversity Measure

Existing measures have limitations in evaluating personalized recommendation diversity. Therefore, we proposed a personalized diversity measure in this section.

Table 2. The movielens 100k dataset (ML-100K) and the million song dataset (MSD)

Data	Stat.					
	#users	#items	#ratings	#genres	Range	Sparsity
ML-100K	943	1,682	100,000	18	1–5	6.30%
MSD	1,217	2,051	88,078	15	1–225	3.53%

4.1 Limitations of Existing Diversity Measures

Our goals are to recommend items that (I) cover a user's interested genres, and (II) have a genre distribution satisfying the user's preference for different genres. Existing diversity measures cannot serve our goals: (1) α-nDCG [4] does not model user preferences (or intent probabilities). (2) IA measures [1] tend to favor the major interests and ignore the minor interests of a user [12]. (3) One of the goals of D$\sharp$-measures [12] is to recommend items that cover as many genres (or intents) as possible, but not to optimize toward individual user's preference.

4.2 Formulation of Personalized Diversity Measure

To overcome these limitations, we propose a personalized diversity measure. Our measure is motivated by α-nDCG [4], which discounts the gain of redundant items by a constant $\alpha \in [0, 1]$. Users often have varied preferences for different genres. A constant cannot model such variances. Intuitively, redundancy under more preferred genres is better than redundancy under less preferred genres.

Let $J_g(r) = 1$ if the item at rank r is labeled with genre g and $J_g(r) = 0$ otherwise, and $C_g(r) = \sum_{k=1}^{r} J_g(k)$. Based on the preference of user u $\{p_g^u\}_{g \in \mathcal{G}}$, we define the *personalized novelty-biased gain* (PNG) for the item at rank r as:

$$PNG(r) = \sum_{g \in \{g\}} h(r) \cdot J_g(r)(p_g^u)^{C_g(r-1)} \tag{4}$$

Here, $h(r) = (2^r - 1)/2^{r_{max}}$. PNG models the marginal gain of an item after a user has seen previous items. We define *p-nDCG* at cutoff C as:

$$p\text{-}nDCG@C = \frac{\sum_{r=1}^{C} PNG(r)/\log(r+1)}{\sum_{r=1}^{C} PNG^*(r)/\log(r+1)} \tag{5}$$

Here, PNG^* is PNG of the ideal list built by Algorithm 2. The algorithm iteratively selects the item that maximizes the p-nDCG score of current item list based on the true ratings and user preferences.

Theoretical Analysis. p-nDCG is effective in capturing user preferences: item lists with a high p-nDCG score tend to contain more items with more preferred genres. To see this, we analyze the ideal list under p-nDCG. If genre g is under-represented in the list, i.e. p_g is high while C_g is low, the PNG for a relevant item with genre g will be large. This makes p-nDCG select more relevant items with genre g as next items. The selection process reaches an equilibrium when each genre is represented according to user preferences:

$$(p_{g_1})^{C_{g_1}} \equiv (p_{g_2})^{C_{g_2}} \ (g_1, g_2 \in \mathcal{G}) \quad \Rightarrow \quad C_g \propto \log(p_g) \ (g \in \mathcal{G}) \qquad (6)$$

The ideal list is effective in reflecting user preferences: the number of items with genre g (C_g) is positively correlated with the preference for genre g (p_g) in the list. This is a desired property for personalized recommendation diversification [18]: each genre needs to be represented according to user preferences in an item list.

5   Experiments

We experiment with the movielens 100k dataset (ML-100K) [6] and the million song dataset (MSD) [2]. ML-100K is a movie rating dataset. It contains 100,000 ratings on 1,682 movies from 943 users. MSD contains music play counts. We use a subset of MSD containing the playing counts of the songs associated properly to one of the predefined genres. This subset contains 88,078 playing counts on 2,051 songs from 1,217 users. The two datasets are summarized in Table 2.

We try both formulations of user preferences and obtain similar results in the experiments. We only show the results using the frequency-based user preference due to the page limit. We use normalized discounted cumulative gain (nDCG), α-nDCG ($\alpha = 0.5$) [4], and the proposed p-nDCG to evaluate algorithm performances. All these measures are computed at cutoff $C = 10$.

5.1   Experiments on Algorithms

The compared methods include **MF** [7], **MMR** [3], **PM-2** [5], and **LTR-N** [21]. We use 5-fold cross validation to tune parameters for all algorithms.

Effects of Parameters. In Fig. 2, we present the effects of tuning (1) D varied from 0.01 to 100, and (2) S varied from 0 to 25. We apply the z-normalization method to amplify the effects. The proposed PDA-JO performs best when $(D, S) = (1, 10)$ on ML-100K and $(D, S) = (10, 5)$ on MSD. Figure 2(a) and (c) show the effects of D on ML-100K ($S = 10$) and MSD ($S = 15$). The performance of PDA-JO increases with the growth of D ($0.1 \leqslant D \leqslant 10$), after which the performance decreases under α-nDCG and p-nDCG. This is because: (1) If D is small, PDA-JO is biased toward rating prediction and disregard diverse ranking, which will degrade the performance under diverse measures. (2) If D is large, rating prediction is less accurate, which will in turn degrade the performance because diverse ranking relies on rating prediction (see Eq. 2). The influence of D is stable when $1 \leqslant D \leqslant 10$. Figure 2(b) and (d) show the effects of S on ML-100K ($D = 1$) and MSD ($D = 10$). The proposed PDA-JO performs

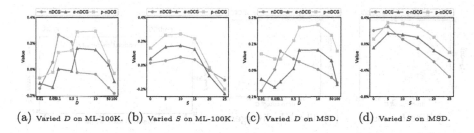

(a) Varied D on ML-100K.   (b) Varied S on ML-100K.   (c) Varied D on MSD.   (d) Varied S on MSD.

Fig. 2. Performances of PDA-JO with varied parameters after z-normalization.

Table 3. Performance comparison of algorithms on ML-100K and MSD. For MMR and PM-2, the subscript is the parameter achieving the best score on validation set.

Method	Performance on MK-100K			Performance on MSD		
	nDCG	α-nDCG	p-nDCG	nDCG	α-nDCG	p-nDCG
MF	0.7206	0.6035	0.5799	0.6061	0.4728	0.5001
$MMR_{0.7}$	0.6944	0.6206 (2.82%)	0.6172 (6.44%)	0.6081	0.4803 (1.58%)	0.5068 (1.33%)
$PM-2_{0.5}$	0.6829	0.6759 (11.98%)	0.6525 (12.53%)	0.5895	0.4954 (4.77%)	0.5179 (3.54%)
LTR-N	0.7301	0.7134 (18.21%)	0.7017 (21.00%)	0.6230	0.4997 (5.70%)	0.5246 (4.89%)
PDA-GR	0.7283	0.7782 (28.93%)	0.7690 (32.61%)	0.6295	0.5430 (14.85%)	0.5665 (13.27%)
PDA-JO	**0.7417**	**0.7846** (29.99%)	**0.7778** (34.13%)	**0.6309**	**0.5579** (18.00%)	**0.5808** (16.14%)

better as S increases ($0 \leqslant S \leqslant 10$), but a performance decrease occurs when $S \geqslant 15$. The overall difference when varying D and S is less than 0.8%, which indicates that PDA-JO is a robust framework.

Comparison of Algorithms. Table 3 compares the performances of all algorithms on ML-100K and MSD. The proposed PDA-JO performs best on both datasets under all three measures. The improvement of PDA-JO over baseline methods is significant based on two-tailed paired t-test. We compare all methods in the following aspects: (1) Personalized diversification methods (PDA-GR and PDA-JO) outperform non-personalized diversification methods (MMR, PM-2, and LTR-N) on all three measures. (2) Heuristic-based methods (MMR and PM-2) sacrifice relevance to boost diversity, while learning-based methods (LTR-N and PDA-JO) can improve both relevance and diversity. (3) PDA-JO is consistently better than PDA-GR for all measures on both datasets.

5.2   Experiments on Measures

We compare the ideal lists of p-nDCG and α-nDCG ($\alpha = 0.5$) on ML-100K as follows: (1) For each user, we randomly split ratings into a training set (80%) and a test set (20%). We also use time-based split (the most current 20% are used for testing), and the results are similar; (2) We use the training set to build the ideal list of p-nDCG (α-nDCG) by Algorithm 2. Here, the user preferences used to compute the p-nDCG score are obtained using the training set.

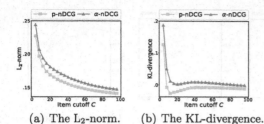

(a) The L_2-norm.   (b) The KL-divergence.

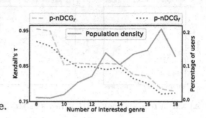

Fig. 3. The distance between the ground-truth user preferences and the genre distribution of the top-C ranked items by p-nDCG (α-nDCG), where the item cutoff $C \in [5, 100]$.

Fig. 4. Rank correlation (Kendall's τ) with α-nDCG, where p-nDCG$_f$ uses the frequency-based user preference while p-nDCG$_r$ uses the rating-based user preference.

Satisfying User Preferences. We show that the ideal list of p-nDCG is more effective than α-nDCG in reflecting user preferences. We compute genre distribution P_p (P_α) of p-nDCG (α-nDCG) by applying user preference formulations to the top-C ranked items in the ideal list. The ground-truth user preference P^* is obtained using the test set. We compute the distance between P^* and P_p (P_α) using KL-divergence or L_2-norm, and average all distances across users. We plot the average distance against item cutoff in Fig. 3. We find that compared with α-nDCG, the genre distribution of the top-C ranked items by p-nDCG consistently better satisfy user preferences, especially when cutoff C is small.

Rank Correlation. We use Kendall's τ to measure rank correlation between the ideal lists of p-nDCG and α-nDCG. The results of averaging Kendall's τ over the users who are interested in the same number of genres are shown in Fig. 4. We find that as the number of interested genres increases, the rank correlation decreases. This is because when a user's interested genres are of the same interest to the user, p-nDCG reduces to α-nDCG. As the number of interested genres grows, the probability that a user has the same preference for different genres decreases. This causes p-nDCG and α-nDCG to produce less similar item lists.

Case Study. We use a real user on ML-100K to illustrate the advantage of p-nDCG in Table 4. The ground-truth column (user preferences) is computed by applying the frequency-based user preference to the test set. We find that: (1) In terms of genre ranking, p-nDCG is more consistent (Kendall's $\tau = 0.89$) with the user preferences than α-nDCG (Kendall's $\tau = 0.39$); (2) The genre distribution of recommended items by p-nDCG is closer (L_2-norm $= 0.20$ using the frequency-based user preference) to the user preferences than α-nDCG (L_2-norm $= 0.29$ using the frequency-based user preference).

6 Related Work

Before receiving attentions in recommendation systems (RS), the problem of diversity is studied in information retrieval (IR) [1,3–5,12,21]. One difference between IR work and our work is that there is ground-truth for test item genres in our work (e.g., ML-100K provides the genre information of movies), but there is no such ground-truth for test document intents (analogous to item genres) in IR work. We explicitly incorporate such genre information into the diverse ranking model, which makes even the naive method effective. Another difference is that the embedding is trainable in our work, but the embedding is not trainable in IR work (it is pre-computed and fixed as relevance features) [21].

Diversity Measures. Several diversity measures are proposed in IR to evaluate the diversity [19]. They are not designed to evaluate the personalized diversity as discussed in Sect. 4.1. In RS, Smyth and McClave [13] define the dissimilarity-based diversity, i.e., the average dissimilarity between all pairs of the recommended items. Vargas et al. argue that the dissimilarity-based diversity is less likely to be perceived as diverse by users than the genre diversity [18]. They propose a Binomial framework to evaluate the genre diversity. The Binomial framework cannot evaluate the relevance (random recommendations may achieve high scores under this framework) and does not model the position of relevant item in an item list. It differs from our measure which evaluates both relevance and diversity and models the relevant item position.

Related Algorithms. Diversification algorithms can be categorized into heuristic-based and learning-based. Heuristic-based methods use some heuristic rules to re-rank the candidate items [3,5,23]. For example, Ziegler et al. propose to select the next item by linearly combining the relevance and the dissimilarity to the selected items based on an intra-list similarity measure [23]. Learning-based methods aim to learn a diverse ranking model from a training set [21]. For example, Xia et al. propose to learn a diverse ranking model by using neural networks to model the marginal novelty of candidate items.

The proposed algorithm is related to model-based collaborative filtering methods, which explain user ratings by factoring the ratings into user embedding and item embedding [7,11]. Our algorithm borrows ideas from learning-to-rank methods [10], which overcome the problems with heuristic predefined ranking function. For example, Tran et. al propose to integrate deep neural networks into the learning-to-rank model [17]. Our algorithm is also related to intent tracking algorithms [14–16,20] in designing highly personalized recommendation systems: we aim to personalize at genre level while intent tracking algorithms personalize at intent level. However, none of these algorithms explicitly consider personalized genre preferences, which is the topic of our work.

Table 4. The ideal list of p-nDCG (α-nDCG) for a real user on ML-100K.

Pos. / Pref.	Top-10 items by α-nDCG										Ct.	Ground-truth		Frequency-based		Rating-based	
	1	2	3	4	5	6	7	8	9	10		Pref.	Rank	Pref.	Rank	Pref.	Rank
Comedy			♦	♦			♦				3	0.3789	1	0.1875	2	0.2000	2
Horror			♦		♦	♦					3	0.2756	2	0.1875	2	0.2000	2
Romance			♦	♦			♦				3	0.2067	3	0.1875	2	0.2000	2
Animation	♦				♦			♦	♦		4	0.0689	4	0.2500	1	0.2154	1
Adventure	♦					♦					2	0.0689	4	0.1250	5	0.1231	5
Thriller		♦									1	0.0011	6	0.0625	6	0.0615	6
Statistics												L_2 (0.29)	τ (0.39)	L_2 (0.26)	τ (0.39)		

Pos. / Pref.	Top-10 items by p-nDCG										Ct.	Ground-truth		Frequency-based		Rating-based	
	1	2	3	4	5	6	7	8	9	10		Pref.	Rank	Pref.	Rank	Pref.	Rank
Comedy	♦				♦			♦	♦		4	0.3789	1	0.2353	1	0.2500	1
Horror	♦	♦						♦		♦	4	0.2756	2	0.2353	1	0.2500	1
Romance	♦				♦			♦			3	0.2067	3	0.1765	3	0.1842	3
Animation				♦		♦				♦	3	0.0689	4	0.1765	3	0.1447	4
Adventure	♦	♦									2	0.0689	4	0.1176	5	0.1184	5
Thriller			♦								1	0.0011	6	0.0588	6	0.0526	6
Statistics												L_2 (0.20)	τ (0.89)	L_2 (0.17)	τ (0.93)		

[1] Diamond ♦ indicates the movie at a certain position is categorized as a certain genre.
[2] L_2 stands for the L_2-norm and τ stands for the Kendall's τ.

7 Conclusion

We studied the problem of personalized recommendation diversification. A personalized diversification algorithm was proposed to incorporate user preferences and jointly optimize both relevance and diversity. To overcome limitations of existing measures, we proposed a personalized diversity measure to evaluate the personalized diversity of recommendations. Experiments using real-world datasets showed that the proposed algorithm outperforms baseline algorithms, including a state-of-the-art leaning-to-rank algorithm. The experiments also validated the effectiveness of the proposed measure in capturing user preferences.

Acknowledgment. This work is supported by Australian Research Council (ARC) Future Fellowships Project FT120100832 and Discovery Project DP180102050.

References

1. Agrawal, R., Gollapudi, S., Halverson, A., Ieong, S.: Diversifying search results. In: WSDM, pp. 5–14. ACM (2009)
2. Bertin-Mahieux, T., Ellis, D.P., Whitman, B., Lamere, P.: The million song dataset. In: ISMIR, vol. 2, p. 10 (2011)
3. Carbonell, J., Goldstein, J.: The use of MMR, diversity-based reranking for reordering documents and producing summaries. In: SIGIR, pp. 335–336 (1998)
4. Clarke, C.L., Kolla, M., Cormack, G.V., Vechtomova, O., Ashkan, A., Büttcher, S., MacKinnon, I.: Novelty and diversity in information retrieval evaluation. In: SIGIR, pp. 659–666. ACM (2008)

5. Dang, V., Croft, W.B.: Diversity by proportionality: an election-based approach to search result diversification. In: SIGIR, pp. 65–74. ACM (2012)
6. Harper, F.M., Konstan, J.A.: The MovieLens datasets: history and context. ACM Trans. Interact. Intell. Syst. (TiiS) **5**(4), 19 (2016)
7. Hu, Y., Koren, Y., Volinsky, C.: Collaborative filtering for implicit feedback datasets. In: ICDM, pp. 263–272. IEEE (2008)
8. Koren, Y.: Factorization meets the neighborhood: a multifaceted collaborative filtering model. In: SIGKDD, pp. 426–434. ACM (2008)
9. Mikolov, T., Sutskever, I., Chen, K., Corrado, G.S., Dean, J.: Distributed representations of words and phrases and their compositionality. In: NIPS (2013)
10. Radlinski, F., Kleinberg, R., Joachims, T.: Learning diverse rankings with multi-armed bandits. In: ICML, pp. 784–791. ACM (2008)
11. Rendle, S., Freudenthaler, C., Gantner, Z., Schmidt-Thieme, L.: BPR: Bayesian personalized ranking from implicit feedback. In: UAI, pp. 452–461 (2009)
12. Sakai, T., Song, R.: Evaluating diversified search results using per-intent graded relevance. In: SIGIR, pp. 1043–1052. ACM (2011)
13. Smyth, B., McClave, P.: Similarity vs. diversity. In: Aha, D.W., Watson, I. (eds.) ICCBR 2001. LNCS (LNAI), vol. 2080, pp. 347–361. Springer, Heidelberg (2001). https://doi.org/10.1007/3-540-44593-5_25
14. Sun, Y., Yuan, N.J., Wang, Y., Xie, X., McDonald, K., Zhang, R.: Contextual intent tracking for personal assistants. In: SIGKDD, pp. 273–282. ACM (2016)
15. Sun, Y., Yuan, N.J., Xie, X., McDonald, K., Zhang, R.: Collaborative nowcasting for contextual recommendation. In: WWW, pp. 1407–1418 (2016)
16. Sun, Y., Yuan, N.J., Xie, X., McDonald, K., Zhang, R.: Collaborative intent prediction with real-time contextual data. TOIS **35**(4), 30:1–30:33 (2017)
17. Tran, T., Phung, D., Venkatesh, S.: Neural choice by elimination via highway networks. In: Cao, H., Li, J., Wang, R. (eds.) PAKDD 2016. LNCS (LNAI), vol. 9794, pp. 15–25. Springer, Cham (2016). https://doi.org/10.1007/978-3-319-42996-0_2
18. Vargas, S., Baltrunas, L., Karatzoglou, A., Castells, P.: Coverage, redundancy and size-awareness in genre diversity for recommender systems. In: RecSys (2014)
19. Wang, X., Wen, J.R., Dou, Z., Sakai, T., Zhang, R.: Search result diversity evaluation based on intent hierarchies. TKDE **30**(1), 156–169 (2018)
20. Wang, Y., Yuan, N.J., Sun, Y., Qin, C., Xie, X.: App download forecasting: an evolutionary hierarchical competition approach. In: IJCAI (2017)
21. Xia, L., Xu, J., Lan, Y., Guo, J., Cheng, X.: Modeling document novelty with neural tensor network for search result diversification. In: SIGIR (2016)
22. Yuan, M., Pavlidis, Y., Jain, M., Caster, K.: Walmart online grocery personalization: behavioral insights and basket recommendations. In: Link, S., Trujillo, J.C. (eds.) ER 2016. LNCS, vol. 9975, pp. 49–64. Springer, Cham (2016). https://doi.org/10.1007/978-3-319-47717-6_5
23. Ziegler, C.N., McNee, S.M., Konstan, J.A., Lausen, G.: Improving recommendation lists through topic diversification. In: WWW, pp. 22–32. ACM (2005)

Personalized Item-of-Interest Recommendation on Storage Constrained Smartphone Based on Word Embedding Quantization

Si-Ying Huang[1,3], Yung-Yu Chen[2], Hung-Yuan Chen[3], Lun-Chi Chen[2], and Yao-Chung Fan[1(✉)]

[1] Department of Computer Science and Engineering,
National Chung Hsing University, Taichung, Taiwan
pe31092168@gmail.com, yfan@nchu.edu.tw
[2] National Center for High-Performance Computing, Hsinchu, Taiwan
denny1232@gmail.com, lunchi0124@gmail.com
[3] Industrial Technology Research Institute, Zhudong, Taiwan
hychen@itri.org.tw

Abstract. In recent years, word embedding models receive tremendous research attentions due to their capability of capturing textual semantics. This study investigates the issue of employing word embedding models into resource-limited smartphones for personalized item recommendation. The challenge lies in that the existing embedding models are often too large to fit into a resource-limited smartphones. One naive idea is to incorporate a secondary storage by residing the model in the secondary storage and processing recommendation with the secondary storage. However, this idea suffers from the burden of additional traffics. To this end, we propose a framework called Word Embedding Quantization (WEQ) that constructs an index upon a given word embedding model and stores the index on the primary storage to enable the use of the word embedding model on smartphones. One challenge for using the index is that the exact user profile is no longer ensured. However, we find that there are opportunities for computing the correct recommendation results by knowing only inexact user profile. In this paper, we propose a series of techniques that leverage the opportunities for computing candidates with the goal of minimizing the accessing cost to a secondary storage. Experiments are made to verify the efficiency of the proposed techniques, which demonstrates the feasibility of the proposed framework.

1 Introduction

Nowadays smartphones have become a ubiquitous medium supporting various forms of functionality and are widely accepted for commons. Users are nearly

S.-Y. Huang and Y.-Y. Chen—These authors contributed equally to the work.
Y.-C. Fan—The person to whom inquiries regarding the paper should be addressed.

© Springer International Publishing AG, part of Springer Nature 2018
D. Phung et al. (Eds.): PAKDD 2018, LNAI 10939, pp. 610–621, 2018.
https://doi.org/10.1007/978-3-319-93040-4_48

with their smartphones 24 h a day. With such an intimacy, a mobile phone has been more than a mini computer for its owner but a personal behavior observer. As a result, mining various data collected from smartphones for user profiling [2,3] has received tremendous interests in the past few years. After the user modeling, smartphones play as an important medium for making personalized item-of-interest recommendation [1,5–8].

In this paper, we focus on employing textual data from smartphones for making a personalized item-of-interest recommendation. Our idea is to treat text data generated by a smartphone user as a user profile. With the user profile, a naive idea for making a personalized recommendation is to define similarity between textual user profiles and candidate items, and then retrieve the most similar items as the predicted items a user may be interested.

For defining similarity, one intuitive idea is to model users' profile and candidate items as Bags-of-Words (BOW) and then define the similarity between two BOWs by *Cosine similarity* or *Jaccard measure*. However, these similarity measures consider only the information at lexical levels; two words with the same meaning will be perpendicular to each other in the metric space. Namely, the words are considered as totally different. For example, "Obama spoke to the media of Illinois" and "President talks to the media in Chicago." have no words in common but are actually with similar semantics.

For addressing this issue, we propose to employ the word embedding models [9,10] to capture the semantics between words. With keyword embedding models, each word is modeled as a vector. An important feature for keyword embedding model is that the keywords sharing common contexts are located in close proximity in the vector space. For example, "CPU" and "RAM" the terms related to personal computers will be close to each other. The same observation holds for "Google", "Facebook", and "Microsoft". For precisely capturing word semantics, keyword embedding models are often trained from a large text corpus. In Fig. 1, we show a visualized results by using multidimensional scaling technique to map the word vector into a two-dimensional space.

While the employment of word embedding model looks promising, we find that the size of word embedded model is often too large to fit into a resource-limited smartphone. For example, for the Chinese Wikipedia dumped in August 2017, there are 740318 unique words. Assume that each word is represented by a 400-dimensional vector and 4 bytes for an entry of a vector. Totally, 1.184 Gigabytes(=740,318*400*4) is required. The employment of models with such sizes in smartphones brings system performance issues and storage concerns.

A naive idea for addressing this problem is to incorporate a client-server model by sending all smartphone data to a server where all computation is performed. However, such idea suffers from the communication cost concerns. Extra network traffics costs for enabling client-server model is required. Existing cloud platforms, e.g. Amazon, charge the server fee according to the amount of network traffics. Furthermore, from user's point of views, data communication always come with costs at data communication fee and battery power consumption.

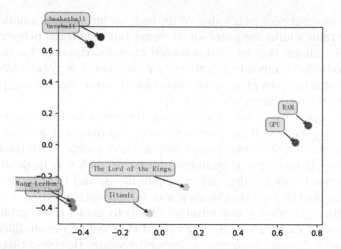

Fig. 1. A word embedding visualization

In this paper, we propose a framework called *Word Embedding Quantization (WEQ)*, which reduces the size of a given word embedding model by quantizing multiple word vectors into bounding areas. Furthermore, we propose techniques for ensuring the correctness of the WEQ employment with respect to the unquantized word embedding model (guaranteeing the recommended items are the same). Our WEQ framework is composed of two components: the offline process of quantizing word embedding model and the online process of computing candidate items based on quantized models. We propose to leverage storage hierarchy on smartphones by storing word embedding models in a layered storage structure; storing quantized model at a primary storage, and the raw word embedding model with the secondary storage. One challenge for using the quantized model is that the exact user profile is no longer ensured. However, we find that there are opportunities for computing the correct recommendation results by knowing only inexact user profile. Accordingly, we propose schemes for efficiently processing candidate computation under the goal of ensuring the result correctness and minimizing the accessing cost to a secondary storage

The contribution of this paper is summarized as follows.

- We propose a WEQ framework for a personalized item-of-interest recommendation based on word embedding model on storage-constrained smartphones.
- Schemes for reducing costs for processing Top-k queries with the quantized word embedding model are designed to ensure the correctness of the recommendation based on the WEQ framework under the goal of minimizing the accessing cost to a secondary storage.
- Extensive experiments with real data sets are conducted. The experiment results show our framework provides nearly three times the performance gain compared with the naive scheme for smartphone personalized item recommendations.

The rest of this paper is organized as follows. Section 2 reviews the existing literature and discuss the position of this paper. Section 3 provides problem formulation for this study. In Sect. 4, we introduce the proposed framework and the techniques for optimizing the secondary storage accessing costs. In Sect. 5 we report experimental results and demonstrate the performance of the proposed framework. Finally, Sect. 6 concludes this paper and provide future research directions.

2   Related Work

Personalized Recommendation on Mobile Devices. Personalized recommendation has been widely studied for various commercial usage. There are also researches on conducting personalized recommendation on mobile devices [1,4,5]. Due to the space limitation, we direct readers to [4] for a complete survey. However, to our best knowledge, the issue of employing word embedding models into resource-limited smartphones remain untouched; most of the existing works are based on traditional BOW models, which fail to capture textual semantics.

Model Compression. As mentioned, the word embedding model is often too large to fit in storage-constrained smartphones. One possible solution is to employ data compression techniques. For example, as proposed in [11], a compression scheme based on neural network and ensemble learning is introduced. Yet another idea is to apply dimension reduction techniques, e.g., low-rank filter banks, as proposed in [12]. These methods may significantly reduce the size of the word embedding model, but they are lossy compression. The recommendation based on the compressed model no longer ensure the correctness of the recommendation results. Compared with the model compression methods, our proposed framework leverages the idea of quantizing the model by grouping word vectors into bounding areas to reduce the size of the model, and propose techniques to guarantee the correctness of the candidate item computation.

Top K. There has been a lot of effort made in establishing safe regions to quickly return the new top-k query with the answers and avoid repetitive calculation [13–15]. For example, in [13], the authors propose region border using linear equations to avoid the computation when there is an update to the position of a top-k query. Also, in [14], the authors propose to determine borders based on the minimum and maximum values from the previous top-k computation. We also direct readers to [15] for a complete survey. However, the existing techniques mainly focus on the issues of reducing computation costs by reusing the results from previous queries, while our work focuses on using bounding areas to reduce the size of word embedding models while guaranteeing the correctness of using only the information from a bounding area to compute Top-k results.

3   Problem Formulation

By employing word embedding model, the problem of personalized items recommendation can be formulated as follow. Assume that we are given a set of words $\{w_1, w_2, ..., w_{|p|}\}$ as a basis for constructing user profile. For each word w_i, we can transform it into a vector $\delta(w_i)$, and then aggregate all $\delta(w_i)$ to form a user profile p. Formally, we get

$$p = \sum_{i=1}^{|p|} \varsigma(w_i) \tag{1}$$

We also model the candidate items in the same manner. That is, for each candidate item, we assume that there is a set of keyword for an item, such as textual item description. We first transform the keywords of an item by the word embedding model and then aggregate all word vectors as the representative for that item. In such a manner, the candidate items can be also formulated as a set of vectors, $A=\{a_1, a_2, ..., a_n\}$. With the given user profile p and the candidate set A, the similarity between p and a can be then computed by $\cos(p, a) = \frac{p \cdot a}{\|p\| \cdot \|a\|}$, and our goal is to compute Top-k similar items with respect to p.

For solving the storage-constrained issue, we propose to leverage storage hierarchy by storing word embedding models in a layered storage structure; storing quantized model at a primary storage, and the raw word embedding model with the secondary storage. As mentioned, the challenge of using the quantized model is that the exact user profile is no longer ensured. However, as will be discussed later, there are opportunities for computing the correct recommendation results by knowing only inexact user profile. Therefore, the research goal of this paper is to minimize the cost of accessing the secondary storage for computing top-k results.

4   The WEQ Framework

4.1   Word Embedding Quantization

Our idea for reducing the size of the model is to quantize a group of words into a bounding area, where bounds the vectors of the words in the group. Thus, the first step is to employ a clustering algorithm to cluster words into groups. After the word clustering, for a given cluster of m words $C=\{w_1, w_2, ..., w_m\}$, we can quantize the word vectors of the words by Algorithm 1. In Example 1, we show a WEQ result example. For ease of discussion, we call the quantization result of a cluster of words as QM (Quantized Mapping) in the following discussion.

Example 1: Assume that we are given three words and their word embedding vectors, i.e., Japan=$< 0.1, 0.17 >$, Tokyo=$< 0.16, 0.1 >$, and Osaka=$< 0.3, 0.2 >$. We have a QM $\{c : (0.186, 0.156), r : 0.122\}$.

There are two things to mention. First, when the number of word vectors (to be quantized together) is large, the space saving is significant. For a given cluster of words, we require only two entries: a center vector c and a radius r to

---

input : $C = \{w_1, ..., w_m\}$;
output: QM: a Quantized Mapping;

1 $c = \frac{1}{m} \sum_{i=1}^{m} \varsigma(w_i)$;
2 $r = 0$;
3 **for** *i=1:m* **do**
4 | $r - \text{MAX}(r, \text{dist}(c, w_i))$
5 **end**
6 $QM = \{Center{:}c, Radius{:}r\}$
7 **return** QM

Algorithm 1: Word Embedding Quantizing Algorithm

---

record the quantized mapping. Second, while the storage cost is saved, the price is that we no longer guarantee the exact vector for individual word; for a word, what we ensure now is an area where its word vector is located.

An important point to indicate is that the employment of WEQ brings imprecise information for user profile modeling. The employment of WEQ makes user profile p into a bounding area. Originally, for every $w_i \in p$, we have its vector v_i through the word embedding model, and use $\sum_{i=1}^{n} \varsigma(w_i)$ as a user profile. However, with the employment of WEQ, for every w_i, we know only its bounding area R_{w_i}. That is, we know only the fact

$$R_{w_i} = \{(c_{w_i} + v), \forall \|v\| \le r_{w_i}\}, \tag{2}$$

where c_{w_i} is the center of QM containing w_i and r_{w_i} is the radius of QM containing w_i.

We note that the bounding area of two different words w_i and w_j shows an additive property. That is, for a given two words $(w_i + w_j)$, the bounding area $R_{w_i + w_j}$ is given by

$$R_{w_i + w_j} = \{(R_{w_i}.c + R_{w_j}.c + v), \forall \|v\| \le R_{w_i}.r + R_{w_j}.r\} \tag{3}$$

As such, a user profile p can be generalized to

$$R_p = \{\sum_{\forall w_i} (R_{w_i}.c + v), \forall \|v\| \le \sum_{\forall i} R_{w_i}.r\} \tag{4}$$

4.2 Bounding Angle Checking Mechanism

With WEQ, we no longer guarantee the exact location of a user profile. Instead, we know only the bounding area R_p. Under such setting, for retrieving top-k items using R_p, there are two cases. First, using R_p produces the same result of using p. That is, no matter where the exact p is located in the bounding area, the same results will be retrieved, and therefore it is safe to processing top-k with R_p. On the other hand, if we can not ensure the same result, we require lookups into a secondary storage (where a complete embedding model is available).

Accordingly, one key issue is that how the correctness of using R_p is checked. In the following discussion, we propose *Bounding Angle Checking Mechanism* (BACM) for this matter. Given a R_p, BACM computes top-k items for p as follows.

- Step 1: Compute top-$(k+1)$ items with respect to $R_p.c$.
- Step 2: Compute a threshold θ_p by $\theta_p = \frac{1}{2}(\theta_{k+1} - \theta_k)$, where θ_k and θ_{k+1} are the distance for the k-th item and the $k+1$-th item with respect to $R_p.c$. This threshold represents the maximal distance between $R_p.c$ and p to have the same top-k results. In Lemma 1, we show the correctness of this observation.
- Step 3: Compute θ_{max} by the following equation

$$\theta_{max} = \arcsin(\frac{r}{\|R_p.c\|}) \tag{5}$$

Note that θ_{max} is the maximal distance between $R_p.c$ and v, $\forall v \in R_p$, as shown in Fig. 2.
- Step 4: If $\theta_{max} \leq \theta_p$, it is safe to use the top-k results based on c. Otherwise, a lookup operation into a complete model is performed to check the real position of p.

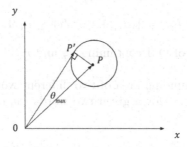

Fig. 2. Maximal distance between a possible p and c

Lemma 1. *For a given R_p, if $\theta_p \leq \frac{1}{2}(\theta_{k+1} - \theta_k)$, the top-k result of p is the same to the top-k result of $R_p.c$, where θ_k denotes the distance between the k-th item and $R_p.c$, θ_{k+1} the distance between the $(k+1)$-th item and $R_p.c$, and θ_p denotes the distance between p and $R_p.c$.*

Proof: Note that as we use $R_p.c$ as a reference for p to compute the top-k results, if we can guarantee the results w.r.t $R_p.c$ is the same as w.r.t. p, we can use the results of $R_p.c$ as a resultant recommendation items for p. Let us consider the worst case that we no longer guarantee the requirement. For a given k-th item and $k + 1$-th item item, if the true position p is close to $k + 1$-th item and away from k-th item, then the correctness may not no longer to be ensured.

Specifically, the worst condition is that if $\theta_k+\theta_p \geq \theta_{k+1}-\theta_p$, we cannot guarantee the correctness of using the result of c. By rewriting the statement, we have

$$\theta_p \leq \frac{1}{2}(\theta_{k+1} - \theta_k) \tag{6}$$

The inequality states that if the condition is violated, it is unsafe to use the results of $R_p.c$.   ∎

In the following, we use a running example to show the process flow of our WEQ framework.

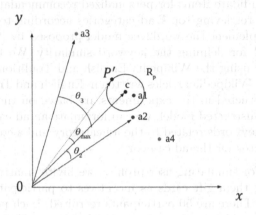

Fig. 3. BACM example

Example 2: Assume that we are given a user profile W={"Japan", "Travel"}, where $QM_{Japan} =< (0.40, 0.2), 0.11 >$ and $QM_{Travel} =< (0.62, 0.9), 0.05 >$, a candidate set A with four items $\{a_1 = (1.13, 1.18), a_2 = (1.13, 0.97), a_3 = (0.40, 1.40), a_4 = (1.42, 0.17) \}$, and top-2 results are asked. Our algorithm works as follows.

- Step 1: by adding the bounding areas of "Japan" and "Travel", we obtain $R_p = \{c : (1.02, 1.1), r : 0.16\}$. Furthermore, compute Top-3 results based on $R_p.c$. Note that $\{a_1, a_2, a_3\}$ is retrieved at this step.
- Step 2: Compute the threshold by the distance between $R_p.c$ and a_2 and $R_p.c$ and a_3. In this case, we have $\theta_p = \frac{1}{2}(\theta_3 - \theta_2) = 10.18°$.
- Step 3: Compute the maximal distance between c and all possible p in R_p. In this case, we have $\theta_{max} = \arcsin(\frac{r}{\|c\|}) = \arcsin(\frac{0.16}{\|(1.02,1.1)\|}) = 6.12°$
- Step 4: Check if it is safe to use only $R_p.c$ by testing $\theta_{max} \leq \theta_p$. In this case, we ensure the same result will be reported, as $\theta_{max} \leq \theta_p$.

As shown in Fig. 3, one can observe that all vectors in R_p are bounded withing the circular area, and the closest possible position denoted as p' for p w.r.t. a_3 is the dashed one in the figure. If this is the case, the distance between p and a_3 is reduced to $\theta_3 - \theta_p$, and the distance between p and a_3 is increased to $\theta_2 + \theta_p$. Accordingly, if $\theta_2 + \theta_p \leq \theta_3 - \theta_p$, we can safely use the results based on R_p.

5   Experiment and Evaluation

5.1   Experimental Settings

In the experiments, we select the personalized advertising application as a testbed for evaluating the performance.

Candidate Items. We simulate the recommendation candidates by employing Google Adwords, which provides a set of words related to some topic, e.g. sports or art topics. We use 100 ad categories from Google Adwords and their keywords to form candidate items for personalized recommendations. We simulate the scenario that retrieving top-k ad categories according to a user profile. In this study, we implement the word2vec model proposed by [9] as a basic word embedding model for defining the keyword similarity. We train the keyword embedding model using the Wikipedia English and Traditional Chinese corpus which contains all Wikipedia articles written in English and Traditional Chinese. The models constructed in the experiments are based on the dump at August 2017. With the constructed model, we can formulate an ad category as a vector by transforming keywords related to the ad category and aggregating them as a representative vector for the ad category.

User Profiles. For simulating user profiles, we invite smartphone users in our campus by giving them gift cards as incentives to participate the data collection experiments. There are 50 participants recruited. Each participant is asked to install our App which implements the Android accessibility service API[1] to capture the texts viewed by the participants. During the data collection process, each participant is asked to use their smartphone for one hour. We use the keyword captured by Accessibility API as the user profiles for evaluating the performance.

We select the most common 100000 words (excluding stop words) from word embedding model learned from the Wikipedia articles as the complete word embedding model. With the words and the mapping, we implement a hierarchical clustering algorithm to form B word clusters to perform word embedding quantization, where B is a given parameter used to simulate the storage size constraint. In this experiment, we use the number of the required lookup operation for computing top-k results using WEQ as our performance metric.

5.2   Performance Results

In this subsection, we present the experimental results for the compared methods under the experimental setting introduced in the last subsection. We compare the proposed method with the scheme that resides word embedding model on a remote server and accesses the required word embedding vectors on demand. We call such a scheme as On-Demand scheme. By the experiment results, we see

---

[1] https://developer.android.com/reference/android/view/accessibility/
AccessibilityEvent.html.

our method requires only 3393 lookup operations, which provides nearly three times the performance gain compared with the On-Demand scheme.

Effects on Varying Parameter. B In Fig. 4(a), we show the costs with respect to various B values. We experimentally set parameter B to 10000, 20000, and 30000 to observe the difference in performance. One can observe that when B is increased, the required look-up operations are reduced. This is an expected result, as when B is large, one can store more information on smartphones and therefore reduce the chances of lookup into the secondary storage.

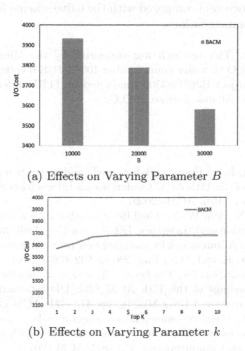

(a) Effects on Varying Parameter B

(b) Effects on Varying Parameter k

Fig. 4. Performance overview

Effects on Varying Parameter. k In Fig. 4(b), we show the costs with respect to various k values, where the x-axis is the value of k and the y-axis is the number of lookup operations required for computing Top-k results. We have two observations for the experiment results. First, one can observe that the cost increases as k increases. The phenomena come from that when k is large, determining true top-k result turns out to be difficult using only the information from the computed bounding area, and therefore a significant number of lookup operations is required to guarantee the correctness of the recommendation results. Second, one can observe that the cost is invariant when $k \geq 4$. The reason for the results is also that when k is large, nearly all information is required to ensure the correctness.

6 Conclusion

In this paper, we propose the WEQ framework for a personalized item-of-interest recommendation based on word embedding model on storage-constrained smartphones. Schemes for reducing costs for processing Top-k queries with the quantized word embedding model are designed to ensure the correctness of the recommendation based on the WEQ framework under the goal of minimizing the accessing cost to a secondary storage. Extensive experiments with real data sets are conducted. The experiment results show our framework provides nearly three times the performance gain compared with the naive scheme for smartphone personalized item recommendations.

Acknowledgement. This research was supported by the Ministry of Science and Technology Taiwan R.O.C. under grant number 106-2221-E-005-082-, and also partially supported by the Project H367B83300 conducted by ITRI under sponsorship of the Ministry of Economic Affairs, Taiwan, R.O.C.

References

1. Penev, A., Wong, R.K.: Framework for timely and accurate ads on mobile devices. In: Proceedings of the 18th ACM Conference on Information and Knowledge Management, pp. 1067–1076. ACM (2009)
2. Zheng, Y., Xie, X., Ma, W.-Y.: GeoLife: a collaborative social networking service among user, location and trajectory. IEEE Data Eng. Bull. **33**(2), 32–39 (2010)
3. Fan, Y.-C., et al.: A framework for enabling user preference profiling through Wi-Fi logs. IEEE Trans. Knowl. Data Eng. **28**(3), 592–603 (2016)
4. Bilenko, M., Richardson, M.: Predictive client-side profiles for personalized advertising. In: Proceedings of the 17th ACM SIGKDD International Conference on Knowledge Discovery and Data Mining, pp. 413–421. ACM (2011)
5. Zhuang, J., et al.: When recommendation meets mobile: contextual and personalized recommendation on the go. In: Proceedings of the 13th International Conference on Ubiquitous Computing, pp. 153–162. ACM (2011)
6. Sarwar, B., et al.: Item-based collaborative filtering recommendation algorithms. In: Proceedings of the 10th International Conference on World Wide Web, pp. 285–295. ACM (2001)
7. Zhu, H., et al.: Mining mobile user preferences for personalized context-aware recommendation. ACM Trans. Intell. Syst. Technol. (TIST) **5**(4), 58 (2015)
8. Yang, X., et al.: On top-k recommendation using social networks. In: Proceedings of the sixth ACM Conference on Recommender Systems, pp. 67–74. ACM (2012)
9. Mikolov, T., et al.: Efficient estimation of word representations in vector space. arXiv preprint arXiv:1301.3781 (2013)
10. Mikolov, T., et al.: Distributed representations of words and phrases and their compositionality. In: Advances in Neural Information Processing Systems, pp. 3111–3119 (2013)
11. Buciluă, C., Caruana, R., Niculescu-Mizil, A.: Model compression. In: Proceedings of the 12th ACM SIGKDD International Conference on Knowledge Discovery and Data Mining, pp. 535–541. ACM (2006)

12. Jaderberg, M., Vedaldi, A., Zisserman, A.: Speeding up convolutional neural networks with low rank expansions. arXiv preprint arXiv:1405.3866 (2014)
13. Zhang, J., Mouratidis, K., Pang, H.: Global immutable region computation. In: Proceedings of the 2014 ACM SIGMOD International Conference on Management of Data, pp. 1151–1162. ACM (2014)
14. Li, Y., et al.: Context-aware advertisement recommendation for high-speed social news feeding. In: 2016 IEEE 32nd International Conference on Data Engineering (ICDE), pp. 505–516. IEEE (2016)
15. Huang, W., et al.: Efficient safe-region construction for moving top-k spatial keyword queries. In: Proceedings of the 21st ACM International Conference on Information and Knowledge Management, pp. 932–941. ACM (2012)

Social Network, Ubiquitous Data and Graph Mining

Topic-Specific Retweet Count Ranking
for Weibo

Hangyu Mao, Yang Xiao, Yuan Wang, Jiakang Wang, and Zhen Xiao[✉]

Department of Computer Science, Peking University, Beijing, China
{mhy,xiaoyang,wangyuan,wangjkcs,xiaozhen}@net.pku.edu.cn

Abstract. In this paper, we study *topic-specific* retweet count ranking problem in Weibo. Two challenges make this task nontrivial. Firstly, traditional methods cannot derive effective feature for tweets, because in topic-specific setting, tweets usually have too many shared contents to distinguish them. We propose a LSTM-embedded autoencoder to generate tweet features with the insight that any different prefixes of tweet text is a possible distinctive feature. Secondly, it is critical to fully catch the meaning of topic in topic-specific setting, but Weibo can provide little information about topic. We leverage real-time news information from Toutiao to enrich the meaning of topic, as more than 85% topics are headline news. We evaluate the proposed components based on ablation methods, and compare the overall solution with a recently-proposed tensor factorization model. Extensive experiments on real Weibo data show the effectiveness and flexibility of our methods.

Keywords: Weibo · Micro-blog · Retweet · Retweet count ranking
Social network

1 Introduction

With the development of micro-blogging services, Weibo, the biggest micro-blogging service in China, has changed the organization of its three major entities (i.e., topic, tweet and user) as shown in Fig. 1. (1) Topics are ranked according to their popularity in the Hot Topic List as shown in the left column. Generally speaking, topic is the group of all tweets sharing the same #topic name#, but it has its own properties such as topic category and topic information. (2) Tweets are divided into common tweets and recommended tweets as show in the right column. Recommended tweets are usually informative and interesting, and they are shown before common tweets in the topic page. (3) Users are encouraged to read tweets in topic pages rather than scattered tweets in their timelines. In a word, topic is becoming the core unit to organize tweets and users in Weibo.

Electronic supplementary material The online version of this chapter (https://doi.org/10.1007/978-3-319-93040-4_49) contains supplementary material, which is available to authorized users.

Fig. 1. The organization of topic, user and tweet in Weibo.

In fact, hot topic is now the main source of page view (PV) and unique visitor (UV) in Weibo. For example, the PV of the topic #Running Man# increases from 13.23 to 42.81 billion after Weibo introduces the "Super Topic" service last year, and the total PV of the top-20 hot topics increases from 127.65 to 361.68 billion. Besides, topic is also beneficial for improving user experience and increasing advertising revenue. On one hand, users can know the detailed and representative information of one topic more easily, because all relevant tweets are grouped and ranked together by topic. On the other hand, advertisers can easily find the target users who browse the specific topic proactively, which means that the advertisements will be more effective.

However, as we know, most users only look through the recommended tweets in the first few pages of a topic. Under this condition, to attract more PV/UV, and further obtain the above mentioned benefits of topic, we need to find out *popular tweets*[1] and make them as the recommended tweets for each topic. In reality, finding out popular tweets from all tweets sharing the same #topic name# is not easy. Currently, the procedure can only be done manually, and it is easy to miss the best popular tweets.

Naturally, in this paper, we try to find out the popular tweets for topics in Weibo automatically (and objectively relative to some metrics)? We formulate this task as a *topic-specific* retweet count ranking problem. Specifically, we predict the retweet count ranking order for all tweets *sharing the same #topic name#, i.e., belonging to the same topic*, and further recommend the highest ranked tweets as the recommended tweets to the corresponding topic (rather than to users). This is why we call our problem *topic-specific* instead of *personalized*. [12,13] have done a lot of work to show that retweet action should be studied at topic level. We refer the readers to their papers for further details.

---

[1] We measure the *popularity* of a tweet by its retweet count. As pointed out by [15, 25,27,28], retweet is the key mechanism for information diffusion on micro-blogging services. A larger retweet count usually means that more users have seen, and will see, the corresponding tweet and topic, and that we will further get more benefits. In fact, researchers often use popular level as the synonym of retweet count [12,14].

More concretely, we propose a Topic-Specific reTweet count Ranking (TSTR) framework to dig this important yet challenging task more deeply. Three main components make up of TSTR framework, making it more applicable in real systems. Firstly, it is impractical to directly deal with the large number of newly-generated tweets in a short time. We propose a Candidate Tweet Filter to filter out unpopular tweets. Secondly, traditional methods cannot derive effective features for tweets, because tweets belonging to the same topic usually have too many shared contents to distinguish them. We propose a LSTM-embedded autoencoder (LSTM-AE) to generate tweet features. This LSTM-AE takes any different prefixes of tweet text as a possible distinctive feature, which makes it very suitable for tweet feature generation. Finally and most importantly, it is critical to fully catch the meaning of topic in topic-specific setting, but we can get little information about topic from Weibo. For example, as shown in Fig. 1, there is only one word (i.e., pay attention) about the most important "Two Sessions" (i.e., NPC and CPPCC) in China. We leverage external real-time news information from Toutiao, the most popular news recommendation platform in China, to enrich the meaning of topic, as we find that more than 85% topics are headline news. We also propose a denoising autoencoder (DAE) to extract topic features from those headline news.

In summary, our contributions are two-fold. (1) This work advances the study of *topic-specific* retweet prediction problem, which has not been well studied like traditional retweet prediction tasks as pointed out by [12,13]. (2) We evaluate the proposed components based on ablation methods, and compare the overall solution with a recently-proposed tensor factorization model. Extensive experiments on real Weibo data show the effectiveness and flexibility of our methods.

2   Related Work

Retweet studies can be roughly divided into retweet analysis and retweet prediction. Retweet analysis aims at understanding why people retweet and which factors impact retweet [1,18,19,26]. Retweet prediction tries to figure out who will retweet a specific tweet or how many times a specific tweet will be retweeted. Our work is an instance of retweet prediction.

Most retweet prediction models formulate retweet count prediction as classification or regression problems [6,14,25,27,28], and only a few researchers study retweet count prediction from a ranking perspective. [17] want to figure out *who* will retweet messages using a Learning-to-Rank framework. They explore a lot of factors, such as retweet history, followers status, followers active time and followers interests, and find that followers who have common interests are more likely to be retweeters. [9] try to answer *who* should share *what*, and extend this problem into two information retrieval scenarios: user ranking and tweet ranking. They propose a Hybrid Factor Non-Negative Matrix Factorization model to estimate each entry of user-tweet matrix. [22] also study both user ranking and tweet ranking. They train a coordinate ascent Learning-to-Rank algorithm to rank the incoming tweets as well as users, and find that tweet-based features have a better predictive ability. [11] study *personalized* tweet ranking according to their probability of being retweeted so that users can find interesting tweets in a short time. They build a user-publisher-tweet graph to re-rank the tweets.

All the above studies focus on user level retweet prediction. Perhaps, [12, 13] are the most relevant studies to ours. The authors also investigate retweet prediction at topic level. They propose a tensor factorization model named V2S to model a set of observed retweet data as a result of three topic-specific factors, i.e., topic virality, user virality and user susceptibility. In their work, they use LDA [4] to generate the latten topics for each tweet. In contrast, topic in our work is directly tagged by users using #topic name#, and all tweets going to be ranked share a same topic.

3   TSTR Framework

3.1   Consideration and Design

There are three major considerations and designs in our system.

The first consideration is that how to deal with the large number of tweets. As we know, a lot of new tweets will be generated in one minute. It is impractical to directly extract features for all those tweets in a short time. Considering that we only care about the popular tweets when we do recommendation, it is natural to propose a Candidate Tweet Filter to filter out unpopular tweets.

The second consideration is that how to derive effective features for tweet. On one hand, tweet is short text with a random length. On the other hand, tweets belonging to the same topic usually have many shared words. Traditional bag-of-word methods and topic model methods such as LDA [4] are not suitable for this task, because those methods suffer from either sparsity or inefficiency for short texts [23]. The recurrent neural network (RNN) may be a better choice, because it can summarize and generate word sequence of arbitrary length and distinguish sequences that have same words but in different orders [8]. So we extend traditional RNN-based encoder-decoder structures and propose a LSTM-AE model with attention mechanism for tweet feature generation.

The third consideration is that how to fully catch the meaning of topic in topic-specific setting. As mentioned in the Introduction, Weibo can provide little information for topics. Fortunately, [16,20] point out that micro-blogging service is more than social network but news media, and over 85% topics are headline news in real world. We also find a similar conclusion in our dataset. So we leverage real-time news information from Toutiao to enrich the meaning of topic. We also propose a DAE model to translate those news information into topic feature.

LSTM-AE and DAE can generate features for tweet and topic, respectively. As for user features, we can crawl them from user database directly. After all features about tweet, topic and user have been generated, we use Tweet Ranker to learn the desired ranking function.

The TSTR framework shown in Fig. 2 summarizes our designs.

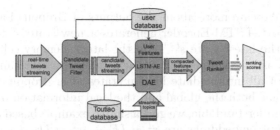

Fig. 2. An illustration of TSTR framework.

3.2   Candidate Tweet Filter

We train a random forest with a dynamic filtering threshold as the Candidate Tweet Filter[2]. Tweet having low popular level with high probability will be filtered out, and others are kept as candidates.

3.3   LSTM-AE

A simplified LSTM-AE structure is shown in Fig. 3(a). LSTM-AE can deal with the random-length property of one tweet and the similar content property of tweets belonging to the same topic, because it is a special RNN. At the same time, it can avoid overfitting because of the added Dropout Layer. Specially, the attention mechanism makes the model focus on more useful parts of the tweet.

The inputs are word embeddings of each tweet. LSTM-AE tries to reconstruct those inputs by minimize the defined loss function in Eq. (1). After training is finished, we extract the outputs of Dropout Layer as tweet features.

Similar RNN-based encoder-decoder structures have been used in other NLP tasks [8,21]. LSTM-AE extends those models in two ways. During training, we extend the loss function. The classical models try to maximize a conditional log-likelihood. Differently, LSTM-AE tries to minimize the mean squared error between (x_n^t, y_n^t) as follows

$$\min_{\theta} \frac{1}{N * T} \sum_{n=1}^{N} \sum_{t=1}^{T} (y_n^t - x_n^t)^2 \tag{1}$$

where x_n^t is the n-th input embedding at time step t, and y_n^t is a function of x_n^t conditioned on model parameter θ. *This difference reflects our insight about how to drive distinctive tweet features in this special topic-specific setting: any different prefixes of tweet text is a possible distinctive feature.* After training is finished, we extend the usage method. The classical models are usually used for generating a target sequence given an input sequence. They care more about the outputs of LSTM-Decoder. Differently, LSTM-AE is used for generating

---

[2] Due to space limitation, we move the features used for building Candidate Tweet Filter into the supplemental material.

tweet features. We care more about the outputs of Dropout Layer. Specifically, at each timestep t, LSTM-Encoder generates a new state s^t for current input x^t, considering the latest state s^{t-1} and the latest memory cell C^{t-1}. The final tweet feature $\overline{X}$ is a concatenation of the last hidden state and the attention-weighted state of all intermediate hidden states after dropout. In this way, we can potentially get both the global and the local information of tweet [21].

To be more understandable, we give a simple example based on tweet text "A like B". The input embeddings are $x^1 = IE_{(A)}$, $x^2 = IE_{(like)}$ and $x^3 = IE_{(B)}$. Because of the memory mechanism of LSTM cell, the hidden state of LSTM-AE can represent the feature embeddings $s^1 = FE_{(A)}$, $s^2 = FE_{(A,like)}$ and $s^3 = FE_{(A,like,B)}$ in some extent. During training, LSTM-AE try to generate output embedding y^t as similar as x^t based on s^t. After training, the concatenation $\overline{X} = [FE_{(A,like,B)}, \alpha_1 FE_{(A)} + \alpha_2 FE_{(A,like)} + \alpha_3 FE_{(A,like,B)}]$ is used as the final tweet feature. As we can see, any prefixes (A), (A,like) and (A,like,B) of tweet text "A like B" is used to generate different features $FE_{(A)}$, $FE_{(A,like)}$ and $FE_{(A,like,B)}$. This is why we say "any different prefixes of tweet text is a possible distinctive feature". This method is useful for our topic-specific application. In contrast, if we only use $FE_{(A,b,C,D,E)}$ as the final tweet feature, it will be too similar with $FE_{(A,B,C,D,E)}$ to distinguish each other, especially when the tweet text is long. Please note that this is very common in topic-specific setting.

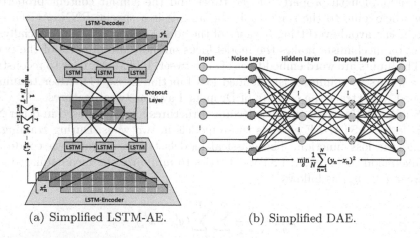

(a) Simplified LSTM-AE.   (b) Simplified DAE.

Fig. 3. Simplified LSTM-AE and DAE structures.

3.4 DAE

As mentioned before, we leverage information from Toutiao to enrich the meaning of topic. Specifically, we use topic as keyword to search Toutiao, and only the returned news headlines are processed. News titles frame the interpretation of the article content and provide the most important information for readers, which have been confirmed by many researchers [2,10]. Note that this method requires a strict time-consistence between topic and the external news.

After we get those news titles, we need to translate them into effective topic features. Our preliminary experiments show that bag-of-word features and topic model features cannot cooperate well with the embedded tweet features. A DAE is proposed to learn embedding features for topic so that both topic features and tweet features have a similar semantic space.

A simplified DAE structure is illustrated in Fig. 3(b). We extract a fixed number of verb and noun phrases with largest $tf\text{-}idf$ values from news title so that the main entity and emotional inclination can be captured on the whole. After that, the corresponding embedding concatenation is used as the input of DAE. As with LSTM-AE, we extract the outputs of Dropout Layer as topic features after training is finished.

Although this structure is common, we have some insights for this design. Firstly, the number of topic is much less than the number of tweet, so a Noise Layer with different noise variances can create more training data. Besides, we also observe that the combination of Noise Layer and Dropout Layer has a lower loss than Noise Layer alone during training. As this simple structure is enough for good results, we leave other novel models for future work.

3.5 Tweet Ranker

In this paper, we use a pair-wise method to train an ensemble of Multiple Additive Regression Tree (MART) as Tweet Ranker, which is similar to LambdaMART [5,7]. The final feature vector is the concatenation of topic features, tweet features and user features[3]. With the supervision of actual retweet count, Tweet Ranker will learn the desired ranking function.

3.6 Chinese Word Embedding

LSTM-AE and DAE assume that we have got the embedding representations of all the Chinese words in topics and tweets. Specifically, we use Neural Probabilistic Language Model [3] to generate those embeddings. Our method is supported by the open source PaddlePaddle[4] deep learning platform, which has more than three million Chinese words as the original corpus.

4 Experiments

Due to space limitation, we move the detailed analyses of experimental data and Candidate Tweet Filter into the supplemental material. However, we would like to highlight some conclusions: (1) the experimental data and the data of the whole Weibo system have a consistent topic distribution; (2) more than 93% unpopular tweets can be filtered out by Candidate Tweet Filter.

To evaluate the ranking results, we adopt 5 widely used metrics following [11,24], i.e., Reciprocal Rank (RR), Precision at k (P@k), Average Precision

---

[3] Due to space limitation, we move the user features into the supplemental material.
[4] http://www.paddlepaddle.org/.

(AP), Normalized Discounted Cumulative Gain at k (NDCG@k), Spearman's Rank Correlation Coefficient (Spearman's ρ). The reported results below contain a "M_" to represent the mean performance on all topics. Some results may also contain a "_#n" to represent that the first n tweets with largest retweet count are relevant. Besides, there are usually 3 to 10 recommended tweets for each topic in Weibo, so we set k and n to 1, 3, 5 and 10 for each group of experiments.

Besides comparing with the recently proposed V2S model [12,13], we also use the following feature sets and their combinations to train ablation models so that we can know the effect of each component in TSTR framework: FC, follower count as feature; UI (User_Info), user features as feature; II (Tweet_Info), original tweet embeddings as feature; TI (Topic_Info), original topic embeddings as feature; II_LSTM, tweet embeddings generated by LSTM-AE as feature; TI_DAE, topic embeddings generated by DAE as feature.

The main experimental results are shown in Fig. 4 and Table 1. We analyze those results from the following five aspects.

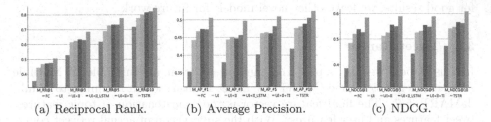

(a) Reciprocal Rank.   (b) Average Precision.   (c) NDCG.

Fig. 4. Results of reciprocal rank, average precision and NDCG.

Table 1. Results of precision and Spearman's ρ.

	FC	UI	UI+II	UI+II_LSTM	UI+II+TI	TSTR
M_P@1_#1	0.25568	0.35795	0.40341	0.40120	0.39205	**0.43713**
M_P@1_#3	0.42614	0.52272	0.55114	0.55689	0.53977	**0.61677**
M_P@1_#5	0.50568	0.59091	0.65909	0.66467	0.65909	**0.71856**
M_P@1_#10	0.61364	0.68750	0.73864	0.76647	0.75000	**0.79042**
M_P@3_#3	0.32765	0.39962	0.39773	0.41517	0.41098	**0.44711**
M_P@3_#5	0.43750	0.53409	0.52083	0.52894	0.55682	**0.58084**
M_P@3_#10	0.55114	0.66856	0.64205	0.65269	0.68561	**0.70259**
M_P@5_#5	0.37273	0.42386	0.41364	0.40838	0.42273	**0.45269**
M_P@5_#10	0.50568	0.59318	0.58068	0.56048	0.59773	**0.62874**
M_P@10_#10	0.38295	0.43750	0.42727	0.41617	0.45114	**0.49042**
Spearman's ρ	0.32757	0.33011	0.37825	0.38225	0.37660	**0.38359**

Ablation Models. (1) FC: all results of FC are much better than random guess. We also use other single feature to do experiments, and we find that the number of follower has a big impact on the prediction. This has been proved by [6,14,15]. (2) UI: we find that ranking according to user features is mostly consistent with ground truth when we only consider information of one entity. In contrast, perhaps due to topic-specific setting, the learned ranking model II is even worse than FC. Detailed results can be found in the supplemental material. [6,9,17,22] use similar baselines, and confirm the importance of user features in their work too. (3) UI+II: we use both user features and tweet features to create a more powerful model. As we can see, most results of UI+II are better than results of UI. In our opinion, feature interaction is the key reason: user features often offer valuable information for tweets. For example, without knowing the user is a gourmet, the II model usually ranks tweets about tourism before tweets about delicacy because tourism is more popular than delicacy in Weibo; and now, the UI+II model can correctly rank tweets about tourism behind tweets about delicacy if those tweets come from a gourmet.

The Proposed LSTM-AE. To test the feature extraction ability of LSTM-AE, we use features from UI+II_LSTM to do an experiment. We find that UI+II_LSTM is better than UI+II and even better than UI+II+TI for metrics such as NDCG. To figure out whether the improvements are statistical generalization, we also apply Student's t-test[5] to the results and find that the corresponding improvements are significant at the level of 0.05. Those results prove that LSTM-AE can generate effective feature for short tweet text with random length, even though tweets belonging to the same topic have similar contents in topic-specific setting.

Hypothesis Testing. To test our hypothesis that real-time news information from Toutiao is potential to boost this retweet count ranking task, we conduct an experiment using features from UI+II+TI and compare it with UI+II. We see that the improvements are marginal, but please note that the corresponding model is only used to test our hypothesis. Considering that the metrics cover a wide range of ranking evaluations and almost all of the results are improved indeed, we conclude that our assumption is reasonable.

The Overall TSTR Model. The TSTR model is build on features from all UI+II_LSTM+TI_DAE. The *improvements* of TSTR compared to other models for different metrics are shown in Table 2. Firstly, all the entries in this table are positive, which indicates that TSTR is a flexible framework to fit different ranking evaluation metrics simultaneously. Those good results may be related to the unbiased feature generation ability of LSTM-AE and DAE. To prove this inference, we train a TSTR model based on NDCG@5 and find that it can achieve similar good results on other metrics. Quantitatively, the average improvements (excluding V2S column) are bigger than 6% for all metrics, and the significant level for metrics such as NDCG and AP can achieve 0.01. Compared to UI+II_LSTM, we can say that TI_DAE is necessary for good results. Compared

---

[5] It can be used to determine if two sets of data are significantly different from each other: https://en.wikipedia.org/wiki/Student%27s_t-test.

to UI+II+TI, we can say that the combination of LSTM-AE and DAE is necessary for good results. Interestingly, the overall ranking improvements (i.e., Spearman's ρ) are much smaller than the top tweets ranking improvements (i.e., other metrics). One possible reason may be that there are many tweets with small retweet count, and it is not easy to fully distinguish them using features generated by LSTM-AE and DAE. This phenomenon means that the proposed TSTR framework is more suitable for applications such as recommendation and hot events detection. In those applications, we care more about the *higher* ranked tweets rather than the ranking of *all* tweets.

Table 2. *Improvements* of TSTR for different metrics compared to UI+II_LSTM, UI+II+TI and V2S models.

	UI+II_LSTM	UI+II+TI	V2S
M_P@1_#1	8.96%	11.50%	18.23%
M_P@1_#3	10.75%	14.27%	8.87%
M_P@1_#5	8.11%	9.02%	5.02%
M_P@1_#10	3.12%	5.39%	3.49%
M_P@3_#3	7.69%	8.79%	9.73%
M_P@3_#5	9.81%	4.31%	6.86%
M_P@3_#10	7.65%	2.48%	3.95%
M_P@5_#5	10.85%	7.09%	6.24%
M_P@5_#10	12.18%	5.19%	3.69%
M_P@10_#10	17.84%	14.24%	3.59%
Ave. Improv.	**9.70%**	**8.23%**	**6.97%**
M_AP_#1	6.72%	7.00%	7.70%
M_AP_#3	11.17%	8.86%	6.11%
M_AP_#5	10.47%	5.76%	4.87%
M_AP_#10	7.53%	3.95%	3.40%
Ave. Improv.	**8.97%**	**6.39%**	**5.52%**
M_RR@1	7.02%	6.47%	7.70%
M_RR@3	8.13%	9.00%	4.26%
M_RR@5	5.95%	5.96%	2.90%
M_RR@10	3.55%	2.66%	1.92%
Ave. Improv.	**6.16%**	**6.02%**	**4.20%**
M_NDCG@1	8.51%	10.95%	5.98%
M_NDCG@3	6.09%	7.18%	5.88%
M_NDCG@5	6.14%	7.43%	5.56%
M_NDCG@10	7.55%	8.33%	4.56%
Ave. Improv.	**7.07%**	**8.47%**	**5.50%**
Spearman's ρ	0.35%	1.86%	0.66%

The Baseline V2S Model. From the average improvement rows in Table 2, we can see that V2S performs better than UI+II_LSTM and UI+II+TI but worse than our TSTR model on the whole. As reported in the original papers, V2S outperforms other state-of-the-art content-based and LDA-based models, so we expect that TSTR could have the same ability too. Fine-grained analyses show that V2S cannot perform well on metrics such as M_P@1_#1, M_P@3_#3, M_AP#1 and M_RR@1, which means that V2S is not suitable for applications such as recommendation and hot events detection where our TSTR model is a better choice as analysed before.

5 Conclusion

In this paper, we leverage real-time news information from Toutiao to improve the topic-specific retweet count ranking task in Weibo. A TSTR framework is proposed to address this important yet challenging problem. A LSTM-AE and a DAE make up of the core part of TSTR framework. LSTM-AE extends traditional RNN-based encoder-decoder models in two ways for generating tweet features. DAE is designed for translating news information into topic features. Extensive experiments on real Weibo data show the effectiveness and flexibility of TSTR framework.

We also provide some useful conclusions. (1) User features are more suitable for this topic-specific ranking task than tweet features. (2) Real-time news information from Toutiao is potential to boost applications in Weibo. (3) The proposed TSTR framework is suitable for applications (e.g., recommendation) caring more about the higher ranked tweets rather than the ranking of all tweets.

We will study the following problems in the future: (1) How to leverage other suitable data for the remaining 15% topics that are not reported in Toutiao. (2) How to use network structure information properly in topic-specific setting.

Acknowledgement. The authors thank the anonymous reviewers for their comments. This work was supported by the National Natural Science Foundation of China under Grant No. 61572044.

References

1. Abdullah, N.A., Nishioka, D., Tanaka, Y., Murayama, Y.: User's action and decision making of retweet messages towards reducing misinformation spread during disaster. J. Inf. Process. **23**(1), 31–40 (2015)
2. Andrew, B.C.: Media-generated shortcuts: do newspaper headlines present another roadblock for low-information rationality? Harv. Int. J. Press/Polit. **12**(2), 24–43 (2007)
3. Bengio, Y., Ducharme, R., Vincent, P., Janvin, C.: A neural probabilistic language model. J. Mach. Learn. Res. **3**, 1137–1155 (2003)
4. Blei, D.M., Ng, A.Y., Jordan, M.I.: Latent Dirichlet allocation. J. Mach. Learn. Res. **3**(January), 993–1022 (2003)

5. Burges, C.J.: From ranknet to lambdarank to lambdamart: an overview. Learning **11**, 23–581 (2010)
6. Can, E.F., Oktay, H., Manmatha, R.: Predicting retweet count using visual cues. In: Proceedings of the 22nd ACM International Conference on Information & Knowledge Management, CIKM 2013, pp. 1481–1484. ACM, New York (2013)
7. Cao, Z., Qin, T., Liu, T.Y., Tsai, M.F., Li, H.: Learning to rank: from pairwise approach to listwise approach. In: Proceedings of the 24th International Conference on Machine Learning, ICML 2007, pp. 129–136. ACM, New York (2007)
8. Cho, K., Van Merriënboer, B., Gulcehre, C., Bahdanau, D., Bougares, F., Schwenk, H., Bengio, Y.: Learning phrase representations using RNN encoder-decoder for statistical machine translation. arXiv preprint arXiv:1406.1078 (2014)
9. Cui, P., Wang, F., Sun, L.: Who should share what?: item-level social influence prediction for users and posts ranking. In: International ACM SIGIR Conference on Research and Development in Information Retrieval, pp. 185–194 (2011)
10. Dor, D.: On newspaper headlines as relevance optimizers. J. Pragmat. **35**(5), 695–721 (2003)
11. Feng, W., Wang, J.: Retweet or not?: personalized tweet re-ranking. In: Proceedings of the Sixth ACM International Conference on Web Search and Data Mining, WSDM 2013, pp. 577–586. ACM, New York (2013)
12. Hoang, T.A., Lim, E.P.: Retweeting: an act of viral users, susceptible users, or viral topics? In: SIAM International Conference on Data Mining (2013)
13. Hoang, T.A., Lim, E.P.: Microblogging content propagation modeling using topic-specific behavioral factors. IEEE Trans. Knowl. Data Eng. **28**(9), 2407–2422 (2016)
14. Hong, L., Dan, O., Davison, B.D.: Predicting popular messages in Twitter. In: Proceedings of the 20th International Conference Companion on World Wide Web, WWW 2011, pp. 57–58. ACM, New York (2011)
15. Jenders, M., Kasneci, G., Naumann, F.: Analyzing and predicting viral tweets. In: Proceedings of the 22nd International Conference on World Wide Web, WWW 2013 Companion, pp. 657–664. ACM, New York (2013)
16. Kwak, H., Lee, C., Park, H., Moon, S.: What is Twitter, a social network or a news media? In: Proceedings of the 19th International Conference on World Wide Web, WWW 2010, pp. 591–600. ACM, New York (2010)
17. Luo, Z., Osborne, M., Tang, J., Wang, T.: Who will retweet me?: finding retweeters in Twitter. In: International ACM SIGIR Conference on Research and Development in Information Retrieval, SIGIR 2013, pp. 869–872. ACM, New York (2013)
18. Macskassy, S.A., Michelson, M.: Why do people retweet? Anti-homophily wins the day! In: ICWSM, pp. 209–216 (2011)
19. Meier, F., Elsweiler, D., Wilson, M.L.: More than liking and bookmarking? Towards understanding Twitter favouriting behaviour. In: ICWSM (2014)
20. Myers, S.A., Zhu, C., Leskovec, J.: Information diffusion and external influence in networks. In: Proceedings of ACM SIGKDD International Conference on Knowledge Discovery and Data Mining, KDD 2012, pp. 33–41. ACM, New York (2012)
21. Shang, L., Lu, Z., Li, H.: Neural responding machine for short-text conversation. arXiv preprint arXiv:1503.02364 (2015)
22. Uysal, I., Croft, W.B.: User oriented tweet ranking: a filtering approach to microblogs. In: Proceedings of the ACM International Conference on Information and Knowledge Management, CIKM 2011, pp. 2261–2264. ACM, New York (2011)
23. Wang, Y., Xiao, Y., Ma, C., Xiao, Z.: Improving users' demographic prediction via the videos they talk about. In: Proceedings of the 2016 Conference on Empirical Methods in Natural Language Processing, Austin, Texas, Association for Computational Linguistics, pp. 1359–1368, November 2016

24. Xiao, Y., Zhao, W.X., Wang, K., Xiao, Z.: Knowledge sharing via social login: exploiting microblogging service for warming up social question answering websites. In: COLING, pp. 656–666. Citeseer (2014)

25. Yang, Z., Guo, J., Cai, K., Tang, J., Li, J., Zhang, L., Su, Z.: Understanding retweeting behaviors in social networks. In: ACM International Conference on Information and Knowledge Management, pp. 1633–1636 (2010)

26. Yuan, N.J., Zhong, Y., Zhang, F., Xie, X., Lin, C.Y., Rui, Y.: Who will reply to/retweet this tweet?: the dynamics of intimacy from online social interactions. In: Proceedings of the Ninth ACM International Conference on Web Search and Data Mining, WSDM 2016, pp. 3–12. ACM, New York (2016)

27. Zhang, Q., Gong, Y., Guo, Y., Huang, X.: Retweet behavior prediction using hierarchical Dirichlet process. In: Twenty-Ninth AAAI Conference on Artificial Intelligence, pp. 403–409 (2015)

28. Zhang, Q., Gong, Y., Wu, J., Huang, H., Huang, X.: Retweet prediction with attention-based deep neural network. In: ACM International on Conference on Information and Knowledge Management, pp. 75–84 (2016)

Motif-Aware Diffusion Network Inference

Qi Tan[1], Yang Liu[1,2], and Jiming Liu[1,2(✉)]

[1] Department of Computer Science, Hong Kong Baptist University,
Hong Kong, China
{csqtan,csygliu,jiming}@comp.hkbu.edu.hk
[2] IRACE Research Center, Hong Kong Baptist University, Shenzhen, China

Abstract. Characterizing and understanding information diffusion over social networks play an important role in various real-world applications. In many scenarios, however, only the states of nodes can be observed while the underlying diffusion networks are unknown. Many methods have therefore been proposed to infer the underlying networks based on node observations. To enhance the inference performance, structural priors of the networks, such as sparsity, scale-free, and community structures, are often incorporated into the learning procedure. As the building blocks of networks, network motifs occur frequently in many social networks, and play an essential role in describing the network structures and functionalities. However, to the best of our knowledge, no existing work exploits this kind of structural primitives in diffusion network inference. In order to address this unexplored yet important issue, in this paper, we propose a novel framework called Motif-Aware Diffusion Network Inference (MADNI), which aims to mine the motif profile from the node observations and infer the underlying network based on the mined motif profile. The mined motif profile and the inferred network are alternately refined until the learning procedure converges. Extensive experiments on both synthetic and real-world datasets validate the effectiveness of the proposed framework.

1 Introduction

Characterizing and understanding information diffusion processes over social networks play an important role in many real-world applications, such as viral marketing [10] and rumor detection [3]. However, in many scenarios, the underlying diffusion networks are hidden [19,20]; what we do have is the states of nodes observed over time. Therefore, inferring the underlying networks based on the observations of node states is of great importance and has received much attention recently [5,19,20].

Utilizing the network structure properties (e.g., community structure [8] and scale-free property [21]) as the prior in the inference procedure has been proved effective in improving the performance of network inference [7,18]. Network motifs, which are regarded as the building blocks of networks [1,17], occur frequently in many real-world networks and play a key role in analyzing the network structure and interpreting the network functionality. For example, as an

© Springer International Publishing AG, part of Springer Nature 2018
D. Phung et al. (Eds.): PAKDD 2018, LNAI 10939, pp. 638–650, 2018.
https://doi.org/10.1007/978-3-319-93040-4_50

important mesoscale structure, motif patterns characterize the local structure of the network connectivity and contribute to network classification [16] and community detection [2]. Moreover, the incoherent feed-forward loop, which is a representative triangle network motif, is commonly found in gene regulation networks and can provide fold-change detection [4].

Although motifs are of great importance in describing the network structures and functionalities, to the best of our knowledge, no existing work exploits this kind of structural primitives in diffusion network inference. In order to address this unexplored yet important issue, we propose a novel framework called Motif-Aware Diffusion Network Inference (MADNI), which takes the network motifs into account when inferring the underlying diffusion networks. Figure 1 schematically illustrates the idea and procedure of the proposed framework.

The contributions of this paper are summarized as follows.

1. We investigate an unexplored yet important issue, i.e., how to integrate motif prior into diffusion network inference.
2. We propose a novel learning framework MADNI to mine the motif profile and incorporate the uncovered motif profile into the network inference procedure.
3. We perform extensive experiments on both synthetic and real-world datasets, showing the effectiveness of the proposed framework.

2 Related Work

The proposed MADNI aims to jointly mine the network motif prior and infer the underlying diffusion network. This section therefore reviews some related works in underlying diffusion network inference and the inference with network priors.

The diffusion network inference problem refers to tracing the diffusion edges based on the observed infection time sequence. Gomez et al. proposed an algorithm NETINF [5] to infer the diffusion edges through maximizing the likelihood of observed infection time by utilizing submodular optimization. To infer the heterogeneous transmission rates and time-varying network, NETRATE [19] and INFOPATH [6] have been proposed respectively. Rong et al. proposed a model-free approach NPDC [20] to utilizes the statistical difference of the infection time intervals between nodes connected with diffusion edges versus those without diffusion edges in network inference. Moreover, Hu et al. proposed a clustering embedded approach CENI [9] to improve the efficiency of network inference by clustering the nodes on the embedded space.

Incorporating the network prior into the learning procedure generally improves the performance of network inference [8,13,21]. Many literatures have focused on inferring the scale-free networks and modular networks with block structure. Liu and Ihler added a log l_1 norm regularization on the estimated graph structure in the Gaussian graphical model learning to encourage the estimated graph becoming scale-free [15]. Liu et al. introduced the weight inverse graph prior to encourage specific node distribution [13]. Hosseini Lee. introduced a block prior to encourage sparse connections between blocks [8].

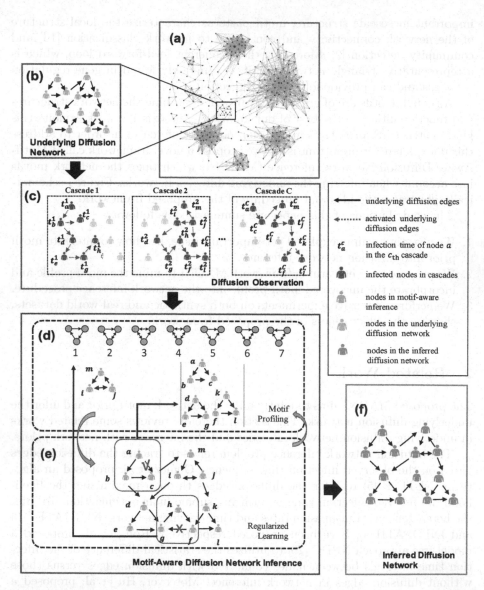

Fig. 1. Schematic illustration of the proposed motif-aware diffusion network inference (MADNI). (a) An example of the diffusion network. (b) An enlarged part of the diffusion network in (a) for algorithm illustration. (c) The observed cascades on the underlying diffusion network shown in (b). For each cascade, only the infection time of the influenced nodes (the red nodes) are observed, such as $\mathbf{t}^1 = \{t_a^1, t_b^1, \cdots\}$. (d) The motif profile is mined from the cascade data by estimating the frequency of various motif patterns in the underlying diffusion network. (e) The underlying diffusion network is learned via motif prior regularized learning. The mined motif profile and the learned network are alternately refined until the inferred network converges. (f) The diffusion network inferred by the MADNI framework. (Color figure online)

However, motif, the structural primitives of many real-world networks, has not been taken into account in underlying diffusion network inference yet. Therefore, in this paper, we attempt to fill this gap by proposing a novel framework to discover the underlying motif patterns and incorporate the uncovered motif patterns into the network inference procedure.

3   Motif-Aware Diffusion Network Inference

In this section, we present our framework, Motif-Aware Diffusion Network Inference (MADNI). First, we provide the necessary notations and formally define the problem. Then we introduce how to estimate the initial structure motif profile from the cascade data and present a novel scheme for inferring the diffusion network with the mined structural motif profile.

3.1   Notations and Problem Formulation

Let $G =< V, E >$ denote the directed diffusion network, where V denotes the set of N nodes (representing the individuals, Blog sites or locations) and E denotes the set of edges (representing the directed influence from one node to another). Generally, the edges in E is represented by the $N \times N$ adjacency matrix A, where an entry (i, j) of A, A_{ij}, is the transmission rate from node i to node j. A subgraph $G_s = (V_s, E_s)$ of G satisfies $V_s \subseteq V$, $E_s \subseteq E$. Two subgraphs $G_s^1 = (V_1, E_1)$ and $G_s^2 = (V_2, E_2)$ are isomorphic if there exists a projection $\varphi : V_1 \rightarrow V_2$ with $(u, v) \in E_1 \leftrightarrow (\varphi(u), \varphi(v)) \in E_2$ for all (u, v).

Motif. A motif pattern with k nodes is a non-isomorphic, connected subgraph frequently appearing in a large network. Figure 1(d) shows all seven close connected triangle motif patterns.

Consider that the diffusion observation U is collected over the underlying diffusion network G and consists of a set of C cascades. Each cascade t^c ($c = 1, ..., C$) is a collection of observed infection time stamps within the population during a time interval of length T and can be represented as an N-dimensional vector $t^c := (t_1^c, \cdots, t_N^c)$, where $t_n^c \in [0, T] \cup \{\infty\}$ indicates the infection time of node n in cascade c. The symbol ∞ labels users that are not infected during observation window $[0, T]$. Given the above diffusion observation O, we aim to infer the underlying relation between nodes on G, i.e., the adjacency matrix A.

3.2   Estimating Motif Pattern from Cascade Data

In this subsection, we introduce a straightforward yet effective approach to estimate the motif frequency from the cascade data, which is expected to be helpful in inferring the underlying diffusion network. When scanning the cascade sequences, we record the occurrence matrix $CO \in R^{N \times N}$ as follows. In each cascade, if $t_i^c + t^W > t_j^c > t_i^c$, where t^W is the time window, $CO_{i,j}$ increases by one. Therefore, $CO_{i,j}$ reflects how many times that node i may influence node

j. We could further take into account the interval between the infection time of node i and that of node j, i.e., $CO_{i,j}$ increases by $\sigma(-(t_j^c - t_i^c))$, where $\sigma(\cdot)$ could be the Exponential function or Rayleigh function [19]. Based on this occurrence matrix, we extract the significant pairwise influences by calculating the score of the edge count: $S_{i,j}^E = \frac{CO_{i,j} - mean(CO)}{std(CO)}$, where $mean(CO)$ and $std(CO)$ denote the mean and standard deviation of all elements in CO, respectively. The edges with low scores are filtered out. Then we could count the subgraph frequency f^m based on the S^E by assuming that there exists an edge $e_{i,j}$ if $S_{i,j}^E$ is nonzero, and further determine the significant motifs by calculating the Z-score of the subgraph as $Q^m = \frac{f^m - mean(\mathcal{F}^m)}{std(\mathcal{F}^m)}$, where $\mathcal{F}^m$ is the m-th motif frequency of a set of samples drawn by randomly shuffling S^E [17]. The m-th motif is significant if Z^m is far above 1 [16]. We denote the procedure of extracting significant motifs as $\mathcal{M} \leftarrow \mathcal{Q}(S^E)$.

3.3   Motif Prior Regularization

Different types of networks exhibit distinct motif frequency profiles [16]. In order to incorporate the motif prior into network inference, we propose an edge-centered regularization to adjust the motif frequency profile of the estimated network. The main idea is that if an edge is forming a high frequent motif, less penalization will be given to it. Let $Z^m \in \Re^{N \times N}$ denote the motif count matrix of G for a certain motif pattern m, where $Z_{i,j}^m$ indicates the number of instances of motif m that containing the edge $(i \rightarrow j)$ [2]. We detect the motif m from the adjacency matrix A and count $Z_{i,j}^m$ for each edge, which is denoted as $Z^m = \mathcal{P}(A, m)$. Furthermore, we have $Z = \sum_{m \in \mathcal{M}} Z^m = \sum_{m \in \mathcal{M}} \mathcal{P}(A, m) = \mathcal{P}(A)$. The motif pattern in $\mathcal{M}$ could be selected as significant motifs detected from the networks or based on prior knowledge. In this paper, the seven close connected triangle motif patterns shown in Fig. 1 are considered since the empirical studies have revealed that these motif patterns appear frequently and play special roles in social networks [14]. Based on the motif count matrix, the reweighted regularization can be constructed for learning the network with significant frequent motif patterns:

$$R(\mathcal{P}(A)) = |M \circ A| = \sum_{i,j=1}^{N} |\frac{A_{i,j}}{Z_{i,j} + 1}| \tag{1}$$

3.4   Learning

We aim to find the diffusion network such that likelihood of diffusion observation is maximized. The likelihood of diffusion observation is calculated as follow.

Pairwise transmission likelihood. With the cascades data, the pairwise transmission likelihood is calculated as follows. Define $f(t_i^c | t_j^c, A_{j,i})$ as the transmission likelihood from node j to node i, which is related to the infection time interval $\Delta t = (t_j^c - t_i^c)$ and the transmission rate $A_{j,i}$. Moreover, a node can only be infected by an infected node. The exponential parametric likelihood model is adopted: $f(t_i^c|t_j^c, A_{j,i}) = A_{j,i} \exp^{-A_{j,i}(t_i^c - t_j^c)}$ if $t_j^c < t_i^c$ and 0 otherwise. The survival likelihood of edge $(j \rightarrow i)$, denoted as $S(t_i^c|t_j^c, A_{j,i})$, is the probability that node i is not infected by node j by time t_i^c, which is calculated as: $S(t_i^c|t_j^c, A_{j,i}) = 1 - F(t_i^c|t_j^c, A_{j,i})$, where $F(t_i^c|t_j^c, A_{j,i})$ is the cumulative function of the transmission likelihood.

Likelihood of a cascade. The likelihood of the observe infections $\hat{t}^c = (t_1^c, \cdots, t_N^c)$ is calculated as:

$$f(\hat{t}^c; A) = \prod_{t_i^c \leq T} \prod_{t_m^c > T} S(T|t_i^c, A_{i,m}) \prod_{k:t_k^c < t_i^c} S(t_i^c|t_k^c, A_{k,i}) \times \sum_{j:t_j^c < t_i^c} H(t_i^c|t_j^c, A_{j,i}),$$

(2)

where $H(t_i^c|t_j^c, A_{j,i})$ is the hazard function: $H(t_i^c|t_j^c, A_{j,i}) = \frac{f(t_i^c|t_j^c, A_{j,i})}{S(t_i^c|t_j^c, A_{j,i})}$.

Network inference. We aim to search A that maximizes the likelihood of cascade observation O. In our framework, the networks are estimated through maximizing the regularized likelihood function as follow.

$$\max_A \left(L(O|A) - R(\mathcal{P}(A)) \right) = \max_A \left(\sum_{c \in C} \log f(t^c, A) - R(\mathcal{P}(A)) \right)$$

(3)

$$s.t. \ A_{j,i} \geq 0, i, j = 1, \cdots, N$$

where $L(O|A)$ is the likelihood function of observation given the network topology and $R(\mathcal{P}(A))$ is the regularization term. We can use ADMM or projected gradient descent [6] to enforce A to be nonnegative. Thus the matrix gradient in terms of A is written as $\frac{\partial L}{\partial A} - M$. The additional computational bounden of adding the regularization term is just the addition of a $N \times N$ matrix. The gradient for edges linking to node k in the cascade where node k is uninfected is

$$\frac{\partial L^c}{\partial A_{j,k}} = T - t_j^c$$

(4)

and the gradient for edges linking to node k in the cascades where node k is infected is:

$$\frac{\partial L^c}{\partial A_{j,k}} = (t_k^c - t_j^c) - \frac{1}{\sum_{l:t_l^c < t_k^c} A_{l,k}}$$

(5)

Summating the above term over all cascades gives the gradient for edges linking to node k. Starting from $Z = \mathcal{P}(CO)$ or a plain prior, i.e., $Z = \mathcal{P}(0^{N \times N})$, we update the network structure and the motif count matrix alternately until the estimated network structure remains unchanged. The detailed procedure of the proposed MADNI framework is provided in Algorithm 1.

3.5   Computational Complexity Analysis

In this subsection, we analyze the computational complexity of the proposed framework. The time cost of occurrence matrix counting is $O(CL)$, where C is the

---

Algorithm 1: Motif-Aware Diffusion Network Inference (MADNI)

Input : The observation O: C cascades $\{t^c = (t_1^c, ..., t_N^c)|_{c=1}^C\}$
Output: The estimated network $\hat{A}$
for $c = 1, \cdots, C$ **do**
    if $t_i^c + t^W > t_j^c > t_i^c$ **then**
        $CO_{i,j} \leftarrow CO_{i,j} + \sigma(-(t_j^c - t_i^c));$ ▷ Construct occurrence matrix
    end
end
for $i, j = 1, \cdots, N$ **do**
    $S_{i,j}^E \leftarrow \frac{CO_{i,j} - mean(CO)}{std(CO)}$;   ▷ Calculate edge significance
end
$\mathcal{M} \leftarrow \mathcal{Q}(S^E)$;   ▷ Initialize candidature motifs set $\mathcal{M}$
$Z \leftarrow \mathcal{P}(S^E, \mathcal{M})$;   ▷ Initialize motif profile
while *not converged* **do**
    $\hat{A} \leftarrow \arg\max_A L(O|A) - R(Z)$;   ▷ Learn diffusion network
    $Z \leftarrow \mathcal{P}(\hat{A}, \mathcal{M})$;   ▷ Update motif profile
end

---

number of cascades and L is the length of a cascade. Assume on average there are K elements in each row of $S^E + (S^E)^T$, then the cost of motif counting in S^E and its random shuffling variants is $O(NK^2)$. For each iteration of diffusion network learning and motif profile updating, the computational demand comes from two parts: motif count matrix calculation and network inference. Assume there are M edges in the graph and the maximum degree is D_{max}. For each edge e, the cost of calculating the motif count C_e is $O(D_{max})$ and thus the cost of calculating the matrix count for all edges is $O(MD_{max})$. The network structure is inferred via an iterative way, in which the complexity in each iteration is $O(CN^2)$. If the maximum number of iterations in network inference problem is N_i, then the complexity of network inference is $O(N_iCN^2)$. As $O(CN_iN^2) > O(MD_{max})$ holds in general, the total computational cost of the proposed framework is $O(CN_iN^2)$.

4 Validations

In this section, we evaluate the performance of our framework in diffusion network inference on both the synthetic and real-world networks, in terms of Precision, Recall and $F1$ score [7].

4.1 Experiments on Synthetic Networks

In this subsection, we evaluate the performance of our framework on synthetic networks and cascades. We first construct the synthetic network with N nodes and then generate C cascades using exponential diffusion model on the network

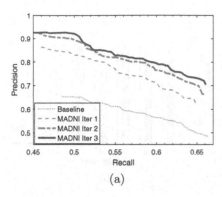
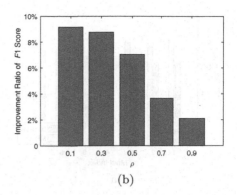

(a)   (b)

Fig. 2. (a) Performance comparison (in terms of Precision and Recall) between MADNI and the baseline. MADNI significantly outperforms the baseline method with only one iteration. The performance of MADNI is further improved after each iteration, and the learning procedure quickly converges in only three iterations. (b) $F1$ improvement ratio of MADNI over the baseline method with varying ρ. MADNI outperforms the baseline even if the target network is close to a random network ($\rho = 0.9$). When the target network becomes more structured, i.e., ρ becomes smaller, the improvement ratio becomes more significant.

as the observation [19]. We choose NETRATE [19] as the baseline method for comparison in our experiments as it explores the global convexity of network inference problem.

Comparison with baseline. We first evaluate the performance of the proposed framework on inferring the motif-dense network, i.e., the network with certain motif patterns occurring frequently. The motif-dense network is generated from a random motif network model, which enumerates the combinations over all nodes and assigns a specific motif to each node combination with probability p. Here we choose the feed-forward loop motif, i.e., Motif 5 illustrated in Fig. 1(d), for our experiment, as it is a commonly observed motif in social networks [16]. Figure 2(a) shows the Precision and Recall of the baseline method and those of MADNI after different number of iterations. It can be seen that the proposed framework achieves significant improvement over the baseline method with only one iteration, which validates the effectiveness of taking the motif into consideration when inferring the structured networks. Furthermore, the performance of the proposed framework can be further improved after each iteration, and the learning procedure quickly converges in only three iterations.

Performance improvement with random edges. After having evaluated the performance of the proposed framework on the motif-dense network, we further test our framework on the networks consist of both significant motifs and random edges. Specifically, we generate the target network from a motif-dense network and a random network. The proportion of random network is indicated by a parameter ρ, where $\rho = 0$ indicates the complete motif-dense network, which is used in our previous experiment; while $\rho = 1$ indicates the complete random

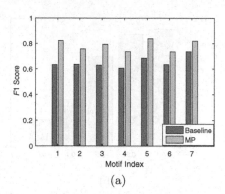
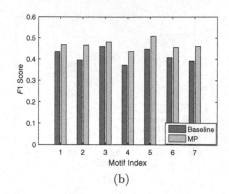

(a)   (b)

Fig. 3. Performance comparison (in terms of $F1$ score) between MADNI and the baseline in (a) Exponential and (b) Rayleigh cascades on the networks with different types of close connected triangle motifs.

network. The $F1$ score improvement ratio of MADNI over the baseline method with varying ρ is shown in Fig. 2(b). Here the improvement ratio is defined as $(F1m - F1_b)/F1_b$, where $F1_m$ is the $F1$ score of MADNI and $F1_b$ is the $F1$ score of the baseline method. It can be observed from the figure that the proposed framework outperforms the baseline even if the target network is close to a random network ($\rho = 0.9$). When the target network becomes more structured, i.e., the ratio of motif-dense network increases, the improvement ratio becomes more significant.

Adaptivity over various motif patterns. In order to show that the performance improvement is independent of the specific motif, we examine the performance of our framework on inferring the networks with different types of frequently occurred motifs. Specifically, we consider all the seven close connected triangle motifs in this experiment. We generate the cascades using the Exponential and Rayleigh cascade models [19] and set $\rho = 0.5$. The comparison results are shown in Fig. 3. MADNI consistently performs better than the baseline over networks with different types of frequently occurred motifs, which demonstrates the adaptivity of the proposed framework over various motif patterns.

4.2   Experiment on a Real-World Network

In this subsection, we evaluate the proposed framework on a real-world network, i.e., an email communication network of an European Research Institute consisting of 320 nodes and 3031 edges [12]. Similar to the synthetic experiments in Sect. 4.1, we generate C ($= 1000, 4000, 10000$) cascades on the network as the observations.

In this experiment, we compare the proposed framework with seven methods: NETINF [5], NETINF with community structure prior, NETINF with scale-free prior, NETRATE [19], NETRATE with community structure prior, NETRATE with scale-free prior, and CENI [9]. Here NETINF and NETRATE are classical

Table 1. Performance comparison (in terms of $F1$ score) between the proposed MADNI-I/MADNI-R methods and seven competing diffusion network inference algorithms in a real-world network experiment. The proposed methods perform the best.

Methods	Number of cascades		
	$C = 1000$	$C = 4000$	$C = 10000$
NETINF	0.5675	0.8040	0.8333
NETINF + Community structure	0.5944	0.8041	0.8363
NETINF + Scale-free	0.6121	0.8044	0.8397
NETRATE	0.6636	0.7900	0.8350
NETRATE + Community structure	0.6385	0.7900	0.8351
NETRATE + Scale-free	0.6426	0.7901	0.8431
CENI	0.3390	0.8058	0.8517
MADNI-I	0.6287	**0.8188**	0.8464
MADNI-R	**0.6685**	0.7998	**0.8600**

network inference methods, community and scale-free structure are representative structural priors, and CENI is a state-of-the-art network inference algorithm. For the proposed framework, NETINF and NETRATE are employed to learn the diffusion network, i.e., maximize $(L(O|A) - R(Z))$, respectively. Therefore, in this experiment, we name the methods generated from the proposed framework as MADNI-I (corresponding to NETINF) and MADNI-R (corresponding to NETRATE), respectively.

Table 1 lists the performance (in terms of $F1$ score) of the proposed MADNI-I/MADNI-R methods and the aforementioned seven competing algorithms under different numbers of cascades. The proposed methods achieve the best performance under all settings of C, indicating that the motif prior is powerful in characterizing the complex structure of real-world networks.

4.3   Experiment on Real-World Cascades

In this subsection, we evaluate the performance of the proposed framework on a real-world information cascade dataset, i.e., MemeTracker dataset [11]. Meme-Tracker collects the quotes and phrases posted by the mass medium and Blog sites. This dataset contains 1.5 million news articles and Blog from August 2008 to May 2009. The articles may include hyperlinks of their sources and thus the information propagation can be tracked by the flow of hyperlinks. A site publishes a piece of information with corresponding hyperlink. Sites receives this piece of information would publish similar information and link to their sources. Thus a collection of hyperlinks with time stamps could be regarded as a hyperlink cascade. We construct the hyperlink cascades from top 500 mass media and Blog sites. The total number of cascades is 11262. We test the performance of the proposed methods as well as NETINF, NETINF+Community

Table 2. Performance comparison (in terms of $F1$ score) between the proposed MADNI-I/MADNI-R methods and five competing diffusion network inference algorithms in a real-world cascade experiment. The proposed MADNI-I performs the best.

Methods	Number of cascades	
	Sub (C = 4000)	All (C = 11262)
NETINF	0.2414	0.3879
NETINF + Community structure	0.2425	0.3460
NETINF + Scale-free	0.2730	0.3800
NETRATE	0.2455	0.2608
CENI	0.2538	0.2873
MADNI-I	**0.2746**	**0.3885**
MADNI-R	0.2472	0.2959

structure, NETINF+Scale-free, NETRATE, and CENI with 4000 cascades and all the 11262 cascades, respectively.

Table 2 shows the $F1$ scores of the proposed methods and five competing algorithms under different numbers of cascades. By modeling the motif-prior in the network inference procedure, the proposed MADNI-I performs better than the other competing algorithms under both settings of C.

5 Conclusion

In this paper, we presented a novel MADNI framework, which mines the motif patterns of the underlying diffusion network and incorporates the uncovered motifs into the network inference procedure via a reweighted motif regularization. By taking the network motifs into consideration, the proposed framework achieves the best performance on both synthetic and real-world datasets.

Future work will be explored from two aspects. First, in the current work, we have only considered the closed triangle motifs in the network inference as they are elementary. In order to better characterize the network structure, more complex motifs such as the higher-order ones should also be taken into account. Second, we will further validate the generalization ability and the flexibility of the proposed framework by incorporating motifs into different baseline methods and comparing the proposed framework to more network inference approaches with various kinds of structural priors.

Acknowledgment. The authors would like to thank the anonymous referees for their valuable comments and helpful suggestions. This work was supported in part by the National Natural Science Foundation of China under Grant 61503317, in part by the grants from the Research Grant Council of Hong Kong SAR under Projects RGC/HKBU12202415 and RGC/HKBU12202417, in part by the the Science and Technology Research and Development Fund of Shenzhen with Project Code

JCYJ20170307161544087, and in part by the Faculty Research Grant of Hong Kong Baptist University (HKBU) under Project FRG2/16-17/032.

References

1. Alon, U.: Network motifs: theory and experimental approaches. Nat. Rev. Genet. **8**(6), 450–461 (2007)
2. Benson, A.R., Gleich, D.F., Leskovec, J.: Higher-order organization of complex networks. Science **353**(6295), 163–166 (2016)
3. Farajtabar, M., Rodriguez, M.G., Zamani, M., Du, N., Zha, H., Song, L.: Back to the past: source identification in diffusion networks from partially observed cascades. In: Proceedings of 18th AISTATS, pp. 232–240 (2015)
4. Goentoro, L., Shoval, O., Kirschner, M.W., Alon, U.: The incoherent feedforward loop can provide fold-change detection in gene regulation. Mol. Cell **36**(5), 894–899 (2009)
5. Gomez Rodriguez, M., Leskovec, J., Krause, A.: Inferring networks of diffusion and influence. In: Proceedings of 16th SIGKDD, pp. 1019–1028 (2010)
6. Gomez Rodriguez, M., Leskovec, J., Schölkopf, B.: Structure and dynamics of information pathways in online media. In: Proceedings of 6th ACM WSDM, pp. 23–32 (2013)
7. He, X., Liu, Y.: Not enough data?: Joint inferring multiple diffusion networks via network generation priors. In: Proceedings of 10th WSDM, pp. 465–474 (2017)
8. Hosseini, S.M.J., Lee, S.I.: Learning sparse gaussian graphical models with overlapping blocks. In: NIPS, vol. 30, pp. 3801–3809 (2016)
9. Hu, Q., Xie, S., Lin, S., Wang, S., Philip, S.Y.: Clustering embedded approaches for efficient information network inference. Data Sci. Eng. **1**(1), 29–40 (2016)
10. Leskovec, J., Adamic, L.A., Huberman, B.A.: The dynamics of viral marketing. ACM Trans. Web **1**(1), 5 (2007)
11. Leskovec, J., Backstrom, L., Kleinberg, J.: Meme-tracking and the dynamics of the news cycle. In: Proceedings of 15th SIGKDD, pp. 497–506 (2009)
12. Leskovec, J., Krevl, A.: SNAP Datasets: Stanford large network dataset collection, June 2014. http://snap.stanford.edu/data
13. Liu, H., Ioannidis, S., Bhagat, S., Chuah, C.N.: Adding structure: social network inference with graph priors. In: 12th ACM International Workshop on Mining and Learning with Graphs (2016)
14. Liu, K., Cheung, W.K., Liu, J.: Detecting multiple stochastic network motifs in network data. Knowl. Inf. Syst. **42**(1), 49–74 (2015)
15. Liu, Q., Ihler, A.: Learning scale free networks by reweighted l1 regularization. In: Proceedings of 14th AISTATS, pp. 40–48 (2011)
16. Milo, R., Itzkovitz, S., Kashtan, N., Levitt, R., Shen-Orr, S., Ayzenshtat, I., Sheffer, M., Alon, U.: Superfamilies of evolved and designed networks. Science **303**(5663), 1538–1542 (2004)
17. Milo, R., Shen-Orr, S., Itzkovitz, S., Kashtan, N., Chklovskii, D., Alon, U.: Network motifs: simple building blocks of complex networks. Science **298**(5594), 824–827 (2002)
18. Mukherjee, S., Speed, T.P.: Network inference using informative priors. PNAS **105**(38), 14313–14318 (2008)

19. Rodriguez, M.G., Balduzzi, D., Schölkopf, B.: Uncovering the temporal dynamics of diffusion networks. In: Proceedings of 28th ICML, pp. 561–568 (2011)
20. Rong, Y., Zhu, Q., Cheng, H.: A model-free approach to infer the diffusion network from event cascade. In: Proceedings of 25th CIKM, pp. 1653–1662 (2016)
21. Tang, Q., Sun, S., Xu, J.: Learning scale-free networks by dynamic node specific degree prior. In: Proceedings of 32nd ICML, pp. 2247–2255 (2015)

Tri-Fly: Distributed Estimation of Global and Local Triangle Counts in Graph Streams

Kijung Shin[1(✉)], Mohammad Hammoud[2], Euiwoong Lee[1], Jinoh Oh[3], and Christos Faloutsos[1]

[1] Carnegie Mellon University, Pittsburgh, USA
{kijungs,euiwoonl,christos}@cs.cmu.edu
[2] Carnegie Mellon University in Qatar, Doha, Qatar
mhhamoud@cmu.edu
[3] Adobe Systems, San Jose, USA
joh@adobe.com

Abstract. Given a graph stream, how can we estimate the number of triangles in it using multiple machines with limited storage?

Counting triangles (i.e., cycles of length three) is a classical graph problem whose importance has been recognized in diverse fields, including data mining, social network analysis, and databases. Recently, for triangle counting in massive graphs, two approaches have been intensively studied. One approach is streaming algorithms, which estimate the count of triangles incrementally in time-evolving graphs or in large graphs only part of which can be stored. The other approach is distributed algorithms for utilizing computational power and storage of multiple machines.

Can we have the best of both worlds? We propose TRI-FLY, the first distributed streaming algorithm for approximate triangle counting. Making one pass over a graph stream, TRI-FLY rapidly and accurately estimates the counts of global triangles and local triangles incident to each node. Compared to state-of-the-art single-machine streaming algorithms, TRI-FLY is (a) **Accurate**: yields up to *4.5× smaller* estimation error, (b) **Fast**: runs up to *8.8× faster* with linear scalability, and (c) **Theoretically sound**: gives unbiased estimates with smaller variances.

Keywords: Graph stream · Triangle counting · Edge sampling

1 Introduction

Counting triangles (i.e., cycles of length three) is a classical graph problem whose importance has been recognized in diverse areas. In data mining, the count of triangles was used for dense subgraph mining [19], spam detection [5], degeneracy estimation [16], and web structure analysis [8]. In social network analysis, many important concepts (e.g., the clustering coefficients and social balance) are based on the count of triangles [20]. In databases, the count of triangles, which measures the degree of transitivity of a relation, can be used for query optimization [4].

© Springer International Publishing AG, part of Springer Nature 2018
D. Phung et al. (Eds.): PAKDD 2018, LNAI 10939, pp. 651–663, 2018.
https://doi.org/10.1007/978-3-319-93040-4_51

Due to this importance, many algorithms have been developed for counting global triangles (i.e., all triangles in a graph) and/or local triangles (i.e., triangles incident to each node in a graph). Especially, for triangle counting in massive graphs, recent work has focused largely on streaming algorithms [2,7,10,11,14, 15] and distributed algorithms [3,12,17].

In a graph stream, where edges are streamed from sources, streaming algorithms [2,7,10,11,14,15] estimate the count of triangles by making one pass over the stream, even when the stream does not fit in the underlying storage. Moreover, since streaming algorithms incrementally update their estimates as each edge arrives, they can naturally be used for maintaining and updating approximate triangle counts in dynamic graphs growing with new edges. However, existing streaming algorithms are designed to run on a single machine and do not utilize multiple machines for faster or more accurate estimation.

On the other hand, distributed algorithms have been employed for utilizing computational and storage resources in distributed-memory [3] and MAPRE-DUCE [12,17] settings. However, they do not provide the advantages of streaming algorithms. That is, they assume that all edges can be stored in the underlying storage and accessed multiple times. Moreover, since they are batch algorithms rather than incremental algorithms, they do not support efficient updates of triangle counts in dynamic graphs growing with new edges.

In this work, we propose TRI-FLY, the first distributed streaming algorithm for approximate counting of global and local triangles. TRI-FLY gives the advantages of both streaming and distributed algorithms, outperforming state-of-the-art single-machine streaming algorithms. Our theoretical and empirical analyses show that TRI-FLY has the following advantages:

- **Accurate:** TRI-FLY produces up to *4.5×* smaller *estimation error* than baselines with similar speeds (Fig. 3).
- **Fast:** TRI-FLY runs in linear time (Fig. 2(c)) up to *8.8× faster* than baselines with similar accuracies (Fig. 3).
- **Theoretically sound:** TRI-FLY gives unbiased estimates with variances inversely proportional to the number of machines (Theorems 1 and 2).

Reproducibility: The code and datasets used in the paper are available at http://www.cs.cmu.edu/~kijungs/codes/trifly/.

2   Related Work

Triangle Counting in Graph Streams. Streaming algorithms estimate the count of triangles by making one pass over a graph stream. Streaming algorithms use sampling because they assume limited storage that may not store all edges. A simple but effective sampling technique is edge sampling. DOULION [18] uniformly samples edges of a large graph, and estimates its global triangle count from that in the sampled graph. MASCOT [11] improves upon DOULION in terms of accuracy by utilizing unsampled edges. Specifically, whenever an edge arrives, MASCOT counts the global and local triangles formed by the incoming edge and

Table 1. Comparison of triangle counting algorithms. Notice that only our proposed algorithm Tri-Fly satisfies all the criteria.

	(Distributed)		(Streaming)		(Proposed) Tri-Fly
	[12,17]	[3]	[2,9,14]	[7,10,11,15]	
Single-pass stream processing			✓	✓	✓
Approximation for large graphs		✓	✓	✓	✓
Global & local triangle counting	✓			✓	✓
Larger data w/more machines	✓	✓			✓
More accurate w/more machines		✓			✓

edges sampled so far, even if the incoming edge is not sampled but discarded. While MASCOT may discard edges even when storage is not full, TRIEST$_\text{IMPR}$ [7] always maintains as many samples as storage allows, leading to higher accuracy. When edges are streamed in the chronological order, WRS [15] improves upon TRIEST$_\text{IMPR}$ in terms of accuracy by exploiting temporal dependencies in the edges. In addition to edge sampling, wedge sampling [9], neighborhood sampling [14], and sample-and-hold [2] were used for global triangle counting, and node coloring [10] was used for local triangle counting. Neighborhood sampling was parallelized in a shared-memory setting where edges arrive in batches, and it was also extended to cases where edges are streamed from multiple sources [13].

Distributed Triangle Counting. Many MAPREDUCE algorithms for exact counts of triangles have been proposed based on the assumption that all edges of the input graph are stored in a distributed file system. The first such algorithm [6] parallelizes node iterator, a serial algorithm for triangle counting. GP [17] divides the input graph into overlapping subgraphs and assigns them to machines, which count the triangles in the assigned subgraphs in parallel. Since the subgraphs are not disjoint, GP produces a large amount of intermediate data, which were reduced in [12]. The idea of dividing the input graph into overlapping subgraphs was used also in a distributed memory setting [3]. These existing distributed algorithms are batch algorithms for static graphs, while we propose incremental algorithms for dynamic graph streams.

The aforementioned streaming algorithms and distributed algorithms are summarized and compared in Table 1.

3   Notations and Problem Definition

3.1   Notations (Table 2)

Consider an undirected graph $\mathcal{G} = (\mathcal{V}, \mathcal{E})$ with the set of nodes $\mathcal{V}$ and the set of edges $\mathcal{E}$. Each unordered pair $(u, v) \in \mathcal{E}$ indicates the edge between two distinct nodes $u, v \in \mathcal{V}$. We denote the set of triangles (i.e., three nodes, every pair of which is connected by an edge) in $\mathcal{G}$ by $\mathcal{T}$ and those with node u by $\mathcal{T}[u] \subset \mathcal{T}$.

Table 2. Table of frequently-used symbols.

	Symbol	Definition
Notations for graph streams (Sect. 3)	$\mathcal{G}^{(t)} = (\mathcal{V}^{(t)}, \mathcal{E}^{(t)})$	Graph at time t
	$e^{(t)}$	Edge that arrives at time t
	(u, v)	Edge between nodes u and v
	t_{uv}	Arrival time of edge (u, v)
	(u, v, w)	Triangle with nodes u, v, and w
	$\mathcal{T}^{(t)}$	Set of global triangles in $\mathcal{G}^{(t)}$
	$\mathcal{T}^{(t)}[u]$	Set of local triangles with node u in $\mathcal{G}^{(t)}$
Notations for algorithm (Sect. 4)	$\mathcal{M}, \mathcal{W}, \mathcal{A}$	Sets of masters, workers, and aggregators
	k	Maximum number of edges stored in each worker
	l_i	Number of edges that worker i has received
	$h : \mathcal{V} \cup \{*\} \rightarrow \mathcal{A}$	Hash function that maps nodes to aggregators
	$\bar{c}$	Estimate of the count of global triangles
	$c[u]$	Estimate of the count of local triangles of node u

We call $\mathcal{T}$ *global triangles* and $\mathcal{T}[u]$ *local triangles* of node u. Each unordered triple $(u, v, w) \in \mathcal{T}$ denotes the triangle with three distinct nodes $u, v, w \in \mathcal{V}$.

Consider a graph stream $(e^{(1)}, e^{(2)}, ...)$ where $e^{(t)}$ denotes the edge that arrives at time $t \in \{1, 2, ...\}$. We use t_{uv} to denote the arrival time of edge (u, v). Let $\mathcal{G}^{(t)} = (\mathcal{V}^{(t)}, \mathcal{E}^{(t)})$ be the graph at time t consisting of the nodes and edges arriving at time t or earlier. Then, $\mathcal{T}^{(t)}$ denotes the set of global triangles in $\mathcal{G}^{(t)}$ and $\mathcal{T}^{(t)}[u] \subset \mathcal{T}^{(t)}$ denotes the set of local triangles of each node $u \in \mathcal{V}^{(t)}$ in $\mathcal{G}^{(t)}$.

3.2   Problem Definition

In this work, we consider the problem of estimating the counts of global and local triangles in a graph stream using multiple machines with limited storage. Specifically, we assume the following realistic conditions:

C1 **No prior knowledge:** no information about the input graph stream (e.g., the number of edges, degree distribution, etc.) is available in advance.

C2 **Shared nothing architecture:** each machine cannot access data stored in the other machines.

C3 **Limited storage:** at most k (≥ 2) edges can be stored in each of n machines, while the number of edges in the input graph stream can be greater than k or even nk.

C4 **Single pass:** edges are processed one by one in their arrival order. Past edges cannot be accessed unless they are stored (in the storage in C3).

Based on these conditions, we define the problem of distributed estimation of global and local triangle counts in a graph stream.

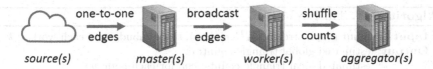

Fig. 1. Flow of data in TRI-FLY.

Problem 1 (Distributed Estimation of Triangle Counts in a Graph Stream).

(1) **Given:** a graph stream $(e^{(1)}, e^{(2)}, ...)$, and n distributed storages each of which can store up to k (≥ 2) edges
(2) **Minimize:** the estimation errors of global triangle count $|\mathcal{T}^{(t)}|$ and local triangle counts $\{|\mathcal{T}^{(t)}[u]|\}_{u \in \mathcal{V}^{(t)}}$ for each time $t \in \{1, 2, ...\}$.

Instead of minimizing a specific measure of estimation error, we use a general approach of simultaneously reducing bias and variance to reduce various measures of estimation error robustly.

4   Proposed Method: Tri-Fly

We propose TRI-FLY, a distributed streaming algorithm for approximate triangle counting. We first present the overview of TRI-FLY. Then, we discuss its details. Lastly, we provide theoretical analyses on its accuracy and complexity.

4.1   Overview (Fig. 1)

Figure 1 shows the flow of data in TRI-FLY. Edges are streamed from sources to masters so that each edge is sent to exactly one master. Each master broadcasts the received edges to the workers. Each worker estimates the global and local triangle counts independently using its local storage. To this end, we adapt TRIEST$_{\text{IMPR}}$, which estimates both global and local triangle counts with no prior knowledge, although any streaming algorithm can be used instead.[1] The counts are shuffled so that the counts of local triangles of each node (or the counts of global triangles) are sent to the same aggregator. The aggregators aggregate the counts and give the final estimates of the counts of global and local triangles.

4.2   Detailed Algorithm (Algorithm 1)

Algorithm 1 describes TRI-FLY. We first define the notations used in it. Then, we explain masters, workers, and aggregators. Lastly, we discuss lazy aggregation.

Notations. We use $\mathcal{M}$, $\mathcal{W}$ and $\mathcal{A}$ to indicate the set of masters, workers and aggregators, respectively. Each worker can store up to k (≥ 2) edges, and $\mathcal{E}_i$ denotes the set of edges currently stored in worker $i \in \mathcal{W}$. We let $\mathcal{G}_i = (\mathcal{V}_i, \mathcal{E}_i)$

---

[1] e.g., WRS [15] can be used instead if edges are streamed in the chronological order.

Algorithm 1. TRI-FLY

    Input   : input graph stream: $(e^{(1)}, e^{(2)}, ...)$, storage budget in each worker: k
    Output: estimated global triangle count: $\bar{c}$
                  estimated local triangle counts: $c[u]$ for each node u
    - Master (each master):

2 **for** *each edge (u, v) from the sources* **do**
3   ⌊ broadcast (u, v) to the workers

    - Worker (each worker $i \in \mathcal{W}$):

5 $l_i \leftarrow 0, \mathcal{E}_i \leftarrow \emptyset$
6 **for** *each edge (u, v) from the masters* **do**
7    $sum \leftarrow 0$
8   **for** *each node $w \in \mathcal{N}_i[u] \cap \mathcal{N}_i[v]$* **do**
9   ⌈ send $(w, 1/(p_i[uvw]))$ to aggregator $h(w)$
10   ⌊ $sum \leftarrow sum + 1/(p_i[uvw])$ ▷ see Eq. (1) for $p_i[uvw]$
11   send $(*, sum)$ to aggregator $h(*)$ ▷ '$*$': key for the global triangle count
12   send (u, sum) to aggregator $h(u)$ and (v, sum) to aggregator $h(v)$
13    $l_i \leftarrow l_i + 1$
14   **if** $|\mathcal{E}_i| < k$ **then** $\mathcal{E}_i \leftarrow \mathcal{E}_i \cup \{(u, v)\}$
15   **else if** *a random number in Bernoulli(k/l_i) is* 1 **then**
16   ⌊ replace a random edge in $\mathcal{E}_i$ with (u, v)

    - Aggregator (each aggregator $j \in \mathcal{A}$):

17 **if** $h(*) = j$ **then** $\bar{c} \leftarrow 0$
18 initialize an empty map c with default value 0
19 **for** *each pair (u, δ) from the workers* **do**
20   **if** $u = *$ **then** $\bar{c} \leftarrow \bar{c} + \delta/|\mathcal{W}|$
21   **else** $c[u] \leftarrow c[u] + \delta/|\mathcal{W}|$

---

be the graph consisting of the edges in $\mathcal{E}_i$. For each node $u \in \mathcal{V}_i$, $\mathcal{N}_i[u]$ indicates the neighbors of u in $\mathcal{G}_i$. We use l_i to denote the number of edges that worker $i \in \mathcal{W}$ has received so far. If $l_i > k$, then $l_i > |\mathcal{E}_i|$ since not all received edges can be stored. We use $h : \mathcal{V}_i \cup \{*\} \rightarrow \mathcal{A}$ to denote a hash function that maps nodes (the keys for local triangle counts) and '$*$' (the key for global triangle counts) to aggregators. Lastly, $\bar{c}$ denotes the estimate of the count of global triangles, and for each node u, $c[u]$ denotes the estimate of the count of local triangles of u.

Masters (lines 2–3). Whenever each master receives an edge from the sources, the master broadcasts the edge to the workers.

Workers (lines 5–16). Each worker independently estimates the global and local triangle counts, and shuffles the counts across the aggregators. Note that the workers use different random seeds and thus shuffle different counts. Each worker $i \in \mathcal{W}$ starts with an empty storage (i.e., $\mathcal{E}_i = \emptyset$) (line 5). Whenever it receives an edge (u, v) from a master (line 6), the worker counts the triangles composed of (u, v) and two edges in its local storage; and sends the counts to the corresponding aggregators using hash function h (lines 7–12). Then, the worker samples (u,v) in its local storage with non-zero probability (lines 13–16). Below, we explain in detail how each worker samples edges and counts triangles.

For sampling (lines 13–16), each worker $i \in \mathcal{W}$ first increases l_i, the number of edges that it has received, by one (line 13). If its local storage is not full (i.e., $|\mathcal{E}_i| < k$), the worker always stores (u,v) by adding (u,v) to $\mathcal{E}_i$ (line 14). If the local storage is full (i.e., $|\mathcal{E}_i| = k$), the worker stores (u,v) with probability k/l_i by replacing a random edge in $\mathcal{E}_i$ with (u,v) (lines 15–16). This is the standard reservoir sampling, which guarantees that each pair of the l_i edges is sampled (i.e., included in $\mathcal{E}_i$) with the equal probability $\min\left(1, \frac{k(k-1)}{l_i(l_i-1)}\right)$.

For counting (lines 7–12), each worker $i \in \mathcal{W}$ finds the common neighbors of nodes u and v in graph $\mathcal{G}_i$ (i.e., the graph consisting of the edges $\mathcal{E}_i$ in its local storage) (line 8). Each common neighbor w indicates the existence of the triangle (u,v,w). Thus, for each common neighbor w, the worker increases the global triangle count and the local triangle counts of nodes u, v, and w by sending the increases to the corresponding aggregators (lines 9, 11, and 12). The amount of increase in the counts is $1/(p_i[uvw])$ for each triangle (u,v,w), where

$$p_i[uvw] := \min\left(1, \frac{k(k-1)}{l_i(l_i-1)}\right) \tag{1}$$

is the probability that triangle (u,v,w) is discovered by worker i (i.e., both (v,w) and (w,u) are in $\mathcal{E}_i$ when worker i receives (u,v)), as explained above. Increasing counts by $1/(p_i[uvw])$ guarantees that the expected amount of the increase sent from each worker is exactly $1(= p_i[uvw] \times 1/(p_i[uvw]) + (1 - p_i[uvw]) \times 0)$ for each triangle, enabling TRI-FLY to give unbiased estimates (see Theorem 1).

Aggregators (lines 17–21). Each aggregator maintains and updates the triangle counts assigned by the hash function h. That is, aggregator $j \subset \mathcal{A}$ maintains the estimate $c[u]$ of the count of local triangles of node u if $h(u) = j$. Likewise, aggregator $j \in \mathcal{A}$ maintains the estimate $\bar{c}$ of the count of global triangles if $h(*) = j$. Specifically, each aggregator increases the estimates by $1/|\mathcal{W}|$ of what it receives, averaging the increases sent from the workers (lines 20–21).

Lazy Aggregation (Optional). In Algorithm 1, each worker sends the increase of the local triangle count of node w to the corresponding aggregator whenever it discovers each triangle (u,v,w) (line 9). Likewise, each worker sends the updates of the global triangle count and the local triangle counts of nodes u and v to the corresponding aggregators whenever it processes each edge (u,v) (lines 11–12). In cases where this eager aggregation is not needed, we can reduce the amount of shuffled data by employing lazy aggregation. That is, counts can be aggregated locally in each worker until they are queried. If queried, the counts are sent to and aggregated in the aggregators and removed from the workers.

4.3   Bias and Variance Analyses

We analyze the biases and variances of the estimates given by TRI-FLY. The biases and variances determine the errors of the estimates. For the analyses,

Table 3. Time and space complexities of Tri-Fly for processing the first t edges in the input graph stream.

	Masters (total)	Workers (each)	Workers (total)	Aggregators (total)
Time	$O(t \cdot |\mathcal{W}|)$	$O(t \cdot \min(t, k))$	$O(|\mathcal{W}| \cdot t \cdot \min(t, k))$	$O(|\mathcal{W}| \cdot t \cdot \min(t, k))$*
Space	$O(|\mathcal{M}|)$	$O(\min(t, k))$	$O(|\mathcal{W}| \cdot \min(t, k))$	$O(|\mathcal{V}^{(t)}|)$

*Can be reduced by lazy aggregation

let $\mathcal{G}^{(t)} = (\mathcal{V}^{(t)}, \mathcal{E}^{(t)})$ be the graph with the edges arriving at time t or earlier. We define $\bar{c}^{(t)}$ as $\bar{c}$ in the aggregator $h(*)$ after edge $e^{(t)}$ is processed. Likewise, for each node $u \in \mathcal{V}^{(t)}$, let $c^{(t)}[u]$ be $c[u]$ in the aggregator $h(u)$ after $e^{(t)}$ is processed. Then, $\bar{c}^{(t)}$ is an estimate of $|\mathcal{T}^{(t)}|$, the global triangle count in $\mathcal{G}^{(t)}$, and each $c^{(t)}[u]$ is an estimate of $|\mathcal{T}^{(t)}[u]|$, the local triangle count of u in $\mathcal{G}^{(t)}$.

We first prove the unbiasedness of Tri-Fly, formalized in Theorem 1.

Theorem 1 (Unbiasedness of Tri-Fly). *At any time, the expected values of the estimates given by* Tri-Fly *are equal to the true global and local triangle counts. That is, in Algorithm 1,*

$$\mathbb{E}[\bar{c}^{(t)}] = |\mathcal{T}^{(t)}|, \ \forall t \geq 1, \ and \ \mathbb{E}[c^{(t)}[u]] = |\mathcal{T}^{(t)}[u]|, \ \forall u \in \mathcal{V}^{(t)}, \ \forall t \geq 1.$$

Proof Sketch. Consider a triangle $(u, v, w) \in \mathcal{T}^{(t)}$. Let $d_i[uvw]$ be the contribution of (u, v, w) to $\bar{c}^{(t)}$ by each worker $i \in \mathcal{W}$. Then, by the definition of $p_i[uvw]$, and lines 11 and 20 of Algorithm 1, $d_i[uvw] = 1/(|\mathcal{W}| \cdot p_i[uvw])$ with probability $p_i[uvw]$, and $d_i[uvw] = 0$ with probability $(1 - p_i[uvw])$. Therefore, $\mathbb{E}[d_i[uvw]] = 1/|\mathcal{W}|$. Then, $\mathbb{E}[d_i[uvw]] = 1/|\mathcal{W}|$ and linearity of expectation imply

$$\mathbb{E}[\bar{c}^{(t)}] = \mathbb{E}\left[\sum_{i \in \mathcal{W}} \sum_{(u,v,w) \in \mathcal{T}^{(t)}} d_i[uvw] \right] = \sum_{i \in \mathcal{W}} \sum_{(u,v,w) \in \mathcal{T}^{(t)}} \mathbb{E}[d_i[uvw]] = |\mathcal{T}^{(t)}|, \forall t \geq 1.$$

See [1] for a full proof with the unbiasedness of the other estimates. ∎

Theorem 2 presents the result of our variance analysis given in the supplementary document [1]. The variance of each $c^{(t)}[u]$ can be analyzed in the same manner considering only the triangles with node u.

Theorem 2 (Variance of Tri-Fly). *The variance of the estimate $\bar{c}^{(t)}$ in* Tri-Fly *is inversely proportional to the number of workers. Let $r^{(t)}$ be the number of triangle pairs in $\mathcal{T}^{(t)}$ where (a) an edge is shared and (b) the shared edge is not last to arrive in any of the two triangles. Let $z^{(t)}$ be $\max\left(0, |\mathcal{T}^{(t)}|\left(\frac{(t-1)(t-2)}{k(k-1)} - 1\right)\right.$ $\left. + r^{(t)}\left(\frac{t-1-k}{k}\right)\right)$. Then, Eq. (2) holds in Algorithm 1.*

$$Var[\bar{c}^{(t)}] \leq \frac{z^{(t)}}{|\mathcal{W}|}, \ \forall t \geq 1. \tag{2}$$

Proof Sketch. For each worker $i \in \mathcal{W}$, let $\bar{c}_i^{(t)}$ be the global triangle count sent from the worker by time t. Then, $\bar{c}^{(t)} = \sum_{i \in \mathcal{W}} \bar{c}_i^{(t)} / |\mathcal{W}|$ (line 20 of Algorithm 1). Equation (2) follows from $Var[\bar{c}_i^{(t)}] \leq z^{(t)}$ for each $i \in \mathcal{W}$ (Lemma 1 in [1]) and independence between $\bar{c}_i^{(t)}$ and $\bar{c}_j^{(t)}$ for $i \neq j$. See [1] for a full proof. ∎

Table 4. Summary of real-world and synthetic graph streams.

Name	# Nodes	# Edges	Summary
BerkStan	685, 230	6, 649, 470	Web
Patent	3, 774, 768	16, 518, 947	Citation
Flickr	2, 302, 925	22, 838, 276	Friendship
FriendSter	65, 608, 366	1, 806, 067, 135	Friendship
Random (800 GB)	1, 000, 000	1, 000, 000, 000 − 100, 000, 000, 000	Synthetic

4.4 Time and Space Complexity Analyses

We summarize the time and space complexities of TRI-FLY in Table 3. Detailed analyses with proofs are given in the supplementary document [1]. Notice that, with a fixed storage budget k, the time complexity of TRI-FLY is linear in the number of edges, as confirmed empirically in Sect. 5.2. The results also suggest that reducing storage budget k and using more masters and aggregators need to be considered if the input stream is too fast to be processed.

5 Experiments

In this section, we conduct experiments to answer the following questions:

- **Q1. Illustration of Theorems**: Does TRI-FLY give unbiased estimates? How rapidly do their variances drop as the number of workers is scaled up? How does TRI-FLY scale with the size of the input stream?
- **Q2. Performance**: How fast and accurate is TRI-FLY compared to the best single-machine streaming algorithms?

5.1 Experimental Settings

Machines: All experiments were conducted on a cluster of 40 machines with 3.47 GHz Intel Xeon X5690 CPUs and 32 GB RAM.

Datasets: The graph datasets used in the paper are summarized in Table 4. The self loops, duplicated edges, and the directions of edges were ignored.

Implementations: We implemented TRI-FLY, TRIEST$_{IMPR}$ [7] (single-machine) and MASCOT [11] (single-machine) in C++ and MPICH 3.1. In them, sampled edges were stored in main memory in the adjacency list format. For TRI-FLY, we used 1 master and 1 aggregator. We used lazy aggregation (see the last paragraph of Sect. 4.2) and aggregated all counts once at the end of each input stream. We simulated graph streams by streaming edges in a random order from the disk of machines that host the master of TRI-FLY or single-machine algorithms.

Evaluation Metrics: We evaluated the accuracy of each algorithm at the end of each input stream. Let x be the true global triangle count, and $\hat{x}$ be its estimate

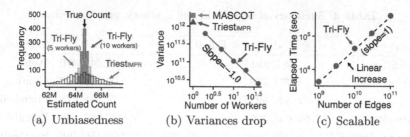

Fig. 2. Theoretical properties of Tri-Fly. (a) TRI-FLY gives unbiased estimates. (b) The variances of the estimates drop inversely proportional to the number of workers. (c) TRI-FLY scales linearly with the size of the input stream.

obtained by an evaluated algorithm. Likewise, for each node $u \in \mathcal{V}$, let $x[u]$ be the true local triangle count of u and $\hat{x}[u]$ be its estimate. We used *global error*, defined as $\frac{|x-\hat{x}|}{1+x}$, and *RMSE*, defined as $\sqrt{\frac{1}{|\mathcal{V}|}\sum_{u \in \mathcal{V}}(x[u]-\hat{x}[u])^2}$, to evaluate the accuracy of global triangle counting and local triangle counting, respectively.

5.2   Q1. Illustration of Our Theorems (Fig. 2)

Illustration of Unbiasedness (Theorem 1). Figure 2(a) shows the distributions of 1,000 estimates of the global triangle count in the BerkStan dataset obtained by TRI-FLY and TRIEST$_{\text{IMPR}}$. We set storage budget k so that each worker stored up to 5% of the edges. The averages of the estimates given by TRI-FLY were close to the true triangle count, as expected from Theorem 1.

Illustration of Variance Decrease (Theorem 2). Under the same experimental settings, Fig. 2(b) shows the variances of the estimates of the global triangle count obtained by different algorithms. We measured the sample variance of 1,000 estimates in each setting. The variance in TRI-FLY dropped inversely proportional to the number of workers, as expected in Theorem 2.

Illustration of Linear Scalability (Sect. 4.4). We measured the running time of TRI-FLY while varying the size of the input stream. We used 30 workers and fixed the storage budget k to 10^7. To measure the scalability independently of the speed of input streams, we measured the time taken to process edges, ignoring the time taken by the master to wait for the arrival of edges in input streams. Figure 2(c) shows the results with random graph streams with 1 million nodes and different numbers of edges. The largest one was **800 GB** with **100 billion edges**. TRI-FLY scaled linearly with the size of the input stream, as expected in Sect. 4.4. We obtained similar results when graph streams with realistic structure were used (see the supplementary document [1]).

5.3   Q2. Performance (Fig. 3)

Since TRI-FLY is the first distributed streaming algorithm for triangle counting, there is **no direct competitor** of TRI-FLY. As baselines, we used TRIEST$_{\text{IMPR}}$

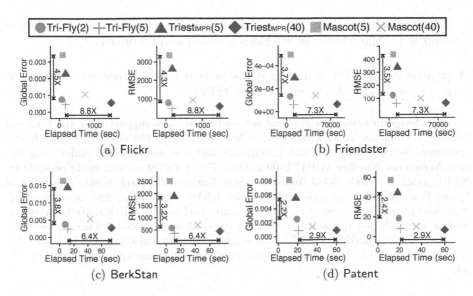

Fig. 3. Tri-Fly achieves both speed and accuracy. In each plot, points represent the speeds and errors of different algorithms (the numbers in parentheses indicate the percentage of edges that can be stored in each worker). TRI-FLY was up to **4.5× more accurate** than single-machine baselines with similar speeds, and it was up to **8.8× faster** than those with similar accuracies.

[7] and MASCOT [11], which are the state-of-the-art single-machine streaming algorithms estimating both global and local triangle counts (see Table 1).

We measured the speeds and accuracies of the considered algorithms with different storage budgets. To compare their speeds independently of the speed of input streams, we measured the time taken by each algorithm to process edges, ignoring the time taken to wait for the arrival of edges in input streams. All evaluation metrics and running times were averaged over 10 trials in the Friendster dataset and over 100 trials in the other datasets.

As seen in Fig. 3, TRI-FLY showed the best performance in every dataset. Specifically, TRI-FLY was up to **8.8×** **faster** than baselines with similar accuracies. In terms of global error and RMSE, TRI-FLY was up to **4.5× and 4.3× more accurate** than baselines with similar speeds, respectively.

6   Conclusion

In this work, we propose TRI-FLY, the first distributed streaming algorithm estimating the counts of global and local triangles with the following strengths:

- **Accurate:** TRI-FLY yields up to *4.5×* and *4.3× smaller estimation errors* for global and local triangle counts than similarly fast baselines (Fig. 3).
- **Fast:** TRI-FLY is up to *8.8× faster* than similarly accurate baselines (Fig. 3). TRI-FLY scales linearly with the size of the stream (Fig. 2(c)).

- **Theoretically sound:** TRI-FLY yields unbiased estimates whose variances drop as the the number of machines is scaled up (Theorems 1 and 2).

Reproducibility: The code and datasets used in the paper are available at http://www.cs.cmu.edu/~kijungs/codes/trifly/.

Acknowledgements. This material is based upon work supported by the National Science Foundation under Grants No. CNS-1314632 and IIS-1408924. Research was sponsored by the Army Research Laboratory and was accomplished under Cooperative Agreement Number W911NF-09-2-0053. This publication was made possible by NPRP grant # 7-1330-2-483 from the Qatar National Research Fund (a member of Qatar Foundation). Shin was supported by KFAS Scholarship. Any opinions, findings, and conclusions or recommendations expressed in this material are those of the author(s) and do not necessarily reflect the views of the National Science Foundation, or other funding parties. The U.S. Government is authorized to reproduce and distribute reprints for Government purposes notwithstanding any copyright notation hereon.

References

1. Supplementary document (2018). http://www.cs.cmu.edu/~kijungs/codes/trifly/supple.pdf
2. Ahmed, N.K., Duffield, N., Neville, J., Kompella, R.: Graph sample and hold: a framework for big-graph analytics. In: KDD (2014)
3. Arifuzzaman, S., Khan, M., Marathe, M.: PATRIC: a parallel algorithm for counting triangles in massive networks. In: CIKM (2013)
4. Bar-Yossef, Z., Kumar, R., Sivakumar, D.: Reductions in streaming algorithms, with an application to counting triangles in graphs. In: SODA (2002)
5. Becchetti, L., Boldi, P., Castillo, C., Gionis, A.: Efficient algorithms for large-scale local triangle counting. TKDD **4**(3), 13 (2010)
6. Cohen, J.: Graph twiddling in a MapReduce world. Comput. Sci. Eng. **11**(4), 29–41 (2009)
7. De Stefani, L., Epasto, A., Riondato, M., Upfal, E.: TRIÈST: counting local and global triangles in fully-dynamic streams with fixed memory size. In: KDD (2016)
8. Eckmann, J.P., Moses, E.: Curvature of co-links uncovers hidden thematic layers in the world wide web. PNAS **99**(9), 5825–5829 (2002)
9. Jha, M., Seshadhri, C., Pinar, A.: A space efficient streaming algorithm for triangle counting using the birthday paradox. In: KDD (2013)
10. Kutzkov, K., Pagh, R.: On the streaming complexity of computing local clustering coefficients. In: WSDM (2013)
11. Lim, Y., Kang, U.: MASCOT: memory-efficient and accurate sampling for counting local triangles in graph streams. In: KDD (2015)
12. Park, H.M., Myaeng, S.H., Kang, U.: PTE: enumerating trillion triangles on distributed systems. In: KDD (2016)
13. Pavan, A., Tangwongan, K., Tirthapura, S.: Parallel and distributed triangle counting on graph streams. Technical report, IBM (2013)
14. Pavan, A., Tangwongsan, K., Tirthapura, S., Wu, K.L.: Counting and sampling triangles from a graph stream. PVLDB **6**(14), 1870–1881 (2013)

15. Shin, K.: WRS: waiting room sampling for accurate triangle counting in real graph streams. In: ICDM (2017)
16. Shin, K., Eliassi-Rad, T., Faloutsos, C.: Patterns and anomalies in k-cores of real-world graphs with applications. Knowl. Inf. Syst. **54**(3), 677–710 (2018)
17. Suri, S., Vassilvitskii, S.: Counting triangles and the curse of the last reducer. In: WWW (2011)
18. Tsourakakis, C.E., Kang, U., Miller, G.L., Faloutsos, C.: DOULION: counting triangles in massive graphs with a coin. In: KDD (2009)
19. Wang, J., Cheng, J.: Truss decomposition in massive networks. PVLDB **5**(9), 812–823 (2012)
20. Wasserman, S., Faust, K.: Social Network Analysis: Methods and Applications, vol. 8. Cambridge University Press, Cambridge (1994)

WFSM-MaxPWS: An Efficient Approach for Mining Weighted Frequent Subgraphs from Edge-Weighted Graph Databases

Md. Ashraful Islam[1], Chowdhury Farhan Ahmed[1], Carson K. Leung[2]([✉]) [iD],
and Calvin S. H. Hoi[2]

[1] Department of Computer Science and Engineering,
University of Dhaka, Dhaka, Bangladesh
csedu.ashraful@gmail.com, farhan@du.ac.bd
[2] Department of Computer Science, University of Manitoba, Winnipeg, MB, Canada
{kleung,hoic}@cs.umanitoba.ca

Abstract. Weighted frequent subgraph mining comes with an inherent challenge—namely, weighted support does not support the downward closure property, which is often used in mining algorithms for reducing the search space. Although this challenge attracted attention from several researchers, most existing works in this field use either affinity based pruning or alternative anti-monotonic weighting technique for subgraphs other than average edge-weight. In this paper, we propose an efficient weighted frequent subgraph mining algorithm called WFSM-MaxPWS. Our algorithm uses the MaxPWS pruning technique, which significantly reduces search space without changing subgraph weighting scheme while ensuring completeness. Our evaluation results on three different graph datasets with two different weight distributions (normal and negative exponential) showed that our WFSM-MaxPWS algorithm led to significant runtime improvement over the existing MaxW pruning technique (which is a concept for weighted pattern mining in computing subgraph weight by taking average of edge weights).

1 Introduction

As *frequent pattern mining* has been an appealing area of data mining, many algorithms have been developed for general pattern mining [5,12,13] to specific pattern mining (e.g., mining sequential patterns, weighted patterns, web access sequences, data streams). To deal with complex data, graph mining [3,7] has emerged as an inevitable area. When compared to unweighted graphs, *weighted graphs* have strong representational power. *Weighted frequent subgraphs* describe underlying graph database more accurately, and thus contribute greatly in areas like feature extraction for graph classification, association rule mining, graph clustering. For example, it is impossible to identify sophisticated metamorphic malwares using signature based approaches. Runtime behavior, though very accurate, is hardly used due to its slower detection rate. As an alternative, malware call-graph analysis has been shown to be very effective. However, insertion

© Springer International Publishing AG, part of Springer Nature 2018
D. Phung et al. (Eds.): PAKDD 2018, LNAI 10939, pp. 664–676, 2018.
https://doi.org/10.1007/978-3-319-93040-4_52

of huge benign codes in malwares reduces the chance of identifying comparatively less frequent suspicious subgraphs when using traditional subgraph mining algorithms. So, the malware detection task can take advantage of weighted frequent subgraph mining by giving higher weights to those call sequences.

Frequent subgraph mining is a tedious task partially because of subgraph isomorphism checking and exponential growth of candidate patterns. Luckily, canonical ordering of graphs and anti-monotonicity of downward closure property have made frequent subgraph mining a feasible task. However, for weighted frequent subgraph mining, the downward closure property no longer holds. Existing algorithms handle general weighted frequent itemset mining in FP-tree based mining algorithms by considering GMaxW (weight of maximum weighted item from the initial global FP-tree) and then recursively considering LMaxW (weight of maximum weighted item in local conditional FP-trees) as the maximum possible itemset weight for pruning [1]. The completeness is ensured by sorting itemsets in weight ascending order. This approach generates a moderate number of candidate patterns because recursively lighter weights are considered for pruning. However, as MaxW is usually much heavier than the actual average weight of subgraph, a huge number of unnecessary candidate subgraphs are generated, and thus leading to long runtime. Although attempts to reduce the number of generated candidate sets have been made, most of the existing approaches either use affinity based pruning (which usually imposes extra conditions to measure interestingness of a subgraph) or other alternative weighting techniques (which redefines subgraph weights such that the downward closure property is satisfied).

In this paper, we aim to reduce the number of generated candidate sets via our non-trivial adoption of affinity-based conditions and subgraph-weighting techniques. The work is inspired by our observation that, when considering MaxW as the maximum possible weight of extended subgraph, those already-seen average weights of the subgraph are often ignored. So, we decided to make good use of (i) these already-seen weights of a subgraph and (ii) some statistical information about the dataset to calculate the maximum possible weight and frequency for extensions of subgraph up to many edges. Consequently, the **Maximum Possible Weighted Support (MaxPWS)** can be computed for a subgraph. Moreover, by making an intelligent change in canonical ordering for weighted subgraphs, the magnitude of MaxPWS can further be brought closer to the actual weighted support, which then leads to safe and effective pruning of unnecessary candidate subgraphs. Hence, *our key contributions* of this paper are as follows:

- a tighter pruning technique **MaxPWS** for weighted subgraph mining, and
- a canonical ordering for weighted subgraphs that makes MaxPWS tighter,
- a weighted frequent subgraph mining algorithm called **WFSM-MaxPWS**, which uses the MaxPWS pruning technique.

The remainder of this paper is organized as follows. The next section presents background and related works. Section 3 describes our MaxPWS pruning technique, canonical order modification, and WFSM-MaxPWS algorithm. Experimental results and conclusions are given in Sects. 4 and 5, respectively.

2   Background and Related Works

Let us give some background information about **weighted graph mining**, which aims to find weighted frequent subgraph. An *edge-weighted graph G* is a collection of nodes V, edges E, together with a mapping between edge-set E and weights. For a function $\boldsymbol{W}(\boldsymbol{e})$ that returns the weight of edge $\boldsymbol{e}$, the weight of a subgraph $\boldsymbol{g}$ can be defined as: $W(g) = \frac{1}{n}\sum_{i=1}^{n} W(e_i)$, where each e_i is an edge of g. Given a graph database GDB of weighted graphs and a minimum weighted support threshold τ, a subgraph g is said to be *weighted frequent* if $wsup(g) \geq \tau$ where $wsup(g) = W(g) \times sup(g)$. In this paper, we focus on the condition that "all edges with same edge label and end-point node label have the same weight".

As the base for our proposed MaxPWS pruning technique, the **gSpan algorithm** [16] represents each subgraph by using DFScode, and it uses an extended tuple comparison rule to rank the DFScode of a subgraph by following the rightmost path extension. Among several isomorphism of a subgraph, the one with the lowest rank is said to be *canonical*. In this paper, we have modified such an extended tuple comparison rule for *weighted* graphs. A challenge is that such a modification affects the rank and changes the canonical DFScode of a subgraph.

In terms of **related works**, both GWF-mining and CWF-mining algorithms [15]—as extensions of utility based itemset mining (which considers non-weighted support)—mine internally and externally weighted graphs respectively by considering external weighted frequency for complex data. Along this direction, further extensions include closed and maximal subgraph mining [14]. In contrast, we consider individual subgraph weight calculated from edge weights and non-weighted frequency.

Eichinger et al. [4] showed that frequent subgraph mining task yields more precise results when considering weight-based constraints. As these weight-based constraints are not anti-monotonic, their algorithm returns approximate and thus *incomplete* results. In contrast, our proposed algorithm is *complete*.

Yang et al. [17] performed weighted subgraph mining on *single individual* weighted graphs. In contrast, our proposed algorithm focuses on mining weighted frequent patterns from a *set of weighted graphs*.

Three subgraph weighting techniques—namely, Average Total Weighting (ATW), Affinity Weighting (AW), and Utility Based Weighting (UBW) were proposed [9] and adapted for *longitudinal social network data* [8]. Among them, ATW requires redefining the subgraph weight as a ratio between total graph database weight and subgraph support set weight. Although ATW is anti-monotonic, it fails to discriminate between two subgraphs having the same support set. AW prunes a subgraph if its edges fail to satisfy some weight correlation condition, whereas UBW discards a subgraph if its weight-share falls below a weight-share threshold λ. On the contrary, our MaxPWS does not require redefining subgraph weight as it takes an average of edge weights. Hence, it can discriminate between two subgraphs even if they have exactly same support set. Moreover, MaxPWS can be applied to any affinity or utility conditions.

Lee and Yun [10] applied weighted-support affinity to reduce search space. Length-decreasing support constraints [11] were used as a weighted smallest-valid extension for frequent graph mining because graph patterns extracted from a given graph database can have various features (e.g., different pattern lengths). Moreover, a distributed approach of weighted frequent subgraph mining using ATW weighting technique [2] was proposed. Another approach [6] finds regular patterns from weighted-directed dynamic graphs with jitter. These four algorithms focus on *special or specific* weighted frequent subgraph mining tasks. In contrast, our algorithm focuses on the *general* weighted pattern mining task.

3   Our Proposed Algorithm

To reduce candidate subgraph generation, we propose (i) a tight pruning condition **MaxPWS** for weighted frequent subgraph mining and (ii) an algorithm called **WFSM-MaxPWS** for **W**eighted **F**requent **S**ubgraph **M**ining by using the MaxPWS condition. In this section, we first show our proposed canonical ordering for edge-weighted graphs, and then discuss details about MaxPWS.

3.1   WFSM-MaxPWS Canonical Ordering of Subgraph

For edge-weighted graphs, we add weight-property in the extended edge-tuple representation used in gSpan. Consequently, the new extended tuple for an edge (u, v) is of the following form as a 6-tuplet:

$$\langle dis_u, dis_v, L(u), L(v), L(u,v), W(u,v) \rangle,$$

where $W(u, v)$ is the weight of edge (u, v). To give a rank of a DFScode of subgraph, the WFSM-MaxPWS canonical order gives the highest priority to (dis_u, dis_v) as in gSpan. The second highest priority is given to edge weight $W(u, v)$. The higher the weight, the smaller is the tuple for the same discovery times. The third priority is given to a lexicographic comparison on node and edge label trio. To elaborate, let

$$t_1 = \langle v_i, v_j, L(v_i), L(v_j), L(v_i, v_j), W(v_i, v_j) \rangle; \text{ and}$$
$$t_2 = \langle v_x, v_y, L(v_x), L(v_y), L(v_x, v_y), W(v_x, v_y) \rangle.$$

Then, $t_1 < t_2$ if and only if

1. $(v_i, v_j) <_e (v_x, v_y)$; or
2. $(v_i, v_j) = (v_x, v_y)$ and $W(v_i, v_j) > W(v_x, v_y)$; or
3. $(v_i, v_j) = (v_x, v_y)$ and $W(v_i, v_j) = (v_x, v_y)$ and
   $\langle L(v_i), L(v_j), L(v_i, v_j) \rangle <_l \langle L(v_x), L(v_y), L(v_x, v_y) \rangle.$

Here, $<_e$ is an ordering on edge, and $<_l$ is an ordering on vertex and edge labels. Note that $<_l$ follows lexicographic order. The edge order $(<_e)$ rule is derived from the *rightmost path extension* sequence. The edge that extended earlier is smaller.

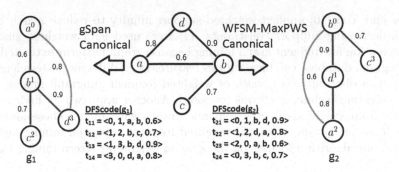

Fig. 1. Canonical code comparison with gSpan (without edge label)

Figure 1 shows a comparison between the canonical representations in gSpan [16] and our WFSM-MaxPWS algorithm. For simplicity and readability, tuples are shown as 5-tuplets by omitting the edge labels $L(u, v)$. As gSpan puts an edge ordering during the mining time of endpoint nodes and compares the (node labels-edge label)-trio in lexicographic order, tuple t_{11} becomes the smallest for graph in the figure. On the other hand, after performing edge ordering during the node mining time, the second importance of our WFSM-MaxPWS is put on edge weight. The *higher the weight, the smaller is the tuple.* As the first tuple of any DFScode has node discovery pair $(0, 1)$, the highest weighted edge would be the smallest. So, WFSM-MaxPWS would consider tuple $\langle 0, 1, b, d, 0.9 \rangle$ as the smallest. DFScode with other edges as first tuple would not be canonical.

Lemma 1. *No tuple can have a weight higher than the weight of the first tuple in a canonical WFSM-MaxPWS DFScode.*

Proof. (By induction on edge count m) For the base case (when $m = 1$), first tuple is the only tuple. Hence, there is no other tuple with a heavier weight.

For the inductive step, let us assume that the first tuple in DFScode C_x for $1 < m \leq x$ is $t_{x0} = \langle 0, 1, L(v_1), L(v_2), L(v_1, v_2), W(v_1, v_2) \rangle$. If we extend the subgraph with a tuple having a weight lower or equal to the first tuple, then the condition continues to hold. However, if we extend with a higher weighted tuple $t_{xk} = \langle *, *, L(u_1), L(u_2), L(u_1, u_2), W(u_1, u_2) \rangle$, then there exists a new possible DFScode C_y having $t_{y0} = \langle 0, 1, L(u_1), L(u_2), L(u_1, u_2), W(u_1, u_2) \rangle$ as first tuple. Here, $t_{x0} > t_{y0}$ implies that $Rank(C_y) < Rank(\{C_x, t_{xk}\})$. So, $\{C_x, t_{xk}\}$ cannot be canonical. Thus, we cannot extend a subgraph with a higher weighted tuple while preserving canonicity. □

3.2 MaxPWS Pruning Technique

We divide the entire graph database into partitions $p_1, p_2, p_3, \ldots, p_x, \ldots, p_M$. Partition p_1 is the set of graphs in the database having the minimum number of edges; partition p_M is the collection of graphs having the maximum number of edges. Graphs in the same partition have the same number of edges. Each graph

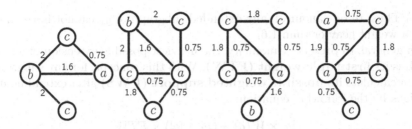

Fig. 2. Sample graph database GDB

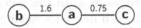

Fig. 3. Sample subgraph g_x

in partition p_x contain at least one more edge than any graphs in partition p_{x-1} and contain at least one fewer edge than graphs in partition p_{x+1}. We called these partition-collection **Edge-Class (EC)**. For the sample graph database in Fig. 2, there are three partitions. Hence, $EC = \{p_1 = \{G_1\}_4; \; p_2 = \{G_3, G_4\}_6; \; p_3 = \{G_2\}_7\}$, where the subscript after each curly brackets indicates the edge count of the partition. For any subgraph of consideration, we calculate its *occurrence list*, which is a collection of subsets from each Edge-Class partition entry where each member graph of the collection is a superset for that subgraph.

Definition 1. Let g be a subgraph. Its **occurrence list (OL)** is defined as $OL(g) = \{q_1, q_2, q_3, \ldots\}$ where $\forall q_i \in OL(g)[q_i \subseteq p_i$ and g is a subgraph of each member of $q_i]$. We call q_i an *occurrence list member (OLM)*.

For example, for the database in Fig. 2, the occurrence list for subgraph g_x shown in Fig. 3 is $OL(g_x) = \{q_1 = \{G_1\}_4; \; q_2 = \{G_3\}_6; \; q_3 = \{G_2\}_7\}$.

We can also calculate the **possible occurrence list (pol)** and **maximum possible frequency (mpf)** for a subgraph based on its OL for extension of that subgraph up to a different edge count. For example, from the OL of g_x, if we extend the subgraph g_x up to 4 edges, then the corresponding *pol* of the extended subgraph $g_{x\to4}$ is still $pol(g_{x\to4}) = \{\{G_1\}_4; \{G_3\}_6; \{G_2\}_7\}$. Similarly, the mpf of $g_{x\to4}$ becomes $mpf(g_{x\to4}) = 3$. However, if we extend g_x beyond 4 edges, graph G_1 can no longer be a part of its OL because it contains only 4 edges in total. For extension up to 6 edges, $pol(g_{x\to6}) = \{\{G_3\}_6; \{G_2\}_7\}$ and $mpf(g_{x\to6}) = 2$. Similarly, for extension up to 7 edge, $pol(g_{x\to7}) = \{\{G_2\}_7\}$ and $mpf(g_{x\to7}) = 1$. Now, the heaviest possible extension of g_x up to 4, 6 and 7 edges can contain 2, 4 and 5 more MaxW-weight edges, respectively. For the database in Fig. 2, it is 2.0 (edge b–c). According to Lemma 1, this extension can be canonical if and only if the first tuple in the canonical DFScode of g_x has a weight at least 2.0. However, canonical DFScode of g_x is $\{\langle 0, 1, a, b, 1.6 \rangle, \langle 0, 2, a, c, 0.75 \rangle\}$. (For simplicity, we omitted edge labels here.) Thus, to be canonical, any extended subgraph of g_x cannot have an edge with a weight heavier than 1.6.

In general, to preserve canonicity, the heaviest possible extension of any subgraph is its **first tuple weight (FTW)**. With this concept, for any subgraph g, we can calculate its **possible weighted support (PWS)** after extending up to m-edges by the following equation:

$$PWS(g_{\rightarrow m}) = \frac{|g| \times W(g) + (m - |g|) \times FTW}{m} \times mpf(g_{\rightarrow m}), \qquad (1)$$

where $|g|$ is the number of edges in g. In this calculation, g is assumed to be extended up to m edges with each new edge having the FTW weight to ensure that this imaginary extended supergraph has the maximum possible weight w.r.t. the graph database (because extension with a heavier weighted edge would not be canonical). We calculate MaxPWS by taking the maximum among all PWS values for g. If MaxPWS fails to satisfy the minimum weighted support threshold τ, then we can safely prune g. Otherwise, we need to extend g because its extended subgraph have potential to be weighted-frequent.

For example, consider subgraph g_x in Fig. 3 with $W(g_x) = \frac{1.6+0.75}{2} = 1.175$. For graph database in Fig. 2, PWS values are as follows:

- $PWS(g_{x \rightarrow 4}) = \frac{2 \times 1.175 + (4-2) \times 1.6}{4} \times 3 = 1.3875 \times 3 = 4.1625$
- $PWS(g_{x \rightarrow 6}) = \frac{2 \times 1.175 + (6-2) \times 1.6}{6} \times 2 \approx 1.4583 \times 2 = 2.9166$
- $PWS(g_{x \rightarrow 7}) = \frac{2 \times 1.175 + (7-2) \times 1.6}{7} \times 1 \approx 1.4786 \times 1 = 1.4786$

Hence, $MaxPWS(g_x) = 4.1625$. In contrast, the MaxW measure for g_x is $2 \times 3 = 6$ (because $MaxW = 2$ and frequency $= 3$), which is greater than MaxPWS.

Note that, if we do not use the modified canonical ordering for weighted graphs as proposed in Sect. 3.1, then the PWS calculation would have to use MaxW (instead of FTW as shown in Eq. (2)):

$$PWS(g_{\rightarrow m}) = \frac{|g| \times W(g) + (m - |g|) \times MaxW}{m} \times mpf(g_{\rightarrow m}) \qquad (2)$$

The resulting algorithm (which uses MaxW) is called **MaxPWS-gSpan algorithm**, which can be considered as a variant of our WFSM-MaxPWS algorithm (which uses FTW).

Lemma 2. *MaxPWS-measure is anti-monotonic.*

Proof. (By contradiction) Suppose that MaxPWS is not anti-monotonic. Then, there exists an extended subgraph g_x of g for which $MaxPWS(g_x) \geq \tau$ even though $MaxPWS(g) < \tau$. If $|g_x| - |g| = k$, then

$$W(g_x) \times |g_x| \leq W(g) \times |g| + k \times FTW. \qquad (3)$$

If MaxPWS of g_x occurs in m-edge extension, then $PWS(g_{x \rightarrow m}) > PWS(g_{\rightarrow m})$. This means $\frac{|g_x| \times W(g_x) + (m - |g_x|) \times FTW}{m} \times mpf(g_{x \rightarrow m}) > \frac{|g| \times W(g) + (m-|g|) \times FTW}{m} \times$

$mpf(g_{\to m})$. Due to the downward closure property of frequency, $mpf(g_{x\to m})$ must be less than or equal to $mpf(g_{\to m})$. So, by removing the mpf terms from both side, $|g_x| \times W(g_x) + (m - |g_x|) \times FTW > |g| \times W(g) + (m - |g|) \times FTW$. By using Eq. (3) and $|g_x| = |g| + k$, we get $W(g) \times |g| + k \times FTW + (m - |g| - k) \times FTW > |g| \times W(g) + (m - |g|) \times FTW$. Consequently, $k \times FTW + (m - |g| - k) \times FTW > (m - |g|) \times FTW$, and thus $(m - |g|) \times FTW > (m - |g|) \times FTW$, which is impossible. Thus, $MaxPWS(g_x)$ cannot be greater than $MaxPWS(g)$. So, MaxPWS-measure is anti-monotonic. □

Corollary 1. *If $MaxPWS(g) < \tau$, then g has no potential weighted frequent extension.* □

Corollary 2. *Due to MaxPWS-measure $\leq$ MaxW-measure, MaxPWS pruning technique prunes more unnecessary patterns.* □

3.3   The WFSM-MaxPWS Algorithm

A pseudocode for WFSM-MaxPWS is shown in Algorithm 1. Here, our WFSM-MaxPWS algorithm takes the following four input parameters: (i) the canonical DFScode of a graph C, (ii) graph database D, (iii) weighted support threshold τ, and (iv) occurrence list OL_C of C. With the initial call, $C = \emptyset$ and $OL_C = EC$ (edge class). WFSM-MaxPWS puts all frequent weighted subgraphs in a *result* set.

Algorithm 1. Algorithm *WFSM-MaxPWS*

1: **procedure** *WFSM-MaxPWS*(C, D, τ, OL_C)
2:    $OL\_vec =$ rightmost-path-extension(C, OL_C, D)
3:    **for each** $(t, OL_t) \in OL\_vec$ **do**
4:        $C' = C \cup t$
5:        **if** *IS_CANONICAL*(C') = false **then** continue.
6:        Set $sup_{C'} = 0$ and $MaxPWS = 0$
7:        **for each** $q_i \in OL_t$ in reverse order (last to first) **do**
8:            $m_i =$ edge count of q_i
9:            $sup_{C'} = sup_{C'} + |q_i|$ //cardinality of q_i
10:            $PWS_i = \frac{|C'| \times W(C') + (m_i - |C'|) \times FTW}{m_i} \times sup_{C'}$
11:            **if** $MaxPWS < PWS_i$ **then** $MaxPWS = PWS_i$
12:        $wsup_{C'} = W(C') \times sup_{C'}$
13:        **if** $wsup_{C'} \geq \tau$ **then**
14:            $result = result \cup C'$ //Enlist C' as frequent weighted subgraph
15:            *WFSM-MaxPWS*(C', D, τ, OL_t)
16:        **else if** $MaxPWS \geq \tau$ **then** *WFSM-MaxPWS*(C', D, τ, OL_t)

The function "rightmost-path-extension" in line 2 enumerates all possible extensions on rightmost path of the given DFScode and returns a vector of those extensions as new edge tuples and their corresponding occurrence list OL_vec.

Each entry in OL_vec is then checked for canonicity (lines 4 & 5). The support of code C' is updated while iterating through each entry in OL (line 9). Simultaneously, possible weighted support from the highest to the lowest $OLM(q_i)$ is calculated to find $MaxPWS$ (lines 8–11). The actual weighted support $wsup$ of the code/subgraph is calculated in line 11. If it satisfies weighted support threshold condition (line 13), then it is enlisted as a frequent weighted subgraph (line 14) and sent for further extension (line 15). Otherwise, if $MaxPWS$ satisfies weighted support threshold condition, though it will not be enlisted as frequent weighted subgraph, then it will be sent to the WFSM-MaxPWS algorithm for further extension (line 16) as its extended graph still has a chance to be frequent weighted subgraph.

4   Experimental Results

To evaluate the performance of our proposed algorithm WFSM-MaxPWS, we conducted several experiments on a PC with an Intel Core i3-2100 CPU at 3.10 GHz and 4 GB RAM running MS Windows 10 operating system. We analyze performance of WFSM-MaxPWS w.r.t. runtime, search-space reduction efficiency, and memory requirement. For comparison, we used MaxW-gSpan (which uses MaxW pruning technique with gSpan) as the baseline algorithm. Both WFSM-MaxPWS and MaxW-gSpan were implemented in Python.

Regarding the test datasets, several graphs datasets[1] were selected from Pub-Chem[2], which provides information on biological activities of small molecules and contains the bioassay records for anti-cancer screen tests with different cancer cell lines. Each dataset captures a certain type of cancer screen with the outcome active or inactive. In particular, we used **MCF-7**, **P388** and **Yeast** datasets. Since these datasets come with no weights, we added weights according to two different weight distributions—namely, *normal* and *negative exponential*. Dataset statistics after adding weight is given in Table 1.

Table 1. Datasets

Dataset	#graphs	Distinct #edges	Avg #edges	Distribution	MinW	MaxW
MCF-7	2,293	54	36	normal ($\mu = 0.5$, $\sigma = 0.07$)	0.13	0.86
				negExpo (f = 1)	0.07	0.98
P388	2,297	64	30	normal ($\mu = 10$, $\sigma = 1.5$)	3.16	16.84
				negExpo (f = 18)	1.55	16.86
Yeast	9,567	125	26	normal ($\mu = 2$, $\sigma = 0.3$)	0.46	3.55
				negExpo (f = 3.6)	0.23	3.53

---

[1] http://www.cs.ucsb.edu/~xyan/dataset.htm.
[2] https://pubchem.ncbi.nlm.nih.gov/.

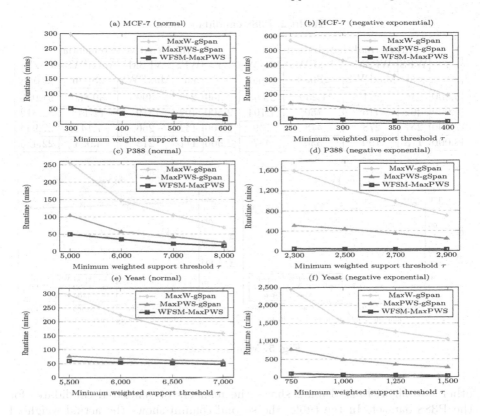

Fig. 4. Runtime

To analyze the performance of our proposed algorithm WFSM-MaxPWS (which uses Eq. (1) for the PWS calculation), we compared it with MaxPWS-gSpan (which uses Eq. (2) for the PWS calculation) and the baseline MaxW-gSpan algorithm (which simply uses the MaxW pruning technique) by using the three datasets in Table 1. We examined the following aspects: (i) runtime (ii) search-space reduction, and (iii) memory usage.

Figure 4 shows the **runtimes** on three datasets (each with two distributions). Here, both our WFSM-MaxPWS and MaxPWS-gSpan algorithms ran significantly faster than MaxW-gSpan. Between the former two, WFSM-MaxPWS ran faster than MaxPWS-gSpan, and the margin was wider for datasets having negative exponential weight distributions because WFSM-MaxPWS takes full advantage of distribution by considering FTW (instead of MaxW) in the MaxPWS calculation. When the weights follows positive exponential, WFSM-MaxPWS still ran faster than MaxPWS-gSpan or MaxW-gSpan, though the gap between the latter two was smaller.

A reason behind the runtime improvement of WFSM-MaxPWS over the other two algorithms is its efficiency in **search-space reduction**. Specifically, WFSM-MaxPWS generates a very small number of candidates when compared with the

Table 2. P388 candidate count

(a) Normal distribution					(b) Negative exponential			
τ	WFS cnt	WFSM-MaxPWS	MaxPWS-gSpan	MaxW-gSpan	τ	WFS cnt	WFSM-MaxPWS	MaxPWS-gSpan	MaxW-gSpan
5,000	616	711	1,259	2,053	2,300	248	399	30,869	91,053
6,000	362	415	750	1,104	2,500	192	319	12,148	38,312
7,000	257	278	478	739	2,700	142	266	7,145	27,944
8,000	173	199	330	487	2,900	111	212	4,953	22,455

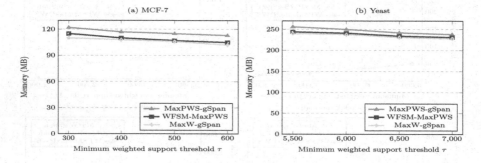

Fig. 5. Memory usage

other two algorithms. Table 2 shows the numbers of generated candidates for the P388 dataset. In the table, the second column shows the actual weighted frequent subgraph count (WFS cnt). The third to fifth columns show candidate counts for WFSM-MaxPWS, MaxPWS-gSpan and MaxW-gSpan. Our WFSM-MaxPWS generated the smallest number of candidates when compared with the other two algorithms, especially with negative exponential distribution. Such a reduction in search space effectively reduced runtime.

As for **memory usage**, both WFSM-MaxPWS and MaxPWS-gSpan required just slightly more memory than MaxW-gSpan due to the occurrence list storage. As shown in Fig. 5 on both MCF-7 and Yeast datasets, the slight increase in memory requirement was insignificant, especially when compared with fruitful benefits of reduction in both runtime and the number of generated candidates.

5   Conclusions

In this paper, we proposed a weighted frequent subgraph mining algorithm called WFSM-MaxPWS, which uses MaxPWS-measure to reduce search-space. Max-PWS is calculated using a modified canonical ordering for weighted graphs to achieve smallest possible upper bound of maximum possible weighted support for any extensions of a particular subgraph along with ensuring no loss of weighted frequent subgraph patterns. Experimental results and comparative analysis on three real datasets with normal and negative exponential distribution show that our WFSM-MaxPWS algorithm outperforms the existing MaxW-gSpan algorithm w.r.t. runtime and reduction in the number of generated candidates. Moreover, our modified canonical ordering for weighted graphs facilitates Max-PWS calculation to achieve even better performance. This concept of modified canonical ordering and MaxPWS pruning have potential to be further utilized in uncertain or utility-based graph databases.

Acknowledgement. This project is partially supported by NSERC (Canada) and U. Manitoba.

References

1. Ahmed, C.F., Tanbeer, S.K., Jeong, B.S., Lee, Y.K., Choi, H.J.: Single-pass incremental and interactive mining for weighted frequent patterns. Expert Syst. Appl. **39**(9), 7976–7994 (2012)
2. Babu, N., John, A.: A distributed approach to weighted frequent subgraph mining. In: ICETT 2016. IEEE (2016). https://doi.org/10.1109/ICETT.2016.7873705
3. Cheng, Z., Flouvat, F., Selmaoui-Folcher, N.: Mining recurrent patterns in a dynamic attributed graph. In: Kim, J., Shim, K., Cao, L., Lee, J.-G., Lin, X., Moon, Y.-S. (eds.) PAKDD 2017. LNCS (LNAI), vol. 10235, pp. 631–643. Springer, Cham (2017). https://doi.org/10.1007/978-3-319-57529-2_49
4. Eichinger, F., Huber, M., Böhm, K.: On the usefulness of weight-based constraints in frequent subgraph mining. In: Bramer, M., Petridis, M., Hopgood, A. (eds.) Research and Development in Intelligent Systems, vol. XXVII, pp. 65–78. Springer, London (2011). https://doi.org/10.1007/978-0-85729-130-1_5
5. Fariha, A., Ahmed, C.F., Leung, C.K.-S., Abdullah, S.M., Cao, L.: Mining frequent patterns from human interactions in meetings using directed acyclic graphs. In: Pei, J., Tseng, V.S., Cao, L., Motoda, H., Xu, G. (eds.) PAKDD 2013. LNCS (LNAI), vol. 7818, pp. 38–49. Springer, Heidelberg (2013). https://doi.org/10.1007/978-3-642-37453-1_4
6. Gupta, A., Thakur, H., Gupta, T., Yadav, S.: Regular pattern mining (with jitter) on weighted-directed dynamic graphs. JESTEC **12**(2), 349–364 (2017)
7. Huang, Z., Ye, Y., Li, X., Liu, F., Chen, H.: Joint weighted nonnegative matrix factorization for mining attributed graphs. In: Kim, J., Shim, K., Cao, L., Lee, J.-G., Lin, X., Moon, Y.-S. (eds.) PAKDD 2017. LNCS (LNAI), vol. 10234, pp. 368–380. Springer, Cham (2017). https://doi.org/10.1007/978-3-319-57454-7_29
8. Jiang, C., Coenen, F., Zito, M.: Finding frequent subgraphs in longitudinal social network data using a weighted graph mining approach. In: Cao, L., Feng, Y., Zhong, J. (eds.) ADMA 2010. LNCS (LNAI), vol. 6440, pp. 405–416. Springer, Heidelberg (2010). https://doi.org/10.1007/978-3-642-17316-5_39
9. Jiang, C., Coenen, F., Zito, M.: Frequent sub-graph mining on edge weighted graphs. In: Pedersen, T.B., Mohania, M.K., Tjoa, A.M. (eds.) DaWaK 2010. LNCS, vol. 6263, pp. 77–88. Springer, Heidelberg (2010). https://doi.org/10.1007/978-3-642-15105-7_7
10. Lee, G., Yun, U.: Mining strongly correlated sub-graph patterns by considering weight and support constraints. IJMUE **8**(1), 197–206 (2013)

11. Lee, G., Yun, U., Kim, D.: A weight-based approach: frequent graph pattern mining with length-decreasing support constraints using weighted smallest valid extension. Adv. Sci. Lett. **22**(9), 2480–2484 (2016)
12. Leung, C.K.-S., Mateo, M.A.F., Brajczuk, D.A.: A tree-based approach for frequent pattern mining from uncertain data. In: Washio, T., Suzuki, E., Ting, K.M., Inokuchi, A. (eds.) PAKDD 2008. LNCS (LNAI), vol. 5012, pp. 653–661. Springer, Heidelberg (2008). https://doi.org/10.1007/978-3-540-68125-0_61
13. Leung, C.K.-S., Tanbeer, S.K.: PUF-tree: a compact tree structure for frequent pattern mining of uncertain data. In: Pei, J., Tseng, V.S., Cao, L., Motoda, H., Xu, G. (eds.) PAKDD 2013. LNCS (LNAI), vol. 7818, pp. 13–25. Springer, Heidelberg (2013). https://doi.org/10.1007/978-3-642-37453-1_2
14. Ozaki, T., Etoh, M.: Closed and maximal subgraph mining in internally and externally weighted graph databases. In: IEEE AINA 2011 Workshops, pp. 626–631 (2011)
15. Shinoda, M., Ozaki, T., Ohkawa, T.: Weighted frequent subgraph mining in weighted graph databases. In: IEEE ICDM 2009 Workshops, pp. 58–63 (2009)
16. Yan, X., Han, J.: gSpan: graph-based substructure pattern mining. In: IEEE ICDM 2002, pp. 721–724 (2002)
17. Yang, J., Su, W., Li, S., Dalkilic, M.M.: WIGM: discovery of subgraph patterns in a large weighted graph. In: SIAM SDM 2012, pp. 1083–1094 (2012)

A Game-Theoretic Adversarial Approach to Dynamic Network Prediction

Jia Li$^{(\boxtimes)}$, Brian Ziebart, and Tanya Berger-Wolf

Department of Computer Science, University of Illinois at Chicago, Chicago, USA
{jli213,bziebart,tanyabw}@uic.edu

Abstract. Predicting the evolution of a dynamic network—the addition of new edges and the removal of existing edges—is challenging. In part, this is because: (1) networks are often noisy; (2) various performance measures emphasize different aspects of prediction; and (3) it is not clear which network features are useful for prediction. To address these challenges, we develop a novel framework for robust dynamic network prediction using an adversarial formulation that leverages both edge-based and global network features to make predictions. We conduct experiments on five distinct dynamic network datasets to show the superiority of our approach compared to state-of-the-art methods.

1 Introduction

Dynamic networks are a powerful tool for understanding complex systems over time, with applications in social science, business, neuroscience, and many other fields. Predicting future network structure enables hypothesis testing of the network evolution and interventions before an event occurs (e.g., preventing disease spread). Despite the importance, learning to predict network changes over time is a challenging task. While existing network prediction methods [6,9,17] often succeed in specific settings, they may be ill-suited to address the variety of networks encountered in the general prediction task. The difficulty of designing a general method for dynamic network prediction is evident in the paucity of methods and the modest improvements in performance over the last decade.

One challenge is that dynamic networks are noisy objects and network data typically suffers both from temporal and topological errors [13]. This is further complicated by various evaluation measures used to emphasize different qualities of network prediction for different applications. For example, when supporting resource allocation decisions (e.g., cellphone tower placement), the recall matters more since it relates to the quality of service. In product recommendation, new product purchases have different values than repeat purchases when creating advertising strategies. It is difficult to design a general method that can optimize these different measures. Moreover, there is often a mismatch between the desired metrics and the one that is de facto being optimized. There are a wide range of potentially useful features for network prediction, including general patterns, such as triangle closure (two people are more likely to become friends if they have

© Springer International Publishing AG, part of Springer Nature 2018
D. Phung et al. (Eds.): PAKDD 2018, LNAI 10939, pp. 677–688, 2018.
https://doi.org/10.1007/978-3-319-93040-4_53

many common friends) [2], edge-specific patterns (edge *e* shows up every three time steps; or A frequently calls B after calling C or D), or intricate dependency structures. Most methods focus on a subset of these feature types, while in reality a mix of those explains the topology of most networks.

To address these challenges, we formulate the prediction task as an adversarial game between a *predictor* player, optimizing for predictive performance, and an *adversarial* player, providing worst case approximation of training data, while constrained to match important properties. This creates network predictions that are robust to worst-case uncertainty that arises due to noise and limited observational data. While existing approaches often sacrifice predictive performance in favor of scalability, we focus on improving performance and defer improving the scalability of our approach to future work. Our formulation produces temporal exponential random graph models (TERGM) [12] when logloss is used. However, our model generalizes beyond this special case, with the advantages of: **flexible cost metrics** that can be tuned towards end-use application performance measures; **edge-based features** that represent either general or edge-specific patterns; and **global summary features** that model the dependency structures while allowing model learning to be a convex optimization problem. We demonstrate the benefits of our approach on five dynamic network prediction tasks with comparison to state-of-the-art methods.

2   Related Work

Holme and Saramäki [13] provide an excellent survey of dynamic network definitions, modeling, and analysis methods. There are two main types of networks with topologies that change over time. In *evolving networks*, nodes and edges are added but not deleted (*e.g.,* citation networks). Thus, predictions are needed for new nodes and added edges (known as *link prediction* [21,32,33]). *Dynamic networks* are more general, with node and edge additions and deletions. In this paper, we focus on predicting additions and removals of edges in the *dynamic network*, essentially predicting the *change* in the network its truly dynamic part.

Dynamic network prediction seeks to predict the future structure of the network, given past network structure (and possibly attribute information). The attribute information, which typically improves prediction accuracy, is quite often not available. For the purely topological approaches, some use generative models, such as preferential attachment [3]. Matrix and tensor factorization techniques [9] are used for making predictions. The structured prediction [17] makes prediction using frequent subgraphs. A hybrid of preferential attachment and ARIMA model is proposed in [14]. Early warning subgraphs and network's vector space embedding are used as features to predict recurrent subgraphs in [23]. There are also supervised learning methods that use topological features to rank list of nodes and the edges around them [21,26]. Separate models for edge formation and dissolution are used by [31]. There are common problems with these approaches. Fundamentally, they do not explicitly define a loss to optimize. Although different measures (e.g. AUC, precision, recall) are used for evaluation, it is not clear whether the methods optimize these measures. We focus on F1-score, which balances between the precision and recall and follow [24] by optimizing cost-sensitive error to approximate the F1-score to avoid the complexity of optimizing a multivariate loss [30].

3 Adversarial Dynamic Network Prediction

We seek a robust method for the general changing network prediction problem:

Definition 1. *The **dynamic network change prediction task** is: given previous networks $\mathcal{G} = (\tilde{G}_1, \ldots, \tilde{G}_{t-1})$, provide a prediction $\hat{G}_t$ for the actual network $\tilde{G}_t$ at timestep t that minimizes $loss(\hat{G}_t, \tilde{G}_t)$, using available training network time series from empirical distribution $\tilde{P}(\tilde{G}_1, \ldots, \tilde{G}_t)$.*

Each network is defined by a static set of nodes V and undirected edges E_t that vary over time: $G_t = (V, E_t)$. For compactness, we let $\mathcal{G}$ represent the history of networks, $\tilde{G}_1, \ldots, \tilde{G}_{t-1}$, when predicting G_t. We denote the existence of an edge from node i to j at time t as $e_{i,j,t} = 1$ and absence as $e_{i,j,t} = 0$. We use $\hat{G}$ for predicted network, $\tilde{G}$ for empirical network, and an adversarially-chosen approximation of the network, $\check{G}_t$. We denote the corresponding edges with these networks using variables $\check{e}_{i,j,t}$ and predicted edges are denoted as $\hat{e}_{i,j,t}$. When testing, the predicted timestep t' is different from any training timestep t.

Loss functions of interest are defined in terms of the differences between edge predictions $\hat{G}_t$ and the actual network $\tilde{G}_t$: $loss(\hat{G}_t, \tilde{G}_t)$. Graph edit distance $loss_{edit}(\hat{G}_t, \tilde{G}_t) = \sum_{1 \leq i < j \leq |V|} I(\hat{e}_{i,j,t} \neq \tilde{e}_{i,j,t})$, is a loss function that additively measures the edge disagreement. However, since networks are often sparse, trivial predictors (e.g., predicting no edges) perform well despite being uninformative. Another loss is to weigh each type of error separately so that errors that are rare can be penalized more heavily. There are two types of changes that may arise in a dynamic network: addition of an edge (start: the beginning of an edge existence) and removal of an existing edge (end: the end of an edge existence). There are four types of errors: a missed start (MS) or a missed end (ME) or falsely predicted as a wrong start (WS) or a wrong end (WE). A loss function that weighs each of these errors differently is: $loss(\tilde{G}_t, \hat{G}_t, \mathcal{G}) =$

$$\lambda_1 \, MS(\tilde{G}_t, \hat{G}_t, \mathcal{G}) + \lambda_2 \, ME(\tilde{G}_t, \hat{G}_t, \mathcal{G}) + \lambda_3 \, WS(\tilde{G}_t, \hat{G}_t, \mathcal{G}) + WE(\tilde{G}_t, \hat{G}_t, \mathcal{G}).$$

3.1 Adversarial Prediction Formulation

Recently, adversarial methods have been successful in problems such as classification [4], sequence tagging [20] and information retrieval [30]. These methods significantly improve predictive performance by optimizing desired loss functions. However, previous adversarial approaches only focus on data with little or no structure. Our model leverages structure in networks while maintaining the robustness guarantee. We formulate our prediction model as a two-player zero-sum game with the adversarial player maximizing the loss while predictor player minimizing the loss. For an unweighted network with three nodes, the payoffs of this game using the graph edit distance as the loss function are shown in Table 1.

The predictor player chooses a distribution $\hat{P}(\hat{G}_t|\mathcal{G})$ over the rows while the adversary controls another distribution $\check{P}(\check{G}_t|\mathcal{G})$ over the columns. Such distributions represent the probabilities of each possible network (pure strategy). The goal of the adversary is to provide the worst case approximation of the training data. However, it cannot be arbitrarily random as this would not reflect meaninful structure of the observed graph dynamics, so the adversary is constrained to match

Table 1. Lagrangian-augmented payoffs (losses) using edit distance for the conditional network prediction game with three edges. Each predicted edge is either absent (0) or present (1), corresponding to 2^3 pure strategies for each player.

	000	001	010	011	100	101	110	111
000	0	1	1	2	1	2	2	3
001	1	0	2	1	2	1	3	2
010	1	2	0	1	2	3	1	2
011	2	1	1	0	3	2	2	1
100	1	2	2	3	0	1	1	2
101	2	1	3	2	1	0	2	1
110	2	3	1	2	1	2	0	1
111	3	2	2	1	2	1	1	0

empirical statistics of the training data, as shown formally in Eq. (2). We consider generic features Φ that constrain the adversary in this section, and discuss specific features in next section The general form of our adversarial formulation defining games of this sort is:

$$\min_{\hat{P}(\hat{G}_t|\mathcal{G})} \max_{\check{P}(\check{G}_t|\mathcal{G})} \mathbb{E}_{\tilde{P}(\mathcal{G})\check{P}(\check{G}_t|\mathcal{G})\hat{P}(\hat{G}_t|\mathcal{G})}[\text{loss}(\hat{G}_t,\check{G}_t)] \tag{1}$$

$$\text{subject to: } \mathbb{E}_{\tilde{P}(\mathcal{G})\check{P}(\check{G}_t|\mathcal{G})}[\Phi(\mathcal{G},\check{G}_t)] = \mathbb{E}_{\tilde{P}(\mathcal{G},\tilde{G}_t)}[\Phi(\mathcal{G},\tilde{G}_t)]. \tag{2}$$

We replace the constraints with Lagrangian potential terms parameterized by θ using the standard method of Lagrangian multipliers , yielding:

$$\min_{\theta} \mathbb{E}_{\tilde{P}(\mathcal{G},\tilde{G}_t)}\left[-\theta \cdot \Phi(\mathcal{G},\tilde{G}_t) + \underbrace{\min_{\hat{P}(\hat{G}_t|\mathcal{G})} \max_{\check{P}(\check{G}_t|\mathcal{G})} \mathbb{E}_{\check{P}(\check{G}_t|\mathcal{G})\hat{P}(\hat{G}_t|\mathcal{G})}[\text{loss}(\hat{G}_t,\check{G}_t) + \theta \cdot \Phi(\mathcal{G},\check{G}_t)]}_{\text{Lagrangian-augmented zero-sum game}}\right].$$

Each network prediction game is augmented with the Lagrangian potential terms and can then be solved within the expectation of this unconstrained optimization problem. The augmented game is shown in Table 2 with the Lagrangian poten-

Table 2. The Lagrangian-augmented payoffs (losses) for the game in Table 1.

	000	001	010	011	100	101	110	111
000	$0+\psi_{000}$	$1+\psi_{001}$	$1+\psi_{010}$	$2+\psi_{011}$	$1+\psi_{100}$	$2+\psi_{101}$	$2+\psi_{110}$	$3+\psi_{111}$
001	$1+\psi_{000}$	$0+\psi_{001}$	$2+\psi_{010}$	$1+\psi_{011}$	$2+\psi_{100}$	$1+\psi_{101}$	$3+\psi_{110}$	$2+\psi_{111}$
010	$1+\psi_{000}$	$2+\psi_{001}$	$0+\psi_{010}$	$1+\psi_{011}$	$2+\psi_{100}$	$3+\psi_{101}$	$1+\psi_{110}$	$2+\psi_{111}$
011	$2+\psi_{000}$	$1+\psi_{001}$	$1+\psi_{010}$	$0+\psi_{011}$	$3+\psi_{100}$	$2+\psi_{101}$	$2+\psi_{110}$	$1+\psi_{111}$
100	$1+\psi_{000}$	$2+\psi_{001}$	$2+\psi_{010}$	$3+\psi_{011}$	$0+\psi_{100}$	$1+\psi_{101}$	$1+\psi_{110}$	$2+\psi_{111}$
101	$2+\psi_{000}$	$1+\psi_{001}$	$3+\psi_{010}$	$2+\psi_{011}$	$1+\psi_{100}$	$0+\psi_{101}$	$2+\psi_{110}$	$1+\psi_{111}$
110	$2+\psi_{000}$	$3+\psi_{001}$	$1+\psi_{010}$	$2+\psi_{011}$	$1+\psi_{100}$	$2+\psi_{101}$	$0+\psi_{110}$	$1+\psi_{111}$
111	$3+\psi_{000}$	$2+\psi_{001}$	$2+\psi_{010}$	$1+\psi_{011}$	$2+\psi_{100}$	$1+\psi_{101}$	$1+\psi_{110}$	$0+\psi_{111}$

tial term relating to the adversary's choice compactly defined as: $\psi_{\check{G}_t} \triangleq \theta \cdot \Phi(\mathcal{G},\check{G}_t)$. Each cell has a Lagrangian potential term (added to the edit distance) that can be viewed as a motivation for the adversarial player to be similar to the training data. In general, the Nash equilibrium of this game consists of the predictor player's distribution over rows and the adversary's distribution over columns. The Nash equilibrium is defined by neither players unilaterally improving their expected payoffs by switching to another strategy.

There are three main steps for solving the optimization: (1) **Obtaining the adversary's equilibrium distribution** for the Lagrangian-augmented game; (2) Computing expected features under the adversary's equilibrium distribution; and (3) **Updating the Lagrangian parameters** of the game until convergence is reached. If not, repeat from the first step.

Finding the optimal parameters θ is a convex optimization problem. Producing predictions under the adversarial network prediction framework requires solving the zero-sum games between the predictor and the adversary. However, this is computationally difficult to achieve naïvely when the network size is moderately large since the size of the game grows exponentially with the number of edges. We employ the double oracle method [22] to tame this complexity. As detailed in Algorithm 1, we start with a subset of pure strategies for both players (Line 1). The game continues when either has a better strategy to improve its payoff value. We compute the probability distributions of the networks by solving the game (Line 3). Given the probability distribution of the predictor player, we try to find if there is a better pure strategy (network) for the adversarial player that will increase the game value (Line 4). If there is, it will be added to the pure strategies for the adversarial player. Similarly for predictor player (Line 5–6). We repeat the process until there is no more strategy added for both players. The equilibrium produced by this algorithm is guaranteed to be an equilibrium for the full game [22]. We show methods for efficiently finding best responses for different sets of features in the following section.

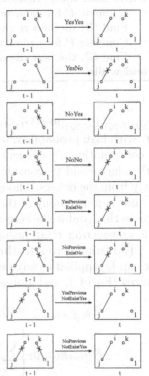

Fig. 1. Features with a line denoting an edge exists, and a line with an X to denote the absence of an edge.

3.2 Performance Guarantees and Features

Employing a worst-case assumption provides robustness guarantees. Since the approximation is worse (by definition) for the predictor, we can be assured that the performance of the predictor on training data will be better than the performance against the adversary: $\mathbb{E}[\text{loss}(\hat{G}_t, G_t)] \leq \mathbb{E}[\text{loss}(\hat{G}_t, \check{G}_t)]$. The constraints in Eq. (2) enforce a degree of similarity between the adversary and the training data distribution, tightening this bound.

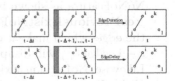

Fig. 2. Edge duration and delay features with edges denoted as in Fig. 1.

Algorithm 1. Double oracle for network edge prediction.

Require: Lagrangian potentials, ψ; loss function loss$(\mathcal{G}, \hat{G}_t, \check{G}_t)$
Ensure: Nash equilibrium, $\hat{P}, \check{P}$
1: Initialize pure strategies for predictor player $\hat{S}$ and adversarial player $\check{S}$
2: **while** game values changing **do**
3:    Compute $\hat{P}$ for game over graphs in $\hat{S}$ and $\check{S}$
4:    Find $\check{G}_t^*$, the best response to $\hat{P}$; $\check{S} \leftarrow \check{S} \cup \check{G}_t^*$
5:    Compute $\check{P}$ for game over graphs in $\hat{S}$ and $\check{S}$
6:    Find $\hat{G}_t^*$, the best response to $\check{P}$; $\hat{S} \leftarrow \hat{S} \cup \hat{G}_t^*$
7: **end while**

$\mathbb{E}_{P(\mathcal{G})(G_t|\mathcal{G})\hat{P}(\hat{G}_t|\mathcal{G})}[\text{loss}(\hat{G}_t, G_t)]$. Our goal is to define a constraint set (features) that, with high probability, will include the true conditional distribution $P(G_t|\mathcal{G})$ in a computationally efficient manner. In the extreme case, features pertaining to single edges could be considered. This allows each edge to be predicted independently from the others. Though computationally efficient, the resulting predictor does not capture many of the dependencies that exist between edges. We choose features that enable the zero-sum network prediction game to be efficiently solved while considering relationships between edges. We constrain our features Φ for the adversarial player to three generic types defined by the network topology that can be applied to any network: edge-specific features (Eq. (3)), generic edge features (Eq. (4)), and global summary features (Eq. (5)):

$$\mathbb{E}_{\check{P}(\mathcal{G})\check{P}(\check{G}_t|\mathcal{G})}[\Phi(\mathcal{G}, \check{e}_{i,j})] = \mathbb{E}_{\check{P}(\mathcal{G},\tilde{G}_t)}[\Phi(\mathcal{G}, \tilde{e}_{i,j})] \tag{3}$$

$$\mathbb{E}_{\check{P}(\mathcal{G})\check{P}(\check{G}_t|\mathcal{G})}\left[\sum_{x,y \in \mathcal{V}} \Phi(\mathcal{G}, \check{e}_{x,y})\right] = \mathbb{E}_{\check{P}(\mathcal{G},\tilde{G}_t)}\left[\sum_{x,y \in \mathcal{V}} \Phi(\mathcal{G}, \tilde{e}_{x,y})\right] \tag{4}$$

$$\mathbb{E}_{\check{P}(\mathcal{G})\check{P}(\check{G}_t|\mathcal{G})}[\Phi(\mathcal{G}, S(\check{G}_t))] = \mathbb{E}_{\check{P}(\mathcal{G},\tilde{G}_t)}[\Phi(\mathcal{G}, S(\tilde{G}_t))]. \tag{5}$$

Edge-specific features (As shown in Eq. 3), relate to one particular edge, $e_{i,j}$. Some of the possible features (shown in Fig. 1) are:

- **YesYesFeature** edge $e_{i,j}$ exists at time t if edge $e_{k,l}$ exists at time $t-1$, $f(e_{k,l,t-1} = 1, e_{i,j,t} = 1)$. Similarly, we define YesNoFeature, NoYesFeature, NoNoFeature as shown in Fig. 1.
- **YesPreviousExistNoFeature:** When edge $e_{i,j}$ exists at time $t-1$, we select the edges which make it disappear at time t. If edge $e_{i,j}$ and $e_{l,k}$ exist at time $t-1$, edge $e_{i,j}$ will disappear at time t, $f(e_{k,l,t-1} = 1, e_{i,j,t-1} = 1, e_{i,j,t} = 0)$. Likewise, NoPreviousExistNoFeature is defined as shown in Fig. 1.
- **DelayFeature:** As the bottom of Fig. 2, edge $e_{i,j}$ exists after Δt timesteps after edge $e_{k,l}$ last exist, $f(e_{k,l,t-\Delta t} = 1, e_{k,l,t-\Delta t+1,\ldots,t-1} = 0, e_{i,j,t} = 1)$.

For all of these features, i, j can be the same as k, l. The $t-1$ can be generalized to $t - \Delta t$. For pairwise edge features such as YesYesFeature, there are m^2 potential features, where m is the total edges. In practice, we constrain such features to the most informative ones which can be selected based on a support frequency threshold. Taking YesYesFeature as an example, the support is defined as the number of times that edge $e_{i,j}$ occurred one time step after edge $e_{k,l}$ exists divided by the number of time steps edge $e_{k,l}$ exist. We select top 10 features for each edge and each edge-specific feature.

Generic edge features (Eq. 4) describe general patterns for all edges in the network. For instance, if an edge exists in the previous timestep, it tends to remain in existence at the current timestep. These patterns are applicable to all edges rather than to specific edges. Our features are:

- **GenericYesYesFeature** if an edge exists in the previous timestep, the same edge continues to exist in the current timestep, $f(e_{x,y,t-1} = 1, e_{x,y,t} = 1)$. The GenericNoNoFeature is similarly defined.
- **PreviousDegreeFeature** is the previous degrees of node x and node y.
- **EdgeDurationFeature:** As shown at the top in Fig. 2, for an edge that exists at timestep t, this feature is based on the number of previous timesteps, up to but not including t, of the edge's continuous existence, $f(e_{i,j,t-\Delta t} = 0, e_{i,j,t-\Delta t+1,...,t-1} = 1, e_{i,j,t} = 1)$.

Additionally, we include Adamic/Adar [2], recency [25], linkratio, and triad collocation elements [31] as features. Most of the generic edge and edge-specific features are interchangeable. When a feature is a general pattern for all edges, we should use it as a generic feature with the advantage of increased number of observation. When the variance of a pattern is high for different edges, we should use specific features. If one classifier is constructed for all the edges in the network, only generic edge features can be employed. On the other hand, if a classifier is built for each edge, the shared patterns of different edges are ignored. In most applications, both generic patterns as well as edge specific patterns govern the evolution of the graph structure. Our framework is flexible incorporating both types of features.

Global summary features are summaries of the network. The previous two types of features consider each edge independently at the predicted timestep. On the contrary, global summary features jointly consider all of the edges so that dependencies of edges are incorporated into the prediction.

- **TotalAddedEdgesFeature** is the total number of edges in existence at the timestep t but do not exist at time $t - 1$.
- **TotalRemovedEdgesFeature** is the number of edges that exist at timestep $t - 1$ but do not exist at time t.

Similarly, we can define other global summary features such as the total number of edges and total number of changed edges. Global features can be defined on a subset of edges such as ones that share the same node attributes. Using multiple global summary features simultaneously requires a way to search for optimal sets of edges for all possible combinations of global summary feature values. When the edge sets affected by the global summary features are non-overlapping, multiple global features, such as TotalAddedEdgesFeature and TotalRemovedEdgesFeature, can be used simultaneously easily. We consider these two global features in our experiment and use them as global summary features.

Algorithm 2. Best response for weighted edge losses

Require: $\hat{P}(\hat{G}_t|\mathcal{G})$, θ, $\varPhi(\mathcal{G}, G_t)$
Ensure: Best Response Network
1: **for** edge $e_{i,j,t}$ **do**
2:     Compute marginal probabilities $\hat{P}(\hat{e}_{i,j,t}|\mathcal{G})$
3:     Compute the cost C of having edge $C(\check{e}_{i,j,t}=1)$ and not having edge $C(\check{e}_{i,j,t}=0)$
4: **end for**
5: Sort edges by $C(\check{e}_{i,j,t}=1) - C(\check{e}_{i,j,t}=0)$
6: $C_{exist} = 0$; $C_{not\_exist} = \sum_{i,j} C(\check{e}_{i,j,t}=0)$
7: $total\_edge = 0$; $maxR = -\infty$
8: **for** edge $\check{e}_{i,j,t}$ in ordered edges **do**
9:     $C_{exist} = C_{exist} + C(\check{e}_{i,j,t}=1)$
10:     $C_{not\_exist} = C_{not\_exist} - C(\check{e}_{i,j,t}=0)$
11:     $total\_edge = total\_edge + 1$
12:     **if** $C_{exist} + C_{not\_exist} + C_{total\_edge} > maxC$ **then**
13:         $maxC = C_{exist} + C_{not\_exist} + C_{total\_edge}$
14:         $m = total\_edge$
15:     **end if**
16: **end for**
17: Add the first m sorted edges to the network.

Algorithm 2 finds best responses for the adversarial player (corresponds to Line 4 of Algorithm 1). We use the total number of edges as an example. Given the distribution of the predictor player, the algorithm computes the marginal probability of each edge (Line 2). Line 3 computes the "cost" (loss) for an edge existing and not existing for the edge generic and specific features. These two types of features can be computed independently for each edge. We sort all the edges according to the cost of including them (Line 5). Since there are also costs for having a different number of total edges, the algorithm considers each possible number of total edges (Line 8–16). When we consider m total edges, we select edges which will increase the cost the most. The total cost includes the costs for all the existing edges (Line 9), for non-existing edges (Line 10), and for having m number of edges. After we go through all the possible number of edges to include, we find the best overall cost and the corresponding number of edge m. After we select the number of edges to add, we add the top edges that will maximize the cost. For other global features, it is similar except different subset of edges are sorted by different criteria.

3.3   Relationship with TERGMs

TERGMs are mostly used for hypothesis testing. They can also be viewed as a network prediction approach with logloss as the loss function. TERGMs can model dependent structures, such as triangles, of the predicted network. TERGM uses pseudo-likelihood approximation or Markov Chain Monte Carlo maximum likelihood estimation. When optimizing logloss, our framework is equivalent to TERGM. However, our model can optimize other loss functions, such as cost-sensitive loss Additionally, our model is able to incorporate global summary features, such as number of total edges, while still obtaining the optimal solution, whereas TERGM does not have the capability to do so without approximation.

4   Experiments

We compare our approach with state-of-the-art baselines on five datasets (whose summary statistics shown in Table 3). Baboon **Proximity** [8,29] and **Follow** are collected using GPS collars on a group of baboons in Kenya. For the proximity network, there is an edge if two individuals are closer than 10 m. **Follow** edges are defined as trajectories of two baboons follow each other (using dynamic time warping). Haggle **Conference** [27] are networks of attendees at IEEE INFOCOM

Table 3. Summary statistics of the datasets.

Name	Nodes	# of training networks	# of testing networks	Time window
Proximity	16	277	70	20 min
Follow	16	277	70	20 min
Conference	41	114	28	30 min
Reality	96	614	157	8 h
Infectious	410	1124	298	20 s

Table 4. Weighted F_1 score.

	SP	CWT	CP	LR	TERGM	Adv
Follow	46.98%	44.07%	44.07%	44.07%	44.07%	**50.85%**
Proximity	16.99%	38.80%	38.80%	38.36%	41.48%	**45.16%**
Conference	29.16%	40.28%	40.24%	40.31%	38.11%	**46.54%**
Reality	43.55%	43.20%	43.08%	42.84%	42.82%	**51.86%**
Infectious	37.52%	39.82%	40.00%	37.06%	-	**41.63%**

2005 conference. **Reality** Mining [10] is a dataset of 96 students and faculty members (nodes) at MIT to study human interactions using smart phones. **Infectious** [15] is an interaction network when people visit the Science Gallery. These datasets are from diverse domains, ranging from human Bluetooth interactions over hours to animal GPS proximity on the order of seconds, which results in vastly different network structures and mechanisms. To choose a proper time window for defining the dynamic network, we employ the TWIN algorithm [7]. For the publicly available datasets, we split the whole sequence of networks based on time since there is one sequence of networks for each dataset. For baboon datasets, we randomly pick eight days as training and the rest as testing.

The state-of-the-art methods we compare with: **Structured Prediction (SP)** [17] uses frequent subgraphs to predict the most likely topology. **Collapsed Weighted Tensor (CWT)** [9] is a matrix factorization approach. **CANDECOMP/PARAFAC (CP)** [9] is based on tensor factorization. We use a Matlab Tensor Toolbox implementation [1,5]. **Logistic Regression (LR)** is a general classifier for all the edges with general edge features. **Temporal Exponential Random Graph Model (TERGM)** which we followed [18] and used bootstrap pseudolikelihood approach in **btergm** package in R [19] for the one-step-ahead prediction. Note that the Infectious network is too large for the tool to handle, so we do not report a result. **Adversarial Prediction**

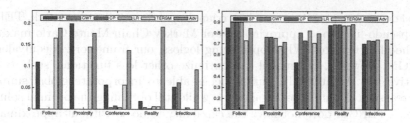

Fig. 3. F1 score for predicting new edges (left) and edge disappearances (right).

(**Adv**), proposed method, obtains each game equilibrium using Gurobi [11] and employs delay features with support of 0.9. We use grid search of a validation set for selecting the cost weights λ. We constrain the pure strategies in the double oracle method to 100 during training.

We use the average of the F_1 score (Table 4) for starting edge prediction and for ending edge prediction weighted by the number of start n_{start} and end occurrences n_{end}: $F_{weighted} = F_{start} * \frac{n_{start}}{n_{start}+n_{end}} + F_{end} * \frac{n_{end}}{n_{start}+n_{end}}$.

Our Adv approach performs better than all the baseline approaches on all datasets. Figure 3(left) shows the F_1 score of predicting new edges. For the CWT and CP methods, their F1-scores for edge addition are very low, since these two approaches predict that nearly no edge exists. Although the two approaches have high F1-scores when predicting edge removal, as shown in Fig. 3(right), the predictions are not very meaningful because predicting all the edges as non-existing is useless for real-world applications. Our Adv approach outperforms the baseline methods on predicting new edges, except for the proximity dataset, in which we are worse than the SP approach. However, the good performance of the SP approach is at the expense of edge removal performance. For predicting edge removals, our method performs better on two out of five datasets and shows competitive performance for the other datasets. Another common approach for evaluation is to use the summary statistics of the network. Table 5 shows the evaluation results of degree distribution correlation computed using GraphCrunch 2 [16]. Although our proposed approach is not designed to optimize global performance measure, our model performs better than the baseline approaches on three datasets. TERGM also performs well on global summary statistic as TERGMs are designed to model networks with similar statistics rather than focusing on

Table 5. Degree distribution correlation.

	SP	CWT	CP	LR	TERGM	Adv
Follow	0.0631	0	0	0	**0.0664**	0.0406
Proximity	−0.0092	0	0	0.0509	0.0286	**0.0962**
Conference	0	−0.2533	0	0.0741	**0.3867**	0.0303
Reality	0.5174	0	0.0033	0.4058	0.3109	**0.5864**
Infectious	0.1215	0.0034	0	0.0034	-	**0.1297**

Table 6. Average prediction time per network (in seconds).

	SP	CWT	CP	LR	TERGM	Adv
Follow	2.4e−3	1.2e−3	100.44	1.1e−3	0.03	0.35
Proximity	0.19	6.1e−4	100.31	7.3e−4	0.05	0.47
Conference	0.03	1.6e−3	358.06	7.4e−2	0.04	0.57
Reality	0.07	0.07	412.81	1.57	0.05	1.86
Infectious	0.09	0.04	1098.10	19.77	-	4.85

specific edges. Table 6 shows the average prediction time per network. Our approach takes a longer time compared to SP, CWT and TERGM, comparable with LR, and much shorter than CP. Since the network is defined in time windows with at least 20 s, Adv can be used for real-time prediction. Though we recognize the scalability limitations of our approach, a number of methods, ranging from basic engineering techniques to stochastic optimization methods, have promising potentials for future work.

5   Conclusions

The variety of network characteristics and applications make dynamic network prediction a challenging task. In this paper, we developed a robust approach to predict edge changes in dynamic networks. By having worst-case approximations of the available training data, it provides robustness guarantee to the inherent uncertainty of reasoning from limited amounts of noisy data, as encountered in many dynamic network settings. As future work, we plan to investigate marginal decompositions of our game [28] to improve scalability.

References

1. Acar, E., Dunlavy, D.M., Kolda, T.G.: A scalable optimization approach for fitting canonical tensor decompositions. J. Chemom. **25**(2), 67–86 (2011)
2. Adamic, L.A., Adar, E.: Friends and neighbors on the web. Soc. Nctw. **25**(3), 211–230 (2003)
3. Albert, R., Barabási, A.L.: Statistical mechanics of complex networks. Rev. Mod. Phys. **74**, 47–97 (2002)
4. Asif, K., Xing, W., Behpour, S., Ziebart, B.D.: Adversarial cost-sensitive classification. In: International Conference on UAI (2015)
5. Bader, B.W., Kolda, T.G., et al.: MATLAB Tensor Toolbox Version 2.6. Available online, February 2015
6. Bringmann, B., Berlingerio, M., Bonchi, F., Gionis, A.: Learning and predicting the evolution of social networks. IEEE Intell. Syst. **25**(4), 26–35 (2010)
7. Caceres, R.S., Berger-Wolf, T., Grossman, R.: Temporal scale of processes in dynamic networks. In: ICDMW, pp. 925–932. IEEE (2011)
8. Crofoot, M.C., Kays, R.W., Wikelski, M.: Data from: shared decision-making drives collective movement in wild baboons (2015)
9. Dunlavy, D.M., Kolda, T.G., Acar, E.: Temporal link prediction using matrix and tensor factorizations. TKDD **5**(2), 10 (2011)
10. Eagle, N., Pentland, A.: Reality mining: sensing complex social systems. Pers. Ubiquit. Comput. **10**(4), 255–268 (2006)
11. Gurobi Optimization, Inc.: Gurobi Optimizer Reference Manual (2015)
12. Hanneke, S., Fu, W., Xing, E.P., et al.: Discrete temporal models of social networks. Electron. J. Stat. **4**, 585–605 (2010)
13. Holme, P., Saramäki, J. (eds.): Temporal Networks. Understanding Complex Systems. Springer, Heidelberg (2013). https://doi.org/10.1007/978-3-642-36461-7
14. Huang, Z., Lin, D.K.: The time-series link prediction problem with applications in communication surveillance. INFORMS J. Comput. **21**(2), 286–303 (2009)

15. Isella, L., Stehlé, J., Barrat, A., Cattuto, C., Pinton, J.F., Van den Broeck, W.: What's in a crowd? Analysis of face-to-face behavioral networks. J. Theor. Biol. **271**(1), 166–180 (2011)
16. Kuchaiev, O., Stevanović, A., Hayes, W., Pržulj, N.: Graphcrunch 2: software tool for network modeling, alignment and clustering. BMC Bioinform. **12**(1), 24 (2011)
17. Lahiri, M., Berger-Wolf, T.Y.: Structure prediction in temporal networks using frequent subgraphs. In: CIDM 2007, pp. 35–42. IEEE (2007)
18. Leifeld, P., Cranmer, S.J., Desmarais, B.A.: Temporal exponential random graph models with btergm: Estimation and bootstrap confidence intervals. J. Stat. Softw. (2015)
19. Leifeld, P., Cranmer, S.J., Desmarais, B.A.: btergm: Temporal exponential random graph models by bootstrapped pseudolikelihood. R Package v. 1.9.0 (2017)
20. Li, J., Asif, K., Wang, H., Ziebart, B.D., Berger-Wolf, T.: Adversarial sequence tagging. In: IJCAI (2016)
21. Liben-Nowell, D., Kleinberg, J.: The link-prediction problem for social networks. J. Am. Soc. Inform. Sci. Technol. **58**(7), 1019–1031 (2007)
22. McMahan, H.B., Gordon, G.J., Blum, A.: Planning in the presence of cost functions controlled by an adversary. In: Proceedings of the ICML, pp. 536–543 (2003)
23. Nagrecha, S., Chawla, N.V., Bunke, H.: Recurrent subgraph prediction. In: Proceedings of the 2015 ASONAM, pp. 416–423. ACM (2015)
24. Parambath, S.P., Usunier, N., Grandvalet, Y.: Optimizing f-measures by cost-sensitive classification. In: Advances in NIPS, pp. 2123–2131 (2014)
25. Potgieter, A., April, K.A., Cooke, R.J., Osunmakinde, I.O.: Temporality in link prediction: understanding social complexity. Emerg.: Complex. Org. **11**(1), 69 (2009)
26. Pujari, M., Kanawati, R.: Link prediction in complex networks by supervised rank aggregation. In: ICTAI 2012, vol. 1, pp. 782–789. IEEE (2012)
27. Scott, J., Gass, R., Crowcroft, J., Hui, P., Diot, C., Chaintreau, A.: CRAWDAD dataset cambridge/haggle (v. 2009-05-29). Downloaded from http://crawdad.org/cambridge/haggle/20090529. Accessed May 2009
28. Shi, Z., Zhang, X., Yu, Y.: Bregman divergence for stochastic variance reduction: saddle-point and adversarial prediction. In: Advances in Neural Information Processing Systems, pp. 6033–6043 (2017)
29. Strandburg-Peshkin, A., Farine, D.R., Couzin, I.D., Crofoot, M.C.: Shared decision-making drives collective movement in wild baboons. Science **348**(6241), 1358–1361 (2015)
30. Wang, H., Xing, W., Asif, K., Ziebart, B.: Adversarial prediction games for multivariate losses. In: Advances in NIPS, pp. 2710–2718 (2015)
31. Yang, Y., Chawla, N.V., Basu, P., Prabhala, B., La Porta, T.: Link prediction in human mobility networks. In: ASONAM, pp. 380–387. IEEE (2013)
32. Yu, W., Cheng, W., Aggarwal, C.C., Chen, H., Wang, W.: Link prediction with spatial and temporal consistency in dynamic networks. In: Proceedings of the International Joint Conference on Artificial Intelligence, pp. 3343–3349. AAAI Press (2017)
33. Zhu, L., Guo, D., Yin, J., Ver Steeg, G., Galstyan, A.: Scalable temporal latent space inference for link prediction in dynamic social networks. IEEE Trans. Knowl. Data Eng. **28**(10), 2765–2777 (2016)

Targeted Influence Minimization in Social Networks

Xinjue Wang[1], Ke Deng[1(✉)], Jianxin Li[2], Jeffery Xu Yu[3], Christian S. Jensen[4], and Xiaochun Yang[5]

[1] RMIT University, Melbourne, Australia
s3308497@student.rmit.edu.au, ke.deng@rmit.edu.au
[2] University of Western Australia, Perth, Australia
jianxin.li@uwa.edu.au
[3] Chinese University of Hong Kong, Hong Kong, China
yu@se.cuhk.edu.hk
[4] Aarhus University, Aarhus, Denmark
csj@cs.aau.dk
[5] Northeastern University, Shenyang, China
yangxc@mail.neu.edu.cn

Abstract. An online social network can be used for the diffusion of malicious information like derogatory rumors, disinformation, hate speech, revenge pornography, etc. This motivates the study of influence minimization that aim to prevent the spread of malicious information. Unlike previous influence minimization work, this study considers the influence minimization in relation to a particular group of social network users, called *targeted influence minimization*. Thus, the objective is to protect a set of users, called *target nodes*, from malicious information originating from another set of users, called *active nodes*. This study also addresses two fundamental, but largely ignored, issues in different influence minimization problems: (i) the impact of a budget on the solution; (ii) robust sampling. To this end, two scenarios are investigated, namely unconstrained and constrained budget. Given an unconstrained budget, we provide an optimal solution; Given a constrained budget, we show the problem is NP-hard and develop a greedy algorithm with an $(1-1/e)$-approximation. More importantly, in order to solve the influence minimization problem in large, real-world social networks, we propose a robust sampling-based solution with a desirable theoretic bound. Extensive experiments using real social network datasets offer insight into the effectiveness and efficiency of the proposed solutions.

1 Introduction

Online social networks can be used for the diffusion of not only positive information such as innovations, news, and novel ideas, but also malicious information such as disinformation and hate speech. Research on maximizing the influence of positive information, called *Influence Maximization*, offers insight to social

© Springer International Publishing AG, part of Springer Nature 2018
D. Phung et al. (Eds.): PAKDD 2018, LNAI 10939, pp. 689–700, 2018.
https://doi.org/10.1007/978-3-319-93040-4_54

network users on how to best propagate the awareness of products and services and has attracted substantial attention [2,4,8,11]. Likewise, the problem of reducing the influence of negative information, called *Influence Minimization*, is also attracting attention [5–7,9,14]. One line of studies on influence minimization aims to find a certain number of edges in social networks such that by deleting these edges, the influence of any information is minimized at the end of the propagation process, no matter which nodes initially have the information [5–7]. An other line of studies assume that a specific set of nodes initially have some information to be spread, The aim is then to delete a certain number of edges or nodes such that the influence of the information is minimized while considering the topics of the information [14] or considering the spread of counter-information from competitors in the same period of time [9,12].

Unlike the above works, we propose, define, and solve a new problem of so-called targeted influence minimization. This problem and its solutions are relevant to many applications. For example, a government agent may want to shield young social network users from pornography or recruitment to terrorism; or a company may initiate a campaign to protect their customers from defamatory information spread by their competitors. The targeted influence minimization problem can be briefly described as follows: given a set of source nodes I with information to be spread and a set of target nodes T in a social network, the aim is to find the minimum set of edges under a budget constraint such that deleting these edges minimizes the influence from I to T. The deletion of an edge (u_1, u_2) can be considered as persuading u_1 does not spread any information to u_2, or u_2 does not accept any information from u_1. Note that T may include all nodes other than I in a social network in the extreme case. Suppose a set of nodes I regularly spread information for business B_1. A competitor B_2 may initiate a campaign to prevent such information from a set of target nodes T, such as the customers of B_2. To do that, it needs find a set of edges under the campaign budget such that these edges will not pass any information related to B_1. As a consequence, the influence from I to T can be reduced to the minimum level.

All existing studies on influence minimization simply assume the budget is insufficient and provide a greedy algorithm. However, this assumption is not always true. We develop an optimal solution to completely block propagated information for the target users if the budget is sufficient. Otherwise, the problem is proved to be NP-hard, and a greedy algorithm is developed. To meet the time requirement in handling large social network data, a novel sampling based solution is provided.

The rest of the paper is organized as follows. We define the problem of targeted influence minimization in Sect. 2, solve the problem when the budget is unconstrained in Sect. 3 and when it is constrained in Sect. 4. Section 5 develops an efficient sampling based solution to enable scalability to large social networks. Finally, we evaluate the effectiveness and efficiency of our proposed solutions using real social network data in Sect. 6 and conclude in Sect. 7.

2   Problem Definition

A social network is modeled as a directed graph $G = (V, E)$, where V is a set of nodes and $E \subseteq V \times V$ is a set of edges. A set of nodes $I \subseteq V$ are called active nodes and have information to be diffused in the social network. Another set of nodes $T \subseteq V$, $I \cap T = \emptyset$, are called target nodes and are the recipients of interest.

2.1   Diffusion Model

We assume the Linear Threshold (LT) diffusion model [4]. Thus, each edge (u, v) comes with a weight $b_{u,v} \in [0, 1]$ to represent the influence u has on v. If a message is from u, the influence of this message on v is added by $b_{u,v}$. If the message is from all neighbors of v, denoted as $Adj(v)$, then influence, $v.inf$ of the message on v is $\sum_{u \in Adj(v)} b_{u,v}$. An activation threshold, $v.\tau$, is associated with v. If $v.inf \geq v.\tau$, v is activated; otherwise, v is not activated.

It has been shown that diffusion in the LT model is equivalent to the process of reachability under random choice of live edges in graph instances [4]. Given a graph $G = (V, E)$, each node $v \in V$ selects at most one of its incoming edges at random, choosing the edge connecting u to v with probability $b_{u,v}$ and not choosing any other edge with probability $1 - \sum_{u \in Adj(v)} b_{u,v}$; the chosen edge is called *live*. After processing each node in V this way, a graph instance G_x containing only the live edges and all the nodes in G is generated. In G_x, suppose a set of nodes I are active initially; an inactive node $u \in V$ ends up as active if and only if G_x contains a path from any node in I to u.

The set of all graph instances that can be generated from G is denoted as χ_G. The influence of I to a set of nodes $T \subseteq (V \setminus I)$ in graph G under the LT diffusion model is defined as follows:

$$\Lambda_G(I, T) = \sum_{G_x \in \chi_G} Prob[G_x] r_{G_x}(I, T), \qquad (1)$$

where $r_{G_x}(I, T)$ is the number of nodes in T reachable from any node in I in graph instance G_x, and $Prob[G_x]$ is the probability of graph instance G_x.

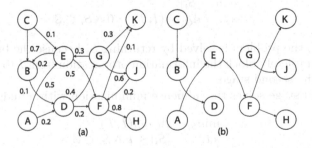

Fig. 1. A social network and an instance graph.

Figure 1(a) illustrates a social network, and an instance graph using the LT diffusion model is shown in Fig. 1(b). The probability of the instance graph is 0.000504. Suppose $I = \{A, B, C\}$ and $T = \{K, J, H\}$. Then K and H are reachable from A, while J is not reachable from any node in I.

2.2 Targeted Influence Minimization

A social network $G = (V, E)$ from which a subset of edges $S \subseteq E$ has been deleted is denoted as $G(S)$.

Definition 1 (Targeted Influence Minimization (TIMin)). *Given a social network $G = (V, E)$, a set of active nodes $I \subseteq V$, a set of target nodes $T \subseteq \{V \backslash I\}$ and a positive real number k as a budget, suppose $\mathbb{S} = \{S_1, S_2, \cdots, S_n\}$ contains all possible sets of edges where $|S_i| \leq k$, $1 \leq i \leq n$;*

- *if there does not exist $S_i \in \mathbb{S}$ such that $\Lambda_{G(S_i)}(I, T) = 0$, TIMin aims to find the set $S_* \in \mathbb{S}$ such that $\Lambda_{G(S_*)}(I, T)$ is minimal;*
- *if a set $S_i \in \mathbb{S}$ exist such that $\Lambda_{G(S_i)}(I, T) = 0$, TIMin aims to find a set $S_* \in \mathbb{S}$ such that $\Lambda_{G(S_*)}(I, T) = 0$ and $|S_*|$ is minimal.*

In the former case, the budget is insufficient to completely block the information propagation from I to T. In the latter case, the budget is sufficient to do so. As an example, consider Fig. 1, where $I = \{A, B, C\}$ and $T = \{K, J, H\}$. A budget $k = 2$ is insufficient to completely block the information propagation from I to T. Thus, TIMin aims to find the set of edges S_* such that $\Lambda_{G(S_*)}(I, T)$ is minimized. Given a budget of $k = 10$, there are many sets of edges that, if deleted, will completely block the information propagation from I to T. In this situation, among all such sets of edges, TIMin aims to find one with the minimum number of edges.

Given active nodes I and target nodes T, we initially need to determine whether the budget k is sufficient or not since this is not known in advance. This leads to the following processing framework.

1. The first stage solves the influence minimization with am unconstrained budget, defined as follows.

$$\min \quad |S_i| \qquad\qquad\qquad (2)$$
$$s.t. \quad \Lambda_{G(S_i)}(I, T) = 0 \wedge S_i \subset \mathbb{S}$$

If $|S_i| \leq k$, the problem is solved by returning S_i because the budget is sufficient to completely block the information propagation from I to T; otherwise, we go to the second stage.

2. The second stage solves the influence minimization with a budget k, defined as follows.

$$\min_{S_i} \quad \Lambda_{G(S_i)}(I, T) \qquad\qquad\qquad (3)$$
$$s.t. \quad |S_i| \leq k \wedge S_i \subset \mathbb{S}$$

3   Budget Unconstrained Solution

We first examine whether the budget is sufficient to completely block the information propagation from I to T. For this purpose, TIMin with unconstrained budget (i.e., $k = \infty$) is solved as a *minimum cut* or *maximum flow* problem. Let s and t be a source node and sink node in a flow network, respectively. In optimization theory, the *max-flow min-cut theorem* states that the maximum amount of flow passing from the source to the sink is equal to the total weight of the edges in the minimum cut, i.e., equal to the smallest total weight of the edges that, if removed, would disconnect the source from the sink [10]. If multiple sources and multiple sinks exist, the problem is transformed into a single-source and single-sink maximum flow problem by adding two new nodes: one connecting all source nodes and the other connecting all sink nodes; the weights of the new edges connected to the two new nodes are ∞.

Lemma 1. *Given a social network $G = (V, E)$, a set of source nodes $I \subseteq V$, and a set of target nodes $T \subseteq \{V \setminus I\}$, the influence minimization is equivalent to the minimum cut problem if budget $k = \infty$.*

The *minimum cut* or *maximum flow* problem is well studied [10]. We adopt Dinic's algorithm to solve this problem [3].

4   Budget Constrained Solution

Theorem 1. *TIMin with an insufficient budget k is NP-hard.*

Due the result in Theorem 1, we provide a greedy algorithm to solve targeted influence minimization with an insufficient budget. The greedy algorithm searches for a set of edges $S \subseteq E$ such that $|S| \leq k$ and the following objective function is maximized.

$$f(S) = \Lambda_G(I, T) - \Lambda_{G(S)}(I, T), \tag{4}$$

where $\Lambda_G(.,.)$ is computed using Eq. 1.

The greedy algorithm proceeds iteratively. Initially, S is empty. In each iteration, it computes the value of each edge e in $G(S)$ as follows.

$$value(e) = \Lambda_{G(S)}(I, T) - \Lambda_{G(S')}(I, T), \tag{5}$$

where $S' = S \cup \{e\}$. The value of e, $value(e)$, is the reduction of influence from I to T with and without e in $G(S)$. Among all edges, the one with the maximum value, say e_*, is deleted. Then e_* is inserted into S, and the remaining budget is decremented by 1. The process terminates when the remaining budget reaches 0. The greedy algorithm is an $(1 - 1/e)$-approximation (≈ 0.632-approximation) since the objective function is non-negative, monotonous, and submodular [5].

5   Sampling-Based Solution

It is prohibitively expensive to directly generate all graph instances and compute the value of each edge in each iteration. Therefore, we devise a sampling-based solution. The solution is inspired by a recent influence maximization study [13], but significant adaptions are required.

Reverse Influence Set (RIS). [13] aim to select at most k nodes with maximum influence in a social network. The method is based on RIS that computes the influence of nodes using graph instances. Specifically, the reverse reachable (RR) node set for each node in each graph instance is generated. Given a node v in graph instance G_x, the RR set contains all nodes in G_x that can reach v. Using the sampling method, a number of nodes are randomly selected from V; the RR set for each node is generated using a randomly selected graph instance. So, a number of random RR sets are obtained. If a node u has a great impact on other nodes, u will have high probability of appearing in the random RR sets. As a result, the problem is transformed to the *maximum coverage problem* of identifying at most k nodes that cover the maximum number of the random RR sets. It has been shown that if the number of random RR sets θ is no less than $(8+2\varepsilon)|V|\frac{\ln|V|+\ln\binom{|V|}{k}+\ln 2}{OPT_k\varepsilon^2}$, then RIS returns an $(1-1/e-\epsilon)$-approximate solution with at least $1-|V|^{-1}$ probability ($\epsilon \in (0,1)$) [1].

5.1   Minimum Influence Path

RIS cannot be applied to our problem without significant modification due to two reasons.

- The random RR set is about node-to-node reachability. In our problem, however, we delete the edges to make reachable-nodes unreachable. While it is straightforward to determine node-to-node reachability, it is more difficult to identify edges the deletion of which makes reachable-nodes unreachable. The reason is that there may be many different paths between two reachable nodes, so deleting an edge does not necessarily block the reachability.
- The random RR set is for the reachability of any node. In our problem, however, only the source nodes I and the target nodes T are relevant.

We propose a novel sampling-based method called *Minimum Influence Path* (MIP) to solve TIMin. The idea is to exploit the fact that each node in a graph instance under the LT diffusion model has at most one incoming edge. Specifically, each node $v \in V$ in the graph instance generation process picks at most one of its incoming edges at random, selecting the edge from $w \in Adj(v)$ with probability $b_{w,v}$, and selecting no edge with probability $1 - \sum_{w \in Adj(v)} b_{w,v}$. Figure 1 (b) shows an example.

As a result, for two nodes v and u, if v is reachable from u in the graph instance, it is easy to observe that the following properties hold: (i) there is one and only one path from u to v in the graph instance, and (ii) the path is acyclic. Therefore, the information propagation from u to v in this graph instance can be blocked by removing any edge in the path. On the other hand, if v is not reachable from u in the graph instance by deleting an edge e, this does not indicate that v is not reachable from u in other graph instances. However, if v is not reachable from u in many graph instances by deleting e, this implies that the information propagation from v to u is less likely to happen even though it is not impossible. So, the problem is to delete those edges that block the paths from source nodes to target nodes are blocked in many graph instances.

On the other hand, if v is not reachable from u in the graph instance, the information propagation is blocked without deleting any edge. This may occur for two reasons. First, v is not reachable from u in graph G. Second, v is not reachable from u in this graph instance. If v is not reachable from u in many graph instances, this implies that the information propagation from v to u is less likely to happen even though it is not impossible.

Given a node in $v \in T$, the *minimum influence path* in a graph instance is the path to v from any node $u \in I$ with the fewest edges. Figure 2(a) shows a graph instance where $I = \{u_1, u_2, u_3\}$ and $T = \{v_1, v_2, v_3, v_4\}$. The minimum influence path from I to each target node is shown in Fig. 2(b). The minimum influence path to v_1 is (e_1, e_2, e_3). Cutting any edge in the minimum influence path will prevent I from reaching v_1 in this graph instance. Intuitively, the edge appearing in more minimum influence paths is more likely to, if deleted, lead to the more influence reduction. In this graph instance, edge e_5 appears in the minimum influence paths of v_2 and v_3 such that deleting e_5 prevents I from reaching two nodes. If deleting e_5 prevents I from reaching many nodes in T in other graph instances, e_5 is likely to be the edge in the solution of MIP.

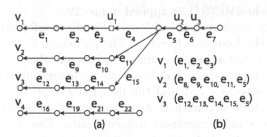

(a)   (b)

Fig. 2. Reverse influence paths.

5.2 Sampling-Based Greedy Algorithm

The pseudo-code of the sampling-based greedy algorithm is presented in Algorithm 1. First, we randomly generate a graph instance in lines 5–7. One node in T is selected randomly in line 8, and the minimum influence path of this node is generated in line 9. This way, θ nodes have been sampled, and the minimum influence path is generated for each of them. Note that a graph instance is more

---

Algorithm 1. Sampling-based Solution

  Input: $G = (V, E)$, I, T, k, θ
  Output: S_*

1 $i \leftarrow 0$
2 $M \leftarrow \emptyset$
3 **while** $i \leq \theta$ **do**
4     $j \leftarrow 0$
     // generate a graph instance i
5   **foreach** $v \in V$ **do**
6   **if** *generateEdge()* **then**
7   randomly select $w \in Adj(v)$ with probability $b_{w,v}$

8     $u \leftarrow$ randomly select a node in V
9     $u.M \leftarrow$ minInfPath(u)
10    $M \leftarrow M \cup u.M$
11    $i \leftarrow i + 1$

12 $S_* \leftarrow$ incrementalMC(M)
13 **return** S_*

---

likely to be selected if the probability of the graph instance is high. If deleting an edge can prevent I from reaching many nodes in T in many graph instances, this edge is more likely to appear in the minimum influence paths. So, the problem is transformed to the *maximum coverage problem* of selecting at most k edges to cover the sampled nodes as many as possible. In our solution, we assume that the specified budget is sufficient, otherwise, the budget unconstrained solution is applied. To this end, the incremental solution of maximum coverage problem, known as incrementalMC(M), is applied in line 12.

The maximum coverage problem is solved using an adapted greedy algorithm that is aware of the budget sufficiency. The pseudo-code is presented in Algorithm 2. The generated minimum influence paths and the corresponding reverse minimum influence paths are used. For each minimum influence path, the algorithm maintains a node $v \in I$ and the list of the edges in the path. For each reverse minimum influence path, it maintains an edge e and a list of the nodes each of which has e in its minimum influence path. The reverse influence minimum paths are constructed while the influence minimum paths are generated (line 2). First, the edge with the longest reverse minimum influence paths is moved to solution S_* (lines 7–8). Then, the nodes in the reverse minimum influence path are processed by finding their minimum influence paths and removing them (line 9); for any edge in the minimum influence paths, its reverse minimum influence paths is found and updated (lines 10–13). The process is repeated until k edges are selected (line 4) or no complete path exists in the remaining influence minimum paths (line 5). The budget sufficiency awareness is implemented by checking whether no complete path exists.

Algorithm 2. Incremental Maximum Coverage

  Input: Mip
  Output: S_*

1   $i = 0$
2   RMip ← construct reverse minimum influence path
3   $S_* ← \emptyset$
4   **while** $i \leq k$ **do**
5   **if** *no complete path in Mip* **then**
6       $\lfloor$ break;
7   rp_e ← the longest path in RMip
8   delete rp_e from RMip
9     $S_* ← S_* \cup e$
10   **foreach** $v \in p_e$ **do**
11   delete path p_v from Mip
12   **foreach** $e \in p_v$ **do**
13        $\lfloor$ rp_e ← delete v from rp_e

14   **return** S_*

Theorem 2. *If $|S_*| \leq k$, the probability that the information propagation from I to T is completely blocked is at most $\frac{1}{n}$; the $|S_*|$ is an $\frac{1}{n}$-approximation of the optimal solution.*

6 Experimental Study

We evaluate the effectiveness and efficiency of our proposed algorithms by comparing with two heuristic algorithms called *Random* and *Weight*. *Random* selects edges randomly until budget k is used. *Weight* selects edges with largest edge weights. Their performance are evaluated in different parameter settings using three real-world networks: *Wiki* with 7,115 nodes and 103,689 edges, *Ego-twitter* with 23,370 nodes and 33,101 edges, and *Epinions* with 75,879 nodes and 508,837 edges. All the three datasets are downloaded from the *Stanford Dataset Collection*[1].

6.1 Evaluation of Effectiveness

Varying k: Figure 3 shows the experimental results when varying k while the source node set I and the target node set T are fixed in size at 500 unless stated otherwise. The source and target nodes are selected randomly.

The study shows that *Greedy* and *Sampling* are able to greatly reduce the influence of I on T for all three datasets given a sufficiently large value of k. When k is above 100, both solutions are able to reduce the influence by up to

---

[1] http://snap.stanford.edu/data/.

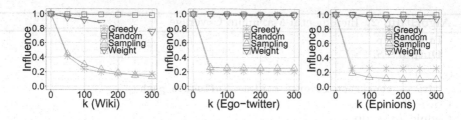

Fig. 3. Remaining influence from I on T when varying k.

80%. Next, *Random* and *Weight* can slightly reduce the influence in *Wiki*. They do not work for *Ego-twitter* and *Epinions*. *Random* and *Weight* cannot block the influence well because the selection of their deleted edges are not relevant to target users. However, this matter is taken into account in *Greedy* and *Sampling*. So the influences minimized by *Greedy* or *Sampling* are always larger than that of *Random* or *Weight*.

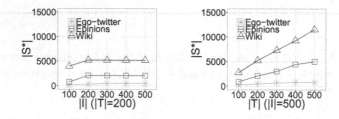

Fig. 4. #edges deleted for unconstrained budget.

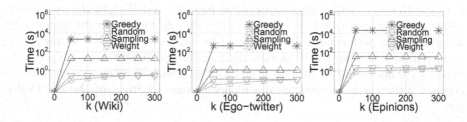

Fig. 5. Time cost when varying k.

Budget Unconstrained Evaluation: As shown in Fig. 4, we can see that the influence from I to T can be blocked completely by deleting a certain number of edges. When $|I| = 500$, $|T| = 100$, it requires 243 edges for *Ego-Twitter*. But more edges must be deleted for *Epinions* and *Wiki* because *Ego-Twitter* dataset is much sparse than *Epinions* or *Wiki* datasets. In order to minimize the influence of I on T in the same parameter settings, it has to delete more edges so that all the paths connecting from I to T can be disconnected. However,

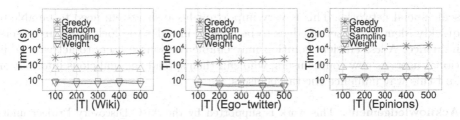

Fig. 6. Time cost when varying $|T|$.

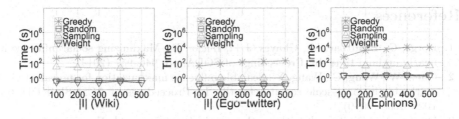

Fig. 7. Time cost when varying I.

when $|T|$ becomes large, it is quite challenging to completely block the initial users' influence on the target users because a large number of edges need to be deleted. Our sampling solution can be applied to block the majority of the influence.

6.2 Evaluation of Efficiency

We evaluate the efficiency of the four solutions when varying k, T, and I. Figures 5, 6 and 7 present the results. Our sampling solution is capable of outperforming the greedy solution by 2 orders of magnitude in all datasets. Both solutions are stable in performance when we increase k. But the time cost of *Greedy* grows with the increase of T or I. Compared with *Greedy* and *Sampling*, *Random* and *Weight* have the best efficiency because their deleted edges can be found without too much computation. But, as we have seen, their lack of effectiveness render them of little use. Therefore, the sampling solution is the best choice for targeted influence minimization in terms of effectiveness and efficiency.

7 Conclusion

In this work, we propose and formalize the problem of targeted influence minimization in social networks that has not previously been studied. We present different solutions that address the computational challenges associated with this problem. We report on empirical studies showing that the proposed solution is capable of quickly blocking 80% or more the influence of source users on target users. The proposed sampling-based solution is efficient when applied to large scale social networks. This is very important because system need to be able to quickly identify the set of edges to be deleted in order to block the source users' influence. A less efficient solution may enable the source users to activate additional users as new source users, who can then spread the malicious information and this way influence the target users.

Acknowledgement. This work is supported by the ARC Discovery Project under grant No. DP160102114.

References

1. Borgs, C., Brautbar, M., Chayes, J.T., Lucier, B.: Maximizing social influence in nearly optimal time. In: Proceedings of SODA, pp. 946–957 (2014)
2. Chen, W., Wang, C., Wang, Y.: Scalable influence maximization for prevalent viral marketing in large-scale social networks. In: Proceedings of ACM SIGKDD, pp. 1029–1038 (2010)
3. Dinitz, Y.: Dinitz' algorithm: the original version and Even's version. In: Goldreich, O., Rosenberg, A.L., Selman, A.L. (eds.) Theoretical Computer Science, Essays in Memory of Shimon Even. LNCS, vol. 3895, pp. 218–240. Springer, Heidelberg (2006). https://doi.org/10.1007/11685654_10
4. Kempe, D., Kleinberg, J., Tardos, É.: Maximizing the spread of influence through a social network. In: Proceedings of ACM SIGKDD, pp. 137–146 (2003)
5. Khalil, E., Dilkina, B., Song, L.: CuttingEdge: influence minimization in networks. In: Proceedings of Workshop on Frontiers of Network Analysis: Methods, Models, and Applications at NIPS (2013)
6. Kimura, M., Saito, K., Motoda, H.: Minimizing the spread of contamination by blocking links in a network. In: Proceedings of AAAI, pp. 1175–1180 (2008)
7. Kimura, M., Saito, K., Nakano, R.: Extracting influential nodes for information diffusion on a social network. In: Proceedings of AAAI, pp. 1371–1376 (2007)
8. Li, Y., Zhang, D., Tan, K.: Real-time targeted influence maximization for online advertisements. PVLDB 8(10), 1070–1081 (2015)
9. Luo, C., Cui, K., Zheng, X., Zeng, D.D.: Time critical disinformation influence minimization in online social networks. In: Proceedings of JISIC, pp. 68–74 (2014)
10. Papadimitriou, C.H., Steiglitz, K.: The max-flow, min-cut theorem. In: Combinatorial Optimization: Algorithms and Complexity, pp. 117–120. Prentice-Hall (1982)
11. Shirazipourazad, S., Bogard, B., Vachhani, H., Sen, A., Horn, P.: Influence propagation in adversarial setting: how to defeat competition with least amount of investment. In: Proceedings of ACM SIGMOD, pp. 585–594 (2012)
12. Song, C., Hsu, W., Lee, M.: Temporal influence blocking: minimizing the effect of misinformation in social networks. In: Proceedings of IEEE ICDE, pp. 847–858 (2017)
13. Tang, Y., Xiao, X., Shi, Y.: Influence maximization: near-optimal time complexity meets practical efficiency. In: Proceedings of ACM SIGMOD, pp. 75–86 (2014)
14. Yao, Q., Shi, R., Zhou, C., Wang, P., Guo, L.: Topic-aware social influence minimization. In: Proceedings of WWW 2015 Companion, no. 1, pp. 139–140 (2015)

Maximizing Social Influence
on Target Users

Yu-Ting Wen[(✉)], Wen-Chih Peng, and Hong-Han Shuai

Department of Computer Science, National Chiao Tung University, Hsinchu, Taiwan
{ytwen,wcpeng}@cs.nctu.edu.tw, hhshuai@nctu.edu.tw

Abstract. Influence maximization has attracted a considerable amount of research work due to the explosive growth in online social networks. Existing studies of influence maximization on social networks aim at deriving a set of users (referred to as seed users) in a social network to maximize the expected number of users influenced by those seed users. However, in some scenarios, such as election campaigns and target audience marketing, the requirement of the influence maximization is to influence a set of specific users. This set of users is defined as the target set of users. In this paper, given a target set of users, we study the Target Influence Maximization (TIM) problem with the purpose of maximizing the number of users within the target set. We particularly focus on two important issues: (1) how to capture the social influence among users, and (2) how to develop an efficient scheme that offers wide influence spread on specified subsets. Experiment results on real-world datasets validate the performance of the solution for TIM using our proposed approaches.

1 Introduction

Through the powerful word-of-mouth effect in social networks, both industrial and academic communities have been prompted to pay close research attention to information dissemination and market recommendation. In influence maximization, the goal is to select a subset of nodes in a social network that maximizes the spread of influence [7]. However, most of the relevant studies have focused on the entire social network, and have neglected the fact that real-world applications such as virtual marketing often want to promote their products to a certain group of customers. For example, a social network system (e.g., Facebook) wants to provide companies (e.g., travel agency) with marketing services by targeting their potential customers, i.e., the *target user set*, who had clicked the ad or had logged into their web-service, to promote their businesses. When a company has a limited budget for the initial "seed users", it is critical to effectively select those users (who may not be in the target user set); the selected users can then influence their friends, their friends' friends and so on, resulting in the maximal influence spread in the target user set.

In this paper, we formulate the *Target Influence Maximization* (TIM) problem that aims to maximize the influences over a target set of users in a social

© Springer International Publishing AG, part of Springer Nature 2018
D. Phung et al. (Eds.): PAKDD 2018, LNAI 10939, pp. 701–712, 2018.
https://doi.org/10.1007/978-3-319-93040-4_55

network, which refers to a subset of the entire social network. Consider an example of social network in Fig. 1, where the target user set $\mathcal{T}$ is $\{0, 3, 7, 9, 10\}$ and the number of seed users k is 2, the goal is to identify the two seeds to influence the users in the target set $\mathcal{T}$. As a result, the optimal seed set is $\{4, 10\}$ which differs from the global influential nodes $\{4, 8\}$. We will explain this example in detail in Sect. 2.

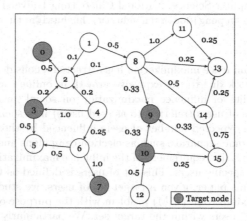

Fig. 1. An example of a target influence social network

There are two main issues to address target influence maximization. The first is to obtain the social influence among users. In real life, although the graph representing the structure of a social network is often explicitly available, the edge weight representing the degree of influence is difficult to determine. Since we focus on the influence spread on a specified subset of the network, we aim to learn more accurate results that consider the propagation trace from real data rather than constant values or intuitively assigned node in-degree numbers. By borrowing some concepts in recent works on modeling the influence flows [5,12,13], we further derive propagation probabilities based on users' historical actions, such as "repost/like/follow" on posts, place visits among location-based social network or membership of groups. The second issue is to efficiently derive the set of k seeds while maximizing the number of influenced users in the specified target user set. Existing efficient influence maximization algorithms [1,2,7] cannot meet the efficiency requirement because prior works still need to enumerate a large number of users as the seed set and measure the corresponding influence spread. Thus, although there is relevant literature on viral marketing focused on certain types of users such as the location-aware [8] and keyword-aware influence maximization problems [9], these algorithms can not be directly applied to the target influence maximization problem. Therefore, TIM calls for effective methods to achieve high performance while not sacrificing much influence spread.

In summary, the contributions of this paper are three-fold:

- We formulate the problem of Target Influence Maximization. We first analyze multiple real-world datasets, with the results showing the similar influence propagation process to neighbor nodes, motivating our algorithm design.
- We quantitatively capture the social influence on a user by leveraging information embedded in the social network. Since both user behavior and contents shared by users should be taken into consideration, the decision making of user behavior is modeled by a probabilistic methodology.
- To speed up the selection of k seeds, we build cluster-based indexes using precomputed information on user behavior clusters to improve the search performance.

2   Problem Formulation

We model a social network as a directed graph $G = (V, E)$, where the vertices in V are users and the edges in E are *followee-follower* relationships. For example, edge (u,v) with weight 1.0 represents that user v has a 100% chance to be influenced (activated) if user u is influenced (activated), where u is the influencer and v is the follower. If a user u is selected as a seed, u becomes active and it will also have the chance to activate its out-neighbors. If u's out-neighbor v becomes active, v will in turn activate the out-neighbors of v to propagate the influence. There are many methods to model this process such as the Independent Cascade (IC) model, the Linear Threshold (LT) model, Diffusion, etc. We adopt the widely-studied IC model [7] with learned edges weights (also known as the Weighted Cascade model) in this paper. Consider a vertex u first becomes active in step t, the influence is independently propagated from vertex u to its currently inactive neighbors $\mathcal{N}_u = \{v_1, v_2, ...\}$. The node u has a single chance to activate each of its inactive neighbors. Node $v_i \in \mathcal{N}_u$ becomes active with the activation probability p_{u,v_i} along the edge (u, v_i) at step $t+1$. No matter whether u succeeds in activating v_i or not, u cannot make any further attempts on v_i. As time unfolds, more and more of the nodes become active, and the process terminates when there is no newly activated vertex.

Formally, given a query $Q = (\mathcal{T}, k)$ with the target user set $\mathcal{T}$ and an integer k. The goal is to find a set of k seeds $\mathcal{S}$ from the graph, i.e., a subset of V with k nodes, to activate the maximum number of vertices in $\mathcal{T}$. The number of activated vertices in $\mathcal{T}$ is called *influence spread*, denoted as $\sigma(\mathcal{S}, \mathcal{T})$. As each vertex has a probability of being activated through multiple vertices, it is necessary to compute the expected number of $\sigma(\mathcal{S}, \mathcal{T})$. The target influence maximization problem is formulated as follows:

Definition 1. *Target Influence Maximization.* *Given a social network G and a query $Q = (\mathcal{T}, k)$, find a k-vertex set $\mathcal{S}^* \in G$, such that for any other k-vertex set $\mathcal{S} \in G$, $\sigma(\mathcal{S}^*, \mathcal{T}) \leq \sigma(\mathcal{S}, \mathcal{T})$. $\mathcal{S}^*$ is called a seed set, and each vertex in $\mathcal{S}^*$ is called a seed.*

Differing from existing influence maximization algorithms [7] which compute the top-k vertices to maximize the influence spread on all vertices, we focus on maximizing the influence spread on a given specific target user set.

3   Probabilistic Social Influence Model

The goal of this section is to predict the expected influence spread of nodes precisely. Rather than assuming edge probabilities to be randomly or to be determined by node degrees, we use real-world propagation trace to simulate the possible world. In other words, we can think of a propagation trace as a possible outcome of a set of probabilistic choices.

Definition 2. *Probabilistic Social Influence Model. Given the social graph $G = (V, E)$ with additional action logs $\mathcal{A}$ as propagation trace, let $\mathcal{X}$ denote the set of all possible worlds. The probabilistic social influence model (PSI) rebuilds G by computing edge weights p of directed links, which represent the probability of activation.*

Then, with the directed and edge-weighted social network $G = (V, E, p)$ independently of the model m chosen, the expected spread $\sigma_m(\mathcal{S})$ can be written as:

$$\sigma_m(\mathcal{S}) = \sum_{\mathcal{X} \in G} Pr[\mathcal{X}] \cdot \sigma_m^{\mathcal{X}}(\mathcal{S}) \tag{1}$$

where $\sigma_m^{\mathcal{X}}(\mathcal{S})$ is the number of nodes reachable from S in the possible world $\mathcal{X}$. The number of possible worlds is clearly exponential. Indeed, computing $\sigma_m(\mathcal{S})$ under the IC models is #P-hard [1,2], and the standard approach [7] tackles influence spread computation from the perspective of Eq. 1: sample a possible world $X \in G$, compute $\sigma_m^{\mathcal{X}}(S)$, and repeat until the number of sampled worlds is large enough. We now develop an alternative approach for computing influence spread, by rewriting Eq. 1, giving a different perspective. Let $path(S, u)$ be an indicator random variable that is 1 if there exists a directed path from the set S to u and 0 otherwise. Moreover let $path_X(S, u)$ denote the outcome of the random variable in a possible world $\mathcal{X} \in G$. Then we have:

$$\sigma_m(\mathcal{S}) = \sum_{u \in V} E[path(\mathcal{S}, u)] = \sum_{u \in V} Pr[path(\mathcal{S}, u) = 1] \tag{2}$$

That is, the expected spread of a set $\mathcal{S}$ is the sum over each node $u \in V$ of the probability of the node u getting activated given that $\mathcal{S}$ is the initial seed set.

The standard approach samples possible worlds from the perspective of Eq. 1. To leverage the available data on real propagation traces, we observe that these traces are similar to possible worlds, except that they are "real available worlds". Thus, in this paper, we approach the computation of influence spread from the perspective of Eq. 2, i.e., we estimate directly $Pr[path(S, u) = 1]$ using the propagation traces that we have in the action log.

3.1 Assigning Direct Weight

In order to estimate $Pr[path(S, u) = 1]$ using available propagation traces, it is natural to interpret such quantity as the fraction of the actions initiated by S that propagated to u, given that S is the seed set. More precisely, we could estimate this probability as

$$\frac{|a \in A|initiate(a, S)\&\exists t : (u, a, t)|}{|a \in A|initiate(a, S)|}$$

where $initiate(a, S)$ is true iff S is precisely the set of $initiators$ of action a. Unfortunately, this approach suffers from a sparsity issue which is intrinsic to the influence maximization problem [10]. If we need to be able to estimate $Pr[path(S, u) = 1]$ for any set S and node u, we will need an enormous number of propagation traces corresponding to various combinations, where each trace has as its $initiator$ set precisely the required node set S. It is clearly impractical to find a real action log where this can be realized. To overcome this obstacle, we propose a different approach to estimating $Pr[path(S, u) = 1]$ by taking a "u-centric" perspective: we assign weights to the possible $influencers$ of a node u whenever u performs an action.

We now revisit the problem of defining the direct weight $\gamma_{v,u}(a)$ given by a node u to a neighbor v for action a. The observation of [5] shows that influence decays over time in an exponential fashion and that some users are more easily influenced than others. Motivated by these ideas, we propose to assign direct weight as:

$$\gamma_{v,u}(a) = \frac{infl(u)}{N_{in}(u, a)} \cdot \exp(-\frac{t(u, a) - t(v, a)}{\tau_{v,u}})$$

Here, $\tau_{v,u}$ is the average time taken for actions to propagate from user v to user u. The exponential term in the equation achieves the desired effect that influence decays over time. $infl(u)$ denotes the user influenceability, that is, how prone the user u is to be influenced by the social context. Precisely, $infl(u)$ is defined as the fraction of actions that u performs under the influence of at least one of its neighbors v, i.e., u performs the action a, such that $t(u, a) - t(v, a) \leq \tau_{v,u}$; note that the direct weight is normalized by $N_{in}(u, a)$, i.e., all neighbors of u that have performed the same action a before u. This is to ensure the sum of direct weights assigned to neighbors of u for action a is at most 1.

3.2 Social Influence Distribution

When a user u performs an action a, we want to give direct influence weight, denoted by $\gamma_{v,u}(a)$, to all $v \in N_{in}(u, a)$. The sum of the direct weights given by a user to its neighbors is under the constraint that no more than 1. We can have various ways of assigning direct weight: for ease of exposition, we assume for the moment to give equal weights to each neighbor v of u, i.e., $\gamma_{v,u}(a) = \frac{1}{d_{in}(u,a)}$ for all $v \in N_{in}(u, a)$. Later we will see a more sophisticated method of assigning direct weight. Intuitively, we also want to distribute influence weight transitively backwards in the propagation graph $G(a)$, such that not only u gives weight to the users $v \in N_{in}(u, a)$, but they in turn pass on the weight to their predecessors in $G(a)$ and so on. This suggests the following definition of total weight given to a user v for influencing u on action a, corresponding to multiple propagation paths:

$$\Gamma_{v,u}(a) = \sum_{w \in N_{in}(u,a)} \Gamma_{v,w}(a) \cdot \gamma_{w,u}(a)$$

where the base of the recursion is $\Gamma_{v,v}(a) = 1$. Sometimes, when the action is clear from the context, we can omit it and simply write $\gamma_{v,u}$ and $\Gamma_{v,u}$. From here on, as a running example, we consider the influence graph in Fig. 1 as the propagation graph $G(a)$ with edges labeled with direct weights $\gamma_{v,u}(a) = 1/d_{in}(u, a)$. For instance, $\Gamma_{4,0} = \Gamma_{4,2} \cdot \gamma_{2,0} = (\Gamma_{4,4} \cdot \gamma_{4,2} + \Gamma_{4,3} \cdot \gamma_{3,2}) \cdot \gamma_{2,0} = (1 \cdot 0.2 + 1 \cdot 0.2) \cdot 0.5 = 0.2$. We next define the total weight given to a set of nodes $S \subseteq V(a)$ for influencing user u on action a as follows:

$$\Gamma_{S,u}(a) = \begin{cases} 1 & \text{if } v \in S; \\ \sum_{w \in N_{in}(u,a)} \Gamma_{S,w}(a) \cdot \gamma_{w,u}(a) & \text{otherwise} \end{cases}$$

Consider again the propagation graph $G(a)$ in Fig. 1. Let $S = \{4, 8\}$. Then, $\Gamma_{S,u}$ is the fraction of flow reaching 0 that flows from either 4 or 8: $\Gamma_{S,0} = \Gamma_{S,2} \cdot \gamma_{2,0} = (1 \cdot 0.2 + 1 \cdot 0.2 + 1 \cdot 0.1) \cdot 0.5 = 0.25$.

3.3 Objective Function

The final question for the PSI model is how to aggregate the influence weight over the whole action log set $\mathcal{A}$. Consider two nodes v and u: the total influence weight given to v by u for all actions in $\mathcal{A}$ is simply obtained by taking the total weight over all actions and normalizing it by the number of actions performed by u (denoted A_u). This is justified by the fact that weights are assigned by u backward to its potential *influencers*. We define:

$$\kappa_{v,u} = \frac{1}{A_u} \sum_{a \in A} \Gamma_{v,u}(a) \tag{3}$$

Intuitively, it denotes the average weight given to v for influencing u over all actions that u performs. Similarly, for the case of a set of nodes $S \in V$, we can define the total influence weight for all the actions in A as $\kappa_{S,u}$.

Note that $\kappa_{S,u}$ corresponds to $Pr[path(\mathcal{S}, u) = 1]$ in Eq. 2. Finally, we define the influence spread $\sigma_{PSI}(S)$ as the total influence weight given to S from the whole social network:

$$\sigma_{PSI}(S) = \sum_{u \in V} \kappa_{S,u} \tag{4}$$

In the spirit of the influence maximization problem, this is the objective function that we want to maximize. The problem is NP-hard and the function $\sigma_{PSI}(\cdot)$ is submodular and monotone, paving the way for an approximation algorithm, which will be presented in next section.

4   Cluster-Based Assembling Method

In the target influence maximization problem, we focus on the specific user set rather than on the whole user graph. In other words, we want to avoid unnecessary computations of $Pr(S \cup \{u\}, T)$, especially for insignificant vertices to the target set. To achieve this goal, a cluster-based assembling approximate framework is proposed.

Table 1. The *influencer* set Σ_v^r and *follower* set Σ_v^e of $v \in V$

v	Σ_v^r	Σ_v^e
0	{0,1,2,3,4,8}	{0}
1	{1}	{0,1,2,8,9,11,13,14,15}
2	{1,2,3,4,8}	{0,2}
3	{3,4}	{0,2,3,5}
4	{4}	{0,2,3,4,5,6}
5	{3,4,5,6}	{5}
6	{4,6}	{5,7}
7	{4,6,7}	{7}
8	{1,8}	{0,2,8,9,11,13,14,15}
9	{1,8,9,10,12,14,15}	{9}
10	{10,12}	{9,10,15}
11	{1,8,11}	{11,13}
12	{12}	{9,10,12,15}
13	{1,8,11,13,14}	{13}
14	{1,8,14}	{9,13,14,15}
15	{1,8,10,12,14,15}	{9,15}

4.1   Cluster Detection Using Influence Behavior

We want to identify a set of vertices (denoted as candidate set C) which includes all possible seeds. Thus we only need to consider the candidate seeds in C to identify the seed set S. Existing algorithms take all users V as the candidate set C, and our goal is to reduce the set as much as possible.

Intuitively, vertices in T will be considered as candidate seeds. In addition, many vertices have influences on the vertices in T, and some of them have large influences and some have small influences. To differentiate them, we want to eliminate those insignificant vertices with small influences. For example, if vertex u's influence on $v \in T$ is smaller than a threshold θ, i.e., $MIP_G(u,v) < \theta$, u is an insignificant vertex for v. If u is an insignificant vertex for every vertex in T, we will not take it as a candidate seed.

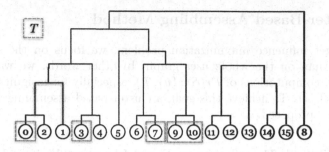

Fig. 2. Hierarchical Tree using the example in Fig. 1 and Table 1

Definition 3. *Candidate Seed.* *Given a query* $Q = (\mathcal{T}, k)$, u *is a candidate seed if* $MIP_G(u, v) \geq \theta$.

Let $\Sigma_v^r = \{u | MIP(u, v) \geq \theta\}$ denotes the *influencer set* of vertex v and $\Sigma_v^e = \{u | MIP(v, u) \geq \theta\}$ denote the *follower set* of vertex v. To support efficient online queries, we precompute these lists for each vertex. Obviously, $C = \cup_{v \in T} \Sigma_v^r$ is a candidate seed set.

To reduce the redundant computation for similar vertices, we build a Hierarchical Tree for vertices in V based on their influence behaviors, and utilize the Hierarchical Tree to compute the set of vertices in the target set $\mathcal{T}$.

Initialization. For each vertex $v \in \mathcal{T}$, we enumerate v's in-neighbors, e.g., u. If $MIP(u, v) \geq \theta$, we add u into C and continue traversing u's in-neighbors. Iteratively we get proper C. For example, Table 1 shows the influencer set and follower set of vertices in Fig. 1. Suppose $\theta = 0.05$. For target user set $\mathcal{T} = \{0, 3, 7, 9, 10\}$, vertices in the query are candidate seeds intuitively. Vertices 1, 2, 4, 5, 6, 8, 12 and 14 are in the candidate set as they have influence on the target user set. Vertices 5, 11 and 13 are candidate seeds as they are not followers of any member of the target user set.

Tree Structure. With the influencer set and the follower set, we can build a Hierarchical Tree using a clustering algorithm to produce hierarchical clusters [4]. The similarity of users' influence behavior allows us to identify communities in the network. The tree shows the order in which the nodes join together to form communities. Each hierarchical tree node D_i represents the user set that has a similar influence spread, which can be utilized to fasten the candidate seed identification (Fig. 2).

4.2 Greedy Algorithm

However, if the query target set is large, it is still expensive to calculate the candidates set C and initial influence $Pr(\{u\}, \mathcal{T})$ for every vertex $u \in C$. Thus, we maintain two indexes to efficiently identify candidate seeds.

Node-Influencer Index. For each Hierarchical Tree node D_i, we precompute the union of influencer sets of vertices in node D_i, denoted by $\Sigma_{D_i}^r = \cup_{v \in D_i} \Sigma_v^r$.

In addition, for each vertex $u \in \Sigma_{D_i}^r$, we also precompute its influence on vertices in D_i, denoted by $Pr(\{u\}, V_{D_i}) = \sum_{v \in D_i} MIP(u, v)$. We use a *node-influencer index* to maintain all the vertices in $\Sigma_{D_i}^r$, with their influences on D_i in decreasing order.

Influencer-Node Index. On the other hand, for each vertex u, we keep a *influencer-node index* of tree nodes whose influencer set contains u, with the corresponding influences.

Therefore, we can devise an assembly-based greedy algorithm to achieve $(1 - 1/e)$ approximate ratio with the indexes. In the online search, we initialize a max-heap H to identify seeds. We first load the node-vertex lists of the tree nodes in $\mathcal{T}$. Then we check the vertices in order and add the vertices with the largest influences into H. We obtain a lower bound B_L for the next seed's influence using the heap and an upper bound B_U for the unvisited vertices' influences using these lists. If $B_L \leq B_U$, the first seed is found; otherwise we add more vertices into the heap. To find the seed from the heap, we need to update their influences. By determining the upper bound of the lists with the influences of the current vertices, we use the expansion-based algorithm [8] to identify the next seed using the heap. Then it takes the influence of the top vertex u of the heap as a lower bound B_L of the next seed. If $B_L \leq B_H$, the top vertex u is the next seed, and the algorithm pops it from the heap and adds it into S. Iteratively we can find top-k seeds.

5 Experiment Evaluation

The goals of our experiments are manifold. At a high level, we evaluate (1) the different models for edge weights with respect to the accuracy of the spread prediction; and (2) the quality and efficiency of seed selection under our approximation algorithms.

5.1 Experiment Setup

Datasets. We use three real-world datasets, namely *Amazon* [10,11], and *Gowalla* [3] and a *Facebook* dataset crawled by ourself for performance evaluation. The details of the three datasets are shown in Table 2. Note that the "product" attribute for the location-based social networks *Gowalla* and *Facebook* represents the "location".

Queries. We randomly generated two types of queries with different target set sizes. (1) Small set queries: the target set contained 100, 200, 500, 1000 and 5000 vertices; (2) Proportional set queries: the target set size is in proportion to the graph node size $|U|$, e.g., the target set size of a 10-percent set query equals $0.1 * |U|$. In the experiments, we report the average performance of 1,000 queries in each query type.

Performance Evaluation. To simulate the ground truth, for each action (check-in in *Gowalla* and *Facebook*; and purchase history in *Amazon*) in the

Table 2. Statistics of datasets

Property		Network		
		Amazon	Gowalla	Facebook
#records	Actions	34,686,770	6,442,890	869,317
#nodes	User product	6,643,669	196,591 1,280,969	29,519
#edge	Friend	39,513	950,327	39,513

test set, we take the set of users that are the first to do the action (e.g., visit a location) among their friends, i.e., the set of "initiators" of the action, to be the seed set. The actual spread is the number of users who performed that action, also considered as influence spread. This allows for a fair comparison of all methods from a neutral standpoint, which is a first in itself.

5.2 Spread Achieved

In this experiment, we compare the influence spread achieved by the seed selection under different target set queries. We adopt **IC** and **LT** described in the previous subsection to show the results of global influence maximization. Considering the compatibility with our problem, we also design two baseline heuristic algorithms to make comparison: PageRank (**PR**) and High Degree (**HD**), which select the top-k nodes with respect to degree and PageRank score respectively, as seeds.

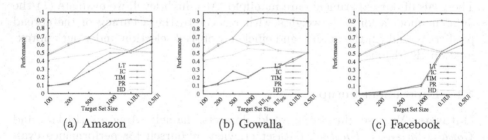

(a) Amazon   (b) Gowalla   (c) Facebook

Fig. 3. Influence spread achieved on different datasets

The experiment results are illustrated in Fig. 3. Note that the size of k is fixed to $|T|/50$ here. Figure 3 presents significant superiority of the **TIM** algorithm over the others when the size of set T is increasing. The global algorithms **IC** and **LT** result in bad performances when $|T|$ is relatively small compared to $|V|$. As $|T|$ increases, the overlap of $|T|$ and $|V|$ also increases, which leads to a larger influenced number among the target user set. The two baseline compared algorithms have similar performance, of which the High Degree performs a little better. The values of online probability and threshold have effect on the final number of influenced target users, but they do not change the overall trend of the experiment data. The **TIM** algorithm outperforms the other algorithms in all cases when the online probability and threshold are different.

In general, the TIM algorithm has an outstanding advantage over the other ones. When the size of set T is large, this advantage is more significant, while in the special case where T is large and online probability is high, the performance of the TIM algorithm is similar with others.

5.3   Running Time

Next, we show the results of the running time taken by the various models. We include two well-recognized greedy algorithms, **CELF++** [6] and **MIA** [1,2], for comparison. Figure 4 reports the results for small set queries. It can be seen that TIM with the cluster-based index (**TIM**) outperforms the other methods.

For the sake of completeness, we also examine the scalability of the TIM algorithm with respect to the size of the action log, in number of tuples. The synthetic data set is generated by randomly choosing propagation traces from the complete dataset action logs and selecting all the corresponding action logs.

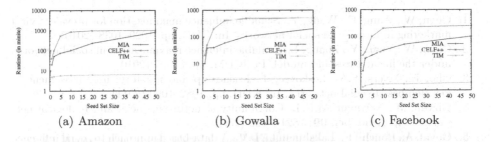

(a) Amazon   (b) Gowalla   (c) Facebook

Fig. 4. Running time comparison

Figure 5 shows the time take by **TIM** to select 50 seeds against the number of tuples used by the three datasets. Most of the time taken by the algorithm is consumed in scanning the action logs.

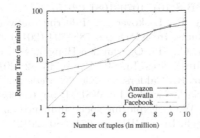

Fig. 5. Running time against the size of action logs

6   Conclusion

This paper presents the target influence maximization problem to select the set of seed users in a social network that maximizes the expected number of target users influenced by the seed users. We first designed a probabilistic model to capture the *social influence* between linked users using action logs. Then, to meet the instant-speed requirement for supporting online queries, a cluster-based index structure and a greedy algorithm were proposed to return the solution with $(1 - 1/e)$ approximation ratio. According to the experiment results, our method achieved high performance while a keeping large influence spread in the target user set, and outperformed the state-of-the-art algorithms for the TIM problem in terms of both effectiveness and efficiency.

Acknowledgments. Wen-Chih Peng was supported in part by MOST Taiwan, Project No. 106-3114-E-009-010 and 106-2628-E-009-008-MY3. In addition, this work is partially funded by Microsoft Research Asia.

References

1. Chen, W., Wang, C., Wang, Y.: Scalable influence maximization for prevalent viral marketing in large-scale social networks. In: KDD, pp. 1029–1038 (2010)
2. Chen, W., Yuan, Y., Zhang, L.: Scalable influence maximization in social networks under the linear threshold model. In: ICDM, pp. 88–97 (2010)
3. Cho, E., Myers, S.A., Leskovec, J.: Friendship and mobility: user movement in location-based social networks. In: KDD, pp. 1082–1090 (2011)
4. Girvan, M., Newman, M.E.J.: Community structure in social and biological networks. Nat. Acad. Sci. **99**, 7821–7826 (2002)
5. Goyal, A., Bonchi, F., Lakshmanan, L.V.: A data-based approach to social influence maximization. VLDB **5**(1), 73–84 (2011)
6. Goyal, A., Lu, W., Lakshmanan, L.V.: CELF++: optimizing the greedy algorithm for influence maximization in social networks. In: WWW, pp. 47–48 (2011)
7. Kempe, D., Kleinberg, J., Tardos, É.: Maximizing the spread of influence through a social network. In: KDD, pp. 137–146 (2003)
8. Li, G., Chen, S., Feng, J., Tan, K.l., Li, W.s.: Efficient location-aware influence maximization. In: SIGMOD, pp. 87–98 (2014)
9. Li, Y., Zhang, D., Tan, K.L.: Real-time targeted influence maximization for online advertisements. VLDB **8**(10), 1070–1081 (2015)
10. McAuley, J., Pandey, R., Leskovec, J.: Inferring networks of substitutable and complementary products. In: KDD, pp. 785–794 (2015)
11. McAuley, J., Targett, C., Shi, Q., van den Hengel, A.: Image-based recommendations on styles and substitutes. In: SIGIR, pp. 43–52 (2015)
12. Saleem, M.A., Kumar, R., Calders, T., Xie, X., Pedersen, T.B.: Location influence in location-based social networks. In: WSDM, pp. 621–630 (2017)
13. Ye, M., Liu, X., Lee, W.C.: Exploring social influence for recommendation: a generative model approach. In: SIGIR, pp. 671–680 (2012)

Team Expansion in Collaborative Environments

Lun Zhao[1], Yuan Yao[1(✉)], Guibing Guo[2], Hanghang Tong[3], Feng Xu[1], and Jian Lu[1]

[1] State Key Laboratory for Novel Software Technology,
Nanjing University, Nanjing, China
zhaolun@smail.nju.edu.cn, {y.yao,xf,lj}@nju.edu.cn
[2] Software College, Northeastern University, Shenyang, China
guogb@swc.neu.edu.cn
[3] Arizona State University, Tempe, USA
hanghang.tong@asu.edu

Abstract. In this paper, we study the team expansion problem in collaborative environments where people collaborate with each other in the form of a team, which might need to be expanded frequently by having additional team members during the course of the project. Intuitively, there are three factors as well as the interactions between them that have a profound impact on the performance of the expanded team, including (1) the task the team is performing, (2) the existing team members, and (3) the new candidate team member. However, the vast majority of the existing work either considers these factors separately, or even ignores some of these factors. In this paper, we propose a neural network based approach TECE to simultaneously model the interactions between the team task, the team members as well as the candidate team members. Experimental evaluations on real-world datasets demonstrate the effectiveness of the proposed approach.

Keywords: Team expansion · Candidate member prediction
Collaborative environments · Neural networks

1 Introduction

In many application domains, people tend to frequently collaborate with others in the form of teams for specific tasks. For example, in open source software community, developers distributed worldwide work with each other by forming developer teams for specific projects; in research community, researchers form research teams and they collaborate with each other for research projects/papers; in the film industry, the crew works together as a team for a film shooting. In this work, we refer to these application domains as *collaborative environments*.

One of the key features in these collaborative environments is *team mobility*, i.e., teams are formed upon specific tasks and individuals can participate

© Springer International Publishing AG, part of Springer Nature 2018
D. Phung et al. (Eds.): PAKDD 2018, LNAI 10939, pp. 713–725, 2018.
https://doi.org/10.1007/978-3-319-93040-4_56

in several teams depending on their own interest and capabilities. In such environments, we usually need to expand the team by adding new members during team formation or when the existing team encounters difficulties on the task. In this paper, we put our focus on the team expansion problem in collaborative environments.

Broadly speaking, team expansion is related to several existing lines of research (please refer to the related work section for more details): (1) people and task matching which searches for an optimal match between people capabilities and task requirements [3,23], (2) recommendation which recommends items to users [11,18], and (3) social proximity analysis which computes the proximities between users [13,21]. However, they all suffer from some limitations for the team expansion problem studied in this paper: (1) people and task matching methods need concrete descriptions about people capabilities and task requirements, while such descriptions are usually unavailable; (2) recommendation methods mainly focus on recommending items for users, while team expansion aims to recommend users for items (i.e., tasks) where the 'chemistry' between existing users and the new user matters; (3) social proximity analysis methods analyze the social connections between users, while the matching between users and tasks are widely ignored.

In this paper, we propose a neural network based approach (TECE) for team expansion in collaborative environments. The basic considerations of TECE are three-fold: (1) no concrete requirements and capabilities for the candidate team member are needed, (2) the candidate should match the task, and (3) the candidate should match the existing team members. To this end, we propose to automatically match the candidate members to both team tasks and existing team members, based on the existing interactions between them. To match a candidate member with the given task, we exploit the collaborative filtering idea from recommender systems by mining the existing interactions between individuals and tasks; to match a candidate member with the existing team members, we take the team leader as a proxy of the team members and incorporate the social connections between the candidate and the team leader into the model. Additionally, we adopt deep models with multiple non-linear neural layers to capture the complex relationships between candidate members, team leaders, and team tasks.

The main contributions of this paper include:

- We formally define the team expansion problem in collaborative environments, which has a wide range of applications.
- We propose the TECE model to solve the team expansion problem. The proposed TECE simultaneously considers three important factors (team task, existing team members, and candidate team member) as well as their interactions.
- Experimental results on two real-world datasets show that the proposed method can outperform several competitors in terms of accurately identifying candidate members. For example, TECE can achieve up to 22.1% improvement compared with its best competitors.

The rest of the paper is organized as follows. Section 2 defines the team expansion problem. Section 3 describes the proposed approach. Section 4 presents the experimental results. Section 5 covers related work, and Sect. 6 concludes.

2   Problem Statement

In this section, we present the problem definition. Without loss of generality, we assume that each team corresponds to a unique task. Therefore, we use t to denote both the team task and the team itself. We assume that there are m teams/tasks and n unique individuals, and we use $\mathcal{T} = \{t_1, t_2, \ldots, t_m\}$ and $\mathcal{I} = \{i_1, i_2, \ldots, i_n\}$ to denote the set of teams and individuals, respectively. The existing interactions between team tasks and individuals are contained in matrix $\mathbf{R}$. For example, $\mathbf{R}(t, i) = 1$ means that individual i belongs or once belonged to team t, and $\mathbf{R}(t, i) = 0$ indicates otherwise. For a specific team t, we use $\mathcal{I}_t$ to denote the existing members in the team. Specially, we assume that there is a team leader $o_t \in \mathcal{I}_t$ (e.g., the owner of the team) in each team. With the above notations, we define the team expansion problem in collaborative environments as follows.

Problem 1. Team Expansion Problem in Collaborative Environments (TECE).

Given: (1) a collection of teams/tasks $\mathcal{T}$ each of which has a team leader o_t,
   (2) a collection of individuals $\mathcal{I}$, (3) the existing interactions $\mathbf{R}$ between
   teams/tasks and individuals, and (4) a team $t_{test} \in \mathcal{T}$ that is about to expand;
Find: the candidate member to join team t_{test}.

The above team expansion problem can be formulated as estimating the fitness scores of unobserved entries in $\mathbf{R}$, which resembles the recommendation problem. However, different from traditional recommendation problem, when recommending a candidate member to a team, we need to pay special attention to the 'chemistry' between the candidate member and existing team members. The goal of team expansion is to generate a ranked list of candidates for the team that is about to expand. The ranked list is determined by the estimated scores of unobserved entries in $\mathbf{R}$. For example, suppose team t is the team to expand. We take team t itself, the team leader o_t, and a candidate team member i as input, and the goal is to learn a mapping function f to obtain the estimated fitness score between candidate i and the team t with leader o_t. Formally, we have $\hat{\mathbf{R}}(t, i) = f(t, o_t, i | \Theta)$, where $\hat{\mathbf{R}}(t, i)$ denotes the estimated fitness score, and Θ contains the model parameters. In the next section, we will show how we construct the mapping function f and learn its parameters.

3   The Proposed Approach

In this section, we present the proposed TECE model, followed by some discussions and generalizations.

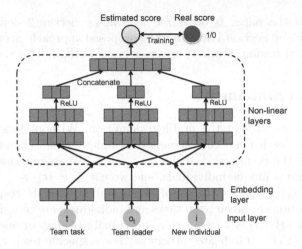

Fig. 1. The overview of TECE.

3.1 The TECE Model

Figure 1 shows the overview of the proposed approach. As we can see, TECE takes (the one-hot encodings of) team task id, team leader id, and candidate member id as input, and embeds each of them into a low-dimensional vector. After that, the resulting embeddings are fed into several non-linear layers to learn the complex interactions between them. To be specific, TECE exploits the collaborative filtering idea from recommender systems to model the interactions between candidate team members and team tasks; it treats the team leader as a proxy of the team members and model the social interactions between candidate members and team leaders; it also models the interactions between team tasks and team members as team state which could impact the ideal candidate member. The last layer contains the final high-level features from the above interactions for predicting the fitness score between the input candidate and the input team (with its team leader). Finally, the output layer produces the estimated fitness score.

Next, we present the details of the TECE model. As mentioned in introduction, team expansion needs to consider the interactions between candidate member, team task, and existing team members. In the following, we first separately consider these interactions, and then describe the neural network based objective functions.

(A) Matching Candidate Team Member with Team Task. For the matching between candidate team member and team task, we first denote the embedding vectors/features of team task and candidate member as $\mathbf{p}_t \in \mathbb{R}^d$ and $\mathbf{q}_i \in \mathbb{R}^d$, respectively, where d is the vector dimensionality. Then, we model the interaction vector $\mathbf{r}_{ti}$ between team task t and candidate i as follows,

$$\mathbf{r}_{ti} = \mathbf{p}_t \odot \mathbf{q}_i \tag{1}$$

where $\odot$ denotes element-wise multiplication. For simplicity, we assume all the embedding vectors are row vectors in this paper.

(B) Matching Candidate Team Member with Existing Team Members (Team Leader). For the matching between candidate member and existing members, we pay special attention to the team leader as he/she plays important roles in building the team spirit and improving team performance [4,19]. Therefore, we use the team leader as a proxy of the team members in this work. Based on this, we model the interactions between team leader o_t and candidate i as the likelihood of their cooperation in the current team. We use c_{ti} to stand for the cooperation likelihood, and it can be computed as

$$c_{ti} = \mathbf{q}_i \odot \mathbf{q}_{o_t} \tag{2}$$

where we still use the element-wise multiplication, and $\mathbf{q}_i$ and $\mathbf{q}_{o_t}$ are the embedding vectors for candidate i and team leader o_t, respectively. Note that team leader is one of the individuals in $\mathcal{I}$.

(C) Team State Building. In addition to matching the candidate with team task and team leader, we consider to build the team state vector which describes the interactions between team task and team leader. Such interactions could impact what is the best way the ideal candidate member should interact with team state. For example, if the current team state is harmonic, the ideal candidate should maintain the existing 'chemistry'; if there is a lack of timely communication in the current team state, the ideal candidate should act as a communication bridge. Specially, we define the team state vector $\mathbf{s}_t$ as

$$\mathbf{s}_t = \mathbf{p}_t \odot \mathbf{q}_{o_t} \tag{3}$$

where $\mathbf{p}_t$ is the embedding vector of team task t, and $\mathbf{q}_{o_t}$ is the embedding vector of team leader o_t.

(D) Single-layer Modeling. Next, based on the above three intermediate vectors, we can build a neural network to compute the fitness score $\hat{\mathbf{R}}(t, i)$. In the simplest case, we can concatenate these vectors and feed the resulting vector into a dense (fully-connected) layer, whose output score $\hat{\mathbf{R}}(t, i)$ can be computed as

$$\hat{\mathbf{R}}(t, i) = f^{out}([\mathbf{r}_{ti}, \mathbf{c}_{ti}, \mathbf{s}_t]\mathbf{x}^T) \tag{4}$$

where $\mathbf{x}$ is the weight vector, and f^{out} can be set as the sigmoid function $f^{out}(x) = \frac{1}{1+e^{-x}}$.

Note that for the network we describe in Eq. (4), it can be seen as a generalization of the traditional collaborative filtering based recommendation. It can be degenerated to traditional collaborative filtering if we take only the $\mathbf{r}_{ti}$ as input, define $\mathbf{x}$ as a row vector of 1s, and define f^{out} as the identify function. In other words, compared to traditional recommendation, we further consider the current team state ($\mathbf{s}_t$) and the interactions between candidate members and team leaders ($\mathbf{c}_{ti}$).

(E) Multi-layer Modeling. For single-layer network, simply using a vector concatenation as final features is insufficient to represent the complex interactions between candidate team members, team leaders, and team tasks. To address this issue, we add multiple non-linear layers to model these interactions. Take the cooperation likelihood vector $\mathbf{c}_{ti}$ as an example, the multi-layer modeling for fitness score $\hat{\mathbf{R}}(t, i)$ can be defined as follows:

$$\mathbf{z}^{(1)} = f^{(1)}(\mathbf{W}^{(1)}\mathbf{c}_{ti} + \mathbf{b}^{(1)})$$
$$\mathbf{z}^{(2)} = f^{(2)}(\mathbf{W}^{(2)}\mathbf{z}^{(1)} + \mathbf{b}^{(2)})$$
$$\cdots$$
$$\mathbf{z}^{(L)} = f^{(L)}(\mathbf{W}^{(L)}\mathbf{z}^{(L-1)} + \mathbf{b}^{(L)})$$
$$\hat{\mathbf{R}}(t, i) = f^{out}(\mathbf{z}^{(L)}\mathbf{x}^T) \tag{5}$$

where L is the layer number, $\mathbf{W}^{(i)}$, $\mathbf{b}^{(i)}$, and $f^{(i)}$ denote the weight matrix, bias vector, and activation function for the corresponding layer, respectively. f^{out} and $\mathbf{x}$ are defined in Eq. (4). For the activation function, we choose the ReLU (Rectified Linear Units) function for $f^{(i)}$. Similarly, we can add multiple non-linear layers for each of the three intermediate vectors as defined in Eqs. (1)–(3), and we may also add multiple non-linear layers for the concatenated vector as defined in Eq. (4).

(F) Objective Function. Finally, we define the objective function to learn the embedding vectors $\mathbf{p}_t$, $\mathbf{q}_i$, and $\mathbf{q}_{o_t}$, as well as the other model parameters Θ. Specially, we adopt the following logistic-like objective function

$$\underset{\mathbf{P},\mathbf{Q},\Theta}{\operatorname{argmax}} \sum_{(t,i)\in\mathbf{R}} \mathbf{R}(t, i) \log(\hat{\mathbf{R}}(t, i)) + (1 - \mathbf{R}(t, i)) \log(1 - \hat{\mathbf{R}}(t, i)) \tag{6}$$

where $\mathbf{P}$ contains the embedding vectors $\mathbf{p}_t$, $\mathbf{Q}$ contains the embedding vectors $\mathbf{q}_i$ and $\mathbf{q}_{o_t}$, $\mathbf{R}(t, i)$ is the real fitness score between team task t and candidate i, and $\hat{\mathbf{R}}(t, i)$ is estimated fitness score by our model.

Here, the ground truth is contained in the existing interaction matrix $\mathbf{R}$ with $\mathbf{R}(t, i) = 1$ as positive label and $\mathbf{R}(t, i) = 0$ as negative label. The remaining problem is to sample the negative (t, i) pairs. Directly choosing all the possible (t, i) pairs is computational expensive (quadratic time complexity). Choosing only the positive labels (i.e., $\mathbf{R}(t, i) = 1$) would lead to trivial solutions (i.e., the feature values towards infinity). In this work, we keep all the positive labels, and randomly sample r (sampling ratio) negative labels (e.g., $\mathbf{R}(t, j) = 0$) in terms of team task t for each positive label. Based on the sampling strategy, a stochastic gradient ascent learning algorithm can be applied for optimization.

3.2 Generalizations and Discussions

The TECE model is open for some reasonable adjustments on the model architecture. Here, we discuss some possible generalizations of the proposed method.

First, the proposed model is flexible with several special cases. For example, we may consider the single-layer objective function as discussed above. Such treatment may improve the training efficiency while lower the prediction accuracy. Besides, for the three inputs, we can delete either the team task or the team leader which will degenerate the model to social proximity model and recommendation model, respectively. We will experimentally evaluate some the above special cases in the next section.

Second, as a common practice in neural networks, we use element-wise multiplication in Eqs. (1)–(3) to model the interactions between team task, team leader, and candidate team member. In addition to this operation, other operations such as inner product, concatenation, average pooling, and max pooling can also be used. In fact, we can simultaneously use element-wise multiplication and concatenation to obtain more intermediate vectors. In this work, we omit such extensions for brevity.

Third, in our model, when matching the candidate member with team members, we take team leader as a proxy. We can also incorporate all the existing team members into the model by average pooling or max pooling. Take average pooling as an example. We can flatten the embedding vectors of existing team members by computing the average vector. We will experimentally evaluate this in the experimental section.

Fourth, in this work, we employ the team-individual interaction history only as input for the team expansion problem. Actually, we can consider much richer information such as the text descriptions of team tasks, member profiles, and temporal effects when such information is available. We leave these extensions as future work.

4 Experiments

4.1 Experimental Setup

Datasets. We conducted experiments on two real datasets: GitHub[1] and DBLP[2]. GitHub is an open-source software development platform. The data contains the information about projects, developers, and the actions from developers to projects. Since there are many toy projects and inactive users, we filter the data by deleting the developers who have contributed to less than five projects, and the projects whose 'star' is no more than five and whose team member number is less than five. DBLP is an open database that collects scientific articles in the field of computer science [20]. We treat each article as a team task, authors as team members, and the first author as the team leader. We use a subset of the whole dataset, and the subset contains the areas of Data Mining, Machine Learning, Database, and Artificial Intelligence. Similar to the processing steps above, we filter out the authors who have published less than three articles, and the articles whose author number is less than three. Overall,

---

[1] http://ghtorrent.org/downloads.html.
[2] https://cn.aminer.org/billboard/citation.

Table 1. Statistics of the datasets.

Dataset	# of team tasks	# of individuals	Avg. members per team	Avg. teams per individual
GitHub	10,505	30,258	31.56	10.96
DBLP	17,838	29,423	3.59	2.18

both software development and research article publication can be seen as collaborative environments where people frequently collaborate with others in the form of teams for specific tasks. Both datasets are publicly available, and the statistics are listed in Table 1.

Compared Methods. We compare TECE with the following methods including two social proximity analysis methods and two recommendation methods.

- Co-rank. It is a heuristic method. The basic idea is to rank the candidate team members based on their cooperation times with the team leader.
- RW [21]. This is a random walk method that computes the proximities between users in a network. We adapt the method to the bipartite network of individuals and tasks.
- BPR [17]. This is a classic recommendation model for the one-class feedback case. It directly optimizes the rankings between an observed feedback and an unobserved feedback.
- NCF [9]. NCF is a recent neural network based model designed for one-class recommender systems. It treats the recommendation problem as a binary classification problem, and adopts neural networks to model the interactions between users and items.

Evaluation Metrics. To evaluate the effectiveness of the compared methods, we adopt the following two widely used evaluation metrics. Specifically, we output the top K candidate members in a ranked list, and compute the HR and nDCG metrics as follows.

$$HR@K = \frac{1}{|T|} \sum_{t=1}^{|T|} hit_t, \qquad nDCG@K = \frac{1}{|T|} \sum_{t=1}^{|T|} \frac{\log 2}{\log(r_t + 1)}$$

where T is the test set of teams, $hit_t \in \{0, 1\}$ is a binary value indicating whether the ground-truth candidate is in the top K list, and $r_t \in \{1, 2, ..., K\}$ is the ranking of the ground-truth candidate in the ranked list. $r_t = 0$ if the candidate is not in the top K list. In this work, we set K to 1, 5, 10, 15, and 20.

Experimental Settings. To setup the experiments, we randomly select one team member in each team as the test set. The training data is used to train the model where a ranked list of possible candidates for each team will be generated. Here, we basically assume that the existing member is a suitable match for the corresponding team task. While this is not always true in reality, we mitigate the

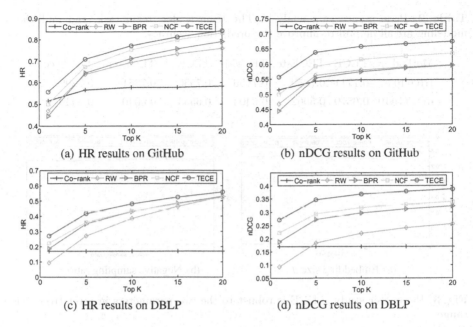

(a) HR results on GitHub

(b) nDCG results on GitHub

(c) HR results on DBLP

(d) nDCG results on DBLP

Fig. 2. Effectiveness comparisons. TECE generally outperforms the compared methods in both HR and NDCG on both datasets.

issue by filtering the datasets and keeping the teams that can be considered as good teams (e.g., software projects that have been starred several times, or scientific papers that have been published in top venues). During the testing stage, since predictions on all the candidates would be time-consuming, we randomly select 100 negative samples for each ground-truth candidate.

For the parameters, we either follow the default setting or set them equally. For example, we set the embedding dimensionality to 32 for all the methods (Co-rank and RW are not applicable). For the two non-linear layers of TECE, we set the embedding size as 16 and 8, respectively.

4.2 Experimental Results

(A) Effectiveness Comparisons. We first compare TECE with the existing methods, and report the results in Fig. 2. First of all, we can observe from the figures that the proposed TECE generally outperforms the compared methods in terms of the two evaluation metrics on both datasets. For example, TECE improves the best competitor (NCF) by up to 11.3% on the GitHub data, and by up to 22.1% on the DBLP data. Basically, TECE is better than NCF and BPR as it further considers the matching between candidate member and team leader; TECE is better than RW and Co-rank as it further considers the matching between candidate member and team task.

Second, we can also observe that the overall results on the GitHub data are better than those on the DBLP datasets. This is due to the fact that the

Table 2. Performance gain analysis. The team leader, team task, and multi-layer modeling are all helpful to improve the prediction accuracy.

Method	TECE	TECE-tc	TECE-lc	TECE-sl	TECE-em	TECE-con
HR@10	0.7711	0.7359	0.7556	0.7550	0.7351	0.7299
nDCG@10	0.6579	0.5962	0.6404	0.6553	0.6040	0.5949

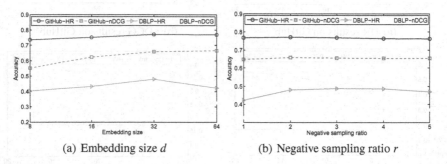

(a) Embedding size d (b) Negative sampling ratio r

Fig. 3. Parameter study. TECE is robust to the two parameters in a relatively wide range.

DBLP data is much sparser, making the problem more challenging. Third, for the Co-rank method, it can identify the candidate team member with a relatively high accuracy at top 1 ($K = 1$), while the accuracy increases slowly as K increases. The reason is that Co-rank can only find the candidates in a local neighborhood (i.e., previously cooperated), while the global perspective of task-candidate matching is ignored.

(B) Performance Gain Analysis. Next, we analyze the proposed TECE by checking the performance of its components and variants. For its components, we delete the team leader, team task, and multi-layer modeling of TECE, and obtain the TECE-tc (task and candidate), TECE-lc (leader and candidate), and TECE-sl (single-layer) method, respectively. For its variants, we consider to substitute team leader with the average-pooling on all the existing team members, and substitute the element-wise product with concatenation. The resulting methods are referred to as TECE-em (existing member) and TECE-con (concatenation), respectively. For brevity, we report the HR@10 and nDCG@10 results on the GitHub data in Table 2.

First, we can observe from the table that all the three components (i.e., team leader, team task, and multi-layer) of TECE are helpful to improve the prediction accuracy. For example, when team leader, team task, and multi-layer modeling are incorporated, the HR@10 performance of TECE improves by 4.8%, 2.1%, and 2.1%, respectively. Second, TECE is better than its two variants TECE-em and TECE-con. The improvement over TECE-em indicates the important role of the team leader, and the improvement over TECE-con indicates that the element-wise multiplication is a better way to model the interactions between team tasks, team leaders, and candidate team members.

(C) Parameter Study. Finally, we conduct a parameter study of the proposed method in terms of the embedding size d and the negative sampling ratio r. The results are shown in Fig. 3 where we still report the HR@10 and nDCG@10 results. As we can see from Fig. 3(a), the prediction accuracy generally improves when the embedding size d grows from 8 to 32. No significant improvement can be observed when d becomes larger. For the negative sampling ratio r, slight improvement can be observed when $r = 3$ for the DBLP data. In general, TECE is robust to the two parameters in a relatively wide range. In this paper, we fix embedding size d to 32 and sampling ratio r to 2 for simplicity.

5 Related Work

In this section, we briefly review the related work including people and task matching, recommender systems, social proximity analysis, etc.

People and Task Matching. In the operations research community, the people and task matching problem has been extensively studied [3,5,23]. Typically, the matching problem is often formulated as an integer linear program, and the goal is to search for an optimal match between people capabilities and task requirements. This line of work needs explicit and concrete descriptions of people capabilities and task requirements, while such descriptions are usually unavailable or inaccurate in many real applications.

Recommender Systems. Team expansion is related to recommender systems. One of the branches in recommender systems takes collaborative filtering as the model basis, and recommends items to users based on the existing interactions/feedback between users and items [11,17]. Later, some researchers further incorporate social connections between users into the model [15,25], and some others adopt deep neural networks. For example, normal deep networks [9], stacked auto-encoder [22], convolutional neural network [10], and recurrent neural network [24] have been used for modeling the user feedback, item content, temporal effect, etc. Different from the existing recommendation methods whose goal is to recommend items to users, team expansion aims to recommend users to items/tasks, where the 'chemistry' between existing users and the candidate user matters.

Social Proximity Analysis. Since the social connections between the candidate member and the existing members matter for the team expansion problem, our work is also related to existing social proximity analysis work [2,14,21]. For example, Tong et al. [21] propose fast random walks based on which the social proximity between two nodes in a network can be computed; Cummings and Kiesler [6] find that prior working experience is the best predictor for the collaborative tie strength; recently, Han and Tang [8] propose the social group invitation problem, and solve the problem from a group evolution viewpoint. This line of work mainly focuses on the proximity analysis between users in the social network, while the matching between individuals and tasks are widely ignored.

Team Formation and Optimization. The team formation or team expansion problem has been studied in the entrepreneurial context [5,7], where interpersonal attraction, knowledge and communication skills make the essential factors for a successful team. In computer science, the team formation problem [1,16] has also been studied. However, existing work still requires the explicit descriptions of task requirements and user skills. The most related work is perhaps the recent work by Li et al. [12]. They propose to reformulate the team expansion problem as a team replacement problem by defining a virtual member with the desired skill set and communication structure, and then replacing this member with a most similar substitute. In contrast to their work, we do not require the descriptions of skill set and communication structure, and our focus is to find a candidate team member by exploiting the interactions between candidates, team leaders, and team tasks.

6 Conclusions

In this paper, we have proposed the team expansion problem in collaborative environments, and proposed a neural network based approach TECE for the problem. The key idea of TECE is to match the candidate team member with both team task and team leader. Additionally, TECE models the non-linear interactions between them via a multi-layer architecture. Experimental evaluations on real-world datasets demonstrate that the proposed approach can outperform several competitors in terms of accurately identifying candidate members. Future directions include exploring richer information such as task descriptions and member profiles for the team expansion problem, and handling the cold-start cases of the team expansion problem.

Acknowledgments. This work is supported by the National Natural Science Foundation of China (No. 61690204, 61672274, 61702252), the National Key Research and Development Program of China (No. 2016YFB1000802), the Fundamental Research Funds for the Central Universities (No. 020214380033), and the Collaborative Innovation Center of Novel Software Technology and Industrialization. Guibing Guo is partially supported by the National Natural Science Foundation for Young Scientists of China (No. 61702084). Hanghang Tong is partially supported by NSF (IIS-1651203, IIS-1715385, CNS-1629888 and IIS-1743040), DTRA (HDTRA1-16-0017), ARO (W911NF-16-1-0168), and gifts from Huawei and Baidu.

References

1. Anagnostopoulos, A., Becchetti, L., Castillo, C., Gionis, A., Leonardi, S.: Online team formation in social networks. In: WWW, pp. 839–848. ACM (2012)
2. Backstrom, L., Leskovec, J.: Supervised random walks: predicting and recommending links in social networks. In: WSDM, pp. 635–644. ACM (2011)
3. Baykasoglu, A., Dereli, T., Das, S.: Project team selection using fuzzy optimization approach. Cybern. Syst.: Int. J. **38**(2), 155–185 (2007)
4. Bradley, J.H., Hebert, F.J.: The effect of personality type on team performance. J. Manag. Dev. **16**(5), 337–353 (1997)

5. Chen, S.J., Lin, L.: Modeling team member characteristics for the formation of a multifunctional team in concurrent engineering. IEEE Trans. Eng. Manag. **51**(2), 111–124 (2004)
6. Cummings, J.N., Kiesler, S.: Who collaborates successfully?: prior experience reduces collaboration barriers in distributed interdisciplinary research. In: CSCW, pp. 437–446. ACM (2008)
7. Forbes, D.P., Borchert, P.S., Zellmer-Bruhn, M.E., Sapienza, H.J.: Entrepreneurial team formation: an exploration of new member addition. Entrep. Theory Pract. **30**(2), 225–248 (2006)
8. Han, Y., Tang, J.: Who to invite next? Predicting invitees of social groups. In: AAAI, pp. 3714–3720 (2017)
9. He, X., Liao, L., Zhang, H., Nie, L., Hu, X., Chua, T.S.: Neural collaborative filtering. In: WWW, pp. 173–182 (2017)
10. Kim, D., Park, C., Oh, J., Lee, S., Yu, H.: Convolutional matrix factorization for document context-aware recommendation. In: RecSys, pp. 233–240. ACM (2016)
11. Koren, Y., Bell, R., Volinsky, C.: Matrix factorization techniques for recommender systems. Computer **42**(8), 30–37 (2009)
12. Li, L., Tong, H., Cao, N., Ehrlich, K., Lin, Y.R., Buchler, N.: Enhancing team composition in professional networks: problem definitions and fast solutions. IEEE Trans. Knowl. Data Eng. **29**(3), 613–626 (2017)
13. Li, L., Yao, Y., Tang, J., Fan, W., Tong, H.: QUINT: on query-specific optimal networks. In: KDD, pp. 985–994. ACM (2016)
14. Liben-Nowell, D., Kleinberg, J.: The link-prediction problem for social networks. J. Assoc. Inf. Sci. Technol. **58**(7), 1019–1031 (2007)
15. Ma, H., Yang, H., Lyu, M.R., King, I.: SoRec: social recommendation using probabilistic matrix factorization. In: CIKM, pp. 931–940. ACM (2008)
16. Rangapuram, S.S., Bühler, T., Hein, M.: Towards realistic team formation in social networks based on densest subgraphs. In: Proceedings of the 22nd International Conference on World Wide Web, pp. 1077–1088. ACM (2013)
17. Rendle, S., Freudenthaler, C., Gantner, Z., Schmidt-Thieme, L.: BPR: Bayesian personalized ranking from implicit feedback. In: UAI, pp. 452–461. AUAI Press (2009)
18. Resnick, P., Varian, H.R.: Recommender systems. Commun. ACM **40**(3), 56–58 (1997)
19. Soomro, A.B., Salleh, N., Mendes, E., Grundy, J., Burch, G., Nordin, A.: The effect of software engineers? Personality traits on team climate and performance: a systematic literature review. Inf. Softw. Technol. **73**, 52–65 (2016)
20. Tang, J., Zhang, J., Yao, L., Li, J., Zhang, L., Su, Z.: Arnetminer: extraction and mining of academic social networks. In: Proceedings of the 14th ACM SIGKDD International Conference on Knowledge Discovery and Data Mining, pp. 990–998. ACM (2008)
21. Tong, H., Faloutsos, C., Pan, J.Y.: Fast random walk with restart and its applications. In: ICDM, pp. 613–622 (2006)
22. Wang, H., Wang, N., Yeung, D.Y.: Collaborative deep learning for recommender systems. In: KDD, pp. 1235–1244. ACM (2015)
23. Wi, H., Oh, S., Mun, J., Jung, M.: A team formation model based on knowledge and collaboration. Expert Syst. Appl. **36**(5), 9121–9134 (2009)
24. Wu, C.Y., Ahmed, A., Beutel, A., Smola, A.J., Jing, H.: Recurrent recommender networks. In: WSDM, pp. 495–503. ACM (2017)
25. Yao, Y., Tong, H., Yan, G., Xu, F., Zhang, X., Szymanski, B.K., Lu, J.: Dual-regularized one-class collaborative filtering. In: CIKM, pp. 759–768. ACM (2014)

HashAlign: Hash-Based Alignment
of Multiple Graphs

Mark Heimann[✉], Wei Lee, Shengjie Pan, Kuan-Yu Chen, and Danai Koutra

Computer Science and Engineering, University of Michigan, Ann Arbor, USA
{mheimann,weile,jessepan,kyuchen,dkoutra}@umich.edu

Abstract. Fusing or aligning two or more networks is a fundamental building block of many graph mining tasks (e.g., recommendation systems, link prediction, collective analysis of networks). Most past work has focused on formulating *pairwise* graph alignment as an optimization problem with varying constraints and relaxations. In this paper, we study the problem of *multiple* graph alignment (collectively aligning multiple graphs at once) and propose HASHALIGN, an efficient and intuitive hash-based framework for network alignment that leverages structural properties and other node and edge attributes (if available) simultaneously. We introduce a new construction of LSH families, as well as robust node and graph features that are tailored for this task. Our method quickly aligns multiple graphs while avoiding the all-pairwise-comparison problem by expressing all alignments in terms of a chosen 'center' graph. Our extensive experiments on synthetic and real networks show that, on average, HASHALIGN is 2× faster and 10 to 20% more accurate than the baselines in *pairwise* alignment, and 2× faster while 50% more accurate in *multiple* graph alignment.

1 Introduction

Much of the data that is generated daily naturally form graphs, such as interactions between users in social media, communication via email or phone calls, question answering in forums, interactions between proteins, and more. Additionally, graphs may be inferred from non-network data [17]. For joint analysis, it is often desirable to fuse multiple graph data sources by finding the corresponding nodes across them. This task, known as graph alignment or matching, is the focus of our work. It is a core graph theoretical problem that has attracted significant interest, both in academia and industry, due to its numerous applications: identifying users in social networks [19], matching similar documents in lingual matching [4], brain graph alignment in neuroscience, protein-protein alignment [4,5], chemical compound comparison, and more.

In many applications, the goal is to align *multiple* (more than two) networks at once. Most existing methods get as input two networks, so they handle multiple network alignment by expensively computing all pairwise alignments. In

---

M. Heimann and W. Lee—These authors contributed equally to this work.

© Springer International Publishing AG, part of Springer Nature 2018
D. Phung et al. (Eds.): PAKDD 2018, LNAI 10939, pp. 726–739, 2018.
https://doi.org/10.1007/978-3-319-93040-4_57

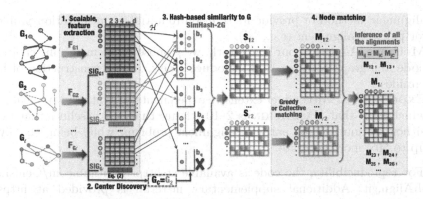

Fig. 1. Overview of proposed approach: HASHALIGN with input l undirected, weighted, attributed graphs (node/edge attributes are denoted with different shades/lines). The framework consists of four parts: (1) scalable, fast, robust, node-ID-invariant feature extraction per graph; (2) 'center' graph discovery, which is $G_C = G_2$ in this example; (3) efficient, hash-based similarity computation, $\mathbf{S}_{iC}$, between each graph G_i and G_C (buckets with red crosses do not contribute any pairwise similarity computations, and thus help with efficiency); and (4) node matching computation to find at most one matching per node in $\mathbf{M}_{ij}$.

this paper, we seek to devise an efficient method that collectively aligns multiple networks and can readily adapt to the existence or not of other node/edge information in addition to the graph topology, without increasing its complexity.

Problem 1 (Multiple Graph Alignment with Side Information). **Given** l graphs, $G_1(\mathcal{V}_1, \mathcal{E}_1), \ldots, G_l(\mathcal{V}_l, \mathcal{E}_l)$, where $\mathcal{V}_i$ and $\mathcal{E}_i$ are the node and edge sets of graph G_i, respectively, *with or without node/edge attributes*, we seek to **find** the correspondence between their nodes efficiently, so that the input graphs are as close to each other as possible.

To solve this problem, we propose HASHALIGN, an unsupervised method that is based on three key ideas: (i) inferring the similarity between nodes in different graphs based on structural properties and node/edge attributes; (ii) leveraging Locality Sensitive Hashing (LSH) [6] to minimize the number of pairwise node comparisons; (iii) choosing a 'center' graph out of l input graphs to which to align all the others, thereby avoiding solving $\binom{l}{2}$ pairwise graph alignment problems (instead solving $l - 1$ alignments and quickly inferring the others by applying simple transformations in the form of sparse matrix multiplications). Figure 1 contains a pictoral overview. Our main contributions are:

- **Flexible Framework.** We propose an efficient and accurate hashing-based family of algorithms, HASHALIGN, which solves the multiple network alignment problem. Our method is general and can readily incorporate any available node and edge attributes. HASHALIGN can be used as a standalone alignment method or provide its solution to initialize optimization problems for pairwise alignment (e.g., [4]).

- **Methods.** As part of our framework, we propose problem-specific choices of node and graph features, and introduce a new, robust construction of hash families.
- **Experiments.** We conduct extensive experiments on synthetic and real data, which show that HASHALIGN is 2–10× faster than the baselines that tackle either the multiple or pairwise alignment problem, while being equally or up to 50% more accurate.

For reproducibility, the code is available at https://github.com/GemsLab/HashAlign.git. Additional supplementary material is provided at https://markheimann.github.io/papers/HashAlign-PAKDD18-full.pdf.

2   Related Work

We review work that is relevant to our problem space and choices of techniques:

Graph Alignment. Scalable methods for pairwise graph alignment include a distributed, belief-propagation-based method for protein alignment [5], a message-passing algorithm for aligning sparse networks when some [4] or all [18] possible matchings are considered, alignment of bipartite networks [13,14], and attributed graph alignment [20]. *Multiple* network alignment, however, poses a further scalability challenge. For instance, the recent optimization-based formulation of [16] solves a bipartite matching problem in $O(n^3)$ time using the Hungarian algorithm. Zhang and Yu [19] introduce the notion of transitivity between graphs to align social networks more scalably with some partial node matchings (anchor links) known *a priori*. Our method HASHALIGN preserves this notion of transitivity for any type of network and requires no anchor links.

Locality-Sensitive Hashing. This technique for efficient similarity search has been used to accelerate the well-known k-nearest neighbor algorithm, often offering theoretical and practical improvements even over sophisticated data structures such as k-d trees [3]. It has also found use in matching problems in other domains, such as ontology matching in information retrieval [8]. In our proposed method, we leverage LSH to efficiently find nodes that are similar. For networks, [12] uses MinHash to find sets of similar nodes in a *single* attributed graph by relying on the adjacency matrix as features, but this is not applicable to the graph alignment setting. Thus, we introduce node-ID invariant representations and adapt LSH to find similarities *across* networks. Our contribution is orthogonal to prior works: a framework for network alignment, HASHALIGN, in which we propose design choices geared toward our specific domain.

3   Proposed Formulation: Two-Graph Alignment

In this section, we first introduce the alignment problem for two graphs. We then describe our proposed approach, and in the next section we extend it to

Table 1. Symbols and definitions. We use bold capital letters for matrices, bold lowercase letters for vectors and normal lowercase letters for scalars.

Symbols	Definitions
$G_p = (\mathcal{V}_p, \mathcal{E}_p)$	Graph p with vertex set $\mathcal{V}_p$ and edge set $\mathcal{E}_p$
$\|\mathcal{V}_p\| = n_p,\ \|\mathcal{E}_p\| = m_p$	Number of nodes and edges in graph G_p, resp.
$s_{G_p}(v)$	$1 \times d_s$ vector of the structural invariants (e.g., PageRank) for node $v \in G_p$
$a_{n\,G_p}(v), a_{e\,G_p}(v)$	$1 \times d_{an}$ vector of node/edge attributes for node $v \in G_p$
$\mathbf{A}_{n\,G_p}, \mathbf{A}_{e\,G_p}$	The stacked node/edge attr. matrices of size $n_p \times d_{an}$ and $n_p \times n_p \times d_{ae}$
$d = d_s + d_{an} + d_{ae}$	Total number of (structural, node, and edge) features
$f_{G_p}(v), \mathbf{F}_{G_p}$	$1 \times d$ all-feature vec. for node $v \in G_p$ and the resp. stacked $n_p \times d$ mat.
SIG_{G_p}	$1 \times 5d$ graph 'signature' vector representing graph G_p
$d(G_i, G_j)$	Distance between graphs G_i and G_j
$\mathbf{S}_{ij}$	Sparse $n_i \times n_j$ similarity matrix between graph G_i and G_j
$\mathbf{M}_{ij}$	$n_i \times n_j$ alignment between graph G_i and G_j
b_i	Bucket i (hashing)
Z	Number of bands (hashing)

the multiple graph alignment problem. Table 1 summarizes the main notations used in our analysis.

3.1 Definition: Relaxed Two-Graph Alignment Problem

The typical graph alignment problem aims to find a one-to-one matching between the nodes of two input graphs. This problem is important, but in many applications it suffices to solve a relaxed version of it: finding *a small set of nodes* that are *likely* to correspond to a given node. Thus, we relax the original alignment problem as follows:

Problem 2 (Relaxed two-graph alignment). Given two graphs, $G_1(\mathcal{V}_1, \mathcal{E}_1)$ and $G_2(\mathcal{V}_2, \mathcal{E}_2)$, which may be (un)directed, (un)weighted and attributed / plain, **we seek to efficiently find** a *sparse*, weighted bipartite graph $G_S = (\mathcal{V}_1 \cup \mathcal{V}_2, \mathcal{E}_S)$ with edges representing potential matching pairs and being weighted by the likelihood of the match:

$$\forall \text{ potential match } (u,v), u \in \mathcal{V}_1, v \in \mathcal{V}_2, \exists e \in \mathcal{E}_S : w_e = sim(u,v)$$

and $|\mathcal{E}_S| < \alpha \cdot \max\{n_1, n_2\}$ where $\alpha \in \mathbb{Z}$ ($\alpha > 1$) controls the density of G_S.

To make sure that nodes are matched only to a few of their closest counterparts, the main requirement in Problem 2 is that G_S is sparse, i.e., $|\mathcal{E}_S| \ll n_1 \times n_2$. Most graph alignment methods find 1-1 matchings between the vertex sets [4,7], and a few approaches relax the requirements of the typical optimization problem to find probabilistic matchings [14], but each method targets a different type of graph (e.g., unipartite, undirected) and most of them rely only on the network structure. In this work we propose a different, intuitive similarity-based approach that encompasses all these settings, leveraging a suitably rich node representation to achieve superior accuracy.

A naive similarity-based method is to: (i) compute all the pairwise similarities between the nodes in G_1 and G_2, and (ii) keep only the edges with similarities greater than a user-specified threshold. Although this approach results in a sparse graph G_S, it has several drawbacks. First, it is computationally expensive, since it computes *all* $n_1 \times n_2$ pairs of similarities and later applies the threshold for edge filtering. Second, the threshold is arbitrary and affects the potential node matchings significantly. Third, it is not clear how to choose the 'right' node representation for similarity computations. Our proposed approach uses hashing to overcome all these issues.

3.2 Node Representation: Handling Node and Edge Attributes

Our framework, HASHALIGN, requires a vector representation of each node. We want these to be comparable across graphs and also leverage node/edge attributes seamlessly (Fig. 2).

We propose to represent each node u with a vector $\mathbf{f}(u)$ of structural features and node/edge attributes (if available). The benefit of this representation is that it can be adjusted to the type of graphs and available information *without any* changes in the problem formulation. Furthermore,

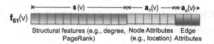

Fig. 2. Proposed feature-based, node-ID invariant representation of vertex v.

it is *node-ID invariant* and can thus be meaningfully compared across graphs. This is not true of representation learning methods like DeepWalk and node2vec, which sample context nodes by their IDs with random walks [9] and thus are not applicable to our multi-network setting [10]. Specifically, in Step 1 of our framework (Fig. 1), we concatenate d_s structural features, d_{an} node attributes, and d_{ae} edge attributes:

- **Structural features** $s \in \mathbb{R}^{1 \times d_s}$. Examples include the so-called local features (e.g., degree variants) and egonet features. The egonet of node u is defined as the induced subgraph of u and its neighbors, and structural features specific to the egonet include its number of edges, its degree, and more. In addition to these features, we also consider features that combine locality with globality, such as PageRank and various types of centrality. We choose specific structural features that are most robust to noise (Sect. 5).
- **Node attributes** $a_n \in \mathbb{R}^{1 \times d_{an}}$. If a graph contains node attributes, the node feature vectors $\mathbf{f}$ can be extended to include those. Numerical features can be simply concatenated with the structural features, while categorical attributes can be incorporated by using 1-hot encoding and concatenated to the previously formed feature vector.
- **Edge attributes** $a_e \in \mathbb{R}^{1 \times d_{ae}}$. We propose converting numerical edge features to node attributes by applying an aggregate function $\xi : \mathcal{E}^{deg_u} \to \mathbb{R}$ (where the domain is the set of edges incident to $u \in \mathcal{V}$ and deg_u is its degree). Examples for $\xi()$ include sum, average, standard deviation, etc. For

categorical features, we propose to encode the distribution of values per feature. For example, if a feature has q possible values, then q entries with their frequencies will be concatenated with the previous features.

3.3 Proposed Hashing-Based Computation of Potential Matchings

Now we have, for each node u in graph G_p, a real-valued vector $\mathbf{f}_{G_p}(u) \in \mathbb{R}^d$ constructed as described in Sect. 3.2. We propose to use Locality Sensitive Hashing (LSH) [6] to find a small number of potential matchings between nodes across graphs (i.e., nodes with high similarity) scalably, without computing all pairs of $n_1 \times n_2$ similarities. In a nutshell, given a similarity function, LSH reduces the dimensionality of high-dimensional data while preserving their local similarities; that is, it efficiently maps similar data points (in our case, nodes) to the same buckets with high probability. Our proposed hashing approach for alignment takes as input the $\mathbf{F}_{G_1} \in \mathbb{R}^{n_1 \times d}$ feature matrix (with row-wise node representations) for G_1 and $\mathbf{F}_{G_2} \in \mathbb{R}^{n_2 \times d}$ for G_2, and hashes them row-wise using an LSH family $\mathcal{H}$.

Definition 1 (LSH-2G). *Given $\mathcal{V}_1$ and $\mathcal{V}_2$, the nodes in graph G_1 and G_2 respectively, along with a similarity function $\phi : \mathbb{R}^d \times \mathbb{R}^d \rightarrow [0,1]$, $\mathcal{H}$ is an LSH-2G family of hash functions such that the probability of two nodes $u, v \in \mathcal{V}_1$ hashing to the same bucket is equal to their similarity, and additionally the probability of two nodes $u \in \mathcal{V}_1$ and $v \in \mathcal{V}_2$ hashing to the same bucket is equal to their similarity: $Pr[h(u) = h(v)] = \phi(\mathbf{f}_{G_1}(u), \mathbf{f}_{G_2}(v))$.*

We propose an LSH-2G family based on the standard measure of cosine similarity (with Euclidean distance in the supplementary material.) We introduce SimHash-2G, a modified version of SimHash [3] that is based on LSH-2G described above. SimHash-2G chooses K randomly generated column vectors $\{\mathbf{r}_1, \ldots, \mathbf{r}_K\} \subset \mathbb{R}^d$ that follow the standard Gaussian distribution (i.e., K random hyperplanes). The LSH-2G family consists of K hash functions: $h_k(u) = sign(\mathbf{f}_{G_p}(u) \cdot \mathbf{r}_k)$. Each of these projects node u on either side of the random hyperplane $\mathbf{r}_k$ (positive or negative sign). For random hyperplane k, the probability of two nodes $u \in \mathcal{V}_1$ and $v \in \mathcal{V}_2$ being mapped to the same bucket is $Pr[h_k(u) = h_k(v)] = 1 - \frac{\theta_{uv}}{\pi}$, where $\theta_{uv} = cos^{-1} \frac{\mathbf{f}_{G_1}(u)\mathbf{f}_{G_2}(v)}{\|\mathbf{f}_{G_1}(u)\|_2 \|\mathbf{f}_{G_2}(v)\|_2}$. The angle $1 - \frac{\theta_{uv}}{\pi}$ captures the proximity of u and v. SimHash-2G computes only the similarity for pairs of nodes according to our proposed SKD-construction (see below).

If a hash function $h_i \in \mathcal{H}$ maps two nodes to the same bucket, that indicates that they *could* be similar, but there is some probability of error. The technique of amplification creates a new LSH family $\mathcal{G}$ with hash function g defined over the functions in $\mathcal{H} = \{h_1, h_2, \ldots, h_K\}$, in order to reduce that probability of error. A standard technique is AND-construction where $g(u) = g(v) \implies \forall i\ h_i(u) = h_i(v)$.

However, the AND-construction is too strict and may lead to many false negatives when finding node matchings. To ameliorate that we use the banding technique: (i) we split each feature vector into Z equal bands, and (ii) per band z, we apply a corresponding LSH-2G family $\mathcal{H}_z$ using AND-construction. In each band, a node can fall into only one bucket, and thus collides with nodes in that same bucket (potential matchings). To handle the observed *skewed* distribution of nodes to buckets and guarantee that each node will have some potential matchings, we introduce the notion of 'importance' of a node collision within a band and propose the SKD-construction.

Definition 2 (Importance σ_{tot} of node collision). *Given two nodes $u \in \mathcal{V}_1$ and $v \in \mathcal{V}_2$, and an LSH-2G family $\mathcal{H} = \{h_1, \ldots, h_K\}$ s.t. $\forall j \; h_j(u) = h_j(v)$ (i.e., both nodes are mapped to bucket $b_\mathcal{H}$), we define the importance of their collision based on $\mathcal{H}$ as the inverse of the size of the corresponding bucket: $\sigma_\mathcal{H}(u, v) = \frac{1}{|b_\mathcal{H}|}$. The total importance score of a node pair collision over all bands and their corresponding LSH families $\mathcal{H}' = \{\mathcal{H}_1, \ldots \mathcal{H}_Z\}$ is defined as: $\sigma_{tot}(u, v) = \sum_{\mathcal{H} \in \mathcal{H}'} \sigma_\mathcal{H}(u, v) \cdot \mathbb{1}_{h_j(u) = h_j(v), \forall h_j \in \mathcal{H}}$.*

Intuitively, the importance of a collision is higher if a few nodes are mapped to a bucket, as the bucket has higher discriminative power. The notion of importance tackles the skewness that we observe in the mapped nodes in the graph alignment setting. Based on this definition, we propose the SKD-construction (where SKD stands for SKeweD).

Definition 3 (SKD-construction). *Given $u \in \mathcal{V}_1$ and $v \in \mathcal{V}_2$, and LSH-2G families $\mathcal{H}' = \{\mathcal{H}_1, \ldots, \mathcal{H}_Z\}$, a new family $\mathcal{G}$ with hash function g is based on SKD-construction:*

$$g(u) = g(v) \implies \sigma_{tot}(u, v) \in TOP_\alpha(u),$$

where $TOP_\alpha(u)$ is the set of top-α total importance scores $\sigma_{tot}(u, v')$ for $v' \in \mathcal{V}_2$, and α is the small factor that controls the density of G_S in Problem 2.

Intuitively, SKD-construction computes the pairwise similarities of nodes that collide often (but not always, like AND-construction) and have important collisions that manage to distinguish the nodes (i.e., it penalizes functions that lead to skewed results).

3.4 From Similarities to Matchings

As shown in Step 3 of Fig. 1, the hashing approach that we introduced returns a small number of high similarities between the nodes of graphs G_1 and G_2, giving us an $n_1 \times n_2$ sparse matrix $\mathbf{S}$ with node similarities. Here we provide ways to use the similarity information in $\mathbf{S}$ to find the node matchings or correspondences $\mathbf{M} \in \mathbb{Z}^{n_1 \times n_2}$:

- **Greedy matching:** Assuming that the higher the similarity score, the more likely two nodes are to match [14,20], we can greedily make independent decisions for the best match of each node in G_1 through a function $\chi : \mathcal{V}_1 \to \mathcal{V}_2$

s.t. $\chi(u) = \text{argmax}_v\{\mathbf{S}_{uv}\}$. Since nodes are matched independently, this is very efficient and parallelizable, but may match more than one node in graph G_1 to the same node in G_2. It is a preferred method for very large networks or networks of different sizes, and also when multiple potential matchings are desired. In the latter case, it can be trivially extended by updating the function $\chi()$ to return more top potential matchings (instead of only the best one).

- **Collective matching:** An alternative is to leverage existing approaches that find 1-to-1 matchings collectively, given a similarity matrix $\mathbf{S}$. In Sect. 5 we consider scalable options for doing so and study their trade-offs.

4 HashAlign: Multiple Graph Alignment

In this section, we extend our HashAlign framework to multiple graph alignment, extending the formal definition of the relaxed 2-graph alignment problem.

Problem 3 (Relaxed multiple graph alignment). Given a set of graphs, $\mathcal{G} = \{G_1(\mathcal{V}_1, \mathcal{E}_1), \ldots, G_l(\mathcal{V}_l, \mathcal{E}_l)\}$, which may be (un)directed, (un)weighted and attributed / plain, we seek to efficiently **find** a sparse, weighted bipartite graph $G_{Sij} = (\mathcal{V}_i \cup \mathcal{V}_j, \mathcal{E}_{Sij})$ for each pair of graphs $<G_i, G_j>$, s.t. $\mathcal{E}_{Sij}$ has the potential matching pairs between their vertex sets and the weights describe how likely the nodes are to match.

Efficient Computation. The key insight to reduce computation is to use one of the l graphs as the 'baseline' graph G_C and align the remaining $l-1$ graphs with it in parallel. This approach avoids computing $O(l^2)$ pairwise graph alignments, instead leading to $l-1$ matching matrices $\mathbf{M}_{2C}, \ldots, \mathbf{M}_{lC}$ (w.l.o.g. we choose graph $G_C = G_1$ in our notation, but we will explain next the choice of G_C). Inspired by the idea of transitivity [19], which requires node matching consistency between pairs of graphs (Fig. 3), we effi-

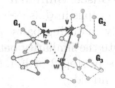

Fig. 3. Node matching consistency: If $u = v$ and $v = w$, then u should match to w (by transitivity).

ciently infer the remaining matching matrices $\mathbf{M}_{ij}$ (where $i, j \neq C$) via sparse matrix multiplications (Step 4 in Fig. 1): $\mathbf{M}_{ij} = \mathbf{M}_{iC} \cdot \mathbf{M}_{jC}^T$.

Choice of G_C. To reduce the induced alignment errors and their propagation to the inferred matchings, we propose the 'center' graph (i.e., the graph in $\mathcal{G}$ with the minimum total distance from the remaining graphs) as the baseline graph G_C (Step 2 in Fig. 1):

$$\text{argmin}_C \sum_j d(G_C, G_j) = \sum_j \|\text{SIG}_{G_C} - \text{SIG}_{G_j}\|_2,$$

where $d(G_C, G_j)$ is the distance between G_C and G_j, and SIG is a graph 'signature', which we create by applying an aggregate feature function $\xi()$ over a

---

Algorithm 1. HASHALIGN

Input: (1) $\mathcal{G}=\{G_1, G_2, \cdots, G_l\}$; (2) [OPT] Per graph i, node/edge attr.
$\mathbf{A}_{n_{G_i}}^{n \times d_{a_n}} / \mathbf{A}_{e_{G_i}}^{n \times n \times d_{a_e}}$
Output: A set of matching matrices $\{\mathbf{M}_{ij}\}$ for $i,j \in \{1, \ldots, l\}$
1: /* **STEPS 1&2**: NODE REPRESENTATION AND CENTER DISCOVERY */
2: **For** $G \in \mathcal{G}$ **do**
3:      $\mathbf{F}_G = \text{extractFeatures}(G, \mathbf{A}_G^{n \times d_{a_n}}, \mathbf{A}_G^{n \times n \times d_{a_e}})$ ▷ Sec. 3.2
4: $G_C = \text{findCenter}(\xi(\mathbf{F}_{G1}), \xi(\mathbf{F}_{G2}), \cdots, \xi(\mathbf{F}_{Gl}))$ ▷ Eq. (4) & aggregate function
     $\xi()$=SIG
5: /* **STEP 3**: HASH-BASED SIMILARITY (assuming q buckets in total) */
6: $\{b_1, \ldots, b_q\} = \text{SimHash-2G}(\mathbf{F}_{G1}, \ldots, \mathbf{F}_{Gl})$ ▷ Sec. 3.3 (or EDHash-2G in Appendix B)
7: $\{\mathbf{S}_{1C}, \mathbf{S}_{2C}, \cdots, \mathbf{S}_{lC}\} = \text{computeSparseSimilarities}(b_1, \ldots, b_q)$ ▷ SKD-construction
8: /* **STEP 4**: NODE MATCHING */
9: $\{\mathbf{M}_{1C}, \mathbf{M}_{2C}, \cdots, \mathbf{M}_{lC}\} = \text{GREEDY or COLLECTIVE}(\mathbf{S}_{1C}, \mathbf{S}_{2C}, \cdots, \mathbf{S}_{lC})$ ▷ Sec. 3.4
10: **For** $i,j \in \{1, \ldots, l\}$ **do**
11:      $\mathbf{M}_{ij} = \mathbf{M}_{iC} \times \mathbf{M}_{jC}^T$ ▷ Sec. 4

---

graph's nodes. In our work, we use the mean, median, standard deviation, skewness, and kurtosis of each of the d features, giving us a $5d$-dimensional vector (shown in Step 1 of Fig. 1). Intuitively and empirically, the center graph being as close as possible the other graphs can make the center-based alignments more precise.

Hash-Based Similarity. After hashing all the feature-based node vectors of all the graphs in $\mathcal{G}$ as described in Sect. 3.3, we compute the similarity scores for possibly matching pairs of nodes according to the SKD-construction. We only compute the similarity between nodes in the center graph (right hand-side in the buckets in step 2 of Fig. 1) and nodes in the *peripheral*, or non-center, graphs (left hand-side).

Putting Everything Together. We propose HASHALIGN, a fast, hash-based, multiple graph alignment approach, which is described at a high level in Algorithm 1 (and pictorially in Fig. 1, where $Z = 1$ for simplicity.) It consists of four main steps: (i) node representation, (ii) 'center' graph identification, (iii) hash-based similarity, and (iv) node matching. In line 7 of Algorithm 1, SimHash-2G is applied to l graphs in parallel.

Computational Complexity of HASHALIGN. Our framework makes two main substitutions for computational savings. First, it replaces full pairwise similarity computations that are quadratic in the number of nodes with hashing in only $O(K \cdot n_p \cdot d)$ time for graph G_p with n_p nodes, if we use K hash functions on d-dimensional feature vectors. Second, it replaces all $\binom{l}{2}$ pairwise network alignments with only $l-1$ pairwise network alignments to a center graph (chosen in $O(l^2 \cdot d)$ time), inferring the remainder with sparse matrix multiplications. More details are given in the supplementary material.

5 Experimental Analysis

In this section, we seek to answer the following questions: (1) How robust is our framework compared to baselines for different levels of noise in the graphs (both in the structure and node/edge attributes)? (2) How could HASHALIGN help existing alignment methods perform better and how could these help our method? (3) How do our methods scale when aligning multiple graphs collectively? We answer these questions on three datasets, described in Table 2. We also include additional experiments, such as a *sensitivity analysis* of HASHALIGN to different parameters, in the supplementary material.

Baselines. We consider 3 baseline methods commonly used in the literature: NetAlign [4], Final [20], and IsoRank [18]. We compare their performance against our method, HASHALIGN, where we infer alignments greedily from the hashing-based node similarities. The baselines accept a matrix $\mathbf{L}$ representing prior alignment information between the nodes of the original graphs. By default, we provide a thresholded similarity matrix based on the node attributes to assure good performance based on the attribute information, even for the baselines that are not formulated for it (NetAlign and IsoRank). We also consider two variants of HASHALIGN, namely HASHALIGN-NA and HASHALIGN-FN, which respectively use NetAlign and Final to infer alignments from the node similarities as the final step of HASHALIGN (Sect. 3.4).

Data. We evaluate our proposed algorithms on three datasets along with the synthetic data that we generated from them (via permutations and added noise, as in [14,20]). Formally, given a graph G_1 with adjacency matrix $\mathbf{A}$, we create a noisy graph G_2 with matrix $\mathbf{B} = \mathbf{PAP}^\top$ (i.e., a permutation of itself), where $\mathbf{P}$ is a randomly generated permutation matrix (i.e., with one nonzero entry per row/column). Synthetic noise is applied to both graph structure and labels throughout our experiments to simulate real-world scenarios where the graphs are matchable but different. The noise level p indicates that with probability p, Gaussian noise with $std = 1$ is added to an edge weight; a binary edge label is flipped; or a categorical node/edge label value is changed.

Table 2. Description of real datasets.

Datasets	# Nodes	# Edges	Graph type	Labels	Description
Connectome [1]	941	9,622	Undirected	-	fMRI-inferred graphs
E-mail [2]	1,133	5,451	Undirected	5	Email communications
DBLP [20]	42,252	210,320	Undirected	1	Coauthorship network

Evaluation Metric. Following the literature, we compute the alignment accuracy as $\frac{\text{\# correct matchings}}{\text{\# total matchings}}$, where the total number of matchings between G_i and G_j is equal to the minimum number of nodes between the two graphs, $min\{n_i, n_j\}$.

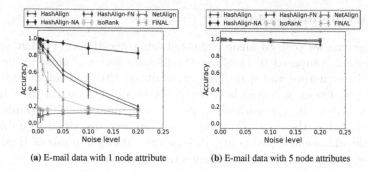

(a) E-mail data with 1 node attribute   (b) E-mail data with 5 node attributes

Fig. 4. E-mail dataset (experimental results on other datasets are similar): Effectiveness w.r.t. noise on both attributes and the graph structure. Methods based on the HASHALIGN framework achieve highest accuracy, particularly with limited node attribute information.

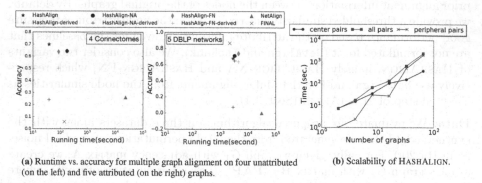

(a) Runtime vs. accuracy for multiple graph alignment on four unattributed   (b) Scalability of HASHALIGN.
(on the left) and five attributed (on the right) graphs.

Fig. 5. (a) HASHALIGN has stable efficiency across different kinds of networks. Final is fast, but its accuracy is subject to whether node/edge labels exist. (b) HASHALIGN scales linearly in terms of alignment with the center graph.

Experimental Setup. We used the following structural attributes: degree, betweenness centrality, PageRank, egonet degree, average neighbor degree, and egonet connectivity. We chose attributes that are robust to noise (graphs to be aligned are often seen as noisy permutations of each other). More details are given in the supplementary material.

To test the ability of HASHALIGN to incorporate different kinds of features, we also generate synthetic node/edge attributes, if none are available in the real data. For each noise level p, we generate 3 pairs of graphs and report the average accuracy (along with a 95% confidence interval). HASHALIGN is implemented in Python2.7, and the structural feature extraction is based on SNAP [15]. We ran the experiments on Intel(R) Xeon(R) CPU E5 @ 3.50 GHz and 256 GB RAM.

Two-Graph Alignment. We aligned pairs of graphs on all the datasets (with G_1 the real graph and G_2 its noisy permutation at noise level p, generated as described above) and got consistent results. Only the result from the E-mail network is shown for brevity. With only 1 binary attribute (Fig. 4a), `NetAlign` and `IsoRank` perform poorly because the similarity matrix **L** built using just 1 attribute is not informative enough. However, these methods work significantly better in our framework, and the gap between HASHALIGN variants and others grows as noise levels increase. In Fig. 4b, HASHALIGN-NA achieves perfect results in the presence of 5 node attributes, though all methods perform essentially perfectly with abundant node attribute information.

Multiple Graph Alignment. We evaluate HASHALIGN against other methods for multiple graph matching on two datasets: five connectome networks [1] without any labels, and four DBLP co-author networks extracted from the whole DBLP dataset following the settings in [20] with one categorical label (the most frequent conference that an author attends.) Both experiments are conducted with $p = 2\%$ noise.

Figure 5a shows how different methods perform in terms of efficiency and accuracy. When there is no label information to help guide the alignment process (i.e., in the case of connectomes), HASHALIGN achieves best accuracy with short running time for peripheral-center graph pairs alignment, followed by HASHALIGN-FN and HASHALIGN-NA. As for the DBLP networks, since the label with 29 *distinct* values is very discriminative, `Final` can achieve very good efficiency, while HASHALIGN and its variants also have comparable performance. However without node labels, `Final` matches less than 10% of all node pairs, which can be boosted to over 60% if we feed it the hash-based similarity matrices of HASHALIGN. Pairwise graph alignment is the most computationally expensive for large numbers of graphs (see Fig. 5b), but for fewer graphs of the sizes in Fig. 5a, computing all pairwise alignments yields the highest accuracy, and is thus our recommendation if computational resources are not an issue. However, center graph alignment (the '*derived*' versions of HASHALIGN variants) often still outperforms the baselines, and in some cases matches the accuracy of the full pairwise comparisons (e.g., HASHALIGN-NA on the connectome data.)

These results clearly show that HASHALIGN leads to significant improvement over existing methods with regard to both accuracy and runtime. In summary, we see that on average, HASHALIGN (including its variants) are 2× faster and 10 to 20% more accurate than the baselines in pairwise alignment, and 2× faster while up to 50% more accurate in multiple graph alignment. However, these existing methods may have their place within our framework (see Step 4 of Fig. 1), where they may be used to accurately infer alignments from the hashing-based node similarities.

We also verify that the proposed method scales as the number of graphs grows by generating up to 64 synthetic graphs from the aforementioned connectome network with $p = 0.02$ noise, $Z = 2$ and $K = 40$. As shown in Fig. 5b, HASHALIGN's runtime scales linearly in terms of alignment with the center graph. The runtime for peripheral graph alignments (i.e., w/o the center graph) using sparse matrix multiplication scales subquadratically, as the slope indicates. We omitted the runtime for feature extraction as it is linear on the number of graphs, and does not contribute much to the runtime.

6   Conclusions

We study the problem of multiple graph alignment and propose HASHALIGN, an intuitive, fast and effective similarity-based approach that readily handles any type of input graph. Our method adapts LSH to graph alignment, with a new construction technique and an appropriate node-ID-invariant node representation for this task. Leveraging the rule of matching transitivity, it scales up to many graphs while avoiding solving the expensive alignment task for each pair of graphs separately. Our experiments on real data (incl. sensitivity analysis in the supplementary material) show that HASHALIGN can stand alone as a multi-network alignment tool or be combined with existing methods that require a small set of possible matchings as input. In most cases, it is more accurate, more robust to noise, and/or faster than the baselines. Our work suggests that hashing is a promising direction for scaling up network alignment. Future work could include extending HASHALIGN to use learned node representations specifically designed for multi-network problems, as in the very recent work of [11]. Here one challenge would be devising suitable graph signatures for efficient multiple graph alignment.

Acknowledgements. This material is based upon work supported in part by the National Science Foundation under Grant No. IIS 1743088, and the University of Michigan. Any opinions, findings, and conclusions or recommendations expressed in this material are those of the author(s) and do not necessarily reflect the views of the National Science Foundation or other funding parties. The U.S. Government is authorized to reproduce and distribute reprints for Government purposes notwithstanding any copyright notation here on.

References

1. COBRE (2012). http://fcon_1000.projects.nitrc.org/indi/retro/cobre.html
2. Konect: Koblenz network collection (2016). http://konect.uni-koblenz.de/networks/
3. Andoni, A., Indyk, P.: Near-optimal hashing algorithms for approximate nearest neighbor in high dimensions. In: FOCS. IEEE (2006)
4. Bayati, M., Gleich, D.F., Saberi, A., Wang, Y.: Message-passing algorithms for sparse network alignment. ACM TKDD **7**(1), 3:1–3:31 (2013)
5. Bradde, S., Braunstein, A., Mahmoudi, H., Tria, F., Weigt, M., Zecchina, R.: Aligning graphs and finding substructures by a cavity approach. Europhys. Lett. **89**, 37009 (2010)
6. Datar, M., Immorlica, N., Indyk, P., Mirrokni, V.S.: Locality-sensitive hashing scheme based on p-stable distributions. In: SCG, pp. 253–262. ACM (2004)
7. Ding, C.H.Q., Li, T., Jordan, M.I.: Nonnegative matrix factorization for combinatorial optimization: spectral clustering, graph matching, and clique finding. In: ICDM (2008)
8. Duan, S., Fokoue, A., Hassanzadeh, O., Kementsietsidis, A., Srinivas, K., Ward, M.J.: Instance-based matching of large ontologies using locality-sensitive hashing. In: Cudré-Mauroux, P., Heflin, J., Sirin, E., Tudorache, T., Euzenat, J., Hauswirth,

M., Parreira, J.X., Hendler, J., Schreiber, G., Bernstein, A., Blomqvist, E. (eds.) ISWC 2012. LNCS, vol. 7649, pp. 49–64. Springer, Heidelberg (2012). https://doi. org/10.1007/978-3-642-35176-1_4

9. Goyal, P., Ferrara, E.: Graph embedding techniques, applications, and performance: a survey. arXiv preprint arXiv:1705.02801 (2017)

10. Heimann, M., Koutra, D.: On generalizing neural node embedding methods to multi-network problems. In: KDD MLG Workshop (2017)

11. Heimann, M., Shen, H., Koutra, D.: Node representation learning for multiple networks: The case of graph alignment. arXiv preprint arXiv:1802.06257 (2018)

12. Khan, K.U., Nawaz, W., Lee, Y.K.: Set-based unified approach for attributed graph summarization. In: IEEE BDCC, December 2014. https://doi.org/10.1109/ BDCloud.2014.108

13. Koutra, D., Faloutsos, C.: Individual and collective graph mining: principles, algorithms, and applications. Synth. Lect. Data Min. Knowl. Discov. 9(2), 1–206 (2017)

14. Koutra, D., Tong, H., Lubensky, D.: Big-align: fast bipartite graph alignment. In: ICDM. IEEE (2013)

15. Leskovec, J., Sosič, R.: SNAP: a general-purpose network analysis and graph-mining library. ACM TIST 8(1), 1 (2016)

16. Malmi, E., Chawla, S., Gionis, A.: Lagrangian relaxations for multiple network alignment. Data Mining Knowl. Discov. 31, 1–28 (2017)

17. Safavi, T., Sripada, C., Koutra, D.: Scalable hashing-based network discovery. In: ICDM. IEEE (2017)

18. Singh, R., Xu, J., Berger, B.: Global alignment of multiple protein interaction networks with application to functional orthology detection. PNAS 105(35), 12763–12768 (2008)

19. Zhang, J., Yu, P.S.: Multiple anonymized social networks alignment. In: ICDM. IEEE (2015)

20. Zhang, S., Tong, H.: Final: Fast attributed network alignment. In: KDD. ACM (2016)

Evaluating and Analyzing Reliability over Decentralized and Complex Networks

Jaron Mar[1], Jiamou Liu[1](✉), Yanni Tang[2], Wu Chen[3](✉), and Tianyi Sun[1]

[1] The University of Auckland, Auckland, New Zealand
jiamou.liu@auckland.ac.nz
[2] Southwest University, Chongqing, China
[3] Institute of Logic and Intelligence, and College of Computer and Information Science, Southwest University, Chongqing, China
chenwu@swu.edu.cn

Abstract. In an increasingly interconnected and distributed world, the ability to ensure communications becomes pivotal in day-to-day operations. Given a network whose edges are prone to failures and disruptions, reliability captures the probability that traffic will reach a target location by traversing edges starting from a given source. This paper investigates reliability in decentralized and complex networks. To evaluate reliability, we introduce a multi-agent method that involves pathfinding agents to reduce the graph. Performance of this method is tested on scale-free and small-world networks as well as real-world spatial networks. We also investigate reliability score which aims to rank the capability of nodes in terms of traffic dissemination traffic across all nodes. Analysis over spatial networks indicates that the reliability score correlates with central and sub-central regions in a geographical region.

Keywords: Network reliability · Spatial network analysis
Transportation networks · Monte-Carlo method

1 Introduction

In an increasingly interconnected and distributed world, the ability to ensure communications becomes pivotal in day-to-day operations. This raises the issue of reliability, which captures the extent to which an individual is able to communicate consistently in the presence of uncertainty. Take, as an example, the case of a wireless mesh network [4]. As the wireless links between devices are subject to random failures, it is crucial to measure the capability of a device in sending or receiving messages [12]. It is important to evaluate this capability in a network context, as this may reveal hidden patterns, differentiate between reliable/unreliable regions, facilitate optimization and improve network robustness.

This paper investigates reliability from a structural perspective. Abstractly, a complex system, e.g. communication, physical or social networks, consists of many components or *nodes*, whose functions are characterized by their mutual

© Springer International Publishing AG, part of Springer Nature 2018
D. Phung et al. (Eds.): PAKDD 2018, LNAI 10939, pp. 740–751, 2018.
https://doi.org/10.1007/978-3-319-93040-4_58

interactions. As links between nodes are prone to random error and disruption, we use a *stochastic binary system* to model uncertainty, where a link is either *up* or *down* [18]. *Two-terminal reliability* measures the probability that a message (or any other item) starting from a particular node reaches a target along links of the network [14]. The sum of two-terminal reliability over all possible targets from a specific source indicates the "positional reliability" held by the source.

Positional reliability reveals important insights. Extending beyond applications in wireless mesh network discussed above, consider a transportation network. Here, failures of edges imply blockage of traffic along stretches of roads due to incidents or blockage. A node with low reliability implies a location that is more vulnerable in terms of transporting goods across the network [15]. A third potential application involves the social network of employees in a company. Edges represent communication channels between people. Unexpected disruptions may occur that affect communication between two individuals. In this scenario, a node having low reliability implies its uncertainty to deliver messages across the entire organization, and thus corresponds to a form of *structural holes*, which is crucial in social network analysis [10]. A node with high reliability, on the other hand, refers to influential individuals in the organization [17].

Contributions. This paper has three main contributions: (1) Monte Carlo approaches have been a common approach to evaluate two-terminal reliability [16]. As this method is confronted with high computational costs over large networks, we propose a new method to approximate reliability by deploying multiple agents who traverse asynchronously within the network, pruning unnecessary parts of the network away. This is also useful for decentralized networks where no global information is stored. (2) Building on notions in reliability theory, we investigate the reliability score of nodes in a network. This rank uncovers significant individuals in the network structure and reveals new insights. (3) The tools explored in the paper can be applied to the analysis of spatial networks, and evaluate reliability over geographical locations. Our analysis reveals that in general, regions of high reliability correspond to areas of high economic and social activities.

2   Main Definitions and Related Works

By a *network*, we mean a graph $G = (V, E)$ where V is the node set and E is the (undirected) edge set; no multi-edge nor self-loop is present. A *path* in the network is a sequence of edges $v_0 v_1, \ldots, v_{k-1} v_k$; the path is *simple* if $v_0, \ldots, v_k$ are pairwise distinct. A *subgraph* is a graph $G' = (V, E')$ where $E' \subseteq E$. We use $S(G)$ to denote the set of all subgraphs of G.

Edges represent two-way links that are subject to random failures; at any time instance, the network G presents itself in the form of a subgraph $G' = (V, E')$ where $E' \subseteq E$ contains all edges that are *up* and edges in $E \setminus E'$ are *down*. More abstractly, a *stochastic binary system (SBS)* defined on G assumes that each edge $e \in E$ is up with a probability $p_e \in [0, 1]$; use $\boldsymbol{p}$ to denote the vector $(p_e)_{e \in E}$. The SBS specifies a probability distribution over $S(G)$: For any subset $E' \subseteq E$, the probability of the subgraph $G' = (V, E')$ is given by $\prod_{e \in E'} p_e \cdot \prod_{e \notin E'} (1 - p_e)$. For subgraph $G' = (V, E')$ and nodes $s, t \in V$, set $\Phi(G', s, t) = 1$ if G' contains a path between s and t; and $\Phi(G', s, t) = 0$ otherwise.

Definition 1 [7]. *Given nodes s and t in V, the* two-terminal reliability *between s and t is the probability that a path exists from s to t along which every edge is up, i.e., it is $R(G, s, t, \boldsymbol{p}) = \Pr(\Phi(G', s, t) = 1)$.*

For a fixed source s, assume that the target t is chosen uniformly randomly. The expected rate of successful communication reflects the reliability of s, i.e., to what extent we expect a message from s may reach a randomly chosen node.

Definition 2. *Define $R_s(t) = R(G, s, t, \boldsymbol{p})$ for all $t \in V$ where $t \neq s$. The reliability score of a node $s \in V$ is $\mathsf{rel}(s) = \mathbb{E}[R_s]$, in other words,*

$$\mathsf{rel}(s) = \left(\sum \{ R(G, s, t, \boldsymbol{p}) \mid t \neq s \} \right) / (|V| - 1)$$

Related Works. Exact evaluations of network reliability include min-cut or Boolean algebraic approaches, which are only suitable for small networks or networks of specific types (e.g. parallel-series networks) due to high complexity [25]. Computing two-terminal reliability over a network is NP-hard in general [6]. In light of its hardness, Monte Carlo methods were commonly used [16]. More recently, methods for approximating network reliability emerged which utilized optimization techniques such as neural networks [28], ant colony [23], and binary decision diagrams [27]. We point out that this paper differs from the works above: While these earlier works focus on the ability of a network to surpass random failures and disruptions to maintain consistent performance, this paper aims to evaluate and rank individual nodes in the network. A failure to communicate between two nodes does not mean that the system "fails", and thus, the goal is not to predict how likely the network prevents failures, but rather, to give a quantitative comparison on the "strength" of nodes with respect to dispersing traffic.

Mathematical models of networks allow rigorous simulation and analysis of a range of phenomena, such as information diffusion [17], community structures [21], and error-resilience [5]. Tools such as centralities are used to study power and influence in organizational networks [20], network integration [22,24], and potential repercussions of tie breakages [13,19]. Reliability is also important to spatial networks analysis. Chassin and Posse in [11] modeled cascading failures of an electric grid using a scale-free model. Neumayer and Modiano in [26] studied the reliability of optical fiber networks that are laid out along physical terrestrial paths. The authors of [9] proposed a challenge to enrich network science with reliability analytics. This paper echoes this call and investigates a range of real-world spatial networks.

Algorithm 1. ReliabEstimate($V, E, \boldsymbol{p}, s, t$)

INPUT: Graph (V, E), $\boldsymbol{p}$, nodes s, t
 1: stack $:=$ new stack with element (s, null)
 2: **while** stack not empty **do**
 3:     $(v, e) :=$ stack.pop()
 4:     **if** e is down **then continue**
 5:     **else if** the state of e is not determined
 6:         Sample e's state (up/down) with probability p_e; **continue** if e is down.
 7:     **if** v is not seen **then**
 8:         set v as seen
 9:         **if** $v = t$ **then return** Success
10:         **for all** incident edges vu **do**
11:             **if** vu is up **then** add (u, vu) to stack
12: **return** Failure

---

3 Agent-Based Reliability Estimation

3.1 Estimating Two-Terminal Reliability

The crude Monte Carlo method takes K samples (which are subgraphs of G) and evaluates the frequency of samples that contain a path between s and t, and outputs it as an estimation of $R := R(G, s, t, \boldsymbol{p})$. The variance of this method is $R(1 - R)/K$. The method takes time $O(|E|)$ per sample.

In a decentralized network, one may deploy independent threads, i.e., "agents" who traverse the graph starting from s, each maintaining their own state of the graph and terminating successfully if they reach t or unsuccessfully if they can not explore the graph further due to the edges being down or all possible edges have been traversed. The type of traversal performed by the agents may affect the running time on different graphs structures. Algorithm 1 describes the operation of such an agent who traverses the graph depth-first; a breadth-first agent could be defined with minor adaptations.

Efficiency is improved by the ReliabEstimate algorithm as an agent may not sample the entire graph. We now aim to further reduce computational cost.

Definition 3. *An edge e is called (s, t)-critical if it appears on a simple path between s and t. The (s, t)-critical subgraph is $G\lceil_{s,t} = (V_{s,t}, E_{s,t})$ where $V_{s,t}$ contains all end points of critical edges and $E_{s,t}$ is the set of critical edges.*

Theorem 1. *For a graph $G = (V, E)$, probability vector $\boldsymbol{p}$ and $s, t \in V$, we have $R(G, s, t, \boldsymbol{p}) = R(G\lceil_{s,t}, s, t, \boldsymbol{p}')$ where $\boldsymbol{p}' = (p_e)_{e \in E'}$.*

Proof. Recall that $S(G\lceil_{s,t})$ denotes the set of subgraphs of $G\lceil_{s,t}$. For a sample $G' \in S(G\lceil_{s,t})$ taken from the probability distribution defined by $\boldsymbol{p}\lceil E_{s,t}$ on $S(G\lceil_{s,t})$, let $\Phi'(G', s, t) = 1$ if G' contains a path between s and t; and 0 otherwise. $\Pr(\Phi'(G', s, t) = 1) = \sum_{\substack{E' \subseteq E_{s,t} \\ \Phi'((V, E'), s, t) = 1}} \left(\prod_{e \in E'} p_e \cdot \prod_{e \notin E'} (1 - p_e) \right)$.

As $\displaystyle\sum_{F'\subseteq E\backslash E_{s,t}}\left(\prod_{e'\in F'}p_{e'}\cdot\prod_{e'\notin F'}(1-p_{e'})\right) = 1$, $\Pr(\Phi'(G',s,t)=1)$ equals to

$$\sum_{\substack{E'\subseteq E_{s,t}\\ \Phi'((V,E'),s,t)=1}}\left(\prod_{e\in E'}p_e\cdot\prod_{e\notin E'}(1-p_e)\right)\cdot\sum_{F'\subseteq E\backslash E_{s,t}}\left(\prod_{e'\in F'}p_e\cdot\prod_{e'\notin F'}(1-p_e)\right),\ \text{which}$$

is $\displaystyle\sum_{\substack{E'\subseteq E_{s,t}\\ \Phi'((V,E'),s,t)=1}}\sum_{F'\subseteq E\backslash E_{s,t}}\left(\prod_{e\in E'}p_e\cdot\prod_{e\notin E'}(1-p_e)\cdot\prod_{e'\in F'}p_e\cdot\prod_{e'\notin F'}(1-p_e)\right)$ and is

thus equal to $\displaystyle\sum_{\substack{G'=(V,E')\in S(G)\\ \Phi(G',s,t)=1}}\left(\prod_{e\in E'}p_e\cdot\prod_{e\notin E'}(1-p_e)\right) = R(G,s,t,\boldsymbol{p}).$ $\qquad\square$

Therefore to estimate $R(G,s,t,\boldsymbol{p})$, it is sufficient to apply the ReliabEstimate algorithm on the critical subgraph of G. The next goal is to compute the critical subgraph $G\!\upharpoonright_{s,t}$ from a given G. Our method relies on two procedures: The first is a traverse algorithm where multiple agents traverse along edges and learn about the graph topology collectively; see Algorithm 2 for the operation of a single agent. The second procedure uses the results of the first procedure and backtracks from t to build up the critical subgraph; see Algorithm 3.

Algorithm 2. GraphReduction-Traverse(G,s,t)

INPUT: (V,E), nodes s,t
 1: **procedure** AGENTTRAVERSE(Agent A, Node u)
 2:     **while** stack is not empty **do**   ▷ stack is shared among all agents
 3:         $(v,e) :=$ stack.pop()
 4:         **if** both e and v are labeled unseen **then**
 5:             Label v and e as seen (which are made available to other agents)
 6:             $U := \varnothing$ ▷ U stores unseen edges
 7:             **for all** edges $vu \in E$ **do**
 8:                 **if** vu is not seen and $u \neq s$ **then**
 9:                     $U := U \cup \{vu\}$ and stack.push(u,vu)
10:             **if** $U = \varnothing$ **then** call TERMINATE(A,v)
11:             **else if** $|U| > 1$ **then** set OrnPath $:= (v)$, and call TERMINATE(A,v)
12:             Pick edge vv' from U ▷ the next edge
13:             Append v' to OrnPath
14:             Call TERMINATE(A,t) if $v' = t$; otherwise, stack.push(v',vv').
15:         **else**
16:             Call TERMINATE(A,v)
17: **procedure** TERMINATE(AGENT A, NODE v)
18:     store OrnPath into OP(v) at the node v
19:     initialize OrnPath as an empty list
20:     **if** stack is not empty **then** start new traversal
21:     **else** wait for stack to become non-empty to start new traversal

---

Agents in the traversal procedure share a common stack which contains nodes from which a traversal starts. An edge (or node) is labeled as either seen or unseen, depending on whether it has been traversed/seen. This label is visible by all agents, and thus an edge is used at most once. When traversing the network, an agent starts from the popped node v, and follows an outward direction away from s. The agent terminates when: it just traversed an edge that had already been used, or it reaches a dead end, or all outgoing neighbors have been exhausted, or it has reached t; The condition at Line 11 is a special termination condition to speed up the backtrack algorithm by making the paths the agents traverse and saves as disjoint as possible. We give a conceptual overview:

Definition 4. *An* oriented path *is a sequence* $(v_0, v_1, \ldots, v_k)$ *such that each* $v_i v_{i+1}$ *is an edge; it is* simple *if no node is repeated; it is in a valid orientation if it is a subsequence of a simple oriented path from s to v_k.*

The algorithm partitions the graph into a set of disjoint oriented paths, each in a valid orientation. Any oriented path is stored in its end point v_k. Fig. 1 illustrates a simple example to make clear of the notions above. Here the source s corresponds to A (in green) and the target t is H (in red). Traversed paths by different agents are shown with arrows of various colors and the number on each edge indicates the step the agent would make. The set of oriented paths $OP(v)$ stored on node v are indicated beside each node v.

The backtrack procedure iteratively adds nodes into the critical subgraph $G\lceil_{s,t}$. The target t is the first added node as any agent that terminated at t would have stored the path it took in the set $OP(t)$. Any nodes (and edges) along this path will be added to the critical subgraph. We then continue the backtrack by checking paths that the agents traversed through which terminated at these added nodes. In a distributed implementation the backtrack algorithm can start once any agent reaches t where it is possible to have multiple agents working on the backtracking as well. As each edge is traversed at most once, the total running time among all agents in the graph reduction process is $O(|V| + |E|)$ which may be reduced further if agents run in parallel.

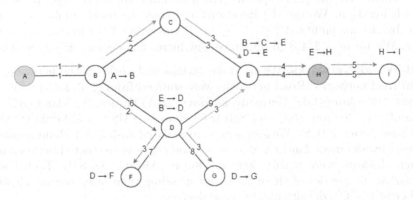

Fig. 1. The resulting oriented paths of the traverse algorithm. (Color figure online)

Algorithm 3. GraphReduction-Backtrack(G, s, t)

INPUT A (possibly $\varnothing$) set of oriented path $OP(u)$ stored in each $u \in V$
OUTPUT Critical subgraph $G\restriction_{s,t}$
1: Initialize backtrack as a new queue containing t
2: **while** backtrack not empty **do**
3:     $v :=$ backtrack.dequeue()
4:     **if** $v \neq$ BTseen and $v \neq s$ **then**
5:         Set BTseen $:= v$
6:         **if** $OP(v) \neq \varnothing$ **then**
7:             add all nodes and edges along all oriented paths in $OP(v)$ to G'
8:             enqueue all nodes along this oriented path to backtrack
9: **return** $G' = (V', E')$

3.2 Evaluation of the Graph Reduction Algorithm

Our next goal is to evaluate the performance of graph reduction. The critical subgraph is expected to be significantly smaller than the original graph. Our experiments use a variety of generated and real-world graphs. The generated networks rely on two well-known random graph models: Watts-Strogatz's small world model (WS) that are characterized by high clustering coefficients and small average distances between nodes [29], as well as Barabási-Albert's scale-free model (BA) that are characterized by power-law degree distributions [8].

WS Model. The small-world networks set a rewiring probability β, which to emulate real-world graphs lies between 0.01 and 0.1, and a degree k of each node [29]. The higher β gets the less the graph will represent a regular lattice and the higher k gets, the higher the average degree will be. The analysis of reliability would not be as interesting on a regular lattice which should exhibit high reliability and minimal reduction due to the non-existence of bridges. Hence we set $\beta = 0.1$ while k at 2, 4 and 6.

BA Model. The scale-free networks were generated using a preferential attachment scheme. An integer m specifies the maximum number of edges to be added at each iteration. We use $|V|$ iterations and thus the total number of edges is given by the inequality $|V| \leq |E| \leq m \cdot |V|$ as every iteration we add 1 to m edges. We use $m = 2, 4, 8$ to illustrate how increasing density affects results.

Real-World Networks. We also take various real world spatial networks, e.g., urban road networks. Road networks were sourced from [1–3] for the following regions: Oldenburg (OD) Germany, Le Havre (HA) France, Auckland (AK) New Zealand, San Joaquin (SJ) and San (SF) Francisco USA, California (CA) and USA road network (US). We also generate emulated ad-hoc Auckland region networks networks using land zones as nodes and edges were created between nodes if their distance were within 300 and 400 m (AK300, AK400). Table 1 shows the various properties of these networks including their size, average clustering coefficient (ACC), density and average degree.

Table 1. Properties of generated and real-world spatial networks.

Network	#Node	#Edge	ACC	Density	Average degree
WS-k2	1000	1000	0.00	0.00200	2.00
WS-k4	1000	2000	0.371	0.00400	4.00
WS-k6	1000	3000	0.451	0.00601	6.00
BA-m2	1000	1514	0.0282	0.00302	3.02
BA-m4	1000	2483	0.0269	0.00495	4.96
BA-m8	1000	4467	0.0524	0.00891	8.92
AK300	3986	25544	0.593	0.00322	12.8
AK400	3986	42936	0.625	0.00541	21.5
AK	46588	77498	0.363	7.14E-05	3.33
OD	6105	7029	0.0108	3.77E-04	2.30
LH	11734	15135	0.0444	2.20E-04	2.58
CA	21048	21693	7.13E-05	9.79E-05	2.06
SJ	18263	23797	0.0188	1.43E-04	2.61
US	175813	179102	0.000249	1.16E-05	2.04
SF	174956	221802	0.0203	1.45E-05	2.54

Experiment 1 [Expected Graph Reduction]. Our goal is to evaluate the size of critical subgraphs for any nodes s and t. Given $G = (V, E)$ and $G\rceil_{s,t} = (V_{s,t}, E_{s,t})$ where $s, t \in V$, *node and edge* (s, t)-*reduction* refer to $\sum_{s \neq t \in V}(|V| - |V_{s,t}|)/n(n-1)$ and $\sum_{s \neq t \in V}(|E| - |E_{s,t}|)/n(n-1)$, respectively. We ran the graph reduction algorithm on all pairs (s, t) in the generated networks to find an estimated reduction. As shown in the table on the right of Fig. 2, BA networks tend to contain dense cores that form a Hamiltonian cycle. As a Hamiltonian cycle is a closed simple walk through every node, any critical subgraph will contain all nodes. For small-world networks, as k increases, the graph tends to contain a Hamiltonian cycle. As m and k increase, less reduction is achieved. The WS network with $k = 2$ has a large number of bridges which results in a large reduction. For all pairs of nodes in the graph of size 1000 and $k = 2$, all pairs lead to at least 80% reduction. Most real-world networks achieve significant size reduction in the critical subgraph.

Experiment 2 [Accuracy and Performance]. This experiment evaluates accuracy and time taken between the algorithms with/without graph reduction. Both algorithms deploy agents running in parallel. Based on 10000 randomly selected pairs of nodes (s, t), a baseline is established using the Monte Carlo algorithm with 10000 samples followed by running both algorithms with the sample sizes 10, 100, 500, 1000, 2500 and 5000. Another goal is to compare performances between depth-first and breadth-first agents. Thus we evaluate the crude Monte Carlo, GraphReduction using DFS, and the same algorithm using BFS. The results are visualized in two plots, one shows average difference given

Small World |V|=1000, k=2, Edge reduction   Estimated average graph reduction.

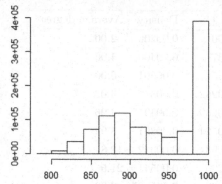

Graph	Node Reduction	Edge Reduction
WS-k2	937	937
WS-k4	0	0
WS-k6	0	0
BA-m2	369	371
BA-m4	164	164
BA-m8	88	88
AK300	166	153
AK400	77	88
OD	1767	1861
CA	2472	2474
SJ	4917	5037
AK	37938	61445
LH	2699	2815
US	51578	51669
SF	56761	58493

Fig. 2. (Left) Reduction on the WS graphs with 1000 nodes and $k = 2$. Each bar represents the number of node pairs (s, t) such that $G\lceil_{s,t}$ results in a node reduction that falls within a given interval. (Right) The table indicates estimated average graph reduction for the networks.

by (*baseline reliability – sample size reliability*) for each algorithm and the other plots the average relative speedup compared to the baseline times calculated by (*mean baseline time – mean sample time*) ÷ *mean baseline time*. Results are shown in Fig. 3. Agent-based evaluation approximates reliability with good accuracy. Expectedly, as the number of samples grows, the difference in reliability converges towards 0. In all networks, our agent-based algorithm performs faster than the standard Monte Carlo method with more apparent differences as the number of samples grows. Depth-first agents perform slightly better on WS and real-world networks and breadth-first slightly better on BA graphs.

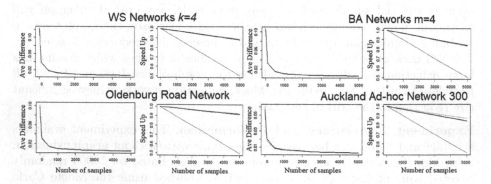

Fig. 3. Average reliability difference and speed up results. Red line is the crude Monte Carlo method (baseline); blue is the BFS traversal, and green is the DFS traversal. (Color figure online)

Reliability Score Distributions. The reliability scores rel(s) of nodes reveal a form of their relative positional advantage. We analyze the distribution of this score in spatial networks. All road networks consistently follow a right-skewed distribution with long-tail; See Fig. 4. The low overall reliability score can be explained by the sparse roads surrounding the dense clusters of the important urban centres. Figure 5 visualizes 6 spatial networks and illustrates reliability using a gradient in which red, orange and green represent low, middle and high

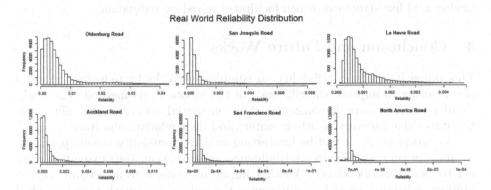

Fig. 4. Distribution of reliability scores of nodes in the real-world spatial networks.

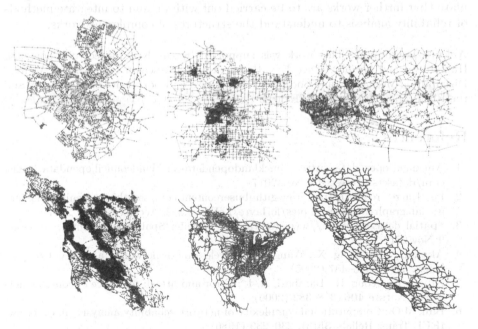

Fig. 5. Visualizations of real-world road networks of Oldenburg (top-left), San Joaquin (top-mid), Le Havre (top-right), San Francisco (bottom-left), USA (bottom-mid), California (bottom-right). Green color indicates nodes of high reliability score. (Color figure online)

scores, respectively. In general, clusters that comprise of nodes with high reliability score also correlate to areas of high importance to the regions. The urban centers in all of the graphs clearly stand out where the varying degrees of greenness among different clusters represent their levels of importance. A general pattern is that smaller such clusters tend to have smaller reliability scores. For example, for San Francisco Bay, one can identify prevalently green areas at San Francisco city, Palo Alto, San Jose, Oakland and Berkeley, all major centers in the Bay area. One can also observe that clusters with the highest reliability having grid like structures which facilitates to reduce redundancy.

4   Conclusion and Future Works

This paper investigates reliability on complex networks by reducing the graph size in order to improve the efficiency of the Monte Carlo method. The analysis of reliability score demonstrates that, on spatial networks, reliability score facilitates the discovery of urban centers and their relative importance.

As future works, it will be interesting to derive reliability-based approaches to analyze network clusters and influence maximization, thus extending existing tools in network science. Another important topic is to investigate ways to improve reliability score for (sub)graphs through add/strengthening ties which may lead to applications in communication, social and spatial networks. We speculate that further works are to be carried out with an aim to integrate methods of reliability analysis to understand the structures of complex networks.

Acknowledgement. This work was supported by the Key Project of Chongqing Humanities and Social Science Key Research Base: Research on Coalition Welfare Distribution Mechanism and Social Cohesion Based on Cooperative Game Theory and the Ratification Number is 18SKB047.

References

1. Auckland open data. https://aucklandopendata-aucklandcouncil.opendata.arcgis.com/datasets. Accessed 9 Nov 2017
2. Le Havre road. https://raw.githubusercontent.com/samthiriot/genlab/master/genlab.graphstream.examples/leHavre/LeHavre.dgs. Accessed 9 Nov 2017
3. Spatial dataset. https://www.cs.utah.edu/~lifeifei/SpatialDataset.htm. Accessed 9 Nov 2017
4. Akyildiz, I.F., Wang, X., Wang, W.: Wireless mesh networks: a survey. Comput. Netw. **47**(4), 445–487 (2005)
5. Albert, R., Jeong, H., Barabási, A.-L.: Error and attack tolerance of complex networks. Nature **406**, 378–382 (2000)
6. Ball, M.O.: Computational complexity of network reliability analysis: an overview. IEEE Trans. Reliab. **35**(3), 230–239 (1986)
7. Ball, M.O., Colbourn, C.J., Provan, J.S.: Network reliability. In: Handbooks in Operations Research and Management Science, vol. 7, pp. 673–762 (1995)
8. Barabási, A.-L., Albert, R.: Emergence of scaling in random networks. Science **286**(5439), 509–512 (1999)

9. Bobbio, A., Ferraris, C., Terruggia, R.: New challenges in network reliability analysis. In: CNIP, vol. 6, pp. 554–564 (2006)
10. Burt, R.S.: Structural Holes: The Social Structure of Competition. Harvard University Press, Cambridge (2009)
11. Chassin, D.P., Posse, C.: Evaluating north american electric grid reliability using the barabási-albert network model. Phys. A: Stat. Mech. Appl. **355**(2), 667–677 (2005)
12. Chen, F., Liu, J., Li, Z., Wang, Y.: Routing with uncertainty in wireless mesh networks. In: IWQoS 2010, pp. 1–5. IEEE (2010)
13. Csima, B.F., Khoussainov, B., Liu, J.: Computable categoricity of graphs with finite components. In: Beckmann, A., Dimitracopoulos, C., Löwe, B. (eds.) CiE 2008. LNCS, vol. 5028, pp. 139–148. Springer, Heidelberg (2008). https://doi.org/10.1007/978-3-540-69407-6_15
14. Gertsbakh, I.B., Shpungin, Y.: Models of Network Reliability: Analysis, Combinatorics, and Monte Carlo. CRC Press, Boca Raton (2016)
15. Guimera, R., Mossa, S., Turtschi, A., Amaral, L.N.: The worldwide air transportation network: anomalous centrality, community structure, and cities' global roles. Proc. Nat. Acad. Sci. **102**(22), 7794–7799 (2005)
16. Kamat, S.J., Riley, M.W.: Determination of reliability using event-based monte carlo simulation. IEEE Trans. Reliab. **24**(1), 73–75 (1975)
17. Kempe, D., Kleinberg, J., Tardos, É.: Maximizing the spread of influence through a social network. In: Proceedings of SIGKDD 2003, pp. 137–146. ACM (2003)
18. Lin, M.-S.: An efficient algorithm for computing the reliability of stochastic binary systems. IEICE Trans. Inf. Syst. **87**(3), 745–750 (2004)
19. Liu, J., Li, L., Russell, K.: What becomes of the broken hearted?: an agent-based approach to self-evaluation, interpersonal loss, and suicide ideation. In: Proceedings of AAMAS 2017 (2017)
20. Liu, J., Moskvina, A.: Hierarchies, ties and power in organizational networks: model and analysis. Soc. Netw. Anal. Mining **6**(1), 106 (2016)
21. Liu, J., Wei, Z.: Community detection based on graph dynamical systems with asynchronous runs. In: 2014 Second International Symposium on Computing and Networking (CANDAR). IEEE (2014)
22. Liu, J., Wei, Z.: Network, popularity and social cohesion: a game-theoretic approach. In: AAAI (2017)
23. Meziane, R., Massim, Y., Zeblah, A., Ghoraf, A., Rahli, R.: Reliability optimization using ant colony algorithm under performance and cost constraints. Electr. Power Syst. Res. **76**(1), 1–8 (2005)
24. Moskvina, A., Liu, J.: Togetherness: an algorithmic approach to network integration. In: 2016 IEEE/ACM International Conference on Advances in Social Networks Analysis and Mining (ASONAM). IEEE (2016)
25. Nelson, A.C., Batts, J.R., Beadles, R.L.: A computer program for approximating system reliability. IEEE Trans. Rel. **19**(2), 61–65 (1970)
26. Neumayer, S., Modiano, E.: Network reliability under geographically correlated line and disk failure models. Comput. Netw. **94**, 14–28 (2016)
27. Sebastio, S., Trivedi, K.S., Wang, D., Yin, X.: Fast computation of bounds for two-terminal network reliability. Eur. J. Oper. Res. **238**(3), 810–823 (2014)
28. Srivaree-ratana, C., Konak, A., Smith, A.E.: Estimation of all-terminal network reliability using an artificial neural network. Comput. Oper. Res. **29**(7), 849–868 (2002)
29. Watts, D.J., Strogatz, S.H.: Collective dynamics of 'small-world' networks. Nature **393**(6684), 440 (1998)

Efficient Exact and Approximate Algorithms for Computing Betweenness Centrality in Directed Graphs

Mostafa Haghir Chehreghani$^{(\boxtimes)}$, Albert Bifet, and Talel Abdessalem

LTCI, Télécom ParisTech, Université Paris-Saclay, Paris, France
mostafa.chehreghani@gmail.com,
{albert.bifet,talel.abdessalem}@telecom-paristech.fr

Abstract. In this paper, first given a directed network G and a vertex $r \in V(G)$, we propose a new exact algorithm to compute betweenness score of r. Our algorithm pre-computes a set $\mathcal{RF}(r)$, which is used to prune a huge amount of computations that do not contribute in the betweenness score of r. Then, for the cases where $\mathcal{RF}(r)$ is large, we present a randomized algorithm that samples from $\mathcal{RF}(r)$ and performs computations for only the sampled elements. We show that this algorithm provides an (ϵ, δ)-approximation of the betweenness score of r. Finally, we empirically evaluate our algorithms and show that they significantly outperform the most efficient existing algorithms, in terms of both running time and accuracy. Our experiments also show that our proposed algorithms can effectively compute betweenness scores of all vertices in a set of vertices.

Keywords: Directed graphs · Betweenness centrality
Exact algorithm · Approximate algorithm

1 Introduction

Graphs are an important tool to model data in different domains, including social networks, bioinformatics, road networks and the world wide web. A property seen in most of these real-world networks is that the ties between vertices do not always represent reciprocal relations [18]. As a result, the networks formed in these domains are *directed graphs* where any edge has a direction and the edges are not always symmetric. *Centrality* is a structural property of vertices (or edges) in the network that quantifies their relative importance. A well-known and widely-used centrality notion is *betweenness centrality*. Freeman [10] defined and used it to measure the control of a human over the communications among others in a social network.

Although there exist polynomial time and space algorithms for betweenness centrality computation, they are expensive in practice. The most efficient existing exact method is Brandes's algorithm [3] whose time complexity is $O(nm)$

© Springer International Publishing AG, part of Springer Nature 2018
D. Phung et al. (Eds.): PAKDD 2018, LNAI 10939, pp. 752–764, 2018.
https://doi.org/10.1007/978-3-319-93040-4_59

for unweighted graphs and $O(nm + n^2 \log n)$ for weighted graphs with positive weights (n and m are the number of vertices and the number of edges of the network, respectively). This means this algorithm is not applicable, even for mid-size networks. However, there are observations that may improve computation of betweenness centrality in practice. In several applications it is sufficient to compute betweenness score of only one or a few vertices. For instance, this index might be computed for only core vertices of communities in social/information networks or for only hubs in communication networks. Another example, discussed in [1], is handling cascading failures. It has been shown that the failure of a vertex with a higher betweenness score may cause greater collapse of the network [21]. Therefore, failed vertices should be recovered in the order of their betweenness scores. This means it is required to compute betweenness scores of only failed vertices, that are a very small subset of all vertices. Note that these vertices are not necessarily those that have the highest betweenness scores in the network. Hence, algorithms that identify vertices with the highest betweenness scores [19] are not applicable.

In the current paper, we exploit this observation to design more effective exact and approximate algorithms for computing betweenness centrality in large directed graphs. Our algorithms are based on computing the set of *reachable vertices* for a given vertex r. On the one hand, this set can be computed very efficiently. On the other hand, it indicates the potential source vertices whose dependency scores on r are non-zero, as a result, it helps us to avoid a huge amount of computations that do not contribute in the betweenness score of r. In this paper, our key contributions are as follows.

- Given a directed graph G (with n vertices and m edges) and a vertex $r \in V(G)$, we present an efficient exact algorithm to compute betweenness score of r. The algorithm is based on pre computing the set *reachable vertices* of r, denoted by $\mathcal{RF}(r)$. $\mathcal{RF}(r)$ can be computed in $O(m)$ time for both unweighted graphs and weighted graphs with positive weights. Time complexity of the whole exact algorithm depends on the size of $\mathcal{RF}(r)$ and it is respectively $O(m \cdot |\mathcal{RF}(r)|)$ and $O(m \cdot |\mathcal{RF}(r)| + n \log n \cdot |\mathcal{RF}(r)|)$ for unweighted graphs and weighted graphs with positive weights. $|\mathcal{RF}(r)|$ is bounded from above by n and in most cases, it can be considered as a small constant (see Sect. 5). Hence, in many cases, time complexity of our proposed exact algorithm for unweighted graphs is linear, in terms of m, and it is $O(m + n \log n)$ for weighted graphs with positive weights.
- In the cases where $\mathcal{RF}(r)$ is large, our exact algorithm might be intractable in practice. To address this issue, we present a randomized algorithm that samples elements from $\mathcal{RF}(r)$ and performs computations for only the sampled elements. We show that this algorithm provides an (ϵ, δ)-approximation of the betweenness score of r.
- In order to evaluate the empirical efficiency of our proposed algorithms, we perform extensive experiments over several real-world datasets. Our experiments show that our algorithms significantly outperform the most efficient existing algorithms, in terms of both running time and accuracy. While our

algorithm is intuitively designed to estimate betweenness score of only one vertex, our experiments reveal that it can efficiently compute betweenness scores of all vertices in a given set.

2 Preliminaries

We assume that the reader is familiar with basic concepts in graph theory. Throughout the paper, G refers to a graph (network). For simplicity, we assume that G is a directed, connected and loop-free graph without multi-edges. Throughout the paper, we assume that G is an unweighted graph, unless it is explicitly mentioned that G is weighted. $V(G)$ and $E(G)$ refer to the set of vertices and the set of edges of G, respectively. We use n and m to refer to $|V(G)|$ and $|E(G)|$, respectively. For a vertex $v \in V(G)$, the number of head ends adjacent to v is called its *in degree*, and the number of tail ends adjacent to v is called its *out degree*. A *shortest path* from $u \in V(G)$ to $v \in V(G)$ is a path whose length is minimum, among all paths from u to v. For two vertices $u, v \in V(G)$, if G is unweighted, by $d(u, v)$ we denote the length (the number of edges) of a shortest path connecting u to v. If G is weighted, $d(u, v)$ denotes the sum of the weights of the edges of a shortest path connecting u to v. By definition, $d(u, u) = 0$. Note that in directed graphs, $d(u, v)$ is not necessarily equal to $d(v, u)$. For $s, t \in V(G)$, σ_{st} denotes the number of shortest paths between s and t, and $\sigma_{st}(v)$ denotes the number of shortest paths between s and t that also pass through v. *Betweenness centrality* of a vertex v is defined as: $BC(v) = \sum_{s,t \in V(G) \setminus \{v\}} \frac{\sigma_{st}(v)}{\sigma_{st}}$. A notion which is widely used for counting the number of shortest paths in a graph is the directed acyclic graph (DAG) containing all shortest paths starting from a vertex s (see e.g., [3]). In this paper, we refer to it as *shortest-path-DAG*, or *SPD* in short, rooted at s. Brandes [3] introduced the notion of the *dependency score* of a vertex $s \in V(G)$ on a vertex $v \in V(G) \setminus \{s\}$, which is defined as $\delta_{s\bullet}(v) = \sum_{t \in V(G) \setminus \{v,s\}} \delta_{st}(v)$, where $\delta_{st}(v) = \frac{\sigma_{st}(v)}{\sigma_{st}}$. We have: $BC(v) = \sum_{s \in V(G) \setminus \{v\}} \delta_{s\bullet}(v)$.

3 Related Work

Brandes [3] introduced an efficient algorithm for computing betweenness centrality of a vertex, which is performed in $O(nm)$ and $O(nm + n^2 \log n)$ times for unweighted and weighted networks with positive weights, respectively. Çatalyürek et al. [5] presented the *compression* and *shattering* techniques to improve the efficiency of Brandes's algorithm for large graphs. Kang et al. [13] presented betweenness centrality indices suitable for very large networks. The authors of [6,9] respectively studied *group betweenness* and *co-betweenness*, the two natural extensions of betweenness to sets of vertices. Brandes and Pich [4] proposed an approximate algorithm based on selecting k source vertices and computing dependency scores of them on the other vertices in the graph. Chehreghani [7] proposed a non-uniform sampler for unbiased estimation of the betweenness score of a single vertex. Similar to [4,7], our proposed algorithms are (source) vertex sampler. However, the sampling strategy we use is novel and is based on computing the set $\mathcal{RF}$. Riondato and Kornaropoulos [19] and Riondato and Upfal [20] presented shortest path samplers for estimating betweenness centrality of all vertices or the k vertices that have the highest betweenness scores in a graph. Finally, Borassi and Natale [2] presented the KADABRA algorithm, which uses balanced bidirectional BFS (bb-BFS) to sample shortest paths. There are in the literature several algorithms for computing betweenness centrality in dynamic graphs (see e.g., [11]). An overview of other (distance-based) centrality indices can be found in [8].

4   Computing Betweenness Centrality in Directed Graphs

In this section, we present our exact and approximate algorithms for computing betweenness centrality of a given vertex r in a large directed graph. First in Sect. 4.1, we introduce *reachable vertices* and show that they are sufficient to compute betweenness score of r. Then in Sects. 4.2 and 4.3, we respectively present our exact and approximate algorithms.

4.1   Reachable Vertices

Let G be a directed graph and $r \in V(G)$. Suppose that we want to compute betweenness score of r. To do so, as Brandes algorithm [3] suggests, for each vertex $s \in V(G)$, we may form the SPD rooted at s and compute the dependency score of s on r. Betweenness score of r will be the sum of all the dependency scores. However, it is possible that in a directed graph for many vertices s, there is no path from s to r and as a result, dependency score of s on r is 0. An example of this situation is depicted in Fig. 1(a). In the graph of this figure, suppose that we want to compute betweenness score of vertex r_1. If we form the SPD rooted at v_1, after visiting the parts of the graph indicated by hachures, we find out that there is no shortest path from v_1 to r_1 and hence, $\delta_{v_1 \bullet}(r_1)$ is 0. The same holds for all vertices in the hachured part of the graph, i.e., dependency scores of these vertices on r_1 are 0. The question arising here is that whether there exists an efficient way to detect the vertices whose dependency scores on r are 0 (so that we can avoid forming SPDs rooted at them)? In the rest of this section, we try to answer this question. We first introduce a (usually small) subset of vertices, called *reachable vertices* and denoted with $RF(r)$, that are sufficient to compute betweenness score of r. Then, we discuss how this set can be computed efficiently.

Definition 1. *Let G be a directed graph and $r, v \in V(G)$. We say r is reachable from v if there is a (directed) path from v to r. The set of vertices that r is reachable from them is denoted by $RF(r)$.*

Proposition 1. *Let G be a directed graph and $r \in V(G)$. If out degree of r is 0, $BC(r)$ is 0, too. Otherwise, we have: $BC(r) = \sum_{v \in RF(r)} \delta_{v \bullet}(r)$.*

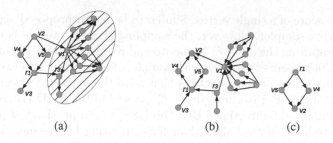

Fig. 1. In Fig. 1(a), the dependency scores of the vertices in the hachured part of the graph (and also v_3) on r_1 is 0. The graph of Fig. 1(b) presents the reverse graph of the graph of Fig. 1(a). Figure 1(c) shows how $\mathcal{RF}(r_1)$ is computed.

Proposition 1 suggests that for computing betweenness score of r, we first check whether *out degree* of r is greater than 0 and if so, we compute $RF(r)$. Betweenness score of r is exactly computed using Proposition 1.

If $RF(r)$ is already known, this procedure can significantly improve computation of betweenness centrality of r. The reason is that, as our experiments show, in real-world directed networks $RF(r)$ is usually significantly smaller than $V(G)$. However, computing $RF(r)$ can be computationally expensive as in the worst case, it requires the same amount of time as computing betweenness score of r. This motivates us to try to define a set $\mathcal{RF}(r)$ that satisfies the following properties: (i) $RF(r) \subseteq \mathcal{RF}(r)$ and (ii) $\mathcal{RF}(r)$ can be computed effectively in a time much faster than computing $BC(r)$. Condition (i) implies that each vertex $v \in V(G)$ whose dependency score on r is greater than 0, belongs to $\mathcal{RF}(r)$ and as a result, $BC(r) = \sum_{v \in \mathcal{RF}(r)} \delta_{v\bullet}(r)$. In the following, we present a definition of $\mathcal{RF}(r)$ and a simple and efficient algorithm to compute it.

Definition 2. *Let G be a directed graph. Reverse graph of G, denoted by $R(G)$, is a directed graph such that: (i) $V(R(G)) = V(G)$, and (ii) $(u, v) \in E(R(G))$ if and only if $(v, u) \in E(G)$.*

For example, the graph of Fig. 1(b) presents the *reverse graph* of the graph of Fig. 1(a).

Definition 3. *Let G be a directed graph and $r \in V(G)$. We define $\mathcal{RF}(G)$ as the set that contains any vertex v such that there is a path from r to v in $R(G)$.*

Proposition 2. *Let G be a directed graph and $r \in V(G)$. We have: $RF(r) = \mathcal{RF}(r)$.*

An advantage of the above definition of $\mathcal{RF}(r)$ is that it can be efficiently computed as follows: (i) first, by flipping the direction of the edges of G, $R(G)$ is constructed, (ii) then, if G is weighted, the weights of the edges are ignored, (iii) finally, a breadth first search (BFS) or a depth-first search (DFS) on $R(G)$ starting from r is performed. All the vertices that are met during the BFS (or DFS), except r, are added to $\mathcal{RF}(r)$. In fact, while in $RF(r)$ we require to solve the multi-source shortest path problem (MSSP), in $\mathcal{RF}(r)$ this is reduced to the single-source shortest path problem (SSSP), which can be addressed much faster. Figure 1 shows an example of this procedure, where in order to compute $\mathcal{RF}(r_1)$, we first generate $R(G)$ (Fig. 1(b)) and then, we run a BFS (or DFS) starting from r_1 (Fig. 1(c)). The set of vertices that are met during the traversal except r_1, i.e., vertices v_2, v_4 and v_5, form $\mathcal{RF}(r_1)$.

For a vertex $r \in V(G)$, each of the steps of the procedure of computing $\mathcal{RF}(r)$, for both unweighted graphs and weighted graphs, can be computed in $O(m)$ time. Hence, time complexity of the procedure of computing $\mathcal{RF}(r)$ for both unweighted graphs and weighted graphs is $O(m)$. Therefore, $\mathcal{RF}(r)$ can be computed in a time much faster than computing betweenness score of r. Furthermore, Proposition 2 says that $\mathcal{RF}(r)$ contains all the members of $RF(r)$. These mean both the afore-mentioned conditions are satisfied.

4.2 The Exact Algorithm

In this section, using the notions and definitions presented in Sect. 4.1, we propose an effective algorithm to compute exact betweenness score of a given vertex r in a directed graph G.

Algorithm 1 presents the high level pseudo code of the E-BCD algorithm proposed for computing exact betweenness score of r in G. After checking whether or not *out degree* of r is 0, the algorithm follows two main steps: (i) computing $\mathcal{RF}(G)$ (Lines 12–16 of Algorithm 1), where we use the procedure described in Sect. 4.1 to compute $\mathcal{RF}(r)$; and (ii) computing $BC(r)$ (Lines 12–16 of Algorithm 1), where for each vertex $v \in \mathcal{RF}(r)$, we form the SPD rooted at v and compute the dependency score of v on the other vertices and add the value of $\delta_{v\bullet}(r)$ to the betweenness score of r. Note that if G is weighted, while in the first step the weights of its edges are ignored, in the second step and during forming SPDs and computing dependency scores, we take the weights into account. Note also that in Algorithm 1, after computing $\mathcal{RF}(r)$, techniques proposed to improve exact betweenness centrality computation, such *compression* and *shattering* [5], can be used to improve the efficiency of the second step. This means the algorithm proposed here is orthogonal to the techniques such as shattering and compression and therefore, they can be merged.

Complexity Analysis. On the one hand, time complexity of the first step is $O(m)$. On the other hand, time complexity of each iteration in Lines 13–16 is $O(m)$ for unweighted graphs and $O(m+n \log n)$ for weighted graphs with positive weights. As a result, time complexity of E-BCD is $O(m|\mathcal{RF}(G)|)$ for unweighted graphs and $O(m|\mathcal{RF}(G)| + n \log n|\mathcal{RF}(G)|)$ for weighted graphs with positive weights.

4.3 The Approximate Algorithm

For a vertex $r \in V(G)$, $\mathcal{RF}(r)$ is always smaller than n and as our experiments (reported in Sect. 5) show, the difference is usually significant. Therefore, E-BCD is usually significantly more efficient than the existing exact algorithms such as

Algorithm 1. The algorithm of computing exact betweenness score in directed graphs.

1: E-BCD
2: **Input.** A directed network G and a vertex $r \in V(G)$.
3: **Output.** Betweenness score of r.
4: **if** *out degree* of r is 0 **then**
5:     **return** 0.
6: **end if**
7: $\mathcal{RF}(r) \leftarrow \emptyset$.
8: $R(G) \leftarrow$ compute the reverse graph of G.
9: If G is weighted, ignore the weights of the edges of $R(G)$.
10: Perform a BFS or DFS on $R(G)$ starting from r.
11: Add to $\mathcal{RF}(r)$ all the visited vertices, except r.
12: $bc \leftarrow 0$.
13: **for all** vertices $v \in \mathcal{RF}(G)$ **do**
14:     Form the SPD rooted at v and compute the dependency scores of v on the other vertices.
15:     $bc \leftarrow bc + \delta_{v\bullet}(r)$.
16: **end for**
17: **return** bc.

Brandes's algorithm [3]. However, in some cases, the size of $\mathcal{RF}(r)$ can be large (see again Sect. 5). To make the algorithm tractable for the cases where $\mathcal{RF}(r)$ is large, in this section we propose a randomized algorithm that picks some elements of $\mathcal{RF}(r)$ uniformly at random and only processes these vertices.

Algorithm 2 shows the high level pseudo code of our randomized algorithm, called A-BCD. Similar to E-BCD, A-BCD first computes $\mathcal{RF}(r)$. Then, at each iteration t $(1 \leq t \leq T)$, it picks a vertex v_t from $\mathcal{RF}(r)$ uniformly at random, forms the SPD rooted at v_t and computes $\delta_{v_t\bullet}(r)$. In the end, betweenness of r is estimated as the sum of the computed dependency scores on r multiply by $\frac{|\mathcal{RF}(r)|}{T}$. Time complexity of Algorithm 2 can be analyzed in a way similar to Algorithm 1. In the following, we use Hoeffding's inequality [12], to derive an error bound for bc. First in Proportion 3, we prove that in Algorithm 2 the expected value of bc is $BC(r)$. Then in Proportion 4, we present the error bound.

Proposition 3. *In Algorithm 2, we have:* $\mathbb{E}[bc] = BC(r)$.

Proposition 4. *In Algorithm 2, let K be the maximum dependency score that a vertex may have on r. For a given $\epsilon \in \mathbb{R}^+$, we have:*

$$\mathbb{P}[|BC(r) - bc| > \epsilon] \leq 2 \exp\left(-2T \cdot \left(\frac{\epsilon}{K \cdot |\mathcal{RF}(r)|}\right)^2\right). \tag{1}$$

Inequality 1 says that for given values $\epsilon \in \mathbb{R}^+$ and $\delta \in (0, 1)$, if T is chosen such that $T \geq \frac{\ln\left(\frac{2}{\delta}\right) \cdot K^2 \cdot |\mathcal{RF}(r)|^2}{2\epsilon^2}$, Algorithm 2 estimates betweenness score of r within an additive error ϵ with probability at least $1 - \delta$. The difference between

---

Algorithm 2. The algorithm of computing exact betweenness score in directed graphs.

1: A-BCD
2: **Input.** A network G, a vertex $r \in V(G)$ and the number of samples T.
3: **Output.** Estimated betweenness score of r.
4: **if** *out degree* of r is 0 **then**
5:     **return** 0.
6: **end if**
7: $\mathcal{RF}(r) \leftarrow \emptyset$.
8: $R(G) \leftarrow$ compute the reverse graph of G.
9: If G is weighted, ignore the weights of the edges of $R(G)$.
10: Perform a BFS or DFS on $R(G)$ starting from r.
11: Add to $\mathcal{RF}(r)$ all visited vertices, except r.
12: $bc \leftarrow 0$.
13: **for all** $t = 1$ **to** T **do**
14:     Select a vertex $v_t \in \mathcal{RF}(r)$ uniformly at random.
15:     Form the SPD rooted at v_t and compute dependency scores of v_t on the other vertices.
16:     $bc \leftarrow bc + \frac{\delta_{v_t \bullet}(r) \cdot |\mathcal{RF}(r)|}{T}$.
17: **end for**
18: **return** bc.

---

the number of samples presented in Proposition 4 and the number of samples required by the methods that uniformly sample from the set of all vertices (e.g., [4]) is that in the later case, the lower bound on the number of samples is a function of n^2, instead of $|\mathcal{RF}(r)|^2$. As we will see in Sect. 5, for most of the vertices, $|\mathcal{RF}(r)| \ll n$. In addition, $|\mathcal{RF}(r)| \ll n$ implies that the error bound presented in Proposition 4 is considerably better than the error bound of shortest path sampling algorithms such as [17].

5   Experimental Results

We perform extensive experiments on several real-world networks to assess the quantitative and qualitative behavior of our proposed exact and approximate algorithms. We test the algorithms over several real-world datasets from different domains, including the *amazon* product co-purchasing network [14], the *com-dblp* co-authorship network [22], the *com-amazon* network [22] the *p2p-Gnutella31* peer-to-peer network [16], the *slashdot* technology-related news network [15] and the *soc-sign-epinions* who-trust-whom online social network [15]. All the networks are treated as directed graphs. For a vertex $r \in V(G)$, its empirical approximation error is defined as: $Error(v) = \frac{|App(v) - BC(v)|}{BC(v)} \times 100$, where $App(v)$ is the calculated approximate score.

As mentioned before, for a directed graph G and a vertex $r \in V(G)$, both of our proposed exact and approximate algorithms first compute $\mathcal{RF}(r)$, which can be done very effectively. Then, based on the size of $\mathcal{RF}(r)$, someone may decide

Table 1. Empirical evaluation of BCD against KADABRA for randomly chosen vertices. Values of δ and ϵ are 0.1 and 0.01, respectively. All the reported times are in seconds. The number of samples in A-BCD is 1000.

Dataset	Random vertices		KADABRA			BCD			
	$BC(r)$	$\frac{\lceil \mathcal{RF}(r) \rceil}{n}$	#samples	Time	Error (%)	E/A	Time	Time$_{\mathcal{RF}}$	Error (%)
Amazon	19613.1	0.1800	16739	19.14	100	A	2.60	0.26	0.26
	87523.6	0.0005			100	E	0.67	0.29	0
	35752.6	0.0020			100	E	1.26	0.29	0
	10449.4	0.00001			100	E	0.11	0.30	0
	1837.58	0.0001			100	E	0.17	0.30	0
Com-amazon	1486.8	0.00003	15036	27.70	100	E	0.14	0.27	0
	364	0.000008			100	E	0.12	0.27	0
	11	0.00004			100	E	0.15	0.27	0
	1701.51	0.0018			100	E	1.41	0.28	0
	139	0.00004			100	E	0.15	0.27	0
Com-dblp	10153	0.0065	17873	26.14	100	A	5.74	0.26	1.10
	34326.5	0.00003			100	E	0.13	0.27	0
	232994	0.00006			100	E	0.21	0.27	0
	1957.93	0.0002			100	E	0.48	0.27	0
	303543	0.0001			100	E	0.53	0.29	0
Email-EuAll	1869.16	0.000008	17066	16.01	100	E	0.03	0.08	0
	2269.29	0.0002			100	E	0.14	0.08	0
	241434	0.0942			100	A	1.88	0.07	1.72
	3	0.000008			100	E	0.03	0.07	0
	503650	0.4966			100	A	1.78	0.08	3.59
P2p-Gnutella31	12655.2	0.00003	16401	6.88	100	E	0.03	0.04	0
	3538.79	0.0027			100	E	0.95	0.04	0
	27824.9	0.00004			100	E	0.03	0.04	0
	6175.2	0.3857			100	A	2.44	0.06	11.31
	4582130	0.00004			100	E	0.02	0.04	0
Slashdot0902	15940.9	0.0002	17421	7.95	100	E	0.17	0.16	0
	15891.7	0.00003			100	E	0.06	0.15	0
	21744	0.00003			100	E	0.05	0.15	0
	43067	0.0044			100	E	2.30	0.17	0
	6165.01	0.00002			100	E	0.05	0.15	0
Soc-sign-epinions	2352.43	0.2760	19099	11.28	100	A	4.57	0.17	55.34
	9198.78	0.0198			100	A	4.60	0.15	18.48
	75201.9	0.0002			100	E	0.24	0.14	0
	8802	0.0002			100	E	0.19	0.14	0
	8052	0.00002			100	E	0.04	0.14	0
Web-NotreDame	140	0.00002	19908	27.29	100	E	0.08	0.25	0
	9003.53	0.0024			100	E	1.84	0.25	0
	4212.33	0.0001			100	E	0.18	0.25	0
	2157.42	0.00009			100	E	0.14	0.25	0
	3079.5	0.0003			100	E	0.35	0.25	0

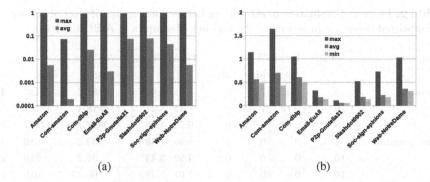

(a)   (b)

Fig. 2. Figure 2(a) shows $|\mathcal{RF}|$ divided by n and Fig. 2(b) presents time to compute $\mathcal{RF}$.

to use either the exact algorithm or the approximate algorithm. Hence in our experiments, we follow the following procedure. First, we compute $\mathcal{RF}(r)$. Then, if $|\mathcal{RF}(r)| \leq \tau$, we run E-BCD; otherwise, we run A-BCD with τ as the number of samples. We refer to this procedure as BCD. The value of τ depends on the amount of time someone wants to spend for computing betweenness centrality. In our experiments reported here, we set τ to 1000. We compare our method against the most efficient existing algorithm for approximating betweenness centrality, which is KADABRA [2].

The efficiency of BCD depends on the size and computation time of $\mathcal{RF}$; if for many vertices these values are very small, BCD can estimate betweenness scores of these vertices very efficiently. To evaluate this, over all the networks, we measure maximum, average and minimum $\mathcal{RF}$ sizes (divided by n) and their computation times. The results are reported in Fig. 2. Figure 2(a) shows that while maximum size of $\mathcal{RF}$ of a vertex in a network is very close to n, its average size is much smaller! Note that in Fig. 2(a) the vertical axis is in the logarithmic scale and since the minimum size is always 0, it is omitted. Figure 2(b) shows that in most cases, $\mathcal{RF}$ can be computed within less than 1 s and it is always computed within less than 2 s!

Table 1 reports the results of our first experiments. For KADABRA, we have set ϵ and δ to 0.01 and 0.1, respectively[1]. Over each dataset, we choose 5 vertices at random and run our algorithm for any of these vertices. In the column "A/E" of this table, "A" means that $\mathcal{RF}$ is larger than 1000, therefore, the approximate algorithm has been employed, and "E" means that the computed score by our proposed algorithm is exact (hence, the approximation error is 0). For the BCD algorithm, we measure both "Time" and "Time$_{\mathcal{RF}}$", where "Time$_{\mathcal{RF}}$" is the time of computing $\mathcal{RF}$ and "Time" is the running time of the other parts of the

---

[1] For given values of ϵ and δ, KADABRA computes the *normalized betweenness* of the vertices of the graph within an error ϵ with a probability at least $1 - \delta$. The *normalized betweenness* of a vertex is its betweenness score divided by $n \cdot (n - 1)$. Therefore, we multiply the scores computed by KADABRA by $n \cdot (n - 1)$.

Table 2. Empirical evaluation of estimating betweenness scores of a set of vertices. All the reported times are in seconds. The number of samples in A-BCD is 1000.

Dataset	Set size	Error (%)			Time	Time$_{\mathcal{RF}}$	$\mathcal{RF}$ size		
		Avg.	Max.	Min.			Avg.	Max.	Min.
Amazon	5	1.47	7.10	0	4.81	1.44	9581.6	47187	4
	10	0.73	7.10	0	7.42	3.21	4818.4	47187	1
	15	0.88	7.10	0	9.74	4.98	3497.798	47187	1
Com-amazon	5	0	0	0	1.98	1.36	132.2	616	3
	10	0	0	0	4.92	3.43	91.2	616	2
	15	0	0	0	7.07	5.48	65.93	616	1
Com-dblp	5	0.22	1.10	0	7.09	1.36	447.8	2092	11
	10	3.47	19.45	0	20.71	3.08	24483.6	227218	1
	15	2.32	19.45	0	28.81	4.92	21351.33	227218	1
Email-EuAll	5	1.06	3.59	0	3.86	0.38	26584.6	111674	2
	10	1.39	7.95	0	9.76	0.78	19020.9	111674	2
	15	0.93	7.95	0	13.52	1.27	12742.8	111674	2
P2p-Gnutella31	5	2.26	11.31	0	3.47	0.22	4864.2	24141	2
	10	7.26	39.17	0	23.09	0.46	5493.6	24141	2
	15	6.79	39.17	0	33.27	0.72	8637.73	28122	2
Slashdot0902	5	0	0	0	2.62	0.78	79.6	369	2
	10	5.04	50.48	0	11.37	1.38	3784.3	26802	1
	15	4.92	50.48	0	14.93	1.99	6662.86	62089	1
Soc-sign-epinions	5	13.37	48.37	0	9.64	0.74	7817.2	36393	3
	10	9.68	48.37	0	17.71	1.52	20302.7	109520	1
	15	9.38	48.37	0	28.46	2.28	15538.86	109520	1
Web-NotreDame	5	0	0	0	2.58	1.25	200.6	797	9
	10	0	0	0	6.89	2.44	231.5	1092	9
	15	0.03	0.30	0	13.16	3.62	414.46	2610	1

algorithm. The total running time of BCD is the sum of "Time" and "Time$_{\mathcal{RF}}$". As can be seen in Table 1, for most of the randomly picked up vertices, $\mathcal{RF}$ is very small and it can be computed very efficiently. This gives exact results in a very short time, less than 3 s in total. In all these cases, KADABRA, while spends considerably more time, simply estimates the scores as 0 (therefore, we consider its approximation error as 100%). The randomly picked up vertices belong to the different ranges of betweenness scores, including high, medium and low.

In the experiments reported in Table 1, BCD is used to estimate betweenness score of only one vertex. However, in practice it might be required to estimate betweenness scores of a given set of vertices. How efficient is BCD in this setting? To answer this question, we select a random set of vertices and run BCD for each vertex in the set. The results are reported in Table 2, where the set contains 5, 10 or 15 vertices. Over all the datasets and for each set of vertices, we report the average, maximum and minimum errors of the vertices. For all the datasets, minimum error is always 0. In Table 2, "Time$_{\mathcal{RF}}$" is the total time of computing $\mathcal{RF}$ of all the vertices in the set and "Time" is the total time of the other steps of computing betweenness scores of all the vertices in the set. Therefore, the total running time of BCD for a given dataset and a given set is the sum of "Time" and "Time$_{\mathcal{RF}}$". The results presented in Table 2 reveal that for estimating betweenness scores of a set of vertices, BCD significantly outperforms KADABRA. While in most cases the total running time of BCD is less than the running time of KADABRA (even when the size of the set is 15), BCD gives much more accurate results. In particular, over datasets such as *amazon, com-amazon, email-EuAll* and *web-NotreDame*, even for the sets of size 15, BCD is faster than KADABRA and it always produces much more accurate results.

6   Conclusion

In this paper, first given a directed network G and a vertex $r \in V(G)$, we proposed a new exact algorithm to compute betweenness score of r. Our algorithm computes a set $\mathcal{RF}(r)$, which is used to prune a huge amount of computations that do not contribute in the betweenness score of r. Then, for the cases where $\mathcal{RF}(r)$ is large, we presented a randomized algorithm that samples from $\mathcal{RF}(r)$ and performs computations for only the sampled elements. Finally, we empirically evaluated our algorithms and showed that in most cases, they significantly outperform the most efficient existing randomized algorithms, in terms of both running time and accuracy.

Acknowledgements. This work has been supported by the ANR project IDOLE.

References

1. Agarwal, M., Singh, R.R., Chaudhary, S., Iyengar, S.: Betweenness ordering problem: an efficient non-uniform sampling technique for large graphs. CoRR abs/1409.6470 (2014)
2. Borassi, M., Natale, E.: KADABRA is an adaptive algorithm for betweenness via random approximation. In: 24th European Symposium on Algorithms, pp. 1–18 (2016)
3. Brandes, U.: A faster algorithm for betweenness centrality. J. Math. Sociol. **25**(2), 163–177 (2001)
4. Brandes, U., Pich, C.: Centrality estimation in large networks. Int. J. Bifurcat. Chaos **17**(7), 303–318 (2007)
5. Çatalyürek, Ü.V., Kaya, K., Sariyüce, A.E., Saule, E.: Shattering and compressing networks for betweenness centrality. In: SDM, pp. 686–694 (2013)
6. Chehreghani, M.H.: Effective co-betweenness centrality computation. In: Seventh ACM International Conference on Web Search and Data Mining, pp. 423–432 (2014)
7. Chehreghani, M.H.: An efficient algorithm for approximate betweenness centrality computation. Comput. J. **57**(9), 1371–1382 (2014)
8. Chehreghani, M.H., Bifet, A., Abdessalem, T.: Discriminative distance-based network indices with application to link prediction. Comput. J. (2018, to appear)

9. Everett, M., Borgatti, S.: The centrality of groups and classes. J. Math. Sociol. **23**(3), 181–201 (1999)
10. Freeman, L.C.: A set of measures of centrality based upon betweenness, sociometry. Soc. Netw. **40**, 35–41 (1977)
11. Hayashi, T., Akiba, T., Yoshida, Y.: Fully dynamic betweenness centrality maintenance on massive networks. Proc. VLDB Endowment **9**(2), 48–59 (2015)
12. Hoeffding, W.: Probability inequalities for sums of bounded random variables. J. Am. Stat. Assoc. **58**(301), 13–30 (1963)
13. Kang, U., Papadimitriou, S., Sun, J., Tong, H.: Centralities in large networks: algorithms and observations. In: SDM, pp. 119–130 (2011)
14. Leskovec, J., Adamic, L.A., Huberman, B.A.: The dynamics of viral marketing. ACM Trans. Web (TWEB) **1**(1) (2007)
15. Leskovec, J., Huttenlocher, D.P., Kleinberg, J.M.: Signed networks in social media. In: CHI, pp. 1361–1370 (2010)
16. Leskovec, J., Kleinberg, J.M., Faloutsos, C.: Graph evolution: densification and shrinking diameters. ACM Trans. Knowl. Discov. Data **1**(1) (2007)
17. Mahmoody, A., Tsourakakis, C.E., Upfal, E.: Scalable betweenness centrality maximization via sampling. In: KDD, pp. 1765–1773 (2016)
18. Newman, M.E.J.: The structure and function of complex networks. SIAM Rev. **45**, 167–256 (2003)
19. Riondato, M., Kornaropoulos, E.M.: Fast approximation of betweenness centrality through sampling. Data Mining Knowl. Discov. **30**(2), 438–475 (2016)
20. Riondato, M., Upfal, E.: ABRA: approximating betweenness centrality in static and dynamic graphs with Rademacher averages. In: KDD, pp. 1145–1154 (2016)
21. Stergiopoulos, G., Kotzanikolaou, P., Theocharidou, M., Gritzalis, D.: Risk mitigation strategies for critical infrastructures based on graph centrality analysis. Int. J. Crit. Infrastruct. Protect. **10**, 34–44 (2015)
22. Yang, J., Leskovec, J.: Defining and evaluating network communities based on ground-truth. In: IEEE ICDM, pp. 745–754 (2012)

Forecasting Bitcoin Price with Graph Chainlets

Cuneyt G. Akcora(✉), Asim Kumer Dey, Yulia R. Gel,
and Murat Kantarcioglu

University of Texas at Dallas, Richardson, USA
{cuneyt.akcora,adey,ygl,muratk}@utdallas.edu

Abstract. Over the last couple of years, Bitcoin cryptocurrency and
the Blockchain technology that forms the basis of Bitcoin have witnessed
a flood of attention. In contrast to fiat currencies used worldwide, the
Bitcoin distributed ledger is publicly available by design. This facilitates
observing all financial interactions on the network, and analyzing how
the network evolves in time. We introduce a novel concept of chainlets,
or Bitcoin subgraphs, which allows us to evaluate the local topological
structure of the Bitcoin graph over time. Furthermore, we assess the role
of chainlets on Bitcoin price formation and dynamics. We investigate the
predictive Granger causality of chainlets and identify certain types of
chainlets that exhibit the highest predictive influence on Bitcoin price
and investment risk.

1 Introduction

Bitcoin cryptocurrency [17] has seen tremendous interest and has achieved sky-
rocketing adoption over the last couple of years. The bitcoin phenomenon is due
not only to revolutionizing online payments but also to a big number of applications the underlying blockchain technology has witnessed in various domains [21].

One interesting aspect of Bitcoin is that a distributed ledger (i.e., blockchain)
is maintained by all the participants to verify the authenticity of each Bitcoin
transaction. The existence of such a distributed ledger creates unique opportunities with respect to graph analysis. Already, different applications have used
the distributed ledger and the Bitcoin graph information to track sex trafficking
[19] and money laundering activity [16].

We believe that the Bitcoin graph can be used for interesting off-the-beaten
track applications. For instance, in most stock analysis platforms, the market
trend is usually predicted by using historical prices and other financial and economic indicators only, without accounting for financial network structure effects.
Since we can observe the complete Bitcoin graph, a natural question to ask is
whether the *local graph structure* impacts the price of an asset (e.g., Bitcoin). In
other domains, local higher-order structures of complex networks, or multiple-
node subgraphs, are found to be an indispensable tool for analysis of network
organization beyond the trivial scale of individual vertices and edges. The core

© Springer International Publishing AG, part of Springer Nature 2018
D. Phung et al. (Eds.): PAKDD 2018, LNAI 10939, pp. 765–776, 2018.
https://doi.org/10.1007/978-3-319-93040-4_60

idea is that if a particular subgraph occurs more or less frequently than the expected baseline occurrence, then such a subgraph is likely to play an important role in network functionality.

Furthermore, structural properties of multiple complex networks can be compared in terms of their (dis)similarities in subgraph patterns. The role of small subgraphs, or network *motifs* and *graphlets*, in organization of complex systems has been first discussed in conjunction with the assessment of stability and robustness of biological networks [15], and later have been studied in a variety of contexts, from social networks to power grids (for overviews see [1] and references therein). Most recently, network motifs are shown to provide an invaluable insight into analysis of functionality and early warning stability indicators in financial networks [9]. However, compared to biological networks, motif-induced inference in financial systems is still an emerging field, and there yet exist no studies on the role of motifs in the analysis of blockchain.

To our knowledge, we are the first to address the impact of local topological structures/motifs on Bitcoin price. We can summarize our contributions as follows:

- We introduce and formalize the notion of *chainlet* motifs to understand the impact of local topological structures on Bitcoin price dynamics.
- We develop techniques to understand which local topological structures (i.e., chainlets) have a higher impact on the price dynamics and use those "important" chainlets for price prediction.
- We compare our techniques to the state of art time series analysis approaches and show that employing chainlets leads to more competitive price prediction mechanisms.

The remainder of this paper is organized as follows: In Sect. 2, we discuss the related work. In Sect. 3, we formally define chainlets using a generalized heterogeneous graph model. In Sect. 4 we compare the price prediction models that use chainlets to other existing models to see the impact of chainlets on price. Finally, in Sect. 5, we conclude with the summary of our results.

2   Related Work

Since the seminal Bitcoin paper [17] in 2008, digital coins [21] have been the most prominent Blockchain applications. Among these, Bitcoin has been the main focus of Blockchain analysis (see [2] for a review).

The earliest studies focused on the transaction graph to locate the coins used in illegal activities, such as money laundering and blackmailing [3,18], which is known as the taint analysis [5]. Moser et al. [16] analyzed the opportunities and limitations of anti-money laundering on Bitcoin by looking at how successive transactions are used to transfer money.

The Bitcoin network itself has also been studied from multiple aspects. For instance, [4] analyzed centralities, and [13] found that since 2010 the Bitcoin network can be considered a scale-free network. Furthermore, [12] tracked the evolution of the Bitcoin transaction network, and modeled degree distributions with power-laws. Although these studies analyzed the Bitcoin graphs, the primary focus was on global graph characteristics. In turn, our *chainlet analysis* sheds light onto local topological structures of Bitcoin and their role on price formation.

A number of recent studies show the utility of global graph features to predict the price [7,11,14]. For instance, [20] analyzed the predictive effects of average balance, clustering coefficient, and number of new edges on the Bitcoin price. Two network flow measures were recently proposed by [23] to quantify the dynamics of the Bitcoin transaction network and to assess the relationship between flow complexity and Bitcoin market variables. Furthermore, [14] identified 16 features for 30, 60 or 120 min intervals and used Random Forest models to predict the price. The core idea behind all these approaches is to extract certain global network features and to employ them for predictions. On the other hand, chainlets provide a finer grained insight at the network transactions. In practice, chainlets can be used to refine the above-mentioned models, so that features are computed on selected subgraphs only. Furthermore, network flows can be detailed in terms of successive chainlets.

3 Methodology

The Bitcoin graph has three main components: *addresses, transactions* and *blocks*. A transaction is a transfer of bitcoins from input addresses to output addresses. Figure 1 shows such a network for 4 transactions and 13 addresses.

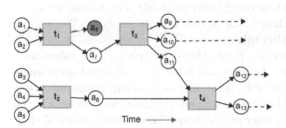

Fig. 1. A transaction-address graph representation of the Bitcoin network. Addresses and transactions are shown with circles and rectangles, respectively. An edge indicates a transfer of coins. The coins at address a_6 are unspent.

Our Bitcoin data come from the official Bitcoin software; we installed the Bitcoin core wallet[1] and had the wallet download the entire Bitcoin history from 2009 to 2018. Afterwards, we parsed the Bitcoin blockchain files, and extracted blocks, transactions and addresses. The source code of our Spark project is available on our Github repository.[2]

We model the Bitcoin graph as the following heterogeneous network with two node types: addresses and transactions.

The Bitcoin Graph Model. The Bitcoin network is a directed graph $\mathcal{G} = (V, E, B)$ where V is a set of vertices, and $E \subseteq V \times V$ is a set of edges. $B =$

---

[1] https://bitcoin.org/en/download.
[2] https://github.com/cakcora/coinworks.

{**Address, Transaction**} represents the set of vertex types. For any vertex $u \in V$, it has a vertex type $\phi(u) \in B$. For each edge $e_{u,v} \in E$ between adjacent nodes u and v, we have $\phi(u) \neq \phi(v)$, and either $\phi(u) = \{$**Transaction**$\}$ or $\phi(v) = \{$**Transaction**$\}$. That is, an edge $e \in E$ represents a coin transfer between an address node and a transaction node. This heterogeneous graph model subsumes the homogeneous case (i.e., $|B| = 1$), where only transaction or address nodes are used, and edges link vertices of the same type. In this paper, we focus on the case where each address node is linked (i.e., input or output address of a transaction) via a transaction node to another address node.

We emphasize three graph rules that shape the actual Bitcoin graph. First, input coins from multiple transactions can be merged and spent in a single transaction (as in transaction t_4 in Fig. 1). Second, in a Bitcoin transaction the input-output address mappings are not explicitly recorded. For instance, consider the transaction t_1 in Fig. 1. The output to address a_6 may come from either a_1 or a_2. Third, coins from multiple input transactions can be spent separately, but those received from one transaction must all be spent in a single transaction. Any amount that is not transferred is considered to be the transaction fee, and gets collected by the miner who creates the block. For this reason, unless it specifies itself as output address again, an address cannot transfer some bitcoins from a previous transaction and keep the change. As a community practice, this address reuse is discouraged, hence most nodes appear in the graph two times; once when they receive coins and once when they spend it. See [2] for a detailed graph representation of Blockchain.

Blocks order transactions in time, whereas each transaction with its input and output nodes represents an immutable decision that is encoded as a subgraph on the Bitcoin network. Rather than using individual edges or nodes, we chose to use this subgraph as the building block in our Bitcoin analysis. We use the term **chainlet** to refer to such subgraphs.

Our choice is due to two reasons. First, the subgraph can be taken as a single data unit because inclusion of nodes and edges in it is based on a single decision. As a transaction is immutable, joint inclusion of input/output nodes in its subgraph cannot be changed afterwards. This is unlike the case on a social network where nodes can become closer on the graph because of actions of their neighbors. Second, we argue and prove that subgraphs have distinct shapes that reflect their role in the network, and we can aggregate these roles to analyze network dynamics.

3.1 Graph Chainlets

We introduce the concept of k-chainlets to assess local higher order topological structure of the Bitcoin graph.

The k-Chainlet Model. A Bitcoin subgraph $\mathcal{G}' = (V', E', B)$ is a *subgraph* of $\mathcal{G}$, if $V' \subseteq V$ and $E' \subseteq E$. If $\mathcal{G}' = (V', E', B)$ is a subgraph of $\mathcal{G}$ and E' contains all edges $e_{u,v} \in E$ such that $(u, v) \in V'$, then G' is called an *induced* subgraph of G. Two graphs $\mathcal{G}' = (V', E', B)$ and $\mathcal{G}'' = (V'', E'', B)$ are called *isomorphic* if there exists a bijection $h : V' \rightarrow V''$ such that all node pairs u, v of G' are adjacent in G' if and only if u and v are adjacent in G''.

Let k-chainlet $\mathcal{G}_k = (V_k, E_k, B)$ be a subgraph of $\mathcal{G}$ with k nodes of type {**Transaction**}. If there exists an iso- morphism between $\mathcal{G}_k$ and $\mathcal{G}'$, $\mathcal{G}' \in \mathcal{G}$, we say that there exists an *occurrence*, or *embedding* of $\mathcal{G}_k$ in $\mathcal{G}$. If a $\mathcal{G}_k$ occurs more/less frequently than expected by chance, it is called a blockchain k-

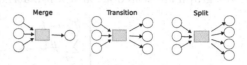

Fig. 2. Merge ($\mathbb{C}_{3\rightarrow 1}$), Transition ($\mathbb{C}_{3\rightarrow 3}$) and Split ($\mathbb{C}_{3\rightarrow 4}$) chainlets for 3 inputs.

chainlet. A *k-chainlet signature* $f_\mathcal{G}(\mathcal{G}_k)$ is a number of occurrences of $\mathcal{G}_k$ in $\mathcal{G}$.

We start by focusing on the 1-chainlet signatures and their properties. For simplicity, we refer to *1-chainlets as chainlets*. A natural classification of chainlets can be made in terms of the number of inputs x and outputs y since there is only one transaction involved.

For a chainlet, we denote $\mathbb{C}_{x\rightarrow y}$ if it has x inputs and y outputs. If the branch is merging with other branches, the cor- responding chainlet will have a higher number of inputs, compared to out- puts. We call these **merge** chainlets, i.e., $\mathbb{C}_{x\rightarrow y}$ such that $x > y$, which show an aggregation of coins into fewer addresses. Two other classes of chainlets are **transition** and **split**

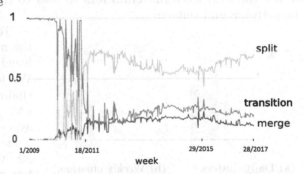

Fig. 3. Percentage of aggregate chainlets in weeks. Splits constitute around 60% of all transactions.

chainlets with $x = y$ and $x < y$, respectively, as shown in Fig. 2. In what follows, we refer to these three chainlet types as the **aggregate chainlets**.

Figure 3 visualizes the percentage of aggregate chainlets in time. For example, the transition chainlets are those $\mathbb{C}_{x\rightarrow x}$ for $x \geq 1$. Figure 3 shows that starting as an unknown project, the Bitcoin network stabilized only after summer 2011. From 2014 and onwards, the split chainlets continued to steadily rise, compared to merge and transition chainlets.

3.2   Clustering Chainlets

The Bitcoin protocol restricts numbers of input and output addresses in a trans- action by putting a limit on the block size (1MB), but the number of inputs and outputs can still reach thousands. As a result, we can have millions of distinct chainlets (e.g., $\mathbb{C}_{1900\rightarrow 200}$, $\mathbb{C}_{1901\rightarrow 200}$ or $\mathbb{C}_{1900\rightarrow 201}$).

We use a matrix representation to model the Bitcoin graph in time with chainlets. For a given time granularity, such as one day, we take snapshots of the Bitcoin network and construct a Bitcoin graph. Chainlet counts obtained from this graph are stored as an $n \times n$-matrix $\mathcal{O}$ such that for $i \leq n, j \leq n$

$$
\mathcal{O}[i,j] = \begin{cases} \#\mathbb{C}_{i \to j} & \text{if } i < n \text{ and } j < n, \\ \sum\limits_{z=n}^{\infty} \#\mathbb{C}_{i \to z} & \text{if } i < n \text{ and } j = n, \\ \sum\limits_{y=n}^{\infty} \#\mathbb{C}_{y \to j} & \text{if } i = n \text{ and } j < n, \\ \sum\limits_{y=n}^{\infty} \sum\limits_{z=n}^{\infty} \#\mathbb{C}_{y \to z} & \text{if } i = n \text{ and } j = n. \end{cases}
$$

In this matrix notation, choosing an n value, e.g., $n = 5$, means that a chainlet with more than 5 inputs/outputs (i.e., $\mathbb{C}_{x \to y}$ s.t., $x \geq 5$ or $y \geq 5$) is recorded in the n-th row or column. That is, we aggregate chainlets with large dimensions that would otherwise fall outside matrix dimensions. In what follows we use the term **extreme chainlets** to refer to these aggregated chainlets on the n-th row and column.

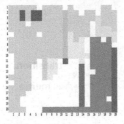   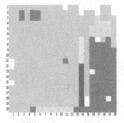

(a) Daily clusters.   (b) Weekly clusters.

Fig. 4. [Color online]. Chainlet clusters with day and week granularities. A chainlet $\mathbb{C}_{x \to y}$ is the intersection cell of the x-th row and y-th column.

To select a suitable value for the matrix dimension n, we analyzed the entire Bitcoin history. We found that % 90.50 of the chainlets have n of 5 (i.e., $\mathbb{C}_{x \to y}$ s.t., $x < 5$ and $y < 5$) in average for daily snapshots. This value reaches % 97.57 for n of 20. We chose to take n of 20, because it can distinguish a sufficiently large number (i.e., 400) of chainlets, and still offers a dense matrix.

With daily and weekly snapshots of the Bitcoin network, we constructed 3.284 and 443 daily and weekly matrices, respectively (with data from 2009 to 2018). Each of the 400 chainlets is represented as a vector of its count in time.

We hierarchically clustered chainlets by using Cosine Similarity [8] over chainlet vectors, and used a similarity cut threshold of 0.7 to create clusters from the hierarchical dendogram. Figure 4 shows the resulting clusters. Cluster memberships are shown with the same color. A white cell denotes a chainlet that constitutes a cluster of its own. In both Fig. 4a and b, higher n values in the right low corner are clustered together, and in the daily clusters extreme chainlets ($\mathbb{C}_{\{x|x>8\} \to 20}$) have their own cluster. An interesting result is that in both matrices extreme chainlets belong to the same clusters with some considerably smaller chainlets such as $\mathbb{C}_{2 \to 3}$, $\mathbb{C}_{3 \to 3}$ and $\mathbb{C}_{2 \to 6}$. In Sect. 4.2 we show that their similarity extends to their impact on price predictions.

4 Experiments

Our experiments first prove the predictive power of chainlets with Granger Causality. We then show how chainlets can be used to predict Bitcoin price.

4.1 Granger Causality

To assess a potential predictive role of chainlets in Bitcoin price formation, we employ a widely adopted econometric concept of Granger causality [6]. The causality test assesses whether one time series is useful in predicting another (see an overview by White et al. [22]). In particular, assume $\mathbf{Y}_t$, $t \in Z^+$ is a $p \times 1$-random vector (e.g., Bitcoin price) and let $\mathcal{F}^t_{(\mathbf{Y})} = \sigma\{\mathbf{Y}_s : s = 0, 1, \ldots, t\}$ denote a σ-algebra generated from all observations of $\mathbf{Y}$ in the market up to time t. Consider a sequence of $(k + 2)$-tuples of random vectors $\{\mathbf{Y}_t, \mathbf{X}_t, \mathbf{Z}^1_t, \ldots, \mathbf{Z}^k_t\}$. For example, in the context of this paper $\mathbf{X}$ can be chainlets and $\mathbf{Z}^1, \ldots, \mathbf{Z}^k$ can be number of transactions. Suppose that for all $h \in Z^+$

$$F_{t+h}\left(\cdot | \mathcal{F}^{t-1}_{(\mathbf{Y},\mathbf{X},\mathbf{Z}^1,\ldots,\mathbf{Z}^k)}\right) = F_{t+h}\left(\cdot | \mathcal{F}^{t-1}_{(\mathbf{Y},\mathbf{Z}^1,\ldots,\mathbf{Z}^k)}\right), \tag{1}$$

where $F_{t+h}\left(\cdot | \mathcal{F}^{t-1}_{(\mathbf{Y},\mathbf{X},\mathbf{Z}^1,\ldots,\mathbf{Z}^k)}\right)$ and $F_{t+h}\left(\cdot | \mathcal{F}^{t-1}_{(\mathbf{Y},\mathbf{Z}^1,\ldots,\mathbf{Z}^k)}\right)$ are conditional distributions of $\mathbf{Y}_{t+h}$, given $\mathbf{Y}_{t-1}, \mathbf{X}_{t-1}, \mathbf{Z}^1_{t-1}, \ldots, \mathbf{Z}^k_{t-1}$ and $\mathbf{Y}_{t-1}, \mathbf{Z}^1_{t-1}, \ldots, \mathbf{Z}^k_{t-1}$, respectively. Then, $\mathbf{X}_{t-1}$ is said *not* to Granger cause (G-cause) $\mathbf{Y}_{t+h}$ with respect to $\mathcal{F}^{t-1}_{(\mathbf{Y},\mathbf{Z}^1,\ldots,\mathbf{Z}^k)}$. Otherwise, $\mathbf{X}$ is said to G-cause $\mathbf{Y}$, which can be denoted by $G_{\mathbf{X} \mapsto \mathbf{Y}}$, where $\mapsto$ represents the direction of causality. Hence, G-causality means that given information on the past of $\mathbf{Y}$ and $\mathbf{Z}^1, \ldots, \mathbf{Z}^k$, the past of $\mathbf{X}$ does not deliver any new information that can be used for predicting $\mathbf{Y}_{t+h}$.

In practice G-causality is typically performed by fitting two linear vector autoregressive (VAR) models of finite order d to $\mathbf{Y}$, with and without $\mathbf{X}$, respectively, and then testing for statistical significance of model coefficients associated with $\mathbf{X}$. Alternatively, we can compare predictive performance of two models (i.e., with and without $\mathbf{X}$), using an F-test, under the null hypothesis of no explanatory power in $\mathbf{X}$. For instance, consider a case of univariate time series y_t, x_t and z_t. To test G-causality of x_t, we compare the fit of the full model $y_t = \alpha_0 + \sum_{k=1}^d \alpha_k y_{t-k} + \sum_{k=1}^d \beta_k x_{t-k} + \sum_{k=1}^d \gamma_k z_{t-k} + e_t$, versus the fit of the reduced model $y_t = \alpha_0 + \sum_{k=1}^d \alpha_k y_{t-k} + \sum_{k=1}^d \beta_k x_{t-k} + \tilde{e}_t$. That is, under the null hypothesis of no predictive effect in x onto y (i.e., x does not G-cause y), $Var(e_t) = Var(\tilde{e}_t)$. If $Var(e_t)$ is (statistically) significantly lower than $Var(\tilde{e}_t)$, then we conclude that x contains additional information that can improve forecasting of y, i.e., $G_{x \mapsto y}$.

Armed with the time series of chainlets, we are now interested in evaluating the potential impact of local graph structures on *future* bitcoin price formation and investment risk. We are primarily interested in two interlinked questions:

1. Do changes in chainlet characteristics exhibit any causal effect on future Bitcoin price and Bitcoin returns?
2. Do chainlets convey some unique information about future Bitcoin prices, given more conventional economic variables and non-network blockchain characteristics?

Table 1 provides summary results of the Granger causality tests for predictive utility of individual/aggregate chainlets, and chainlet clusters[3] in analysis of the Bitcoin price and its log returns (see Fig. 4a for the clusters). Log returns of Bitcoin prices measure the relative change in prices and are defined as $LR_t = \log y_t - \log y_{t-1}$. As a more conventional predictor, we also include the total number of transactions (# of Trans.) into the baseline models. Direction of causality is denoted by $\longmapsto$. Table 1 indicates that individual chainlets, e.g., $\mathbb{C}_{6 \to 1}$, $\mathbb{C}_{1 \to 7}$, $\mathbb{C}_{20 \to 12}$, as well as aggregate chainlets, e.g., split chainlets, have a predictive impact on price formation, and in some cases also exhibit causal linkage with future log returns. Some chainlet clusters have predictive relationship only with Bitcoin price, whereas Cluster 35 G-causes both price and log returns. As expected, total number of transactions also has causality effects on both Bitcoin price and log returns. The G-causality relationships of different chainlets and Bitcoin price indicate that they are likely to contain important predictive information on Bitcoin price formation and volatility.

4.2   Price Prediction

In Sect. 4.1 we show that chainlets G-cause the Bitcoin price and hence, exhibit predictive impact on prices. We are now interested in quantifying the forecasting utility of chainlets. To evaluate the chainlets' predictive power, *we can use any forecasting model and compare predictive performances with and without chainlets.* Typically such a comparative analysis is performed based on the Box-Jenkins (BJ) class of parametric linear models. However, as indicated by [10], more flexible Random Forest (RF) models often tend to outperform the BJ models in their predictive capabilities. In particular, we find that the optimal baseline autoregressive integrated moving average (ARIMA(p, d, q)) models selected by minimizing the Akaike Information criterion (AIC), yield from 0.2% to 40% higher prediction root mean squared error (RMSE) than the RF baseline models.

Here RMSE $= \sqrt{(1/n) \sum_{t=1}^{n} (y_t - \hat{y}_t)^2}$, where y_t is the test set of Bitcoin price and $\hat{y}_t$ is the corresponding predicted value. ARIMA and RF models deliver comparable results, therefore, due to space limitations, we present the comparison study based only on the RF type of models.

---

[3] Some representative chainlets from daily clusters 7, 8, 16 and 35 are $\mathbb{C}_{9 \to 11}$, $\mathbb{C}_{3 \to 17}$, $\mathbb{C}_{8 \to 14}$ and $\mathbb{C}_{1 \to 1}$, respectively.

Table 1. In G-causality, P and LR denote significance in price & log returns, respectively; blank space implies no significance. Confidence level is 95%.

Covariate types	Causality	Outcome with lag effects				
		1	2	3	4	5
# of Trans.	Total # Trans. $\rightarrowtail$ Outcome	LR	LR	P/LR	P/LR	
Aggregate Chainlets	Merge Chainlets $\rightarrowtail$ Outcome	-	-	-	-	-
	Split Chainlets $\rightarrowtail$ Outcome	-	LR	P/LR	P	-
	Trans. Chainlets $\rightarrowtail$ Outcome	-	-	-	-	-
Individual Chainlets	$\mathbb{C}_{1\to7} \rightarrowtail$ Outcome	P	P	P	P	P
	$\mathbb{C}_{6\to1} \rightarrowtail$ Outcome	-	P	P	P	-
	$\mathbb{C}_{3\to3} \rightarrowtail$ Outcome	-	P	P	P	-
Extreme Chainlets	$\mathbb{C}_{20\to2} \rightarrowtail$ Outcome	LR	P/LR	P/LR	P/LR	P
	$\mathbb{C}_{20\to3} \rightarrowtail$ Outcome	P	P	P	P	P
	$\mathbb{C}_{20\to12} \rightarrowtail$ Outcome	P	P	P	P	P
	$\mathbb{C}_{20\to17} \rightarrowtail$ Outcome	-	-	P	P	P
Chainlet Clusters	Cluster 35 $\rightarrowtail$ Outcome	LR	LR	P/LR	P/LR	-
	Cluster 16 $\rightarrowtail$ Outcome	-	LR	-	-	-
	Cluster 8 $\rightarrowtail$ Outcome	-	P	P	P	P
	Cluster 7 $\rightarrowtail$ Outcome	-	P	P	P	P

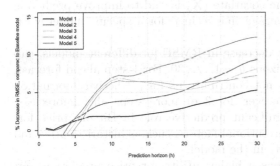

Fig. 5. % Change (decrease) in RMSE compared to the baseline model.

We performed extensive experiments with various chainlets and selected to showcase six of these RF models. Table 2 provides an overview of the constructed models. The baseline model includes only the lagged (past period) values of the Bitcoin price. Other models comprise of lagged prices with different covariates, mainly chainlets or some functions of chainlets such as the mean of all aggregate/split type chainlets and mean of all chainlets in a specific cluster.

In our study each RF model used 500 trees, and sampling all rows of the data set is done with replacement. Number of variables used at each split are, for example, 2, 3 and 4, for Models 1, 2 and 5, respectively.

We continuously change the training data using a sliding window technique, where we choose the window size of 200. That is, at each time step we train our model based on the past 200 values, and armed with this estimated model, we then construct a h step ahead forecast.

Table 2. Model description for Bitcoin price (response) and varying predictors.

Model	Predictors
Baseline M_0	Price lag 1, Price lag 2, Price lag 3
Model 1	Price lag 1, Price lag 2, Price lag 3, # Trans lag 1 , # Trans lag 2, # Trans lag 3
Model 2	Price lag 1, Price lag 2, Price lag 3, Split Pattern lag 1, Split Pattern lag 2, Split Pattern lag 3, Cluster 8 lag 1, Cluster 8 lag 2, Cluster 8 lag 3
Model 3	Price lag 1, Price lag 2, Price lag 3, $\mathbb{C}_{1\to7}$ lag 1, $\mathbb{C}_{1\to7}$ lag 2, $\mathbb{C}_{1\to7}$ lag 3
Model 4	Price lag 1, Price lag 2, Price lag 3, $\mathbb{C}_{1\to7}$ lag 1, $\mathbb{C}_{1\to7}$ lag 2, $\mathbb{C}_{1\to7}$ lag 2, $\mathbb{C}_{6\to1}$ lag 1, $\mathbb{C}_{6\to1}$ lag 2, $\mathbb{C}_{6\to1}$ lag 3
Model 5	Price lag 1, Price lag 2, Price lag 3, $\mathbb{C}_{1\to7}$ lag 1, $\mathbb{C}_{1\to7}$ lag 2, $\mathbb{C}_{1\to7}$ lag 2, $\mathbb{C}_{6\to1}$ lag 1, $\mathbb{C}_{6\to1}$ lag 2, $\mathbb{C}_{6\to1}$ lag 3, $\mathbb{C}_{3\to3}$ lag 1, $\mathbb{C}_{3\to3}$ lag 2, $\mathbb{C}_{3\to3}$ lag 3

Predictive utilities of models in Table 2 over the baseline model can be measured as $\Psi_{(X\to Y)} = \psi(M)/\psi(M_0)$, where ψ is a measure of prediction error, e.g., root mean squared error (RMSE). Here $\psi(M_0)$ is the prediction error of baseline model, where lagged prices are the only predictor; and $\psi(M)$ is the prediction error of a given model, where predictors are lagged prices and other exogenous covariates $(\mathbf{X})$. If $\Psi_{(X\to Y)} < 1$, the covariate $(\mathbf{X})$ is said to improve prediction of Y. We also calculate the percentage change in ψ for a specific model w.r.t. M_0 as $\Delta = \left(1 - \Psi_{(X\to y)}\right)100\%$.

Figure 5 compares the percent decrease in RMSE for different models, calculated for varying prediction horizons $h = 1, \ldots, 30$. For 1-step ahead forecast, chainlets and other covariates do not contribute useful predictive information over history of Bitcoin price. However, for 3 or more steps ahead forecasts, chainlets play an increasingly significant predictive role in Bitcoin price formation, even when other more conventional factors, such as historical price and number of transactions, are already in the model.

Furthermore, some chainlets has a higher utility for price prediction. For example, in Model 5, we observe the highest decrease in RMSE, compared to the baseline model. Models 3 and 4 yield the second highest decrease in RMSE until the forecast horizon h of 20. After h of 20, Model 2 delivers the second highest reduction in RMSE over the baseline model.

Figure 6 compares the observed data with fitted values from baseline model and three other models, i.e., Model 1, 2, and 5. For h of 1, all models deliver similar prediction accuracy and capture the variability of the data very well. Although, as expected, the prediction performance of all models deteriorates as forecasting horizon $h \to \infty$, Models 1, 2, and 5 still yield a noticeably higher predictive accuracy, compared to the baseline model without chainlets.

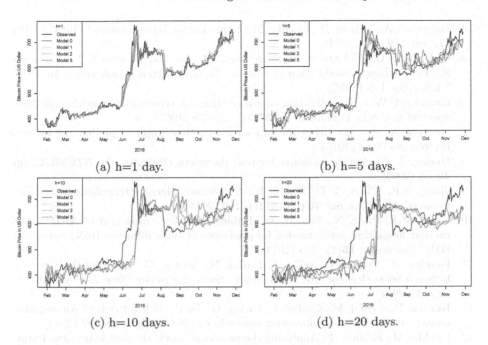

(a) h=1 day.

(b) h=5 days.

(c) h=10 days.

(d) h=20 days.

Fig. 6. [Color Online]. Price prediction for 2016 with 1, 5, 10 for 20 day horizons.

5   Conclusion

We introduce a novel concept of k-chainlets on Bitcoin that expands the ideas of motifs and graphlets to Blockchain graphs. Chainlet analysis provides a deeper insight into local topological properties of the Blockchain and the role of those local higher-order topologies in the Bitcoin price formation. We find that certain types of chainlets have a high predictive utility for Bitcoin prices. Furthermore, extreme chainlets exhibit an important role in the Bitcoin price prediction.

Acknowledgments. This research was supported in part by NIH 1R01HG006844, NSF CNS-1111529, CICI-1547324, IIS-1633331, DMS-1736368 and ARO W911NF-17-1-0356.

References

1. Ahmed, N.K., Neville, J., Rossi, R.A., Duffield, N., Willke, T.L.: Graphlet decomposition: framework, algorithms, and applications. KAIS **50**, 1–32 (2016)
2. Akcora, C.G., Gel, Y.R., Kantarcioglu, M.: Blockchain: a graph primer. arXiv preprint arXiv:1708.08749 (2017)
3. Androulaki, E., Karame, G.O., Roeschlin, M., Scherer, T., Capkun, S.: Evaluating user privacy in bitcoin. In: Sadeghi, A.-R. (ed.) FC 2013. LNCS, vol. 7859, pp. 34–51. Springer, Heidelberg (2013). https://doi.org/10.1007/978-3-642-39884-1_4

4. Baumann, A., Fabian, B., Lischke, M.: Exploring the bitcoin network. In: WEBIST (1), pp. 369–374 (2014)
5. Di Battista, G., Di Donato, V., Patrignani, M., Pizzonia, M., Roselli, V., Tamassia, R.: Bitconeview: visualization of flows in the bitcoin transaction graph. In: IEEE VizSec, pp. 1–8 (2015)
6. Granger, C.W.J.: Investigating causal relations by econometric models and cross-spectral methods. Econometrica **37**(3), 424–438 (1969)
7. Greaves, A., Au, B.: Using the bitcoin transaction graph to predict the price of bitcoin. No Data (2015)
8. Huang, A.: Similarity measures for text document clustering. In: NZCSRSC, pp. 49–56 (2008)
9. Jiang, X.F., Chen, T.T., Zheng, B.: Structure of local interactions in complex financial dynamics. Sci. Rep. **4**(5321), 1–9 (2014)
10. Kane, M.J., Price, N., Scotch, M., Rabinowitz, P.: Comparison of ARIMA and random forest time series models for prediction of avian influenza H5N1 outbreaks. BMC Bioinform. **15**(1), 276 (2014)
11. Kondor, D., Csabai, I., Szüle, J., Pósfai, M., Vattay, G.: Inferring the interplay between network structure and market effects in Bitcoin. New J. Phys. **16**(12), 125003 (2014)
12. Kondor, D., Pósfai, M., Csabai, I., Vattay, G.: Do the rich get richer? An empirical analysis of the Bitcoin transaction network. PLOS One **9**(2), e86197 (2014)
13. Lischke, M., Fabian, B.: Analyzing the bitcoin network: the first four years. Future Internet **8**(1), 7 (2016)
14. Madan, I., Saluja, S., Zhao, A.: Automated bitcoin trading via machine learning algorithms (2015)
15. Milo, R., Shen-Orr, S., Itzkovitz, S., Kashtan, N., Chklovskii, D., Alon, U.: Network motifs: simple building blocks of complex networks. Science **298**(5594), 824–827 (2002)
16. Moser, M., Bohme, R., Breuker, D.: An inquiry into money laundering tools in the bitcoin ecosystem. In: eCRS, pp. 1–14. IEEE (2013)
17. Nakamoto, S.: Bitcoin: a peer-to-peer electronic cash system (2008)
18. Ober, M., Katzenbeisser, S., Hamacher, K.: Structure and anonymity of the bitcoin transaction graph. Future Internet **5**(2), 237–250 (2013)
19. Portnoff, R.S., Huang, D.Y., Doerfler, P., Afroz, S., McCoy, D.: Backpage and bitcoin: uncovering human traffickers. In: SIGKDD, pp. 1595–1604. ACM (2017)
20. Sorgente, M., Cibils, C.: The reaction of a network: exploring the relationship between the Bitcoin network structure and the Bitcoin price. No Data (2014)
21. Tschorsch, F., Scheuermann, B.: Bitcoin and beyond: a technical survey on decentralized digital currencies. IEEE Commun. Surv./Tut. **18**(3), 2084–2123 (2016)
22. White, H., Chalak, K., Lu, X.: Linking granger causality and the pearl causal model with settable systems. In: JMLR, vol. 12, pp. 1–29 (2011)
23. Yang, S.Y., Kim, J.: Bitcoin market return and volatility forecasting using transaction network flow properties. In: IEEE SSCI, pp. 1778–1785 (2015)

Information Propagation Trees
for Protest Event Prediction

Jeffery Ansah[✉], Wei Kang, Lin Liu, Jixue Liu, and Jiuyong Li

School of Information Technology and Mathematical Sciences,
University of South Australia, Adelaide, SA 5095, Australia
jeffery.ansah@mymail.unisa.edu.au,
{wei.kang,lin.liu,jixue.liu,jiuyong.li}@unisa.edu.au

Abstract. Protest event prediction using information propagation from
social media is an important but challenging problem. Despite the
plethora of research, the implicit relationship between social media information propagation and real-world protest events is unknown. Given
some information propagating on social media, how can we tell if a
protest event will occur? What features of information propagation are
useful and how do these features contribute to a pending protest event?
In this paper, we address these questions by presenting a novel formalized
propagation tree model that captures relevant protest information propagating as precursors to protest events. We present a viewpoint of information propagation as trees which captures both temporal and structural
aspects of information propagation. We construct and extract structural
and temporal features daily from propagation trees. We develop a matching scheme that maps daily feature values to protest events. Finally, we
build a robust prediction model that leverages propagation tree features
for protest event prediction. Extensive experiments conducted on Twitter datasets across states in Australia show that our model outperforms
existing state-of-the-art prediction models with an accuracy of up to 89%
and F1-score of 0.84. We also provide insights on the interpretability of
our features to real-world protest events.

Keywords: Propagation trees · Protest · Prediction · Social media
Twitter

1 Introduction

Protest event prediction is an important task with numerous benefits to many
organizations and stakeholders. Early predictions help law enforcement agencies
prepare ahead of anticipated protests, inform tourists of protest-prone zones,
help traffic regulators divert traffic effectively and also help service providers
prioritize on the concerns of citizens.

Social media sites such as Twitter and Facebook have become a major tool
for the organization, mobilization, and coordination of protest events such as
the Arab Springs, Latin America Uprising, Baltimore Riots, 15M, and other

© Springer International Publishing AG, part of Springer Nature 2018
D. Phung et al. (Eds.): PAKDD 2018, LNAI 10939, pp. 777–789, 2018.
https://doi.org/10.1007/978-3-319-93040-4_61

protest events of the like. Existing studies on protest events [4,5] have shown that the ubiquitous nature and increasing use of social media during protest events make it a viable platform for developing techniques to predict future protest events. Recent studies [5,6,8,10,12] have proposed different methods for predicting protest events using social media. These methods generally aim at extracting (e.g., keywords, hashtags, event dates, location mentions), analysing and integrating social media data with appropriate statistical, Natural Language Processing (NLP), and machine learning algorithms to detect or forecast events.

Despite the successes of existing state-of-the-art social media protest prediction approaches, some challenges still exist. For example, text-based approaches such as [5,11] are not designed with information propagation within online communities in mind. Text-based approaches are therefore restricted in their ability to model structural and temporal features of hashtags, users, follower-relationships, retweets etc. from information propagation for protest prediction. Secondly, the size, speed, and complexity of social media posts coupled with the dynamic evolution of online social media information propagation make it computationally challenging to directly observe the implicit relationship between social media posts and real-world protest events. Also, features from information propagation that are highly interpretable for event prediction are unknown. This calls for the need to develop new models that are not restricted by these technical challenges as well as yielding a rich set of interpretable features for protest event prediction without trading off the accuracy of prediction.

In this paper, we address this need by presenting a novel propagation tree model for predicting protest events using information propagation on social media. Our proposed approach can model the relationship between tweets, users, hashtags and follower relationships, and effective in capturing both the temporal and structural features of information propagation for protest event prediction. Given some information propagating on social media about a protest event, how can we tell if a protest event is going to happen? What features of information propagation can give indications of when a protest event is going to occur? We model information propagation as trees and propose a rich set of highly interpretable features that are precursory to protest events to answer these questions.

We first present a formalized model for constructing propagation trees from online conversations on protest events. We continuously extract structural and temporal features from these trees. We develop a feature-to-event mapping scheme that maps temporal and spatial information from ground truth protest events to propagation tree features. We then build a classifier to learn the mapping function for protest event prediction using propagation tree features as inputs. We conduct extensive experiments on real-world datasets to demonstrate the potency of our method over existing state-of-the-art prediction approaches.

More concretely our contributions are summed up as follows:

1. We present a formalized data-driven model for building propagation trees from information propagating on online social networks.

2. We build a prediction model that uses structural and temporal features extracted from propagation trees for protest event prediction. Our proposed model outperforms existing state-of-the-art prediction models.
3. We present a novel set of highly interpretable features that shed light on the implicit relationship between information propagation on social media and real-world protest events.

2   Related Work

Protest event prediction and detection have been an active line of research, presenting advanced techniques that incorporate different statistical and machine learning techniques. Some notable works are underscored in [2,5,6,8,10,12]. A well-known prediction system in this domain is EMBERS [12]. The EMBERS system fuses several models [1,8,12] with data from different sources to forecast protest events in 10 Latin American countries. Earlier work by Kallus [5] presents a random forest protest model that leverages future date heuristics and NLP techniques to extract event type, population, event date from open source data in seven (7) different languages. A Non-Parametric Heterogeneous Graph Scan for disease and protest prediction using Twitter data was proposed by [2].

The advent of social media has facilitated the study of information propagation in social networks. This line of research on information propagation studies [3,4,7,9,13,14] present diverse notions for studying information propagation in online social networks. These studies have shown emerging applications of information propagation in viral systems [13], influence maximization [7], and recommendation networks, etc. However, less effort has been devoted to the application of information propagation to protest event prediction due to the implicit relationship between information propagation on social media and real-world protest events that have not been fully understood.

Our work seeks to combine these two parallel lines of research by modeling information propagation for protest event prediction. Closely related to our work is [1] where the authors used structural features of cascade graphs as input to a regression model to predict protest events in Latin America. However, we present a notion of propagation trees that differs in modeling the dynamics of information propagation. Our approach further incorporates both structural and temporal features of online information propagation for protest event prediction.

3   Preliminaries and Problem Setup

In this section, we present preliminary definitions and also describe the setup of our problem. Throughout this paper, we consider Twitter as our social network to model information propagation on protest events. However, it is worth mentioning that the formalizations can be modified and extended in the context of other social networks and microblogs such as Facebook and Sina Weibo.

3.1 Definitions

Twitter provides a functionality for users to follow other users. This enables users to receive information from those they follow on their timeline. Generally, information propagates on Twitter in the following manner, a Twitter user Alice posts a tweet[1] on a protest event on a given day. Bob, a follower of Alice also posts a tweet on the same protest event after the original post by Alice. As this process continues, information propagates through the network.

Definition 1 (Twitter Follower Network). *Twitter follower network is a directed graph* $\mathbf{G} = \langle \mathbf{V}, \mathbf{E} \rangle$, *where* $\mathbf{V} = \{X_1, X_2, ..., X_N\}$ *is a set of N Twitter users, and* $\mathbf{E} = \{X_i \rightarrow X_j | X_i, X_j \in \mathbf{V}, i \neq j\}$ *is a set of directed edges representing that user X_j is a follower of user X_i on Twitter. Thus information propagates from X_i to X_j if an edge $X_i \rightarrow X_j$ exist in $\mathbf{G}$.*

The graph in Fig. 1 represents a Twitter follower network. A directed edge $\mathbf{A} \rightarrow \mathbf{E}$ shows that user $\mathbf{E}$ follows user $\mathbf{A}$.

Definition 2 (Time Indexed Tweet Series). *Let $\mathcal{C}$ be the tweet corpus posted by users* $\mathbf{V} = \{X_1, X_2, ..., X_N\}$ *in a given day, and $p = (X, c, \tau)$ represent the tweet posted by user $X \in \mathbf{V}$ at time τ with content $c \in \mathcal{C}$. Let $\tau(p)$ denotes the time when p is posted and $\mathcal{U}(p)$ denotes the user of p. A Time Indexed Tweet Series is* $\mathcal{P} = \langle p_1, p_2, ..., p_K \rangle$ *s.t. $\tau(p_i) \leq \tau(p_j)$ if $i \leq j$, where $i, j \in \{1, 2, ..., K\}$ and K is the number of tweets posted on the current day.*

The tweet corpus contains posting time, id's of users who posted the tweets and the content (text) of the post. We assume that the time indexed tweet series are already filtered to obtain relevant tweets on a protest event. Table 1 shows sample tweets that are indexed according to the time of posting.

Definition 3 (Propagation Tree (PT)). *Given a Twitter follower network $\mathbf{G}$ and time indexed tweet series $\mathcal{P}$ for the current day, let τ_i be the time of the first post of X_i in $\mathcal{P}$ and τ_j be the time of the first post of X_j in $\mathcal{P}$. A Propagation Tree* $\mathbf{PT} = \langle \mathbf{V}', \mathbf{E}' \rangle$ *where* $\mathbf{V}' = \{(X_i, \tau_i) \mid X_i \in \mathbf{V}\}$ *is a set of node time pair, and* $\mathbf{E}' = \{(X_i, \tau_i) \rightarrow (X_j, \tau_j) \mid X_i \rightarrow X_j \in \mathbf{E}, \tau_i \leq \tau_j\}$ *is a set of directed edges.*

Propagation trees are constructed using tweets and the Twitter follower relationship. Figure 2 shows sample propagation trees constructed using tweets in Table 1 and the follower network in Fig. 1.

Definition 4 (Propagation Forest). *A propagation forest* $\mathbf{FPT} = \{PT_1, PT_2, ..., PT_M\}$ *is a set of M propagation trees constructed in a day s.t. $M > 1$.*

---

[1] The tweet can be in the form of retweet, @mentions, normal tweet etc.

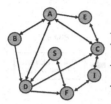

Fig. 1. Twitter follower network

Table 1. Time indexed tweet series

Timestamp	UserID	Tweets
07:00:03	A	We nd our ryts to protest!!! Hit d streets naw,#protest #humanrights
07:14:43	C	#myrights #riots join movement Calln all!!! #killcorruption now
07:34:35	B	RT @XXX show ryts to protest!!!Hit rally now,#protest #humanrights
08:45:42	S	#Unfair taxes we will #protest, destroy and fyt 4 our ryts !!!
10:23:27	F	Gov't dsnt care abt us.. We will giv them nthn bt more #protest
10:24:11	E	#Revolution #protest 4 change, fyt like hell and stop this grt injustice

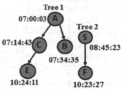

Fig. 2. Propagation trees

3.2 Problem Setup

We use propagation trees to capture how information propagates in a particular community, clique or how online groups discuss a protest event of interest. Usually, large scale protest information propagation among users who are connected by Twitter follower relationship is rare [1]. The basic assumption is that relevant protest information propagating in these groups are precursors to a real protest event that will happen on the ground. Retrospective studies [4] have shown that large growth rates and increasing size of online conversations among users of the same follower network are precursory to pending protest events. We are interested in such temporal and structural precursory features that capture the growth rate, size, duration and other dynamics of information propagation within an online community of Twitter users. These features we posit can offer insights as to if a protest event will occur.

Definition 5 (Problem Definition). *Given a set of protest information propagating on Twitter (Time Indexed Tweet Series) $\mathcal{P}$ and $\mathbf{G}$ on day i, the goal is to predict the occurrence of a future protest event $E_{\Delta i}$ on day Δi, where $\Delta i \in \{i+1, i+2, i+3\}$. Formally, this is formulated as learning a mapping function from propagation tree features $\mathcal{Z}$ to the occurrence of a future protest event on day Δi, i.e., $f : \mathcal{Z} \longrightarrow \{E_{\Delta i}\}$ where $E_{\Delta i} = 1$ if there is a protest event on day Δi and $E_{\Delta i} = 0$ otherwise.*

4 The Propagation Tree Framework

In this section, we present a detailed description of our proposed solution. The diagram in Fig. 3 shows the step-by-step stages of our approach. We first extract relevant online conversations from Twitter to discover information propagation about protest events. This requires that tweets are filtered using a set of protest-related keywords. Words such *protests, riots, unrest, placard, violence, blockade* etc. in tweets, are likely to be associated with protest events.

For every user who posts a tweet, we obtain his/her followers and construct propagation trees (Fig. 3 Phase 1). We then propose a matching scheme to map

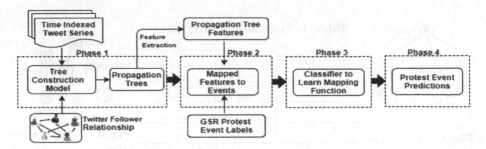

Fig. 3. The overview of our proposed framework.

temporal and structural features of propagation trees to ground truth[2] protest events in Phase 2. In the third phase, we build a classifier to learn the mapping function and output the predictions of our model in Phase 4.

Propagation Tree Construction (Phase 1): The propagation tree framework starts with a data-driven model for building propagation trees. We use Fig. 1, Table 1 and Fig. 2 as reference examples to explain our method of construction propagation trees. We assume the Time Indexed Tweet Series $\mathcal{P}$ used for our formulation are protest-related since we use a protest keyword dictionary to filter out tweets that do not match at least three protest keywords. It is worth mentioning that our notion of propagation trees presented here is expandable to also track hashtags and specific topics propagating in online social networks. We use the following steps to show how propagation trees are constructed following the criteria defined. Given Twitter follower network $\mathbf{G}$, and Time Indexed Tweet Series $\mathcal{P}$, a propagation tree is constructed as follows: A tree starts with a source node representing a user who posted the first tweet on a protest event on a given day. From Table 1, user $\mathbf{A}$ is chosen as the source node of tree 1 (see Fig. 2).

Criterion 1 (Tree Growth). *Given a propagation tree PT_n and a follower network $\mathbf{G} = \{\mathbf{V}, \mathbf{E}\}$. Let (X_m, τ_m) be a new node at time τ_m ($\tau_m > \tau_i \ \forall \ (X_i, \tau_i) \in PT_n$). If there exists a directed edge from X_j to X_m in the follower network $\mathbf{G}$, $X_j \to X_m$, the propagation tree PT_n is grown by adding an edge $X_j \to X_m$ such that $\tau_j \geq \{\tau_i\} \ \forall \ (X_i, \tau_i) \in PT_n$.* From Fig. 1, user $\mathbf{C}$ follows user $\mathbf{A}$. We add node $\mathbf{C}$ and a directed edge $\mathbf{A} \to \mathbf{C}$ to tree 1.

Criterion 2 (Emergence of new Propagation Tree). *Given that PT_n is a propagation tree under construction in the current day with node set $\mathbf{V}'_n$, let $\mathcal{F}(X)$ be the list of followers of X. If X_m has posted a tweet in the current day and; (1) $X_m \notin \mathbf{V}'_n$ (2) $\forall V \in \mathbf{V}'_n$, $X_m \notin \mathcal{F}(V)$, a new tree PT_m is created with the node (X_m, τ_m) as the root.*

After $\mathbf{B}$ is added to tree 1, the next post is $\mathbf{S}$. The follower network shows that user $\mathbf{S}$ does not follow any of the users that have already published a tweet (i.e. $\mathbf{A}, \mathbf{C}, \mathbf{B}$), hence a new PT tree 2 emerges with $\mathbf{S}$ as the root (Fig. 2).

Criterion 3 (Tree and Forest Termination). *Given that PT_n is a propagation tree under construction with a set of nodes $\mathbf{V}'_n$ and (X_i, τ_i) is the last node*

---

[2] The ground truth refers to Gold Standard Record (GSR).

added to PT_n, a propagation tree is terminated if $\forall(X \in \mathbf{V}_n') \nexists \mathcal{F}(X_i)$ who posts a tweet at time $\tau \geq \tau_i$. An **FPT** is terminated after the user of the last tweet is added to the tree construction.

In tree 2 (Fig. 2), after **F** was added to the tree, no follower of **F** posted any tweet in $\mathcal{P}$, so the tree terminates at time *10:23:27*. Also, the forest is terminated after **E** is added to tree 1. Algorithm 1 shows the tree construction process.

Feature Extraction and Event Mapping (Phase 2):

Propagation Tree Features: Temporal and structural features of propagation trees are effective in capturing information propagation in networks [9,13]. In this work, we compute features such as the size, growth rate, duration, etc. of propagation trees. Our features described in Table 2 are effective in capturing how fast a piece of information is spreading within a community, the number of users engaged in that conversation, the rate at which new users join the conversation, the duration etc. The intuition here is that information propagation within online communities on a large scale does not usually occur, and thus an occurrence such phenomenon signals a big event. Trees constructed one or two days prior to a protest event have a short duration, large size and have a sudden surge in the number of users. These characteristics of propagation trees are useful for event prediction task. These features are used as input in our protest event prediction model.

Table 2. Propagation tree features

Features	Description
Size of Largest Tree (LT)	*The tree with the most number of nodes*
Duration of LT	*Time difference between when the root node and last node of LT was created*
Growth Rate of LT	*The ratio of size of LT to the duration of LT*
Forest Tree Size	*Total number of trees in a forest in a given day*
Forest Node Size	*Total number of nodes in a forest in a given day*
Tree Growth Interval	*Average time difference between activation times of the tree root nodes*
Forest Activity Time	*Time interval between the first root node and when the last node was added to forest*
Forest Growth Rate	*The ratio of Forest Node Size to the Forest Duration*
Tree Max. Duration	*The maximum duration of all trees in a given day*
Tree Avg. Duration	*The average duration of all trees in a given day*
Forest Node to Tree Ratio	*The ratio of Forest Node Size to Forest Tree Size*

Feature-to-Event Mapping: Gold Standard Records (GSR) are ground truth data from major online news sources, blogs and articles on real-world protest events compiled by news analysts and domain experts. The GSR contains protest event information such as protest event date, location (by state), and news source (news site). Table 3 shows sample coded GSR events used for our experiments. The event class label of the GSR on day i in a given location l is denoted as $E_{il} \in [0,1]$. The class label has a value of $E_{il} = 1$ if at least one significant protest event was mentioned in the GSR for that location and 0 otherwise. A significant protest event is one that was reported by a major online news site and thus has been recorded in the GSR. We map a set of propagation tree features $\mathcal{Z}_{il}$ extracted on day i for location l, to its corresponding GSR response. Once we obtain this match, we use lead time settings of $\triangle i \in \{i+1, i+2, i+3\}$ days as shown in Fig. 4. We are interested in predicting on each day from $i+1$ to $i+3$ if a significant protest event will occur in a given location.

Event Prediction Classifier (Phase 3): To predict protest events, we treat our prediction task as a binary classification problem. For each day, we predict if a significant protest event will occur in a given state or not. Our Propagation Tree Model (**PTM**) is a Support Vector Machine (SVM) classifier that predicts protest events using features of propagation trees. SVM is efficient in handling model overfitting and also has appropriate kernel functions which yield better performance on both linear and non-linear separable data points.

Algorithm 1. Propagation Tree Construction

1: **procedure** : **Input** $\mathcal{P}$, *follower relationship from* **G**
2: *Select* $\mathcal{U}(p1) \in \mathcal{P}$ *as source node*
3: **for** *every other* $p \in \mathcal{P}$
4: **Check Condition:**
5: **if**: $\mathcal{U}(p)$ *follows any node in an existing tree*
6: $\rightarrow$ Apply **Criterion 1** to grow tree
7: **else:**
8: $\rightarrow$ Apply **Criterion 2**
9: *Exit* **Check Condition** *after all* $p \in \mathcal{P}$ *have been checked*
10: $\rightarrow$ Apply **Criterion 3**
11: **Output** : **PT, FPT**

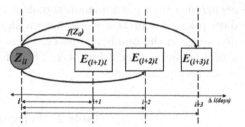

Fig. 4. Feature-to-event mapping

Table 3. Coded GSR events

Date	Country	State	Event headlines	Source
08-05-16	Australia	NSW	*Protesters descend on Newcastle as flotilla to stop coal exports*	XYZ News
05-07-16	Australia	SA	*Coal Mine Workers protest against big pay cut ahead for Fair work hearing*	*XYZ News News-10*

5   Experimental Settings and Results

We present a detailed description of how the experiments were conducted and the results obtained in comparison with existing state-of-the-art models.

5.1   Experiment Setup

Datasets and Preprocessing: To develop and test our methods, we collected over 100 million publicly available tweets published from June 2016 to April 2017 in Australia. As a preprocessing step, the tweets were filtered (using Tweets geolocation normalization and location mention identifiers [11] techniques) to obtain relevant tweets for each state. We built a dictionary of 96 protest-related keywords using words such as *'protest'*, *'unrest'*, *'terror'*, *'action'*, *'placards'* etc. selected by domain experts. For each state, we filtered to obtain a subset of tweets that matched at least three keywords from our dictionary. This is to ensure we obtain only protest-related tweets for our analysis. We then collect the follower list of all users who posted tweets during the period of observation using the Twitter API For each day, we build propagation trees for each state as described in Sects. 3 and 4.

PTM Setup: We adopt the implementation of LIBSVM[3], a library for support vector machines in Weka 3.8 using Radial Basis Function (RBF) as the kernel function. We also applied synthetic minority over-sampling technique SMOTE[4] to handle the issue of unbalanced GSR class labels. We performed 10-fold cross-validation on the training set to choose parameters that yield the highest accuracy. The penalty C and γ are determined using grid search.

It is worth mentioning that we built and tested five other different standard machine learning algorithms (KNN, Random Forest, Naive Bayes, Logistic Regression and Decision Trees) suitable for binary classification problems on our features. While the results of most classifiers were very similar, we present the result[5] of our PTM (using SVM) due to limited space.

Comparison Models: We compare our proposed propagation tree model **PTM** to other existing state-of-the-art prediction models. We followed strictly the implementations described by the authors in their published papers.

GSR Model [1,8] is an autoregressive logistic model that uses lagged values of the GSR on a previous day as a predictor of an event in subsequent days.

Volume Based Model (VBM) [8] is a logistic regression model with LASSO (Least Absolute Shrinkage and Selection Operator) that maps a large set of volume-based features to predict the occurrence of a future protest.

Cascade Graph Model (CGM) [1,12] is a logistic regression model with LASSO that uses structural features of mention-retweet and follower cascade graphs computed daily to forecast the probability of occurrence of a GSR event.

Top Keyword Model (TKM) is our novel variant implementation of the VBM. To build this model, We select the top 14 most occurring keywords in our tweet datasets as our features, computed their volumes daily, and build a k-nearest neighbour (KNN) classifier to predict protest events.

---

[3] http://weka.sourceforge.net/doc.stable/weka/classifiers/functions/LibSVM.html.
[4] http://weka.sourceforge.net/packageMetaData/SMOTE/index.html.
[5] SVM outperforms all other classifiers with best precision, recall and F1-score.

5.2   Performance Metrics

The performance of the various models is evaluated using standard classification metrics. The related performance metrics include; precision, recall (Sensitivity), F1-score, and accuracy. Table 4 shows the description of true positives, false positives, true negatives and false negatives.

5.3   Results and Discussions

Table 5 shows the averaged results of the various models for the three test months (Feb., Mar., and April 2017) using three different lead time settings. For each lead time setting, we run the experiments five times and report the average results. Clearly, our proposed PTM outperforms all the other methods in precision, recall, and F1-score as well as achieving an accuracy of up to 89%. Cascade graph model is the first runner-up most of the time, followed by the GSR model and in some cases the Volume based model. Due to limited space, we only show the distribution of precision, recall and F1-score using a lead time setting of two days for Western Australia (WA) on March 2017 in Fig. 6. Our propagation tree model (**PTM**) also achieves ROC Area Under Curve (AUC) of 0.78 as shown in Fig. 5 clearly showing superiority over the comparison models. The results show that our propagation tree models yield better results than all the other models with an average accuracy of 83%. These results corroborate with earlier work in [1] suggesting that models that capture information propagation are more effective in predicting protest events.

Insights into Propagation Tree Features: Our interest in this work also lies in discovering the usefulness and interpretability of propagation tree features.

To interpret our features, we plot the daily normalized feature distribution in Fig. 7 over one month period for NSW which recorded 11 GSR protest events. From Fig. 7, the normalized distribution shows that the forest size and the size of the largest tree produces strong signals one or two days prior to protest events. We also observed a sudden increase in features such as the growth rate, size of

Table 4. Description of our evaluation metrics

We define accuracy in our context as:

$$Accuracy = \frac{k_{tp}+k_{tn}}{k_{tp}+k_{tn}+k_{fp}+k_{fn}}$$
$$Precision = \frac{k_{tp}}{k_{tp}+k_{fp}},$$
$$Recall = \frac{k_{tp}}{k_{tp}+k_{fn}},$$
$$F_1 = 2 \cdot \frac{(Precision \cdot Recall)}{Precision+Recall}$$

Metric	Description
True positives (k_{tp})	GSR protest days that were correctly predicted as protest event days by the model
True negatives (k_{tn})	GSR no-protest days that were correctly predicted as no-protest event days by the model
False positives (k_{fp})	GSR no-protest event days that were incorrectly predicted as protest events
False negatives (k_{fn})	GSR protest event days that were incorrectly predicted as no-protest event days

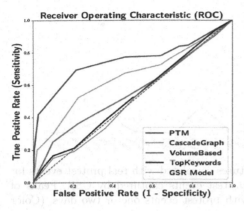

Fig. 5. ROC curves for the various models.

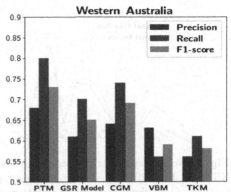

Fig. 6. Precision, recall and F1-score in March 2017 for WA using lead time of two days.

Table 5. Performance comparison of predictive models for South Australia (SA), New South Wales (NSW) and Western Australia (WA). PR = Precision, RE = Recall, Acc.(%) = Accuracy.

State	Model	LeadTime1				LeadTime2				LeadTime3			
		PR	RE	F1	Acc.	PR	RE	F1	Acc.	PR	RE	F1	Acc.
SA	PTM	**0.73**	**0.87**	**0.79**	**86.56**	0.76	**0.87**	**0.84**	**89.43**	**0.79**	**0.85**	**0.81**	**83.82**
	GSR Model	0.61	0.52	0.56	68.81	0.66	0.74	0.70	78.64	0.63	0.78	0.69	71.55
	CascadeGraph	0.70	0.82	0.78	81.40	**0.79**	0.63	0.70	76.52	0.76	0.78	0.79	78.65
	VolumeBased	0.66	0.71	0.68	72.10	0.70	0.72	0.69	73.15	0.76	0.78	0.76	76.32
	TopKeywords	0.69	0.35	0.46	66.35	0.72	0.35	0.48	58.34	0.78	0.75	0.75	76.37
WA	PTM	0.66	**0.82**	**0.73**	**82.94**	0.76	**0.83**	**0.79**	**86.04**	**0.66**	**0.76**	**0.71**	**80.24**
	GSR Model	0.65	0.81	0.71	80.02	0.67	0.82	0.74	81.96	0.65	0.67	0.66	74.28
	CascadeGraph	**0.69**	0.77	0.72	78.10	0.69	0.81	0.74	81.40	0.63	0.78	0.69	78.94
	VolumeBased	0.63	0.67	0.65	67.74	0.76	0.78	0.77	78.64	0.51	0.93	0.66	73.14
	TopKeywords	0.65	0.73	0.68	73.23	0.70	0.72	0.71	72.04	0.71	0.71	0.70	70.90
NSW	PTM	0.60	**0.76**	**0.66**	75.28	**0.76**	**0.85**	**0.80**	**82.99**	0.69	**0.74**	**0.69**	**75.81**
	GSR Model	0.55	0.74	0.63	70.17	0.65	0.80	0.71	70.13	0.61	0.76	0.68	73.64
	CascadeGraph	**0.69**	0.74	0.71	**75.62**	0.59	0.67	0.62	67.91	0.56	0.69	0.61	69.23
	VolumeBased	0.69	0.60	0.63	60.19	0.67	0.68	0.66	67.55	0.57	0.63	0.60	65.12
	TopKeywords	0.67	0.61	0.63	60.78	0.63	0.57	0.59	56.97	0.64	0.60	0.62	60.43

the forest, one or two days before a protest event. The interpretation of this observation is that conversations on protest event are more focused among users of similar follower network groups prior to a protest event. This signifies that features of propagation trees captured from online discussions on protest events are useful antecedents to impending future protest events.

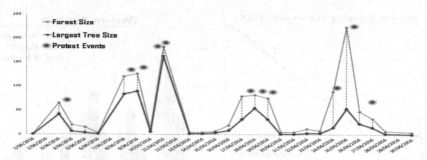

Fig. 7. Normalized distribution of tree features correlated with real protest events for NSW in June 2016. Red dots represent protest events captured by GSR. Peaks of feature values show a strong correlation with protest events one or two days. (Color figure online)

6   Conclusion

In this work, we presented a formalized propagation tree model that captures online social media information as propagation trees. We developed a novel supervised protest prediction approach using interpretable features of propagation trees as precursors to predict impending protest event. We discovered that structural and temporal features extracted from propagation trees are effective precursors to protest events with a lead time of two days. We have shown that our features of propagation trees that capture the social network relationship and timeliness of information propagation incorporated into our model outperforms existing state-of-the-art protest prediction models.

Acknowledgements. We acknowledge Data to Decisions CRC (D2DCRC), Cooperative Research Centres Programme, and the University of South Australia for funding this research. The work has also been partially supported by ARC Discovery project DP170101306.

References

1. Cadena, J., Korkmaz, G., Kuhlman, C.J., Marathe, A., Ramakrishnan, N., Vullikanti, A.: Forecasting social unrest using activity cascades. PloS One **10**(6), e0128879 (2015)
2. Chen, F., Neill, D.B.: Non-parametric scan statistics for event detection and forecasting in heterogeneous social media graphs. In: The 20th ACM SIGKDD, pp. 1166–1175. ACM (2014). https://doi.org/10.1145/2623330.2623619
3. Gomez-Rodriguez, M., Leskovec, J., Krause, A.: Inferring networks of diffusion and influence. ACM TKDD **5**(4), 21 (2012). https://doi.org/10.1145/2086737.2086741
4. Gonzalez-Bailon, S., Borge-Holthoefer, J., Moreno, Y.: Broadcasters and hidden influentials in online protest diffusion. CoRR abs/1203.1868 (2012). http://arxiv.org/abs/1203.1868
5. Kallus, N.: Predicting crowd behavior with big public data. In: the 23rd International Conference on World Wide Web, pp. 625–630. ACM (2014)

6. Kang, W., Chen, J., Li, J., Liu, J., Liu, L., Osborne, G., Lothian, N., Cooper, B., Moschou, T., Neale, G.: Carbon: forecasting civil unrest events by monitoring news and social media. In: Cong, G., Peng, W.-C., Zhang, W.E., Li, C., Sun, A. (eds.) ADMA 2017. LNCS (LNAI), vol. 10604, pp. 859–865. Springer, Cham (2017). https://doi.org/10.1007/978-3-319-69179-4_62
7. Kempe, D., Kleinberg, J., Tardos, É.: Maximizing the spread of influence through a social network. In: 9th ACM SIGKDD, pp. 137–146. ACM (2003)
8. Korkmaz, G., Cadena, J., Kuhlman, C.J., Marathe, A., Vullikanti, A., Ramakrishnan, N.: Combining heterogeneous data sources for civil unrest forecasting. In: 2015 IEEE/ACM (ASONAM), pp. 258–265. IEEE (2015)
9. Krishnan, S., Butler, P., Tandon, R., Leskovec, J., Ramakrishnan, N.: Seeing the forest for the trees: new approaches to forecasting cascades. In: Proceedings of the 8th ACM Conference on Web Science, pp. 249–258. ACM (2016)
10. Muthiah, S., Huang, B., Arredondo, J., Mares, D., Getoor, L., Katz, G., Ramakrishnan, N.: Planned protest modeling in news and social media. In: Proceedings of 29th AAAI Conference on Artificial Intelligence, pp. 3920–3927 (2015)
11. Rahimi, A., Cohn, T., Baldwin, T.: Twitter user geolocation using a unified text and network prediction model. arXiv preprint arXiv:1506.08259 (2015)
12. Ramakrishnan, N., Butler, P., Muthiah, S., Self, N., Khandpur, R., Saraf, P., Wang, W., Cadena, J., Vullikanti, A., Korkmaz, G., et al.: 'Beating the news' with embers: forecasting civil unrest using open source indicators. In: 20th ACM SIGKDD, pp. 1799–1808. ACM (2014)
13. Szabó, G., Huberman, B.A.: Predicting the popularity of online content. Commun. ACM 53(8), 80–88 (2010). https://doi.org/10.1145/1787234.1787254
14. Taxidou, I., Fischer, P.M.: Online analysis of information diffusion in Twitter. In: 23rd ICWWW 2014, pp. 1313–1318. ACM (2014)

Predictive Team Formation Analysis via Feature Representation Learning on Social Networks

Lo Pang-Yun Ting[1], Cheng-Te Li[2,3], and Kun-Ta Chuang[1(✉)]

[1] Department of Computer Science and Information Engineering, National Cheng Kung University, Tainan, Taiwan
lpyting@netdb.csie.ncku.edu.tw , ktchuang@mail.ncku.edu.tw
[2] Institute of Data Science, National Cheng Kung University, Tainan, Taiwan
chengte@mail.ncku.edu.tw
[3] Department of Statistics, National Cheng Kung University, Tainan, Taiwan

Abstract. Team formation is to find a group of experts covering required skills and well collaborating together. Existing studies suffer from two defects: cannot afford flexible designation of team members and do not consider whether the formed team is truly adopted in practice. In this paper, we propose the *Predictive Team Formation* (PTF) problem. PTF provides the flexibility of designated members and delivers the prediction-based formulation to compose the team. We propose two methods by learning the feature representations of experts based on *node2vec* [4]. One is *Biased-n2v* that models the topic bias of each expert in the social network. The other is *Guided-n2v* that refines the transition probabilities between skills and experts to guide the random walk in a heterogeneous graph of expert-expert, expert-skill, and skill-skill. Experiments conducted on DBLP and IMDb datasets exhibit that our methods can significantly outperform the state-of-the-art optimization-based approaches in terms of prediction recall. We also reveal that the designated members with tight social connections can lead to better performance.

1 Introduction

Team formation is one of the essential tasks in social network analysis [7]. Given a set of required skills, the goal of team formation is to select a group of experts possessing the required skills and well collaborating with each other in the underlying social network. A good team to execute a given task needs to ensure that the required skills can be covered by the found team members, and meanwhile the team members can have less cost to communicate with one another. To name a few applications, a project leader plans to organize a team for breast cancer prediction with deep learning techniques. The conductor wants to form a group of musicians being capable of violin, piano, and flute to deliver high-quality performance.

© Springer International Publishing AG, part of Springer Nature 2018
D. Phung et al. (Eds.): PAKDD 2018, LNAI 10939, pp. 790–802, 2018.
https://doi.org/10.1007/978-3-319-93040-4_62

Different optimization methods have been proposed to solve the variants of team formation problems, including affording the team leader [5], allowing some existing designated experts [15], and finding the alternate to replace the unavailable member [12]. These variants can be summarized as: given the set of required skills for a task, before the team is formed, there are no existing members [6,7], a *designated* member (e.g. the leader) [5], few *designated* members [17], and all members except for the unavailable one [12]. However, we think that the problem and the solutions of team formation can be further improved due to two observations. First, while the given task may contain various numbers of designated experts (as elaborated above), none of past methods can simultaneously tackle these variants. Second, although different optimization-based methods have been validated to lead to higher scores of some team goodness measures, it is still unknown that whether the found team members will be accepted (i.e., truly adopted as a team). That says, in practice, it is expected that experts in the formed team need to be adopted and truly work together. But none of previously proposed methods experience such kind of validation.

How can we form the team and ensure the found team members can be truly adopted while allowing the various numbers of designated experts in the query task? This is the first central question this paper aims to answer. The second question lies in *how do the social connections between the designated experts influence the performance of team formation?* To deal with the first question, in this work, we formulate the *Predictive Team Formation* (PTF) problem: given a set of required skills, a set of designated members in a social network, and the team size K, the goal is to find a team of K experts who can truly work together. In other words, we aim at *predicting* the team members based on the given information. To solve the PTF problem, we propose to learn the feature representation (i.e., the embedding vector) of each expert in the social network. The idea is that those who will become team members tend to be close to each other in the feature space that captures how experts co-work and how they adopt skills. The feature representation should be learned by considering their past collaborations in the social network and the skills associated by each expert. Two feature representation learning methods based on *node2vec* [4] are proposed. One is *Biased-n2v*: learning the bias of topic adoption for every expert for *node2vec* random walk sampling in the social network. The other is *Guided-n2v*: extending the transition probabilities between skill and expert nodes in a skill-expert heterogeneous expertise graph for *node2vec*. On the other hand, to answer the second question, our experiments will be conducted by studying various settings of the designated members. The settings include the number of designated experts, the number of team members to be found, and the social density between designated members.

We summarize the contributions of this paper in the following.

- We formulate the problem of *Predictive Team Formation* (PTF). The novelty of PTF is two-fold. One is allowing various numbers of designated expert members in the team. The other is to find the members who can be truly adopted as a team via ground-truth validation. To the best of our knowledge,

we are the pioneer to deliver prediction-based formulation of the team formation problem.

- Technically, we propose two feature representation-based methods to deal
with the PTF problem. One is *Biased-n2v* and the other is *Guided-n2v*. The
former models the topic bias of experts in the social network while the latter
guides the random walk in a heterogeneous expertise graph to improve the
basic *node2vec* framework.
- Experiments conducted on DBLP and IMDb datasets exhibit two findings.
The proposed *Biased-n2v* and *Guided-n2v* can outperform the state-of-the-
art optimization-based and recommendation-based methods. The set of designated experts with higher density values lead to better performance in predictive team formation.

The structure of this paper is presented as follows. We present the problem
statement in Sect. 2, followed by describing the proposed methods in Sect. 3.
The experimental results are exhibited in Sect. 4. Section 5 reviews the relevant
studies, and Sect. 6 concludes this work.

2   Problem Statement

We first describe some preliminary notations. First, a social network is represented by a graph $G = (V, E)$, where V is the node set (i.e., experts) and E is the
edge set (i.e., the collaboration between experts). Each node $u \in V$ is associated
with a set of labels L_u that represent skills. Each edge $e = (v_i, v_j) \in E$ is also
associated with a weight score depicting the communication cost between expert
v_i and v_j. Higher edge weights mean higher communication cost, and thus cannot well collaborate with one another. Second, let $L(S)$ be the set of required
labels covered by a node set S, i.e., $L(S) = L \cap (\cup_{u \in S} L_u)$, where L is the set
of required labels for the given task. Third, a query task of team formation,
denoted by T, consists of a set of required labels L, a set of designated expert
members $S_d \subseteq V$, and K is the team size (i.e., the number of team members),
i.e., $T = \langle L, S_d, K \rangle$, where $S_d = \emptyset$ indicates no designated members are specified.

Predictive Team Formation (PTF). Given a query task of team formation
$T = \langle L, S_d, K \rangle$, and a collection of experts V whose historical collaborations
E can be constructed as a social network $G = (V, E)$, the PTF problem is to
find the set S_r of remaining team members (i.e., $|S_d \cup S_r| = K$) for the task
T in a prediction-based approach, so that ground-truth team members can be
accurately identified, i.e., $S_r = \hat{S} \setminus S_d$, where $\hat{S}$ is the set of ground-truth team
members.

Note that the formulation of existing team formation problems (e.g. [5–7,17]
targets at optimizing a variety of self-defined objective functions (e.g. minimizing
the communication cost among team members [7], minimizing the leader distance
[5], minimizing the coordination cost [1], and maximizing influence-cost ratio [8]).
However, we argue that such optimization-based problem formulation cannot
find the team members who are truly adopted as a team. That says, past studies

do not concern about whether or not the found experts are selected as the team members in the future. Therefore, in our work, we alternatively resort to the prediction-based problem formulation, and expect this can better identify the ground-truth team members. To validate such assumption, we will conduct the experiments to examine the performance of team formation in terms of recall scores.

3 Proposed Methods

In this section, we first describe how to learn the feature representations of experts in a network in Sect. 3.1. Then we present the first method that further considers the topic preference of experts in the learning in Sect. 3.2. Section 3.3 extends the learning in a skill-expert heterogeneous graph.

3.1 Learning Node Representation ($n2v$) for Team Formation

We extend the skip-gram architecture [4,13] from natural language processing (NLP) and node embedding to learn the feature representations of experts in the social network G. In NLP, the skip-gram architecture learns relations between words and their context. Here each node in the network is treated as a word, and some random walk paths are sampled as sentences. We define $N_\chi(v) \subseteq V$ as the neighbor nodes for each node $v \in V$ via a sampling method χ. Here we use a sampling method based on a proposed biased random walk that models the topic preference of experts, which is presented later in Sect. 3.2. The skip-gram model is extended to optimize the log-likelihood of the observed $N_\chi(v)$, conditioned on node v's feature representation $f(v)$ as follows:

$$\max_f \sum_{v \in V} \log P(N_\chi(v) \mid f(v)).$$

To make the optimization process more efficient, we adopt two standard assumptions [4]. First, we assume that, given node v's feature representation, v's neighbor nodes $N_\chi(v)$ can be observed conditionally independent of each other. Then $P(N_\chi(v) \mid f(v))$ can be factorized by the neighbor nodes as follows:

$$P(N_\chi(v) \mid f(v)) = \prod_{n \in N_\chi(v)} P(n \mid f(v)).$$

Second, we assume that any pair of neighboring nodes symmetrically affect each other in the k-dimensional space of feature representation. Therefore, given a node v, the conditional likelihood of every neighbor node $n \in N_\chi(v)$ can be modeled as a *softmax* unit [2] by reversing the previous formula:

$$P(n \mid f(v)) = \frac{\exp(f(n) \cdot f(v))}{\sum_{u \in V} \exp(f(u) \cdot f(v))}.$$

With these assumptions, the objective function can be rewritten as:

$$\max_f \sum_v \left(-\log Z_v + \sum_{n \in N_x(v)} f(n) \cdot f(v) \right),$$

where $Z_v = \sum_{u \in V} \exp(f(v) \cdot f(u))$ can be approximated by negative sampling [14]. In addition, this objective function can be optimized by stochastic gradient descent. Note that to have reliable performance, each node in the graph will be treated as the source of t random walks. These generated $t \times |V|$ random walks will be exploited to learn the node features by stochastic gradient descent. After learning node features, we will use the feature representations of experts to predict the remaining team members.

Team Formation. Given each expert node $v \in V$ has a learned feature representation vector $f(v)$ and the query task $T = \langle L, S_d, K \rangle$, we aim to find the set of remaining team member S_r. The basic idea is that experts with similar feature vectors tend to collaborate with each other. Therefore, by measuring the distance between a candidate expert u and the set S_d of designated experts, and can report the K experts with the lowest distance values as the recommended team members. Specifically, we define the distance $\delta(u, S_d)$ between candidate $u \in V \setminus S_d$ and S_d as:

$$\delta(u, S_d) = \frac{\sum_{v \in S_d} \|f(u) - f(v)\|_2}{|S_d|},$$

where $\|\cdot\|_2$ is the $\mathcal{L}^2$ distance. Those K expert nodes with lowest values $\delta(u, S_d)$ will be recommended as the remaining formed team members. It is worthwhile noticing that if no designated experts are given (i.e., $S_d = \emptyset$), we cannot compute the distance $\delta(u, S_d)$. In such a case, as we also have the feature vector for each required skill label $l \in L$, we alternatively compute the distance $\delta(u, L)$ between candidate expert u and the required label set L. The corresponding intuition is that a proper team member needs to *be accommodated with* the required labels in the feature space.

3.2   The *Biased-n2v* Method

The original skip-gram model (i.e., *n2v*) is to learn the feature vectors f based each node's neighborhood in the network. *node2vec* [4] has proposed a random walk mechanism that fuses the Breadth-first Search and Depth-first Search in the graph. The idea is that nodes possess similar random walk-sampled neighborhood should lead to similar feature vectors. However, such a random walker purely relies on the collaboration network structure. The labels (i.e., skills) associated by each expert node are ignored in the random walk-based neighborhood sampling. Eventually, the skill label knowledge cannot be modeled in the feature representation. For example, suppose that both experts v_i and v_j frequently co-work with experts v_a and v_b. Then v_i and v_j tends to work together if the original $n2v$ is adopted. However, v_i may collaborate with v_a and v_b based on skill l_x while v_j may work with v_a and v_b for skill l_y. We think that a better feature representation learning method should further take nodes' labels and topics into account so that the collaboration preferences of experts can be modeled.

Let $\mathcal{T}$ be the set of topics of expertise in a certain dataset. Each expert can adopt multiple topics. Each skill can also belong to multiple topics. In our experiments in Sect. 4, we consider "conferences" in DBLP data as topics, and "genres" in IMDb data as topics. We first measure the expertise bias of an expert v on a certain expertise topic τ. Let π_τ be the *adoption ratio* of topic τ over all topics, i.e., $\pi_\tau = \frac{a_\tau}{\sum_{\tau' \in \mathcal{T}} a_{\tau'}}$, where a_τ is the number of adoption for topic τ by all experts. We also define an expert v's *adoption contribution* on topic τ (denoted by $\pi_\tau(v)$) as the number of adoption for topic τ by expert v (denoted by $a_\tau(v)$) divided by the number of adoption for topic τ by all experts, i.e., $\pi_\tau(v) = \frac{a_\tau(v)}{\sum_{u \in V} a_\tau(u)}$. By treating π_τ and $\pi_\tau(v)$ as two probability distributions, we propose to exploit the *Kullback-Leibler divergence* to estimate the bias b_v of expert v's topic adoption, given by:

$$b_v = \sum_{\tau \in \mathcal{T}} \pi_\tau(v) \log \frac{\pi_\tau(v)}{\pi_\tau}.$$

A higher score of bias b_v for expert v means that her preference of topic adoption in different teams is far from the general experts' preferences. Therefore, higher bias may lead to less possibility to collaborate with other experts.

To apply the bias of each expert into the learning of feature representation, we aim to refine the edge weights between experts by using the bias values so that the neighborhood sampled by the random walk mechanism can reflect the collaboration bias of experts. Since the bias is defined for each expert, we re-define the edges in the social network as directed ones so that edge weight $w_{i \rightarrow j}$ from node v_i to v_j can differ from that $w_{j \rightarrow i}$ from v_j to v_i. The biased edge weight is defined as: $w_{i \rightarrow j} = w_{ij} \times \frac{1}{b_j}$ (i.e., $w_{j \rightarrow i} = w_{ij} \times \frac{1}{b_i}$, where w_{ij} is the original edge weight in the social network. We obtain edge weights by using Jaccard coefficient: $w_{ij} = \frac{|\mathcal{T}(i) \cap \mathcal{T}(j)|}{|\mathcal{T}(i) \cup \mathcal{T}(j)|}$, where $\mathcal{T}(i)$ is the set of topics that expert i had ever involved. As lower bias scores raise edge weights, and thus boost the possibility of being sampled for learning the feature representation.

3.3   The *Guided-n2v* Method

While *Biased-n2v* attempts to model the bias preference of topic adoption of experts, in which the experts' topics are considered *implicitly* as bias values, we alternatively propose the second method to *explicitly* use the interactions between experts and skills. The basic idea is construct a heterogeneous graph to jointly model the collaborations between experts, the correlations between skills, and the adoption between skills and experts. Then we devise a *guided* random walk mechanism for sampling the neighborhood in such a constructed heterogeneous graph for *node2vec*.

Heterogeneous Expertise Graph. A heterogeneous expertise graph $\mathcal{H} = (\mathcal{V}, \mathcal{E})$ is defined as: its node set is an union of expert set V and label set L, given by $\mathcal{V} = V \cup L$. The edge set of $\mathcal{H}$ is constructed by three parts $\mathcal{E} = E \cup R \cup Q$: (a) E is the edge set E in the social network of experts G, (b) R is the set of *correlation links* between labels, in which each $r_{xy} \in R$ indicates labels l_x and l_y have ever been required by at least one task, and (c) Q is the edge set of *adoption links* between experts and labels, in which each $q_{ix} \in Q$ represents that the label l_x is ever adopted by expert v_i. In addition, each type of edge is associated by a weight value: (a) the weight $w_{i \to j}^E$ for each edge $e_{ij} \in E$ is defined by $w_{ij}^E = \frac{|\mathcal{C}(i) \cap \mathcal{C}(j)|}{|\mathcal{C}(i) \cup \mathcal{C}(j)|}$, where $\mathcal{C}(i)$ is the set of experts who had ever collaborated with expert i, (b) the weight w_{xy}^R for each edge $r_{xy} \in R$ is defined by $w_{xy}^R = \frac{|T(x) \cap T(y)|}{|T(x) \cup T(y)|}$, where $T(x)$ is the set of tasks that require label x, and (c) the weight w_{ix}^Q for each edge $q_{ix} \in Q$ is equally defined as $w_{ix}^Q = 1$.

Given the heterogeneous graph $\mathcal{H}$ is different from the social network G in terms of multiple types of nodes and edges, we need to re-define the mechanism of random walk so that each expert's feature vector can be better learned. The basic idea is that each expert in $\mathcal{H}$ is characterized by not only her past collaborators, but also those who adopt similar skills (even they have never co-worked together). While the former part can be captured by the social links $E \subseteq \mathcal{E}$, the latter needs to be modeled by links in $R \subseteq \mathcal{E}$ and $Q \subseteq \mathcal{E}$. That says, two experts with similar learned feature vectors if they share more collaborators via E and have similar preference of skill adoption via R and E in the heterogeneous graph $\mathcal{H}$. To realize such an idea, we propose to re-define the random walk mechanism so that the neighbors of each expert in $\mathcal{H}$ can be sampled to preserve the her collaborators and the other experts with similar adopted skills.

Since the traversal of the random walker is determined by the transition probabilities between nodes, we re-define the transition probabilities by introducing three parameters λ, μ, and ϕ by following the formulation of the second-order random walk [4]. Consider that the random walk currently arrives at node v from node t, and needs to further surfer to one of v's neighbors x according to the corresponding transition probability $\omega_{v \to x}$. We define $\omega_{v \to x}$ based on the types of nodes v, t and x, which can be divided into two cases. First, if either $\underline{t \in V \text{ and } v \in V}$ or $\underline{t \in L \text{ and } v \in V}$:

$$\omega_{v \to x} = \begin{cases} \lambda \cdot w_{v \to x} & \text{if } d_{tx} = 0 \\ w_{v \to x} & \text{if } d_{tx} = 1 \\ \mu \cdot w_{v \to x} & \text{if } d_{tx} = 2, x \in V \\ \phi \cdot w_{v \to x} & \text{if } d_{tx} = 2, x \in L \end{cases}'$$

where d_{tx} represents the distance of shortest path between node t and x in $\mathcal{H}$. Second, if $\underline{t \in V \text{ and } v \in L}$:

$$\omega_{v \to x} = \begin{cases} 0 & \text{if } d_{tx} = 0 \\ w_{v \to x} & \text{if } d_{tx} \neq 0 \end{cases}.$$

Based on the refined transition probability $\omega_{v \to x}$, *node2vec* is allowed to sample the neighboring nodes and learn the feature representation of each expert v_i in the heterogeneous graph $\mathcal{H}$. Note that the parameters λ, μ, and ϕ control the search strategy towards "breadth-first search" or "depth-first search". As different datasets may have various settings of parameters λ, μ, and ϕ, we will use a validation data subset to the parameter values that lead to better performance.

4   Experimental Results

We conduct experiments to answer three questions. First, we wonder whether the proposed feature representation learning methods can better perform in forming teams, compared with existing optimization-based methods. Second, by varying the number of designated experts and the number of experts to be found, how does the performance of our methods evolve? Third, how does the performance of predictive team formation be affected by the social connections among the designated experts?

Datasets. We employ two datasets for the experiments. The first is DBLP, in which authors are experts, and authors of each paper are considered as a team. By removing stop words, keywords in paper titles are considered as skills. We collect papers published in conferences up to 2017 from a variety of areas: DATABASE = {SIGMOD, VLDB, ICDE, ICDT, EDBT, PODS}, DATA MINING = {KDD, WWW, SDM, PKDD, ICDM}, ARTIFICIAL INTEL-LIGENCE = {ICML, ECML, COLT, UAI}, and THEORY = {SODA, FOCS, STOC, STACS}. We consider authors publishing at least three papers. For an author, skills are those keywords that have been adopted by her at least two times. End up we have 10,724 papers, 3,716 experts, and 827 skills. The second is IMDb. Movies from 2000 to 2009 are collected, in which actors are experts, and actors of each movie are considered as a team. We consider actors involving in at least four movies. By removing stop words, keywords in movie titles are considered as skills. For an actor, skills are those keywords that have been adopted by her at least two times. Totally we have 3,513 movies, 3,716 experts, and 672 skills.

Competing Methods. The main objective of this work is to form the team via prediction. Therefore, we aim at examining whether the performance of feature representation learning is better than existing optimization-based methods. The first two competing methods are *EnhancedSteiner* and *CoverSteiner* algorithms proposed to solve the typical team formation problem [7]. When required skills or experts are given, both algorithms find the team members by minimizing the communication cost between members. *EnhancedSteiner* was validated to out-perform existing baselines. The other method for the experiments is *node2vec* [4], which is one of the state-of-the-art method of feature representation learn-ing. If *node2vec* can outperform optimization-based methods, we can say that feature representation learning can be more suitable to solve the team formation problem. In addition to the original *node2vec* applied on the social network G

of experts, we have *node2vec-h* applied on the heterogeneous expertise graph $\mathcal{H}$. The parameter settings (*i.e.*, dimensions, walks per node, walk length, and so on) used for *node2vec* algorithm is the same as [4]. Last, we would like to understand how the proposed *Biased-n2v* and *Guided-n2v* perform in both datasets. The parameters (λ, μ, ϕ) we use in *Guided-n2v* for DBLP and IMDb datasets are (1, 0.5, 0.2) and (2, 1, 0.5), respectively (these values lead to the best performance).

Evaluation Settings. The experiments consist of three parts. First, we examine the performance by varying the number of team members to be found (i.e., $|S_r| = K - |S_d|$). Second, we report the performance by varying the number of designated experts (i.e., $|S_d|$). Third, to understand how the social structure of designated experts S_d in the query team affects the performance, we present the results by varying the *social density* ψ. The social density of S_d is defined as:

$$\psi(S_d) = \frac{\sum_{u,v \in S_d, u \neq v} c(u,v)}{\binom{|S_d|}{2}},$$

where $c(u,v) = 1$ if node u is connected with node v; $c(u,v) = 0$, otherwise. We consider *Recall* as the evaluation metric, which is defined as: $Recall = \frac{|S_r \cap \hat{S}_r|}{|\hat{S}_r|}$, where S_r is the set of predicted team members and $\hat{S}_r$ is the set of true remaining team members (i.e., ground truth), and $|S_r| = |\hat{S}_r|$. Among all of the teams in each dataset, we use 80% teams for training (i.e., construct the social network and the heterogeneous graph, and learn the feature vectors of experts), and use the other 20% for testing (i.e., serve as the query tasks and compile their sets of required skills). In each of the testing query task, the set of designated experts (if specified) is randomly selected from all team members.

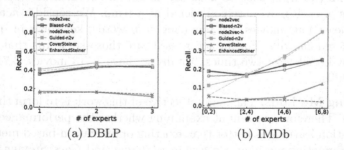

Fig. 1. Results by varying the number of team members to be predicted (i.e., $|S_r|$).

Evaluation Results. The results by varying the number of remaining team members to be predicted are exhibited in Fig. 1, and the results by varying the number of designated experts are shown in Fig. 2. We can have the following findings. First, the approach of feature representation learning (e.g. *node2vec*) significantly outperforms the optimization-based methods *CoverSteiner* and

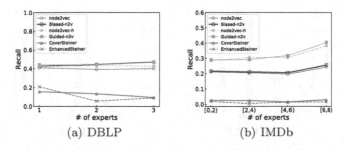

(a) DBLP   (b) IMDb

Fig. 2. Results by varying the number of designated experts (i.e., $|S_d|$).

EnhancedSteiner. Such results imply the existing optimization-based methods fall to identify the experts that will be truly adopted by teams. In addition, the distance measure between candidate members and designated experts in the feature space can to some extent reflect the potential collaboration in team formation. Second, when either the number of remaining team members or the number of designated experts increase, the feature representation learning approaches are able to stably lead to significantly higher Recall values. These results reveal that feature learning can be more effective for larger teams, and more designated experts can provide more clues related to expertise and collaboration for the query task. Last but not least, our extended *Biased-n2v* can outperform *node2vec* in the social network of experts. And our extended *Guided-n2v* is better than *node2vec_h* in the heterogeneous expertise graph. Furthermore, among these feature learning methods, *Guided-n2v* generally leads to the highest Recall values. Such outcomes prove the usefulness of implicitly modeling the preferences of topic adoption for experts (*Biased-n2v* and explicitly learning the expert-expert, skill-expert, and skill-skill interactions in an joint manner *Guided-n2v*). Eventually we recommend using *Guided-n2v* to find the expert members for the proposed predictive team formation problem. The results for different intervals of social density $\psi(S_d)$ by using *Guided-n2v* are exhibited in Fig. 3. We can find that when the social density of the set of designated experts gets higher, the Recall value tends to accordingly get boosted. Such results unveil some insights. If the team formation can be built on some experts who have ever collaborated with each other (e.g. find experienced employees to execute a new project in a software company by providing team members who had ever co-worked, it can benefit in finding the proper remaining team members who can be truly adopted. To better form an effective team, we recommend the team organizers to select the designated experts who have ever collaborated with each other.

5   Related Work

Team formation is proposed to find a set of experts such that not only a set of required skills are covered, but also the communication cost among team members is minimized [7]. There are a series of follow-up extensions considering

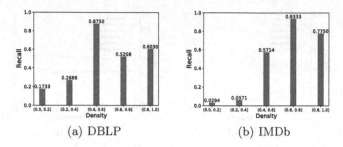

(a) DBLP   (b) IMDb

Fig. 3. Results of different social density $\psi(S_d)$ for the set of designated experts by *Guided-n2v*.

various real scenarios: jointly finding a team leaders and forming the team [5], simultaneously tackling multiple sets of required skills [1], specifying the number of experts for each skill [10], allowing geographical and team-size constraints [15], imposing swarm-based optimization [3], and recommending other individuals to replace some of existing team members [12]. *Community Search* [16] alternatively finds a densely-connected subgraph based on a set of given nodes, instead of required skills. We claim that existing studies found team members are not well proven to be truly adopted by the team organizers. We think a good team formation algorithm should not only recommend team members but also ensure them to be adopted. This is the essential difference of past studies from the present work. Recently, the team member replacement [12] is proposed to find a suitable alternate to replace the existing team member who is no longer available.

Social event organization aims at composing a group of persons that satisfying various kinds of event requirements. Socio-spatial Group Query [19] is to find a group of persons who are not only geographically close to each other but also acquainted with each other. A follow-up work [9] recommends a group of users satisfying required labels and being acquainted with each other for an event host. SEO [11] further composes multiple event groups simultaneously. Marketing effect maximization [20] aims to find a set of nodes that are geographically close to the event location and attract more those users satisfying event themes. The bottleneck-aware social event arrangement (BSEA) [18] further considers social influence to recommend events for users.

6 Conclusions

In this paper, we answer two questions. The first is how to find the team members of a query task with required skills and various numbers of designated experts and ensure the found team members will be truly adopted by the team. We show that feature representation learning approaches can better predict the team members, compared with existing optimization-based team formation methods. The extended *Biased-n2v* and *Guided-n2v* methods can further lead to higher Recall values, compared with the *node2vec* method. Second, we unfold how the social connections among designated experts influence the performance of predictive team formation. We find that if the designated experts tightly connect with each other, the performance of identifying the true team members get boosted. We suggest team organizers to provide experts who have ever co-worked so that the predicted team members can better fit the requirement of executing the task.

Acknowledgment. This paper was supported in part by Ministry of Science and Technology, R.O.C., under Contract 105-2221-E-006-140-MY2, 106-3114-E-006-002, 107-2636-E-006-002, and 106-2628-E-006-005-MY3, and by Academia Sinica under AS-107-TP-A05.

References

1. Anagnostopoulos, A., Becchetti, L., Castillo, C., Gionis, A., Leonardi, S.: Online team formation in social networks. In: Proceedings of ACM WWW (2012)
2. Anzai, Y.: Pattern Recognition and Machine Learning. Elsevier, New York (2012)
3. Basiri, J., Taghiyareh, F., Ghorbani, A.: Collaborative team formation using brain drain optimization: a practical and effective solution. World Wide Web J. (2017)
4. Grover, A., Leskovec, J.: Node2vec: scalable feature learning for networks. In: Proceedings of ACM SIGKDD (2016)
5. Kargar, M., Zihayat, M., An, A.: Discovering top-k teams of experts with/without a leader in social networks. In: Proceedings of ACM CIKM (2011)
6. Kargar, M., Zihayat, M., An, A.: Finding affordable and collaborative teams from a network of experts. In: Proceedings of SDM (2013)
7. Lappas, T., Liu, K., Terzi, E.: Finding a team of experts in social networks. In: Proceedings of ACM SIGKDD (2009)
8. Li, C.T., Huang, M.Y., Yan, R.: Team formation with influence maximization for influential event organization on social networks. World Wide Web J. (2017)
9. Li, C.T., Shan, M.K.: Composing activity groups in social networks. In: Proceedings of ACM CIKM (2012)
10. Li, C.T., Shan, M.K., Lin, S.D.: On team formation with expertise query in collaborative social networks. KAIS (2015)
11. Li, K., Lu, W., Bhagat, S., Lakshmanan, L.V., Yu, C.: On social event organization. In: Proceedings of the 20th ACM SIGKDD (2014)
12. Li, L., Tong, H., Cao, N., Ehrlich, K., Lin, Y.R., Buchler, N.: Replacing the irreplaceable: fast algorithms for team member recommendation. In: Proceedings of ACM WWW (2015)
13. Mikolov, T., Chen, K., Corrado, G., Dean, J.: Efficient estimation of word representations in vector space. In: Proceedings of Workshop at ICLR (2013)
14. Mikolov, T., Sutskever, I., Chen, K., Corrado, G., Dean, J.: Distributed representations of words and phrases and their compositionality. In: Proceedings of NIPS (2013)
15. Rangapuram, S.S., Bühler, T., Hein, M.: Towards realistic team formation in social networks based on densest subgraphs. In: Proceedings of ACM WWW (2013)
16. Sozio, M., Gionis, A.: The community-search problem and how to plan a successful cocktail party. In: Proceedings of ACM SIGKDD (2010)
17. Teng, Y.C., Wang, J.Z., Huang, J.L.: Team formation with the communication load constraint in social networks. In: Peng, W.C. et al. (eds.) PAKDD 2014. LNCS, vol. 8643, pp. 125–136. Springer, Heidelberg (2014). https://doi.org/10.1007/978-3-319-13186-3_12

18. Tong, Y., She, J., Meng, R.: Bottleneck-aware arrangement over event-based social networks: the max-min approach. World Wide Web J. (2016)
19. Yang, D.N., Shen, C.Y., Lee, W.C., Chen, M.S.: On socio-spatial group query for location-based social networks. In: Proceedings of ACM SIGKDD (2012)
20. Yu, Z., Zhang, D., Yu, Z., Yang, D.: Participant selection for offline event marketing leveraging location-based social networks. IEEE Trans. Syst. Man Cybern.: Syst. (2015)

Leveraging Local Interactions for Geolocating Social Media Users

Mohammad Ebrahimi[1,2(✉)], Elaheh ShafieiBavani[1,2], Raymond Wong[1,2], and Fang Chen[1,2]

[1] School of Computer Science and Engineering, University of New South Wales, Sydney, NSW 2052, Australia
{mohammade,elahehs,wong,fang}@cse.unsw.edu.au
[2] Data61-CSIRO, Sydney, Australia

Abstract. Predicting the geolocation of social media users is one of the core tasks in many applications, such as rapid disaster response, targeted advertisement, and recommending local events. In this paper, we introduce a new approach for user geolocation that unifies users' social relationships, textual content, and metadata. Our two key contributions are as follows: (1) We leverage semantic similarity between users' posts to predict their geographic proximity. To achieve this, we train and utilize a powerful word embedding model over millions of tweets. (2) To deal with isolated users in the social graph, we utilize a stacking-based learning approach to predict users' locations based on their tweets' textual content and metadata. Evaluation on three standard Twitter benchmark datasets shows that our approach outperforms state-of-the-art user geolocation methods.

Keywords: Geolocation · Twitter · Local intreractions

1 Introduction

Associating data with the particular geolocation from which it originated creates a powerful tool for different applications such as rapid disaster response, opinion analysis, and recommender systems [1,21]. However, only a small amount of social media data has been geolocation-annotated; for example, less than 1% of Twitter data has geo-coordinates provided [18], even though the platform supports geolocation metadata. Hence, recent work has focused on automatic geolocation inference (geoinference) of social media posts or users. *User geolocation* is the task of predicting the primary (or *"home"*) location of a user from available sources of information, such as text posted by that individual, or network relationships with other individuals [12]. Geolocation methods usually train a model on the small set of users whose locations are known, and predict locations of other users using the resulting model. These models broadly fall into three categories: text-based [8,31], network-based [18], and hybrid methods that combine both text and network models [6,7,28,30].

© Springer International Publishing AG, part of Springer Nature 2018
D. Phung et al. (Eds.): PAKDD 2018, LNAI 10939, pp. 803–815, 2018.
https://doi.org/10.1007/978-3-319-93040-4_63

Herein, we propose a hybrid method to tackle user geolocation prediction in Twitter. Our main contributions can be summarized as follows: (1) We leverage semantic textual similarity between users' posts to predict which individuals in the ego network may be most proximate. The main idea is that users who live nearby are more likely to influence each other and discuss more local topics. Detecting these *local interactions* improves the accuracy of the label propagation algorithm used for geolocation inference. To achieve this, we train and utilize a powerful word embedding model over 235 million tweets. (2) To deal with isolated users in the social graph, we employ a stacking-based learning approach to estimate users' locations based on their tweets' textual content and metadata. (3) We show that including these methods in a hybrid geoinference approach achieves superior results than state-of-the-art methods.

The rest of the paper is organized as follows. We review the previous works in Sect. 2. Utilized data is described in Sect. 3. Section 4 explains the proposed approach. The experimental results are given in Sect. 5, and finally, we conclude the paper and outline possible future work in Sect. 6.

2 Related Work

2.1 Text-Based Methods

Text-based methods utilize the geographical bias of language use in social media for geolocation. These methods have widely used probability distributions of words over locations. Maximum likelihood estimation approaches [3] and language modeling approaches minimizing KL-divergence [31] have succeeded in predicting user locations using word distributions. Topic modeling approaches to extract latent topics with geographical regions [8,15] have also been explored considering word distributions. Supervised learning methods with word features are also popular in text-based geoinference. Multinomial Naive Bayes [11,12], logistic regression [12,34], hierarchical logistic regression [34], and multi-layer neural network with stacked denoising autoencoder [22] have realized geolocation prediction from text. A semi-supervised machine learning approach by [2] has also been produced using a sparse-coding and dictionary learning. In [16], a kernel-based method has been used to smooth linguistic features over very small grid sizes and consequently alleviate data sparseness. More recently, a neural network-based geolocation approach has been proposed in [29]. The authors used the parameters of the hidden layer of the neural network as word and phrase embeddings, and performed a nearest neighbor search on a sample of city names and dialect terms. While having good results, text-based approaches are often limited to those users who generate text that contains geographic references [18].

2.2 Network-Based Methods

Network-based methods rely on the geospatial homophily of interactions between users. An early work by [5] proposed an approach in which the location of a given user is inferred by simply taking the most-frequently seen location among its social network. In [18], the idea of location inference has been extended as label propagation over some form of friendship graph by interpreting location labels spatially. Locations are then inferred using an iterative, multi-pass procedure. This method has been further extended by [4] to take into account edge weights in the social network, and to limit the propagation of noisy locations. They weigh locations as a function of how many times users interacted there, hence favoring locations of friends with evidence of a close relationship. However, the number of interactions between users does not necessarily correlate with their location proximity. In [33], a collective geographical embedding has been proposed for geoinference. The main limitation of network-based models is that they completely fail to geolocate users who are not connected to geolocated components of the graph (i.e., isolated users).

2.3 Hybrid Methods

Li et al. [20] proposed a geolocation method by integrating both friendship and content information in a probabilistic model. Rahimi et al. [30] showed that geolocation predictions from text can be used as a back-off for disconnected users in a network-based approach. In [28], a hybrid approach has been proposed by propagating information on a graph built from user mentions in Twitter messages, together with dongle nodes corresponding to the results of a text-based geolocation method. Rahimi et al. [29] have also proposed a text geoloation method based on neural network and incorporated it into their network-based approach [28]. Metadata such as location fields have also been useful as effective clues to predict geolocation [14]. Different geoinference approaches have been proposed to consider text and metadata information simultaneously. Combinatory approaches such as dynamically weighted ensemble method [23], stacking [12], ensemble learning method [17], and average pooling with a neural network [25], have strengthened geolocation prediction.

3 Data

We have used three benchmark Twitter user geolocation datasets in our experiments: **(1) TwUs** [31], **(2) TwWorld** [11], and **(3) WNUT** [13]. These datasets have been used widely for training and evaluation of geolocation models. They are all pre-partitioned into training, development and test sets. Table 1 summarizes descriptive statistics for the three datasets.

4 The Proposed Approach

Figure 1 illustrates an overview of the proposed approach. We first construct a social graph as a representation of users' social relationships (Sect. 4.1). A clustering-based approach is used to refine the social graph by filtering highly-mentioned users (celebrities). We then propose a new method based on word

Table 1. Datasets Details

Dataset	Scope	Tweets	Users	Train	Test	Dev
TwUs	United States	38M	450K	430K	10K	10K
TwWorld	World	12M	1.4M	1.38M	10K	10K
WNUT	World	13M	1.02M	1M	10K	10K

embeddings to detect local interactions and leverage them to improve geoinference performance (Sect. 4.2). To this end, for each pair of connected users in the graph, we estimate their proximity by computing their tweets' semantic similarity, and weigh the corresponding edge with the estimated similarity score. To deal with isolated users in the social graph, we employ a back-off strategy to take advantages of other sources that might contain location information (Sect. 4.3). For this purpose, we apply an ensemble learning approach over users' tweets and metadata to estimate the locations of test users. Finally, we run a label propagation algorithm over the social graph to infer the locations of users with unknown locations (Sect. 4.4).

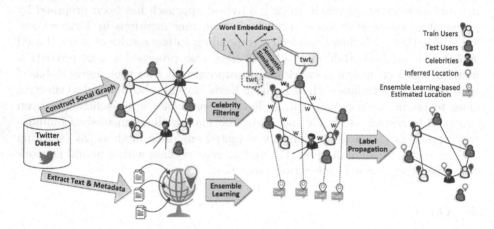

Fig. 1. Overview of the proposed approach

4.1   Construction of Social Graph

We build an undirected graph from interactions among Twitter users based on @-mentions in their tweets. In this graph, nodes are all users in a dataset (train and test), as well as other external users mentioned in their tweets and undirected edges are created between two users if either user mentioned the other[1].

---

[1] We consider uni-directional mentions, since bi-directional mentions are too rare to be useful in the datasets used in our experiments [30].

Ebrahimi et al. [6,7] showed that pruning the social graph by filtering celebrities (i.e., users that are mentioned by more than T distinct users) effectively decreases the propagation of noisy locations, and consequently improves the performance of geoinference. Following [6,7], we utilize a density-based clustering algorithm, DBSCAN [9], to cluster celebrity's geolocated mentioners based on their geographical coordinates. If the algorithm outputs only one cluster containing more than the predefined proportion (δ) of total geolocated mentioners, the celebrity is considered as a Local one. Otherwise, it will be considered as a Global celebrity. To construct a refined social graph, we remove Global celebrities and preserve Local ones as useful location indicators.

DBSCAN requires two parameters: ε *(eps)* and *MinPts*. We set parameter ε to 70 for TwUs, 130 for TwWORLD, and 80 for WNUT. Following [6,7], we set the *MinPts* dynamically to $\eta = 30$ percent of total number of points, and parameter δ to 0.8. Celebrity Threshold (T) was set to 5 for TwUs and WNUT, and 15 for TwWORLD. All parameters were chosen using grid search with development sets of the three datasets.

4.2 Predicting Geographical Proximity from Linguistic Similarity

Previously, [18] employed two traditional similarity metrics to predict users' geographical proximity, however, they reported weak correlations for the surface-level metrics. This was our motivation for a deeper analysis and comparing semantic content of tweets to predict local interactions.

Recent developments in distributed semantic representations (e.g., [24]), also called word embeddings, have been shown to be highly effective in measuring semantic similarity between vocabulary terms. Word embeddings techniques assign each term a low-dimensional (comparing to the size of vocabulary) vector in a semantic vector space. In this space, close vectors are supposed to demonstrate high semantic or syntactic similarity between the corresponding words. We utilize word2vec [24] as a successful implementation of word embeddings that learns a vector representation for each word using a shallow neural network language model. Specifically, it uses a neural network architecture (the skip-gram model) that consists of an input layer, a projection layer, and an output layer to predict nearby words. Each word vector is trained to maximize the log probability of neighboring words in a corpus, i.e., given a sequence of words $w_1, \ldots, w_T$,

$$\frac{1}{T} \sum_{t=1}^{T} \sum_{j \in nb(t)} \log p(w_j | w_t) \tag{1}$$

where $nb(t)$ is the set of neighboring words of word w_t and $p(w_j|w_t)$ is the hierarchical softmax of the associated word vectors $\vec{w}_j$ and $\vec{w}_t$.

The training on very large datasets allows the model to learn complex word relationships such as $king - m\vec{a}n + wom\vec{a}n \approx qu\vec{e}en$ [24] and $Sydney - Austr\vec{a}lia + Germ\vec{a}ny \approx Berlin$. We have used a large corpus of Twitter microposts (235M tweets) to infer the word embeddings (Sect. 5.1).

In order to measure semantic similarity between tweets based on word embeddings, we employ the Word Mover's Distance (WMD) algorithm [19]. This algorithm has been proposed to accurately estimate the similarity degree between a pair of documents, and shown highly efficient in measuring the short-text semantic similarity [26]. Adopting to our problem, the WMD formulates the dissimilarity degree between a pair of tweets, T and T', by calculating the minimum amount of summing up individual distances (travel costs) that embedded words in T need to travel to reach the embedded words in T':

$$WMD(T,T') = \min_{F \geq 0} \sum_{w_i \in T} \sum_{w_j \in T'} F_{w_i w_j} \times d(w_i, w_j)$$

$$\text{subject to,} \sum_{w_i \in T} F_{w_i w_j} = \frac{c(w_j, T')}{|T'|}, \forall w_j \in T', \sum_{w_j \in T'} F_{w_i w_j} = \frac{c(w_i, T)}{|T|}, \forall w_i \in T$$

(2)

where F is a flow matrix which indicates how much probability mass should flow (or travel) from word w_i in T to word w_j in T', and vice versa; $c(w_i, T)$ denotes the occurrence frequency of the word w_i in the tweet T. $d(w_i, w_j)$ denotes the individual distance (or travel cost) between a pair of words w_i and w_j corresponding to their learned word embeddings $\vec{w}_i$ and $\vec{w}_j$: $d(w_i, w_j) = \|\vec{w}_i - \vec{w}_j\|_2$. Having normalized Word Mover's Distance between two tweets T and T', we compute the semantic textual similarity (STS) as: $STS_{wmd}(T, T') = 1 - WMD(T, T')$. Consequently, each edge in the social graph is weighted by the semantic textual similarity of tweets posted by connecting users.

4.3 Predicting Locations of Isolated Users

As reported by [30], many test users are not transitively connected to any training node. It results in the label propagation failing to assign a location to isolated users. This usually happens when users do not use @-mentions, or when a set of nodes constitutes a disconnected component of the graph [30].

To alleviate this issue, we take advantages of both tweets' textual content and user-declared metadata to estimate the location of each test user. The predicted location is then used as an initial estimation during label propagation. To this end, we use the stacking approach proposed by [12] to combine together the tweet text (TEXT) and metadata fields: description (DESC), user-declared location (LOC) and user-declared time zone (TZ). The stacking approach consists of the three following steps: First, a multinomial naive Bayes base classifier ($L0$) is trained for each data type (i.e., TEXT, DESC, LOC, and TZ). Next, the outputs of the four classifiers on the training set are obtained using 10-fold cross validation. Lastly, a meta-classifier ($L1$ classifier) is trained over the base classifiers, using l_2-regularized logistic regression.

The stacking-based estimated location is attached as dongle node [30] to each test user in the social graph. The dongle nodes are treated in the same way as other labeled nodes (i.e., the training nodes). This iteratively adjusts the locations based on both the known training users and predicted test users. In such a way, the inferred locations of test users will better match neighboring users in their sub-graph, or in the case of disconnected nodes, will retain their initial classification estimate.

4.4 Label Propagation with Modified Adsorption

We utilize Modified Adsorption [32] as our label propagation algorithm, since it allows different levels of influence between prior/known labels and propagated label distributions. Modified Adsorption is a graph-based semi-supervised learning algorithm which has been used for open domain class-instance acquisition. It computes a soft assignment of labels to the nodes of a graph $G = (V, E, W)$, where V is the set of nodes with $|V| = n$, E is the set of edges, and W is an edge weight matrix. Out of the $n_l + n_u = n$ nodes in G, we have prior knowledge of labels for n_l nodes (training users), while the remaining n_u nodes are unlabeled (test users). Assume C is the set of labels, with $|C| = m$ representing the total number of labels. Y is the $n \times m$ matrix storing training label information. The l_{th} element of the vector Y_v encodes the prior knowledge for vertex v. Another vector, $\hat{Y} \in \mathbb{R}_+$, is the output of the algorithm (estimated label distribution for the nodes), using similar semantics as Y. The goal of Modified Adsorption is to compute $\hat{Y}$ such that the following objective function is minimized:

$$C(\hat{Y}) = \sum_l \left[\mu_1 (Y_l - \hat{Y}_l)^T S(Y_l - \hat{Y}) + \mu_2 \hat{Y}_l^T L \hat{Y}_l + \mu_3 \|\hat{Y}_l - R_l\|_2^2 \right] \qquad (3)$$

where μ_1, μ_2, and μ_3 are hyperparameters; L is the Laplacian of an undirected graph derived from G, but with revised edge weights; and R is an $n \times m$ matrix of per-node label prior, if any, with R_l representing the l_{th} column of R. S is a diagonal binary matrix indicating if a node is labelled or not.

In our experiments, we set the label confidence for training and test users to 1.0 and 0, respectively. For each Local celebrity, we initialize its location to the weighted median latitude and weighted median longitude of all its geolocated mentioners [6,7]. We set the label confidence for Local celebrites to 0.6, so that their label can be changed over the propagation process. Training users along with Local celebrities with their corresponding labels confidences are added to the seed set. We set μ_1, μ_2, and μ_3 to 0.9, 0.15, and 0, respectively. It should be noted that optimal values of μ_1, μ_2, and μ_3 have been chosen using a grid search with development sets of the three datasets. Finally, we run the Modified Adsorption algorithm iteratively until convergence, which usually occurs at or before 10 iterations.

5 Experimental Results

5.1 Experiment Setting

To infer the word embeddings, we have used tweets from training sets of TwUs (38M tweets), WNUT (9M tweets) and TwWORLD-Ex (an extended version of

TwWORLD comprising 188M tweets [11]). The resulting dataset contains 235M raw tweets. In the preprocessing step, we used replacement tokens for URLs, mentions and numbers. We did not replace hashtags as doing so experimentally demonstrated to decrease the accuracy.

To construct look-up table containing per-word feature vectors, we used the Python *gensim*[2] library which wraps the original Google Word2Vec C code. The model was trained using the Skip-gram architecture and negative sampling ($k = 5$) for five iterations, with a context window of 3 and subsampling factor of 0.001. It is noteworthy that to be part of the vocabulary, words should occur at least five times in the corpus. The final word2vec model has a vocabulary of 3M words and word representations of dimensionality 400. We chose word embeddings of size 400 because smaller embeddings experimentally showed to capture not as much detail and resulted in a lower accuracy. Larger word embeddings, on the other hand, made the model too complex to train.

5.2   Evaluation Metrics

We evaluate our approach in the following three commonly used metrics for user geolocation: **Acc@161**: The percentage of predicted locations which are within a 161 km (100 mile) radius of the actual location [3], as a proxy for accuracy within a metro area; **Mean**: The mean error (distance from the predicted location to the actual location) in km [8]; and **Median**: The median value (in km) of error distances in predictions [8].

5.3   Results

Tables 2 and 3 present the geolocation inference results over the three datasets. The results show that our proposed hybrid method (GEOCELWE-STACK) achieves the best performance in terms of all evaluation metrics. Our network-based method (GEOCELWE), on the other hand, outperforms existing network-based and text-based methods.

To evaluate the contribution of different components, we compare the median errors for variants of our approach. As can be seen in Fig. 2, the performance deteriorates the most when we exclude the stacking-based component (GEOCELWE). The performance further drops when we eliminate local interaction detection component, and use the number of mentions in tweets as edge weight instead of utilizing tweets' semantic similarities (GEOCEL). The main reason is that users' location proximity is much better correlated with their tweets' semantic similarity than the number of mentions. The performance also declines when we exclude celebrity filtering component and remove all celebrities from the social graph (GEO). It confirms the importance of keeping Local celebrities and propagating their locations through the social graph. Even so, our analysis show that in some cases, ordinary users who have mentioned by lots of their friends have been wrongly identified as Local celebrities. One potential approach

---

[2] https://radimrehurek.com/gensim/.

Table 2. Performance of Text-based (TB), Network-based (NB), and Hybrid (Hyb) geolocation methods over TwUs and TwWorld datasets ("-" signifies that no results were published for the given dataset).

	Category	TwUs			TwWorld		
		Acc@161	Mean	Median	Acc@161	Mean	Median
WB-Uniform [34]	TB	49	703	170	32	1714	490
WB-KDTree [34]	TB	48	686	191	31	1669	509
MDN-Shared [27]	TB	42	655	216	-	-	-
MLP+KDTree [29]	TB	54	562	121	35	1456	406
MLP+K-Mean [29]	TB	55	581	91	36	1417	373
LP-Rahimi [30]	NB	37	747	431	-	-	-
MadCelB [28]	NB	54	709	117	-	-	-
MadCelW [28]	NB	54	705	116	45	2525	279
GeoCel [7]	NB	59	546	83	48	2027	215
LP-LR [30]	Hyb	50	620	157	-	-	-
MadCelW-LR [28]	Hyb	60	529	78	53	1403	111
MadCelW-MLP [29]	Hyb	61	515	77	53	1280	104
GeoCel-BK [7]	Hyb	66	438	56	54	1216	95
GeoCelWe	NB	60	472	67	49	1892	197
GeoCelWe-Stack	Hyb	**69**	**371**	**43**	**59**	**394**	**31**

Table 3. Geolocation performance over WNUT dataset. We have also reported the *Accuracy* to make our results comparable with the state-of-the-art methods.

	Category	WNUT			
		Accuracy	Acc@161	Mean	Median
FujiXerox [25]	Hyb	47.6	-	1122	16.1
CSIRO [17]	Hyb	52.6	-	1929	21.7
GeoCelWe	NB	42.4	56	1853	98
GeoCelWe-Stack	Hyb	**55.7**	**70**	**1018**	0

to distinguish such users from Local celebrities is to look at bi-directional mentions as an indication of "real friendship". However, the utilized datasets do not include enough bi-directional mentions to be useful for this purpose.

It can be concluded that: (1) Using semantic textual similarity of tweets makes our approach able to detect and leverage local interactions in user's ego network. It results in improving the accuracy of label propagation algorithm and ultimately the quality of geoinference. (2) Utilizing the stacking-based approach to estimate the locations of isolated users can cover the limitations of network-based approach and significantly improve the prediction performance.

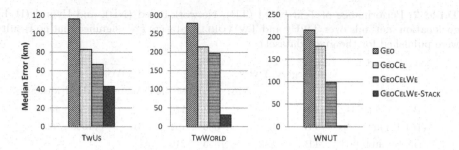

Fig. 2. Comparing the effect of different components on geoinference performance

5.4 Other Textual Similarity Measures

As a part of our experiments, we have used alternative textual similarity methods to test whether they can improve users' proximity prediction for geolocation. To this end, we have developed the following metrics to measure the semantic similarity of tweets.

1. **Cosine Similarity (CoSim) between TF-IDF Vectors**
2. **Jensen-Shannon Divergence (JSD) between Probability Distributions**
3. **WMD with Word Embeddings Trained on Google News[3] (WE-GN)**
   For the last metric, we replace our word embedding model trained on Twitter data (WE-TW) with the pre-trained Google News word2vec model, and use WMD algorithm to compute the semantic similarity of tweets. Since this model has been learned over formal text, tweet content is normalized before measuring textual similarity. To this end, we utilize the Twitter normalization lexicon [10] to replace abbreviations and slangs with the correct versions.

Figure 3 shows the performance of our geoinference approach (in terms of median error over the development sets of the three datasets), using different similarity metrics. From the comparison results, we make two observations as follows. Firstly, similarity measures based on word embeddings, specially when trained on Twitter data, consistently outperform other metrics. Secondly, comparing word embedding models, our trained model (WE-TW) is capable of measuring geographic closeness of tweets much better than Google News model (WE-GN). One possible reason is the potential differences between genres in Twitter data and news text. Moreover, as reported in several social media dialect studies [12], slangs (particularly regional slangs) and abbreviations provide cues about authors' geographic locations. Given that, this is likely that normalizing tweets eliminates useful information about geographic proximity.

---

[3] https://code.google.com/archive/p/word2vec/.

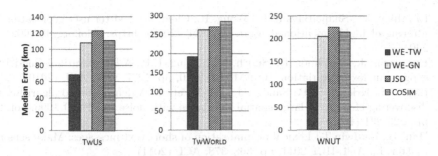

Fig. 3. Comparing geoinference performance using different similarity metrics

6 Conclusion and Future Work

In this paper, we have proposed a hybrid method to infer the locations of social media users. We have made the following contributions: (1) We have leveraged semantic similarity between tweets to predict users' local interactions. To achieve this, we have trained a powerful word embedding model over 235 million tweets. (2) To cover the limitation of network-based methods in geolocating isolated users, we have used an ensemble learning approach to take advantages of tweets' textual content and metadata. (3) We have conducted comprehensive experiments on three standard Twitter datasets, and demonstrated that our method outperforms state-of-the-art geolocation methods.

As future work, we plan to look at more efficient ways of utilizing metadata, textual, and network information in a joint model, and leveraging temporal data (e.g., the time that a user typically posts a tweet) to improve user geoinference.

References

1. Ashktorab, Z., Brown, C., Nandi, M., Culotta, A.: Tweedr: mining twitter to inform disaster response. In: ISCRAM 2014 (2014)
2. Cha, M., Gwon, Y., Kung, H.T.: Twitter geolocation and regional classification via sparse coding. In: ICWSM 2015, pp. 582–585 (2015)
3. Cheng, Z., Caverlee, J., Lee, K.: You are where you tweet: a content-based approach to geo-locating twitter users. In: CIKM 2010, pp. 759–768. ACM (2010)
4. Compton, R., Jurgens, D., Allen, D.: Geotagging one hundred million twitter accounts with total variation minimization. In: BigData 2014, pp. 393–401. IEEE (2014)
5. Davis Jr., C.A., Pappa, G.L., Rocha de Oliveira, D.R., Arcanjo, F.L.: Inferring the location of twitter messages based on user relationships. Trans. GIS **15**(6), 735–751 (2011)
6. Ebrahimi, M., ShafieiBavani, E., Wong, R., Chen, F.: Exploring celebrities on inferring user geolocation in twitter. In: Kim, J., Shim, K., Cao, L., Lee, J.-G., Lin, X., Moon, Y.-S. (eds.) PAKDD 2017. LNCS (LNAI), vol. 10234, pp. 395–406. Springer, Cham (2017). https://doi.org/10.1007/978-3-319-57454-7_31

7. Ebrahimi, M., ShafieiBavani, E., Wong, R., Chen, F.: Twitter user geolocation by filtering of highly mentioned users. JASIST (2018). https://doi.org/10.1002/asi.24011

8. Eisenstein, J., O'Connor, B., Smith, N.A., Xing, E.P.: A latent variable model for geographic lexical variation. In: EMNLP 2010, pp. 1277–1287. ACL (2010)

9. Ester, M., Kriegel, H.-P., Sander, J., Xu, X., et al.: A density-based algorithm for discovering clusters in large spatial databases with noise. In: KDD 1996, vol. 96, pp. 226–231 (1996)

10. Han, B., Baldwin,T.: Lexical normalisation of short text messages: Makn sens a# twitter. In: ACL-HLT 2011, pp. 368–378. ACL (2011)

11. Han, B., Cook, P., Baldwin, T.: Geolocation prediction in social media data by finding location indicative words. In: COLING 2012, pp. 1045–1062 (2012)

12. Han, B., Cook, P., Baldwin, T.: Text-based twitter user geolocation prediction. Artif. Intell. Res. **49**, 451–500 (2014)

13. Han, B., Hugo, A., Rahimi, A., Derczynski, L., Baldwin, T.: Twitter geolocation prediction shared task of the 2016 workshop on noisy user-generated text. In: WNUT 2016, pp. 213–217 (2016)

14. Hecht, B., Hong, L., Suh, B., Chi, E.H.: Tweets from Justin Bieber's heart: the dynamics of the location field in user profiles. In: ACM SIGCHI 2011, pp. 237–246. ACM (2011)

15. Hong, L., Ahmed, A., Gurumurthy, S., Smola, A.J., Tsioutsiouliklis, K.: Discovering geographical topics in the twitter stream. In: WWW 2012, pp. 769–778. ACM (2012)

16. Hulden, M., Silfverberg, M., Francom, J.: Kernel density estimation for text-based geolocation. In: AAAI 2015, pp. 145–150 (2015)

17. Jayasinghe, G., Jin, B., Mchugh, J., Robinson, B., Wan, S.: Csiro data61 at the WNUT geo shared task. In: WNUT 2016, pp. 218–226 (2016)

18. Jurgens, D.: That's what friends are for: inferring location in online social media platforms based on social relationships. In: ICWSM 2013, vol. 13, pp. 273–282 (2013)

19. Kusner, M.J., Sun, Y., Kolkin, N.I., Weinberger, K.Q.: From word embeddings to document distances. In: ICML 2015, pp. 957–966 (2015)

20. Li, R., Wang, S., Deng, H., Wang, R., Chang, K.C.-C.: Towards social user profiling: unified and discriminative influence model for inferring home locations. In: SIGKDD 2012, pp. 1023–1031. ACM (2012)

21. Lian, D., Ge, Y., Zhang, F., Yuan, N.J., Xie, X., Zhou, T., Rui, Y.: Content-aware collaborative filtering for location recommendation based on human mobility data. In: ICDM 2015, pp. 261–270. IEEE (2015)

22. Liu, J., Inkpen, D.: Estimating user location in social media with stacked denoising auto-encoders. In: NAACL-HLT 2015, pp. 201–210 (2015)

23. Mahmud, J., Nichols, J., Drews, C.: Where is this tweet from? inferring home locations of twitter users. In: ICWSM 2012, vol. 12, pp. 511–514 (2012)

24. Mikolov, T., Sutskever, I., Chen, K., Corrado, G.S., Dean, J.: Distributed representations of words and phrases and their compositionality. In: NIPS 2013, pp. 3111–3119 (2013)

25. Miura, Y., Taniguchi, M., Taniguchi, T., Ohkuma, T.: A simple scalable neural networks based model for geolocation prediction in twitter. In: WNUT 2016, pp. 235–239 (2016)

26. Qiang, J., Chen, P., Wang, T., Wu, X.: Topic modeling over short texts by incorporating word embeddings. In: Kim, J., Shim, K., Cao, L., Lee, J.-G., Lin, X., Moon, Y.-S. (eds.) PAKDD 2017. LNCS (LNAI), vol. 10235, pp. 363–374. Springer, Cham (2017). https://doi.org/10.1007/978-3-319-57529-2_29

27. Rahimi, A., Baldwin, T., Cohn, T.: Continuous representation of location for geolocation and lexical dialectology using mixture density networks. In: EMNLP 2017, pp. 167–176. ACL (2017)

28. Rahimi, A., Cohn, T., Baldwin, T.: Twitter user geolocation using a unified text and network prediction model. In: ACL-IJCNLP 2015, pp. 630–636. ACL (2015)

29. Rahimi, A., Cohn, T., Baldwin, T.: A neural model for user geolocation and lexical dialectology. In: ACL 2017 (2017)

30. Rahimi, A., Vu, D., Cohn, T., Baldwin, T.: Exploiting text and network context for geolocation of social media users. In: NAACL-HLT 2015, pp. 1362–1367. ACL (2015)

31. Roller, S., Speriosu, M., Rallapalli, S., Wing, B., Baldridge, J.: Supervised text-based geolocation using language models on an adaptive grid. In: EMNLP-CONLL 2012, pp. 1500–1510. ACL (2012)

32. Talukdar, P.P., Crammer, K.: New regularized algorithms for transductive learning. In: Buntine, W., Grobelnik, M., Mladenić, D., Shawe-Taylor, J. (eds.) ECML PKDD 2009. LNCS (LNAI), vol. 5782, pp. 442–457. Springer, Heidelberg (2009). https://doi.org/10.1007/978-3-642-04174-7_29

33. Wang, F., Lu, C.-T., Qu, Y., Yu, P.S.: Collective geographical embedding for geolocating social network users. In: Kim, J., Shim, K., Cao, L., Lee, J.-G., Lin, X., Moon, Y.-S. (eds.) PAKDD 2017. LNCS (LNAI), vol. 10234, pp. 599–611. Springer, Cham (2017). https://doi.org/10.1007/978-3-319-57454-7_47

34. Wing, B., Baldridge, J.: Hierarchical discriminative classification for text-based geolocation. In: EMNLP 2014, pp. 336–348. ACL (2014)

Utilizing Sequences of Touch Gestures for User Verification on Mobile Devices

Liron Ben Kimon$^{(\boxtimes)}$, Yisroel Mirsky, Lior Rokach, and Bracha Shapira

Department of Software and Information Systems Engineering, Ben-Gurion University of the Negev, Beer Sheva, Israel {benkimol, yisroel}@post.bgu.ac.il, {liorrk, bshapira}@bgu.ac.il

Abstract. Smartphones have become ubiquitous in our daily lives; they are used for a wide range of tasks and store increasing amounts of personal data. To minimize risk and prevent misuse of this data by unauthorized users, access must be restricted to verified users. Current classification-based methods for gesture-based user verification only consider single gestures, and not sequences. In this paper, we present a method which utilizes information from sequences of touchscreen gestures, and the context in which the gestures were made using only basic touch features. To evaluate our approach, we built an application which records all the necessary data from the device (touch and contextual sensors which do not consume significant battery life). Using XGBoost on the collected data, we were able to classify between a legitimate user and the population of illegitimate users (imposters) with an average equal error rate (EER) of 4.78% and an average area under the curve (AUC) of 98.15%. Our method demonstrates that by considering only basic touch features and utilizing sequences of gestures, as opposed to individual gestures, the accuracy of the verification process improves significantly.

Keywords: Continuous user verification · Mobile · Security
Touchscreen gestures · Sequence recognition · Context · Behavioral models
XGBoost

1 Introduction

Smartphones have become increasingly popular on a global scale. Large parts of our personal lives, such as personal SMS messages, emails, and credit card numbers, are accessible from our mobile devices. During 2013, 3.1 million American consumers (one of out of every ten smartphone users) were victims of phone theft, out of which 68% were unable to recover their data after the theft occurred [1]. Moreover, sometimes theft occurs from the victim's hand (i.e., a device 'snatch') [2]. Since many users do not lock their devices due to inconvenience [3], smartphone theft may result in unauthorized access to a user's private and sensitive information. The conventional solution is point authentication. There is a wide range of point authentication methods, but each has its weaknesses. For example, password authentication does not adequately protect

© Springer International Publishing AG, part of Springer Nature 2018
D. Phung et al. (Eds.): PAKDD 2018, LNAI 10939, pp. 816–828, 2018.
https://doi.org/10.1007/978-3-319-93040-4_64

smartphones from attackers, because users often choose simple, easy to guess passwords [4, 5]. Pattern swipe unlock leaves oily residues or smudges on the screen of the mobile device allowing attackers to deduce the pattern via a 'smudge-attack' by holding the screen up to a light or using high-resolution photography [6]. The following are the necessary steps for continuous user verification: (1) continuously identify the user, (2) detect when the user is not an authorized user, and finally (3) lock the device and notify the owner when unauthorized usage has been detected. Previous studies considered single gestures when training their machine learning algorithm. However, intuitively, a sequence of gestures may capture a user's behavior and intent better than a single gesture. With this understanding, we propose a novel technique for capturing the user's personal behavior. Specifically, we utilize a combination of strong features to learn a model. The features relate to the following two categories (1) **Gesture Trace Features.** The sequence of gestures a user performs implicitly captures the user's behavior and intentions. Whenever a gesture is performed, we extracted features that capture the present gesture, and aggregated features that capture the last few gestures which the user made. For the touch features we used only basic features, without doing any heavy manipulations on the data. The aggregated features represent the user's implicit context and aim to capture the user's behavior, (2) **Contextual Features.** To help segregate concepts, for each gesture made we also provide explicit contextual features. Specifically, an indication of the app presently on the foreground, the user's physical activity (e.g., walking, driving), and the device's screen orientation. The way a person touches his/her smartphone screen is not only unique to the user but is also unique under different contexts [9]. The combination of gesture trace features and explicit contextual features can greatly improve the accuracy of continuous verification. Therefore, the contributions of this paper is threefold: (1) We introduce the idea of including information from gesture traces to model the user's behavior. To implement this concept, we propose features which can be extracted from a trace, and determine the minimum required length needed for a trace, (2) We propose a novel combination of features that do not drastically consume battery life: touch gesture traces, and explicit contextual information, (3) We evaluate the proposed method on real data collected from 20 volunteers who used the provided device for two weeks each while most of the other works test their approaches on volunteers in a controlled lab environment where the natural behavior of the user cannot be modeled. This paper completes a work-in-progress published by the authors in [10]. The rest of this paper is organized as follows. Section 2 reviews previous work in the field. Section 3 presents the attack scenario considered in this paper. Section 4 describes the proposed method. Section 5 describes the evaluation setup, and Sect. 6 provides the results. Finally, in Sect. 7 we conclude with a discussion of results.

2 Related Work

There have been many previous studies on continuous user verification for mobile devices. A good survey of techniques can be found in [11]. We briefly review here techniques involving touch gestures. These works in general follow the same approach:

(1) Capture sensor data from a single gesture, (2) extract features from the gesture and other sensors during that interval, and (3) apply a machine learning model to determine if the gesture belongs to the authorized user. These works advance continuous user verification by either proposing a new improved way of representing the gestures as a feature vector or new modeling of the feature vectors to identify the true user. Tao Feng, et al. [7] collected specific touch gesture information, including the finger motion speed, direction, the pressure at each sampled touch point, the gesture length and touch curvature (the slope of the gesture). This was done for a single gesture at a time. In [12], the authors collected the context of the application running in order to improve user authentication. In contrast to our work, they did not utilize other explicit contexts available from the device (e.g., user's physical activity). To the best of our knowledge, our work is the first to utilize the information gathered from sequences of touch gestures (what we refer to as a gesture trace). In this paper we demonstrate that better performance is achievable by considering the last few gestures performed, as opposed to just the present gesture. We also consider button presses made during these traces (such as back and home button). This temporal contextual information captures the user's activity, and behavior. Most of previous state-of-the-art papers classify the gestures into types such as tap, scroll, zoom, etc. In this paper we used basic raw data such as X and Y coordinates, without specifying the type of the gesture in order to emphasize the improvement of the accuracy by utilizing sequences of gestures. In addition, most state-of-the-art papers did not examine real life scenarios and focused only on predefined tasks [9], in this paper we used real life scenarios without any constraints on the user's behavior. Although the authors in [12, 13] have already considered using contextual features surrounding a touch gesture, the authors either built separate classifiers for each suggested modality [15], or excluded strong contextual features [8, 13, 14]. We believe that there is a strong connection between the modalities and therefore it is logical to include all the data in a single model, and we consider contextual features such as current application, user's physical activity, power consumption of the device and button's clicks.

3 Attack Scenario

The purpose of this research is to protect the phone from usage which can cause the owner considerable damage. We assume the following attack scenario: (1) the victim left his/her device unattended, (2) the device was stolen by the attacker, and then (3) the attacker attempts to perform malicious activities (send an email or instant message using the victim's name, browse and possibly copy photos or videos, etc.). In general, our proposed method of continuous verification can defend against other attack scenarios with our loss of generality. Moreover, it is important to note that we assume these activities can only be achieved by physically interacting with the screen. Since our approach relies on obtaining a trace of gestures it is important to note the number of gestures an attacker would need to perform to reach the goals above. To estimate the number of gestures needed, we had five volunteers perform each of the above attacks on somebody else's phone. The volunteers were not informed of the purpose the experiment. The volunteers were experienced with the victim's platform so there was no bias on navigation, only layout. From this experiment we found that the uninformed thief performs at least 12–30 touch gestures to attain his/her goal. We also observed that users tend to perform quite a lot of brief gestures when scrolling, or searching for an icon. In attacks that involve typing, the number of gestures were considerably more.

4 Suggested Approach

In this paper, we propose a novel verification method which continuously verifies users on mobile devices by monitoring and analyzing touchscreen gestures. To describe the approach, we first provide the following definitions: (1) **Gesture** - The interaction a user makes on the device screen, starting from the moment the finger touches the screen, until the finger has been lifted, (2) **Gesture Trace** - The gestures that have been made prior to the current gesture (within some time), (3) **Context of a Gesture** - The state of the device (soft sensing) and the state of the user (hard sensing), while the gesture was made. A binary classification model M is trained where class '0' is the authorized user, and class '1' is a collection of other unauthorized users. In other words, a feature vector x provided to M has a label '0' if the instance that the vector represents belongs to the target user, or '1' if it belongs to a random stranger, obtained from another similar model device.

Continuous verification is provided by performing the following for every gesture of the user;

Data Acquisition:

1. Acquire the raw sensor values for the gesture, power consumption of the device, screen orientation, current application on the foreground, and the user's physical activity.
2. Store this sensor data in a local event database.

Feature Extraction:

3. Extract n_g features which describe the current gesture.
4. Extract n_c features which describe the context of the current gesture.
5. Extract n_h aggregated features from the trace of gestures made within a time interval.
6. Form the feature vector $\bar{x}$ with the $n = n_g + n_c + n_h$ features collected above, and store it in the local database.
7. Form $\bar{x}^T$ by concatenating the last T feature vectors $(\bar{x})$ made.

User Verification:

8. Use the classification model M to determine if $\bar{x}^T$ belongs to the authorized user. If it does, then add $\bar{x}^T$ the training repository (and M), otherwise lock the device and notify the owner by email.

To provide this service, an agent runs in the background of the device. Both the storage of the training repository, and the induction of the machine learning model M are offloaded to a remote server in the cloud. When there are enough instances in the remote repository, the server induces a new model M based on the instances from this user, and instances from other random users. The model is then pushed to the agent running on the device. In order to support concept drifts, the model keeps updating when new data arrives. A detailed description of these steps is provided in the sections that follow.

4.1 Data Acquisition

The data acquisition phase consists of acquiring information that is generated on the device. Some of this data is stored temporarily on the device in a local event database that is later used for extracting the features which are based on recent/passed activity. Every gesture begins with the act of touching the screen and ends with lifting the finger from the device. One gesture can produce one or more sensor records, depending on the length of the gesture. Each sample consists of four values: timestamp, X coordinate (XC), Y coordinate (YC), and the area of the screen covered by the finger (AR). The sensor records from each gesture is stored in the local event database and discarded after sometime. As mentioned above, we decided to use only those basic touch features instead of classifying the gestures into types (which may improve the results) because the main goal of this paper is to introduce the influence of using a sequence of gestures on the accuracy. Along with the touch sensor data, contextual sensors are recorded as well. These contextual sensors fall into one of two categories:

Hard sensing – Provides additional data about the state of the user:

- **User activity** – the locomotion of the user, using the Android ActivityManager API. There are four predefined options: tilting, still, in vehicle, and on foot.
- **Soft sensing** – Provides additional data about the state of the device:
- **Current application** – the application running on the foreground when the gesture was made.
- **Screen orientation** – the orientation of the screen when the gesture was made. There are three possible values: portrait, landscape, and reverse landscape.
- **Power consumption of the device** – the power consumption of the device while the gesture was made. We used the built-in voltage and current sensors, available via the operating system's API, every 30 s.
- **Button pressed** – events where the user presses one of the following buttons: back, menu, short click on home button, long click on home button, volume up, volume down, screen on, and screen off.

Table 1. The features extracted for each touch gesture

	The features extracted	n
Current Gesture Data $\vec{x}_g$	• X, Y coordinates at the beginning and at the end of the gesture • Area of the finger at the beginning and at the end of the gesture • Segment features – the μ and σ of XC,YC,VX, VY, and AR • Number of samples in the entire gesture • Duration of the gesture	38
Contextual Data $\vec{x}_c$	• Application running in the foreground • User activity • Screen orientation	21
Gesture Trace Data $\vec{x}_t$	• Touch history – μ and σ of the touch features during I • Buttons – total number of times each button was clicked during I • Power consumption – μ and σ of the current and voltage slopes during I • Number of gestures over I	89

4.2 Feature Extraction

Whenever a gesture is made, and the raw sensor data has been collected, feature extraction is performed to create three feature vectors. These vectors are $\vec{x}_g, \vec{x}_c$, and $\vec{x}_h$, which are the features describing the current gesture, the features describing the current context, and the features describing the gesture history, respectively. Table 1 contains a brief description of each of these feature vectors. Once the features are collected $\vec{x} = \vec{x}_g$, $\vec{x}_c$, $\vec{x}_h$ is they are added to a local database. Let T be a set parameter which is the sequence length of the last gestures made, which are to be analyzed. The final feature vector presented to the machine learning classifier M, is the concatenation of the last T feature vectors that represent the recent sequence of gestures made. Concretely, $\vec{x}^T = \vec{x}_t, \vec{x}_{t-1}, \ldots, \vec{x}_{t-T}$ totaling $T * (n_g + n_c + n_h)$ features. $\vec{x}^T$ is also sent to the remote server (which maintains the model M).

Gesture Features. As mentioned in Sect. 4.1, each touch gesture consists of a variable number of sensor records (across the gesture itself). The following section describes how we extracted the n_g features for $\vec{x}_g$. First we calculate the direction vector of the coordinates – the difference between the values of the X and Y coordinates of every sensor record in the gesture, and the previous record in the same gesture. The first instance in the gesture gets a value of 0 for both features. We refer to these two features as VX and VY. Next, in order to summarize the gesture as a constant number of features, we segmented the gesture over the time plane as follows: Let k be the number of equal sized partitions to make out of each gesture. When a gesture is recorded, the samples are partitioned into k sets, for which we aggregate relevant statistics. For every set we calculate the average and the standard deviation of each of the following sensor records: XC, YC, VX, VY, and AR. Finally, we extract the following 6 features: the X and Y coordinates and the area of the screen covered by the finger at the beginning and at the end of the gesture, the number of samples in the entire trace, and the duration of the trace. Every feature vector $\vec{x}_g$ is stored in the local event database. Feature vectors older than 30 s are deleted.

Contextual Features. To create $\vec{x}_c$, the following features were extracted: the application running on the foreground, the user's activity (e.g., walking, standing still...), and the screen orientation. Because of the vast number of possible applications, we only considered the 14 most popular apps. To represent the above features from these apps we used a binary representation. For example, the feature 'FacebookONScreen' will receive the value 1 if Facebook is running on the foreground and 0 otherwise. Therefore, 14 features were extracted for the applications, 4 features for the user's activity, and 3 features for the screen orientation.

Gesture Trace Features. We define the trace interval I as the time span starting from I_s seconds ago until the present time (for example $I_s = 30$).

The feature vector $\vec{x}_h$ is created in the following manner: First the feature vectors $\vec{x}_h$ that were created during I, not including the current $\vec{x}_h$, are retrieved from the local event database. We calculate the average and standard deviation of the features in these vectors collectively to produce the first n_g features. Next, we make a feature for each button type which captures the number of time each respective button was pressed during I. Finally, we calculate the slope of the current and the voltage for each two consecutive values over I. Using these slopes, we add features which represented their average and standard deviation. The slopes were calculated as follows:

$$slope_{current} = \frac{c2 - c1}{t2 - t1}, \tag{1}$$

$$slope_{voltage} = \frac{v2 - v1}{t2 - t1} \tag{2}$$

Finally, the last feature of the $\vec{x}_h$ is the total number of gestures that occurred over *I*.

4.3   Verification

Once $\vec{x}^T$ has been attained, it is passed to the machine learning model M (stored locally). Using M it is possible to verify whether the user is authorized. $\vec{x}^T$ has many features, yet the machine learning model needs to be robust and efficient to execute. Therefore, we opted a decision tree ensemble as our classifier, using one of the following machine learning algorithms: XGBoost (extreme gradient boosting) and (2) Random Forest. Although the classification model M is stored and executed locally on the device, M is trained, and maintained off-site on a remote server. The remote server receives and stores the feature vectors from the participating users. Let U be the collection of users who are a part of the system. Let $u_i \in U$ be the user of the present device, and M_{u_i} be that user's model. The server trains and updates M_{u_i} as a one-vs-all classifier in the following way: All feature vectors $\vec{x}^T$, which were generated by u_i, are given the label '0'. Then a random set of users from U are selected such that each

Table 2. The average EER results for each number of segments tested

EER	Segments
0.08417	2
0.08340	3
0.08395	4

Table 3. The average EER results for each number of applications tested

EER	Applications
0.08592	6
0.08484	8
0.08340	10
0.08329	12
0.08291	**14**
0.08321	16

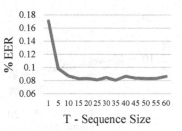

Fig. 1. Comparison of different sequence sizes

selected user $u_j \neq u_i$. Next, all feature vectors x^T produced by the selected users, are given the label '1'. Finally, both sets of labeled instances are joined to form a single training set, and the model M_{u_i} is induced.

5 Experiment Setup

5.1 Data Collection

In order to collect the data, we developed an application that extracts all of the necessary raw features from the mobile device (touch and contextual data). The application had no GUI and ran in the background without interrupting regular device usage. The application was responsible for detecting the touch events that were produced by the device, and their context (both hard and soft sensing). The application was installed on rooted Galaxy S4 smartphones given to 20 volunteers (eight males and twelve females, aged 18–40). The volunteers used the phone as their own personal phone for two weeks. All of the participants had experience with some types of smartphone and were familiar with smartphone use. In order to collect accurate unbiased data, the participants had complete freedom to whatever they wanted to do with the device: download applications, change settings, etc. During our evaluation we compared pairs of users where one participant served as the owner of the phone with another participant who served as the thief. Therefore, it was critical to maintain uniformity in the users' measurements, and all volunteers were given a device with the same configuration: Samsung Galaxy S4 – Android KitKat version 4.4.2. Each volunteer used the device for two weeks, during which the data was recorded by our application. The raw data was downloaded from the devices when the participants returned their smartphones at the end of the experiment.

5.2 Evaluation Metrics

The area under the curve (AUC) metric was used to evaluate verification performance, since it is known as the most appropriate method to evaluate a binary classification model [16], especially in order to avoid the prior probability effect on the accuracy

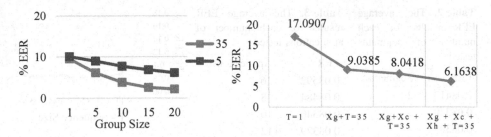

Fig. 2. Different sized groups for sequence sizes of 5 and 35 instances

Fig. 3. Verification improvement when considering different features and sequence size

[17]. The AUC demonstrates the trade-off between the true positive rate (TPR) and false positive rate (FPR). We also computed the equal error rate (EER) metric, which is the rate at which the false acceptance rate (FAR) and false rejection rate (FRR) are equal. The lower the equal error rate value, the higher the accuracy of the verification system. The FAR and FRR are calculating as $FAR = \frac{\#verified\ imposters}{\#imposters}$ and $FRR = \frac{\#rejected\ genuine}{\#genuine}$. To calculate the EER, we first computed the predicted probability of each data point belonging to each class. Then, in order to vary the threshold value, at the particular threshold value the corresponding FAR and FRR are derived. All FAR and FRR pairs are used to plot the receiver operating characteristic curve (ROC). The corresponding value of the point at which FAR and FRR are equal is the EER. The EER is well known metric, and previous state-of-the-art papers uses this metric in order to present the results of their method.

5.3 Hyper Parameter Tuning

As described in Sect. 4.2 there are three parameters which our approach requires: the sequence length T, the aggregation interval I_s, and the number of segments for each gesture k. We performed a series of preliminary experiments in order to tune the model and set the best values for these hyper parameters. For T we examined values in the range of 1–60, for I_s we examined two values 30 and 60 s, and for k we examined the values 2, 3, and 4. In addition to these hyper parameters, we also examined the number of most popular applications which should be considered in the contextual features (6–16 apps at a time). For the tuning, we used the XGB classifier, as it is empirically known to run faster and more accurate than other classifiers. We ran the algorithm on each possible pair of our 20 participants ($\frac{20!}{(20-2)!(2!)} = 190$ combinations). 90% of each participant's data was used for training, and the rest was used for testing the algorithm. We split the data according to the time of the events to maintain the consistency of the sequences and simulate real world scenarios. For each run we calculated the EER and then calculated the average EER result over all runs. In each experiment we manipulated one parameter and fixed the others. Figure 1 shows that the optimal T is between 5 and 35 instances (gestures). Table 2 shows that a k of 3 produced the best results. Finally, Table 3 shows that good results are achieved by considering the top 14 apps. We also found that $I_s = 30$ seconds provided the best results.

Table 4. Average EER and AUC results for both insider and outsider attack scenarios

	EER	AUC
Outsider attack	6.82%	96.25%
Insider attack	4.78%	98.15%

6 Experiments and Results

Using the hyperparameters tuned as described above, we performed experiments to evaluate the accuracy of classifying the instance $\bar{x}^T$ into authorized or non-authorized user (authorized users has the label '0', and unauthorized users has the label '1'). As described in Sect. 4, our model is trained on data taken from many different users. Specifically, instances labeled as the positive class are taken from the smartphone's owner, and instances labeled as the negative class are taken from other users (possibly from the same company/organization). Therefore, in our evaluations, we examine two possible scenarios: (1) **Outsider Attack Scenario** – The scenario where the attacker does not belong to the same organization as the target user, and thus the model was not trained using the attacker's data. For this scenario, we included the attacker's data in the test set and not in the training set. (2) **Insider Attack Scenario** – The scenario where the attacker is an individual within the same organization as the target user, and the model was trained on using the attacker's data. This scenario represents the case where the attacker performs an 'insider attack' on a colleague's smartphone, from within the same organization. For this scenario, we include some of the attacker's data in the training set. In order to simulate the insider and outsider attack scenarios, we conducted the following experiment. For the insider attack scenario, for every participant that we wanted to verify, we randomly choose 15 other participants on which the model was trained (we set the number of participants to 15 empirically after trying other numbers). We split the data from these participants, so that 90% of the data was used to train the model, and the other 10% formed the test set. For the outsider attack scenario, we randomly chose two other participants on which the model was not trained (non-verified users) and used 10% of their data as the test set to simulate attackers.

6.1 Outsider and Insider Attack Scenario

The following is our analysis under the outsider attack scenario. In order to calculate the EER, we predicted the user's probability for each instance (aggregation of gestures). We calculated the EER for each user, as well as the average EER. In spite of the fact that the best results were obtained for a sequence size of 35, it could be argued that 35 gestures are too many for a theft scenario, since one may want to identify that a device was stolen much earlier. Therefore, we report results for sequence sizes of 5 and 35 instances. In addition, instead of calculating the probability for each instance, we calculated the average probability for a group of instances. This was done because sometimes one misclassified instance can harm the EER results, and calculation of the average probability can mitigate this error. We used five different sized groups of instances: 1, 5, 10, 15 and 20. The results in Fig. 2 demonstrates that when the group of instances is larger, the results are more accurate. In addition, a sequence size of 35 instances results in more improved performance than a sequence size of five instances. Thus, there is a trade-off between the sequence size, the size of the group, and the response time. It is intuitive that in cases in which there is more information, the results will be more accurate, but we aim to verify thieves as soon as possible. The results show that a sequence size of 35 instances and a group size of five achieved an average EER of 6.11%. Nearly the same result was achieved with a sequence size of five instances and a group size of 20 instances (average EER of 6.12%). The best EER result obtained was 2.19%, which was achieved when considering a sequence size of 35 instances and a group size of 20 instances. In addition, we wanted to compare the XGB classifier's performance to the performance of the random forest classifier for the outsider attack scenario. We performed the same experiment, testing both the insider and outsider attacks. In both algorithms we tested a sequence size of 35 instances and the five group sizes previously mentioned. The results show that for every case, the XGB classifier outperformed the random forest classifier. Finally, we calculated the average EER and AUC results for both insider and outsider attack scenarios. The results were calculated based on a sequence size of 5 instances and a group of 15 instances when calculating the prediction. We were able to achieve an EER of 4.78% and AUC of 98.15% for the insider attack scenario, versus an EER of 6.82% and AUC of 95.55% for the outsider attack scenario, as presented in Table 4. The following is our analysis under the contaminated model scenario (insider attack scenario). Results in Fig. 3 demonstrate that adding our proposed features and considering a sequence of gestures, it is possible to improve the verification accuracy. Considering only the touch features, without taking into account context or gesture trace features, the average EER was 9.03%. After considering the context of the gestures, the results improved, and an average EER of 8.04% was attained; similarly, after considering the gesture trace features, the results improved, and an average EER of 6.16% was attained. The influence of the sequence size is also shown in Fig. 3. The average EER obtained when considering a sequence size of one instance, i.e., no sequence at all, was 17.09%, and after concatenating 35 instances the result improved dramatically to 8.04%. As mentioned earlier, the optimum sequence size is between 5 and 35.

6.2   Response Time and Comparison with State-of-the-Art

We found that the optimum trace length T is 35. From our data collection experiment, we observed that a user performs 35 gestures in 13.8 s on average, with standard deviation of 25 s. From our trials described in Sect. 3, it can be assumed that an uninformed attacker will perform at least 35 gestures to achieve his/her goal. This is because the attacker does not know the layout and the location of every application and file. Although $T = 35$ is the optimal value, a user can set $T = 5$ and still obtain good results (as shown in Fig. 1). Moreover, the verification time (feature extraction and model execution) is negligible. We found that it takes only a few micro seconds to perform the classification task. Therefore, with respect to the attack scenario in Sect. 3, the attacker should be detected before the attack is complete. In [13] the authors achieved ERRs ranging from 6.33% to 15.4%. In [8] the authors achieved an EER between 0% and 4%, depend on the predefined task. In [9] the authors achieved EERs ranging from 6.1% to 6.9%. Although our method only achieved EERs ranging from 4.78% to 6.82%, all of the studies above were performed in a controlled lab setting where typically participants performed a set of pre-defined tasks, sometimes repeatedly. In our study we gave the volunteers full control over their devices, and they used the devices as their own personal device for two weeks. In addition, most state-of-the-are papers utilize the type of gesture in order to build a model. We used only basic touch features, and still we were able to achieve good results without doing any complicated computing on the data, which may cost us an expensive time when building the model and detecting a thief. Therefore, our method arguably performs better than state-of-the-art touch-based user authentication, while using sensors that do not drastically consume battery life.

7 Conclusion

We presented the design, implementation, and evaluation of a user verification method for mobile devices using touchscreen gestures, the context under which the touch was made, and the trace of the last few gestures made. For our evaluations, we developed an Android application and experimented with real subjects for several weeks each. The results show that by combining our proposed modalities and sequence history, the EER can be reduced by over 3%.

References

1. Lookout Blog: Phone Theft in America: What really happens when your phone gets grabbed (2014). https://blog.lookout.com/blog/2014/05/07/phone-theft-in-america/
2. Lee, A.: A Thief Snatched My iPhone (2014). http://readwrite.com/2014/06/23/iphone-smartphone-theft-crime/
3. Consumer Report: Smart phone thefts rose to 3.1 million in 2013 (2014). http://www.consumerreports.org/cro/news/2014/04/smart-phone-thefts-rose-to-3-1-million-last-year/index.htm
4. Clarke, N.L., Furnell, S.M.: Authentication of users on mobile telephones - a survey of attitudes and practices. Comput. Secur. 24(7), 519–527 (2005)
5. Vance, A.: If your password is 123456, just make it hackme. N.Y. Times 20, A1 (2010)
6. Aviv, A.J., Gibson, K., Mossop, E., Blaze, M., Smith, J.M.: Smudge attacks on smartphone touch screens. In: USENIX Conference on Offensive Technology, pp. 1–7 (2010)
7. Feng, T., Liu, Z., Kwon, K.A., Shi, W., Carbunar, B., Jiang, Y., Nguyen, N.: Continuous mobile authentication using touchscreen gestures. In: 2012 IEEE International Conference on Technologies for Homeland Security, HST 2012, pp. 451–456 (2012)
8. Frank, M., et al.: Touchalytics: on the applicability of touchscreen input as a behavioral biometric for continuous authentication. IEEE Trans. Inf. Forensics Secur. 8(1), 136–148 (2013)
9. Murmuria, R., Stavrou, A., Barbará, D., Fleck, D.: Continuous authentication on mobile devices using power consumption, touch gestures and physical movement of users. In: Bos, H., Monrose, F. (eds.) RAID 2015. LNCS, vol. 9404, pp. 405–424. Springer, Cham (2015). https://doi.org/10.1007/978-3-319-26362-5_19

10. Ben Kimon, L., et al.: User verification on mobile devices using sequences of touch gestures. In: Proceedings of the 25th Conference on User Modeling, Adaptation and Personalization. ACM (2017)
11. Patel, V.M., Chellappa, R., Chandra, D., Barbello, B.: Continuous user authentication on mobile devices: recent progress and remaining challenges. IEEE Sig. Process. Mag. **33**(4), 49–61 (2016)
12. Feng, T., Yang, J., Yan, Z., Tapia, E.M., Shi, W.: TIPS: context-aware implicit user identification using touch screen in uncontrolled environments. In: Proceedings of the 15th Workshop on Mobile Computing Systems and Applications, pp. 9:1–9:6 (2014)
13. Zhao, X., Feng, T., Shi, W.: Continuous mobile authentication using a novel graphic touch gesture feature. In: IEEE 6th International Conference on Biometrics: Theory, Applications and Systems, BTAS 2013 (2013)
14. Zhao, X., Feng, T., Shi, W., Kakadiaris, I.A.: Mobile user authentication using statistical touch dynamics images. IEEE Trans. Inf. Forensics Secur. **9**(11), 1780–1789 (2014)
15. Shi, W., Yang, J., Jiang, Y., Yang, F., Xiong, Y.: SenGuard: passive user identification on smartphones using multiple sensors. In: International Conference on Wireless and Mobile Computing, Networking and Communications, pp. 141–148 (2011)
16. Huang, J., Ling, C.X.: Using AUC and accuracy in evaluating learning algorithms. IEEE Trans. Knowl. Data Eng. **17**(3), 299–310 (2005)
17. Rokach, L., Maimom, O.: Data mining with decision trees: theory and applications (2007)

Author Index

Printed in the United States
By Bookmasters